Probability and Its Applications

Published in association with the Applied Probability Trust

Editors: J. Gani, C.C. Heyde, P. Jagers, T.G. Kurtz

Probability and Its Applications

Anderson: Continuous-Time Markov Chains.
Azencott/Dacunha-Castelle: Series of Irregular Observations.
Bass: Diffusions and Elliptic Operators.
Bass: Probabilistic Techniques in Analysis.
Chen: Eigenvalues, Inequalities, and Ergodic Theory
Choi: ARMA Model Identification.
Daley/Vere-Jones: An Introduction to the Theory of Point Processes.
   Volume I: Elementary Theory and Methods, Second Edition.
de la Peña/Giné: Decoupling: From Dependence to Independence.
Del Moral: Feynman-Kac Formulae: Genealogical and Interacting Particle Systems with
   Applications.
Durrett: Probability Models for DNA Sequence Evolution.
Galambos/Simonelli: Bonferroni-type Inequalities with Applications.
Gani (Editor): The Craft of Probabilistic Modelling.
Grandell: Aspects of Risk Theory.
Gut: Stopped Random Walks.
Guyon: Random Fields on a Network.
Kallenberg: Foundations of Modern Probability, Second Edition.
Last/Brandt: Marked Point Processes on the Real Line.
Leadbetter/Lindgren/Rootzén: Extremes and Related Properties of Random Sequences
   and Processes.
Molchanov: Theory of Random Sets.
Nualart: The Malliavin Calculus and Related Topics.
Rachev/Rüschendorf: Mass Transportation Problems. Volume I: Theory.
Rachev/Rüschendofr: Mass Transportation Problems. Volume II: Applications.
Resnick: Extreme Values, Regular Variation and Point Processes.
Shedler: Regeneration and Networks of Queues.
Silvestrov: Limit Theorems for Randomly Stopped Stochastic Processes.
Thorisson: Coupling, Stationarity, and Regeneration.
Todorovic: An Introduction to Stochastic Processes and Their Applications.

Ilya Molchanov

Theory of Random Sets

With 33 Figures

Ilya Molchanov
Department of Mathematical Statistics and Actuarial Science, University of Berne,
CH-3012 Berne, Switzerland

Series Editors
J. Gani
Stochastic Analysis Group CMA
Australian National University
Canberra ACT 0200
Australia

C.C. Heyde
Stochastic Analysis Group, CMA
Australian National University
Canberra ACT 0200
Australia

P. Jagers
Mathematical Statistics
Chalmers University of Technology
SE-412 96 Göteborg
Sweden

T.G. Kurtz
Department of Mathematics
University of Wisconsin
480 Lincoln Drive
Madison, WI 53706
USA

Mathematics Subject Classification (2000): 60-02, 60D05, 06B35, 26A15, 26A24, 28B20, 31C15, 43A05, 43A35, 49J53, 52A22, 52A30, 54C60, 54C65, 54F05, 60E07, 60G70, 60F05, 60H25, 60J99, 62M30, 90C15, 91B72, 93E20

British Library Cataloguing in Publication Data
Molchanov, Ilya S., 1962–
  Theory of random sets. — (Probability and its applications)
  1. Random sets 2. Stochastic geometry
  I. Title
  519.2

Library of Congress Cataloging-in-Publication Data
CIP data available.

Apart from any fair dealing for the purposes of research or private study, or criticism or review, as permitted under the Copyright, Designs and Patents Act 1988, this publication may only be reproduced, stored or transmitted, in any form or by any means, with the prior permission in writing of the publishers, or in the case of reprographic reproduction in accordance with the terms of licences issued by the Copyright Licensing Agency. Enquiries concerning reproduction outside those terms should be sent to the publishers.

ISBN: 978-1-84996-949-9
Springer Science+Business Media
springeronline.com

© Springer-Verlag London Limited 2005
Softcover reprint of the hardcover 1st edition 2005

The use of registered names, trademarks, etc., in this publication does not imply, even in the absence of a specific statement, that such names are exempt from the relevant laws and regulations and therefore free for general use.

The publisher makes no representation, express or implied, with regard to the accuracy of the information contained in this book and cannot accept any legal responsibility or liability for any errors or omissions that may be made.

Typesetting: Camera-ready by author.
12/3830-543210 Printed on acid-free paper   SPIN 10997451

To my mother

Preface

History

The studies of random geometrical objects go back to the famous Buffon needle problem. Similar to the ideas of Geometric Probabilities that can be traced back to the first results in probability theory, the concept of a random set was mentioned for the first time together with the mathematical foundations of Probability Theory. A.N. Kolmogorov [321, p. 46] wrote in 1933:

> Let G be a measurable region of the plane whose shape depends on chance; in other words, let us assign to every elementary event ξ of a field of probability a definite measurable plane region G. We shall denote by J the area of the region G and by $\mathbf{P}(x, y)$ the probability that the point (x, y) belongs to the region G. Then
> $$\mathbf{E}(J) = \iint \mathbf{P}(x, y) \mathrm{d}x \mathrm{d}y.$$

One can notice that this is the formulation of Robbins' theorem and $\mathbf{P}(x, y)$ is the coverage function of the random set G.

The further progress in the theory of random sets relied on the developments in the following areas:

- studies of random elements in abstract spaces, for example groups and algebras, see Grenander [210];
- the general theory of stochastic processes, see Dellacherie [131];
- advances in image analysis and microscopy that required a satisfactory mathematical theory of distributions for binary images (or random sets), see Serra [532].

The mathematical theory of random sets can be traced back to the book by Matheron [381]. G. Matheron formulated the exact definition of a random closed set and developed the relevant techniques that enriched the convex geometry and laid out the foundations of mathematical morphology. Broadly speaking, the convex geometry contribution concerned properties of functionals of random sets, while the morphological part concentrated on operations with the sets themselves.

The relationship between random sets and convex geometry later on has been thoroughly explored within the stochastic geometry literature, see, e.g. Weil and Wieacker [607]. Within the stochastic geometry, random sets represent one type of objects along with point processes, random tessellations, etc., see Stoyan, Kendall and Mecke [544]. The main techniques stem from convex and integral geometry, see Schneider [520] and Schneider and Weil[523].

The mathematical morphology part of G. Matheron's book gave rise to numerous applications in image processing (Dougherty [146]) and abstract studies of operations with sets, often in the framework of the lattice theory (Heijmans [228]).

Since 1975 when G. Matheron's book [381] was published, the theory of random sets has enjoyed substantial developments. D.G. Kendall's seminal paper [295] already contained the first steps into many directions such as lattices, weak convergence, spectral representation, infinite divisibility. Many of these concepts have been elaborated later on in connection to the relevant ideas in pure mathematics. This made many of the concepts and notation used in [295] obsolete, so that we will follow the modern terminology that fits better into the system developed by G. Matheron; most of his notation was taken as the basis for the current text.

The modern directions in random sets theory concern

- relationships to the theories of semigroups and continuous lattices;
- properties of capacities;
- limit theorems for Minkowski sums and relevant techniques from probabilities in Banach spaces;
- limit theorems for unions of random sets, which are related to the theory of extreme values;
- stochastic optimisation ideas in relation to random sets that appear as epigraphs of random functions;
- studies of properties of level sets and excursions of stochastic processes.

These directions constitute the main core of this book which aims to cast the random sets theory in the conventional probabilistic framework that involves distributional properties, limit theorems and the relevant analytical tools.

Central topics of the book

The whole story told in this book concentrates on several important concepts in the theory of random sets.

The first concept is the *capacity functional* that determines the distribution of a random closed set in a locally compact Hausdorff separable space. It is related to positive definite functions on semigroups and lattices. Unlike probability measures, the capacity functional is non-additive. The studies of non-additive measures are abundant, especially, in view of applications to game theory, where the non-additive measure determines the gain attained by a coalition of players. The capacity functional can be used to characterise the weak convergence of random sets and some properties of their distributions. In particular, this concerns unions of random closed sets, where the regular variation property of the capacity functional is of primary importance. It is possible to consider random capacities that unify the concepts of a random closed set and a random upper semicontinuous function. However, the capacity functional does not help to deal with a number of other issues, for instance to define the expectation of a random closed set.

Here the leading role is taken over by the concept of a *selection*, which is a (single-valued) random element that almost surely belongs to a random set. In this framework it is convenient to view a random closed set as a multifunction (or set-valued function) on a probability space and use the well-developed machinery of set-valued analysis. It is possible to find a countable family of selections that completely fills the random closed set and is called its *Castaing representation*. By taking expectations of integrable selections, one defines the *selection expectation* of a random closed set. However, the families of all selections are very rich even for simple random sets.

Fortunately, it is possible to overcome this difficulty by using the concept of the *support function*. The selection expectation of a random set defined of a non-atomic probability space is always convex and can be alternatively defined by taking the expectation of the support function. The *Minkowski sum* of random sets is defined as the set of sums of all their points or all their selections and can be equivalently formalised using the arithmetic sum of the support functions. Therefore, limit theorems for Minkowski sums of random sets can be derived from the existing results in Banach spaces, since the family of support functions can be embedded into a Banach space. The support function concept establishes numerous links to convex geometry ideas. It also makes it possible to study set-valued processes, e.g. set-valued martingales and set-valued shot-noise.

Important examples of random closed sets appear as *epigraphs* of random lower semicontinuous functions. Viewing the epigraphs as random closed sets makes it possible to obtain results for lower semicontinuous functions under the weakest possible conditions. In particular, this concerns the convergence of minimum values and minimisers, which is the subject of stochastic optimisation theory.

It is possible to consider the family of closed sets as both a semigroup and a lattice. Therefore, random closed sets are simply a special case of general lattice- or semigroup-valued random elements. The concept of probability measure on a *lattice* is indispensable in the modern theory of random sets.

It is convenient to work with random *closed* sets, which is the typical setting in this book, although in some places we mention random open sets and random Borel sets.

Plan

Since the concept of a set is central for mathematics, the book is highly interdisciplinary and aims to unite a number of mathematical theories and concepts: capacities, convex geometry, set-valued analysis, topology, harmonic analysis on semigroups, continuous lattices, non-additive measures and upper/lower probabilities, limit theorems in Banach spaces, general theory of stochastic processes, extreme values, stochastic optimisation, point processes and random measures.

The book starts with a definition of random closed sets. The space $\mathbb{E}$ which random sets belong to, is very often assumed to be locally compact Hausdorff with a countable base. The Euclidean space $\mathbb{R}^d$ is a generic example (apart from rare moments when $\mathbb{E}$ is a line). Often we switch to the more general case of $\mathbb{E}$ being a Polish space or Banach space (if a linear structure is essential). Then the Choquet theorem concerning the existence of random sets distributions is proved and relationships with set-valued analysis (or multifunctions) and lattices are explained. The rest of Chapter 1 relies on the concept of the capacity functional. First it highlights relationships between capacity functionals and properties of random sets, then develops some analytic theory, convergence concepts, applications to point processes and random capacities and finally explains various interpretations for capacities that stem from game theory, imprecise probabilities and robust statistics.

Chapter 2 concerns expectation concepts for random closed sets. The main part is devoted to the selection (or Aumann) expectation that is based on the idea of the selection. Chapter 3 continues this topic by dealing with Minkowski sums of random sets. The dual representation of the selection expectation – as a set of expectations of all selections and as the expectation of the support function – makes it possible to refer to limit theorems in Banach spaces in order to prove the corresponding results for random closed sets. The generality of presentation varies in order to explain which properties of the carrier space $\mathbb{E}$ are essential for particular results.

The scheme of unions for random sets is closely related to extremes of random variables and further generalisations for pointwise extremes of stochastic processes. Chapter 4 describes the main results for the union scheme and explains the background ideas that mostly stem from the studies of lattice-valued random elements.

Chapter 5 is devoted to links between random sets and stochastic processes. On the one hand, this concerns set-valued processes that develop in time, in particular, set-valued martingales. On the other hand, the subject matter concerns random sets interpretations of conventional stochastic processes, where random sets appear as graphs, level sets or epigraphs (hypographs).

The Appendices summarise the necessary mathematical background that is normally scattered between various texts. There is an extensive bibliography and a detailed subject index.

Several areas that are related to random sets are only mentioned in brief. For instance, these areas include the theory of set-indexed processes, where random sets appear as stopping times (or stopping sets), excursions of random fields and potential theory for Markov processes that provides further examples of capacities related to hitting times and paths of stochastic processes.

It is planned that a companion volume to this book will concern models of random sets (germ-grain models, random fractals, growth processes, etc), convex geometry techniques, statistical inference for stationary and compact random sets and related modelling issues in image analysis.

Conventions

The numbering follows a two-digit pattern, where the first digit is the section number of the current chapter. When referring to results from other chapters, we add the chapter number using the three digit numbering scheme. When referring to the Appendices, the first digit is a letter that designates the particular appendix. The statements in theorems and propositions are mostly numbered by Roman numbers, while the conditions usually follow Arabic numeration.

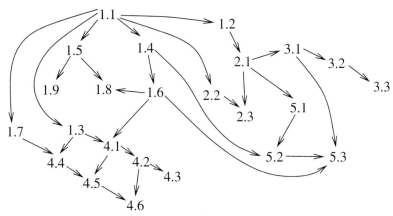

A rough dependence guide between the sections.

Although the main concepts in this book are used throughout the whole presentation, it is anticipated that a reader will be able to read the book from the middle. The concepts are often restated and notation is set to be as consistent as possible taking into account various conventions within a number of mathematical areas that build up this book.

The problems scattered through the text are essentially open, meaning that their solutions are currently not known to the author.

The supporting information (e.g. bibliographies) for this book is available through Springer WEB site or from

http://www.cx.unibe.ch/~ilya/rsbook/index.html

Acknowledgements

G. Matheron's book *Random Sets and Integral Geometry* [381] accompanied me throughout my whole life in mathematics since 1981 where I first saw its Russian translation (published in 1978). Then I became fascinated in this cocktail of techniques from topology, convex geometry and probability theory that essentially makes up the theory of random sets.

This book project has spanned my work and life in four different countries: Germany, the Netherlands, Scotland and Switzerland. I would like to thank people of all these and many other countries who supported me at various stages of my work and from whom I had a chance to learn. In particular, I would like to thank Dietrich Stoyan who, a while ago, encouraged me to start writing this book and to my colleagues in Bern for a wonderful working and living environment that helped me to finish this project in a final spurt.

I am grateful to the creators of XEmacs software which was absolutely indispensable during my work on this large LaTeX project and to the staff of Springer who helped me to complete this work.

Finally, I would like to thank my family and close friends for being always ready to help whatever happens.

Bern, Switzerland *Ilya Molchanov*
April 2004

Contents

1 Random Closed Sets and Capacity Functionals 1
  1   The Choquet theorem 1
      1.1   Set-valued random elements 1
      1.2   Capacity functionals 4
      1.3   Proofs of the Choquet theorem 13
      1.4   Separating classes 18
      1.5   Random compact sets 20
      1.6   Further functionals related to random sets 22
  2   Measurability and selections 25
      2.1   Multifunctions in metric spaces 25
      2.2   Selections of random closed sets 31
      2.3   Measurability of set-theoretic operations 37
      2.4   Random closed sets in Polish spaces 40
      2.5   Non-closed random sets 41
  3   Lattice-theoretic framework 42
      3.1   Basic constructions 42
      3.2   Existence of measures on partially ordered sets 43
      3.3   Locally finite measures on posets 46
      3.4   Existence of random sets distributions 47
  4   Capacity functionals and properties of random closed sets 49
      4.1   Invariance and stationarity 49
      4.2   Separable random sets and inclusion functionals 51
      4.3   Regenerative events 56
      4.4   Robbins' theorem 59
      4.5   Hausdorff dimension 60
      4.6   Random open sets 63
      4.7   C-additive capacities and random convex sets 64
      4.8   Comparison of random sets 67
  5   Calculus with capacities 70
      5.1   Choquet integral 70
      5.2   The Radon–Nikodym theorem for capacities 72

		5.3	Dominating probability measures	75
		5.4	Carathéodory's extension	77
		5.5	Derivatives of capacities	79
	6	Convergence		84
		6.1	Weak convergence	84
		6.2	Convergence almost surely and in probability	90
		6.3	Probability metrics	93
	7	Random sets and hitting processes		97
		7.1	Hitting processes	97
		7.2	Trapping systems	99
		7.3	Distributions of random convex sets	102
	8	Point processes and random measures		105
		8.1	Random sets and point processes	105
		8.2	A representation of random sets as point processes	112
		8.3	Random sets and random measures	115
		8.4	Random capacities	117
		8.5	Robbin's theorem for random capacities	119
	9	Various interpretations of capacities		124
		9.1	Non-additive measures	124
		9.2	Belief functions	127
		9.3	Upper and lower probabilities	129
		9.4	Capacities in robust statistics	132
	Notes to Chapter 1			134
2	**Expectations of Random Sets**			**145**
	1	The selection expectation		145
		1.1	Integrable selections	145
		1.2	The selection expectation	150
		1.3	Applications to characterisation of distributions	160
		1.4	Variants of the selection expectation	161
		1.5	Convergence of the selection expectations	165
		1.6	Conditional expectation	170
	2	Further definitions of expectations		174
		2.1	Linearisation approach	174
		2.2	The Vorob'ev expectation	176
		2.3	Distance average	178
		2.4	Radius-vector expectation	182
	3	Expectations on lattices and in metric spaces		183
		3.1	Evaluations and expectations on lattices	183
		3.2	Fréchet expectation	184
		3.3	Expectations of Doss and Herer	186
		3.4	Properties of expectations	190
	Notes to Chapter 2			190

3 Minkowski Addition ... 195
1 Strong law of large numbers for random sets ... 195
- 1.1 Minkowski sums of deterministic sets ... 195
- 1.2 Strong law of large numbers ... 198
- 1.3 Applications of the strong law of large numbers ... 200
- 1.4 Non-identically distributed summands ... 206
- 1.5 Non-compact summands ... 209

2 The central limit theorem ... 213
- 2.1 A central limit theorem for Minkowski averages ... 213
- 2.2 Gaussian random sets ... 218
- 2.3 Stable random compact sets ... 220
- 2.4 Minkowski infinitely divisible random compact sets ... 221

3 Further results related to Minkowski sums ... 223
- 3.1 Law of iterated logarithm ... 223
- 3.2 Three series theorem ... 224
- 3.3 Komlós theorem ... 226
- 3.4 Renewal theorems for random convex compact sets ... 226
- 3.5 Ergodic theorems ... 230
- 3.6 Large deviation estimates ... 232
- 3.7 Convergence of functionals ... 233
- 3.8 Convergence of random broken lines ... 234
- 3.9 An application to allocation problem ... 235
- 3.10 Infinite divisibility in positive convex cones ... 236

Notes to Chapter 3 ... 237

4 Unions of Random Sets ... 241
1 Union-infinite-divisibility and union-stability ... 241
- 1.1 Extreme values: a reminder ... 241
- 1.2 Infinite divisibility for unions ... 242
- 1.3 Union-stable random sets ... 247
- 1.4 Other normalisations ... 253
- 1.5 Infinite divisibility of lattice-valued random elements ... 258

2 Weak convergence of normalised unions ... 262
- 2.1 Sufficient conditions ... 262
- 2.2 Necessary conditions ... 265
- 2.3 Scheme of series for unions of random closed sets ... 269

3 Convergence with probability 1 ... 270
- 3.1 Regularly varying capacities ... 270
- 3.2 Almost sure convergence of scaled unions ... 272
- 3.3 Stability and relative stability of unions ... 275
- 3.4 Functionals of unions ... 277

4 Convex hulls ... 278
- 4.1 Infinite divisibility with respect to convex hulls ... 278
- 4.2 Convex-stable sets ... 281
- 4.3 Convergence of normalised convex hulls ... 284

XVI   Contents

      5   Unions and convex hulls of random functions 286
          5.1   Random points 286
          5.2   Multivalued mappings 288
      6   Probability metrics method 293
          6.1   Inequalities between metrics 293
          6.2   Ideal metrics and their applications 295
    Notes to Chapter 4 299

5   **Random Sets and Random Functions**........................... 303
      1   Random multivalued functions 303
          1.1   Multivalued martingales 303
          1.2   Set-valued random processes 312
          1.3   Random functions with stochastic domains.............. 319
      2   Levels and excursion sets of random functions 322
          2.1   Excursions of random fields 322
          2.2   Random subsets of the positive half-line and filtrations 325
          2.3   Level sets of strong Markov processes.................. 329
          2.4   Set-valued stopping times and set-indexed martingales 334
      3   Semicontinuous random functions 336
          3.1   Epigraphs of random functions and epiconvergence 336
          3.2   Stochastic optimisation 348
          3.3   Epigraphs and extremal processes 353
          3.4   Increasing set-valued processes of excursion sets 361
          3.5   Strong law of large numbers for epigraphical sums 363
          3.6   Level sums of random upper semicontinuous functions..... 366
          3.7   Graphical convergence of random functions 369
    Notes to Chapter 5 378

    Appendices.. 387
      A   Topological spaces and linear spaces 387
      B   Space of closed sets....................................... 398
      C   Compact sets and the Hausdorff metric 402
      D   Multifunctions and semicontinuity 409
      E   Measures, probabilities and capacities 412
      F   Convex sets 421
      G   Semigroups and harmonic analysis 425
      H   Regular variation 428

References 435

List of Notation 463

Name Index 467

Subject Index... 475

1
Random Closed Sets and Capacity Functionals

1 The Choquet theorem

1.1 Set-valued random elements

As the name suggests, a random set is an object with values being sets, so that the corresponding record space is the space of subsets of a given carrier space. At this stage, a mere definition of a general random element like a random set presents little difficulty as soon as a σ-algebra on the record space is specified. The principal new feature is that random sets may have something inside (different to random variables and random vectors) and the development of this idea is crucial in the studies of random sets. Because the family of all sets is too large, it is usual to consider random closed sets defined as random elements in the space of closed subsets of a certain topological space $\mathbb{E}$. The family of closed subsets of $\mathbb{E}$ is denoted by $\mathcal{F}$, while $\mathcal{K}$ and $\mathcal{G}$ denote respectively the family of all compact and open subsets of $\mathbb{E}$. It is often assumed that $\mathbb{E}$ is a *locally compact Hausdorff second countable* topological space (LCHS space). The Euclidean space $\mathbb{R}^d$ is a generic example of such space $\mathbb{E}$.

Let us fix a complete probability space $(\Omega, \mathfrak{F}, \mathbf{P})$ which will be used throughout to define random elements. It is natural to call an $\mathcal{F}$-valued random element a random closed set. However, one should be more specific about measurability issues, which acquire considerably more importance when studying random elements in complex spaces. In other words, when defining a random element it is necessary to specify which information is available in terms of the observable events from the σ-algebra $\mathfrak{F}$ in Ω. It is essential to ensure that the measurability requirement is restrictive enough to ensure that all functionals of interest become random variables. At the same time, the measurability condition must not be too strict in order to include as many random elements as possible. The following definition describes a rather flexible and useful concept of a random closed set.

Definition 1.1 (Definition of a random closed set). A map $X: \Omega \mapsto \mathcal{F}$ is called a *random closed set* if, for every compact set K in $\mathbb{E}$,

$$\{\omega: X \cap K \neq \emptyset\} \in \mathfrak{F}. \qquad (1.1)$$

Condition (1.1) simply means that observing X one can always say if X hits or misses any given compact set K. In more abstract language, (1.1) says that the map $X\colon \Omega \mapsto \mathcal{F}$ is measurable as a map between the underlying probability space and the space $\mathcal{F}$ equipped with the σ-algebra $\mathfrak{B}(\mathcal{F})$ generated by $\{F \in \mathcal{F}:\ F \cap K \neq \emptyset\}$ for K running through the family $\mathcal{K}$ of compact subsets of $\mathbb{E}$. Note that $\mathfrak{B}(\mathcal{F})$ is called the Effros σ-algebra, which is discussed in greater detail in Section 2.1 for the case of a general Polish space $\mathbb{E}$. As in Appendix B, we write

$$\mathcal{F}_K = \{F \in \mathcal{F}:\ F \cap K \neq \emptyset\}.$$

The σ-algebra generated by $\mathcal{F}_K$ for all K from $\mathcal{K}$ clearly contains

$$\mathcal{F}^K = \{F \in \mathcal{F}:\ F \cap K = \emptyset\}.$$

Furthermore, for every G from the family $\mathcal{G}$ of open sets,

$$\mathcal{F}_G = \{F \in \mathcal{F}:\ F \cap G \neq \emptyset\} = \bigcap_n \mathcal{F}_{K_n},$$

where $\{K_n, n \geq 1\}$ is a sequence of compact sets such that $K_n \uparrow G$ (here the local compactness of $\mathbb{E}$ is essential). Therefore, $\mathcal{F}_G \in \mathfrak{B}(\mathcal{F})$ for all $G \in \mathcal{G}$. It should be noted that the *Fell topology* on $\mathcal{F}$ (discussed in Appendix B) is generated by open sets $\mathcal{F}_G$ for $G \in \mathcal{G}$ and $\mathcal{F}^K$ for $K \in \mathcal{K}$. Therefore, the σ-algebra generated by $\mathcal{F}_K$ for $K \in \mathcal{K}$ coincides with the Borel σ-algebra generated by the Fell topology on $\mathcal{F}$. It is possible to reformulate Definition 1.1 as follows.

Definition 1.1'. A map $X\colon \Omega \mapsto \mathcal{F}$ is called a random closed set if X is measurable with respect to the Borel σ-algebra on $\mathcal{F}$ with respect to the Fell topology, i.e.

$$X^{-1}(\mathcal{X}) = \{\omega:\ X(\omega) \in \mathcal{X}\} \in \mathfrak{F}$$

for each $\mathcal{X} \in \mathfrak{B}(\mathcal{F})$.

Then (1.1) can be formulated as

$$X^{-1}(\mathcal{F}_K) = \{\omega:\ X(\omega) \in \mathcal{F}_K\} \in \mathfrak{F}. \tag{1.2}$$

As in Appendix D, we often write $X^-(K)$ instead of $X^{-1}(\mathcal{F}_K)$. It is easy to see that (1.2) implies the measurability of a number of further events, e.g. $\{X \cap G \neq \emptyset\}$ for every $G \in \mathcal{G}$, $\{X \cap F \neq \emptyset\}$ and $\{X \subset F\}$ for every $F \in \mathcal{F}$.

Since σ-algebra $\mathfrak{B}(\mathcal{F})$ is the Borel σ-algebra with respect to a topology on $\mathcal{F}$, this often leads to the conclusion that $f(X)$ is a random closed set if X is a random closed set and the map $f\colon \mathcal{F} \mapsto \mathcal{F}$ is continuous or semicontinuous (and therefore measurable).

Example 1.2 (Simple examples of random closed sets).
(i) If ξ is a random element in $\mathbb{E}$ (measurable with respect to the Borel σ-algebra on $\mathbb{E}$), then the singleton $X = \{\xi\}$ is a random closed set.

(ii) If ξ is a random variable, then $X = (-\infty, \xi]$ is a random closed set on the line $\mathbb{E} = \mathbb{R}^1$. Indeed, $\{X \cap K \neq \emptyset\} = \{\xi \geq \inf K\}$ is a measurable event for every $K \subset \mathbb{E}$. Along the same line, $X = (-\infty, \xi_1] \times \cdots \times (-\infty, \xi_d]$ is a random closed subset of $\mathbb{R}^d$ if $(\xi_1, \ldots, \xi_n)$ is a d-dimensional random vector.

(iii) If ξ_1, ξ_2, ξ_3 are three random vectors in $\mathbb{R}^d$, then the triangle with vertices ξ_1, ξ_2 and ξ_3 is a random closed set. If ξ is a random vector in $\mathbb{R}^d$ and η is a non-negative random variable, then random ball $B_\eta(\xi)$ of radius η centred at ξ is a random closed set. While it is possible to deduce this directly from Definition 1.1, it is easier to refer to general results established later on in Theorem 2.25.

(iv) Let ζ_x, $x \in \mathbb{E}$, be a real-valued stochastic process on $\mathbb{E}$ with continuous sample paths. Then its level set $X = \{x : \zeta_x = t\}$ is a random closed set for every $t \in \mathbb{R}$. Indeed, $\{X \cap K = \emptyset\} = \{\inf_{x \in K} \zeta_x > t\} \cup \{\sup_{x \in K} \zeta_x < t\}$ is measurable. Similarly, $\{x : \zeta_x \leq t\}$ and $\{x : \zeta_x \geq t\}$ are random closed sets.

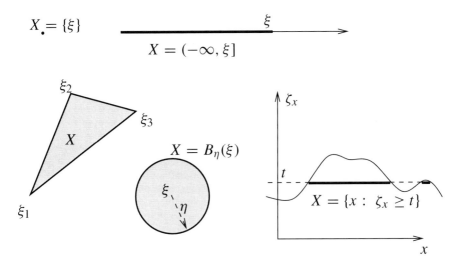

Figure 1.1. Simple examples of random closed sets.

Example 1.3 (Random variables associated with random closed sets).

(i) It is easy to see that the norm $\|X\| = \sup\{\|x\| : x \in X\}$ for a random closed set X in $\mathbb{E} = \mathbb{R}^d$ is a random variable (with possibly infinite values). The event $\{\|X\| > t\}$ means that X hits an open set G being the complement to the closed ball of radius t centred at the origin.

(ii) For every $x \in \mathbb{E}$ the indicator $\mathbf{1}_X(x)$ (equal to 1 if $x \in X$ and to zero otherwise) is a random variable.

(iii) If μ is a locally finite Borel measure on $\mathbb{E}$, then $\mu(X)$ is a random variable. This follows directly from Fubini's theorem since $\mu(X) = \int \mathbf{1}_X(x) \mu(dx)$, see Section 4.4.

If two random closed sets X and Y share the same distribution, then we write $X \stackrel{d}{\sim} Y$. This means that $\mathbf{P}\{X \in \mathcal{X}\} = \mathbf{P}\{Y \in \mathcal{X}\}$ for every measurable family of closed sets $\mathcal{X} \in \mathfrak{B}(\mathcal{F})$.

1.2 Capacity functionals

Definition

The distribution of a random closed set X is determined by $\mathbf{P}(\mathcal{X}) = \mathbf{P}\{X \in \mathcal{X}\}$ for all $\mathcal{X} \in \mathfrak{B}(\mathcal{F})$. The particular choice of $\mathcal{X} = \mathcal{F}_K$ and $\mathbf{P}\{X \in \mathcal{F}_K\} = \mathbf{P}\{X \cap K \neq \emptyset\}$ is useful since the families $\mathcal{F}_K$, $K \in \mathcal{K}$, generate the Borel σ-algebra $\mathfrak{B}(\mathcal{F})$.

Definition 1.4 (Capacity functional). A functional $T_X : \mathcal{K} \mapsto [0, 1]$ given by

$$T_X(K) = \mathbf{P}\{X \cap K \neq \emptyset\}, \quad K \in \mathcal{K}, \tag{1.3}$$

is said to be the *capacity functional* of X. We write $T(K)$ instead of $T_X(K)$ where no ambiguity occurs.

Example 1.5 (Capacity functionals of simple random sets).
  (i) If $X = \{\xi\}$ is a random singleton, then $T_X(K) = \mathbf{P}\{\xi \in K\}$, so that the capacity functional is the probability distribution of ξ.
  (ii) Let $X = \{\xi_1, \xi_2\}$ be the set formed by two independent identically distributed random elements in $\mathbb{E}$. Then $T_X(K) = 1 - (1 - \mathbf{P}\{\xi_1 \in K\})^2$. For instance if ξ_1 and ξ_2 are the numbers shown by two dice, then $T_X(\{6\})$ is the probability that at least one dice shows six.
  (iii) Let $X = (-\infty, \xi]$ be a random closed set in $\mathbb{R}$, where ξ is a random variable. Then $T_X(K) = \mathbf{P}\{\xi > \inf K\}$ for all $K \in \mathcal{K}$.
  (iv) If $X = \{x \in \mathbb{E} : \zeta_x \geq t\}$ for a real-valued sample continuous stochastic process ζ_x, $x \in \mathbb{E}$, then $T_X(K) = \mathbf{P}\{\sup_{x \in K} \zeta_x \geq t\}$.

It follows immediately from the definition of $T = T_X$ that

$$T(\emptyset) = 0, \tag{1.4}$$

and

$$0 \leq T(K) \leq 1, \quad K \in \mathcal{K}. \tag{1.5}$$

Since $\mathcal{F}_{K_n} \downarrow \mathcal{F}_K$ as $K_n \downarrow K$, the continuity property of the probability measure $\mathbf{P}$ implies that T is *upper semicontinuous* (see Proposition D.7), i.e.

$$T(K_n) \downarrow T(K) \quad \text{as } K_n \downarrow K \text{ in } \mathcal{K}. \tag{1.6}$$

Properties (1.4) and (1.6) mean that T is a (topological) precapacity that can be extended to the family of all subsets of $\mathbb{E}$ as described in Appendix E.

It is easy to see that the capacity functional T is *monotone*, i.e.

$$T(K_1) \leq T(K_2) \quad \text{if } K_1 \subset K_2.$$

1 The Choquet theorem

Moreover, T satisfies a stronger monotonicity property described below. With every functional T defined on a family of (compact) sets we can associate the following *successive differences*:

$$\Delta_{K_1} T(K) = T(K) - T(K \cup K_1), \tag{1.7}$$

$$\Delta_{K_n} \cdots \Delta_{K_1} T(K) = \Delta_{K_{n-1}} \cdots \Delta_{K_1} T(K)$$
$$- \Delta_{K_{n-1}} \cdots \Delta_{K_1} T(K \cup K_n), \quad n \geq 2. \tag{1.8}$$

If T from (1.3) is a capacity functional of X, then

$$\Delta_{K_1} T(K) = \mathbf{P}\{X \cap K \neq \emptyset\} - \mathbf{P}\{X \cap (K \cup K_1) \neq \emptyset\}$$
$$= -\mathbf{P}\{X \cap K_1 \neq \emptyset, \ X \cap K = \emptyset\}.$$

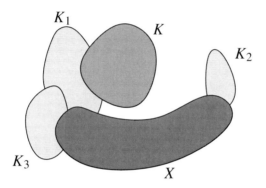

Figure 1.2. Set X from $\mathcal{F}^K_{K_1, K_2, K_3}$.

Applying this argument consecutively yields an important relationship between the higher-order successive differences and the distribution of X

$$-\Delta_{K_n} \cdots \Delta_{K_1} T(K) = \mathbf{P}\{X \cap K = \emptyset, \ X \cap K_i \neq \emptyset, \ i = 1, \ldots, n\}$$
$$= \mathbf{P}\left\{X \in \mathcal{F}^K_{K_1, \ldots, K_n}\right\}, \tag{1.9}$$

where

$$\mathcal{F}^K_{K_1, \ldots, K_n} = \{F \in \mathcal{F}: \ F \cap K = \emptyset, \ F \cap K_1 \neq \emptyset, \ldots, F \cap K_n \neq \emptyset\},$$

see Figure 1.2. In particular, (1.9) implies

$$\Delta_{K_n} \cdots \Delta_{K_1} T(K) \leq 0 \tag{1.10}$$

for all $n \geq 1$ and $K, K_1, \ldots, K_n \in \mathcal{K}$.

Example 1.6 (Higher-order differences).
(i) Let $X = \{\xi\}$ be a random singleton with distribution $\mathbf{P}$. Then

$$-\Delta_{K_n}\cdots\Delta_{K_1}T(K) = \mathbf{P}\left\{\xi \in (K_1 \cap \cdots \cap K_n \cap K^c)\right\}.$$

(ii) Let $X = (-\infty, \xi_1] \times (-\infty, \xi_2]$ be a random closed set in the plane $\mathbb{R}^2$. Then $-\Delta_{\{x\}}T(\{y,z\})$ for $x = (a,c)$, $y = (b,c)$, $z = (a,d)$ is the probability that ξ lies in the rectangle $[a,b) \times [c,d)$, see Figure 1.3.

(iii) Let $X = \{x : \zeta_x \geq 0\}$ for a continuous random function ζ. Then

$$-\Delta_{K_n}\cdots\Delta_{K_1}T(K) = \mathbf{P}\left\{\sup_{x \in K} \zeta_x < 0,\ \sup_{x \in K_i} \zeta_x \geq 0,\ i = 1,\ldots,n\right\}.$$

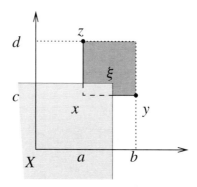

Figure 1.3. Random closed set from Example 1.6(ii).

The properties of the capacity functional T resemble those of the distribution function. The upper semicontinuity property (1.6) is similar to the right-continuity, and (1.10) generalises the monotonicity concept. However, in contrast to measures, functional T is not additive, but only *subadditive*, i.e.

$$T(K_1 \cup K_2) \leq T(K_1) + T(K_2) \tag{1.11}$$

for all compact sets K_1 and K_2.

Example 1.7 (Non-additive capacity functional). If $X = B_r(\xi)$ is the ball of radius r centred at a random point ξ in $\mathbb{R}^d$, then $T_X(K) = \mathbf{P}\{\xi \in K^r\}$, which is not a measure, since the r-envelopes K_1^r and K_2^r are not necessarily disjoint for disjoint K_1 and K_2.

Complete alternation and monotonicity

Because of the importance of properties (1.6) and (1.10) it is natural to consider general functionals on $\mathcal{K}$ that satisfy these properties without immediate reference to distributions of random closed sets. A real-valued functional φ on $\mathcal{K}$ which satisfies (1.4), (1.5), (1.6) and (1.10) is said to be a *capacity functional*. In other words, a capacity functional is a functional on $\mathcal{K}$ which takes values in [0, 1], equals 0 on the empty set and is upper semicontinuous and completely alternating on $\mathcal{K}$. The latter concept is addressed in the following definition.

Definition 1.8 (Completely alternating and completely ∪-monotone functionals). Let $\mathcal{D}$ be a family of sets which is closed under finite unions (so that $M_1 \cup M_2 \in \mathcal{D}$ if $M_1, M_2 \in \mathcal{D}$). A real-valued functional φ defined on $\mathcal{D}$ is said to be
(i) *completely alternating* or *completely ∪-alternating* (notation $\varphi \in \mathsf{A}(\mathcal{D})$ or $\varphi \in \mathsf{A}_\cup(\mathcal{D})$) if
$$\Delta_{K_n} \cdots \Delta_{K_1} \varphi(K) \leq 0, \quad n \geq 1, \ K, K_1, \ldots, K_n \in \mathcal{D}. \quad (1.12)$$

If (1.12) holds for all $n \leq m$, then φ is said to be alternating of degree m (or m-alternating).
(ii) *completely ∪-monotone* (notation $\varphi \in \mathsf{M}_\cup(\mathcal{D})$) if
$$\Delta_{K_n} \cdots \Delta_{K_1} \varphi(K) \geq 0, \quad n \geq 1, \ K, K_1, \ldots, K_n \in \mathcal{D}.$$

As (1.10) shows, the capacity functional T is completely alternating. Definition 1.8 is usually applied to the case when $\mathcal{D} = \mathcal{K}$. It complies with Definition G.5 applied to the semigroup $\mathcal{D}$ with the union being the semigroup operation. Another natural semigroup operation is the intersection of sets, which leads to other (however closely related) concepts of alternating and monotone functionals. Similar to the definition of $\Delta_{K_n} \cdots \Delta_{K_1} \varphi(K)$, we introduce the following successive differences

$$\nabla_{K_1} \varphi(K) = \varphi(K) - \varphi(K \cap K_1), \quad (1.13)$$
$$\nabla_{K_n} \cdots \nabla_{K_1} \varphi(K) = \nabla_{K_{n-1}} \cdots \nabla_{K_1} \varphi(K)$$
$$\quad - \nabla_{K_{n-1}} \cdots \nabla_{K_1} \varphi(K \cap K_n), \quad n \geq 2. \quad (1.14)$$

The following definition is a direct counterpart of Definition 1.8.

Definition 1.9 (Completely ∩-alternating and completely monotone functionals). Let $\mathcal{D}$ be a family of sets which is closed under finite intersections. A real-valued functional φ defined on $\mathcal{D}$ is said to be
(i) *completely ∩-alternating* (notation $\varphi \in \mathsf{A}_\cap(\mathcal{D})$) if
$$\nabla_{K_n} \cdots \nabla_{K_1} \varphi(K) \leq 0, \quad n \geq 1, \ K, K_1, \ldots, K_n \in \mathcal{D};$$

(ii) *completely monotone* or *completely ∩-monotone* (notation $\varphi \in \mathsf{M}(\mathcal{D})$ or $\varphi \in \mathsf{M}_\cap(\mathcal{D})$) if
$$\nabla_{K_n} \cdots \nabla_{K_1} \varphi(K) \geq 0, \quad n \geq 1, \ K, K_1, \ldots, K_n \in \mathcal{D}.$$

When saying that φ is completely alternating we always mean that φ is completely ∪-alternating, while φ being completely monotone means that φ is completely ∩-monotone. For every functional φ on $\mathcal{D}$ with values in [0, 1], its *dual* functional

$$\tilde{\varphi}(K) = 1 - \varphi(K^c), \quad K^c \in \mathcal{D}, \tag{1.15}$$

is defined on the family $\mathcal{D}' = \{K^c : K \in \mathcal{D}\}$ of complements to the sets from $\mathcal{D}$.

Proposition 1.10. *Let $\varphi \colon \mathcal{D} \mapsto [0, 1]$. Then*
(i) *$\varphi \in \mathsf{A}_\cup(\mathcal{D})$ if and only if, for any fixed $L \in \mathcal{D}$,*

$$-\Delta_L \varphi(K) = \varphi(K \cup L) - \varphi(K) \in \mathsf{M}_\cup(\mathcal{D});$$

(ii) *$\varphi \in \mathsf{A}_\cap(\mathcal{D})$ if and only if, for any fixed $L \in \mathcal{D}$,*

$$-\nabla_L \varphi(K) = \varphi(K \cap L) - \varphi(K) \in \mathsf{M}_\cap(\mathcal{D}).$$

(iii) *Let $\varphi \colon \mathcal{D} \mapsto [0, 1]$. Then $\varphi \in \mathsf{A}_\cup(\mathcal{D})$ (respectively $\varphi \in \mathsf{A}_\cap(\mathcal{D})$) if and only functional $\tilde{\varphi}(K) \in \mathsf{M}_\cap(\mathcal{D}')$ (respectively $\tilde{\varphi}(K) \in \mathsf{M}_\cup(\mathcal{D}')$) for the dual functional $\tilde{\varphi}$ on $\mathcal{D}' = \{K^c : K \in \mathcal{D}\}$.*

Proof. (i) It suffices to note that

$$\Delta_{K_n} \cdots \Delta_{K_1}(-\Delta_L \varphi(K)) = -\Delta_L \Delta_{K_n} \cdots \Delta_{K_1} \varphi(K)$$

with a similar relationship valid for the successive differences based on intersections. Statement (ii) is proved similarly. The proof of (iii) is a matter of verification that

$$\Delta_{K_n} \cdots \Delta_{K_1} \tilde{\varphi}(K) = -\nabla_{K_n^c} \cdots \nabla_{K_1^c} \varphi(K^c). \qquad \square$$

Alternation and monotonicity of capacity functionals

Every measure μ is a completely alternating functional, since

$$-\Delta_{K_n} \cdots \Delta_{K_1} \mu(K) = \mu((K_1 \cup \cdots \cup K_n) \setminus K) \geq 0.$$

In particular, $\Delta_{K_1} \mu(K) = -\mu(K_1)$ if K and K_1 are disjoint.

Note that φ is increasing if and only if

$$\Delta_{K_1} \varphi(K) = \varphi(K) - \varphi(K \cup K_1)$$

is non-positive. Furthermore, for $n = 2$,

$$\Delta_{K_2} \Delta_{K_1} \varphi(K) = \varphi(K) - \varphi(K \cup K_1) - \varphi(K \cup K_2) + \varphi(K \cup K_1 \cup K_2).$$

Therefore, (1.12) for $n = 2$ is equivalent to

$$\varphi(K) + \varphi(K \cup K_1 \cup K_2) \leq \varphi(K \cup K_1) + \varphi(K \cup K_2). \tag{1.16}$$

In particular, if $K = \emptyset$ and $\varphi(\emptyset) = 0$, then

$$\varphi(K_1 \cup K_2) \leq \varphi(K_1) + \varphi(K_2), \tag{1.17}$$

meaning that φ is subadditive. Clearly, if $\varphi = \mu$ is a measure, then (1.17) turns into an equality for disjoint K_1 and K_2. For an increasing φ, inequality (1.16) is equivalent to
$$\varphi(K_1 \cap K_2) + \varphi(K_1 \cup K_2) \leq \varphi(K_1) + \varphi(K_2) \tag{1.18}$$
for all K_1 and K_2. A functional φ satisfying (1.18) is called *concave* or strongly subadditive. Functionals satisfying the reverse inequality in (1.18) are called *convex* or strongly superadditive. If only $\Delta_{K_1}\varphi(K)$ and $\Delta_{K_2}\Delta_{K_1}\varphi(K)$ are non-positive, then φ is called 2-alternating. Therefore, φ is 2-alternating if it is both concave and monotone.

According to Definition E.8, a function $\varphi \colon \mathcal{P} \mapsto [-\infty, +\infty]$ on the family $\mathcal{P}$ of all subsets of $\mathbb{E}$ is called a *capacity* (or $\mathcal{K}$-capacity) if it satisfies the following conditions:

(i) $M \subset M'$ implies $\varphi(M) \leq \varphi(M')$;
(ii) $M_n \uparrow M$ implies $\varphi(M_n) \uparrow \varphi(M)$;
(iii) $K_n \downarrow K$ for compact sets K_n, K implies $\varphi(K_n) \downarrow \varphi(K)$.

Definition 1.8 singles out those capacities which are completely alternating or completely monotone. Since the family $\mathcal{K}$ forms a *semigroup* with union being the semigroup operation and the neutral element being the empty set, it is possible to use the results of Appendix G within this context. It follows from Theorem G.6 that each completely alternating capacity is *negative definite* on $\mathcal{K}$. Theorem G.8 states that $\varphi \in \mathsf{A}_\cup(\mathcal{K})$ (respectively $\varphi \in \mathsf{A}_\cap(\mathcal{K})$) if and only if $e^{-t\varphi} \in \mathsf{M}_\cup(\mathcal{K})$ (respectively $e^{-t\varphi} \in \mathsf{M}_\cap(\mathcal{K})$) for all $t > 0$. Let us formulate one particularly important corollary of this fact.

Proposition 1.11. *If φ is a completely alternating non-negative capacity with possibly infinite values, then $T(K) = 1 - e^{-\varphi(K)}$ is a completely alternating capacity with values in $[0, 1]$.*

Proposition 1.11 is often used to construct a capacity functional from a completely alternating upper semicontinuous capacity that may take values greater than 1. The random closed set with the capacity functional T from Proposition 1.11 is infinite divisible for unions, see Chapter 4.

Extension of capacity functional

As explained in Appendix E, a capacity φ defined on $\mathcal{K}$ can be naturally extended onto the family $\mathcal{P}$ of all subsets of $\mathbb{E}$ keeping alternation or the monotonicity properties enjoyed by φ. In its application to capacity functionals of random closed sets, put
$$T^*(G) = \sup\{T(K) : K \in \mathcal{K}, K \subset G\}, \quad G \in \mathcal{G}, \tag{1.19}$$
and
$$T^*(M) = \inf\{T^*(G) : G \in \mathcal{G}, G \supset M\}, \quad M \in \mathcal{P}. \tag{1.20}$$

Theorem 1.12 (Consistency of extension).
 (i) $T^*(K) = T(K)$ for each $K \in \mathcal{K}$.
 (ii) For each Borel set B,
$$T^*(B) = \sup\{T(K) : K \in \mathcal{K}, K \subset B\}.$$

Proof. The first statement follows from the upper semicontinuity of T. Note that $T^*(K)$ is a limit of $T^*(G_n)$ for a sequence of open sets $G_n \downarrow K$. By choosing $K_n \in \mathcal{K}$ such that $K \subset K_n \subset G_n$ we deduce that $T(K_n) \downarrow T^*(K)$, while at the same time $T(K_n) \downarrow T(K)$ since T is upper semicontinuous. The second statement is a corollary from the more intricate Choquet capacitability theorem, see Theorem E.9. □

Since the extension T^* coincides with T on $\mathcal{K}$, in the following we use the same notation T to denote the extension, i.e. $T(G)$ or $T(B)$ denotes the values of the extended T on arbitrary open set G and Borel set B. Theorem 1.12 and the continuity property of probability measures imply $T(B) = \mathbf{P}\{X \cap B \neq \emptyset\}$ for all Borel B.

The Choquet theorem

Since the σ-algebra $\mathfrak{B}(\mathcal{F})$ is rich, it is difficult to explicitly assign a measure to its elements. Nonetheless, since the σ-algebra $\mathfrak{B}(\mathcal{F})$ is generated by the families $\mathcal{F}_K$, $K \in \mathcal{K}$, it is quite natural to expect that a capacity functional on $\mathcal{K}$ determines uniquely the distribution of a random closed set. The following fundamental theorem singles out upper semicontinuous completely alternating capacities on $\mathcal{K}$ as those which correspond to distributions of random closed sets. The uniqueness part easily follows from the fact that σ-algebra $\mathfrak{B}(\mathcal{F})$ is generated by $\mathcal{F}_K$ for $K \in \mathcal{K}$. It is the existence part that is more complicated. The proof of the Choquet theorem is presented in Section 1.3.

Theorem 1.13 (Choquet theorem). *Let $\mathbb{E}$ be a LCHS space. A functional $T: \mathcal{K} \mapsto [0,1]$ such that $T(\emptyset) = 0$ is the capacity functional of a (necessarily unique) random closed set in $\mathbb{E}$ if and only if T is upper semicontinuous and completely alternating.*

The following results follow from the uniqueness part of the Choquet theorem.

Proposition 1.14. *Let $\mathbb{E}$ be a LCHS space.*
 (i) *The capacity functional T_X of a random closed set X is a probability measure if and only if X is a random singleton.*
 (ii) *T_X is a sub-probability measure (i.e. a measure with the total mass not exceeding 1) if and only if X with probability 1 consists of at most a single point, i.e.* $\mathbf{P}\{\mathrm{card}(X) > 1\} = 0$.
 (iii) *A random closed set X is deterministic if and only if $T_X(K)$ takes only values 0 or 1 for each $K \in \mathcal{K}$.*

Proposition 1.14(iii) (and the uniqueness part of the Choquet theorem) does not hold in an arbitrary (e.g. not locally compact) space $\mathbb{E}$. For instance, if $\mathbb{E} = \mathbb{R}$ with the discrete metric, then compact sets are necessarily finite, so that $T_X(K) = 0$ for each $K \in \mathcal{K}$ if $X = \{\xi\}$ is a random singleton with a non-atomic distribution.

Maxitive capacity functionals

A functional T is said to be *maxitive* if

$$T(K_1 \cup K_2) = \max(T(K_1), T(K_2)) \qquad (1.21)$$

for all compact sets K_1 and K_2. Maxitive functionals arise naturally in the theory of extremal processes, see Norberg [430, 431]. Every sup-measure (defined in Appendix E) is maxitive, while the converse is false since the definition of sup-measures involves taking a supremum over an arbitrary family of sets on the right-hand side of (1.21). If T is maxitive on $\mathcal{K}$, then (1.21) also holds for the extension of T onto the family of open sets and all subsets of $\mathbb{E}$.

Example 1.15 (Maxitive capacity). Define a *maxitive capacity* T by

$$T(K) = \sup\{f(x) : x \in K\}, \qquad (1.22)$$

where $f : \mathbb{E} \mapsto [0, 1]$ is an upper semicontinuous function. Then $T = f^{\vee}$ is the sup-integral of f as defined in Appendix E. This capacity functional T describes the distribution of the random closed set $X = \{x \in \mathbb{E} : f(x) \geq \alpha\}$, where α is a random variable uniformly distributed on $[0, 1]$.

The following proposition shows that Example 1.15 actually describes all maxitive capacities that correspond to distributions of random closed sets. In a sense, the upper semicontinuity assumption makes it possible to move from finite maximum in (1.21) to a general supremum over all singletons.

Proposition 1.16 (Maxitive upper semicontinuous capacities). *If T is a maxitive upper semicontinuous functional with values in $[0, 1]$, then T is given by (1.22) for an upper semicontinuous function $f : \mathbb{E} \mapsto [0, 1]$.*

Proof. Since T is upper semicontinuous, $f(x) = T(\{x\})$ is an upper semicontinuous function and $T(K_n) \downarrow T(\{x\})$ if $K_n \downarrow \{x\}$. This implies that for each $x \in \mathbb{E}$ and any $\varepsilon > 0$ there exists a neighbourhood $G_\varepsilon(x)$ of x such that $T(G_\varepsilon(x)) < f(x) + \varepsilon$. Every $K \in \mathcal{K}$ is covered by $G_\varepsilon(x)$, $x \in K$, so that K has a finite subcover of $G_\varepsilon(x_1), \ldots, G_\varepsilon(x_n)$. Then (1.21) implies

$$T(K) \leq \max(T(G_\varepsilon(x_1)), \ldots, T(G_\varepsilon(x_n))) \leq \max(f(x_1), \ldots, f(x_n)) + \varepsilon,$$

whence (1.22) immediately holds. □

Proposition 1.16 means that together with the upper semicontinuity assumption, (1.21) implies that T is a sup measure. If (1.21) holds for all K_1 and K_2 from a family of sets $\mathcal{D}$ closed under finite unions, then T is called maxitive on $\mathcal{D}$.

Theorem 1.17 (Complete alternation of a maxitive capacity). *Every functional φ maxitive on a family $\mathcal{D}$ closed under finite unions is completely alternating on $\mathcal{D}$.*

Proof. Consider arbitrary $K, K_1, K_2, \ldots \in \mathcal{D}$. Let us prove by induction that

$$\Delta_{K_n} \cdots \Delta_{K_1} \varphi(K) = \varphi(K) - \varphi(K \cup K_1) \qquad (1.23)$$

given that $\varphi(K_1) = \min(\varphi(K_i), i = 1, \ldots, n)$. This fact is evident for $n = 1$. Assume that $\varphi(K_1) = \min(\varphi(K_i), i = 1, \ldots, n+1)$. Using the induction assumption, it is easy to see that

$$\Delta_{K_{n+1}} \cdots \Delta_{K_1} \varphi(K) = \Delta_{K_n} \cdots \Delta_{K_1} \varphi(K) - \Delta_{K_n} \cdots \Delta_{K_1} \varphi(K \cup K_{n+1})$$
$$= [\varphi(K) - \varphi(K \cup K_1)] - [\varphi(K \cup K_{n+1}) - \varphi(K \cup K_{n+1} \cup K_1)].$$

By the maxitivity assumption and the choice of K_1,

$$\varphi(K \cup K_{n+1}) - \varphi(K \cup K_{n+1} \cup K_1)$$
$$= \max(\varphi(K), \varphi(K_{n+1})) - \max(\varphi(K), \varphi(K_{n+1}), \varphi(K_1)) = 0.$$

Now the monotonicity of φ implies that the left-hand side of (1.23) is non-positive, i.e. φ is completely alternating. □

For example, the *Hausdorff dimension* is a maxitive functional on sets in $\mathbb{R}^d$, and so is completely alternating. However, it is not upper semicontinuous, whence there is no random closed set whose capacity functional is the Hausdorff dimension.

Independence and conditional distributions

Definition 1.18 (Independent random sets). Random closed sets $X_1, \ldots, X_n$ are said to be *independent* if

$$\mathbf{P}\{X_1 \in \mathcal{X}_1, \ldots, X_n \in \mathcal{X}_n\} = \mathbf{P}\{X_1 \in \mathcal{X}_1\} \cdots \mathbf{P}\{X_n \in \mathcal{X}_n\}$$

for all $\mathcal{X}_1, \ldots, \mathcal{X}_n \in \mathfrak{B}(\mathcal{F})$.

The Choquet theorem can be used to characterise independent random closed sets in a LCHS space.

Proposition 1.19. *Random closed sets $X_1, \ldots, X_n$ are independent if and only if*

$$\mathbf{P}\{X_1 \cap K_1 \neq \emptyset, \ldots, X_n \cap K_n \neq \emptyset\} = \prod_{i=1}^{n} T_{X_i}(K_i)$$

for all $K_1, \ldots, K_n \in \mathcal{K}$.

Conditional distributions of random sets can be derived in the same way as conditional distributions of random elements in an abstract measurable space. However, this is not the case for conditional expectation, as the latter refers to a linear structure on the space of sets, see Chapter 2.

If $\mathfrak{H}$ is a sub-σ-algebra of $\mathfrak{F}$, then the conditional probabilities $T_X(K|\mathfrak{H}) = \mathbf{P}\{X \cap K \neq \emptyset | \mathfrak{H}\}$ are defined in the usual way. As noticed in Section 1.4, it suffices to define the capacity functional on a countable family $\mathcal{A}$ of compact sets, which simplifies the measurability issues. The family $T_X(K|\mathfrak{H})$, $K \in \mathcal{A}$, is a random capacity functional that defines the conditional distribution X given $\mathfrak{H}$.

1.3 Proofs of the Choquet theorem

Measure-theoretic proof

The proof given by Matheron [381] is based on the routine application of the measure-theoretic arguments related to extension of measures from algebras to σ-algebras. In fact, the idea goes back to the fundamental paper by Choquet [98] and his theorem on characterisation of positive definite functionals on cones. Here we discuss only sufficiency, since the necessity is evident from the explanations provided in Section 1.2.

Let us start with several auxiliary lemmas. The first two are entirely non-topological and their proofs do not refer to any topological assumption on the carrier space $\mathbb{E}$.

Lemma 1.20. *Let $\mathcal{V}$ be a family of subsets of $\mathbb{E}$ which contains $\emptyset$ and is closed under finite unions. Let $\mathfrak{V}$ be the family which is closed under finite intersections and generated by $\mathcal{F}_V$ and $\mathcal{F}^V$ for $V \in \mathcal{V}$. Then $\mathfrak{V}$ is an algebra and each non-empty $\mathcal{Y} \in \mathfrak{V}$ can be represented as*

$$\mathcal{Y} = \mathcal{F}^V_{V_1,\ldots,V_n} \tag{1.24}$$

for some $n \geq 0$ and $V, V_1, \ldots, V_n \in \mathcal{V}$ with $V_i \not\subset V \cup V_j$ for $i \neq j$ (then (1.24) is said to be a reduced representation of $\mathcal{Y}$). If $\mathcal{Y} = \mathcal{F}^{V'}_{V'_1,\ldots,V'_k}$ is another reduced representation of $\mathcal{Y}$, then $V = V'$, $n = k$, and for each $i \in \{1,\ldots,n\}$ there exists $j_i \in \{1,\ldots,n\}$ such that $V \cup V_i = V \cup V'_{j_i}$.

Proof. The family $\mathfrak{V}$ is closed under finite intersections and $\emptyset = \mathcal{F}_\emptyset \in \mathfrak{V}$. If $\mathcal{Y} \in \mathfrak{V}$, then the complement to $\mathcal{Y}$,

$$\mathcal{F} \setminus \mathcal{Y} = \mathcal{F}_V \cup \mathcal{F}^{V \cup V_1} \cup \mathcal{F}^{V \cup V_2}_{V_1} \cup \cdots \cup \mathcal{F}^{V \cup V_n}_{V_1,\ldots,V_{n-1}},$$

is a finite union of sets from $\mathfrak{V}$. Hence $\mathfrak{V}$ is an algebra.

If $\mathcal{Y}$ satisfies (1.24) with $V_i \subset V \cup V_j$ for some $i \neq j$, then the set V_j can be eliminated without changing $\mathcal{Y}$. Therefore, a reduced representation of $\mathcal{Y}$ exists. Consider two reduced representations of a non-empty $\mathcal{Y}$. Without loss of generality assume that there exists a point $x \in V' \setminus V$. Since $\mathcal{Y} \neq \emptyset$, there exist k points (some of them may be identical) $x_1, \ldots, x_k$ such that $x_j \in V'_j \setminus V'$, $1 \leq j \leq k$ and

$$\{x_1,\ldots,x_k\} \in \mathcal{F}^{V'}_{V'_1,\ldots,V'_k} = \mathcal{Y} = \mathcal{F}^V_{V_1,\ldots,V_n}.$$

Since $x \notin V$, we have $\{x, x_1, \ldots, x_k\} \in \mathcal{F}^V_{V_1,\ldots,V_n}$. At the same time, $x \in V'$, whence $\{x, x_1, \ldots, x_k\} \notin \mathcal{F}^{V'}_{V'_1,\ldots,V'_k}$. The obtained contradiction shows that $V = V'$.

Choose $y \in V_n \setminus V$ and $y_i \in V_i \setminus (V \cup V_n)$, $i = 1, \ldots, n-1$. Since $\{y_1, \ldots, y_{n-1}\} \notin \mathcal{Y}$ and $\{y, y_1, \ldots, y_{n-1}\} \in \mathcal{Y}$, there exists $j_n \in \{1, \ldots, k\}$ such that $y \in V'_{j_n}$ and $y_i \notin V'_{j_n}$ for $i = 1, \ldots, n-1$. For any other point $y' \in V_n \setminus V$ we similarly conclude that $y' \in V'_{j_n}$, whence $V_n \setminus V \subset V'_{j_n}$ and

$$V_n \subset V \cup V'_{j_n}.$$

Using identical arguments in the other direction we obtain $V'_{j_n} \setminus V \subset V_{i_n}$. If $i_n \neq n$, this leads to $V_n \subset V_{i_n} \cup V$ and so contradicts the assumption that $\mathcal{Y}$ has a reduced representation. Thus, $i_n = n$ and $V_n \setminus V = V_{j_n} \setminus V$. The proof is finished by repeating these arguments for every other set V_i, $i = 1, \ldots, n-1$. □

Lemma 1.21. *In the notation of Lemma 1.20, let T be a completely alternating functional on $\mathcal{V}$ such that $T(\emptyset) = 0$, $0 \leq T \leq 1$. Then there exists a unique additive map $\mathbf{P} : \mathfrak{V} \mapsto [0,1]$ such that $\mathbf{P}(\emptyset) = 0$ and $\mathbf{P}(\mathcal{F}_V) = T(V)$ for all $V \in \mathcal{V}$. This map is given by*

$$\mathbf{P}(\mathcal{Y}) = -\Delta_{V_n} \cdots \Delta_{V_1} T(V), \tag{1.25}$$

where $\mathcal{Y} = \mathcal{F}^V_{V_1,\ldots,V_n}$ is any representation of $\mathcal{Y} \in \mathfrak{V}$.

Proof. By the additivity property, we get

$$\mathbf{P}(\mathcal{F}^V_{V_1,\ldots,V_n}) = \mathbf{P}(\mathcal{F}^V_{V_1,\ldots,V_{n-1}}) - \mathbf{P}(\mathcal{F}^{V \cup V_n}_{V_1,\ldots,V_{n-1}}), \tag{1.26}$$

which immediately shows that the only additive extension of $\mathbf{P}(\mathcal{F}_V) = T(V)$ is given by (1.25). It is easy to show that the right-hand side of (1.25) retains its value if any representation of $\mathcal{Y}$ is replaced by its reduced representation. Furthermore,

$$\Delta_{V_n} \cdots \Delta_{V_1} T(V) = \Delta_{V_n \cup V} \cdots \Delta_{V_1 \cup V} T(V),$$

which, together with Lemma 1.20, show that $\mathbf{P}(\mathcal{Y})$ is identical for any reduced representation of $\mathcal{Y}$. The function $\mathbf{P}$ is non-negative since T is completely alternating and $\mathbf{P}(\emptyset) = \mathbf{P}(\mathcal{F}_\emptyset) = T(\emptyset) = 0$. Furthermore, (1.26) implies

$$\mathbf{P}(\mathcal{F}^V_{V_1,\ldots,V_n}) \leq \mathbf{P}(\mathcal{F}^V_{V_1,\ldots,V_{n-1}}) \leq \cdots \leq \mathbf{P}(\mathcal{F}^V) = 1 - T(V) \leq 1.$$

It remains to show that P is additive. Let $\mathcal{Y}$ and $\mathcal{Y}'$ be two disjoint non-empty elements of $\mathfrak{V}$ with the reduced representations

$$\mathcal{Y} = \mathcal{F}^V_{V_1,\ldots,V_n}, \quad \mathcal{Y}' = \mathcal{F}^{V'}_{V'_1,\ldots,V'_k},$$

such that $\mathcal{Y} \cup \mathcal{Y}' \in \mathfrak{V}$. Since

$$\mathcal{Y} \cap \mathcal{Y}' = \mathcal{F}^{V \cup V'}_{V_1,\ldots,V_n,V'_1,\ldots,V'_k} = \emptyset,$$

without loss of generality assume that $V_n \subset V \cup V'$. Since $\mathcal{Y} \cup \mathcal{Y}' \in \mathfrak{V}$, this union itself has a reduced representation

$$\mathcal{Y} \cup \mathcal{Y}' = \mathcal{F}^{V''}_{V''_1,\ldots,V''_m}.$$

If $V = \mathbb{E}$, then $\mathcal{Y} = \{\emptyset\}$ if all subscripts in the representation of $\mathcal{Y}$ are empty, or $\mathcal{Y} = \emptyset$ otherwise, so that the additivity is trivial. Assume that there exists $x \notin V$

and $x_i \in V_i \setminus V$, $i = 1, \ldots, n$. Then $F = \{x, x_1, \ldots, x_n\} \in \mathcal{Y}$. Since $F \in \mathcal{Y} \cup \mathcal{Y}'$, we have $F \cap V'' = \emptyset$, i.e. $x \notin V''$. Therefore, $V'' \subset V$. Similar arguments lead to $V'' \subset V'$, whence
$$V'' \subset (V \cap V').$$

Let us show that $V'' = V$. Assume that there exist points $x \in V \setminus V''$ and $x' \in V' \setminus V''$. Choose points $x_i'' \in V_i'' \setminus V''$ for $i = 1, \ldots, m$. Then $\{x, x', x_1'', \ldots, x_m''\} \in \mathcal{Y} \cup \mathcal{Y}'$, so that $\{x, x'\} \cap V = \emptyset$ or $\{x, x'\} \cap V' = \emptyset$. Since both these statements lead to contradictions, we conclude that $V = V''$ or $V' = V''$. The latter is impossible, since then $V_n \subset V \cup V' = V$ leads to $\mathcal{Y} = \emptyset$. Therefore, $V = V''$, $V \subset V'$ and $V_n \subset V'$.

For each $F \in \mathcal{Y} \cup \mathcal{Y}'$, the condition $F \cap V_n \neq \emptyset$ yields $F \notin \mathcal{Y}'$, while $F \cap V_n = \emptyset$ implies $F \in \mathcal{Y}'$. Thus,
$$\mathcal{Y} = (\mathcal{Y} \cup \mathcal{Y}') \cap \mathcal{F}_{V_n} = \mathcal{F}^V_{V_1'', \ldots, V_m'', V_n},$$
$$\mathcal{Y}' = (\mathcal{Y} \cup \mathcal{Y}') \cap \mathcal{F}^{V_n} = \mathcal{F}^{V \cup V_n}_{V_1'', \ldots, V_m''}.$$

Then
$$-\mathbf{P}(\mathcal{Y}) = \Delta_{V_n} \Delta_{V_m''} \cdots \Delta_{V_1''} T(V)$$
$$= \Delta_{V_m''} \cdots \Delta_{V_1''} T(V) - \Delta_{V_m''} \cdots \Delta_{V_1''} T(V \cup V_n)$$
$$= -\mathbf{P}(\mathcal{Y} \cup \mathcal{Y}') + \mathbf{P}(\mathcal{Y}'),$$

which implies the additivity of $\mathbf{P}$ on $\mathfrak{V}$. □

The following lemma uses the upper semicontinuity assumption on T and the local compactness of $\mathbb{E}$.

Lemma 1.22. *Let T be a completely alternating upper semicontinuous functional on $\mathcal{K}$. By the same letter denote its extension defined by (1.19) and (1.20). Consider any two open sets G and G_0, any $K \in \mathcal{K}$, a sequence $\{K_n, n \geq 1\} \subset \mathcal{K}$ such that $K_n \uparrow G$ and a sequence $\{G_n, n \geq 1\} \subset \mathcal{G}$ such that $G_n \downarrow K$ and $G_n \supset \mathrm{cl}(G_{n+1}) \in \mathcal{K}$ for every $n \geq 1$. Then*
$$T(G_0 \cup K \cup G) = \lim_{n \to \infty} T(G_0 \cup G_n \cup K_n).$$

Proof. Since T is monotone,
$$T(G_0 \cup K \cup K_n) \leq T(G_0 \cup G_n \cup K_n) \leq T(G_0 \cup G_n \cup G).$$

For each open $G' \supset G_0 \cup G \cup K$ we have $G' \supset G_n$ for sufficiently large n. By (1.20), $T(G_0 \cup G_n \cup G) \downarrow T(G_0 \cup G \cup K)$. Similarly, $T(K \cup K_n \cup G_0)$ converges to $T(K \cup G \cup G_0)$, since T is continuous from below. □

Proof of the Choquet theorem. Let $\mathcal{V}$ be the family of sets $V = G \cup K$ where $G \in \mathcal{G}$ and $K \in \mathcal{K}$. It is possible to extend T to a completely alternating capacity on $\mathcal{V}$. By Lemma 1.21, formula (1.25) determines an additive map from $\mathfrak{V}$ to $[0, 1]$. Note that $\mathfrak{V}$ generates the σ-algebra $\mathfrak{B}(\mathcal{F})$. By known results on extensions of measures from an algebra to the corresponding σ-algebra (see Neveu [424, Prop. I.6.2]) it suffices to find a family $\mathfrak{V}' \subset \mathfrak{V}$ which consists of compact sets (in the Fell topology on $\mathcal{F}$) such that for each $\mathcal{Y} \in \mathfrak{V}$

$$\mathbf{P}(\mathcal{Y}) = \sup\{\mathbf{P}(\mathcal{Y}') : \mathcal{Y}' \in \mathfrak{V}'\}. \tag{1.27}$$

Let $\mathfrak{V}'$ consist of $\mathcal{F}_{K_1,\ldots,K_n}^G$, where $n \geq 0$, $G \in \mathcal{G}$ and $K_1, \ldots, K_n \in \mathcal{K}$. Then the elements of $\mathfrak{V}'$ are compact in the Fell topology and $\mathfrak{V}' \subset \mathfrak{V}$. It remains to prove (1.27).

Let $\mathcal{Y} = \mathcal{F}_{V_1,\ldots,V_n}^V \in \mathfrak{V}$ with $V = G_0 \cup K_0$, $G_0 \in \mathcal{G}$ and $K_0 \in \mathcal{K}$. There exists a sequence $\{G_k, k \geq 1\}$ of open sets such that $G_k \downarrow K_0$ and $G_k \supset \operatorname{cl}(G_{k+1}) \in \mathcal{K}$ for all $k \geq 1$. Hence V is a limit of a decreasing sequence of open sets $G_0 \cup G_k$. Similarly, for each $i \in \{1, \ldots, n\}$, V_i can be obtained as a limit of an increasing sequence $\{K_{ik}, k \geq 1\}$ of compact sets. Define

$$\mathcal{Y}_k = \mathcal{F}_{K_{1k},\ldots,K_{nk}}^{G_0 \cup G_k}.$$

Then $\mathcal{Y}_k \in \mathfrak{V}'$ and $\mathcal{Y}_k \uparrow \mathcal{Y}$ as $k \to \infty$. In order to show that $\mathbf{P}(\mathcal{Y}_k) \uparrow \mathbf{P}(\mathcal{Y})$ note that

$$\mathbf{P}(\mathcal{Y}) = -T(V) + \sum_i T(V \cup V_i) - \sum_{i_1 < i_2} T(V \cup V_{i_1} \cup V_{i_2}) + \cdots,$$

$$\mathbf{P}(\mathcal{Y}_k) = -T(G_0 \cup G_k) + \sum_i T(G_0 \cup G_k \cup K_{ik})$$
$$- \sum_{i_1 < i_2} T(G_0 \cup G_k \cup K_{i_1 k} \cup K_{i_2 k}) + \cdots.$$

Since both the sums above are finite and, by Lemma 1.22, each of the summands in the second sum converges to the corresponding summand of the first sum, one has $\mathbf{P}(\mathcal{Y}_k) \uparrow \mathbf{P}(\mathcal{Y})$. □

Harmonic analysis proof

Now we outline a proof which refers to techniques from *harmonic analysis* on semigroups, see Berg, Christensen and Ressel [61]. It is based on Theorem G.10 of Appendix G which characterises positive definite functions on *idempotent* semigroups. The family $\mathcal{K}$ of compact sets is an Abelian semigroup with respect to the union operation. The union operation is idempotent, i.e. $K \cup K = K$. The key idea of the proof is to identify all (continuous in some sense) *semicharacters* on $(\mathcal{K}, \cup)$ as elements of $\mathcal{F}$.

Let $\mathcal{I}$ be the set of all subsemigroups I of $(\mathcal{K}, \cup)$, which satisfy

$$K, L \in I \Rightarrow K \cup L \in I \quad \text{and} \quad K \subset L, L \in I \Rightarrow K \in I. \tag{1.28}$$

Define $\tilde{K} = \{I \in \mathcal{I} : K \in I\}$ and equip $\mathcal{I}$ with the coarsest topology in which the sets $\tilde{K}$ and $\mathcal{I} \setminus \tilde{K}$ are open for all $K \in \mathcal{K}$. Let $\mathbf{1}_I$ denote the map $\mathbf{1}_I(K) = \mathbf{1}_{K \in I}$ from $\mathcal{K}$ into $\{0, 1\}$. Furthermore, $\mathcal{I}_r$ denotes the set of all $I \in \mathcal{I}$ such that $\mathbf{1}_I$ is upper semicontinuous.

Lemma 1.23. For $F \in \mathcal{F}$ let $I_F = \mathcal{K}^F = \{K \in \mathcal{K} : K \cap F = \emptyset\}$. Then $F \mapsto I_F$ is a bijection between $\mathcal{F}$ and $\mathcal{I}_r$ and the inverse mapping is $I \mapsto \mathbb{E} \setminus \cup_{K \in I} \operatorname{Int} K$.

Proof. It is obvious that $I_F \in \mathcal{I}$. If $K \cap F = \emptyset$, then there exists an open neighbourhood of K which does not intersect F, so that the function $\mathbf{1}_{I_F}$ is upper semicontinuous.

Let $I \in \mathcal{I}_r$. With each $K \in I$, the family I contains a neighbourhood of K with a compact closure, whence

$$\bigcup_{K \in I} K = \bigcup_{K \in I} \operatorname{Int} K.$$

Therefore, $F = \mathbb{E} \setminus \cup_{K \in I} \operatorname{Int} K$ is closed and $I \subset I_F$. If $L \in I_F$, then L is covered by $\operatorname{Int} K$, $K \in I$. Therefore, L is covered by a finite number of compact sets from I,

$$L \subset \operatorname{Int} K_1 \cup \cdots \cup \operatorname{Int} K_n \subset K_1 \cup \cdots \cup K_n.$$

Because of (1.28), we see that $L \in I$, hence $I = I_F$.

Finally, if F_1 and F_2 are different closed sets and $x \in F_1 \setminus F_2$, then $\{x\} \in I_{F_2} \setminus I_{F_1}$, so that $I_{F_2} \neq I_{F_1}$ confirming the bijectivity. □

The following proposition strengthens the result of Lemma 1.23.

Proposition 1.24. *The mapping $c \colon \mathcal{I} \mapsto \mathcal{F}$ defined by*

$$c(I) = \mathbb{E} \setminus \bigcup_{K \in I} \operatorname{Int} K$$

is continuous on $\mathcal{I}$ with respect to the Fell topology on $\mathcal{F}$, and maps $\mathcal{I}_r$ bijectively onto $\mathcal{F}$.

Proof. It suffices to prove that $c^{-1}(\mathcal{F}^K)$ and $c^{-1}(\mathcal{F}_G)$ are open in $\mathcal{I}$ for $K \in \mathcal{K}$ and $G \in \mathcal{G}$. Note that $c(I) \cap K = \emptyset$ is equivalent to the existence of $L \in I$ such that $K \subset \operatorname{Int} L$. Therefore,

$$c^{-1}(\mathcal{F}^K) = \bigcup_{L \in \mathcal{K},\, K \subset \operatorname{Int} L} \tilde{L}, \qquad (1.29)$$

which shows that $c^{-1}(\mathcal{F}^K)$ is open in $\mathcal{I}$, since $\tilde{L}$ is open.

Similarly, $c(I) \cap G \neq \emptyset$ is equivalent to the existence of $K \in \mathcal{K} \setminus I$ such that $K \subset G$, hence

$$c^{-1}(\mathcal{F}_G) = \bigcup_{K \in \mathcal{K},\, K \subset G} (\mathcal{I} \setminus \tilde{K})$$

is open in $\mathcal{I}$. □

Proof of the Choquet theorem. Let T be an upper semicontinuous completely alternating capacity such that $T(\emptyset) = 0$ and $0 \leq T \leq 1$. It follows from the upper semicontinuity condition that

$$Q(K) = \sup\{Q(L) : L \in \mathcal{K}, \ K \subset \operatorname{Int} L\},$$

where $Q(K) = 1 - T(K)$.

Note that $\mathcal{I}$ is isomorphic to the set of semicharacters on $(\mathcal{K}, \cup)$, i.e. real-valued maps $\mathcal{K} \mapsto \mathbb{R}$ satisfying $\chi(\emptyset) = 1$, $\chi(K \cup L) = \chi(K)\chi(L)$, see Appendix G. Theorem G.6 implies that the functional $Q = 1 - T$ is positive definite on $\mathcal{K}$, i.e.

$$\sum_{i,j=1}^{n} c_i \bar{c}_j Q(K_i \cup K_j) \geq 0$$

for complex $c_1, \ldots, c_n$, $n \geq 1$, where $\bar{c}_i$ denotes the complex conjugate to c_i. By Theorem G.10, there exists a measure ν on $\mathcal{I}$ such that

$$Q(K) = \nu(\{I \in \mathcal{I} : K \in I\}) = \nu(\tilde{K}).$$

Now (1.29) and the continuity property of Radon measures ($\sup_\alpha \nu(G_\alpha) = \nu(\cup_\alpha G_\alpha)$ for upward filtering family of open sets G_α) yield

$$\nu(\cup_{L \in \mathcal{K}, \ K \subset \operatorname{Int} L} \tilde{L}) = \sup\{\nu(\tilde{L}) : L \in \mathcal{K}, \ K \subset \operatorname{Int} L\} = \nu(c^{-1}(\mathcal{F}^K)).$$

Hence $Q(K) = \mu(\mathcal{F}^K)$, where μ is the image of measure ν under the continuous mapping $c: \mathcal{I} \mapsto \mathcal{F}$. The uniqueness part is straightforward, since the families of sets $\mathcal{F}^K_{K_1,\ldots,K_n}$ generate $\mathfrak{B}(\mathcal{F})$. □

Another proof given by Norberg [432] is based on powerful techniques from the theory of lattices and is also applicable for random closed sets in non-Hausdorff spaces, see Section 3.4.

1.4 Separating classes

The Choquet theorem establishes that a probability measure on $\mathfrak{B}(\mathcal{F})$ can be determined by its values on $\mathcal{F}_K$ for $K \in \mathcal{K}$, i.e. the capacity functional on $\mathcal{K}$. However, the capacity functional defined on the whole family $\mathcal{K}$ of compact sets is still rather difficult to define constructively, because the family $\mathcal{K}$ is too rich and it is generally complicated to check the complete alternation conditions imposed on the capacity functional. Fortunately, in some cases it is possible to reduce the family of compact sets such that an upper semicontinuous completely alternating functional on this family extends to a unique probability measure on $\mathfrak{B}(\mathcal{F})$, i.e. defines a distribution of a unique random closed set. In some cases it is possible to achieve this by considering random closed sets with special realisations, e.g. those which are convex with probability one, see Section 7. Below we discuss possibilities of restricting the capacity functional in such a way that the restriction still determines the distribution of a general random closed set. Recall that

$$\mathfrak{B}_k = \{B \in \mathfrak{B}(\mathbb{E}) : \operatorname{cl} B \in \mathcal{K}\}$$

denotes the family of all *relatively compact* Borel sets in $\mathbb{E}$.

Definition 1.25 (Separating class). A class $\mathcal{A} \subset \mathfrak{B}_k$ is called *separating* if $\emptyset \in \mathcal{A}$ and, for all $K \in \mathcal{K}$ and $G \in \mathcal{G}$ with $K \subset G$, there exists an $A \in \mathcal{A}$ such that $K \subset A \subset G$, see Figure 1.4. A family of sets $\mathcal{A}_0$ is said to be a *pre-separating* class if the family of finite unions of sets from $\mathcal{A}_0$ forms a separating class.

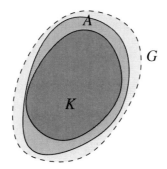

Figure 1.4. Set A from a separating class.

It follows from the topological assumptions on $\mathbb{E}$ (locally compact Hausdorff second countable) that, for each pair $K \subset G$ from Definition 1.25, there exists an open set G_1 with a compact closure such that $K \subset G_1 \subset \overline{G}_1 \subset G$. Definition 1.25 implies the existence of $A \in \mathcal{A}$ such that $K \subset \operatorname{Int} A \subset \overline{A} \subset G$. Since $\mathbb{E}$ is second countable, this means that every separating class includes a countable separating subclass.

Let $\varphi \colon \mathcal{A} \mapsto [0, \infty]$ be an increasing function on a separating class $\mathcal{A}$. Define its *outer extension* φ^- and the *inner extension* φ^0 by

$$\varphi^-(K) = \inf\{\varphi(A) : A \in \mathcal{A}, K \subset \operatorname{Int} A\}, \quad K \in \mathcal{K}, \tag{1.30}$$

$$\varphi^0(G) = \sup\{\varphi(A) : A \in \mathcal{A}, \operatorname{cl} A \subset G\}, \quad G \in \mathcal{G}. \tag{1.31}$$

If φ_1 is the restriction of φ onto a separating subclass $\mathcal{A}_1 \subset \mathcal{A}$, then $\varphi_1^- = \varphi^-$ and $\varphi^0 = \varphi_1^0$. Note also that $\varphi^{--} = \varphi^{0-} = \varphi^-$ and $\varphi^{00} = \varphi^{-0} = \varphi^0$.

Definition 1.26 (Continuity set). A set $B \in \mathfrak{B}_k$ is said to be a *continuity set* of φ if

$$\varphi^0(\operatorname{Int} B) = \varphi^-(\operatorname{cl} B), \tag{1.32}$$

where we allow for $\infty = \infty$. Let $\mathfrak{S}_\varphi$ denote the family of continuity sets for φ.

The family $\mathfrak{S}_\varphi \cap \mathcal{K}$ (and $\mathfrak{S}_\varphi$) is a separating class itself. For each pair $K \subset G$ with $\operatorname{cl} G \in \mathcal{K}$, consider an increasing family of compact sets K_t, $0 \leq t < 1$, such that $K_0 = K$ and $K_t \uparrow G$ as $t \uparrow 1$. Then there are at most a countable number of t

such that $K_t \notin \mathfrak{S}_\varphi$, whence we can pick any other $K_s \in \mathfrak{S}_\varphi$ as a set separating K and G.

If $\mathcal{A} \subset \mathcal{K}$ and φ is upper semicontinuous on $\mathcal{A}$, then $\varphi^- = \varphi$. This is similar to the property of extensions of upper semicontinuous capacities formulated in Theorem 1.12.

Theorem 1.27 (Capacity functional on a separating class). *Let $\mathcal{A}$ be a separating class, closed under finite unions. Suppose that $\varphi \colon \mathcal{A} \mapsto [0,1]$ is completely alternating on $\mathcal{A}$ with $\varphi(\emptyset) = 0$. Then there exists a unique random set X with the capacity functional $T(K) = \varphi^-(K)$ for all $K \in \mathcal{K}$. In particular, if $\mathcal{A} \subset \mathcal{K}$ and φ is an upper semicontinuous completely alternating functional on $\mathcal{A}$, then there exists a unique random closed set with the capacity functional T equal to φ on $\mathcal{A}$.*

Proof. The functional φ^- is completely alternating if φ is too. Furthermore, the functional φ^- is upper semicontinuous, so that the Choquet theorem is applicable to φ^-. □

Important separating classes are the family of $\mathcal{K}_{\mathrm{ub}}$ of all finite unions of balls of positive radii, or the class $\mathcal{K}_{\mathrm{up}}$ of all finite unions of parallelepipeds, see Salinetti and Wets [512] and Lyashenko [367]. Both these classes can be replaced with their countable subfamilies of balls with rational centres and radii and parallelepipeds with rational vertices. The following proposition includes the particular cases mentioned above.

Proposition 1.28. *Let $\mathcal{K}_0$ be the family of the closures for all relatively compact sets from a base of the topology on $\mathbb{E}$. Then $\mathcal{K}_0$ is a separating class and, for each capacity functional T on $\mathcal{K}_0$, there is a unique random closed set X such that $T(K) = \mathbf{P}\{X \cap K \neq \emptyset\}$ for all $K \in \mathcal{K}_0$.*

Example 1.29. Let $\mathbb{E} = \mathbb{R}$ be the real line. Consider the family $\mathcal{K}_0$ that consists of finite unions of closed bounded segments. It follows from Proposition 1.28 that the values of a capacity functional T on $\mathcal{K}_0$ determine uniquely the distribution of a random closed subset of $\mathbb{R}$. It is easy to see that T is upper semicontinuous on $\mathcal{K}_0$ if and only if $T([a,b])$ is right-continuous with respect to b and left-continuous with respect to a.

1.5 Random compact sets

If $\mathbb{E}$ is locally compact, then the family $\mathcal{K}$ of all compact sets is a measurable subclass of $\mathcal{F}$, i.e. $\mathcal{K} \in \mathfrak{B}(\mathcal{F})$. Indeed,

$$\mathcal{K} = \bigcup_{n \geq 1} \{F \in \mathcal{F} : F \subset K_n\},$$

where $\{K_n, n \geq 1\}$ is a sequence of compact sets such that $K_n \uparrow \mathbb{E}$ as $n \to \infty$. Note that $\mathcal{K} \in \mathfrak{B}(\mathcal{F})$ also for a general metric separable space $\mathbb{E}$ with $\mathfrak{B}(\mathcal{F})$ being the Effros σ-algebra (see Definition 2.1), since

$$\mathcal{K} = \bigcap_{m\geq 1}\bigcup_{n\geq 1}\bigcup_{x_1,\ldots,x_n\in\mathbb{Q}} \{F\in\mathcal{F}: F\subset \cup_{i=1}^{n} B_{1/m}(x_i)\},$$

where $\mathbb{Q}$ is a countable dense set in $\mathbb{E}$ (recall that $B_r(x)$ denotes the closed ball of radius r centred at x). The above argument spells out the fact that a closed set is compact if it possesses a finite m^{-1}-net for all $m \geq 1$.

Definition 1.30 (Random compact set). A random closed set X with almost surely compact values (so that $X \in \mathcal{K}$ a.s.) is called a *random compact set*.

Alternatively, it is possible to construct a random compact set directly as a $\mathcal{K}$-valued random element. The myopic topology on $\mathcal{K}$ (or the Hausdorff metric if $\mathbb{E}$ is a metric space) on $\mathcal{K}$ generates the Borel σ-algebra $\mathfrak{B}(\mathcal{K})$ on $\mathcal{K}$ that can be used to define a random compact set as a measurable $\mathcal{K}$-valued map $X\colon \Omega \mapsto \mathcal{K}$, see Appendix C. By Theorem C.5(iii), the σ-algebra $\mathfrak{B}(\mathcal{K})$ is generated by $\{K \in \mathcal{K} : K \cap G \neq \emptyset\}$ for $G \in \mathcal{G}$. If $\mathbb{E}$ is locally compact, then every open set can be approximated by compact sets, whence $\mathfrak{B}(\mathcal{K}) = \mathfrak{B}(\mathcal{F}) \cap \mathcal{K}$, i.e. the Borel σ-algebra on $\mathcal{K}$ coincides with the trace of $\mathfrak{B}(\mathcal{F})$ on $\mathcal{K}$. Therefore, these two natural approaches to define a random compact set produce the same object if $\mathbb{E}$ is locally compact. In a general topological space $\mathbb{E}$, Definition 1.30 is consistently used to define random compact sets. If $\mathcal{K}$ does not belong to $\mathfrak{B}(\mathcal{F})$, the condition $X \in \mathcal{K}$ a.s. is understood as

$$\sup\{\mathbf{P}\{X \in \mathcal{Y}\} : \mathcal{Y} \in \mathfrak{B}(\mathcal{F}), \mathcal{Y} \subset \mathcal{K}\} = 1.$$

The following result is a sort of "tightness" theorem for distributions of random compact sets.

Theorem 1.31 (Tightness for random compact sets). *Let X be a random compact set in a Polish space $\mathbb{E}$. For all $\varepsilon > 0$ there exists $K \in \mathcal{K}$ such that $\mathbf{P}\{X \subset K\} \geq 1-\varepsilon$.*

Proof. Let $\mathbb{Q} = \{x_k, k \geq 1\}$ be a countable dense set in $\mathbb{E}$. Note that

$$\lim_{n\to\infty} \mathbf{P}\left\{X \subset \bigcup_{k=1}^{n} B_{1/m}(x_k)\right\} = 1.$$

Choose $n = n_m$ such that

$$\mathbf{P}\left\{X \subset \bigcup_{k=1}^{n} B_{1/m}(x_k)\right\} \geq 1 - \frac{\varepsilon}{2^m}.$$

Define a compact set K as

$$K = \bigcap_{m\geq 1}\left[\bigcup_{k=1}^{n_m} B_{1/m}(x_k)\right].$$

Then

$$\mathbf{P}\{X \not\subset K\} = \mathbf{P}\left\{X \cap \bigcup_{m \geq 1}\left[\bigcup_{k=1}^{n_m} B_{1/m}(x_k)\right]^c \neq \emptyset\right\}$$

$$\leq \sum_{m \geq 1} \mathbf{P}\left\{X \not\subset \left[\bigcup_{k=1}^{n_m} B_{1/m}(x_k)\right]\right\} \leq \varepsilon \sum_{m \geq 1} 2^{-m} = \varepsilon. \qquad \square$$

1.6 Further functionals related to random sets

Avoidance, containment and inclusion functionals

The capacity functional $T_X(K)$ is defined as the probability that X hits a compact set K and therefore is often called the *hitting* functional of X. Along the same line, it is possible to define further functionals associated with a random closed set X.

Definition 1.32. For a random closed set X,

$$Q_X(K) = \mathbf{P}\{X \cap K = \emptyset\}, \quad K \in \mathcal{K},$$

is said to be the *avoidance* functional,

$$C_X(F) = \mathbf{P}\{X \subset F\}, \quad F \in \mathcal{F},$$

is the *containment* functional, and

$$I_X(K) = \mathbf{P}\{K \subset X\}, \quad K \in \mathcal{K},$$

is the *inclusion* functional.

All these functionals can be extended onto the family of open sets and all sets in the same way at it has been done for the capacity functional by means of (1.19) and (1.20). Let us list several obvious relationships between the introduced functionals

$$Q_X(K) = 1 - T_X(K),$$
$$C_X(F) = Q_X(F^c) = 1 - T_X(F^c),$$
$$I_X(K) = Q_{X^c}(K) = 1 - T_{X^c}(K).$$

The inclusion functional is related to the capacity functional of the complement X^c, the latter being an open random set, see Section 4.6. The avoidance functional is completely ∪-monotone on $\mathcal{K}$, see Definition 1.8. The containment functional is completely ∩-monotone (also called completely monotone) on $\mathcal{F}$, see Definition 1.9. The containment functional defined on open sets is the dual functional to the capacity functional, see (1.15).

Assume that $\mathbb{E}$ is a LCHS space. A simple reformulation of the Choquet theorem shows that the avoidance functional $Q_X(K)$, $K \in \mathcal{K}$, determines uniquely the distribution of X. The same is true for the containment functional $C_X(F)$, $F \in \mathcal{F}$, defined on the family of closed sets. The containment functional $C_X(K)$ restricted to $K \in \mathcal{K}$

may be degenerated for a non-compact X and so does not determine the distribution of X. However, if X is a random compact set, then

$$C_X(F) = \lim_{n\to\infty} C_X(F \cap K_n),$$

where $\{K_n, n \geq 1\}$ is an increasing sequence of compact sets such that $K_n \uparrow \mathbb{E}$. It is worthwhile to mention the following simple fact.

Proposition 1.33 (Containment functional of a random compact set). *The distribution of a random compact set X is uniquely determined by its containment functional $C_X(K)$, $K \in \mathcal{K}$.*

It is often useful to consider the inclusion functional $I_X(L)$ defined on sets L from the family $\mathfrak{I}$ of finite subsets of $\mathbb{E}$. The continuity of probability measures from the above immediately yields

$$I_X(K) = \inf\{I_X(L) : L \in \mathfrak{I}, L \subset K\} \qquad (1.33)$$

first for countable K and then for each $K \in \mathcal{K}$ referring to the separability of $\mathbb{E}$. As (1.33) shows, the inclusion functional $I_X(L)$, $L \in \mathfrak{I}$, can be uniquely extended onto the whole class $\mathcal{K}$. However, the inclusion functional, in general, does not determine uniquely the distribution of X. For example, if $X = \{\xi\}$ is a random singleton with ξ having an absolutely continuous distribution, then $I_X(K)$ vanishes on each non-empty K, see also Section 4.2.

Coverage function and covariance

It is easy to specify relationships between the capacity functional and the inclusion functional on finite sets. First, if $K = \{x\}$ is a singleton, then

$$p_X(x) = T_X(\{x\}) = I_X(\{x\}) \qquad (1.34)$$

is called the *coverage function* of X. The following proposition follows from the upper semicontinuity of the capacity functional.

Proposition 1.34. *The coverage function $p_X(x)$, $x \in \mathbb{E}$, is upper semicontinuous.*

For two-point sets, the corresponding inclusion functional

$$\Sigma_X(x_1, x_2) = I_X(\{x_1, x_2\}) = \mathbf{P}\{\{x_1, x_2\} \subset X\} \qquad (1.35)$$

is called the *covariance function* of X. It is easy to see that the covariance function is positive definite. The covariance can be expressed using the capacity functional on two-point sets and the coverage function as

$$\Sigma_X(x_1, x_2) = p_X(x_1) + p_X(x_2) - T_X(\{x_1, x_2\}).$$

The n-point coverage probabilities can be calculated from the capacity functional using the inclusion-exclusion formula as

$$I_X(\{x_1,\ldots,x_n\}) = \mathbf{P}\{\{x_1,\ldots,x_n\} \subset X\} = -\Delta_{\{x_n\}} \cdots \Delta_{\{x_1\}} T_X(\emptyset).$$

Therefore, the inclusion functional $I_X(L)$, $L \in \mathfrak{J}$, and the capacity functional $T_X(L)$, $L \in \mathfrak{J}$, restricted to the family $\mathfrak{J}$ of finite sets can be expressed from each other by solving systems of linear equations.

By integrating the covariance function of X one obtains

$$K(z) = \int \Sigma_X(x, x+z) dx = \int \mathbf{P}\{\{x, x+z\} \subset X\} dx$$
$$= \mathbf{E}\,\mathrm{mes}(X \cap (X-z)),$$

called the *geometric covariogram* of X.

Covariances of stationary random sets

If X is stationary (see Definition 4.1), then the covariance function is continuous and depends on the difference between the arguments, i.e. $\Sigma_X(x_1, x_2) = \Sigma(x_1 - x_2)$. If X is a stationary isotropic random set, then $\Sigma_X(x_1, x_2)$ depends only on $r = \|x_1 - x_2\|$. Then, in many applications, it is useful to approximate $\Sigma(r)$ by the *exponential covariance function* given by

$$\Sigma(r) = p(1-p)e^{-ar} + p^2,$$

where p is the common value for $p_X(x)$.

The function $\gamma(x) = \Sigma(0) - \Sigma(x)$ is called the *variogram*. Then

$$\gamma(x-y) = \frac{1}{2}[\mathbf{P}\{x \in X, y \notin X\} + \mathbf{P}\{x \notin X, y \in X\}].$$

Open problem 1.35. Characterise those functions that may appear as covariances (or variograms) of stationary random closed sets. It is well known that the properties $\gamma(0) = 0$, $\gamma(-h) = \gamma(h)$ and

$$\sum_{i,j=1}^{n} c_i c_j \gamma(x_i - x_j) \le 0$$

for any $n \ge 1$, $x_1, \ldots, x_n$, and real numbers $c_1, \ldots, c_n$ that sum to 0 (the conditional negative definiteness of γ) single out those functions that appear as variograms of random fields. Further conditions are required to ensure the existence of the indicator random field (or random closed set) with variogram γ. G. Matheron (see Lantuéjoul [344, p. 27]) conjectured that such γ should satisfy

$$\sum_{i,j=1}^{n} c_i c_j \gamma(x_i - x_j) \le 0$$

for any $n \ge 1$, $x_1, \ldots, x_n$, and $c_1, \ldots, c_n \in \{-1, 0, 1\}$ that sum to 1.

The covariance function $\Sigma(\cdot)$ can be used to develop a spectral theory of stationary random closed sets. For a Borel set A in $\mathbb{R}^d$, define its random measure $\mu(A) = \text{mes}_d(A \cap X)$ (see Section 8.3). If

$$\text{Cov}(A) = \int_A \Sigma(x) dx,$$

then

$$\int_{\mathbb{R}^d} \text{mes}_d(A \cap (B+h)) \text{Cov}(dh)$$

is the covariance between $\mu(A)$ and $\mu(B)$. It is shown in Koch, Ohser and Schladitz [320] that there exists a finite measure ν on $\mathbb{R}^d$, called the Bartlett spectrum, such that

$$\int_{\mathbb{R}^d} \hat{\psi}(x) \Sigma(x) dx = \int_{\mathbb{R}^d} \psi(u) \nu(du),$$

where $\hat{\psi}$ is the Fourier transform of a function ψ which decays sufficiently fast.

Möbius inversion

If $\mathbb{E}$ is a finite space, then the distribution of a random closed set is naturally determined by a finite set of probabilities $P_X(F) = \mathbf{P}\{X = F\}$ for all $F \subset \mathbb{E}$. These probabilities can be found from the containment functional by the Möbius inversion formula

$$P_X(F) = \sum_{K \subset F} (-1)^{\text{card}(F \setminus K)} C_X(K), \quad F \subset \mathbb{E}, \tag{1.36}$$

where $\text{card}(\cdot)$ is the cardinality of the corresponding set (note that all closed sets in a finite space are also compact). In the other direction,

$$C_X(F) = \sum_{K \subset F} P_X(K).$$

2 Measurability and selections

2.1 Multifunctions in metric spaces

Effros measurability

Let $(\Omega, \mathfrak{F}, \mathbf{P})$ be a probability space. A map $X : \Omega \mapsto \mathcal{F}$ from Ω into the space $\mathcal{F}$ of closed subsets of $\mathbb{E}$ is called a (closed-valued) *multifunction* or *set-valued function*. As before, $\mathcal{F}$ is the family of closed subsets of $\mathbb{E}$, but now the space $\mathbb{E}$ is assumed to be a *Polish* (complete separable metric) space. We aim to define random closed sets in $\mathbb{E}$ as measurable multifunctions, so that it is vital to introduce the appropriate measurability concept.

Definition 2.1 (Effros measurability). A map $X: \Omega \mapsto \mathcal{F}$ is called *Effros measurable* if
$$X^-(G) = \{\omega: X(\omega) \cap G \neq \emptyset\} \in \mathfrak{F}$$
for each $G \in \mathcal{G}$, i.e. for each open set G. The *Effros σ-algebra* on $\mathcal{F}$ is generated by the families $\mathcal{F}_G$ for all $G \in \mathcal{G}$.

Sometimes, an Effros measurable multifunction is called weakly measurable as opposed to a strongly measurable X which satisfies $X^-(F) = \{\omega: X(\omega) \cap F \neq \emptyset\} \in \mathfrak{F}$ for every closed set F.

It is possible to view a multifunction X as composed of single-valued measurable functions which "fit inside" X. Such functions are called selections of X. Note that an $\mathbb{E}$-valued random element ξ is a measurable map $\xi: \Omega \mapsto \mathbb{E}$ where the measurability is understood with respect to the conventional Borel σ-algebra $\mathfrak{B}$ on $\mathbb{E}$.

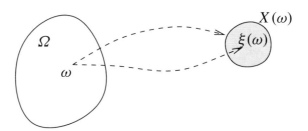

Figure 2.1. A multifunction X and its selection ξ.

Definition 2.2 (Measurable selection). A random element ξ with values in $\mathbb{E}$ is called a (measurable) *selection* of X if $\xi(\omega) \in X(\omega)$ for almost all $\omega \in \Omega$. The family of all selections of X is denoted by $\mathcal{S}(X)$.

Fundamental measurability theorem

If $\mathbb{E}$ is locally compact, then an Effros measurable multifunction is exactly a random closed set as defined in Section 1.1. Indeed, each open set in a locally compact space can be approximated from below by a sequence of compact sets, so that $X^-(G) \in \mathfrak{F}$ for all open G if and only if $X^-(K) \in \mathfrak{F}$ for all $K \in \mathcal{K}$. In a general Polish space other measurability definitions are possible. The following theorem of C. Himmelberg establishes the equivalence of several possible concepts. Its proof can be found in Himmelberg [257] and Castaing and Valadier [91].

Theorem 2.3 (Fundamental measurability theorem for multifunctions). *Let $\mathbb{E}$ be a separable metric space. Consider the following statements.*
(1) $X^-(B) \in \mathfrak{F}$ for every $B \in \mathfrak{B}(\mathbb{E})$.
(2) $X^-(F) \in \mathfrak{F}$ for every $F \in \mathcal{F}$.
(3) $X^-(G) \in \mathfrak{F}$ for every $G \in \mathcal{G}$, i.e. X is Effros measurable.

(4) The distance function $\rho(y, X) = \inf\{\rho(y, x) : x \in X\}$ is a random variable for each $y \in \mathbb{E}$.
(5) There exists a sequence $\{\xi_n\}$ of measurable selections of X such that

$$X = \mathrm{cl}\{\xi_n, \ n \geq 1\}.$$

(6) The graph of X

$$\mathrm{Graph}(X) = \{(\omega, x) \in \Omega \times \mathbb{E} : x \in X(\omega)\}$$

belongs to $\mathfrak{F} \otimes \mathfrak{B}(\mathbb{E})$ (the product σ-algebra of $\mathfrak{F}$ and $\mathfrak{B}(\mathbb{E})$).
Then the following results hold.
(i) $(1) \Rightarrow (2) \Rightarrow (3) \Leftrightarrow (4) \Rightarrow (6)$.
(ii) If $\mathbb{E}$ is a Polish space (i.e. $\mathbb{E}$ is also complete) then $(3) \Leftrightarrow (5)$.
(iii) If $\mathbb{E}$ is a Polish space and the probability space $(\Omega, \mathfrak{F}, \mathbf{P})$ is complete, then (1)–(6) are equivalent.

A measurable map $X \colon \Omega \mapsto \mathcal{F}$ is called a *random closed set* in $\mathbb{E}$. Since we always assume that $\mathbb{E}$ is Polish and the probability space is complete, Theorem 2.3 implies that all listed measurability definitions (1)–(6) are equivalent, so that X may be called measurable if it satisfies any one of them. Unless $\mathbb{E}$ is locally compact, it does not suffice to assume that $X^-(K) = \{\omega : X \cap K \neq \emptyset\} \in \mathfrak{F}$ for every compact set K.

Definition 2.4 (σ-algebra generated by X). The minimal σ-algebra $\mathfrak{F}_X$ generated by a random closed set X is generated by the events $X^-(G) = \{\omega \in \Omega : X(\omega) \cap G \neq \emptyset\}$ for $G \in \mathcal{G}$.

Clearly, $\mathfrak{F}_X$ is the minimal σ-algebra on Ω which ensures that X is Effros measurable. Because of Theorem 2.3, it is possible to generate $\mathfrak{F}_X$ using any of the conditions (1)–(4) or (6) of Theorem 2.3. If $\mathbb{E}$ is locally compact, $\mathfrak{F}_X$ is generated by $X^-(K)$, $K \in \mathcal{K}$.

Measurability of special multifunctions

If X is a random convex weakly compact subset of a Banach space (so that almost all realisations of X are weakly compact convex sets in $\mathbb{E}$), then it is possible to provide a simpler criterion for the measurability of X.

Proposition 2.5 (Measurability of convex-valued multifunction). *If the dual space $\mathbb{E}^*$ is separable, then a weakly compact convex-valued multifunction is measurable if and only if X is scalarly measurable, i.e. the support function of X*

$$h(X, u) = \sup\{\langle x, u \rangle : x \in X\}$$

is a random variable for each continuous linear functional $u \in \mathbb{E}^*$, where $\langle x, u \rangle$ denotes the value of u at x.

Proof. Necessity immediately follows from property (5) of Theorem 2.3, since $h(X, u) = \sup\{\langle \xi_n, u \rangle : n \geq 1\}$.

Sufficiency. Let B_1^* be the unit ball in the dual space $\mathbb{E}^*$. Then, for each $z \in \mathbb{E}$,

$$\rho(z, X) = \inf_{x \in X} \sup_{u \in B_1^*} \langle z - x, u \rangle.$$

Since $x \mapsto \langle z - x, u \rangle$ is a concave function and both X and B_1^* are convex, it is possible to swap inf and sup. Therefore,

$$\rho(z, X) = \sup_{u \in B_1^*} \left[\langle z, u \rangle - \sup_{x \in X} \langle x, u \rangle \right]$$

$$= \sup_{u_n \in B_1^*} [\langle z, u_n \rangle - h(X, u_n)]$$

is measurable, where $\{u_n, \, n \geq 1\}$ is dense in B_1^*. □

The following result provides an especially simple criterion of the measurability for regular closed multifunctions. Recall that a set F is *regular closed* if F coincides with the closure of its interior, i.e. $F = \mathrm{cl}(\mathrm{Int}\, F)$.

Theorem 2.6 (Measurability of regular closed multifunction). *A multifunction X with almost surely regular closed values in a Polish space $\mathbb{E}$ is a random closed set (i.e. X is Effros measurable) if and only if $\{\omega : x \in X(\omega)\} \in \mathfrak{F}$ for every $x \in \mathbb{E}$.*

Proof. Necessity is evident from Theorem 2.3. For sufficiency, fix a countable dense set $\mathbb{Q} = \{x_n, n \geq 1\}$ in $\mathbb{E}$. For each x_n, $A_n = \{\omega : x_n \in X(\omega)\} \in \mathfrak{F}$. Then $\cup_{n \geq 1} A_n = \{\omega : X(\omega) \neq \emptyset\}$ is measurable, so that without loss of generality we may assume that $X(\omega)$ is not empty for all ω.

Define an $\mathbb{E}$-valued random element ξ by putting $\xi(\omega) = x_1$ if $\omega \in A_1$, $\xi(\omega) = x_2$ if $\omega \in A_2 \setminus A_1$, $\xi(\omega) = x_3$ if $\omega \in A_3 \setminus (A_1 \cup A_2)$, etc. Then ξ is a measurable selection of X. Define a countable family of measurable selections as

$$\xi_n(\omega) = \begin{cases} x_n, & \omega \in A_n, \\ \xi(\omega), & \text{otherwise}, \end{cases} \quad n \geq 1.$$

Note that $\mathrm{cl}(X \cap \mathbb{Q}) = X$, since X is regular closed. Then $X = \mathrm{cl}\{\xi_n, \, n \geq 1\}$, whence X is measurable by Theorem 2.3. □

Borel interpretation of Effros σ-algebra

In order to study the convergence of random closed sets, it is essential to know when the Effros σ-algebra coincides with the Borel σ-algebra $\mathfrak{B}(\mathcal{F})$ generated by an appropriate topology on $\mathcal{F}$, i.e.

$$\sigma\{\mathcal{F}_G, \, G \in \mathcal{G}\} = \mathfrak{B}(\mathcal{F}). \tag{2.1}$$

This question has been already addressed in Section 1.1, where it was explained that the Effros σ-algebra is generated by the Fell topology on $\mathcal{F}$ if $\mathbb{E}$ is a locally compact space. The following theorem summarises several important results of this kind for a Polish space $\mathbb{E}$. Recall that the Wijsman topology is a topology of pointwise convergence of distance functions of closed sets, see Appendix B for a survey of various topologies on $\mathcal{F}$.

Theorem 2.7 (Effros σ-algebra and topologies on $\mathcal{F}$).
(i) *If $\mathbb{E}$ is a separable metric space, then the Wijsman topology generates the Borel σ-algebra that fulfils (2.1).*
(ii) *If $\mathbb{E}$ is locally compact, then (2.1) holds if $\mathcal{F}$ is equipped with the Fell topology.*
(iii) *The Effros σ-algebra induced on the family of compact sets $\mathcal{K}$ coincides with the Borel σ-algebra generated by the Hausdorff metric.*

Proof.
(i) First, prove that $\mathcal{F}$ is separable with respect to the Wijsman topology. Let $\mathcal{Q}$ be the family of all finite subsets of a countable dense set $\mathbb{Q}$ in $\mathbb{E}$. For each $F \in \mathcal{F}$ it is possible to find a countable set $F' = \{x_1, x_2, \dots\} \subset F$ such that $F = \mathrm{cl}\, F'$. Let F_n be a set from $\mathcal{Q}$ such that $\rho_H(F_n, F'_n) < n^{-1}$, where $F'_n = \{x_1 \dots, x_n\}$. Then F_n converges to F in Wijsman topology, since

$$\rho(x, F'_n) - n^{-1} \leq \rho(x, F_n) \leq \rho(x, F'_n) + n^{-1}$$

for each $x \in \mathbb{E}$, so that it suffices to notice that $\rho(x, F'_n) \to \rho(x, F)$ as $n \to \infty$.

Since $\mathcal{F}$ is separable, the Borel σ-algebra corresponding to the Wijsman topology is generated by

$$\{F \in \mathcal{F} : |\rho(x, F) - \rho(x, F_0)| < r\} \quad (2.2)$$

for $x \in \mathbb{Q}$, $F_0 \in \mathcal{F}$ and positive rational r. It is easily seen that every set given by (2.2) belongs to the Effros σ-algebra. For the reverse inclusion, referring to the separability of $\mathbb{E}$, it suffices to show that, for each open ball $B_r^o(x_0) = \{x : \rho(x, x_0) < \varepsilon\}$, the set $\{F : F \cap B_r^o(x) \neq \emptyset\}$ belongs to the Borel–Wijsman σ-algebra. This is indeed the case, since

$$\{F : F \cap B_r^o(x) \neq \emptyset\} = \{F : |\rho(x_0, F) - \rho(x_0, \{x_0\})| < r\}.$$

(ii) follows from (i), since the Fell topology generates the same Borel σ-algebra as the Wijsman topology if $\mathbb{E}$ is locally compact, see Theorem B.13(ii).
(iii) is an immediate corollary from Theorem C.5(iii). □

If $\mathbb{E}$ is a Banach space such that the dual space $\mathbb{E}^*$ is separable, then the slice topology (see Beer [56] and Hess [243]) generates the Borel σ-algebra which coincides with the Effros σ-algebra on $\mathcal{F}$. If $\mathbb{E}$ is a reflexive Banach space, then the Mosco and slice topologies coincide, so that the Mosco topology can be used instead, see Beer [56, Cor. 5.4.14].

Approximability of random closed sets

It is possible to work out further measurability properties relying on approximations of random closed sets by random sets with at most a finite number of values.

Definition 2.8 (Simple random sets). A random closed set X is called *simple*, if it assumes at most a finite number of values, so that there exists a finite measurable partition $A_1, \ldots, A_n$ of Ω and sets $F_1, \ldots, F_n \in \mathcal{F}$ such that $X(\omega) = F_i$ for all $\omega \in A_i$, $1 \leq i \leq n$.

It is known (see Appendix B) that the space $\mathcal{F}$ is separable in the Fell topology if $\mathbb{E}$ is LCHS. The separability of $\mathcal{F}$ ensures that in this case each random closed set is an almost sure limit (in the Fell topology) of simple random sets. For a general separable metric space $\mathbb{E}$, this is not always the case.

Definition 2.9 (Approximable random sets). A random closed set X is called *approximable* (with respect to some topology or metric on $\mathcal{F}$) if X is an almost sure limit of a sequence of simple random closed sets (in the chosen topology or metric).

If the convergence in Definition 2.9 is understood with respect to a metric, then the equivalent concept is the total measurability of X, meaning the existence of a sequence of simple random sets that converges to X in probability, see Section 6.2.

Note that approximations of X by simple random sets are always understood with respect to the topology such that the corresponding Borel σ-algebra satisfies (2.1). To be more specific, it is sensible to call X Hausdorff approximable, Wijsman approximable, Mosco approximable, etc. depending on the chosen topology.

Proposition 2.10 (Approximability of random sets). *Assume that (2.1) holds for an appropriate topology on $\mathcal{F}$. Then the approximability property of a random closed set X is equivalent to one of the following conditions.*
  (i) *There exists a subset $\Omega' \subset \Omega$ of a full measure such that $\{X(\omega) : \omega \in \Omega'\}$ is a separable subset of $\mathcal{F}$ with respect to the chosen topology.*
  (ii) *The distribution $\mathbf{P}_X$ induced by $\mathbf{P}$ on $\mathcal{F}$ is a Radon probability measure on $\mathfrak{B}(\mathcal{F})$, i.e. for each $\mathcal{Y} \in \mathfrak{B}(\mathcal{F})$,*

$$\mathbf{P}_X(\mathcal{Y}) = \sup\{\mathbf{P}_X(\mathcal{X}) : \mathcal{X} \text{ is compact in } \mathcal{F}, \ \mathcal{X} \subset \mathcal{Y}\},$$

  where the compactness is understood with respect to the topology satisfying (2.1).

Proof.
(i) follows from a general result on random elements in topological spaces, see Vakhaniya, Tarieladze and Chobanyan [568, Prop. 1.1.9].
(ii) If X is approximable, then there exists a sequence $\{\omega_k, k \geq 1\}$ such that $\{X(\omega_k), k \geq 1\}$ is dense in $\{X(\omega) : \omega \in \Omega \setminus \Omega_0\}$ with $\mathbf{P}(\Omega_0) = 0$. Ulam's theorem [568, Th. 2.3.1] together with (i) complete the proof. □

The following result implies that a measurable random closed (respectively compact) set can be equivalently defined as a Wijsman (respectively Hausdorff) approximable multifunction.

Theorem 2.11 (Wijsman and Hausdorff approximability). *In a Polish space $\mathbb{E}$,*
(i) every random closed set is Wijsman approximable;
(ii) every random compact set is Hausdorff approximable.

Proof.
(i) It is shown in the proof of Theorem 2.7(iii) that the space $\mathcal{F}$ with the Wijsman topology is separable if $\mathbb{E}$ is Polish. Then (i) follows from Proposition 2.10(i).
(ii) It suffices to show that $\mathcal{K}$ is separable with respect to the Hausdorff metric ρ_H. Let $\mathbb{Q}$ be a countable dense set in $\mathbb{E}$. Then the (countable) family $\mathcal{Q}$ of all finite sequences of elements of $\mathbb{Q}$ is dense in $\mathcal{K}$. For this, fix any $K \in \mathcal{K}$ and $\varepsilon > 0$. Then the balls of radius ε centred at the points of $\mathbb{Q}$ cover K. Therefore, K has a finite cover, which means that $\rho(K, Q) < \varepsilon$ for some $Q \in \mathcal{Q}$. □

It should be noted that a bounded random closed set X in a Polish space is not always Hausdorff approximable, even if the realisations of X are almost surely convex.

Example 2.12 (Non-approximable random closed sets). Consider the probability space $\Omega = [0, 1]$ with the Lebesgue σ-algebra and the Lebesgue measure $\mathbf{P}$.
(i) Let $\mathbb{E}$ be the Banach space of real-valued continuous functions on $[0, 1]$ with the uniform norm. Define a multifunction with closed convex values as $X(\omega) = \{x \in \mathbb{E} : \|x\| \leq 1, \ x(\omega) = 0\}$ for $\omega \in \Omega$. Then

$$\rho(x, X(\omega)) = \max(\sup_{t \neq \omega} \max(|x(t)| - 1, 0), \ x(\omega))$$

is a random variable for every $x \in \mathbb{E}$, so that X is a random closed set by Theorem 2.3. However, X cannot be obtained as an almost sure limit in the Hausdorff metric of simple random closed sets, since $\rho_H(X(\omega), X(\omega')) = 1$ for $\omega \neq \omega'$, contrary to Proposition 2.10(i).
(ii) Let $\mathbb{E} = \ell^2$ be the space of square-summable sequences. For each $\omega \in \Omega = [0, 1]$ take its binary expansion $\omega = \sum_{n=1}^{\infty} \omega_n 2^{-n}$ with ω_n equal 0 or 1. Let

$$X(\omega) = \{x \in \ell^2 : \|x\| \leq 1, \ x_n = 0 \text{ for } \omega_n = 0, \ n \geq 1\}.$$

Then X is not Hausdorff approximable, since $\rho_H(X(\omega), X(\omega')) = 1$ for $\omega \neq \omega'$.

2.2 Selections of random closed sets

Fundamental selection theorem

Recall that $\mathcal{S}(X)$ denotes the family of all (measurable) selections of X. The fundamental measurability theorem for multifunctions (Theorem 2.3) implies the following existence theorem for selections.

Theorem 2.13 (Fundamental selection theorem). *If X is an Effros measurable closed-valued almost surely non-empty multifunction in a Polish space $\mathbb{E}$, then $\mathcal{S}(X) \neq \emptyset$.*

The fundamental selection theorem can be proved directly by constructing a sequence of random elements ξ_n with values in a countable dense subset of $\mathbb{E}$ such that $\rho(\xi_n, X) < 2^{-n}$ and $\rho(\xi_n, \xi_{n-1}) < 2^{-n+1}$ for all $n \geq 1$ on a set of full measure. The completeness of $\mathbb{E}$ is crucial to ensure that the sequence of ξ_n possesses an almost sure limit, which becomes the required selection of X. The full proof can be found in Kuratowski and Ryll-Nardzewski [339], Castaing and Valadier [91, Th. III.6] or Aubin and Frankowska [30, Th. 8.1.3].

It should be noted that Theorem 2.3(ii) implies not only the mere existence of selections, but a stronger fact that the selections "fill" X, so that X equals the closure of a countable set of its selections.

Definition 2.14 (Castaing representation). A countable family of selections $\xi_n \in \mathcal{S}(X)$, $n \geq 1$, is said to be the *Castaing representation* of X if $X = \operatorname{cl}\{\xi_n, n \geq 1\}$.

Characterisation of distributions of random sets by their selections

The family of selections $\mathcal{S}(X)$ depends not only on X, but also on the underlying probability space. For instance, a two-point deterministic set $X = \{0, 1\}$ has only two trivial (deterministic) selections if $\mathfrak{F} = \{\emptyset, \Omega\}$ is the trivial σ-algebra, while if $\mathfrak{F}$ is richer, then random variables with possible values 0 and 1 appear as selections. Even if the probability space is fixed and non-atomic, then the situation is not straightforward. The following example describes two identically distributed random closed sets X and Y defined on the same probability space such that $\mathcal{S}(X) \neq \mathcal{S}(Y)$.

Example 2.15 (Identically distributed random sets with different selections). Let $\Omega = [0, 1]$ with the σ-algebra $\mathfrak{F}$ of Lebesgue measurable subsets and the Lebesgue measure $\mathbf{P}$. Define two random closed (even compact) subsets of $\mathbb{E} = \mathbb{R}$ as $X(\omega) = \{-\omega, \omega\}$ and $Y(\omega) = \{-s(\omega), s(\omega)\}$ where $s(\omega) = 2\omega$ if $\omega < 1/2$ and $s(\omega) = 2\omega - 1$ if $\omega \geq 1/2$. It is easy to see that X and Y are identically distributed. However, the selection of Y

$$\eta(\omega) = s(\omega)\mathbf{1}_{[0,1/2)}(\omega) - s(\omega)\mathbf{1}_{[1/2,1]}(\omega)$$

has a distribution which is not shared by any selection of X.

The situation described in Example 2.15 can be explained if one observes that the selection η belongs to the weak closure of $\mathcal{S}(X)$, i.e. η is the weak limit of a sequence $\{\xi_n, n \geq 1\} \subset \mathcal{S}(X)$. Taking the weak closure of the family of random elements is identical to taking the weak closure of the corresponding family of distributions $\{\mathbf{P}_\xi : \xi \in \mathcal{S}(X)\}$. It is well known (see Billingsley [70]) that the weak convergence of random elements (or their distributions) can be metrised by the *Prokhorov metric* given by

$$\mathfrak{p}(\mathbf{P}_1, \mathbf{P}_2) = \inf\{\varepsilon > 0 : \mathbf{P}_1(B) \leq \mathbf{P}_2(B^{\varepsilon-}) + \varepsilon \text{ for all } B \in \mathfrak{B}(\mathbb{E})\} \quad (2.3)$$

Theorem 2.16 (Selections of identically distributed random sets). *Consider two non-atomic probability spaces $(\Omega, \mathfrak{F}, \mathbf{P})$ and $(\Omega', \mathfrak{F}', \mathbf{P}')$ and two random closed sets X and Y in a Polish space $\mathbb{E}$ defined respectively on Ω and Ω'. If X and Y are identically distributed, then the weak closures of $\mathcal{S}(X)$ and $\mathcal{S}(Y)$ coincide.*

Proof. Let $\xi \in \mathcal{S}(X)$ and let $\varepsilon > 0$. We have to find $\eta \in \mathcal{S}(Y)$ such that $\mathfrak{p}(\xi, \eta) < \varepsilon$. It follows from Proposition 2.10(ii) (applied to singletons) that there exists a compact set K such that $\mathbf{P}\{\xi \in B_0\} < \varepsilon$, where $B_0 = \mathbb{E} \setminus K$ is the complement to K. Let $B_1, \ldots, B_m$ be a partition of K into disjoint Borel sets of diameter less than ε. Define $c_i = \mathbf{P}\{\xi \in B_i\}$ and $A_i = Y^-(B_i) = \{\omega \in \Omega' : Y(\omega) \cap B_i \neq \emptyset\}$ for $i = 0, 1, \ldots, m$.

Since X and Y are identically distributed,

$$\mathbf{P}\{X \cap B_I \neq \emptyset\} = \mathbf{P}'\{Y \cap B_I \neq \emptyset\} = \mathbf{P}'(\cup_{i \in I} A_i)$$

for every $I \subset \{0, 1, \ldots, m\}$, where $B_I = \cup_{i \in I} B_i$. Since the B_i's are disjoint,

$$\mathbf{P}\{X \cap B_I \neq \emptyset\} \geq \mathbf{P}\{\xi \in B_I\} = \sum_{i \in I} \mathbf{P}\{\xi \in B_i\} = \sum_{i \in I} c_i.$$

Then

$$\mathbf{P}'(\cup_{i \in I} A_i) = \mathbf{P}\{X \cap B_I \neq \emptyset\} \geq \sum_{i \in I} c_i$$

for each $I \subset \{0, 1, \ldots, m\}$ and

$$\mathbf{P}'(\cup_{i=0}^m A_i) = 1 = \sum_{i=0}^m c_i.$$

By a combinatorial result (Halmos and Vaughan [218], Hart and Kohlberg [223]) which holds for non-atomic probability spaces, there exists a partition $A'_0, A'_1, \ldots, A'_m$ of Ω' such that $A'_i \subset A_i$ and $\mathbf{P}'(A'_i) = c_i$ for $i = 0, 1, \ldots, m$. Define $\eta(\omega)$ for $\omega \in A'_i$ to be a selection of $Y \cap \mathrm{cl}(B_i)$. Then $\eta \in \mathcal{S}(Y)$ and $\mathfrak{p}(\xi, \eta) < \varepsilon$. □

For a σ-algebra $\mathfrak{H} \subset \mathfrak{F}$, let $\mathcal{S}_\mathfrak{H}(X)$ be the family of selections of X that are $\mathfrak{H}$-measurable. The following result states that the weak closed convex hulls of the family of $\mathfrak{F}$-measurable selections of X and $\mathfrak{F}_X$-measurable selections of X coincide, where $\mathfrak{F}_X$ is the σ-algebra generated by X, see Definition 2.4. If the probability space is atomless, then taking convex hulls is no longer necessary.

Theorem 2.17 ($\mathfrak{F}$- and $\mathfrak{F}_X$-measurable selections). *If X is a random closed set in a Polish space $\mathbb{E}$, then the closed convex hulls in the weak topology of $\mathcal{S}(X)$ and $\mathcal{S}_{\mathfrak{F}_X}(X)$ coincide.*

Proof. Although the statement follows from Theorem 2.16, we present here an independent proof. Without loss of generality, assume that X is almost surely non-empty. Since $\mathfrak{F}_X \subset \mathfrak{F}$, it suffices to show that an arbitrary selection $\xi \in \mathcal{S}(X)$ belongs to the closed convex hull of $\mathcal{S}_{\mathfrak{F}_X}(X)$. Let $f\colon \mathbb{E} \mapsto \mathbb{R}$ be a bounded continuous function. Since bounded continuous functions are dual to probability measures on $\mathbb{E}$ we need to show that for every $\varepsilon > 0$ there is $\eta \in \mathcal{S}_{\mathfrak{F}_X}(X)$ such that $\mathbf{E}f(\xi) \leq \mathbf{E}f(\eta) + \varepsilon$.

Note that $Y = \mathrm{cl}\, f(X)$ is a random compact set in $\mathbb{R}^1$ that is also $\mathfrak{F}_X$-measurable. Set $\alpha = \sup Y$. Then $[\alpha - \varepsilon, \alpha] \cap Y$ is an almost surely non-empty and $\mathfrak{F}_X$-measurable random closed set, see Theorem 2.25(iv). By the fundamental selection theorem, it admits a $\mathfrak{F}_X$-measurable selection ζ. Furthermore, a random closed set $Z = X \cap f^{-1}(\{\zeta\})$ is non-empty and so also admits a $\mathfrak{F}_X$-measurable selection η that satisfies $f(\eta) = \zeta$. Finally

$$\mathbf{E}f(\xi) \leq \mathbf{E}\alpha \leq \mathbf{E}\zeta + \varepsilon \leq \mathbf{E}f(\eta) + \varepsilon\,.\qquad\square$$

Proposition 2.18. *Two random closed sets X and Y in a Polish space $\mathbb{E}$ are identically distributed if and only if $\mathcal{S}_{\mathfrak{F}_X}(X) = \mathcal{S}_{\mathfrak{F}_Y}(Y)$.*

Proof. Sufficiency. If X and Y are not identically distributed, then $\mathbf{P}\{X \cap G \neq \emptyset\} \neq \mathbf{P}\{Y \cap G \neq \emptyset\}$ for some $G \in \mathcal{G}$, see Theorem 2.28. Without loss of generality assume that $\mathbf{P}\{X \cap G \neq \emptyset\} > \mathbf{P}\{Y \cap G \neq \emptyset\}$. Using the Castaing representation of X whose members are $\mathfrak{F}_X$-measurable (such a representation exists by Theorem 2.3), one can construct a selection $\xi \in \mathcal{S}_{\mathfrak{F}_X}(X)$ such that $\xi \in G$ whenever $X \cap G \neq \emptyset$. Then for any $\eta \in \mathcal{S}_{\mathfrak{F}_Y}(Y)$ we have

$$\mathbf{P}\{\eta \in G\} \leq \mathbf{P}\{Y \cap G \neq \emptyset\} < \mathbf{P}\{X \cap G \neq \emptyset\} = \mathbf{P}\{\xi \in G\}\,,$$

which shows that no such η shares the distribution with ξ.

Necessity. Consider $\xi \in \mathcal{S}_{\mathfrak{F}_X}(X)$. Then there exists a measurable map $\varphi\colon \mathcal{F} \mapsto \mathbb{E}$ such that $\xi = \varphi(X)$. Therefore, $\eta = \varphi(Y)$ is $\mathfrak{F}_Y$-measurable and has the same distribution as ξ. Furthermore, $\rho(\varphi(X), X)$ has the same distribution as $\rho(\varphi(Y), Y)$. Thus, η is a selection of Y, since $\rho(\varphi(X), X)$ vanishes almost surely. $\square$

Selectionability of distributions

Because the families of selections for identically distributed random sets may be different, it is natural to associate selections with the distribution of X rather than its representation as a multifunction on a particular probability space.

Definition 2.19 (Selectionable distributions). A probability distribution μ on $\mathbb{E}$ is *selectionable* with respect to a distribution ν on $\mathcal{F}$ if there is a random closed set X with the distribution ν and a selection $\xi \in \mathcal{S}(X)$ with distribution μ. The family of all probability measures on $\mathbb{E}$ which are selectionable with respect to a probability measure ν on $\mathcal{F}$ is denoted by $\mathcal{S}(\nu)$.

The following result obtained by Artstein [19] is a corollary from necessary and sufficient conditions for proper matching. If $\mathbb{E}$ is locally compact, it is possible to deduce it from the ordered coupling theorem proved in Section 4.8.

Theorem 2.20 (Selectionability). *A probability distribution μ on a Polish space $\mathbb{E}$ is selectionable with respect to the distribution ν on $\mathcal{F}$ if and only if*

$$\mu(K) \leq \nu(\{F : F \cap K \neq \emptyset\}) \tag{2.4}$$

for all $K \in \mathcal{K}$.

Families of selections

The following results are formulated for random compact sets in Polish spaces. It should be noted that their variants for random closed (non-compact) sets in locally compact spaces are possible to obtain using one-point compactification (see Appendix A) and allowing for a positive mass to be assigned to the compactifying point.

Proposition 2.21 (Selections of random compact sets). *For each probability measure ν on $\mathcal{K}$, the family $\mathcal{S}(\nu)$ is a convex compact set with respect to the weak convergence of measures and their arithmetic addition.*

Proof. If $\mu_1, \mu_2 \in \mathcal{S}(\nu)$, then (2.4) immediately implies that $c\mu_1 + (1-c)\mu_2 \in \mathcal{S}(\nu)$ for every $c \in [0, 1]$. Furthermore, (2.4) can be written for all open F, so that $\mathcal{S}(\nu)$ is closed in the weak topology by Billingsley [70, p. 24, (iv)]. By Proposition 2.10(ii) and Theorem 2.7(iii), there exists a compact set $\mathcal{K}' \subset \mathcal{K}$ such that $\nu(\mathcal{K}') \geq 1 - \varepsilon$ for any fixed $\varepsilon > 0$. The union of all sets from $\mathcal{K}'$ is a compact subset K' of $\mathbb{E}$. By (2.4), $\mu(K') \geq 1 - \varepsilon$, so that $\mathcal{S}(\nu)$ is tight and weakly compact by the Prokhorov theorem, see Billingsley [70, p. 62]. □

Similar to the definition of the Hausdorff metric on the space of compact sets, it is possible to define a distance between compact families of probability measures. Let $\mathbb{M}$ be the metric space of all probability measures on $\mathbb{E}$ with the Prokhorov metric $\mathfrak{p}$. For two compact sets $A_1, A_2 \subset \mathbb{M}$, define

$$\mathfrak{p}_H(A_1, A_2) = \max\left(\sup_{\mu \in A_1} \inf_{\nu \in A_2} \mathfrak{p}(\mu, \nu), \ \sup_{\nu \in A_2} \inf_{\mu \in A_1} \mathfrak{p}(\mu, \nu) \right).$$

The following result is proved by Artstein [16] similar to Theorem 2.16.

Theorem 2.22 (Continuity for families of selections). *The function $\nu \mapsto \mathcal{S}(\nu)$ is $\mathfrak{p}_H$-continuous, i.e. if $\nu_n \to \nu$ weakly on $\mathcal{K}$, then $\mathfrak{p}_H(\mathcal{S}(\nu_n), \mathcal{S}(\nu)) \to 0$ as $n \to \infty$.*

Theorem 2.22, in particular, immediately implies Theorem 2.16 for random compact sets, noticing that the Hausdorff metric does not distinguish between the sets and their closures. Moreover, if ν_n weakly converges to ν on $\mathcal{K}$, then the weak closure of $\mathcal{S}(\nu_n)$ converges to the weak closure of $\mathcal{S}(\nu)$.

Steiner point and selection operators

If X is an almost surely non-empty random convex compact set in $\mathbb{R}^d$ (see Definition 4.32), then its particularly important selection is given by the *Steiner point* $\mathbf{s}(X)$ defined in Appendix F. As shown in Dentcheva [136], the Steiner point possesses a number of useful smoothness properties as a function of X. For instance, the inequality $\|\mathbf{s}(K) - \mathbf{s}(L)\| \leq d\rho_H(K, L)$ for each $K, L \in \mathcal{K}$ implies that the Steiner point is a Lipschitz function on $\mathcal{K}$ with respect to the Hausdorff metric. It is shown by Aubin and Frankowska [30, Th. 9.4.1] that the Steiner point can be equivalently defined as

$$\mathbf{s}(X) = \frac{1}{\varkappa_d} \int_{B_1} m(H(X, u)) du, \qquad (2.5)$$

where B_1 is the unit ball in $\mathbb{R}^d$, $H(X, u) = \{x \in X : h(X, u) = \langle x, u \rangle\}$ is the u-face of X (or the subdifferential $\partial h(X, u)$ of the support function as defined in Appendix F) and $m(H(X, u))$ is the point in $H(X, u)$ with the smallest norm. It follows from Theorem 2.27 (see also (2.6)) that $m(H(X, u))$ is a random element in $\mathbb{E}$. Formula (2.5) can be amended to define a *generalised* Steiner point as

$$\mathbf{s}_\mu(X) = \int_{B_1} m(H(X, u)) \mu(du),$$

where μ is a probability measure on B_1. A convexity argument implies that $\mathbf{s}_\mu(X)$ is a selection of X. As shown by Dentcheva [136, Lemma 5.4], $\mathbf{s}_\mu(X)$ are dense in X for all probability measures μ absolutely continuous with respect to the Lebesgue measure.

The (generalised) Steiner points are particular examples of maps from $\operatorname{co} \mathcal{K}'$ into $\mathbb{E}$ which are continuous with respect to the Hausdorff metric. Recall that $\mathcal{K}'$ (respectively $\mathcal{F}'$) denote the families of non-void compact (respectively closed) sets, while $\operatorname{co} \mathcal{K}'$ and $\operatorname{co} \mathcal{F}'$ are formed by convex sets from the corresponding families. Rephrasing the concept of a selection, an Effros measurable map $\mathfrak{f}: \mathcal{F}' \mapsto \mathbb{E}$ is called a selection *operator* if $\mathfrak{f}(F) \in F$ for every non-empty closed set F.

Proposition 2.23 (Castaing representation using selection operators). *There exists a sequence of selection operators* $\{\mathfrak{f}_n, n \geq 1\}$ *such that* $F = \operatorname{cl}\{\mathfrak{f}_n(F), n \geq 1\}$ *for every* $F \in \mathcal{F}'$.

Proof. Consider the multifunction defined on $\mathcal{F}$ by $I(F) = F$ for all closed F. This is a measurable map with respect to the Effros σ-algebra. Now the Castaing representation of I provides the required family of selection operators. □

A selection operator $\mathfrak{f}$ is continuous if it is continuous in the Wijsman topology. The existence of a continuous selection operator on $\operatorname{co} \mathcal{F}'$ for a separable Banach space $\mathbb{E}$ is shown in Gao and Zhang [187].

Open problem 2.24. Find an explicit construction of the Castaing representation for non-convex random closed sets that may be similar to the representation using weighted Steiner points in the convex case.

It is possible to define a *tangent cone* to a random closed set as the limit of $(X - \xi)/t$ as $t \downarrow 0$, where ξ is a selection of X. The limit is considered in the Fell topology if $\mathbb{E} = \mathbb{R}^d$ and in the Mosco sense if $\mathbb{E}$ is a Banach space.

2.3 Measurability of set-theoretic operations

Set-theoretic operations

Because we always assume that $\mathbb{E}$ is Polish and the probability space is complete, Theorem 2.3 provides a number of equivalent definitions of measurable multifunctions which help to prove the measurability of operations with random closed sets. The Minkowski (elementwise) addition that appears in part (v) of the following theorem is defined in Appendix A and the limits for sequences of sets in (viii) are defined in Appendix B.

Theorem 2.25 (Measurability of set-theoretic operations). *If X is a random closed set in a Polish space $\mathbb{E}$, then the following multifunctions are random closed sets:*
   (i) $\overline{\mathrm{co}}\,(X)$, *the closed convex hull of X;*
  (ii) αX *if α is a random variable;*
 (iii) $\mathrm{cl}(X^c)$, *the closed complement to X, $\mathrm{cl}(\mathrm{Int}(X))$, the closure of the interior of X, and ∂X, the boundary of X.*
If X and Y are two random closed sets, then
  (iv) $X \cup Y$ *and $X \cap Y$ are random closed sets;*
   (v) $\mathrm{cl}(X + Y)$ *is a random closed set (if $\mathbb{E}$ is a Banach space);*
  (vi) *if both X and Y are bounded, then $\rho_\mathrm{H}(X, Y)$ is a random variable.*
If $\{X_n,\ n \geq 1\}$ is a sequence of random closed sets, then
 (vii) $\mathrm{cl}(\cup_{n \geq 1} X_n)$ *and $\cap_{n \geq 1} X_n$ are random closed sets;*
(viii) $\limsup X_n$ *and $\liminf X_n$ are random closed sets.*

Proof.
   (i) Without loss of generality assume that $X \neq \emptyset$ a.s. Consider the Castaing representation $\{\xi_n,\ n \geq 1\}$ of X. Then the countable family of convex combinations of $\{\xi_n,\ n \geq 1\}$ with rational coefficients is dense in $\overline{\mathrm{co}}\,(X)$, so that $\overline{\mathrm{co}}\,(X)$ admits its Castaing representation and, therefore, is measurable.
  (ii) follows immediately from the fact that $\{\alpha\xi_n,\ n \geq 1\}$ is the Castaing representation of αX.
 (iii) For every $G \in \mathcal{G}$, $\{\mathrm{cl}(X^c) \cap G = \emptyset\} = \{G \subset X\}$, so it suffices to show that the latter event is measurable. Let $\{F_n,\ n \geq 1\}$ be an increasing sequence of closed sets such that $F_n \uparrow G$. Then $\{G \subset X\} = \cap_{n \geq 1}\{F_n \subset X\}$, so that it suffices to show that $\{F \subset X\}$ is measurable for every $F \in \mathcal{F}$. Since there exists a countable set of points $\{x_k,\ k \geq 1\}$ which are dense in F, $\{F \subset X\} = \cap_{k \geq 1}\{x_k \in X\} \in \mathfrak{F}$.

Furthermore, cl(Int X) is measurable, since cl(Int X) = cl(Y^c) for Y = cl(X^c). The boundary of X can be represented as $\partial X = X \cap \mathrm{cl}(X^c)$, so that the measurability of ∂X would follow from (iv).

(iv) is a particular case of (vii) to be proved later on.

(v,vi) If $\{\xi_n, n \geq 1\}$ and $\{\eta_m, m \geq 1\}$ are the Castaing representations of X and Y respectively, then $\{\xi_n + \eta_m, n, m \geq 1\}$ is the Castaing representation of cl($X + Y$), whence cl($X + Y$) is measurable. Furthermore,

$$\rho_H(X, Y) = \rho_H(\{\xi_n, n \geq 1\}, \{\eta_m, m \geq 1\})$$

is measurable.

(vii) If $G \in \mathcal{G}$, then

$$\{\mathrm{cl}(\cup_{n \geq 1} X_n) \cap G \neq \emptyset\} = \bigcup_{n \geq 1} \{X_n \cap G \neq \emptyset\} \in \mathfrak{F},$$

which confirms the measurability of cl($\cup_{n \geq 1} X_n$). To show the measurability of countable intersections observe that

$$\mathrm{Graph}(\cap_{n \geq 1} X_n) = \bigcap_{n \geq 1} \mathrm{Graph}(X_n),$$

so that $\cap_{n \geq 1} X_n$ is measurable by Theorem 2.3.

(viii) Note that $X_n^\varepsilon = \{x : \rho(x, X_n) \leq \varepsilon\}$ is a random closed set, since its graph

$$\mathrm{Graph}(X_n^\varepsilon) = \{(\omega, x) : \rho(x, X_n) \leq \varepsilon\}$$

is measurable, since $\rho(\cdot, \cdot)$ is a $(\mathfrak{F} \otimes \mathfrak{B}(\mathbb{E}), \mathfrak{B}(\mathbb{R}))$-measurable function. Now (viii) follows from (vii) taking into account that

$$\liminf_{n \to \infty} X_n = \bigcap_{k \geq 1} \mathrm{cl}\left(\bigcup_{m \geq 1} \bigcap_{n \geq m} X_n^{1/k}\right),$$

$$\limsup_{n \to \infty} X_n = \bigcap_{m \geq 1} \mathrm{cl}\left(\bigcup_{n \geq m} X_n\right). \qquad \square$$

Inverse function and infimum

It is possible to formulate several results on inverse functions in the language of random closed sets.

Theorem 2.26 (Random inverse functions). *Let X and Y be random closed sets in Polish spaces $\mathbb{E}$ and $\mathbb{E}'$ respectively. Let $\zeta_x = \zeta_x(\omega)$, $x \in \mathbb{E}$, be an almost surely continuous $\mathbb{E}'$-valued stochastic process. Then $Z = \{x \in X : \zeta_x \in Y\}$ is a random closed set.*

If $\zeta_X \cap Y \neq \emptyset$ a.s. (where $\zeta_X = \{\zeta_x : x \in X\}$ is the image of X under ζ), then there exists an $\mathbb{E}$-valued random element ξ such that $\zeta_\xi \in Y$ a.s. In particular, if $Y = \{\eta\}$ is a singleton and $\eta \in \zeta_X$ a.s., then there exists an $\mathbb{E}$-valued random element ξ such that $\eta = \zeta_\xi$ a.s.

Proof. Note that $\varphi(\omega, x) = (\omega, \zeta_x(\omega))$ is measurable with respect to the product σ-algebra $\mathfrak{F} \otimes \mathfrak{B}(\mathbb{E})$, whence $\{(\omega, x) : \varphi(\omega, x) \in B\} \in \mathfrak{F} \otimes \mathfrak{B}(\mathbb{E})$ for every $B \in \mathfrak{B}(\mathbb{E}')$. The proof is finished by observing that

$$\text{Graph}(Z) = \text{Graph}(X) \cap \varphi^{-1}(\text{Graph}(Y))$$

is a measurable subset of $\Omega \times \mathbb{E}$. □

The following result concerns the measurability of infimum taken over a random closed set, see Figure 2.2. It can be easily proved using the Castaing representation for its first statement and referring to Theorem 2.26 for the second one.

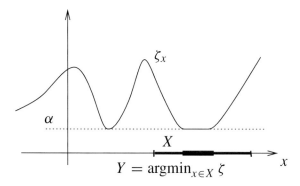

Figure 2.2. The minimum α and argmin inside X of a random function ζ.

Theorem 2.27 (Measurability of infimum). *Let X be an almost surely non-empty random closed set in Polish space $\mathbb{E}$ and let ζ_x be an almost surely continuous stochastic process with values in $\mathbb{R}$. Then*
(i) *the infimum of ζ_x over $x \in X$,*

$$\alpha = \inf_{x \in X} \zeta_x,$$

is a random variable;
(ii) *the set of minimum points,*

$$Y = \{x \in X : \zeta_x = \alpha\} = \text{argmin}_{x \in X} \zeta,$$

is a random closed set.
In particular, this implies that the *support function*

$$h(X, u) = \sup\{\langle x, u \rangle : x \in X\}, \quad u \in \mathbb{E}^*,$$

is a random variable if $\mathbb{E}$ is a Banach space. Furthermore, the α-envelope of X

$$X^\alpha = \{x : \rho(x, X) \leq \alpha\}$$

is a random closed set if α is a non-negative random variable. If ξ is an $\mathbb{E}$-valued random element, then $\rho(\xi, X) = \inf\{\rho(\xi, x) : x \in X\}$ is a random variable; the metric projection of ξ onto X

$$\text{proj}_X(\xi) = \{x \in X : \rho(x, \xi) = \rho(\xi, X)\} \tag{2.6}$$

is a random closed set.

2.4 Random closed sets in Polish spaces

The distribution of a random closed set in a Polish space $\mathbb{E}$ is determined by a probability measure on the Effros σ-algebra. The following theorem provides a uniqueness result.

Theorem 2.28 (Equality in distribution). *Let X and Y be two random closed sets in a Polish space $\mathbb{E}$. Then the following statements are equivalent.*
  (i) X *and* Y *are identically distributed.*
  (ii) For each open set G, $\mathbf{P}\{X \cap G \neq \emptyset\} = \mathbf{P}\{Y \cap G \neq \emptyset\}$.
  (iii) For each finite set $x_1, \ldots, x_n \in \mathbb{E}$ the random vectors $(\rho(x_1, X), \ldots, \rho(x_n, X))$ and $(\rho(x_1, Y), \ldots, \rho(x_n, Y))$ are identically distributed.

Proof. (i) and (ii) are equivalent, since the Effros σ-algebra is generated by the events that appear in (ii). Furthermore, (i) and (iii) are equivalent, since, by Theorem 2.7(i) the Effros σ-algebra coincides with the Borel σ-algebra generated by the Wijsman topology. Therefore, $\rho(x, X)$, $x \in \mathbb{E}$, is a continuous stochastic process whose finite-dimensional distributions are uniquely identified by (iii). Note that $\rho(x, X)$, $x \in \mathbb{E}$, is called the distance function of X, see Definition 2.2.5. □

The capacity functional of a random closed set X on the family of open sets

$$T_X(G) = \mathbf{P}\{X \cap G \neq \emptyset\}, \quad G \in \mathcal{G},$$

or its counterpart $T_X(F)$, $F \in \mathcal{F}$, on the family of closed sets uniquely identify the distribution of X. It is easily seen that T is a completely alternating capacity on $\mathcal{G}$ or $\mathcal{F}$. Its semicontinuity property causes potential difficulties. For example, $T(F_n) \downarrow T(F)$ does not necessarily follow from $F_n \downarrow F$. Note however, that the capacity functional on the family of compact sets no longer determines uniquely the distribution of X if $\mathbb{E}$ is not locally compact.

The Choquet theorem (for a locally compact space $\mathbb{E}$) implies that for every completely alternating upper semicontinuous capacity T with values in $[0, 1]$ there is a unique random closed set X having T as its capacity functional. Unfortunately, the corresponding existence result for random closed sets in general Polish spaces is not known. It should be noted that it is not possible to deduce it immediately from the theory outlined later on in Section 3, since the corresponding lattice fails to have a second countable Scott topology.

Nguyen and Nguyen [428] showed that the Choquet theorem does not hold on any non-compact Polish space under the assumptions that T is completely alternating on $\mathcal{G}$ and satisfies

$$T(G_n) \uparrow T(G) \text{ if } G_n \uparrow G \text{ in the Hausdorff metric}. \tag{2.7}$$

If $\mathbb{E}$ is the unit ball in a Hilbert space, then the *Kuratowski measure of non-compactness* $T(G) = \text{nc}(G)$, see Appendix A, is a functional which is completely alternating, satisfies (2.7) but at the same time violates the σ-subadditivity property, so that there exists an open set G such that

$$T(G) \not\leq \sum_{n=1}^{\infty} T(G_n) \tag{2.8}$$

for a decomposition $G = \cup_n G_n$ into the union of open sets. This means that no random closed set can have T as its capacity functional.

Another counterexample can be provided as follows. Let $T(G) = 0$ if G is bounded and 1 otherwise. Then T is maxitive, hence is completely alternating. It also satisfies (2.7). However, (2.8) does not hold for $G_n = \cup_{k=1}^{n}\{G'_k\}$ with G'_k being a bounded neighbourhood of x_k from a countable set $\{x_n, n \geq 1\}$ in $\mathbb{E}$ leading to violation of (2.8). Therefore, (2.7) is too weak even for a locally compact space E.

Using the inclusion functional of X defined on finite sets, it is easy to come up with a consistent family of finite-dimensional distributions for the indicator random function $\mathbf{1}_X(x)$. However extra conditions are needed to ensure that this becomes an indicator function of a closed set. The inclusion functional also ceases to be informative if the coverage function $p_X(x) = \mathbf{P}\{x \in X\}$ vanishes.

Open problem 2.29. Generalise the Choquet theorem for random closed sets in general Polish spaces.

2.5 Non-closed random sets

A measurable subset Y of $\Omega \times \mathbb{E}$ equipped with the product σ-algebra $\mathfrak{F} \otimes \mathfrak{B}(\mathbb{E})$ is called a *random set*. Every such Y can be equivalently defined as a multifunction

$$X(\omega) = \{x \in \mathbb{E} : (\omega, x) \in Y\},$$

so that X is also called a random set. Then $X: \Omega \mapsto \mathfrak{B}$ is a set-valued function with values being Borel subsets of $\mathbb{E}$ and $Y = \text{Graph}(X)$ becomes the graph of X. Sometimes, X is called a *graph-measurable* random set to stress the fact that the graph of X is a measurable set in the product space. However, X no longer has closed values whence the fundamental measurability theorem is not applicable to confirm the equivalence of various measurability definitions.

A set-valued function $X: \Omega \mapsto \mathfrak{B}$ is called *Borel measurable* if $\{\omega : X(\omega) \cap B \neq \emptyset\} \in \mathfrak{F}$ for all $B \in \mathfrak{B}$. Because the probability space is assumed to be complete, every graph-measurable random set X is also Borel measurable. The inverse implication is in general wrong.

The family $\mathfrak{B}(\mathbb{R}^d)$ with the metric given by the Lebesgue measure of the symmetric difference is a Polish space, so that a random Borel set in $\mathbb{R}^d$ can be defined as a random element with respect to the corresponding Borel σ-algebra, see Straka and Štěpán [548].

Alternatively, it is possible to define a general random set X as a random *indicator* function $\zeta(x) = \mathbf{1}_X(x)$, so that X is a random set if $\zeta(x)$ is a stochastic process on $\mathbb{E}$. Then the distribution of ζ can be determined by its finite-dimensional distributions. However, a number of interesting random closed sets lead to non-separable random functions whose finite-dimensional distributions are degenerated, see Section 4.2. A particularly important example of random open sets will be considered in Section 4.6.

3 Lattice-theoretic framework

3.1 Basic constructions

Because of the partial order on the family of closed sets, it is quite natural to treat the family of sets as a lattice. Below we will provide a brief summary of basic concepts from the theory of continuous lattices.

Let $\mathbb{L}$ be a *partially ordered set* (poset). A non-empty set $D \subset \mathbb{L}$ is called directed if $x, y \in D$ implies that $x \leq z$ and $y \leq z$ for some $z \in D$. Assume that each directed set D has supremum $\vee D$ (then $\mathbb{L}$ is called up-complete). A poset $\mathbb{L}$ is called a *semilattice* if each non-empty finite set has an infimum and is called a *lattice* if each non-empty finite set has both infimum and supremum. Furthermore, $\mathbb{L}$ is a *complete* lattice if every subset of $\mathbb{L}$ has a supremum and an infimum.

We say that x is way below y and write $x \ll y$ if $y \leq \vee D$ for every directed set D implies $x \leq z$ for some $z \in D$. Further we always assume that $\mathbb{L}$ is *continuous*, i.e. for each $x \in \mathbb{L}$, the set $\{y \in \mathbb{L} : y \ll x\}$ is directed with supremum x. A set $U \subset \mathbb{L}$ is called *Scott open* if U is an *upper set*, that is

$$\uparrow x = \{y \in \mathbb{L} : x \leq y\} \subset U \quad \text{for all } x \in U,$$

and if $x \in U$ implies the existence of some $y \in U$ with $y \ll x$. The collection of all Scott open sets is a topology on $\mathbb{L}$ which is denoted by Scott($\mathbb{L}$). A base of the Scott topology on $\mathbb{L}$ consists of $\{y \in \mathbb{L} : x \ll y\}$.

A set $F \subset \mathbb{L}$ is a *filter* if it is a non-empty upper set such that for each pair $x, y \in F$ there exists $z \in F$ satisfying $z \leq x$ and $z \leq y$. Denote by $\mathcal{L} = \text{OFilt}(\mathbb{L})$ the collection of all Scott open filters on $\mathbb{L}$. Note that $\mathcal{L}$ is a continuous poset itself, so that it is possible to define the topology Scott($\mathcal{L}$) and the family OFilt($\mathcal{L}$). The mapping

$$x \mapsto \mathcal{L}_x = \{F \in \mathcal{L} : x \in F\}, \quad x \in \mathbb{L}, \tag{3.1}$$

is an isomorphism between $\mathbb{L}$ and $\mathcal{L}$, which is called the *Lawson duality*. Let $\sigma(\mathbb{L})$ be the minimal σ-algebra on $\mathbb{L}$ generated by the family $\mathcal{L}$. A subset $\mathbb{Q} \subset \mathbb{L}$ is called separating if $x \ll y$ implies that $x \leq q \leq y$ for some $q \in \mathbb{Q}$.

Example 3.1 (Real line). Let $\mathbb{L} = [0, \infty)$ be the real half-line with the conventional order. It is a lattice and it becomes a complete lattice if the infinity is included, i.e. $\mathbb{L} = [0, \infty]$. Furthermore, $x \ll y$ if and only if $x < y$. Scott open sets are (x, ∞) for $x \geq 0$. These sets form the family OFilt($\mathcal{L}$), which is also a lattice. The Lawson duality maps x to the family of sets (y, ∞) with $y < x$, alternatively $x \mapsto [0, y)$.

Proposition 3.2 (see Norberg [432]). *Let $\mathbb{L}$ be a continuous poset. The following four conditions are equivalent:*
 (i) Scott($\mathbb{L}$) is second countable;
 (ii) $\mathbb{L}$ contains a countable separating set;
 (iii) $\mathcal{L}$ contains a countable separating set;
 (iv) Scott($\mathcal{L}$) is second countable.

Further, $\mathbb{L}$ is assumed to have a *second countable* Scott topology implying that $\mathbb{L}$ contains a countable separating subset $\mathbb{Q}$.

3.2 Existence of measures on partially ordered sets

Consider a non-negative mapping λ defined on a collection $\mathbb{F}_c$ of filters in a poset $\mathbb{L}$. Assume that $\mathbb{F}_c$ is closed under non-empty countable intersections and

$$\lambda(K) = \lim_{n \to \infty} \lambda(K_n) \quad \text{if } K_n \downarrow K \in \mathbb{F}_c, \ K_n \in \mathbb{F}_c, \ n \geq 1. \tag{3.2}$$

Assume that
$$\nabla_{K_n} \cdots \nabla_{K_1} \lambda(K) \geq 0 \tag{3.3}$$

for all $n \geq 1$ and $K, K_1, \ldots, K_n \in \mathbb{F}_c$, see (1.13) and (1.14) for the definition of the successive differences ∇. According to Definition 1.9, (3.3) means that λ is completely monotone (more exactly, completely $\cap$-monotone) on $\mathbb{F}_c$.

Fix a collection $\mathbb{F}_o$ of filters on $\mathbb{L}$, which is closed under finite non-empty intersections and is dual to $\mathbb{F}_c$ in the following sense.

Assumptions 3.3.
 (i) Each $K \in \mathbb{F}_c$ is the limit of a decreasing sequence of sets from $\mathbb{F}_o$ and each $G \in \mathbb{F}_o$ is the limit of an increasing sequence of sets from $\mathbb{F}_c$.
 (ii) If $K \subset G$ and $K_n \uparrow G$, where $K, K_1, K_2, \ldots \in \mathbb{F}_c$ and $G \in \mathbb{F}_o$, then $K \subset K_n$ for some n.
 (iii) If $K_n \downarrow K \subset \cup_{n \geq 1} G_n$ with $K, K_1, K_2, \ldots \in \mathbb{F}_c$ and $G_1, G_2, \ldots \in \mathbb{F}_o$, then $K_m \subset \cup_{n \leq m} G_n$ for some m.
 (iv) For each $G \in \mathbb{F}_o$ there exists $K \in \mathbb{F}_c$ such that $G \subset K$.

Although generic elements of $\mathbb{F}_c$ and $\mathbb{F}_o$ are denoted by K and G respectively, the letters K and G do not stand in this context for generic compact and open sets.

Theorem 3.4 (Measures on posets). *If λ satisfies (3.2) and (3.3) and Assumptions 3.3 hold, then λ extends to a measure on the minimal σ-algebra over $\mathbb{F}_c$. This extension is unique if $\mathbb{L} = \cup_{n \geq 1} K_n$ for some $K_1, K_2, \ldots \in \mathbb{F}_c$.*

Example 3.5 (Measure on $\mathbb{R}$). Let $\mathbb{L} = \mathbb{R}$ and let $\mathbb{F}_c$ (respectively $\mathbb{F}_o$) be the family of sets $(-\infty, x]$ (respectively $(-\infty, x)$) for $x \in \mathbb{R}$. Then $\lambda(K)$, $K \in \mathbb{F}_c$ is a function of the right end-point x of K. The condition on λ means that λ is a right-continuous non-decreasing function, which generates a measure on $\mathbb{R}$.

Example 3.6 (Measure on $\mathcal{F}$). Let $\mathbb{L}$ be the family $\mathcal{F}$ of closed sets in a LCHS space $\mathbb{E}$. Furthermore, let $\mathbb{F}_c$ consist of $\mathcal{F}^G$ for $G \in \mathcal{G}$ and let $\mathbb{F}_o = \{\mathcal{F}^K, K \in \mathcal{K}\}$. Note that $\mathcal{F}^{G_1} \cap \mathcal{F}^{G_2} = \mathcal{F}^{G_1 \cup G_2}$ and $\mathcal{F}^{G_n} \downarrow \mathcal{F}^G$ if and only if $G_n \uparrow G$. If $\lambda(\mathcal{F}^G) = Q(G)$, then

$$\nabla_{\mathcal{F}^{G_1}} \lambda(\mathcal{F}^G) = Q(G) - Q(G \cup G_1) = \Delta_{K_1} Q(K),$$

$$\nabla_{\mathcal{F}^{G_n}} \cdots \nabla_{\mathcal{F}^{G_1}} \lambda(\mathcal{F}^G) = \Delta_{G_n} \cdots \Delta_{G_1} Q(G),$$

and (3.3) means that $T = 1 - Q$ is a completely alternating functional on $\mathcal{G}$. The chosen families $\mathbb{F}_c$ and $\mathbb{F}_o$ satisfy Assumptions 3.3 and $\mathcal{F} = \cup_{n \geq 1} \mathcal{F}^{G_n}$ for a sequence of open sets $\{G_n, n \geq 1\}$, so that Theorem 3.4 is applicable in the current framework and yields a variant of the Choquet theorem for capacity functionals defined on open subsets of $\mathbb{E}$.

The proof of Theorem 3.4 is based on a series of intermediate results, which are similar to the measure-theoretic proof of the Choquet theorem in Section 1.3. Below we will only outline the lemmas and propositions that constitute the proof, referring to Norberg [432] for the full details.

Introduce the family

$$\mathcal{V} = \{K \cap G : K \in \mathbb{F}_c, G \in \mathbb{F}_o\}$$

which is similar to the family $\mathcal{V}$ in Lemma 1.20. Following the same line, define

$$\mathfrak{V} = \{V \setminus (V_1 \cup \cdots \cup V_n) : V, V_1, \ldots, V_n \in \mathcal{V}, n \geq 0\}.$$

In the context of Example 3.6, the family $\mathfrak{V}$ consists of elements of the type $\mathcal{F}^{K \cup G}_{K_1 \cup G_1, \ldots, K_n \cup G_n}$, where $K, K_1, \ldots, K_n \in \mathcal{K}$ and $G, G_1, \ldots, G_n \in \mathcal{G}$'

Lemma 3.7. *The family $\mathfrak{V}$ is a semiring (i.e. it is closed under finite intersections and the set-difference of any two elements is a union of a finite number of disjoint elements from $\mathfrak{V}$); it is also an algebra if $\mathbb{L} \in \mathcal{V}$.*

Any representation $Y = V \setminus (V_1 \cup \cdots \cup V_n)$ of a non-empty member $Y \in \mathfrak{V}$ is said to be reduced if $n = 0$ or if $V_i \subset V$ for all $i = 1, \ldots, n$ and $V_i = V_j$ whenever $V_i \subset V_j$. The following proposition generalises Lemma 1.20.

Proposition 3.8. *Every non-empty $Y \in \mathfrak{V}$ has a reduced representation. If $V \setminus (V_1 \cup \cdots \cup V_n)$ and $V' \setminus (V'_1 \cup \cdots \cup V'_m)$ are two reduced representations of Y, then $V = V'$, $m = n$ and $V_i = V'_i$, $1 \leq i \leq n$.*

Extend λ onto $\mathcal{V}$ by approximating each element of $\mathcal{V}$ from below using elements of $\mathbb{F}_c$

$$\lambda(V) = \sup\{\lambda(K) : K \in \mathbb{F}_c, K \subset V\}, \quad V \in \mathcal{V}.$$

Lemma 3.9. *Whenever $V, V_1, V_2 \in \mathcal{V}$, we have $\lambda(V \cap V_1) \leq \lambda(V_1)$ and*

$$\lambda(V \cap V_1) + \lambda(V \cap V_2) \leq \lambda(V) + \lambda(V \cap V_1 \cap V_2).$$

Lemma 3.10. *Let $V_i, V_i' \in \mathcal{V}$ for $1 \leq i \leq n$. If $V_i' \subset V_i$ for all i, then*

$$\lambda(\cap_{i=1}^n V_i) - \lambda(\cap_{i=1}^n V_i') \leq \sum_{i=1}^n (\lambda(V_i) - \lambda(V_i')).$$

This leads to the following statements which are similar to Lemma 1.22.

Proposition 3.11.
(i) *Let $V, V_1, V_2, \ldots \in \mathcal{V}$. If $V_n \downarrow V$, then*

$$\lambda(V) = \lim_{n \to \infty} \lambda(V_n).$$

(ii) *If $V \in \mathcal{V}$ and $K_n \uparrow G$ for $G \in \mathbb{F}_o$, $K_1, K_2, \ldots \in \mathbb{F}_c$, then*

$$\lambda(V \cap G) = \lim_{n \to \infty} \lambda(V \cap K_n).$$

(iii) *Let $V \in \mathcal{V}$. If $K_n \uparrow G$ and $G_n \downarrow K$ with $K, K_1, K_2, \ldots \in \mathbb{F}_c$ and $G, G_1, G_2, \ldots \in \mathbb{F}_o$, then*

$$\lambda(V \cap K \cap G) = \lim_{n \to \infty} \lambda(V \cap K_n \cap G_n).$$

Define the higher order differences for the function λ defined on $\mathcal{V}$ in exactly the same way as was done in (1.13) and (1.14) on the family $\mathbb{F}_c$. Since $\nabla_{V_n} \cdots \nabla_{V_1} \lambda(V)$ does not depend on the order of the subscripts $V_n, \ldots, V_1$, it is convenient to write $\nabla_\mathbf{V} \lambda(F)$ instead of $\nabla_{V_n} \cdots \Delta_{V_1} \lambda(F)$ with $\mathbf{V} = \{V_1, \ldots, V_n\}$. For convenience, put $\nabla_\emptyset \lambda(F) = \lambda(F)$.

Lemma 3.12. *If $V \setminus (V_1 \cup \cdots \cup V_n)$ and $V' \setminus (V_1' \cup \cdots \cup V_m')$ are two (not necessarily reduced) representations of $Y \in \mathfrak{V}$, then*

$$\nabla_{V_n} \cdots \nabla_{V_1} \lambda(V) = \nabla_{V_m'} \cdots \nabla_{V_1'} \lambda(V').$$

Lemma 3.12 shows that the value of $\nabla_\mathbf{V} \lambda(V)$ only depends on the member $Y = V \setminus (V_1 \cup \cdots \cup V_n)$, but not on its representation. Therefore, we can extend λ to the semiring $\mathfrak{V}$ by putting

$$\lambda(V \setminus \cup \mathbf{V}) = \nabla_\mathbf{V} \lambda(F), \quad V \in \mathcal{V}, \, \mathbf{V} \subset \mathcal{V} \text{ finite}. \tag{3.4}$$

Proposition 3.13. *The mapping λ from (3.4) is additive and non-negative on $\mathfrak{V}$.*

The additivity is proved similarly to Lemma 1.21, while the non-negativity can be obtained by approximating elements of $\mathcal{V}$ with elements of $\mathbb{F}_c$. The last steps of the proof aim to show that λ can be extended to a measure on the σ-algebra generated by $\mathcal{V}$. First, extend λ onto the ring $\tilde{\mathfrak{V}}$ generated by $\mathfrak{V}$ (so that $\tilde{\mathfrak{V}}$ is closed under finite intersections and set-differences). It suffices to show that λ is countably subadditive, i.e.

$$\lambda(R_n) \downarrow 0 \quad \text{if} \quad R_n \downarrow \emptyset, R_1, R_2, \ldots \in \tilde{\mathfrak{V}}. \tag{3.5}$$

Define

$$\mathcal{Y} = \{K \setminus \cup \mathbf{G} : K \in \mathbb{F}_c, \ \mathbf{G} \subset \mathbb{F}_o \text{ finite}\}.$$

Then whenever $\cap_{n \geq 1} Y_n = \emptyset$ for some $Y_1, Y_2, \ldots \in \mathcal{Y}$, we have $\cap_{n \leq m} Y_n = \emptyset$ for some finite m.

Proposition 3.14. *For each $R \in \tilde{\mathfrak{V}}$ we have*

$$\lambda(R) = \sup\{\lambda(\cup_{i=1}^n Y_i) : n \geq 1, \ Y_i \in \mathcal{Y}, \ Y_i \subset R, \ 1 \leq i \leq n\}.$$

Finally, it is possible to show that (3.5) holds, whence λ extends to a measure on the σ-algebra generated by $\mathfrak{V}$. This σ-algebra coincides with the σ-algebra generated by $\mathbb{F}_c$. If $\mathbb{L} = \cup_n K_n$ with $K_1, K_2, \ldots \in \mathbb{F}_c$ and μ is another function satisfying $\mu(K) = \lambda(K)$ for all $K \in \mathbb{F}_c$, then $\mu(R) = \lambda(R)$ for all $R \in \tilde{\mathfrak{V}}$, whence $\mu = \lambda$.

3.3 Locally finite measures on posets

A measure λ on $\mathbb{L}$ is said to be locally finite if it is locally finite with respect to the Scott topology, which is equivalent to the condition

$$\lambda(\uparrow x) < \infty, \quad x \in \mathbb{L}.$$

The following result follows from Theorem 3.4 for $\mathbb{F}_c = \{\uparrow x : x \in \mathbb{L}\}$ and $\mathbb{F}_o = \{F \in \mathcal{L} : F \ll \mathbb{L}\}$.

Theorem 3.15 (Locally finite measure on $\mathbb{L}$). *Let $\mathbb{L}$ be a continuous lattice and assume Scott($\mathbb{L}$) to be second countable. Then $\Lambda(x) = \lambda(\uparrow x)$, $x \in \mathbb{L}$, defines a bijection between the family of locally finite measures λ on $\mathbb{L}$ and the family of mappings $\Lambda: \mathbb{L} \mapsto \mathbb{R}_+$ satisfying*

$$\Lambda(x) = \lim_{n \to \infty} \Lambda(x_n), \quad x_n \uparrow x, \ x, x_1, x_2, \ldots \in \mathbb{L}, \tag{3.6}$$

and

$$\nabla_{x_n} \cdots \nabla_{x_1} \Lambda(x) \geq 0, \quad n \geq 1, \ x, x_1, x_2, \ldots, x_n \in \mathbb{L}, \tag{3.7}$$

where

$$\nabla_{x_1} \Lambda(x) = \Lambda(x) - \Lambda(x \vee x_1),$$

$$\nabla_{x_n} \cdots \nabla_{x_1} \Lambda(x) = \nabla_{x_{n-1}} \cdots \nabla_{x_1} \Lambda(x) - \nabla_{x_{n-1}} \cdots \nabla_{x_1} \Lambda(x \vee x_n), \quad n \geq 2.$$

Note that $(\uparrow x_1) \cap (\uparrow x_2) = \uparrow (x_1 \vee x_2)$, so the above definition of the successive differences complies with (1.13) and (1.14). Condition (3.7) means that Λ is a completely monotone function on the semigroup $(\mathbb{L}, \vee)$ (see Appendix G), i.e. $\Lambda \in \mathsf{M}(\mathbb{L})$.

Example 3.16 (Distributions of random variables). The real line is a lattice which is not continuous, since it is not up-complete. But $(-\infty, \infty]$ is a continuous lattice and also $[-\infty, \infty)$ under the reverse order. With this reverse order, $\uparrow x = [-\infty, x]$. Thus, as a special case of Theorem 3.15, one obtains the well known correspondence between locally finite measures on the extended real line and increasing right-continuous functions.

Whenever $F \in \mathcal{L}$, there are $x_1, x_2, \ldots \in \mathbb{L}$ such that $(\uparrow x_n) \uparrow F$ as $n \to \infty$. By the continuity of measures we obtain

$$\lambda(F) = \sup_{x \in F} \lambda(\uparrow x), \quad F \in \mathcal{L},$$

and the dual relationship

$$\lambda(\uparrow x) = \inf_{F \ni x} \lambda(F), \quad x \in \mathbb{L}.$$

The following theorem establishes the existence of locally finite measures on $\mathcal{L}$. Note that $\mathcal{L}_x$ is defined in (3.1).

Theorem 3.17 (Locally finite measure on $\mathcal{L}$). *Let $\mathbb{L}$ be a continuous semi-lattice with a top $\mathbb{I}$ and a second countable Scott topology. The formula $\lambda(\mathcal{L}_x) = \Lambda(x)$, $x \in \mathbb{L}$, defines a bijection between the family of locally finite measures λ on $\mathcal{L}$ and the family of mappings $\Lambda \colon \mathbb{L} \mapsto [0, \infty]$ which are finite on $\mathbb{L}_o = \{x \in \mathbb{L} : x \ll \mathbb{I}\}$ and satisfy (3.6) and (3.7), where the latter holds for all $x, x_1, \ldots, x_n \in \mathbb{L}_o$.*

Because of the Lawson duality between $\mathbb{L}$ and $\mathcal{L}$, we arrive at the following conclusion.

Corollary 3.18 (Measures on $\mathbb{L}$ and functionals on $\mathcal{L}$). *Let $\mathbb{L}$ be a continuous semi-lattice with a top $\mathbb{I}$ and a second countable Scott topology. The formula*

$$Q(F) = \lambda(F), \quad F \in \mathcal{L},$$

defines a bijection between the family of locally finite measures λ on $\mathbb{L}$ and the family of mappings $Q \colon \mathcal{L} \mapsto [0, \infty]$ which are finite on $\mathcal{L}_o = \{F \in \mathcal{L} : F \ll \mathbb{L}\}$ and satisfy

$$Q(F) = \lim_{n \to \infty} Q(F_n), \quad F_n \uparrow F, \ F, F_1, F_2, \ldots \in \mathcal{L}, \tag{3.8}$$

and completely monotone on $\mathcal{L}_o$, i.e.

$$\nabla_{F_n} \cdots \nabla_{F_1} Q(F) \geq 0, \quad n \geq 1, \ F, F_1, \ldots, F_n \in \mathcal{L}_o. \tag{3.9}$$

3.4 Existence of random sets distributions

It turns out that the abstract results of the previous section can be used to prove the Choquet theorem and to extend it for random closed sets in non-Hausdorff spaces.

First, assume that $\mathbb{E}$ is a LCHS (and so metrisable) space. Consider the lattice of all closed subsets of $\mathbb{E}$ with the reversed inclusion order, so that $F_1 \leq F_2$ means $F_1 \supset F_2$. Note that $\uparrow M = \{F \in \mathcal{F}: F \subset M\}$ and $F_1 \ll F_2$ means $\text{Int } F_1 \supset F_2$.

It is important to show that the elements of $\text{OFilt}(\mathcal{F})$ are exactly the families

$$\mathcal{F}^K = \{F \in \mathcal{F}: F \cap K = \emptyset\}, \quad K \in \mathcal{K}.$$

It is easy to see that $\mathcal{F}^K \in \text{OFilt}(\mathcal{F})$. In the other direction, $\mathcal{U} \in \text{OFilt}(\mathcal{F})$ implies that $\mathcal{U} = \cup_{n \geq 1}(\uparrow F_n)$ where $F_n \in \mathcal{F}$ and such that $F_{n+1} \ll F_n$ for all $n \geq 1$, see Norberg [432]. Therefore,

$$\mathcal{U} = \bigcup_{n \geq 1} \{F \in \mathcal{F}: F \subset F_n\} = \bigcup_{n \geq 1} \{F \in \mathcal{F}: F \subset \text{Int } F_n\}$$
$$= \{F \in \mathcal{F}: F \cap M = \emptyset\}$$
$$= \mathcal{F}^M,$$

where M is a closed set being the complement to $\cup_{n \geq 1}(\text{Int } F_n)$. Let us show that M is a compact set. Define

$$F = \bigcap_{i=1}^{\infty} D_n,$$

where

$$D_n = \bigcap_{i=1}^{n} (\mathbb{E} \setminus (M \cap K_i)^{\varepsilon_n}), \tag{3.10}$$

K_i, $i \geq 1$, is a monotone sequence of compact sets such that $\cup K_i = \mathbb{E}$, $\varepsilon_n \downarrow 0$ is a sequence of positive numbers and $(M \cap K_i)^{\varepsilon_n}$ is the ε_n-envelope of $M \cap K_i$. Assume that M is unbounded. Since $\mathcal{F}^M$ is open in Scott topology, for each $F \in \mathcal{F}^M$ there exists $F' \in \mathcal{F}^M$ such that $F' \ll F$. The latter means that $F \supset \cap \mathcal{D}$ for directed $\mathcal{D}$ implies that $F' \supset D$ for some $D \in \mathcal{D}$. However, the latter is not possible for $\mathcal{D} = \{D_n, n \geq 1\}$ defined in (3.10), since

$$F' \supset \bigcap_{i=1}^{n} (\mathbb{E} \setminus (M \cap K_i)^{\varepsilon_n})$$

implies $F' \cap M \neq \emptyset$. Therefore, the family $\mathcal{L} = \text{OFilt}(\mathcal{F})$ can be identified with $\mathcal{K}$ and the Choquet theorem follows from Corollary 3.18.

The lattice framework makes it possible to relax the topological conditions imposed on $\mathbb{E}$ and prove the Choquet theorem for non-Hausdorff topological spaces. Let $\mathbb{E}$ be a space endowed with a second countable topology $\mathcal{G}$ which is also a continuous lattice. As before, $\mathcal{F}$ is the collection of all closed sets, but $\mathcal{K}$ now denotes the family of all compact and saturated sets. A set K is called *saturated* if it is equal to the intersection of all open sets $G \supset K$. Note that we do not assume any of the separation properties for $\mathbb{E}$. If $\mathbb{E}$ is a T_1 space (so that all singletons are closed), then each subset of $\mathbb{E}$ is saturated.

A non-empty $F \in \mathcal{F}$ is said to be *irreducible* if $F \subset F_1 \cup F_2$ for two closed sets F_1 and F_2 implies that $F \subset F_1$ or $F \subset F_2$. Assume that $\mathbb{E}$ is *sober*, which means every irreducible set is the topological closure of a singleton corresponding to a unique element of $\mathbb{E}$. All Hausdorff spaces are sober. The relevance of this concept to the lattice framework is explained by the fact that any continuous poset endowed with its Scott topology is sober.

Consider a lattice $\mathbb{L}$ being the family $\mathcal{F}$ of all closed sets in $\mathbb{E}$ with the reverse inclusion, so that the empty set is the top of $\mathbb{L}$. Then $\mathbb{L}$ is continuous and has a second countable Scott topology. The family $\mathcal{K}$ is a continuous semi-lattice under the same order. If $K \in \mathcal{K}$, then $\mathcal{F}^K$ belongs to the family OFilt($\mathbb{L}$) of open filters of $\mathbb{L} = \mathcal{F}$. The mapping $K \mapsto \mathcal{F}^K$, $K \in \mathcal{K}$, is an isomorphism between $\mathcal{K}$ and OFilt($\mathcal{F}$), which shows that Scott($\mathcal{K}$) is second countable.

Corollary 3.18 applied to $\mathbb{L} = \mathcal{F}$ establishes a correspondence $Q(K) = \lambda(\mathcal{F}^K)$ between a function Q that satisfies (3.8) and (3.9) and a locally finite measure λ on $\mathcal{F}$. The corresponding measure λ is a probability measure if $Q(\emptyset) = 1$. In view of Example 3.6, the above conditions are equivalent to the fact that $T = 1 - Q$ is a capacity functional. Therefore, Corollary 3.18 yields an extension of the Choquet theorem for random closed sets in a second countable sober space $\mathbb{E}$ having a continuous topology (that also implies that $\mathbb{E}$ is locally compact).

4 Capacity functionals and properties of random closed sets

4.1 Invariance and stationarity

Stationary and isotropic random sets

If g is any function (or transformation) acting on $\mathbb{E}$, then the distribution of X is said to be *invariant* with respect to g (or g-invariant) if $X \stackrel{d}{\sim} g(X)$, i.e. X and its image under g are identically distributed. If X is g-invariant for each g from a group of transformations G acting on X, then x is called G-invariant. Particularly important cases appear if $\mathbb{E} = \mathbb{R}^d$ is the Euclidean space and G is either the group of translations on $\mathbb{R}^d$ or the group of all rotations or the group of all rigid motions.

Definition 4.1 (Stationary and isotropic random sets).
(i) A random closed set X in $\mathbb{R}^d$ is called *stationary* if

$$X \stackrel{d}{\sim} (X + a) \tag{4.1}$$

for all $a \in \mathbb{R}^d$, i.e. the distribution of X is invariant under all non-random translations. If (4.1) holds for all a from a linear subspace $\mathbb{H} \subset \mathbb{R}^d$, then X is called $\mathbb{H}$-stationary.

(ii) A random closed set X in $\mathbb{R}^d$ is called *isotropic* if $X \stackrel{d}{\sim} (gX)$ for each deterministic rotation g.

Proposition 4.2 (Stationarity implies unboundedness). *A stationary almost surely non-empty random closed set X in $\mathbb{R}^d$ is unbounded with probability 1 and $\overline{\mathrm{co}}\,(X) = \mathbb{R}^d$ almost surely for the closed convex hull $\overline{\mathrm{co}}\,(X)$ of X.*

Proof. Since X is almost surely non-empty, its support function $h(X, u)$ does not take the value $-\infty$. The stationarity of X implies $h(X, u) \stackrel{d}{\sim} (h(X, u) + \langle a, u \rangle)$ for all $a \in \mathbb{R}^d$. Putting $a = u$ shows that $h(X, u)$ is infinite with probability one for all $u \neq 0$. Applying this argument for a countable dense set of u on the unit sphere yields that $\overline{\mathrm{co}}\,(X) = \mathbb{R}^d$ a.s., whence X is also almost surely unbounded. □

Proposition 4.2 implies that a stationary convex set is either empty almost surely or is almost surely equal to the whole space.

The following proposition follows immediately from the Choquet theorem.

Proposition 4.3 (Invariance properties of the capacity functional).
(i) *A random closed set X is stationary if and only if its capacity functional is translation invariant, i.e. $T_X(K + a) = T_X(K)$ for all $K \in \mathcal{K}$ and $a \in \mathbb{R}^d$.*
(ii) *A random closed set X is isotropic if and only if its capacity functional is rotation invariant, i.e. $T_X(gK) = T_X(K)$ for all $K \in \mathcal{K}$ and all rotations g.*

Stationary random sets on the line

Consider a random closed set X on the real line $\mathbb{E} = \mathbb{R}$. Then X is stationary if and only if $T_X(K) = T_X(K + a)$ for all $a \in \mathbb{R}$ and $K \in \mathcal{K}$. It is possible to relax this condition by imposing it for some subfamilies of compact sets K. For instance, X is said to be *first-order stationary* if $T_X([x, y]) = T_X([x + a, y + a])$ for all $x, y, a \in \mathbb{R}$; X is *second-order stationary* if $T_X(K) = T_X(K + a)$ for all $a \in \mathbb{R}$ and K being unions of two segments, etc.

Proposition 4.4. *Let T be a completely alternating functional on the family $\mathcal{K}_0$ of finite unions of segments in $\mathbb{R}$. If T is first-order stationary and $a(x) = T([0, x])$ is right continuous at $x = 0$, then T is upper semicontinuous on $\mathcal{K}_0$.*

Proof. It was noticed in Example 1.29 that it suffices to show that $T([x, y])$ is right-continuous with respect to y and left-continuous with respect to x. Because of the first-order stationarity, $T([x, y]) = T([0, y - x])$, so that it suffices to show that the function $a(x) = T([0, x])$ is right-continuous for all $x \geq 0$. The 2-alternation property (1.16) of T applied for $K = \{0\}$, $K_1 = [0, u]$ and $K_2 = [-x, 0]$ implies

$$a(u) - a(0) \geq a(x + u) - a(x), \quad u, x \geq 0,$$

whence a is right-continuous for all $x \geq 0$. □

The concept of first- and second-order stationary sets in $\mathbb{R}$ can be extended to the higher-dimensional case and any family of sets $\mathcal{M} \subset \mathcal{K}$, so that X is called *nth-order stationary* on $\mathcal{M}$ if $T_X((K_1 \cup \cdots \cup K_n) + a) = T_X(K_1 \cup \cdots \cup K_n)$ for all

$K_1, \ldots, K_n \in \mathcal{M}$ and $a \in \mathbb{R}^d$. Often $\mathcal{M}$ is chosen to be a pre-separating class, for example, the family of all balls in $\mathbb{R}^d$.

An important case appears if $\mathcal{M}$ is a family of singletons. Then X is first (respectively second) order stationary if its indicator function is the first (respectively second) order stationary random field. If ζ_x is a strictly stationary continuous random field on $\mathbb{R}^d$, then the corresponding level set $X = \{x : \zeta_x = t\}$ is a stationary random closed set in $\mathbb{R}^d$ for every $t \in \mathbb{R}$.

A random closed set X is said to be *quasi-stationary* if $\mathbf{P}\{X \in \mathcal{Y}\} = 0$ for any $\mathcal{Y} \in \mathfrak{B}(\mathcal{F})$ implies $\mathbf{P}\{(X+a) \in \mathcal{Y}\} = 0$ for each $a \in \mathbb{R}^d$. This property can be equivalently reformulated by using higher-order differences (1.8) and choosing $\mathcal{Y} = \mathcal{F}_{K_1,\ldots,K_n}^K$.

Self-similarity

A random closed set X in $\mathbb{R}^d$ is said to be *self-similar* if X coincides in distribution with cX for every $c > 0$. This is the case if and only if the capacity functional satisfies $T(cK) = T(K)$ for every $K \in \mathcal{K}$ and $c > 0$.

Example 4.5 (Self-similar sets).
 (i) Let $\mathbb{C}$ be a deterministic cone in $\mathbb{R}^d$. If X is a random (not necessarily isotropic) rotation of $\mathbb{C}$, then X is self-similar. This is also the case if all realisations of X are cones.
 (ii) Let $X = \{t \geq 0 : w_t = 0\}$ be the set of zeroes for the Wiener process w_t. Then X is self-similar, although it is not a cone itself. Section 5.2.1.
 (iii) The measure Λ on $\mathfrak{B}(\mathbb{R}^d)$ with density $\lambda(x) = \|x\|^{-d}$ satisfies $\Lambda(cK) = \Lambda(K)$ for every $c > 0$ and $K \in \mathcal{K}$. The capacity functional $T(K) = 1 - e^{-\Lambda(K)}$ defines a self-similar random closed set which is the Poisson random set in $\mathbb{R}^d$, see Definition 8.8.

If X is self-similar and a.s. non-empty, then 0 belongs to X almost surely. Therefore, a non-trivial self-similar set cannot be stationary.

Proposition 4.6 (Logarithm of self-similar set). *If X is a self-similar random closed set in $(0, \infty)$, then its logarithm $Y = \{\log(x) : x \in X\}$ is a stationary random closed set in $\mathbb{R}$.*

Proof. For every $K \in \mathcal{K}(\mathbb{R})$ and $a \in \mathbb{R}$, the set $(K + a)$ hits Y if and only if $e^a e^K$ hits X, where $e^K = \{e^t : t \in K\}$. By the self-similarity of X,

$$\mathbf{P}\left\{e^a e^K \cap X \neq \emptyset\right\} = \mathbf{P}\left\{e^K \cap X \neq \emptyset\right\} = \mathbf{P}\{K \cap Y \neq \emptyset\}. \qquad \square$$

4.2 Separable random sets and inclusion functionals

Finite-dimensional distributions of the indicator process

Consider a random function ζ_x, $x \in \mathbb{E}$, with the only possible values being 0 or 1. Then ζ_x is an indicator $\mathbf{1}_Z(x)$ of a (not necessarily closed) set $Z \subset \mathbb{E}$. By Kolmogorov's extension theorem, the distribution of ζ_x is determined by the finite-dimensional distributions

$$\mathbf{P}\{\zeta_x = 1, x \in K, \; \zeta_y = 0, y \in L\} = \mathbf{P}\{K \subset Z, \; L \cap Z = \emptyset\}$$

for K and L from the family $\mathfrak{J}$ of finite subsets of $\mathbb{E}$ (note that the empty set is considered to be an element of $\mathfrak{J}$). The above probabilities can be expressed in terms of the hitting functional of Z

$$T_Z(K) = \mathbf{P}\{Z \cap K \neq \emptyset\}, \quad K \in \mathfrak{J},$$

since

$$\mathbf{P}\{L \cap Z = \emptyset\} = 1 - T_Z(L),$$
$$\mathbf{P}\{x \in Z, \; L \cap Z = \emptyset\} = T_Z(L \cup \{x\}) - T_Z(L),$$
$$\mathbf{P}\{(K \cup \{x\}) \subset Z, \; L \cap Z = \emptyset\} = \mathbf{P}\{K \subset Z, \; L \cap Z = \emptyset\}$$
$$- \mathbf{P}\{K \subset Z, \; (L \cup \{x\}) \cap Z = \emptyset\}.$$

The family of finite-dimensional distributions of the indicator function is consistent if and only if $T_Z(K)$ is a completely alternating capacity on $\mathfrak{J}$. Therefore, every completely alternating capacity T on the family $\mathfrak{J}$ with values in $[0, 1]$ defines uniquely the distribution of a random indicator function on an arbitrary space $\mathbb{E}$. However, this indicator function may correspond to a non-closed random set.

Proposition 4.7 (Extension of capacity on finite sets). *Let $\mathbb{E}$ be a LCHS space. A completely alternating functional $T : \mathfrak{J} \mapsto [0, 1]$ satisfying $T(\emptyset) = 0$ is the capacity functional of a random closed set if and only if T is upper semicontinuous on $\mathfrak{J}$, where the latter is equipped with the topology induced by the myopic topology on the family $\mathcal{K}$ of compact sets.*

Proof. We have to prove *sufficiency* only. Extend T onto the family $\mathcal{G}$ of open sets and then onto the family of compact sets by

$$T^*(G) = \sup\{T(L) : L \subset G, \; L \in \mathfrak{J}\}, \quad G \in \mathcal{G}, \tag{4.2}$$
$$T^*(K) = \inf\{T^*(G) : K \subset G, \; G \in \mathcal{G}\}, \quad K \in \mathcal{K}. \tag{4.3}$$

Then $T^*(K)$ becomes a completely alternating upper semicontinuous functional on $\mathcal{K}$, so that the Choquet theorem implies the existence of a random closed set with the capacity functional T^*. It remains to show that T^* coincides with T on $\mathfrak{J}$.

Let $G_n \downarrow L$ and $T^*(G_n) \downarrow T^*(L)$ as $n \to \infty$. Then there is a sequence $\{L_n, n \geq 1\} \subset \mathfrak{J}$ such that $T(L_n) \downarrow T^*(L)$ and $L_n \downarrow L$ as $n \to \infty$. Since L_n converges to L in the myopic topology on $\mathfrak{J}$, the upper semicontinuity of T implies $\limsup T(L_n) \leq T(L)$. Hence $T^*(L) \leq T(L)$. The proof finishes by combining this fact with the obvious inequality $T^*(L) \geq T(L)$. □

Separability

It is essential to note that Proposition 4.7 establishes the existence but not the uniqueness of a random closed set with the capacity functional defined on finite sets. Different random closed sets may share the same capacity functional restricted onto $\mathfrak{I}$.

The simplest example is provided by a random singleton $X = \{\xi\}$ with ξ having an absolutely continuous distribution in $\mathbb{R}^d$. Then X hits every finite set with probability zero, so that the capacity functional restricted on $\mathfrak{I}$ is indistinguishable with the capacity functional of the almost surely empty set. If A is a closed set in $\mathbb{R}^d$ with a zero Lebesgue measure, then $X = A + \xi$ is also indistinguishable from the empty set if the capacity functional is restricted on the family $\mathfrak{I}$. This implies that for any other random closed set Y

$$\mathbf{P}\{Y \cap L \neq \emptyset\} = \mathbf{P}\{(Y \cup X) \cap L \neq \emptyset\}, \quad L \in \mathfrak{I},$$

meaning that Y and $Y \cup X$ have identical capacity functionals restricted on $\mathfrak{I}$.

Definition 4.8 (Separability and separant). *A random closed set X is said to be* separable *if there exists a countable dense set $\mathbb{Q} \subset \mathbb{E}$ such that X almost surely coincides with $\mathrm{cl}(X \cap \mathbb{Q})$. Every such $\mathbb{Q}$ is called a* separant *of X.*

Definition 4.8 relies on the fact that $\mathrm{cl}(X \cap \mathbb{Q})$ is a random closed set. Indeed, for every open G,

$$\{\mathrm{cl}(X \cap \mathbb{Q}) \cap G = \emptyset\} = \{(X \cap \mathbb{Q}) \cap G = \emptyset\} = \{X \cap B = \emptyset\} \in \mathfrak{F},$$

where $B = \mathbb{Q} \cap G$, so that the conclusion follows from the fundamental measurability theorem. Clearly, $X = \{\xi\}$ and $X = A + \xi$ are not separable if ξ has an absolutely continuous distribution and the Lebesgue measure of A vanishes. The following theorem establishes the existence of a separable random closed set determined by a completely alternating functional on $\mathfrak{I}$.

Theorem 4.9 (Distribution of a separable random set). *Assume that $\mathbb{E}$ is a LCHS space. Let T be a completely alternating functional on $\mathfrak{I}$ with values in $[0, 1]$ such that $T(\emptyset) = 0$ and let T^* be the extension of T defined by (4.2) and (4.3).*

(i) *There exists a random closed set X such that $T_X(K) = T^*(K)$ for all $K \in \mathcal{K}$ and T_X is the smallest capacity functional such that $T(L) \leq T_X(L)$ for all $L \in \mathfrak{I}$.*

(ii) *The random closed set X is separable. If $\mathbb{Q}$ is its separant, then $X = \mathrm{cl}(Z \cap \mathbb{Q})$ a.s. for the random (not necessarily closed) set Z determined by the values of T on $\mathfrak{I}$.*

(iii) *A random closed set X such that $T_X(L) = T(L)$ for all $L \in \mathfrak{I}$ exists if and only if*

$$T^*(\{x\}) = T(\{x\}), \quad x \in \mathbb{E}. \tag{4.4}$$

In this case $\mathrm{cl}(X \cap \mathbb{Q})$ is the unique separable random closed set whose capacity functional coincides with T on $\mathfrak{I}$.

Proof.
(i) The existence directly follows from the Choquet theorem. If T^{**} is a capacity functional such that $T^{**}(L) \geq T(L)$ for all $L \in \mathfrak{I}$, then

$$T^{**}(G) \geq \sup\{T^{**}(L) : L \subset G, \ L \in \mathfrak{I}\}$$
$$\geq \sup\{T(L) : L \subset G, \ L \in \mathfrak{I}\} = T^*(G).$$

Thus, $T^{**}(K) \geq T^*(K)$ for all $K \in \mathcal{K}$.

(ii) Let $\mathcal{G}_0$ be a countable base of the topology on $\mathbb{E}$. For every $G \in \mathcal{G}_0$ define Q_G to be the union of sets $\{L_n, \ n \geq 1\} \subset \mathfrak{I}$ chosen so that $L_n \subset G$ and $T(L_n) \uparrow T^*(G)$ as $n \to \infty$. Let $\mathbb{Q}_0$ be the union of sets Q_G for $G \in \mathcal{G}_0$. Then

$$T^*(G) = \mathbf{P}\{Z \cap \mathbb{Q}_0 \cap G \neq \emptyset\}, \quad G \in \mathcal{G}_0, \tag{4.5}$$

noticing that $\{Z \cap \mathbb{Q}_0 \cap G \neq \emptyset\}$ is a measurable event since $\mathbb{Q}_0$ is countable. By approximating any $G \in \mathcal{G}$ with unions of elements of $\mathcal{G}_0$ it may be shown that (4.5) holds for all $G \in \mathcal{G}$. Since

$$\mathbf{P}\{Z \cap \mathbb{Q}_0 \cap G \neq \emptyset\} = \mathbf{P}\{\mathrm{cl}(Z \cap \mathbb{Q}_0) \cap G \neq \emptyset\}, \quad G \in \mathcal{G},$$

the random closed set $\mathrm{cl}(Z \cap \mathbb{Q}_0)$ has the capacity functional $T^*(G)$ on the family of open sets. Therefore, X coincides in distribution with $\mathrm{cl}(Z \cap \mathbb{Q}_0)$. The set $\mathbb{Q}_0$ may be taken as a separant of X, since $\mathrm{cl}(Z \cap \mathbb{Q}_0) \cap \mathbb{Q}_0 \supset Z \cap \mathbb{Q}_0$.

If $\mathbb{Q}$ is another separant for X, then $\mathbb{Q} \cup \mathbb{Q}_0$ is also a separant for X, whence

$$T^*(G) = \mathbf{P}\{Z \cap \mathbb{Q} \cap G \neq \emptyset\}$$
$$= \mathbf{P}\{Z \cap \mathbb{Q}_0 \cap G \neq \emptyset\}$$
$$= \mathbf{P}\{Z \cap (\mathbb{Q}_0 \cup \mathbb{Q}) \cap G \neq \emptyset\}, \quad G \in \mathcal{G}.$$

By pairwise inclusions of the events involved in the above chain of equalities, we deduce that $\mathrm{cl}(Z \cap \mathbb{Q}) = \mathrm{cl}(Z \cap \mathbb{Q}_0) = X$ a.s.

(iii) Since $\mathbb{Q} \cup L$ is a separant for Z,

$$\{x \in Z\} \subset \{x \in \mathrm{cl}(Z \cap (\mathbb{Q} \cup \{x\}))\} = \{x \in \mathrm{cl}(Z \cap \mathbb{Q})\},$$

where the equality is understood as equivalence up to a set of probability zero. By (4.4), $\mathbf{P}\{x \in \mathrm{cl}(Z \cap \mathbb{Q})\} = \mathbf{P}\{x \in Z\}$. Therefore, events $\{x \in Z\}$ and $\{x \in \mathrm{cl}(Z \cap \mathbb{Q})\}$ coincide up to a set of probability zero, whence $T^*(L) = T(L)$ for all $L \in \mathfrak{I}$. The uniqueness follows from (i). □

The separability concept of random closed sets is similar to the concept of the separability for stochastic processes, see Doob [142] and Gihman and Skorohod [193]. The random closed set X is separable if and only if its indicator function is a separable stochastic process. This means that separable random sets can be explored through their indicator functions. Furthermore, the separability assumption allows us to treat non-closed random sets using methods typical in the theory of stochastic processes. Quite differently from the theory of stochastic processes where the separability is a usual assumption, many interesting random closed sets are not separable. For instance, such a simple random set like a random singleton $X = \{\xi\}$ is not separable and corresponds to a non-separable indicator function if ξ has an absolutely continuous distribution.

Since distributions of separable random closed sets are uniquely identified by the values of their capacity functionals on the family of finite sets, Theorem 4.9(iii) can be reformulated using inclusion functionals.

Proposition 4.10 (Inclusion functional for separable random sets). *A distribution of a separable random closed set X is uniquely determined by its inclusion functional $I_X(L) = \mathbf{P}\{L \subset X\}$ for $L \in \mathfrak{I}$.*

P-continuity

The following definition formulates a continuity assumption related to the capacity functional restricted onto $\mathfrak{I}$. Recall that Proposition 1.34 establishes the upper semi-continuity of the coverage function $p_X(x) = T_X(\{x\})$, $x \in \mathbb{E}$.

Definition 4.11 (P-continuity). A random closed set X is called **P**-*continuous* at $x_0 \in \mathbb{E}$ if $p_X(x) = \mathbf{P}\{x \in X\}$ is continuous at x_0 as a function of x. Furthermore, X is called **P**-continuous if it is **P**-continuous at every $x_0 \in \mathbb{E}$.

Proposition 4.12 (P-continuity and separability).
  (i) *A random closed set X is **P**-continuous at x_0 if and only if*
$$\lim_{x \to x_0} \mathbf{P}\{x_0 \in X, \ x \notin X\} = 0, \quad (4.6)$$
  or, equivalently,
$$\lim_{x \to x_0} \mathbf{P}\{x_0 \notin X, \ x \in X\} = 0.$$
 (ii) *If a separable random closed set X is **P**-continuous, then every countable dense set $\mathbb{Q} \subset \mathbb{E}$ is a separant for X.*
 (iii) *If X is **P**-continuous, then for every two countable dense sets $\mathbb{Q}$ and $\mathbb{Q}'$ we have $\mathrm{cl}(X \cap \mathbb{Q}) = \mathrm{cl}(X \cap \mathbb{Q}')$ a.s.*

Proof.
(i) Clearly,
$$\mathbf{P}\{x_0 \in X, \ x \notin X\} = T_X(\{x_0, x\}) - T_X(\{x\}).$$
Since T is upper semicontinuous, $\limsup_{x \to x_0} T(\{x_0, x\}) \leq T(\{x_0\})$, whence the **P**-continuity at x_0 is equivalent to (4.6).
(ii) Note that (4.6) means that the indicator function of X is a stochastically continuous random function, see Gihman and Skorohod [193]. It is well known that every separable stochastically continuous random function has any countable dense set as its separant.
(iii) By Doob's theorem, the indicator function of X is stochastically equivalent to a separable random function. This separable function is the indicator of the random closed set $\mathrm{cl}(X \cap \mathbb{Q})$, where the choice of $\mathbb{Q}$ is immaterial because of (ii). □

A.s.-continuous random sets

Definition 4.13 (a.s.-continuity). A random closed set X is called *a.s.-continuous* if $\mathbf{P}\{x \in \partial X\} = 0$ for every $x \in \mathbb{E}$.

Note that ∂X (the boundary of X) is a random closed set by Theorem 2.25(iii). If reformulated for the indicator function of X, Definition 4.13 means that the indicator function has a discontinuity at any given point with probability zero. The property of X being a.s.-continuous can be verified using a restriction of the capacity functional onto the family of finite sets.

Proposition 4.14 (a.s.-continuity and separability).
(i) A random closed set X is a.s.-continuous if and only if X is **P**-continuous and $\mathrm{cl}(X \cap \mathbb{Q})$ is a.s.-continuous for some countable dense set $\mathbb{Q} \subset \mathbb{E}$.
(ii) If X is an a.s.-continuous random closed set, then $\mathrm{cl}(\mathrm{Int}(X)) = \mathrm{cl}(X \cap \mathbb{Q})$ a.s. for every countable dense set $\mathbb{Q}$ and $\mathrm{cl}(\mathrm{Int}(X))$ is the random closed set that appears in Theorem 4.9(i) with the distribution derived from $T_X(L)$ for $L \in \mathfrak{I}$. If, in addition, X is separable, then $X = \mathrm{cl}(\mathrm{Int}\, X)$ almost surely, i.e. X is regular closed.

Proof.
(i) If X is a.s.-continuous, then X is **P**-continuous since

$$\{x \in \partial X\} \supset \bigcap_{n \geq 1} \bigcup_{m \geq n} \{x \in X,\ y_m \notin X\}$$

for all sequences $y_n \to x$, whence (4.6) holds. If $Y = \mathrm{cl}(X \cap \mathbb{Q})$, then

$$\partial Y \subset \partial X \subset (\partial Y) \cup (Y^c \cap X),$$

whence Y is a.s.-continuous.

If X is **P**-continuous, then the capacity functionals of X and Y coincide on $\mathfrak{I}$. Therefore, $\mathbf{P}\{x \in (\partial Y) \cup (Y^c \cap X)\} = 0$ for all $x \in \mathbb{E}$, whence $\mathbf{P}\{x \in \partial X\} = \mathbf{P}\{x \in \partial Y\} = 0$.
(ii) If X is a.s.-continuous, then $(X \cap \mathbb{Q}) \subset \mathrm{Int}\, X$ a.s. for every countable dense set $\mathbb{Q}$. Therefore, $\mathrm{cl}(X \cap \mathbb{Q}) \subset \mathrm{cl}(\mathrm{Int}(X))$ a.s. The evident reverse inclusion completes the proof. □

4.3 Regenerative events

p-function

For random closed sets on the line, the calculation of the inclusion functional on the family of finite sets can be considerably simplified by imposing a regenerative property.

Definition 4.15 (Regenerative event). A separable random closed set X on the positive half line such that X contains the origin almost surely, i.e. $0 \in X$ a.s., is said to be a *regenerative event* (or regenerative phenomenon) if

$$I_X(\{t_1,\ldots,t_n\}) = I_X(\{t_1\})I_X(\{t_2 - t_1\}) \cdots I_X(\{t_n - t_{n-1}\})$$

for all $n \geq 2$ and $0 < t_1 < \cdots < t_n$, where I_X is the inclusion functional of X.

Proposition 4.10 implies that the distribution of a regenerative event is determined by its coverage function

$$p(t) = \mathbf{P}\{t \in X\}, \quad t \geq 0, \tag{4.7}$$

which is called the *p-function* of X. All possible p-functions can be characterised by their monotonicity property that is similar to the monotonicity properties of capacity functionals. By the usual inclusion-exclusion formula using p one obtains the avoidance functional $Q(\{t_1,\ldots,t_n\})$ for all $t_1,\ldots,t_n \geq 0$. This observation easily leads to the following result.

Proposition 4.16. *A real valued function $p(t)$, $t > 0$, is a p-function of a regenerative event if and only if the avoidance functional $Q(\{t_1,\ldots,t_n\})$ corresponding to p is non-negative and $Q(\{t_1,\ldots,t_n\}) \geq Q(\{t_1,\ldots,t_n,t_{n+1}\})$ for all $t_1 < \cdots < t_n < t_{n+1}$.*

Note that $p(0) = 1$ by the imposed assumption that X contains the origin almost surely. A p-function and the corresponding regenerative event are said to be *standard* if $p(t) \to 1$ as $t \downarrow 0$. The inequalities for Q from Proposition 4.16 can be restated for the p-function, which leads to the following results proved in Kingman [308].

Theorem 4.17 (Properties of p-function). *Every standard p-function is positive, uniformly continuous on $\mathbb{R}_+$ and of bounded variation in every finite interval. The limit*

$$q = \lim_{t \downarrow 0} t^{-1}(1 - p(t)) \tag{4.8}$$

exists in $[0, \infty]$ and $q = 0$ if and only if p identically equals to 1.

Theorem 4.18 (Laplace transform of p-function). *For every standard p-function p, there exists a unique positive measure μ on $(0, \infty]$ with*

$$\int_{(0,\infty]} (1 - e^{-t})\mu(dt) < \infty, \tag{4.9}$$

such that the Laplace transform $r(\theta) = \int_0^\infty e^{-\theta t} p(t) dt$ satisfies

$$\frac{1}{r(\theta)} = \theta + \int_{(0,\infty]} (1 - e^{-\theta t})\mu(dt) \tag{4.10}$$

for all θ with a positive real part. Equation (4.10) establishes a one-to-one correspondence between standard p-functions and positive measures μ satisfying (4.9).

Lévy measure and subordinator

The measure μ that appears in (4.10) is called the Lévy measure of the corresponding regenerative event. If q from (4.8) is finite, the regenerative event is called *stable* as opposed to an *instantaneous* X which has $q = \infty$. For a stable X, $\mu((0, x]) = qF(x)$, where F is the cumulative distribution function of a strictly positive random variable. Then X is the so-called *alternating renewal process*, i.e. X is the union of exponentially distributed segments with mean length $1/q$ separated by segments of random length with distribution F (all lengths are mutually independent).

If p is a standard p-function, then $\lim_{t\to\infty} p(t) = (1+a)^{-1}$, where $a = \int_{(0,\infty]} x\mu(dx)$. If $\mu(\{\infty\}) > 0$, then X is bounded almost surely and is called transient. If $a = \infty$, then X is called null, while $a < \infty$ identifies a positive X.

Example 4.19 (Alternating renewal process).
(i) If μ is the measure of mass λ concentrated at $\{1\}$, then X is the union of exponentially distributed segments with mean λ^{-1} separated by unit gaps. Then (4.10) implies $r(\theta) = (\lambda + \theta - \lambda e^{-\theta})^{-1}$.
(ii) If μ has the density $\lambda e^{-\lambda t}$, then X is the union of exponentially distributed segments with mean 1 separated by exponentially distributed gaps with mean $1/\lambda$. Then $r(\theta) = (\lambda + \theta)\theta^{-1}(\theta + \lambda + 1)^{-1}$.

Each standard regenerative event X can be represented as the image $\{\zeta(x) : x \geq 0\}$ of an increasing process with independent increments ζ (called *subordinator*) starting from zero and whose cumulant is

$$\mathbf{E}e^{-\theta\zeta(x)} = \exp\left\{-x\left(\theta + \int_{(0,\infty]} (1-e^{-\theta t})\mu(dt)\right)\right\}.$$

As will be shown in Section 5.2.3, this relationship to subordinators holds for the even more general case of strong Markov (or regenerative) random sets. This general case includes also non-separable sets whose distributions are not necessarily determined by their p-functions.

For standard regenerative events the avoidance functional on intervals can be expressed using the p-function and the Lévy measure μ that appears in (4.10).

Proposition 4.20 (Avoidance functional of a regenerative event). *For every standard regenerative event with p-function p and Lévy measure μ,*

$$Q_X((t,s)) = \mathbf{P}\{X \cap [t,s) = \emptyset\}$$
$$= \int_0^t p(v)\mu([s-v,\infty])dv, \quad 0 < t < s < \infty, \qquad (4.11)$$

and

$$\mu([s,\infty]) = \lim_{\varepsilon\downarrow 0} \varepsilon^{-1} Q_X((\varepsilon, s)), \quad s > 0. \qquad (4.12)$$

4 Capacity functionals and properties of random closed sets

Proof. The p-function p is the density of the potential measure U of the subordinator $\zeta(x)$, i.e.
$$U(A) = \mathbf{E}\int_0^\infty \mathbf{1}_A(\zeta(x))dx.$$

If $T(t) = \inf\{x \geq 0 : \zeta(x) > t\}$ is the first passage time strictly above t, then
$$Q_X((t,s)) = \mathbf{P}\{\zeta(T(t)-) \leq t,\ \zeta(T(t)) \geq s\}.$$

It follows from Bertoin [65, Prop. 2, p. 76] that the latter probability is given by
$$\int_0^t \int_s^\infty U(dv)\mu(dz-y) = \int_0^t p(v)\mu([s-v,\infty])dv.$$

Finally, (4.12) can be easily derived by passing to the limit in (4.11). □

Related results will be discussed in greater generality in Section 5.2.3.

4.4 Robbins' theorem

In many cases the capacity functional of X can be used to evaluate the expectation of $\mu(X)$, where μ is a measure on $\mathbb{E}$. The key point is to observe that
$$\mu(X) = \int_\mathbb{E} \mathbf{1}_X(x)\mu(dx),$$

see also Example 1.3(iii). If μ is locally finite, then Fubini's theorem yields the following result.

Theorem 4.21 (Robbins' theorem). *Let X be a random closed set in a Polish space $\mathbb{E}$. If μ is a locally finite measure on Borel sets, then $\mu(X)$ is a random variable and*
$$\mathbf{E}\mu(X) = \int_\mathbb{E} \mathbf{P}\{x \in X\}\,\mu(dx) \tag{4.13}$$

in the sense that if one side is finite then so is the other and they are equal.

Proof. It suffices to show that $\mathbf{1}_F(x) \colon \mathbb{E} \times \mathcal{F} \mapsto \{0,1\}$ is jointly measurable with respect to $\mathfrak{B}(\mathbb{E}) \otimes \mathfrak{B}(\mathcal{F})$, namely,
$$\{(x,F) : \mathbf{1}_F(x) = 0\} = \{(x,F) : x \notin F\}$$
$$= \bigcup_{G \in \mathcal{G}_0} \{(x,F) : x \in G,\ F \cap G = \emptyset\}$$
$$= \bigcup_{G \in \mathcal{G}_0} (G \times \mathcal{F}^G) \in \mathfrak{B}(\mathbb{E}) \otimes \mathfrak{B}(\mathcal{F}),$$

where $\mathcal{G}_0$ is a countable base of topology on $\mathbb{E}$. □

Apart from calculating the expected value of $\mu(X)$, Theorem 4.21 can sometimes be used in the other direction to deduce that $p_X(x)$ vanishes if $\mathbf{E}\mu(X) = 0$ for a sufficiently rich family of measures μ. An easy generalisation of (4.13) for higher-order moments is

$$\mathbf{E}(\mu(X)^k) = \int \cdots \int \mathbf{P}\{\{x_1,\ldots,x_k\} \subset X\}\, dx_1 \cdots dx_k, \qquad (4.14)$$

i.e. the higher-order moments of $\mu(X)$ can be obtained as multiple integrals of the inclusion functional $I_X(\{x_1,\ldots,x_k\})$ of X. An extension of Robbins' theorem for random capacities is discussed in Section 8.5.

Robbins' theorem does not hold for measures which are not locally finite; indeed neither does Fubini's theorem, see e.g. Mattila [382, p. 14]. For example, if μ is the counting measure and $X = \{\xi\}$ is a random singleton with ξ having an absolutely continuous distribution, the left-hand side of (4.13) equals 1, while $\mathbf{P}\{x \in X\} = 0$ for all x, whence the right-hand side vanishes. Similar examples can be easily constructed for X being a point process or any other random set with a vanishing coverage function and non-vanishing $\mu(X)$. For instance, X may be a random collection of curves with $\mu(X)$ being the total curve length. Therefore, Robbins' theorem does not apply to many interesting geometric measures, in particular, the Hausdorff measures. Even the measurability of $\mu(X)$ may fail as the following example shows.

Example 4.22 (Non-measurable $\mu(X)$). Let $g \colon \mathbb{R}^d \mapsto [1, \infty)$ be a non-measurable function. Define a measure on $\mathbb{R}^d$ by

$$\mu(K) = \begin{cases} \sum_{x \in K} g(x) & \text{if } K \text{ is finite}, \\ \infty & \text{otherwise}. \end{cases}$$

Then it is not true that $\mu(X)$ is a random variable for every random closed set X. For example, if $X = \{\xi\}$ is a random singleton, then $\mu(X) = g(\xi)$ is not necessarily a random variable.

4.5 Hausdorff dimension

Bounds on Hausdorff dimension using intersection probabilities

By Robbins' theorem, a random closed set X in $\mathbb{R}^d$ has a positive Lebesgue measure if and only if its coverage function $p(x) = \mathbf{P}\{x \in X\}$ is positive on a set of a positive measure. Then the Hausdorff dimension of X is d. If X is a random singleton (and so has the Hausdorff dimension zero), then $T(K)$ is positive if K has a non-empty interior that includes a part of the support of X. In general, the Hausdorff dimension of X may be assessed by considering the capacity functional on some specially designed sets.

For $n \geq 1$, split the unit cube $[0,1]^d$ in $\mathbb{R}^d$ into 2nd dyadic half-open cubes with side length 2^{-n}. Let $Z^n(p)$ be the union of such cubes where each cube has the probability $p \in (0,1)$ of being included independently of other cubes at any other stage n. Define

4 Capacity functionals and properties of random closed sets　　61

$$Z(p) = \bigcap_{n \geq 0} Z^n(p).$$

The following result shows that the Hausdorff dimension of set K can be explored by considering the intersection of K with $Z(p)$.

Proposition 4.23 (Intersections with $Z(p)$). *Let $\alpha > 0$ and let K be a closed subset of $[0,1]^d$.*
  (i) *If K intersects $Z = Z(2^{-\alpha})$ with positive probability, then $\dim_H(K) \geq \alpha$.*
  (ii) *If X is a random subset of $[0,1]^d$ and X intersects the independent random set $Z = Z(2^{-\alpha})$ with positive probability, then $\dim_H(X) \geq \alpha$ with positive probability.*

Proof.
(i) Let $b = \mathbf{P}\{K \cap Z \neq \emptyset\} > 0$. Then, for any collection of sets A_i formed as unions of dyadic cubes such that $K \subset \cup A_i$, we have

$$b \leq \sum_j \mathbf{P}\{A_j \cap Z \neq \emptyset\} \leq \sum_j \mathbf{P}\{A_j \cap Z^{n_j}(2^{-\alpha}) \neq \emptyset\} \leq \sum_j \operatorname{diam}(A_j)^\alpha,$$

where n_j is defined so that A_j is composed of the cubes of side length 2^{-n_j}. The Carathéodory construction defined in Example E.1 applied for the family $\mathcal{M}$ of dyadic cubes yields the so-called net measure $\mathcal{N}^\alpha(K)$. It is known (see Mattila [382, Sec. 5.2]) that

$$\mathcal{H}^\alpha(K) \leq \mathcal{N}^\alpha(K) \leq 4^\alpha d^{\alpha/2} \mathcal{H}^\alpha(K).$$

If $\dim_H K = \beta < \alpha$, then there exists a collection of dyadic cubes covering K such that $\sum_j \operatorname{diam}(A_j)^\beta < b$, contrary to the assumption. Therefore, $\dim_H K \geq \alpha$.
(ii) By taking conditional expectation and using (i) for a deterministic set K one obtains that

$$0 < \mathbf{P}\{X \cap Z \neq \emptyset\}$$
$$= \mathbf{E}(\mathbf{P}\{X \cap Z \neq \emptyset \mid Z\})$$
$$= \mathbf{E}(\mathbf{1}_{\dim_H X \geq \alpha} \mathbf{P}\{X \cap Z \neq \emptyset \mid Z\})$$
$$\leq \mathbf{E}(\mathbf{1}_{\dim_H X \geq \alpha}) = \mathbf{P}\{\dim_H X \geq \alpha\}. \qquad \square$$

Theorem 4.24 (Lower bound on the Hausdorff dimension). *If a random closed set X has $T_X(K) > 0$ for all deterministic sets K with $\dim_H K > \beta$, then $\dim_H X \geq d - \beta$ with positive probability.*

Proof. Let Z_1 and Z_2 be two independent random sets such that $Z_1 \stackrel{d}{\sim} Z(2^{-(d-\beta)})$ and $Z_2 \stackrel{d}{\sim} Z(2^{-(\beta-\varepsilon)})$. Then

$$Z_1 \cap Z_2 \stackrel{d}{\sim} Z(2^{-(d-\varepsilon)}).$$

By the theory of branching processes, $Z(2^{-(d-\varepsilon)})$ is non-empty with positive probability, since the number of daughter cubes at every step has the expected value $2^{\varepsilon} > 1$. By Proposition 4.23(ii) and letting $\varepsilon \to 0$, one obtains that $\dim_H Z(2^{-(d-\beta)}) \geq \beta$ with positive probability.

The condition of the theorem implies that $\mathbf{P}\{X \cap Z(2^{-(d-\beta-\varepsilon)}) \neq \emptyset\} > 0$ for $\varepsilon > 0$ and $\dim_H X \geq d - \beta$ with positive probability by Proposition 4.23(ii). □

Intersection-equivalence and capacity-equivalence

It is well known that two functions are equivalent at x if their ratio is bounded away from zero and infinity in the neighbourhood of x. It is possible to extend this concept for random (closed) sets using their capacity functionals.

Definition 4.25 (Intersection-equivalent random sets). Two random closed sets X and Y in a Polish space $\mathbb{E}$ are *intersection-equivalent* in the open set G, if there exist constants $c_1, c_2 > 0$ such that for any closed set $F \subset G$,

$$c_1 T_Y(F) \leq T_X(F) \leq c_2 T_Y(F). \tag{4.15}$$

Note that (4.15) means that the ratio $T_X(F)/T_Y(F)$ is bounded above and below by positive constants that do not depend on $F \subset G$. Two random singletons $X = \{\xi\}$ and $Y = \{\eta\}$ with absolutely continuous distributions are intersection-equivalent if the ratio of the densities of ξ and η is bounded away from zero and infinity.

Let cap_f denote the capacity obtained by (E.7) with the kernel

$$k(x, y) = f(\|x - y\|)$$

for a decreasing function $f : \mathbb{R}_+ \mapsto [0, \infty]$.

Definition 4.26 (Capacity equivalence). Two (random) sets X and Y are *capacity equivalent* if there exist positive constants c_1 and c_2 such that

$$c_1 \text{cap}_f(Y) \leq \text{cap}_f(X) \leq c_2 \text{cap}_f(Y)$$

for all decreasing functions f.

The following results are proved by Pemantle, Peres and Shapiro [451].

Theorem 4.27 (Random sets generated by Brownian motion).
(i) The trace of spatial Brownian motion in $\mathbb{R}^d$ with $d \geq 3$ is capacity-equivalent to the unit square $[0, 1]^2$.
(ii) The zero-set $X = \{t \in [0, 1] : w_t = 0\}$ for the Wiener process w_t is capacity-equivalent to the middle-$\frac{1}{2}$ Cantor set K that consists of all points in $[0, 1]$ that have only digits 0 and 3 in their 4-adic expansions.

4.6 Random open sets

It is natural to define random *open* sets as complements to random closed sets, so that $Y: \Omega \mapsto \mathcal{G}$ is called a random open set if its complement $X = Y^c$ is a random closed set. Since $\{Y^c \cap F = \emptyset\} = \{F \subset Y\}$, Y is a random open set if and only if $\{F \subset Y\}$ is a measurable event for every $F \in \mathcal{F}$. Theorem 2.28 implies that the distribution of Y is uniquely determined by its inclusion functional

$$I_Y(F) = \mathbf{P}\{F \subset Y\}, \quad F \in \mathcal{F}.$$

Proposition 4.28 (Closure and interior of random sets). *Assume that $\mathbb{E}$ is LCHS.*
(i) *If Y is a random open set, then its closure $X = \mathrm{cl}(Y)$ is a random closed set.*
(ii) *If X is a random closed set, then its interior $Y = \mathrm{Int}(X)$ is a random open set.*

Proof.
(i) It suffices to note that $\{\mathrm{cl}(Y) \cap G = \emptyset\} = \{Y \cap G = \emptyset\} = \{G \subset Y^c\}$ is a measurable event for every open set G.
(ii) For every $F \in \mathcal{F}$, $\{F \subset Y\} = \{F \cap \mathrm{cl}(X^c) = \emptyset\}$, so that the statement follows from Theorem 2.25(iii). □

The fact that $\mathrm{cl}(Y)$ is a random closed set for an open Y does not imply that Y is a random open set. Proposition 4.28 justifies correctness of the following definition.

Definition 4.29 (Regular closed random set). A random closed set X is called *regular closed* if $X = \mathrm{cl}(\mathrm{Int}(X))$ a.s. A random open set Y is called *regular open* if $Y = \mathrm{Int}(\mathrm{cl}(Y))$ a.s.

Theorem 2.6 implies that X is a random regular closed set if X takes values in the family of regular closed sets and $\{x \in X\}$ is a measurable event for every $x \in \mathbb{E}$. It follows from Theorem 4.9 that the distribution of every regular closed set X is uniquely determined by its inclusion functional I_X defined on the family $\mathfrak{J}$ of finite sets.

It should be noted that for a general open random set Y the inclusion functional cannot be extended uniquely from the family $\mathfrak{J}$ of finite sets onto $\mathcal{K}$. For instance, if $Y = X^c$ where $X = \{\xi\}$ is a random singleton with non-atomic ξ, then $I_Y(L) = 1$ for every finite set L. This situation differs from the case of random closed sets, see (1.33). The following result follows from Theorem 4.9.

Proposition 4.30 (Inclusion functional of random open set). *Let $\mathbb{E}$ be a LCHS space. For every functional $I: \mathfrak{J} \mapsto [0, 1]$ such that $1 - I$ is a capacity functional on $\mathfrak{J}$ there is a unique random open set Y such that $\mathbf{P}\{L \subset Y\} = I(L)$ for all $L \in \mathfrak{J}$ and Y^c is a separable random closed set.*

In particular, if Y is a random convex open set in $\mathbb{R}^d$ (i.e. $Y \in \mathrm{co}\,\mathcal{G}$ a.s), then its complement is a separable random closed set. This yields the following corollary.

Corollary 4.31 (Distribution of random open convex set). *Let I be a functional on the family $\mathfrak{J}$ of finite sets in $\mathbb{R}^d$ such that $I(\emptyset) = 1$, $I(L) \in [0, 1]$ for all $L \in \mathfrak{J}$,*

$1 - I$ is a completely alternating functional on $\mathfrak{J}$, $I(L_n) \to I(L)$ if $\operatorname{co}(L_n) \downarrow \operatorname{co}(L)$ and $I(L) = I(L \cup \{x\})$ if $\operatorname{co}(L) = \operatorname{co}(L \cup \{x\})$. Then there exists a unique random convex open set Y with the inclusion functional I.

4.7 C-additive capacities and random convex sets

C-additivity

Using capacity functionals, it is possible to provide necessary and sufficient conditions for a random closed set X to be almost surely convex. Denote by $\operatorname{co} \mathcal{F}$ (respectively $\operatorname{co} \mathcal{K}$) the family of convex closed (respectively compact) sets in a Banach space $\mathbb{E}$. The families $\operatorname{co} \mathcal{F}$ and $\operatorname{co} \mathcal{K}$ are Effros measurable, for instance,

$$\mathcal{F} \setminus (\operatorname{co} \mathcal{F}) = \bigcup_{x,y \in \mathbb{Q}} \bigcup_{c \in \mathbb{Q}_1} \{F \in \mathcal{F} : \{x, y\} \subset F, \ (cx + (1-c)y) \notin F\}, \quad (4.16)$$

where $[x, y]$ denotes the segment with end-points x and y, $\mathbb{Q}$ is a countable dense set in $\mathbb{R}^d$ and $\mathbb{Q}_1$ is the set of rational points in $[0, 1]$.

Definition 4.32 (Convex random set). A random closed (respectively compact) set X is called random *convex* closed set (respectively random *convex* compact set) if $X \in \operatorname{co} \mathcal{F}$ (respectively $X \in \operatorname{co} \mathcal{K}$) a.s.

To characterise random convex sets in terms of their capacity functionals we make use of the Choquet theorem and so need to assume that $\mathbb{E}$ is locally compact. Without loss of generality consider $\mathbb{E} = \mathbb{R}^d$.

Definition 4.33 (Separated sets). Two compact sets K_1 and K_2 in $\mathbb{R}^d$ are said to be separated by a compact set K if, for every $x_1 \in K_1$ and $x_2 \in K_2$, there is $c \in [0, 1]$ such that $cx_1 + (1-c)x_2 \in K$.

It is easy to see that every two convex compact sets K_1 and K_2 such that $K_1 \cup K_2$ is also convex are separated by $L = K_1 \cap K_2$, see Figure 4.1. The following definition strengthens the subadditive property (1.16) of general capacities.

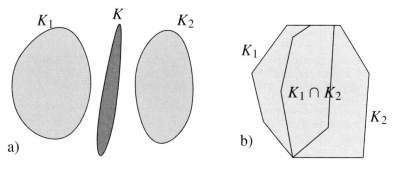

Figure 4.1. a) K_1 and K_2 separated by K; b) K_1 and K_2 separated by $K_1 \cap K_2$.

Definition 4.34 (C-additive capacity).
(i) A capacity φ is said to be *C-additive* on $\mathcal{K}$ if
$$\varphi(K) + \varphi(K \cup K_1 \cup K_2) = \varphi(K \cup K_1) + \varphi(K \cup K_2) \qquad (4.17)$$
for each $K_1, K_2 \in \mathcal{K}$ that are separated by $K \in \mathcal{K}$.
(ii) A capacity φ is said to be C-additive on co $\mathcal{K}$ if
$$\varphi(K_1 \cap K_2) + \varphi(K_1 \cup K_2) = \varphi(K_1) + \varphi(K_2) \qquad (4.18)$$
for all $K_1, K_2 \in \text{co}\,\mathcal{K}$ such that $K_1 \cup K_2 \in \text{co}\,\mathcal{K}$.

Every measure is C-additive with (4.17) valid for disjoint K_1 and K_2 without assuming that K separates K_1 and K_2.

Theorem 4.35 (Convexity and C-additivity). *Let T_X be the capacity functional of a random closed set X in $\mathbb{R}^d$. Then the following statements are equivalent.*
(i) *X is almost surely convex.*
(ii) *T_X is C-additive on $\mathcal{K}$.*
(iii) *T_X is C-additive on co $\mathcal{K}$.*

Proof. If X is a.s. convex, then, for every K_1 and K_2 separated by K, Definition 4.34(i) implies that
$$0 = \mathbf{P}\{X \cap K = \emptyset,\ X \cap K_1 \neq \emptyset,\ X \cap K_2 \neq \emptyset\} = -\Delta_{K_2}\Delta_{K_1} T_X(K),$$
whence T_X satisfies (4.17) so that (ii) follows from (i). Furthermore, (iii) follows from (ii) since K_1 and K_2 are separated by $K_1 \cap K_2$.

It remains to prove that the C-additivity of T_X on co $\mathcal{K}$ implies that X is a.s. convex. For every two points $x, y \in \mathbb{R}^d$ and $z = cx + (1-c)y$ with $c \in [0, 1]$,
$$\mathbf{P}\{\{x, y\} \subset X, z \notin X\} = T_X(K_1) + T_X(K_2) - T_X(K_1 \cap K_2) - T_X(K_1 \cup K_2)$$
$$= 0,$$
which is easily seen by applying (4.18) to $K_1 = [x, z]$ and $K_2 = [z, y]$. Therefore, the probabilities of all events from the countable union in the right-hand side of (4.16) vanish. □

Open problem 4.36. Characterise, in terms of the capacity functional T_X, the property that random closed set X almost surely belongs to the convex ring, i.e. almost all realisations of X are finite unions of convex compact sets.

Open problem 4.37. Apart from the characterisation of convexity it is very difficult to characterise geometric properties of X using its capacity functional. Is it possible to characterise connectivity properties of X using T_X? In particular, this concerns the existence of unbounded connected components, the question typical in continuum percolation theory, see Meester and Roy [387].

Semi-Markov random sets

Definition 4.38 (Semi-Markov random set). A random closed set X in $\mathbb{R}^d$ is said to be *semi-Markov* if its avoidance functional Q_X satisfies

$$Q_X(K \cup K_1 \cup K_2) Q_X(K) = Q_X(K \cup K_1) Q_X(K \cup K_2)$$

for all $K_1, K_2 \in \mathcal{K}$ that are separated by $K \in \mathcal{K}$.

The above definition is equivalent to the property that

$$\Psi_X(K) = -\log(Q_X(K))$$

is a C-additive functional on $\mathcal{K}$. This property will be further discussed in Section 4.1.2 in relation to the union infinite divisibility concept.

Proposition 4.39. *A random closed set X is semi-Markov if and only if $X \cap K_1$ and $X \cap K_2$ are conditionally independent given $\{X \cap K = \emptyset\}$ for K that separates K_1 and K_2.*

Proof. Let K' and K'' be two compact sets. Then

$$\mathbf{P}\left\{(X \cap K_1) \cap K' = \emptyset, (X \cap K_2) \cap K'' = \emptyset \,\middle|\, X \cap K = \emptyset\right\}$$
$$= Q_X((K_1 \cap K') \cup (K_2 \cap K'') \cup K)/Q_X(K)$$
$$= Q_X(K_1 \cap K') Q_X(K_2 \cap K''),$$

since $(K_1 \cap K')$ and $(K_2 \cap K'')$ are also separated by K. Proposition 1.19 (formulated for the avoidance functional) finishes the proof. □

Let X be a stationary semi-Markov random closed set. For every unit vector u and $t_1, t_2 > 0$, Definition 4.38 applied to $K_1 = [0, t_1 u]$, $K_2 = [t_1 u, (t_1 + t_2) u]$ and $K = \{t_1 u\}$ implies that

$$q(u, t_1 + t_2) q_0 = q(u, t_1) q(u, t_2),$$

where $q(u, t) = Q_X([0, tu])$ and $q_0 = Q_X(\{0\})$. Since $q(u, t)$ is monotone with respect to t,

$$q(u, t) = q_0 e^{-\theta(u)t}, \quad t \geq 0,$$

where $\theta(u) = \theta(-u)$ for all u. On the line (if $d = 1$) or for isotropic X, $q(u, t) = q(t)$ does not depend on u and satisfies $q(t) = q_0 e^{-\theta t}$, $t \geq 0$.

Examples of semi-Markov stationary random sets on the line are provided by *alternating renewal processes* given by the union of disjoint random segments of i.i.d. lengths separated by a sequence of i.i.d. exponentially distributed gaps, cf Example 4.19. In particular, if the segments shrink to points, then we obtain a stationary Poisson point process on the real line. Note also that a regenerative event is the complement of a semi-Markov (non-stationary) set.

Open problem 4.40. Characterise semi-Markov sets in $\mathbb{R}^d$. Under an additional assumption of the infinite divisibility for unions, such sets have been characterised in Matheron [381, Th. 5.4.1].

4.8 Comparison of random sets

Comparison of random elements

Let $\mathbb{L}$ be a Polish space partially ordered by a relation $\leq$, so that the set $\{(x, y) \in \mathbb{L} \times \mathbb{L} : x \leq y\}$ is closed in the product topology on $\mathbb{L} \times \mathbb{L}$. A function $f : \mathbb{L} \mapsto \mathbb{R}$ is called *increasing* if $x \leq y$ implies $f(x) \leq f(y)$. A subset $A \subset \mathbb{L}$ is *increasing* if its indicator is an increasing function, i.e. $x \in A$ and $x \leq y$ together imply $y \in A$. The family of all bounded increasing functions (respectively sets) on $\mathbb{L}$ is denoted by $\mathcal{J}(\mathbb{L})$ (respectively $\mathcal{J}_0(\mathbb{L})$). In the lattice theory increasing sets are called upper sets.

Consider two $\mathbb{L}$-valued random elements ξ_1 and ξ_2 which correspond to the probability measures $\mathbf{P}_1$ and $\mathbf{P}_2$ on the family $\mathfrak{B}(\mathbb{L})$ of Borel subsets of $\mathbb{L}$. We say that ξ_1 (or $\mathbf{P}_1$) is *stochastically smaller* than ξ_2 (or $\mathbf{P}_2$) and write $\xi_1 \leq_{st} \xi_2$ (or $\mathbf{P}_1 \leq_{st} \mathbf{P}_2$) if $\mathbf{E}f(\xi_1) \leq \mathbf{E}f(\xi_2)$ for every $f \in \mathcal{J}(\mathbb{L})$. This is equivalent to the requirement that $\mathbf{P}_1(A) \leq \mathbf{P}_2(A)$ for every $A \in \mathcal{J}_0(\mathbb{L})$.

Theorem 4.41 (Stochastic order for probability measures). *The following conditions are equivalent for any two probability measures $\mathbf{P}_1$ and $\mathbf{P}_2$ on $\mathfrak{B}(\mathbb{L})$.*
(1) $\mathbf{P}_1 \leq_{st} \mathbf{P}_2$.
(2) *There exists a probability measure on $\{(x, y) \in \mathbb{L} \times \mathbb{L} : x \leq y\}$ with the first marginal $\mathbf{P}_1$ and the second marginal $\mathbf{P}_2$.*
(3) *There exists a real-valued random variable α and two measurable functions f and g mapping $\mathbb{R}$ into $\mathbb{L}$ with $f \leq g$ such that the distribution of $f(\alpha)$ is $\mathbf{P}_1$ and that of $g(\alpha)$ is $\mathbf{P}_2$.*
(4) *There exist two $\mathbb{L}$-valued random elements $\tilde{\xi}_1$ and $\tilde{\xi}_2$ (providing an ordered coupling of ξ_1 and ξ_2) such that $\tilde{\xi}_1 \leq \tilde{\xi}_2$ a.s. and the distribution of $\tilde{\xi}_i$ is $\mathbf{P}_i$, $i = 1, 2$.*
(5) $\mathbf{P}_1(A) \leq \mathbf{P}_2(A)$ *for all closed $A \in \mathcal{J}_0(\mathbb{L})$.*

Stochastic order for random sets

In the case of random closed sets, $\mathbb{L}$ becomes the family $\mathcal{F} = \mathcal{F}(\mathbb{E})$ of all closed subsets of $\mathbb{E}$ ordered by inclusion. Assume that $\mathbb{E}$ is a LCHS space. Then Theorem B.2(iii) implies that $\mathcal{F}$ is a Polish space in the Fell topology. It is easy to see that $\{(F_1, F_2) : F_1 \subset F_2\}$ is closed in the product Fell topology on $\mathcal{F} \times \mathcal{F}$. Theorem 4.41 is therefore applicable, so that a random closed set X with distribution $\mathbf{P}_1$ is stochastically smaller than a random closed set Y with distribution $\mathbf{P}_2$ if and only if $\mathbf{P}_1(\mathcal{Y}) \leq \mathbf{P}_2(\mathcal{Y})$ for every $\mathcal{Y} \in \mathcal{J}_0(\mathcal{F})$, such that, with every $F \in \mathcal{Y}$, the family $\mathcal{Y}$ contains all closed sets $F' \supset F$. In terms of X and Y this condition can be formulated as

$$\mathbf{P}\{X \in \mathcal{Y}\} \leq \mathbf{P}\{Y \in \mathcal{Y}\} \qquad (4.19)$$

for every increasing family $\mathcal{Y}$ of closed sets. An example of such family is $\mathcal{Y} = \mathcal{F}_K$ for any $K \in \mathcal{K}$. Then (4.19) implies

68   1 Random Closed Sets and Capacity Functionals

$$T_X(K) \leq T_Y(K), \quad K \in \mathcal{K}, \tag{4.20}$$

where T_X and T_Y are the capacity functionals of X and Y respectively. By similar arguments,

$$\mathbf{P}\{X \cap K_1 \neq \emptyset, \ldots, X_n \cap K_n \neq \emptyset\} \leq \mathbf{P}\{Y \cap K_1 \neq \emptyset, \ldots, Y_n \cap K_n \neq \emptyset\} \tag{4.21}$$

for every $n \geq 1$ and all $K_1, \ldots, K_n \in \mathcal{K}$. Using the successive differences introduced in (1.7) and (1.8), it is possible to rewrite (4.21) as

$$\Delta_{K_n} \cdots \Delta_{K_1} T_X(\emptyset) \geq \Delta_{K_n} \cdots \Delta_{K_1} T_Y(\emptyset). \tag{4.22}$$

It should be noted that (4.22) is obtained by specialising (4.19) for $\mathcal{Y} = \mathcal{F}_{K_1,\ldots,K_n}$. Although these families of closed sets do not exhaust all possible choices of $\mathcal{Y} \in \mathcal{J}_0(\mathcal{F})$, the following result confirms that (4.22) is a sufficient condition for (4.19).

Theorem 4.42 (Stochastic order for random closed sets). *Let $\mathbb{E}$ be a LCHS space. For two random closed sets X and Y, X is stochastically smaller than Y if and only if (4.21) or (4.22) holds for every $n \geq 1$ and all compact (or, equivalently, all open, or all closed) sets $K_1, \ldots, K_n$.*

Proof. The necessity of (4.21) follows from (4.19) for all compact sets $K_1, \ldots, K_n$ and furthermore for all closed/open sets by approximations.

A family $\mathcal{Y} \subset \mathcal{F}$ is called *decreasing* if $F \in \mathcal{Y}$ and $F' \subset F$ imply $F' \in \mathcal{Y}$. It suffices to prove that (4.21) implies $\mathbf{P}\{X \notin \mathcal{Y}\} \leq \mathbf{P}\{Y \notin \mathcal{Y}\}$ for every decreasing family $\mathcal{Y}$ which is open in the Fell topology.

It is useful to consider the family $\mathcal{F}$ endowed with decreasing topology having the base $\mathcal{F}^K$, $K \in \mathcal{K}$. Let us show that a decreasing family $\mathcal{Y}$ is open in decreasing topology if $F_n \downarrow F$ for any $F_n, F \in \mathcal{Y}$ implies $F_n \in \mathcal{Y}$ for some n. Suppose that F does not belong to the interior of $\mathcal{Y}$. Then $F \cap K \neq \emptyset$ if $\mathcal{F}^K \subset \mathcal{Y}$ for $K \in \mathcal{K}$. For every $s \notin F$ pick $G_s \in \mathcal{G}$ and $K_s \in \mathcal{K}$ such that $s \in G_s \subset K_s \subset F^c$. Because of the second countability property, $F^c = \cup_{n \geq 1} G_{s_n}$ for a sequence $\{s_n, n \geq 1\}$. Then $\cap_{i=1}^n G_{s_i}^c \downarrow F$, so that by the assumed condition on $\mathcal{Y}$, $\cap_{i=1}^n G_{s_i}^c \in \mathcal{Y}$ for some n. Furthermore, $K = \cap_{i=1}^n K_{s_i} \in \mathcal{K}$ and $F \cap K = \emptyset$. If a closed set H does not intersect K, then $H \subset \cap_{i=1}^n G_{s_i}^c$, so that $H \in \mathcal{Y}$, since $\mathcal{Y}$ is decreasing. Thus, $\mathcal{F}^K \subset \mathcal{Y}$ and $F \cap K = \emptyset$ contrary to the assumption.

The decreasing topology is second countable. To show this, consider a countable base $\mathcal{G}_0$ of $\mathcal{G}$ which consists of open sets with compact closures and note that $F \cap K = \emptyset$ implies $K \subset F^c$, so that a finite collection of $G_i \in \mathcal{G}_0$ satisfies $K \subset \cup_i G_i \subset \cup_i \text{cl}(G_i) \subset F^c$. Thus, $F \cap \cup_i \text{cl}(G_i) = \emptyset$, i.e. a countable base of the decreasing topology is composed of the families $\mathcal{F}^K$ for K being finite unions of sets from $\mathcal{G}_0$.

Consider a decreasing set $\mathcal{Y}$ open in the Fell topology. Let $F_n, F \in \mathcal{Y}$, $n \geq 1$, with $F_n \downarrow F$. Then F_n converges to F in the Fell topology, so that $F_n \in \mathcal{Y}$ for sufficiently large n (since $\mathcal{Y}$ is open). By the above arguments, $\mathcal{Y}$ is open in the decreasing topology. Since the decreasing topology is second countable, $\mathcal{Y} = \cup_{n \geq 1} \mathcal{F}^{K_n}$ for some $K_1, K_2, \ldots \in \mathcal{K}$. Now (4.21) implies

$$\mathbf{P}\{X \not\subset \mathcal{Y}\} = \lim_n \mathbf{P}(\cap_{i=1}^n \{X \cap K_i \neq \emptyset\})$$
$$\leq \lim_n \mathbf{P}(\cap_{i=1}^n \{Y \cap K_i \neq \emptyset\}) = \mathbf{P}\{Y \not\subset \mathcal{Y}\} \,,$$

which finishes the proof. □

Application to selections

If (4.20) holds, then X is said to be *smaller in capacity* than Y. It should be noted that (4.20) is not sufficient in general to deduce that X is stochastically smaller than Y.

Example 4.43 (Smaller in capacity does not imply stochastically smaller).
(i) Consider a two-points space $\mathbb{E} = \{a, b\}$ with the discrete topology (so that all subsets are open). Let X be empty with probability $2/3$ and $X = \{a, b\}$ otherwise. A random closed set Y takes values $\{a\}$ and $\{b\}$ with probabilities $3/8$ each and the value $\{a, b\}$ with probability $1/4$. Then (4.20) holds for all K, whereas $\mathbf{P}\{X = \{a, b\}\} > \mathbf{P}\{Y = \{a, b\}\}$ which is impossible if X is stochastically smaller than Y.
(ii) Consider a random closed set X on the line which is equal to $[1/3, 2/3]$ with probability $1/2$ and to $[0, 1]$ otherwise. Let Y take values $[0, 2/3]$ and $[1/3, 1]$ with probabilities $1/2$. Then X is smaller in capacity than Y. Let $f(B) = 1$ if $B \supset [0, 1]$ and $f(B) = 0$ otherwise. Then $\mathbf{E} f(X) = 1/2$ is strictly greater than $\mathbf{E} f(Y) = 0$, so that X is not stochastically smaller than Y.

However, (4.20) does imply that $X \leq_{st} Y$ if X is a singleton as the following result shows.

Corollary 4.44 (Selectionable distributions). *Assume $\mathbb{E}$ is LCHS. An $\mathbb{E}$-valued random element ξ is stochastically smaller than a random closed set Y if and only if*

$$\mathbf{P}\{\xi \in G\} \leq \mathbf{P}\{Y \cap G \neq \emptyset\} = T_Y(G) \qquad (4.23)$$

for all $G \in \mathcal{G}$ (alternatively, open sets G can be replaced by compact sets $K \in \mathcal{K}$ or closed sets $F \in \mathcal{F}$).

Proof. Since necessity is trivial, we have to prove sufficiency only. Put $X = \{\xi\}$ (the proof also works for a non-Hausdorff sober space $\mathbb{E}$ with X given by the closure of $\{\xi\}$). By (4.23),

$$\mathbf{P}\{X \cap G_1 \neq \emptyset, \dots, X \cap G_n \neq \emptyset\} = \mathbf{P}\{\xi \in (\cap G_i)\}$$
$$\leq \mathbf{P}\{Y \cap (\cap G_i) \neq \emptyset\}$$
$$\leq \mathbf{P}\{Y \cap G_1 \neq \emptyset, \dots, Y \cap G_n \neq \emptyset\} \,.$$

By Theorem 4.42, $X \leq_{st} Y$, which immediately yields the required statement. □

Since $\{Y \subset F\} = \{Y \cap F^c = \emptyset\}$ for $F \in \mathcal{F}$, it is easy to see that (4.23) is equivalent to the condition $\mathbf{P}\{Y \subset F\} \leq \mathbf{P}\{\xi \in F\}$ for all $F \in \mathcal{F}$. The fact that $\{\xi\} \leq_{st} Y$ means that there exists a random element $\hat{\xi} \stackrel{d}{\sim} \xi$ and a random closed set $\hat{Y} \stackrel{d}{\sim} Y$ such that $\hat{\xi}$ is a selection of $\hat{Y}$. Then the distribution $\mathbf{P}$ of ξ is called Y-selectionable, see Section 2.2. An application of Theorem 3 from Strassen [550] shows that if $\mathbf{P}$ is dominated by T_X, then

$$P(A) = \mathbf{E}\mu(A), \qquad (4.24)$$

where μ is a random measure supported by X. For instance, μ may be chosen to be an atomic measure concentrated at a selection of X, so that $\mathbf{P}$ becomes the distribution of the corresponding selection.

5 Calculus with capacities

5.1 Choquet integral

Definition and basic properties

Consider a function f that maps $\mathbb{E}$ into $\mathbb{R}_+ = [0, \infty)$. If φ is a functional defined on subsets of $\mathbb{E}$ such that $\varphi(\{x : f \geq t\})$ is well defined for every $t > 0$, then the *Choquet integral* of f with respect to φ is defined as

$$\int f \, d\varphi = \int_0^\infty \varphi(\{x : f \geq t\}) dt. \qquad (5.1)$$

This integral can be restricted to a subset $M \subset \mathbb{E}$ as

$$\int_M f \, d\varphi = \int f \mathbf{1}_M \, d\varphi = \int_0^\infty \varphi(\{x \in M : f \geq t\}) dt.$$

In particular, the definition of the Choquet integral is applicable if f is a measurable function and φ is one of functionals determined by a random closed set X, e.g. the capacity functional T_X or the containment functional C_X.

Theorem 5.1 (Choquet integral with respect to distributions of random sets). *Let X be an almost surely non-empty random closed set. For every measurable non-negative function f,*

$$\int f \, dT_X = \mathbf{E} \sup f(X), \qquad (5.2)$$

$$\int f \, dC_X = \mathbf{E} \inf f(X), \qquad (5.3)$$

where $f(X) = \{f(x) : x \in X\}$. If X may be empty with positive probability, then (5.2) holds with $\sup \emptyset = 0$.

Proof. The proof follows from Fubini's theorem, since

$$\int f \, dT_X = \int_0^\infty T_X(\{x : f(x) \geq t\}) \, dt$$

$$= \int_0^\infty \mathbf{P}\{\alpha \geq t\} \, dt$$

$$= \mathbf{E}\alpha,$$

where $\alpha = \sup\{f(x) : x \in X\}$. The second statement can be proved similarly. □

Proposition 5.2 (Properties of Choquet integral). *Consider non-negative functions f and g for which the Choquet integral (5.1) is defined for a subadditive functional φ. Then*
(i) *for every $c \geq 0$, $\int (cf) d\varphi = c \int f d\varphi$;*
(ii) *for every $a \geq 0$, $\int (f + a) d\varphi = a + \int f d\varphi$;*
(iii) *$\int (f + g) d\varphi \leq \int f d\varphi + \int g d\varphi$.*

Proof. For $\varphi = T_X$, the proofs immediately follow from Theorem 5.1. The general case can be easily handled using the subadditivity of φ, see [135]. □

Since

$$\int f \, dC_X \leq \int f \, dT_X \quad (5.4)$$

with a *superlinear* functional of f in the left and a *sublinear* functional in the right-hand side, the sandwich theorem (Theorem A.5) implies that there exists a linear functional of f sandwiched between the two sides of (5.4). This linear functional can be represented as an integral $\int f d\mu$, where the corresponding probability measure μ can be identified as the distribution of a selection of X.

Example 5.3 (Choquet integral with respect to sup-measure). Let $\varphi = f^\vee$ be the sup-measure generated by an upper semicontinuous function $f : \mathbb{E} \mapsto [0, 1]$. The corresponding random closed set X appears as $\{x : f(x) \geq \alpha\}$ for α uniformly distributed on $[0, 1]$, see Example 1.15. By Theorem 5.1, $\int g d\varphi = \mathbf{E} \sup\{g(x) : f(x) \geq \alpha\}$. In particular, $\int_M f d\varphi = \varphi(M)^2$ and $\int_{\{y\}} g d\varphi = g(y) f(y)$.

Comonotonic additivity

It is easy to check either directly or with the help of Theorem 5.1 that if φ is a measure, then $\int f d\varphi$ coincides with the usual definition of the Lebesgue integral. In difference to the Lebesgue integral, the Choquet integral is not additive in general. However, its additivity property can be verified if the integrated functions are comonotonic as described in the following definition.

Definition 5.4 (Comonotonic functions). Real-valued functions f and g are called *comonotonic* if $(f(x) - f(x'))(g(x) - g(x')) \geq 0$ for all $x, x' \in \mathbb{E}$. Furthermore, f and g are strongly comonotonic if, for all $x, x' \in \mathbb{E}$, $f(x) < f(x')$ if and only if $g(x) < g(x')$.

The following proposition is easy to prove for $\varphi = T_X$ by using Theorem 5.1 and the fact that

$$\sup\{af(x) + bg(x) : x \in X\} = a \sup f(X) + b \sup g(X)$$

if f and g are comonotonic. The case of a general φ is considered in Dellacherie [130] and Denneberg [135].

Theorem 5.5 (Comonotonic additivity). *For every two comonotonic functions f and g and every $a, b > 0$*

$$\int (af + bg) \, d\varphi = a \int f \, d\varphi + b \int g \, d\varphi$$

holds for every functional φ.

The comonotonicity establishes an equivalence relationship on the family of functions, so that a finite collection of functions is comonotonic if and only if all functions are pairwise comonotonic.

If $\varphi(\mathbb{E}) = 1$, the Choquet integral can be consistently defined for not necessarily non-negative functions as

$$\int f \, d\varphi = \int_0^\infty \varphi(\{x : f(x) \geq t\}) \, dt - \int_{-\infty}^0 [1 - \varphi(\{x : f(x) \geq t\})] \, dt.$$

This integral is called the *upper* integral, while the *lower* integral is defined as

$$(L) \int f \, d\varphi = \int_0^\infty [1 - \varphi(\{x : f(x) \leq t\})] \, dt - \int_{-\infty}^0 \varphi(\{x : f(x) \leq t\}) \, dt.$$

It is easy to see that the upper integral with respect to the capacity functional T_X coincides with the lower integral with respect to the containment functional, the property shared by each pair of dual capacities. Similarly to Theorem 5.1, it is easy to see that if X is almost surely non-empty, then

$$\int f \, dT_X = \mathbf{E} |\sup f(X)|, \quad \int f \, dC_X = \mathbf{E} |\inf f(X)|.$$

5.2 The Radon–Nikodym theorem for capacities

Indefinite Choquet integral

Observe that the Choquet integral of f with respect to a capacity φ yields a new capacity as

$$\psi(K) = \int_K f \, d\varphi, \quad K \in \mathcal{K}. \tag{5.5}$$

Then ψ is said to be an *indefinite integral* of φ and the function f is called the *Radon–Nikodym derivative* of ψ with respect to φ, i.e.

$$f(x) = \frac{d\psi}{d\varphi}(x), \quad x \in \mathbb{E}.$$

Proposition 5.6 (Alternation and semicontinuity). *The degree of alternation (monotonicity) of ψ defined by (5.5) is not less than the analogous degree of φ. In particular if φ is completely alternating (monotonic), then so is ψ. The capacity ψ is upper semicontinuous if both f and φ are upper semicontinuous.*

Proof. It is easy to see that

$$\Delta_{K_n} \cdots \Delta_{K_1} \psi(K) = \int_0^\infty \Delta_{K_n} \cdots \Delta_{K_1} \varphi(\{x \in K : f(x) \geq t\}) \, dt,$$

whence ψ is alternating (monotonic) of a certain degree if φ is. The upper semicontinuity of ψ follows from the monotone convergence theorem. □

In particular, if $\varphi = T_X$ and $\mathbf{E} \sup f(X) \leq 1$, then

$$T_Y(K) = \int_K f \, dT_X = \mathbf{E} f^\vee(X \cap K) \tag{5.6}$$

is a capacity functional of a certain random closed set Y. Therefore, the Choquet integral provides a way of introducing new capacity functionals of random closed sets.

Example 5.7 (Capacity functional defined by indefinite integrals). Let T_Y be defined by (5.6).
  (i) If $f(x) = \mathbf{1}_F(x)$ for $F \in \mathcal{F}$, then $T_Y(K) = \mathbf{P}\{X \cap K \cap F \neq \emptyset\}$ is the capacity functional of $Y = X \cap F$.
  (ii) Let $f(x) = \mathbf{1}_{F_1}(x) + p\mathbf{1}_{F_2}(x)$ with $p \in (0, 1)$ and disjoint closed sets F_1 and F_2. Then Y is the union of $X \cap F_1$ and the random set that equals $X \cap F_2$ with probability p independently of X and is empty otherwise.
  (iii) Let $f(x) = e^{-\|x\|}$ on $\mathbb{E} = \mathbb{R}^d$. Then Y is the intersection of X and independent random ball B_ξ centred at the origin with the exponentially distributed radius of mean 1. Indeed,

$$\mathbf{P}\{X \cap K \cap B_\xi \neq \emptyset\} = \mathbf{P}\{\xi \geq \rho(0, X \cap K)\}$$
$$= \mathbf{E} e^{-\rho(0, X \cap K)} = \mathbf{E} f^\vee(X \cap K).$$

(iv) Let $X = \{\xi\}$ be a random singleton. If $\sup f = 1$, then

$$T_Y(K) = \mathbf{E}(f(\xi)\mathbf{1}_K(\xi))$$

is a probability measure, i.e. Y is a random singleton too.

(v) Let X be a stationary Poisson point process in $\mathbb{R}^d$ with intensity λ, see Definition 8.7. Then

$$\mathbf{E}f^{\vee}(X \cap K) = \int_0^{\sup f} \left(1 - e^{-\operatorname{mes}(\{x \in K: f(x) \geq s\})}\right) dt.$$

Absolute continuity

If the capacity functionals of X and Y are related by (5.6), then $T_Y(K) = 0$ given that $T_X(K) = 0$. This is a particular case of absolute continuity of capacities as defined below.

Definition 5.8 (Absolutely continuous capacities). A capacity ψ is *absolutely continuous* with respect to φ (notation $\psi \ll \varphi$) if, for every $K \in \mathcal{K}$, $\psi(K) = 0$ provided $\varphi(K) = 0$.

While absolute continuity of measures implies the existence of the corresponding Radon–Nikodym derivative, this is no longer true for capacities. To see this, consider the case of a finite $\mathbb{E}$ of cardinality n. Then f is determined by n numbers, which clearly do not suffice to define uniquely a capacity on subsets of $\mathbb{E}$ that may need up to $2^n - 1$ numbers to be completely determined.

Strong decomposition and Radon–Nikodym theorem

Let us formulate here a general *Radon–Nikodym theorem* for capacities φ and ψ that are monotone, subadditive and continuous from below. The pair (φ, ψ) is said to have a *strong decomposition property* if, for every $t \geq 0$, there exists a measurable set A_t such that the following conditions hold

$$t(\psi(A) - \psi(B)) \leq \varphi(A) - \varphi(B) \quad \text{if } B \subset A \subset A_t, \tag{5.7}$$
$$t(\psi(A) - \psi(A \cap A_t)) \geq \varphi(A) - \varphi(A \cap A_t) \quad \text{for all } A. \tag{5.8}$$

Every two measures (μ, ν) possess a strong decomposition property and A_t can be derived from the Hahn decomposition of the signed measure $t\nu - \mu$. The strong decomposition property can be formulated as follows.

Definition 5.9 (Strong decomposition property). The pair (φ, ψ) has a *strong decomposition property* if, for every $t > 0$, there exists a set A_t such that, for $w_t = t\psi - \varphi$, the set function $w_t(A_t \cup \cdot)$ is non-decreasing and $w_t(A_t \cap \cdot)$ is non-increasing.

The following result is proved by Graf [208].

Theorem 5.10 (Radon–Nikodym theorem for capacities). *For every two capacities φ and ψ, ψ is an indefinite integral of φ if and only if (φ, ψ) has the strong decomposition property and $\psi \ll \varphi$.*

Open problem 5.11. Consider a random closed set with the capacity functional T. For a fixed $L \in \mathcal{K}$ define $T_L(K) = \mathbf{P}\{X \cap K \neq \emptyset,\ X \cap L \neq \emptyset\}$, $K \in \mathcal{K}$. It is evident that $T_L \ll T$. Does the pair (T, T_L) possess a strong decomposition property and, if yes, what is the corresponding Radon–Nikodym derivative?

Open problem 5.12. Interpret the conditions of the Radon–Nikodym theorem for capacities (Theorem 5.10) for completely alternating capacities that correspond to distributions of random closed sets. As a first step, note that (5.7) and (5.8) written for $\psi = T_X$ and $\varphi = T_Y$ mean that $t\mathbf{P}_X(\mathcal{F}_A^B) \leq \mathbf{P}_Y(\mathcal{F}_A^B)$ if $B \subset A \subset A_t$ and $t\mathbf{P}_X(\mathcal{F}_A^{A \cap A_t}) \geq \mathbf{P}_Y(\mathcal{F}_A^{A \cap A_t})$ for all A.

5.3 Dominating probability measures

Upper probability

Let $\mathbb{E}$ be a LCHS space. Corollary 4.44 says that if the capacity functional T_X of a non-empty random closed set X dominates a probability measure μ, then X possesses a selection with distribution μ. Let $\mathbb{P}_X$ be the family of all probability measures μ that are dominated by T_X, i.e. $\mu(K) \leq T_X(K)$ for each $K \in \mathcal{K}$. Note that $\mathbb{P}_X$ can be alternatively defined as the family of all measures that dominate the containment functional C_X, implying that all measures from $\mathbb{P}_X$ are "sandwiched" between C_X and T_X. The following result establishes that the capacity functional is *upper probability*, i.e. it equals the upper envelope of all probability measures that it dominates.

Theorem 5.13 (Capacity functional as upper probability). *For every almost surely non-empty random closed set X in a LCHS space,*

$$T_X(K) = \sup\{\mu(K) : \mu \in \mathbb{P}_X\}, \quad K \in \mathcal{K}.$$

Proof. Consider an arbitrary $K \in \mathcal{K}$. Let ξ be a selection of $X \cap K$ if $X \cap K \neq \emptyset$ and otherwise let ξ be equal to any other selection of $X \setminus K$ (note that $X \setminus K$ is closed if $X \cap K = \emptyset$). Then $\xi \in K$ if and only if $X \cap K \neq \emptyset$. If μ is the distribution of ξ, then μ is dominated by T_X and $\mu(K) = T_X(K)$. □

Proposition 5.14. *For every bounded upper semicontinuous non-negative function f on $\mathbb{E}$ and a capacity functional T there exists a probability measure μ such that $T(\{x : f(x) \geq t\}) = \mu(\{x : f(x) \geq t\})$ for all $t \geq 0$.*

Proof. Consider a random closed set X with the capacity functional T. The statement is easily seen by identifying the required μ with the distribution of a selection $\xi \in \mathcal{S}(X)$ that satisfies $f(\xi) = \sup f(X)$. □

The capacity T is said to be *dichotomous* if, for each compact set K and each $\varepsilon > 0$, there are disjoint compact sets K_1 and K_2 contained in K such that $T(K_i) \geq T(K) - \varepsilon$. It is known that the Newton capacity is dichotomous with $\varepsilon = 0$. Also it is easy to see that the maxitive capacity $f^\vee$ is dichotomous if f is continuous near its maximum point. Note that a probability measure is not dichotomous.

The capacity functional T_X is called *equalised* if $(C_X(K) + T_X(K))/2$ is a probability measure. If $X \neq \emptyset$ a.s., Theorem 5.1 implies that T_X is equalised if and only if
$$I(f) = \mathbf{E}(\sup f(X) + \inf f(X))$$
is a linear functional of f. This is the case if $X = \{\xi, \eta\}$ consists of at most two points, since
$$\mathbf{E}[\max(f(\xi), f(\eta)) + \min(f(\xi), f(\eta))] = \mathbf{E}[f(\xi) + f(\eta)]$$
is a linear functional of f.

Finite space case

For capacities on finite spaces, the Möbius inversion can be used to obtain a sufficient domination condition.

Proposition 5.15 (Dominating measures on finite spaces). *Assume that $\mathbb{E}$ is finite. Let C_X be a containment functional and let P_X be its Möbius inverse defined by (1.36). Then every measure μ satisfying*
$$\mu(\{x\}) = \sum_{B \ni x} \lambda(B, x) P_X(B)$$
dominates C_X. Here $\lambda(B, x)$ is any non-negative function defined for $x \in B \subset \mathbb{E}$ satisfying
$$\sum_{B \ni x} \lambda(B, x) = 1, \quad B \subset \mathbb{E}.$$

The following result provides a characterisation of equalised capacity functionals for the case of finite $\mathbb{E}$.

Proposition 5.16 (Equalised capacity functionals on finite space). *Let $\mathbb{E}$ be a finite space. Then T_X is equalised if and only if the cardinality of X is at most 2.*

Proof. For every $L \subset \mathbb{E}$ denote $P(L) = \mathbf{P}\{X = L\}$. Then $C(K) = \sum_{L \subset K} P(L)$ and $T(K) = \sum_{K \cap L \neq \emptyset} P(L)$. The equalising condition yields
$$0 = \sum_{x \in \mathbb{E}} (C(\{x\}) + T(\{x\})) - 2$$
$$= \sum_{x \in \mathbb{E}} P(\{x\}) + \sum_{x \in \mathbb{E}} \sum_{L \ni x} P(L) - 2 \sum_{L \subset \mathbb{E}} P(L).$$

Rearrangement of the terms verifies that

$$\sum_{x \in \mathbb{E}} \sum_{L \ni x} P(L) = \sum_{k=1}^{\operatorname{card}(\mathbb{E})} k \sum_{\operatorname{card}(L)=k} P(L)$$

$$= \sum_{k=3}^{\operatorname{card}(\mathbb{E})} k \sum_{\operatorname{card}(L)=k} P(L) + 2 \sum_{\operatorname{card}(L)=2} P(L) + \sum_{x \in \mathbb{E}} P(\{x\}).$$

Therefore,

$$0 = \sum_{k=3}^{\operatorname{card}(\mathbb{E})} (k-2) \sum_{\operatorname{card}(L)=k} P(L),$$

meaning that $P(L) = 0$ whenever $\operatorname{card}(L) > 2$. □

Open problem 5.17. Characterise equalised capacity functionals on a non-finite space $\mathbb{E}$.

5.4 Carathéodory's extension

Carathéodory's construction described in Appendix E makes it possible to construct a measure $\bar{\varphi}$ from a given capacity φ on a family of sets. Fix a family of sets $\mathcal{M}$ that contains all open balls and is used to construct $\bar{\varphi}$ as described in (E.1) and (E.2). Henceforth we assume that $\mathbb{E} = \mathbb{R}^d$.

Proposition 5.18 (Subadditivity and extension of a measure).
  (i) If φ is subadditive on $\mathcal{M}$, i.e. $\varphi(M_1 \cup M_2) \leq \varphi(M_1) + \varphi(M_2)$ for all $M_1, M_2 \in \mathcal{M}$, then $\varphi \leq \bar{\varphi}$ on $\mathcal{M}$.
  (ii) If φ is the restriction to $\mathcal{M}$ of a locally finite measure μ, then $\bar{\varphi} = \mu$.

Proof.
(i) follows from the fact that Carathéodory's construction preserves the monotonicity property.
(ii) Clearly, $\bar{\varphi} \geq \mu$ and $\bar{\varphi}(M) = \mu(M)$ for every $M \in \mathcal{M}$. Let G be a relatively compact open set, so that $\mu(G) < \infty$ and $\bar{\varphi}(G) < \infty$ also. By the Besicovitch covering theorem (see Federer [167, Th. 2.8.15]), there is a countable disjoint collection of balls $\{B_n, n \geq 1\}$ contained in G such that $\bar{\varphi}(C) = 0$ for $C = G \setminus \bigcup_{n \geq 1} B_n$. Hence $\mu(C) = 0$. But then

$$\bar{\varphi}(G) = \sum_{n \geq 1} \bar{\varphi}(B_n) = \sum_{n \geq 1} \mu(B_n) = \mu(G) - \mu(C) = \mu(A).$$

Thus, μ agrees with $\bar{\varphi}$ on relatively compact open sets, and hence on Borel sets. □

In view of applications discussed later on in Section 8.5, consider Carathéodory's extension $\overline{\upsilon \varphi}$ of the product of two capacities υ and φ. Write $f^\vee$ for a sup-measure generated by a function f.

78   1 Random Closed Sets and Capacity Functionals

Theorem 5.19 (Extension of the product). Suppose $\bar{\varphi} = \mu$ is a locally finite measure.

(i) Assume that υ is a uniformly bounded capacity. Then $f(x) = \upsilon(\{x\})$ is an upper semicontinuous function and, for all Borel A,

$$\overline{\upsilon\varphi}(A) = \int_A f \, d\mu \, .$$

(ii) Let $f \colon \mathbb{R}^d \mapsto \mathbb{R}_+$ be a bounded upper semicontinuous function such that $\int_K f \, d\mu < \infty$ for all compact sets K. Then, for all Borel A,

$$\overline{f^\vee \varphi}(A) = \int_A f \, d\mu \, . \tag{5.9}$$

Proof. We prove only (ii); statement (i) is similar. Write $\nu = \overline{f^\vee \varphi}$ and $\eta(A) = \int_A f \, d\mu$. Suppose A is Borel with $\eta(A) < \infty$. Choosing $\varepsilon > 0$, partition A into a finite number of disjoint Borel sets B_i such that

$$\sum_i a_i \mathbf{1}_{B_i} \leq f \mathbf{1}_A \leq \sum_i a_i \mathbf{1}_{B_i} + \varepsilon \tag{5.10}$$

for some $a_i \in \mathbb{R}$. Integrating over B_i yields that

$$a_i \mu(B_i) \leq \eta(B_i) \leq (a_i + \varepsilon)\mu(B_i) \, .$$

Since η is bounded on compact sets, it is also a locally finite measure and there exist open sets $G_i \supset B_i$ such that

$$\eta(G_i) < (1 + \varepsilon)\eta(B_i) \, , \quad \mu(G_i) < (1 + \varepsilon)\mu(B_i) \, .$$

By the Besicovitch covering theorem there are open balls $C_{ni} \subset G_i$ covering ν-almost all of G_i such that $\text{diam}(C_{ni}) < \delta$ and $f(x_{ni}) \leq f^\vee(C_{ni}) \leq f(x_{ni}) + \varepsilon$, where x_{ni} is the centre of C_{ni}.

By (5.10) we have $a_i \leq f(x_{ni}) \leq a_i + \varepsilon$ so that $a_i \leq f^\vee(C_{ni}) \leq a_i + 2\varepsilon$. Hence for each i

$$\overline{(f^\vee \varphi)}_\delta(B_i) \leq \sum_{n \geq 1} f^\vee(C_{ni}) \varphi(C_{ni}) \leq (a_i + 2\varepsilon) \sum_{n \geq 1} \varphi(C_{ni})$$

giving

$$\nu(B_i) \leq (a_i + 2\varepsilon) \sup_{\delta > 0} \overline{\varphi}_\delta(G_i) = (a_i + 2\varepsilon)\mu(G_i) < (a_i + 2\varepsilon)(1 + \varepsilon)\mu(B_i) \, ,$$

so that

$$\nu(A) \leq \sum_i (a_i + 2\varepsilon)(1 + \varepsilon)\mu(B_i) \, .$$

Since ε was arbitrary we have $\nu(A) \leq \eta(A)$.

Next we prove that (5.9) holds for each upper semicontinuous step-function $f = \sum \alpha_i \mathbf{1}_{B_i}$, where $\alpha_i > 0$ for all $i = 1, \ldots, k$ and $B_1, \ldots, B_k$ are disjoint Borel sets. Since the B_i are disjoint we have $\overline{f^\vee \varphi}(A) = \sum_i \overline{f^\vee \varphi}(A \cap B_i)$. For any δ-cover $\{C_n\}$ of $A \cap B_i$, without loss of generality discarding sets C_n which do not intersect $A \cap B_i$, we have $f^\vee(C_n) \geq \alpha_i$, so that

$$(\overline{f^\vee \varphi})_\delta (A \cap B_i) \geq \alpha_i \overline{\varphi}_\delta (A \cap B_i).$$

Therefore, $\overline{f^\vee \varphi}(A \cap B_i) \geq \alpha_i \mu(A \cap B_i)$. Finally,

$$\overline{f^\vee \varphi}(A) \geq \sum \overline{f^\vee \varphi}(A \cap B_i) \geq \sum \alpha_i \mu(A \cap B_i) = \int_{A \cap (\cup B_i)} f \, d\mu = \int_A f \, d\mu.$$

From this we obtain (5.9) for the step-function f.

Now approximate f by μ-integrable upper semicontinuous step-functions f'_n such that $f'_n(x) \uparrow f(x)$ as $n \to \infty$ for all $x \in A$. By the monotone convergence theorem,

$$\nu(A) = \overline{f^\vee \varphi}(A) \geq \overline{(f'_n)^\vee \varphi}(A) = \int_A f'_n \, d\mu \to \int_A f \, d\mu = \eta(A)$$

so that $\nu = \eta$. $\square$

The following proposition concerns one particularly important case of φ being the *indicator capacity* $\varphi(K) = \mathbf{1}_{F \cap K \neq \emptyset}$ generated by a closed set F. If $\bar{\upsilon}$ is a locally finite (Radon) measure, then Equation (5.11) below is the special case of (5.9) with f being the indicator of F. A simple argument based on checking that any δ-cover of $F \cap K$ can be extended to a δ-cover of K without increasing the sum $\sum \varphi(M_n) \upsilon(M_n)$ shows that the result holds without assuming that $\bar{\upsilon}$ is a locally finite measure.

Proposition 5.20. *For a closed set $F \subset \mathbb{R}^d$ let $\varphi(K) = \mathbf{1}_{F \cap K \neq \emptyset}$ be the indicator capacity. Then, for every set function υ, $\overline{\varphi \upsilon}$ is the restriction of $\bar{\upsilon}$ to F, i.e.*

$$\overline{\varphi \upsilon}(K) = \bar{\upsilon}(F \cap K). \tag{5.11}$$

5.5 Derivatives of capacities

Definition

The definition of the derivative for capacities relies on the vague convergence concept for capacities defined in Appendix E. Assume throughout that $\mathbb{E} = \mathbb{R}^d$.

Definition 5.21 (Derivative for capacities). *A capacity φ is said to be differentiable at $K \in \mathcal{K}$ if, for some $\alpha > 0$ (called the exponent of the derivative), the capacity*

$$\frac{\varphi(K + (tL \cup \{0\})) - \varphi(K)}{t^\alpha}, \quad L \in \mathcal{K},$$

converges vaguely as a function of L to $d_L\varphi(K)$ as $t \downarrow 0$. The limit $d_L\varphi(K)$ (called the *derivative* of φ at K) may be infinite, but it is assumed that $0 < d_L\varphi(K) < \infty$ for at least one $L \in \mathcal{K}$.

If $K = \{x\}$ is a singleton, write $d_L\varphi(x)$ instead of $d_L\varphi(\{x\})$. The differential of a measure μ is defined by Faro, Navarro and Sancho [166] as the weak limit of $\mu(x + tL)/t^\alpha$ as $t \downarrow 0$. In our terms this corresponds to $d_L\mu(\{x\})$, since a differentiable measure satisfies $\mu(\{x\}) = 0$.

Example 5.22 (Derivative at singletons). If φ does not charge singletons (i.e. $\varphi(\{x\}) = 0$ for all x), is homogeneous and translation invariant, then $d_L\varphi(x) = \varphi(L)$. If μ is a measure $\mathbb{R}^d$ with density p with respect to the Lebesgue measure, then $d_L\mu(x) = p(x)\text{mes}(L)$. Similarly,

$$d_L\mu(\{x_1, \ldots, x_n\}) = \text{mes}(L) \sum_{i=1}^{n} p(x_i). \qquad (5.12)$$

Derivatives of capacity functionals

If $\varphi = T$ is the capacity functional of a random closed set X, then $d_L T(K)$ is completely alternating as a function of L and is upper semicontinuous as a vague limit of upper semicontinuous capacities. By Proposition 1.11,

$$T^K(L) = 1 - \exp\{-d_L T(K)\}, \quad L \in \mathcal{K}, \qquad (5.13)$$

is a capacity functional of a random closed set. Considered as a function of L, the derivative $d_L T(K)$ is homogeneous of order α, that is $d_{cL} T(K) = c^\alpha d_L T(K)$ for all $c > 0$. In the theory of random sets homogeneous capacities arise naturally as probability distributions of union-stable random closed sets, see Section 4.1.3. For example, the derivative in (5.12) corresponds to the Poisson point process with intensity $\sum p(x_i)$.

If L contains the origin, then $d_L T(K)$ appears as the normalised limit of

$$T((K + tL) \cup L) - T(K) = \mathbf{P}\{X \cap K = \emptyset, \ X \cap (K + tL) \neq \emptyset\}.$$

The event in the right-hand side means that X hits a neighbourhood of K while not touching K itself. Define $Z_n = X_1 \cup \cdots \cup X_n$ for i.i.d. random closed sets $X_1, X_2, \ldots$ with the capacity functional T that is differentiable at $K \in \mathcal{K}$. Then, for each $L \in \mathcal{K}$ with $0 \in L$,

$$\mathbf{P}\{Z_n \cap (K + n^{-1/\alpha}L) \neq \emptyset \mid Z_n \cap K = \emptyset\}$$

$$= 1 - \left[\mathbf{P}\left\{X_1 \cap (K + n^{-1/\alpha}L) = \emptyset \mid X_1 \cap K = \emptyset\right\}\right]^n$$

$$= 1 - \left[1 - \frac{T(K + n^{-1/\alpha}L) - T(K)}{1 - T(K)}\right]^n$$

converges vaguely as $n \to \infty$ to

$$\tilde{T}(K) = 1 - \exp\left\{-\frac{d_L T(K)}{1 - T(K)}\right\}.$$

Derivative of the Lebesgue measure

Assume that $\mu = \text{mes}$ is the Lebesgue measure in $\mathbb{R}^d$.

Theorem 5.23 (Derivative of the Lebesgue measure). *If K is a regular closed convex compact set, then*

$$d_L \mu(K) = \int_{\mathbb{S}^{d-1}} h(L, u) S_{d-1}(K, du), \quad L \in \mathcal{K}, \tag{5.14}$$

where $h(L, u)$ is the support function of L and $S_{d-1}(K, du)$ is the area measure of K, see Appendix F.

Lemma 5.24. *If K is a regular closed convex compact set, then, for each $L \in \mathcal{K}$,*

$$d_L \mu(K) = d_{\text{co}(L)} \mu(K). \tag{5.15}$$

Proof. Without loss of generality assume that $0 \in L$. If $L = \{0, x\}$, then

$$\mu(K + t\,\text{co}(L)) - \mu(K + tL) = o(t) \quad \text{as } t \downarrow 0.$$

The same argument implies (5.15) for each finite L. A general $L \in \mathcal{K}$ can be approximated from the above by a sequence $\{L_n, n \geq 1\}$ of polyhedrons, such that $\text{co}(L_n) = \text{co}(F_n)$ for some finite set $F_n \subset L_n$, $n \geq 1$. Then $d_{L_n} \mu(K) = d_{\text{co}(L_n)} \mu(K)$ and (5.15) follows from the upper semicontinuity of $d_L \mu(K)$ with respect to L. □

Proof of Theorem 5.23. By Lemma 5.24 it suffices to assume that $L \in \text{co}\,\mathcal{K}$. The translative integral formula (see Schneider [520, Eq. (4.5.32)]) yields

$$\mu(K + tL) = \mu(tL) + \sum_{k=1}^{d-1} V_k(K, tL) + \mu(K).$$

The functionals $V_k(\cdot, tL)$ and $V_k(K, \cdot)$ are additive; the first is homogeneous of degree k, while the second is homogeneous of degree $(d - k)$. The proof is finished by noticing that the functional $V_{d-1}(K, L)$ equals the right-hand side of (5.14). □

Example 5.25 (Derivative of sup-measure). Assume that $\varphi = f^\vee$ for an upper semicontinuous function f. The set

$$\text{argmax}_K f = \{x \in K : f(x) = f^\vee(K)\}$$

is not empty, since f is upper semicontinuous. If there exists a point $x \in \text{argmax}_K f$ such that $x \in \text{Int}\,K$, then $x \in K + tL$ for all sufficiently small t, whence $d_L \varphi(K) = 0$. To exclude this trivial case, assume that $\text{argmax}_K f$ is a subset of ∂K.

Assume that f is continuous in a neighbourhood of K and continuously differentiable in a neighbourhood of $\text{argmax}_K f$. The derivative of the sup-measure (with the exponent $\alpha = 1$) is given by

$$d_L\varphi(K) = \sup_{x \in \operatorname{argmax}_K f} h(L, f'(x)). \tag{5.16}$$

This is easily seen by using the Taylor expansion for $f(y)$ with $y \in x + tL$ and $x \in \operatorname{argmax}_K f$. The random set Z with the capacity functional (5.13) is the union of half-spaces

$$Z = \bigcup_{x \in \operatorname{argmax}_K f} \{z : \langle z, f'(x) \rangle \geq \xi\},$$

where ξ has the exponential distribution with mean 1.

Union of independent random sets

If X_1 and X_2 are independent random closed sets, then

$$T_{X_1 \cup X_2}(K) = 1 - (1 - T_{X_1}(K))(1 - T_{X_2}(K)).$$

If the capacity functionals T_{X_1} and T_{X_2} are differentiable at K with the same exponent α, then

$$d_L T_{X_1 \cup X_2}(K) = (1 - T_{X_2}(K)) d_L T_{X_1}(K) + (1 - T_{X_1}(K)) d_L T_{X_2}(K).$$

Example 5.26. Let $X = \{\xi, \eta\}$ where ξ and η are independent random points with distributions $\mathbf{P}_\xi$ and $\mathbf{P}_\eta$. Then

$$d_L T_X(K) = \mathbf{P}\{\eta \notin K\} d_L \mathbf{P}_\xi(K) + \mathbf{P}\{\xi \notin K\} d_L \mathbf{P}_\eta(K).$$

If ξ and η have absolutely continuous distributions with densities p_ξ and p_η, then $d_L T_X(x) = (p_\xi(x) + p_\eta(x)) \operatorname{mes}(L)$.

Differentiation of the Choquet integral

Below we will find a derivative of the capacity given by the Choquet integral. For a capacity φ, we write $\varphi(x)$ instead of $\varphi(\{x\})$.

Theorem 5.27 (Derivative of Choquet integral at singleton). *Let*

$$\psi(K) = \int_K f \, d\varphi, \quad K \in \mathcal{K},$$

for a continuous non-negative function f and a capacity φ, which is differentiable at $\{x\}$ with exponent α.
  (i) *If $\varphi(x) = 0$ and/or f is Lipschitz of order $\beta > \alpha$ in a neighbourhood of x, then ψ is differentiable at $\{x\}$ with exponent α and*

$$d_L \psi(x) = f(x) d_L \varphi(x).$$

5 Calculus with capacities

(ii) If $\alpha \geq 1$, $\varphi(x) > 0$, f is continuously differentiable in a neighbourhood of x and φ is upper semicontinuous at $\{x\}$, then ψ is differentiable at $\{x\}$ with exponent 1 and

$$d_L\psi(x) = \begin{cases} f(x)d_L\varphi(x) + \varphi(x)h(L, f'(x)), & \alpha = 1, \\ \varphi(x)h(L, f'(x)), & \alpha > 1. \end{cases}$$

Proof. By the definition of the Choquet integral,

$$t^{-\alpha}\left[\int_{x+tL} f d\varphi - \int_{\{x\}} d\varphi\right] = t^{-\alpha}\int_0^\infty [\varphi((x+tL)\cap F_s) - \varphi(\{x\}\cap F_s)]ds$$

$$= t^{-\alpha}I_1 + t^{-\alpha}I_2 + t^{-\alpha}I_3,$$

where $F_s = \{x : f(x) \geq s\}$ and

$$t^{-\alpha}I_1 = t^{-\alpha}\int_0^{\inf f(x+tL)} [\varphi(x+tL) - \varphi(x)]ds$$

$$= \inf f(x+tL)\frac{\varphi(x+tL) - \varphi(x)}{t^\alpha} \to f(x)d_L\varphi(x),$$

$$t^{-\alpha}I_2 = t^{-\alpha}\int_{\inf f(x+tL)}^{f(x)} [\varphi((x+tL)\cap F_s) - \varphi(x)]ds$$

$$\leq \frac{\varphi(x+tL) - \varphi(x)}{t^\alpha}[f(x) - \inf f(x+tL)] \to 0 \quad \text{as } t \downarrow 0,$$

and

$$t^{-\alpha}I_3 = t^{-\alpha}\int_{f(x)}^{\sup f(x+tL)} \varphi((x+tL)\cap F_s)ds.$$

(i) If $\varphi(x) = 0$, then

$$\frac{\varphi((x+tL)\cap F_s)}{t^\alpha} \leq \frac{\varphi(x+tL) - \varphi(x)}{t^\alpha} \to d_L\varphi(x),$$

whence $t^{-\alpha}I_3$ converges to zero. This holds also if f is Lipschitz, since

$$I_3 \leq \varphi(x+tL)(\sup f(x+tL) - f(x)).$$

(ii) In this case $t^{-1}I_1$ converges to zero if $\alpha > 1$ and to $f(x)d_L\varphi(x)$ if $\alpha = 1$. Furthermore,

$$\varphi(x)\frac{\sup f(x+tL) - f(x)}{t} \leq t^{-1}I_3 \leq \varphi(x+tL)\frac{\sup f(x+tL) - f(x)}{t}.$$

Both sides converge to $\varphi(x)h(L, f'(x))$. □

Example 5.28. If $\varphi = \mu$ is the Lebesgue measure and $\psi(K) = \int_K f \, d\mu$ (in this case the Choquet integral coincides with the Lebesgue integral), Theorem 5.27 implies that $d_L \psi(x) = f(x)\mu(L)$. Assume now that $\varphi(K) = \mu(K^r)$ for fixed $r > 0$, where K^r is the r-envelope of K. By (5.14),

$$d_L \varphi(x) = d_L \mu(B_r(0)) = r^{d-1} \int_{\mathbb{S}^{d-1}} h(L, u) \mathcal{H}^{d-1}(du) = \frac{1}{2} r^{d-1} \omega_d \mathbf{b}(L),$$

where $\mathbf{b}(L)$ is the mean width of L. If $\psi = \int f \, d\varphi$, then Theorem 5.27(ii) implies that

$$d_L \psi(x) = \frac{1}{2} r^{d-1} \omega_d \mathbf{b}(L) f(x) + r^d \varkappa_d h(L, f'(x)).$$

If $d = 2$, the corresponding random closed set with the capacity functional (5.13) is the union of two independent random sets: the half-space $\{z : r^d \varkappa_d \langle z, f'(x) \rangle \geq \xi\}$ with the exponentially distributed ξ of mean 1 and the other being the stationary isotropic Poisson line process with intensity $\pi f(x) r^{d-1}$, see Stoyan, Kendall and Mecke [544, p. 250].

Corollary 5.29 (Radon–Nikodym derivative). *If ψ is the Choquet integral of a differentiable capacity φ and $\varphi(x) = 0$ for all x, then, for each $L \in \mathcal{K}$ with $0 \in L$,*

$$\psi(K) = \int_K \frac{d_L \psi(x)}{d_L \varphi(x)} d\varphi, \quad K \in \mathcal{K},$$

where the function $d_L \psi(x)/d_L \varphi(x)$ is independent of L and yields the Radon–Nikodym derivative of ψ with respect to φ.

Note that Corollary 5.29 is trivial if φ and ψ are measures.

6 Convergence

6.1 Weak convergence

Continuity sets

The weak convergence of random closed sets is a special case of the weak convergence of probability measures, since a random closed set is a particular case of a general random element and can be associated with a probability measure on $\mathfrak{B}(\mathcal{F})$.

Definition 6.1 (Weak convergence). A sequence of random closed sets $\{X_n, n \geq 1\}$ is said to converge *weakly* (or converge *in distribution*) to a random closed set X with distribution $\mathbf{P}$ (notation $X_n \xrightarrow{d} X$) if the corresponding probability measures $\{\mathbf{P}_n, n \geq 1\}$ converge weakly to $\mathbf{P}$, i.e.

$$\mathbf{P}_n(\mathcal{Y}) \to \mathbf{P}(\mathcal{Y}) \quad \text{as } n \to \infty \tag{6.1}$$

for each $\mathcal{Y} \in \mathfrak{B}(\mathcal{F})$ such that $\mathbf{P}(\partial \mathcal{Y}) = 0$, where the boundary of $\mathcal{Y}$ is defined with respect to a topology on $\mathcal{F}$ that generates the Effros σ-algebra.

If $\mathbb{E}$ is a LCHS space, then the boundary of $\mathcal{Y}$ in Definition 6.1 is taken in the Fell topology. Since in this case the family $\mathcal{F}$ of closed sets is compact (see Theorem B.2(i)), no tightness conditions are needed for the weak convergence of random closed sets in LCHS spaces, i.e. all families of distributions of random closed sets are relatively compact. This fact can be formulated as follows.

Theorem 6.2 (Helly theorem for random sets). *If $\mathbb{E}$ is a LCHS space, then every sequence $\{X_n, n \geq 1\}$ of random closed sets has a weakly convergent subsequence.*

It is difficult to check (6.1) for all $\mathcal{Y}$ from $\mathfrak{B}(\mathcal{F})$. The first natural step is to use $\mathcal{Y} = \mathcal{F}_K$ for K running through $\mathcal{K}$.

Lemma 6.3. *Let $\mathbb{E}$ be a LCHS space. For each $K \in \mathcal{K}$,*

$$\mathbf{P}(\mathcal{F}_K) = \mathbf{P}(\mathcal{F}_{\mathrm{Int}\,K}) \qquad (6.2)$$

implies $\mathbf{P}(\partial \mathcal{F}_K) = 0$.

Proof. Let us show that the interior of $\mathcal{F}_K$ in the Fell topology contains $\mathcal{F}_{\mathrm{Int}\,K}$. If $F \cap \mathrm{Int}\,K \neq \emptyset$ and F_n Painlevé–Kuratowski converges to F, then $F_n \cap \mathrm{Int}\,K \neq \emptyset$ for all sufficiently large n, see Corollary B.7. It suffices to note that that the Fell topology coincides with the Painlevé–Kuratowski convergence if $\mathbb{E}$ is LCHS, see Theorem B.6. Since $\mathcal{F}_K$ is closed, $\partial \mathcal{F}_K \subset \mathcal{F}_K \setminus \mathcal{F}_{\mathrm{Int}\,K}$, so that $\mathbf{P}(\partial \mathcal{F}_K) = 0$ if (6.2) holds. □

Note that (6.2) is equivalent to $\mathbf{P}\{X \cap K \neq \emptyset, \ X \cap \mathrm{Int}\,K = \emptyset\} = 0$ or

$$T_X(K) = T_X(\mathrm{Int}\,K)$$

for the corresponding random closed set X, where $T_X(\mathrm{Int}\,K)$ is defined using (1.19). The following definition is a special case of Definition 1.26.

Definition 6.4 (Continuity family). The family of relatively compact Borel sets B satisfying

$$T_X(\mathrm{cl}\,B) = T_X(\mathrm{Int}\,B)$$

is called the *continuity family* of X and denoted by $\mathfrak{S}_{T_X}$ or $\mathfrak{S}_X$.

It is shown by Molchanov [394] that $\mathfrak{S}_X$ contains all regular closed compact sets if X is stationary.

Pointwise convergence of capacity functionals

It is straightforward to deduce from Lemma 6.3 that $X_n \xrightarrow{d} X$ yields $T_{X_n}(K) \to T_X(K)$ as $n \to \infty$ for every compact set $K \in \mathfrak{S}_X$. The following theorem characterises the weak convergence of random closed sets in terms of the pointwise convergence of capacity functionals.

Theorem 6.5 (Convergence of capacity functionals). *A sequence of random closed sets $\{X_n, n \geq 1\}$ in a LCHS space converges weakly to a random closed set X if and only if*

$$T_{X_n}(K) \to T_X(K) \quad \text{as } n \to \infty \tag{6.3}$$

for each $K \in \mathfrak{S}_X \cap \mathcal{K}$.

It is possible to prove Theorem 6.5 directly by first showing that $T_{X_n}(G) \to T_X(G)$ for all open sets G such that $T_X(G) = T_X(\text{cl } G)$, then deducing from this convergence of probability measures on the families $\mathcal{Y} = \mathcal{F}_{G_1,\ldots,G_n}^K$ with $\mathbf{P}(\partial \mathcal{Y}) = 0$ and finally referring to Theorem 2.2 of Billingsley [70]. Alternatively, Theorem 6.5 can be obtained as a particular case of Theorem 6.8 below.

Example 6.6 (Convergence of random singletons). If $X_n = \{\xi_n\}$, $n \geq 1$, then the weak convergence of X_n is equivalent to the weak convergence of ξ_n in the conventional sense, see Billingsley [70]. Then $\mathbf{P}\{\xi_n \in K\} = T_{X_n}(K)$, so that (6.3) is read as the weak convergence of the sequence $\{\xi_n, n \geq 1\}$.

Example 6.7 (Convergence of random balls). Random balls $X_n = B_{\eta_n}(\xi_n)$, $n \geq 1$, converge weakly if (η_n, ξ_n) converge weakly as random elements in the product space $\mathbb{R}_+ \times \mathbb{E}$. Indeed, $X_n \cap K \neq \emptyset$ if and only if

$$(\eta_n, \xi_n) \in F = \bigcup_{r \geq 0} (\{r\} \times K^r)$$

with F being a closed set. Since the map $B_r(x) \mapsto (r, x)$ is a continuous bijection between the family of balls and $\mathbb{R}_+ \times \mathbb{E}$, $X_n \xrightarrow{d} X$ implies that (η_n, ξ_n) converges weakly as $n \to \infty$.

Convergence determining classes

The following important theorem relies on Definition 1.25 of the separating class and refers to the notation introduced in Section 1.4. Its formulation involves the capacity functionals extended to the family of all subsets of $\mathbb{E}$ by means of (1.19) and (1.20).

Theorem 6.8 (Characterisation of weak convergence). *A sequence of random closed sets $\{X_n, n \geq 1\}$ in a LCHS space converges weakly to a random closed set X if there exists a separating class $\mathcal{A}$ and an increasing set function $\varphi \colon \mathcal{A} \mapsto [0, 1]$ such that*

$$\varphi^0(\text{Int } B) \leq \liminf_n T_{X_n}(B) \leq \limsup_n T_{X_n}(B) \leq \varphi^-(\text{cl } A) \tag{6.4}$$

for all $A \in \mathcal{A}$. Then $T_X(K) = \varphi^-(K)$ for all $K \in \mathcal{K}$ and $T_X(G) = \varphi^0(G)$ for all $G \in \mathcal{G}$. If $\mathcal{A} \subset \mathfrak{S}_\varphi$, then $T_X(B) = \varphi(B)$ for all $B \in \mathcal{A}$.

Proof. Fix $K \in \mathcal{K}$ and choose a sequence $\{B_m, m \geq 1\} \subset \mathcal{A}$ such that $B_m \downarrow K$ and $K \subset \text{cl } B_{m+1} \subset \text{Int } B_m$ for all $m \geq 1$. By (6.4) and (1.30),

$$\limsup_{n\to\infty} T_{X_n}(K) \le \limsup_{n\to\infty} T_{X_n}(B_n) \le \varphi^-(\operatorname{cl} B_m) \to \varphi^-(K).$$

A similar argument yields

$$\liminf_{n\to\infty} T_{X_n}(K) \ge \liminf_{n\to\infty} T_{X_n}(\operatorname{Int} K) \ge \varphi^0(\operatorname{Int} K).$$

Therefore,

$$\lim_{n\to\infty} T_{X_n}(K) = \varphi^-(K)$$

for all $K \in \mathcal{K} \cap \mathfrak{S}_\varphi$. If $\{X_{n(k)}, k \ge 1\}$ is a subsequence of $\{X_n, n \ge 1\}$, then, by Theorem 6.2, it has a subsequence that converges weakly to a random closed set X. Then $T_X(K) = \varphi^-(K)$ for all K from $\mathcal{K} \cap \mathfrak{S}_\varphi \cap \mathfrak{S}_X$. Every $K \in \mathcal{K}$ can be approximated from the above by a sequence $\{K_n, n \ge 1\} \subset \mathcal{K} \cap \mathfrak{S}_\varphi \cap \mathfrak{S}_X$ such that $\operatorname{Int} K_n \downarrow K$. Since $T_X^- = T_X$ by the semicontinuity of T_X and $\varphi^{--} = \varphi^-$, we obtain that $T_X(K) = \varphi^-(K)$ for all $K \in \mathcal{K}$. The same argument shows that $T_{X'}(K) = \varphi^-(K)$ for every possible weak limit X' of $\{X_{n(k)}, k \ge 1\}$. The Choquet theorem implies $X \stackrel{d}{\sim} X'$. It follows from Billingsley [70, Th. 2.3] that $X_n \stackrel{d}{\to} X$. Similar arguments can be used to show that $T_X(G) = \varphi^0(G)$ for all $G \in \mathcal{G}$, whence $\mathfrak{S}_X = \mathfrak{S}_\varphi$. □

Corollary 6.9 (Sufficient condition for weak convergence). *Let $\mathcal{A}$ be a separating class in a LCHS space. If $T_{X_n}(B) \to T_X(B)$ as $n \to \infty$ for all $B \in \mathcal{A} \cap \mathfrak{S}_X$, then $X_n \stackrel{d}{\to} X$ as $n \to \infty$.*

Corollary 6.9 implies Theorem 6.5 for $\mathcal{A} = \mathcal{K}$. Other typically used separating classes are the class of finite unions of balls of positive radii (or the countable class of finite unions of balls with rational midpoints and positive rational radii) and the class of finite unions of parallelepipeds. These classes are called *convergence determining*. In general, a family $\mathcal{M} \subset \mathcal{K}$ is said to determine the weak convergence if the pointwise convergence of the capacity functionals on $\mathcal{M} \cap \mathfrak{S}_T$ yields the weak convergence of distributions for the corresponding random closed sets.

Convergence of Choquet integrals and selections

It is well known that the weak convergence of random variables is characterised by the convergence of expectations for every bounded continuous function of the variables. A parallel result holds for random closed sets. Recall that $\mathbf{E} f^\vee(X) = \mathbf{E} \sup f(X)$ for a non-negative measurable function f equals the Choquet integral of f with respect to T_X, see Theorem 5.1.

Proposition 6.10 (Convergence of Choquet integrals). *A sequence $\{X_n, n \ge 1\}$ of random closed sets converges weakly to a random closed set X if and only if $\mathbf{E} f^\vee(X_n)$ converges to $\mathbf{E} f^\vee(X)$ for every continuous function $f : \mathbb{E} \mapsto \mathbb{R}$ with a bounded support.*

Proof. It suffices to consider non-negative functions f. Denote $F_s = \{x : f(x) \geq s\}$. Note that $T_X(F_s) \neq T_X(\operatorname{Int} F_s)$ for at most a countable set of s. Therefore, $T_{X_n}(F_s) \to T_X(s)$ for almost all $s > 0$, whence the convergence of the Choquet integrals easily follows. The inverse implication follows from the fact that the indicator function $g(x) = \mathbf{1}_K(x)$ can be approximated from below and from above by continuous functions with bounded supports. □

The weak convergence of random closed sets in a general Polish space $\mathbb{E}$ with a metric ρ can be characterised in terms of the weak convergence of their distance functions. In line with Theorem 2.28, a sequence $\{X_n, n \geq 1\}$ of random closed sets converges weakly to X if and only if the finite-dimensional distributions of the process $\rho(x, X_n)$ converge to the finite-dimensional distributions of $\rho(x, X)$, $x \in \mathbb{E}$, see Salinetti and Wets [512].

The weak convergence of random convex closed sets implies the convergence of their selections.

Proposition 6.11. *Let $\{X_n, n \geq 1\}$ be a sequence of almost surely non-empty random convex closed sets in a separable Banach space, such that $X_n \xrightarrow{d} X$. Then X is a random convex closed set and there exists a sequence of selections $\xi_n \in \mathcal{S}(X_n)$, $n \geq 1$, such that $\xi_n \xrightarrow{d} \xi$ with $\xi \in \mathcal{S}(X)$.*

Proof. Since the family $\operatorname{co} \mathcal{F}'$ of non-empty convex closed sets is closed in $\mathcal{F}$,

$$\mathbf{P}\{X \in \operatorname{co} \mathcal{F}'\} \geq \limsup_{n\to\infty} \mathbf{P}\{X_n \in \operatorname{co} \mathcal{F}'\} = 1.$$

Let $\mathfrak{f}$ be a continuous selection operator on $\operatorname{co} \mathcal{F}'$, see Section 2.2. Define selections of X and X_n by $\xi = \mathfrak{f}(X)$ and $\xi_n = \mathfrak{f}(X_n)$, $n \geq 1$. If g is a bounded continuous function on $\mathbb{E}$, then $g(\mathfrak{f}(F))$ is a continuous real-valued function on $\operatorname{co} \mathcal{F}'$. Now $X_n \xrightarrow{d} X$ implies $\mathbf{E}g(\mathfrak{f}(X_n)) \to \mathbf{E}g(\mathfrak{f}(X))$, which means $\mathbf{E}g(\xi_n) \to \mathbf{E}g(\xi)$, i.e. $\xi_n \xrightarrow{d} \xi$. □

Open problem 6.12. Does $X_n \xrightarrow{d} X$ as $n \to \infty$ for not necessarily convex random closed sets imply that there is a sequence of selection $\xi_n \in \mathcal{S}(X_n)$ and $\xi \in \mathcal{S}(X)$ such that $\xi_n \xrightarrow{d} \xi$? By Proposition 6.11, this holds for random convex sets.

Proposition 6.13 (Convergence of support functions). *Let $\{X_n, n \geq 1\}$ be a sequence of almost surely non-empty random convex compact sets. Then X_n converges weakly to a random convex compact set X if and only if the finite-dimensional distributions of $h(X_n, \cdot)$ converge to those of $h(X, \cdot)$ and $\sup_n \mathbf{P}\{\|X_n\| \geq c\} \to 0$ as $c \to \infty$.*

Proof. It suffices to show that the imposed conditions imply the weak convergence of the support functions in the space of continuous functions on the unit sphere. The corresponding tightness condition (see Billingsley [70]) requires

$$\lim_{\delta\downarrow 0}\limsup_{n\to\infty}\mathbf{P}\left\{\sup_{\|u-v\|\leq\delta}|h(X_n,u)-h(X_n,v)|\geq\varepsilon\right\}=0.$$

The inequality for the support functions (F.4) yields

$$\mathbf{P}\left\{\sup_{\|u-v\|\leq\delta}|h(X_n,u)-h(X_n,v)|\geq\varepsilon\right\}\leq\mathbf{P}\{\|X_n\|\geq\varepsilon/\delta\},$$

whence the condition of the theorem implies that the sequence $h(X_n,\cdot)$ is tight. □

Convergence to a singleton

In optimisation problems it is often possible to assume that a sequence of random closed sets converges to a random closed set X which is either empty or consists of a single point, i.e. $\mathbf{P}\{\mathrm{card}(X)>1\}=0$. The following theorem deals with the convergence of selections for such sequences. Note that a random element ξ is said to be a *generalised selection* of X if $\xi\in X$ a.s. on the event $\{X\neq\emptyset\}$. Recall that a sequence of random elements $\{\xi_n,n\geq 1\}$ is *tight* if for all $\varepsilon>0$ there is a compact set $K\subset\mathbb{E}$ such that $\mathbf{P}\{\xi_n\in K\}\geq 1-\varepsilon$ for all n.

Theorem 6.14 (Weak convergence to a singleton). *Let $\{X_n,n\geq 1\}$ be a sequence of random closed sets in a Polish space $\mathbb{E}$. Assume that X_n weakly converges to a random closed set X such that $\mathrm{card}(X)\leq 1$ a.s. and $\mathbf{P}\{X_n\neq\emptyset\}\to\mathbf{P}\{X\neq\emptyset\}$. Let ξ_n and ξ be generalised selections of X_n and X respectively. Then ξ_n converges in distribution to ξ if at least one of the following conditions holds*
  (i) *the sequence $\{\xi_n,n\geq 1\}$ is tight;*
  (ii) $\mathbf{P}\{\mathrm{card}(X_n)>1\}\to 0.$

Proof.
  (i) Assume without loss of generality that X is almost surely non-empty. By Theorem E.6, it suffices to show that $\limsup\mathbf{P}\{\xi_n\in F\}\leq\mathbf{P}\{\xi\in F\}$ for all $F\in\mathcal{F}$. Fix $\varepsilon>0$ and let K be a compact set such that $\mathbf{P}\{\xi_n\in K\}\geq 1-\varepsilon$ for all $n\geq 1$. Then

$$\mathbf{P}\{\xi_n\in F\}-\varepsilon\leq\mathbf{P}\{\xi_n\in(F\cap K)\}\leq\mathbf{P}\{X_n\cap(F\cap K)\neq\emptyset\}.$$

Since $\mathcal{F}_{F\cap K}$ is closed in $\mathcal{F}$ and $X_n\xrightarrow{d}X$,

$$\limsup_{n\to\infty}\mathbf{P}\{X_n\cap(F\cap K)\neq\emptyset\}\leq\mathbf{P}\{X\cap(F\cap K)\neq\emptyset\}.$$

Since X is assumed to be a.s. non-empty, $X=\{\xi\}$ is a singleton, whence

$$\limsup_{n\to\infty}\mathbf{P}\{\xi_n\in F\}\leq T_X(F\cap K)+\varepsilon=\mathbf{P}\{\xi\in F\}+\varepsilon,$$

so letting $\varepsilon\downarrow 0$ finishes the proof.
  (ii) For every $F\in\mathfrak{S}_X$, one has

$$\mathbf{P}\{X_n\cap F\neq\emptyset,\,\mathrm{card}(X_n)=1\}\leq\mathbf{P}\{\xi_n\in F\}\leq\mathbf{P}\{X_n\cap F\neq\emptyset\},$$

so that the required weak convergence follows from the imposed conditions. □

Semi-differentiability

In sensitivity studies of optimisation problems it is essential to be able to deduce the weak convergence of set-valued functions from the weak convergence of their arguments. Let $F\colon \mathbb{R}^m \mapsto \mathcal{F}(\mathbb{R}^d)$ be a multifunction measurable with respect to the Borel σ-algebra on $\mathbb{R}^m$ and the Effros σ-algebra on $\mathcal{F}(\mathbb{R}^d)$, see Appendix D. If ξ is a random vector in $\mathbb{R}^m$ with distribution **P**, then $F(\xi)$ is a random closed set in $\mathbb{R}^d$. Let **P** be a probability measure on $\mathbb{R}^m$. The multifunction F is called **P**-*a.s. semi-differentiable* at $z_0 \in \mathbb{R}^m$ relative to $x_0 \in F(z_0)$ if there exists a multifunction $F'_{z_0,x_0}\colon \mathbb{R}^m \mapsto \mathcal{F}(\mathbb{R}^d)$ such that

$$F'_{z_0,x_0}(z) = \lim_{t\downarrow 0,\, z' \to z} t^{-1}[F(z_0 + tz') - x_0]$$

holds for all points z except those in a set of **P**-measure zero. The following result follows directly from the continuous mapping theorem, see Billingsley [70, Th. 5.5].

Theorem 6.15 (Weak convergence of semi-differentiable multifunctions). *Let $\{\xi_n, n \geq 1\}$ be a sequence of random vectors in $\mathbb{R}^m$ such that $a_n^{-1}(\xi_n - z_0)$ converges in distribution to a random vector ξ with distribution **P**, where $\{a_n, n \geq 1\}$ is a sequence of positive normalising constants and z_0 is a non-random point in $\mathbb{R}^m$. If F is **P**-a.s. semi-differentiable at z_0 relative to a point $x_0 \in F(z_0)$, then $a_n^{-1}(F(\xi_n) - x_0)$ converges in distribution to $F'_{z_0,x_0}(\xi)$.*

If $F(x) = \{f(x)\}$ is a single valued function, Theorem 6.15 implies that $a_n^{-1}(f(\xi_n) - f(z_0))$ converges in distribution to $f'(z_0)\xi$.

6.2 Convergence almost surely and in probability

Definition

It is easy to define the almost sure convergence of random closed sets using one of the topologies on $\mathcal{F}$ described in Appendix B. For example, $X_n \xrightarrow{\text{PK}} X$ a.s. if $X_n(\omega) \xrightarrow{\text{PK}} X(\omega)$ for almost all $\omega \in \Omega$. If $\mathbb{E}$ is locally compact, this convergence is equivalent to the almost sure convergence in the Fell topology. In this case the indication of the topology is usually omitted and we write $X_n \to X$ a.s. The almost sure convergence of random compact sets is usually defined with respect to the Hausdorff metric as $\rho_H(X_n, X) \to 0$ a.s.

As a consequence of a general property of probability measures in topological spaces, the almost sure convergence of random closed sets (in the Fell topology) implies their weak convergence. On the other hand, a weakly convergent sequence of random closed sets can be realised on a single probability space as an almost surely convergent sequence, see Wichura [610].

Example 6.16 (A.s. convergence of convex hulls). Let K be a convex compact set in $\mathbb{R}^d$ with sufficiently smooth boundary ∂K. Choose n independent random points uniformly distributed in K and denote by P_n their convex hull. Then P_n is a *random* *polyhedron*, such that $P_n \to K$ almost surely as $n \to \infty$. Since the classical paper by Rényi and Sulanke [478], the rate of convergence for various functionals of P_n (e.g. its area $\mathrm{mes}_2(P_n)$) and the corresponding limit theorems have been the focus of attention of many probabilists. For instance, if $d = 2$ and the curvature $k(x)$ does not vanish for all $x \in \partial K$, then

$$\lim_{n\to\infty} n^{2/3}[\mathrm{mes}_2(K) - \mathbf{E}\,\mathrm{mes}_2(P_n)] = (2/3)^{1/3}\,\Gamma(5/3)\,\mathrm{mes}_2(K)^{2/3}\int_{\partial K} k^{1/3}(s)\mathrm{d}s\,,$$

see Schneider [519]. Further results in this direction can be found in Bräker and Hsing [76], Groeneboom [212] and McClure and Vitale [385].

Deterministic limits

Deriving the almost sure convergence of random closed sets in the Fell topology involves checking the conditions (F1) and (F2) of Corollary B.7. These conditions can be reformulated for the sets K and G from some countable subfamilies of $\mathcal{K}$ and $\mathcal{G}$ and then applied for a sequence of random sets with a non-random limit.

Proposition 6.17 (A.s. convergence to deterministic limit). *A sequence $X_n, n \geq 1$, of random closed sets in a LCHS space a.s. converges to a deterministic closed set F if and only if the following conditions hold.*

(R1) *If $K \cap F = \emptyset$ for $K \in \mathcal{K}$, then*

$$\mathbf{P}\{X_n \cap K \neq \emptyset \ i.o.\} = \mathbf{P}\left\{\bigcap_{n=1}^{\infty}\bigcup_{m=n}^{\infty}\{X_n \cap K \neq \emptyset\}\right\} = 0\,,$$

where "i.o." means "infinitely often".
(R2) *If $G \cap F \neq \emptyset$ for $G \in \mathcal{G}$, then*

$$\mathbf{P}\{X_n \cap G = \emptyset \ i.o.\} = \mathbf{P}\left\{\bigcap_{n=1}^{\infty}\bigcup_{m=n}^{\infty}\{X_n \cap G = \emptyset\}\right\} = 0\,.$$

These conditions can be relaxed by replacing $\mathcal{K}$ in (R1) with a separating class $\mathcal{A} \subset \mathcal{K}$ and $\mathcal{G}$ in (R2) with $\mathcal{A}' = \{\mathrm{Int}\,K : K \in \mathcal{A}\}$.

Proposition C.10 together with a usual separability argument based on choosing a countable dense set yields the following result concerning the almost sure convergence of random compact sets.

Proposition 6.18 (A.s. convergence of random compact sets). *Let $\mathcal{V}$ be a closed subset of $\mathcal{K}$ and let $\{X_n, n \geq 1\}$ be a sequence of $\mathcal{V}$-valued random sets such that $\mathrm{cl}(\cup_n X_n)$ is compact almost surely. If $d_\mathrm{H}(X_n, V)$ (see (C.2)) a.s. converges for each $V \in \mathcal{V}$, then $\{X_n, n \geq 1\}$ converges a.s. in the Hausdorff metric.*

In a Banach space it is possible to define weak and strong almost sure limits of a sequence of random closed sets. Hiai [253] showed that if $\sup_{n\geq 1}\|X_n\| < \infty$ a.s. in a reflexive space $\mathbb{E}$, then there exists a random closed set X such that $X = \mathrm{w\!-\!lim\,sup}\,X_n$ a.s.

Convergence in probability

In order to define the convergence of random closed sets in probability, it is necessary to assume that $\mathbb{E}$ is a metric space. Recall that $F^{\varepsilon-}$ is the open ε-envelope of F.

Definition 6.19 (Convergence in probability). A sequence $\{X_n, n \geq 1\}$ is said to converge *in probability* if, for every $\varepsilon > 0$ and $K \in \mathcal{K}$,

$$\mathbf{P}\{[(X_n \setminus X^{\varepsilon-}) \cup (X \setminus X_n^{\varepsilon-})] \cap K \neq \emptyset\} \to 0 \quad \text{as } n \to \infty. \tag{6.5}$$

For brevity, it is sensible to denote

$$Y_{\varepsilon,n} = (X_n \setminus X^{\varepsilon-}) \cup (X \setminus X_n^{\varepsilon-})$$

so that (6.5) means that $Y_{\varepsilon,n} \xrightarrow{d} \emptyset$ for each $\varepsilon > 0$.

Lemma 6.20 (Convergent subsequences). *If the random closed sets $\{X_n, n \geq 1\}$ converge in probability to X, then there exists a subsequence $\{n(i), i \geq 1\}$ such that $X_{n(i)} \to X$ almost surely as $i \to \infty$.*

Proof. Choose a sequence $\{(\varepsilon_i, K_i)\} \subset (\mathbb{R}_+ \times \mathcal{K})$ such that $\varepsilon_i \downarrow 0$, $\sum_{i=1}^{\infty} \varepsilon_i < \infty$ and $K_i \uparrow \mathbb{E}$. By (6.5), for every $i \geq 1$ it is possible to find an integer $n(i)$ such that the sequence $\{n(i), i \geq 1\}$ is strictly increasing and $\mathbf{P}\{Y_{\varepsilon_i,n} \cap K_i \neq \emptyset\} \leq \varepsilon_i$ for all $n \geq n(i)$. Then

$$\sum_{i=1}^{\infty} \mathbf{P}\{Y_{\varepsilon_i,n(i)} \cap K_i \neq \emptyset\} \leq \sum_{i=1}^{\infty} \varepsilon_i < \infty$$

and, by the Borel–Cantelli lemma, $Y_{\varepsilon_i,n(i)} \cap K_i \neq \emptyset$ at most a finite number of times. For every $\varepsilon > 0$ and $K \in \mathcal{K}$, we get $Y_{\varepsilon,n(i)} \subset Y_{\varepsilon_i,n(i)}$ and $K \subset K_i$ for sufficiently large i. Therefore,

$$\lim_{k \to \infty} \mathbf{P}(\cup_{i=k}^{\infty} \{Y_{\varepsilon,n(i)} \cap K \neq \emptyset\}) = 0.$$

This implies $Y_{\varepsilon,n(i)} \to \emptyset$ a.s., whence $X_{n(i)} \to X$ a.s. by Proposition B.3. □

Theorem 6.21. *Let $\tilde{\rho}$ be any metric on $\mathcal{F}$ that is compatible with the Fell topology. If X and X_n, $n \geq 1$, are random closed sets, then the following statements are equivalent:*
  (i) $X_n \to X$ *in probability;*
  (ii) $\tilde{\rho}(X_n, X) \to 0$ *in probability;*
  (iii) *every subsequence of $\{X_n, n \geq 1\}$ contains a further subsequence that converges to X almost surely.*

Proof. The implication (i)⇒(ii) follows from Lemma 6.20. Condition (iii) means that any subsequence of random variables $\{\tilde{\rho}(X_n, X), n \geq 1\}$ contains a further subsequence converging almost surely to zero. This fact for real-valued random variables is equivalent to (ii).

The implication (ii)⇒(i) is easy to prove by assuming that (i) does not hold, so there is a subsequence $\{X_{n(i)}, i \geq 1\}$ such that $\mathbf{P}\{Y_{\varepsilon,n(i)} \cap K \neq \emptyset\} > \varepsilon$ for some fixed $\varepsilon > 0$ and $K \in \mathcal{K}$. Now (ii) implies $X_{n(i(k))} \to X$ a.s. for a further subsequence. By Proposition B.3, $Y_{\varepsilon,n(i(k))} \to \emptyset$ a.s. contrary to the assumption. □

Corollary 6.22. *If $X_n \to X$ in probability, then $X_n \xrightarrow{d} X$ as $n \to \infty$.*

Proof. For every bounded continuous function $g \colon \mathcal{F} \mapsto \mathbb{R}$, the dominated convergence theorem and Theorem 6.21 imply that every subsequence of $\{\mathbf{E}g(X_n), n \geq 1\}$ contains a further subsequence converging to $\mathbf{E}g(X)$, whence $\mathbf{E}g(X_n)$ converges to $\mathbf{E}g(X)$. □

6.3 Probability metrics

Probability metrics in general spaces

In this section we discuss probability metrics in the space of random closed sets distributions which generalise well known concepts of the uniform distance and the Lévy distance between distributions of random variables and the Prokhorov metric for random elements in metric spaces.

Definition 6.23 (Probability metric). A *probability metric* $\mathfrak{m}(\xi, \eta)$ is a numerical function on the space of distributions of random elements, which satisfies the following conditions

$$\mathfrak{m}(\xi, \eta) = 0 \quad \text{implies} \quad \mathbf{P}\{\xi = \eta\} = 1,$$
$$\mathfrak{m}(\xi, \eta) = \mathfrak{m}(\eta, \xi),$$
$$\mathfrak{m}(\xi, \eta) \leq \mathfrak{m}(\xi, \zeta) + \mathfrak{m}(\zeta, \eta),$$

for all random elements ξ, η and ζ.

Since a random compact set is a $\mathcal{K}$-valued random element, probability metrics for random compact sets can be defined by specialising general metrics for the case of random elements in the space $\mathcal{K}$ equipped with the Hausdorff metric ρ_H. For instance, the Prokhorov metric uses only the metric structure of the carrier space (see Rachev [470, p. 30]) and can be defined for random compact sets as

$$\mathfrak{p}(X, Y) = \inf\{\varepsilon > 0 \colon \mathbf{P}\{X \in \mathcal{Y}\} \leq \mathbf{P}\{Y \in \mathcal{Y}^\varepsilon\} + \varepsilon, \; \mathcal{Y} \in \mathfrak{B}(\mathcal{K})\},$$

where $\mathcal{Y}^\varepsilon$ is the ε-neighbourhood of $\mathcal{Y} \subset \mathcal{K}$ in the Hausdorff metric. Another metric can be defined as

$$K_H(X, Y) = \inf\{\varepsilon > 0 \colon \mathbf{P}\{\rho_H(X, Y) > \varepsilon\} < \varepsilon\},$$

where X and Y are random compact sets. This metric K_H metrises the convergence of random compact sets in probability with respect to the Hausdorff metric. An analogue of the so-called "engineering" metric (see Rachev [470, p. 5]) is defined as

$$I_H(X, Y) = \mathbf{E}\rho_H(X, Y).$$

The above mentioned metrics are *compound* [470, p. 39], i.e. their values depend on the joint distributions of X and Y as opposed to *simple* metrics that depend only on their marginal distributions. It is well known that simple metrics are more convenient, since they can be naturally applied to limit theorems. However, many interesting simple metrics for random variables are defined by means of their densities or characteristic functions, which are quite difficult to extend for random closed sets.

Probability metrics based on Castaing representation

Another possible approach to define probability metrics for random sets relies on their Castaing representation using selections. If m is a probability metric for random elements in $\mathbb{E}$, then a metric

$$\mathfrak{m}_H(X, Y) = \max \left\{ \sup_{\xi \in \mathcal{S}(X)} \inf_{\eta \in \mathcal{S}(Y)} \mathfrak{m}(\xi, \eta), \sup_{\eta \in \mathcal{S}(Y)} \inf_{\xi \in \mathcal{S}(X)} \mathfrak{m}(\xi, \eta) \right\}$$

for a.s. non-empty random closed sets X and Y is introduced in the same way as the Hausdorff metric is constructed from a metric on $\mathbb{E}$.

Example 6.24 (Engineering metric). Let $\mathbb{E} = \mathbb{R}^d$ and choose m to be the simple "engineering" metric on the space of integrable random vectors, i.e. $\mathfrak{m}(\xi, \eta) = \rho(\mathbf{E}\xi, \mathbf{E}\eta)$. Assume that both X and Y have at least one integrable selection. Then

$$\mathfrak{m}_H(X, Y) = \max \left\{ \sup_{x \in \mathbf{E}X} \inf_{y \in \mathbf{E}Y} \rho(x, y), \sup_{y \in \mathbf{E}Y} \inf_{x \in \mathbf{E}X} \rho(x, y) \right\}$$
$$= \rho_H(\mathbf{E}X, \mathbf{E}Y),$$

where $\mathbf{E}X$ (respectively $\mathbf{E}Y$) is the set of expectations of all integrable selections of X (respectively Y). The set $\mathbf{E}X$ is the selection expectation of X, which will be studied in detail in Section 2.1.2. Thus, $\mathfrak{m}_H(X, Y)$ is the Hausdorff distance between the selection expectations of X and Y. Unfortunately, for a more complicated metric m the evaluation of $\mathfrak{m}_H$ for random sets is very difficult, since the family of selections is rich even for simple random sets.

Probability metrics based on capacity functionals

A useful generalisation of classical probability metrics can be obtained by replacing distribution functions in their definitions with capacity functionals. The *uniform distance* between the random closed sets X and Y is defined as

$$\mathfrak{u}(X, Y; \mathcal{A}) = \sup\{|T_X(K) - T_Y(K)| : K \in \mathcal{A}\}, \tag{6.6}$$

where $\mathcal{A}$ is a subclass of $\mathcal{K}$. The *Lévy metric* is defined as follows

$$\mathfrak{L}(X, Y; \mathcal{A}) = \inf \{r > 0 :$$
$$T_X(K) \leq T_Y(K^r) + r, \ T_Y(K) \leq T_X(K^r) + r, \ K \in \mathcal{A}\}, \quad (6.7)$$

where K^r is the r-envelope of K. Hereafter we omit $\mathcal{A}$ if $\mathcal{A} = \mathcal{K}$, i.e. $\mathfrak{u}(X, Y) = \mathfrak{u}(X, Y; \mathcal{K})$ and $\mathfrak{L}(X, Y) = \mathfrak{L}(X, Y; \mathcal{K})$ etc.

Example 6.25. Let $X = \{\xi\}$ and $Y = \{\eta\}$ be random singletons. Then $\mathfrak{L}(X, Y)$ is the Lévy–Prokhorov distance between ξ and η (see Zolotarev [630]) and $\mathfrak{u}(X, Y)$ coincides with the total variation distance between the distributions of ξ and η.

Weak convergence

The following result shows that the Lévy metric determines the weak convergence of random sets.

Theorem 6.26 (Lévy metric and weak convergence). *Let $\mathcal{A} \subset \mathcal{K}$ be a separating class. A sequence $\{X_n, n \geq 1\}$ of random closed sets converges weakly to a random closed set X if and only if $\mathfrak{L}(X_n, X; \mathcal{A}(K_0)) \to 0$ as $n \to \infty$ for each $K_0 \in \mathcal{K}$, where $\mathcal{A}(K_0) = \{K \in \mathcal{A} : K \subset K_0\}$.*

Proof. Sufficiency. Let $\mathfrak{L}(X_n, X; \mathcal{A}(K_0)) \to 0$ as $n \to \infty$. It follows from (6.7) that, for $K \in \mathcal{A}(K_0) \cap \mathfrak{S}_X$,
$$T_X(K) \leq T_{X_n}(K^{\varepsilon_n}) + \varepsilon_n \text{ and } T_{X_n}(K) \leq T_X(K^{\varepsilon_n}) + \varepsilon_n, \ n \geq 1, \quad (6.8)$$
where $\varepsilon_n \downarrow 0$ as $n \to \infty$. Assume that $T_X(K) > 0$, whence $\operatorname{Int} K \neq \emptyset$. Since $\mathcal{A}$ is a separating class, there exists a sequence $\{K_n, n \geq 1\} \subset \mathcal{A}$ such that
$$T_X(K_n) \uparrow T_X(\operatorname{Int} K) = T_X(K) \quad (6.9)$$
and $K_n^{\varepsilon_n} \subset K$ for all $n \geq 1$. Since (6.8) holds on $\mathcal{A}(K_0)$,
$$T_{X_n}(K_n) \leq T_{X_n}(K_n^{\varepsilon_n}) + \varepsilon_n \leq T_{X_n}(K) + \varepsilon_n.$$
Thus,
$$T_X(K) - \varepsilon_n - (T_X(K) - T_X(K_n)) \leq T_{X_n}(K) \leq T_X(K) + \varepsilon_n$$
$$+ \left(T_X(K^{\varepsilon_n}) - T_X(K)\right).$$
The upper semicontinuity of T_X and (6.9) yield that $T_{X_n}(K) \to T_X(K)$ as $n \to \infty$.

Necessity. If $X_n \overset{d}{\to} X$, then $T_{X_n}(K) \to T_X(K)$ for each $K \in \mathcal{A} \cap \mathfrak{S}_X$. Let $\varepsilon > 0$ and $K_0 \in \mathcal{K}$ be specified. Consider compact sets $K_1, \ldots, K_m$, which form an ε-net of $\mathcal{A}(K_0)$ in the Hausdorff metric. It is easy to show that $K_i^{r_i}$ belongs to $\mathfrak{S}_X$ for some $r_i \in [\varepsilon, 2\varepsilon]$, $1 \leq i \leq m$. It follows from the pointwise convergence of the capacity functionals on $K_i^{r_i}$ that for a certain integer n_0 and every $n \geq n_0$,
$$|T_{X_n}(K_i^{r_i}) - T_X(K_i^{r_i})| \leq \varepsilon, \quad 1 \leq i \leq m.$$

Let K_j be the nearest neighbour of an arbitrary $K \in \mathcal{A}$ from the chosen ε-net. Then, for all $n \geq n_0$,

$$T_{X_n}(K) \leq T_{X_n}(K_j^\varepsilon) \leq T_{X_n}(K_j^{r_j}) \leq T_X(K_j^{r_j}) + \varepsilon \leq T_X(K^{3\varepsilon}) + 3\varepsilon. \quad (6.10)$$

Similarly,

$$T_X(K) \leq T_{X_n}(K^{3\varepsilon}) + 3\varepsilon. \quad (6.11)$$

Thus, $\mathfrak{L}(X_n, X; \mathcal{A}(K_0)) \leq 3\varepsilon$. Letting ε go to zero proves the necessity. □

Corollary 6.27. *A sequence of random closed sets $\{X_n, n \geq 1\}$ converges weakly to a random compact set X if and only if $\mathfrak{L}(X_n, X) \to 0$ as $n \to \infty$.*

Proof. Sufficiency immediately follows from Theorem 6.26.
Necessity. Let $\{K_n, n \geq 1\}$ be an increasing sequence of compact sets, such that $K_n \uparrow \mathbb{R}^d$ as $n \to \infty$. Then $T_X(\mathbb{R}^d) - T_X(K_n) < \varepsilon$ for a certain n. It is easy to show that $K' = K_n^\delta$ belongs to $\mathfrak{S}_X$ for some $\delta > 0$, whence

$$T_{X_n}(\mathbb{R}^d) - T_{X_n}(K') < \varepsilon$$

for sufficiently large n. Inequalities (6.10) and (6.11) hold for each $K \subset K'$. If $K \not\subset K'$, then

$$T_{X_n}(K) \leq T_{X_n}(K \cap K') + \varepsilon \leq T_X(K^{3\varepsilon}) + 4\varepsilon$$

and

$$T_X(K) \leq T_{X_n}(K^{3\varepsilon}) + 4\varepsilon.$$

Hence $\mathfrak{L}(X_n, X) \to 0$ as $n \to \infty$. □

Uniform convergence

The convergence in the uniform metric does not follow in general from the pointwise convergence of capacity functionals. The following result shows that the uniform convergence is related to the weak convergence of inner envelopes of random sets. For every set F and $\delta > 0$, $F^{-\delta} = \{x : B_\delta(x) \subset F\}$ is the inner parallel set of F. Note that the weak convergence of outer envelopes X_n^δ to X^δ follows from the weak convergence of the corresponding random closed sets.

Proposition 6.28 (Uniform convergence and inner envelopes). *Let $X_n^{-\delta} \xrightarrow{d} X^{-\delta}$ as $n \to \infty$ for every $\delta \geq 0$, where X is an almost surely regular closed random set such that $\mathbf{P}\{\text{Int } X \cap K \neq \emptyset\} = T_X(K)$ for each $K \in \mathcal{K}$. Then $\mathfrak{u}(X_n, X, \mathcal{K}(K_0)) \to 0$ for every $K_0 \in \mathcal{K}$.*

Proof. Without loss of generality assume that $\mathbb{E}$ is compact and $K_0 = \mathbb{E}$. It is easy to see that T_X is continuous in the Hausdorff metric on $\mathcal{K}$ and so $\mathfrak{S}_X = \mathcal{K}$. For every $\delta > 0$ fix a finite δ-net $\mathcal{N}_\delta = \{K_1, \ldots, K_{m(\delta)}\}$ of $\mathcal{K}$ in the Hausdorff metric. For every $K \in \mathcal{K}$ denote by $\mathcal{N}_\delta(K)$ the element of $\mathcal{N}_\delta$ closest to K. Then

$$T_{X_n}(K) - T_X(K) \le \max_{1 \le i \le m} |T_{X_n}(K_i^\delta) - T_X(K_i^\delta)| + \sup_{K \in \mathcal{K}} (T_X(K^\delta) - T_X(K)).$$

The first term converges to zero as $n \to \infty$, while the second one can be made arbitrarily small by the choice of δ. Furthermore,

$$T_{X_n}(K) - T_X(K) \ge T_{X_n^{-2\delta}}(\mathcal{N}_\delta(K)^\delta) - T_X(K).$$

The right-hand side is smaller in absolute value than

$$\sup_{K \in \mathcal{N}_\delta} |T_{X_n^{-2\delta}}(K^\delta) - T_{X^{-2\delta}}(K^\delta)| + \sup_{K \in \mathcal{N}_\delta} [T_X(K^\delta) - T_{X^{-2\delta}}(K)|]$$

$$+ \sup_{K \in \mathcal{K}} [T_X(K^{2\delta}) - T_X(K)].$$

The first term converges to zero by the convergence assumption, the second converges to zero by the assumptions imposed on X, while the third term converges to zero by the continuity of T_X. □

Further probability metrics for random closed sets can be defined using the following idea. Let $\mathcal{H}$ be a family of functions that map $\mathcal{F}$ into $\mathbb{R}$. For $h \in \mathcal{H}$ write $h^\vee$ for the sup-integral of h. Put

$$\mathfrak{m}_\mathcal{H}(X, Y) = \sup_{h \in \mathcal{H}} |\mathbf{E} h^\vee(X) - \mathbf{E} h^\vee(Y)|. \tag{6.12}$$

If the family $\mathcal{H}$ contains all indicators, then $\mathfrak{m}_\mathcal{H}$ is a probability metric. If $h = \mathbf{1}_K$ is an indicator function, then $\mathbf{E} h^\vee(X) - \mathbf{E} h^\vee(Y) = T_X(K) - T_Y(K)$. In general, $\mathfrak{m}_\mathcal{H}$ is a probability metric if the family $\mathcal{H}$ is so rich that the values $\mathbf{E} h^\vee(X)$ for $h \in \mathcal{H}$ determine uniquely the distribution of X. Since $\mathbf{E} h^\vee(X)$ equals the Choquet integral $\int h \, dT_X$, the metric defined by (6.12) is a generalisation of the integral metric from Müller [417].

7 Random sets and hitting processes

7.1 Hitting processes

As we have seen, the distribution of a random closed set in a LCHS space is uniquely determined by the hitting probabilities of compact sets or open sets. This means that a random closed set X is identifiable by its *hitting process*

$$\zeta(G) = \mathbf{1}_{X \cap G \ne \emptyset}, \quad G \in \mathcal{G},$$

considered to be a random function on $\mathcal{G}$, or the process $\zeta(K) = \mathbf{1}_{X \cap K \ne \emptyset}$ being a random function on $\mathcal{K}$, see Figure 7.1. Note that

$$\mathbf{E}\zeta(K) = T_X(K) = \mathbf{P}\{X \cap K \ne \emptyset\}.$$

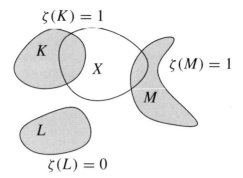

Figure 7.1. Hitting process ζ generated by X.

The process ζ can be extended from its values on a separating class $\mathcal{A}$ using (1.30) and (1.31). The following result deals with the extension of hitting processes and is similar to Theorem 1.27 that concerns extensions of capacity functionals defined on separating classes. The notation ζ^- and ζ^0 also appears in Theorem 1.27.

Theorem 7.1 (Extension of the hitting process). Let $\mathcal{A} \subset \mathfrak{B}_k$ be separating and closed under finite union class and let $\zeta(A)$, $A \in \mathcal{A}$, be an increasing $\{0, 1\}$-valued random process satisfying $\zeta(\emptyset) = 0$ and

$$\zeta(A_1 \cup A_2) = \max(\zeta(A_1), \zeta(A_2)) \quad \text{a.s. for all } A_1, A_2 \in \mathcal{A}. \tag{7.1}$$

Then there exists a random closed set X such that with probability one $\mathbf{1}_{X \cap K \neq \emptyset} = \zeta^-(K)$ for every $K \in \mathcal{K}$ and $\mathbf{1}_{X \cap G \neq \emptyset} = \zeta^0(G)$ for every $G \in \mathcal{G}$. If also

$$\zeta^0(\operatorname{Int} A) = \zeta^-(\operatorname{cl} A) \quad \text{a.s., } A \in \mathcal{A}, \tag{7.2}$$

then the hitting process generated by X is an extension of ζ. If $\mathcal{A} \subset \mathcal{K}$, then (7.2) may be replaced by $\zeta(A) = \zeta^-(A)$ a.s. for all $A \in \mathcal{A}$, and if $\mathcal{A} \subset \mathcal{G}$, then both (7.1) and (7.2) may be replaced by

$$\zeta(\cup_{n \geq 1} A_n) = \sup_{n \geq 1} \zeta(A_n) \quad \text{a.s.} \tag{7.3}$$

for all sequences $\{A_n, n \geq 1\} \subset \mathcal{A}$ such that $\cup_{n \geq 1} A_n \in \mathcal{A}$.

Proof. It is easy to see that $\zeta^0(G_1 \cup G_2) = \max(\zeta^0(G_1), \zeta^0(G_2))$ a.s. for all $G_1, G_2 \in \mathcal{G}$. The same holds simultaneously for all G_1, G_2 from a countable separating class $\mathcal{G}_0 \subset \mathcal{G}$.

If $\{x_1, \ldots, x_n\} \subset K$, then K is covered by $A_1, \ldots, A_n$ where $x_i \in \operatorname{Int} A_i$ and $A_i \in \mathcal{A}$ for all i. Therefore,

$$\zeta^-(K) = \sup_{x \in K} \zeta^-(\{x\}), \quad K \in \mathcal{K}.$$

If $X = \{x \in \mathbb{E} : \zeta^-(\{x\}) = 1\}$, then the hitting process generated by X coincides with ζ^- on $\mathcal{K}$, i.e. $\mathbf{1}_{X \cap K \neq \emptyset} = \zeta^-(K)$ for all $K \in \mathcal{K}$. Therefore, the hitting process coincides with ζ^0 on $\mathcal{G}$.

If $A \in \mathcal{A}$ satisfies $\zeta^0(\text{Int } A) = \zeta^-(\text{cl } A)$, then $\mathbf{1}_{X \cap \text{Int } A \neq \emptyset} = \mathbf{1}_{X \cap \text{cl } A \neq \emptyset}$. The last two assertions are easy to prove noticing that (7.3) implies (7.1) and $\zeta(A) = \zeta^0(\text{Int } A)$ for all $A \in \mathcal{A}$. □

7.2 Trapping systems

$\mathcal{T}$-closure

A hitting process defined on a rich family of sets (e.g. on a separating class) is a rather complicated object. It is quite natural to attempt to restrict it to a smaller family of sets, even at the cost of an incomplete characterisation of the sets that generate this hitting process.

Let us fix a family $\mathcal{T}$ of Borel sets called a *trapping system*. The only assumptions on $\mathcal{T}$ are that every set $A \in \mathcal{T}$ is non-empty and all traps cover $\mathbb{E}$. Assume that every set $F \subset \mathbb{E}$ is accessible only through the knowledge of whether or not F hits sets from the trapping system $\mathcal{T}$. In other words, instead of F we observe its hitting process $\mathbf{1}_{F \cap K \neq \emptyset}$ for $K \in \mathcal{T}$. With every set $F \subset \mathbb{E}$ we associate its $\mathcal{T}$-*closure* $\text{cl}(F; \mathcal{T})$ defined as the intersection of all sets A^c such that $A \cap F = \emptyset$ and $A \in \mathcal{T}$.

It is easy to see that $F \subset \text{cl}(F; \mathcal{T})$ and the $\mathcal{T}$-closure is a monotone idempotent operation, i.e. $\text{cl}(F_1; \mathcal{T}) \subset \text{cl}(F_2; \mathcal{T})$ for $F_1 \subset F_2$ and $\text{cl}(\text{cl}(F; \mathcal{T}); \mathcal{T}) = \text{cl}(F; \mathcal{T})$. A set F is called $\mathcal{T}$-closed if $F = \text{cl}(F; \mathcal{T})$. However, unlike the topological closure, the $\mathcal{T}$-closure does not distribute over intersections or finite unions.

Example 7.2 (Trapping by disks). In $\mathbb{R}^2$ consider the family $\mathcal{T}$ of all disks of radius r. If $r \to \infty$, then the corresponding $\mathcal{T}$-closure coincides with the convex hull operation. Otherwise, the $\mathcal{T}$-closure of F is a subset of the convex hull of F, which is obtained by rolling a disk of radius r outside F, see Figure 7.2.

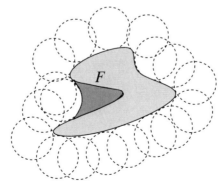

Figure 7.2. $\mathcal{T}$-closure of F using the family of disks in $\mathbb{R}^2$.

Incidence functions

A zero-one function f over $\mathcal{T}$ is called a *strong incidence function* if $A \subset \cup_{i \in I} A_i$ and $f(A) = 1$ imply $f(A_i)$ for some $i \in I$ from an arbitrary family of traps A_i, $i \in I$, and a trap $A \in \mathcal{T}$. A *weak incidence function* satisfies the above requirement for finite collections A_i, $i \in I$, only. The hitting process $\mathbf{1}_{M \cap A \neq \emptyset}$, $A \in \mathcal{T}$, generated by any set M is a strong incidence function.

Proposition 7.3 (Strong incidence functions). *A zero-one function f over a trapping system $\mathcal{T}$ is a strong incidence function if and only if f is the hitting process on $\mathcal{T}$ generated by the set*

$$F = \bigcap_{A \in \mathcal{T}, \, f(A)=0} A^c.$$

This is the unique $\mathcal{T}$-closed set which generates the hitting process f.

Proof. For arbitrary $A \in \mathcal{T}$, $f(A) = 0$ implies $A \cap F = \emptyset$. If $A \cap F = \emptyset$, then A is covered by the union of sets $A' \in \mathcal{T}$ with $f(A') = 0$, whence $f(A) = 0$ by the condition imposed on strong incidence functions. Thus, f is a hitting process generated by F. It is easy to see that F is $\mathcal{T}$-closed. Any two $\mathcal{T}$-closed sets are identical if and only if they avoid the same traps, which immediately implies the uniqueness. □

The above result makes it possible to treat $\mathcal{T}$-closed sets as strong incidence functions on the corresponding trapping system $\mathcal{T}$. If $\mathcal{T}$ is the family $\mathcal{G}$ of all open sets, then the $\mathcal{T}$-closure becomes the topological closure and the family of $\mathcal{T}$-closed sets coincides with the family $\mathcal{F}$ of all closed sets. The flexibility associated with a choice of trapping system makes it possible to adjust the trapping system $\mathcal{T}$ in order to obtain a rather general family of subsets of $\mathbb{E}$ as the family of $\mathcal{T}$-closed sets. The following easy result formalises such a choice.

Proposition 7.4 ($\mathcal{T}$-closed sets). *A system $\mathcal{Z}$ of subsets of $\mathbb{E}$ is identifiable with the system of $\mathcal{T}$-closed sets for a suitably chosen trapping system $\mathcal{T}$ if and only if $\mathcal{Z}$ contains $\emptyset$, $\mathbb{E}$ and is closed under arbitrary intersections. The corresponding trapping system may be obtained as the family of all sets F^c where $F \in \mathcal{Z}$ and $F \neq \mathbb{E}$.*

It suffices to take $\mathcal{T}$ that consists of all the sets F^c, $F \in \mathcal{Z}'$, where $\mathcal{Z}'$ is a subclass of $\mathcal{Z}$ such that every $F \in \mathcal{Z}$ ($F \neq \mathbb{E}$) can be obtained as an intersection of sets from $\mathcal{Z}'$. If the conditions on $\mathcal{Z}$ imposed in Proposition 7.4 are not satisfied, then it is still possible to come up with a trapping system which yields a family of $\mathcal{T}$-closed sets larger than $\mathcal{Z}$. Each $F \in \mathcal{Z}$ is $\mathcal{T}$-closed for each trapping system $\mathcal{T}$ such that each element of $\mathcal{Z}$ can be expressed as an intersection of complements of sets from $\mathcal{T}$.

Random weak and strong incidence functions

Let $\mathcal{T}_\cup$ consist of the empty set and all finite unions of traps from $\mathcal{T}$. Any incidence function can be naturally extended onto the family $\mathcal{T}_\cup$ as $f(\cup A_i) = \max_i(f(A_i))$.

Since the definition of weak incidence functions involves values of the function on any finite collection of traps, it is possible to define a *random weak incidence function* as a stochastic process ζ defined on $\mathcal{T}_\cup$ such that $\zeta(\cup A_i) = \max_i(\zeta(A_i))$ a.s. for all finite collections $A_i \in \mathcal{T}$. Then $T(A) = \mathbf{P}\{\zeta(A) = 1\}$ for $A \in \mathcal{T}_\cup$ determines the finite-dimensional distributions of $\zeta(A)$, $A \in \mathcal{T}_\cup$, as

$$\mathbf{P}\{\zeta(A) = 0, \zeta(A_1) = 1, \ldots, \zeta(A_n) = 1\} = -\Delta_{A_n} \cdots \Delta_{A_1} T(A),$$

see (1.9). Therefore, the distribution of a random weak incidence function can be defined by a completely alternating functional on $\mathcal{T}_\cup$ with values in $[0, 1]$.

However, *random strong incidence functions* are more important from the point of view of their possible interpretation as random $\mathcal{T}$-closed sets, see Proposition 7.3. Their definition involves arbitrary collections of traps, which calls for arguments similar to the separability concept in the studies of stochastic processes. In order to be able to construct a random strong incidence function the trapping system ought to satisfy some separability assumptions.

Definition 7.5 (c-traps and trapping space). A trap A is called a *c-trap*, when every covering by traps of $\mathrm{cl}(A; \mathcal{T})$ can be reduced to a finite subcovering. The pair $(\mathbb{E}, \mathcal{T})$ is called a *trapping space* if $\mathcal{T}$ is a trapping system and also
  (i) with every trap A we can associate a countable system of subtraps of A in such a way that all possible subtraps of A can be obtained as unions of some traps from the chosen countable system;
  (ii) if $x \in A \in \mathcal{T}$, then x belongs to a c-trap whose $\mathcal{T}$-closure is contained in A.

If $\mathcal{T}$ is the system of all open sets, then the above requirements are clearly satisfied if $\mathbb{E}$ is a LCHS space. The c-traps are relatively compact open sets and the countable system in Definition 7.5(i) is the base of the topology.

Theorem 7.6 (Random strong incidence functions). *Let $(\mathbb{E}, \mathcal{T})$ be a trapping space. A completely alternating function T on $\mathcal{T}_\cup$ with values in $[0, 1]$ such that $T(\emptyset) = 0$ corresponds to a random strong incidence function (or, equivalently a random $\mathcal{T}$-closed set) if and only if T is continuous from below, i.e., for every trap $A \in \mathcal{T}$, $T(A)$ equals the supremum of $T(\cup_{i=1}^n A_i)$ for every $n \geq 1$ and all c-traps $A_1, \ldots, A_n$ such that $\cup_{i=1}^n A_i \subset A$.*

Theorem 7.6 can be interpreted from the general point of view of measures on lattices described in Section 3. If the corresponding lattice operation is defined by $F_1 \vee F_2 = \mathrm{cl}(F_1 \cap F_2; \mathcal{T})$, then the conditions on the trapping space ensure that the corresponding Scott topology is second countable, see Proposition 3.2. No other topological assumptions are made on $\mathbb{E}$.

Random closed sets with special realisations

If random closed sets take values in a subfamily $\mathcal{Z}$ of the family $\mathcal{F}$ of all closed sets, then this additional information can be used to reduce the family of compact sets needed to define the capacity functional and still ensure the uniqueness. In this context it is desirable to construct the probability measure on the whole σ-algebra $\mathfrak{B}(\mathcal{F})$. The following theorem provides alternative conditions that make it possible.

Theorem 7.7 (Distributions of random closed sets with restriction on realisations). *Let $\mathbb{E}$ be a LCHS space. Consider $\mathcal{Z} \subset \mathcal{F}$ and $\mathcal{T} \subset \mathcal{K}$ and suppose that the following conditions hold.*

(i) *$\mathcal{T}$ is closed with respect to finite unions.*
(ii) *There exists a countable subclass $\mathcal{G}' \subset \mathcal{G}$ such that any $K \in \mathcal{T}$ is the limit of a decreasing sequence of sets from $\mathcal{G}'$ and also any $G \in \mathcal{G}'$ is the limit of an increasing sequence from $\mathcal{T}$.*
(iii) *For any $G \in \mathcal{G}' \cup \{\emptyset\}$ and $K_1, \ldots, K_n \in \mathcal{T}, n \geq 0$, the family*

$$\mathcal{F}^G_{K_1,\ldots,K_n} \cap \mathcal{Z}$$

is non-empty, provided $K_i \setminus G \neq \emptyset$ for all $1 \leq i \leq n$.
(iv) *The σ-algebra $\sigma(\mathcal{Z})$ generated by*

$$\left\{ \mathcal{F}^K_{G_1,\ldots,G_n} \cap \mathcal{Z} : K \in \mathcal{T} \cup \{\emptyset\}, G_i \in \mathcal{G}', 1 \leq i \leq n \right\}$$

coincides with the σ-algebra $\mathfrak{B}(\mathcal{F}) \cap \mathcal{Z} = \{\mathcal{Y} \cap \mathcal{Z} : \mathcal{Y} \in \mathfrak{B}(\mathcal{F})\}$ induced by $\mathfrak{B}(\mathcal{F})$ on $\mathcal{Z}$.

Let $\tilde{\mathcal{Z}}$ be the closure of $\mathcal{Z}$ in the Fell topology. If $T : \mathcal{T} \mapsto [0, 1]$ is a completely alternating upper semicontinuous functional with $T(\emptyset) = 0$, then there is a (necessary unique) probability $\mathbf{P}$ on $\sigma(\mathcal{Z})$ satisfying $\mathbf{P}\{\mathcal{F}_K \cap \tilde{\mathcal{Z}}\} = T(K)$ for all $K \in \mathcal{T}$.

7.3 Distributions of random convex sets

Containment functional of random convex compact sets

In the following assume that $\mathbb{E}$ is the finite-dimensional Euclidean space $\mathbb{R}^d$. A random closed set X in $\mathbb{R}^d$ is said to be *convex* if its realisations are almost surely convex, i.e. X belongs to $\mathrm{co}\,\mathcal{F}$ almost surely, see Definition 4.32. Of course, the Choquet theorem implies that the distribution of each random convex closed set X in $\mathbb{R}^d$ is determined by the corresponding capacity functional. However, there is a more economical way to define distributions of random convex compact sets.

Theorem 7.8 (Distribution of convex compact sets). *The distribution of a random convex compact set X in $\mathbb{R}^d$ is determined uniquely by the values of the containment functional $C_X(K) = \mathbf{P}\{X \subset K\}$ for $K \in \mathrm{co}\,\mathcal{K}$. Moreover, it suffices to consider all K being convex polytopes.*

Proof. By Proposition 2.5, X is a random convex compact set in $\mathbb{R}^d$ if and only if the support function $h(X, u)$ is a random variable for every u from the unit sphere $\mathbb{S}^{d-1}$. The finite-dimensional distributions of the stochastic process $h(X, u), u \in \mathbb{S}^{d-1}$, are uniquely determined by the containment functional of X. For example,

$$\mathbf{P}\{h(X,u) \le t\} = \lim_{n\to\infty} C_X(\mathbb{H}_u^-(t) \cap B_n),$$

where $\mathbb{H}_u^-(t) = \{x : \langle x, u \rangle \le t\}$ and $\{B_n, n \ge 1\}$ is an increasing sequence of balls of radius n centred at the origin. □

Theorem 7.8 can be obtained using lattice-theoretic arguments described in Section 3.4. If X is a non-convex random compact set, then the containment functional $C_X(K)$, $K \in \text{co}\,\mathcal{K}$, does determine the distribution of $\text{co}(X)$. The containment functional can be extended onto the family $\text{co}\,\mathcal{F}$ of convex closed sets by

$$C_X(F) = \mathbf{P}\{X \subset F\}, \quad F \in \text{co}\,\mathcal{F}.$$

The containment functional is a completely monotone capacity on $\text{co}\,\mathcal{F}$, see Definition 1.9. In other words, it satisfies the following conditions:

(I1) C_X is upper semicontinuous, i.e. $C_X(F_n) \downarrow C_X(F)$ if $F_n \downarrow F$ as $n \to \infty$ for $F, F_1, F_2, \ldots \in \text{co}\,\mathcal{F}$.
(I2) The recurrently defined functionals $\nabla_{F_n} \cdots \nabla_{F_1} C_X(F)$ are non-negative for all $F_1, \ldots, F_n \in \text{co}\,\mathcal{F}$ and $n \ge 1$, see (1.14).

Note that

$$\nabla_{F_n} \cdots \nabla_{F_1} C_X(F) = \mathbf{P}\{X \subset F, X \not\subset F_i, 1 \le i \le n\}.$$

Unbounded random convex closed sets

Although it is tempting to extend Theorem 7.8 for a not necessarily compact random closed set X by considering its containment functional $C_X(F)$ for $F \in \text{co}\,\mathcal{F}$, the following example shows that the distribution of a general non-compact convex random closed set cannot be determined by its containment functional on $\text{co}\,\mathcal{F}$.

Example 7.9 (Unbounded random convex closed set). Let X be the half-space which touches the unit ball B_1 at a random point uniformly distributed on its boundary. Then $C_X(F) = 0$ for each $F \in \text{co}\,\mathcal{F}$, $F \ne \mathbb{R}^d$, so that this containment functional does not determine the distribution of X.

Now consider a special family of random convex closed (not necessarily compact) sets whose distributions are characterised by the containment functionals. Let X be a random convex closed set which is not necessarily compact. Define

$$L_X = \{u \in \mathbb{S}^{d-1} : h(X, u) < \infty \text{ a.s.}\}, \tag{7.4}$$

where $h(X, u)$ is the support function of X. Then L_X is a convex subset of $\mathbb{S}^{d-1}$, where the convexity means that the cone generated by L_X is convex. Note that L_X is not necessarily closed or open. For a convex set $L \subset \mathbb{S}^{d-1}$, define a family of convex closed sets as

$$\mathcal{C}(L) = \{F \in \text{co}\,\mathcal{F} : h(F, u) = \infty, u \notin L, h(F, u) < \infty, u \in L\}. \tag{7.5}$$

Clearly, if $X \in \mathcal{C}(L)$ a.s., then $L_X = L$. In the following we consider random sets with values in $\mathcal{C}(L)$. Consider a σ-algebra $\sigma(L)$ on $\mathcal{C}(L)$ generated by the families $\{F \in \mathcal{C}(L) : F \subset G\}$ for all open convex G. Since the same σ-algebra is generated if G is taken from the family of all open half-spaces, X is a $\sigma(L)$-measurable random element if and only if $h(X, u)$, $u \in L$, is a random function.

Proposition 7.10. *For each convex set* $L \subset \mathbb{S}^{d-1}$, $\sigma(L) = \mathfrak{B}(\mathcal{F}) \cap \mathcal{C}(L)$, *i.e. the σ-algebra $\sigma(L)$ coincides with the σ-algebra induced by $\mathfrak{B}(\mathcal{F})$ on the family $\mathcal{C}(L)$.*

Proof. One-side inclusion $\sigma(L) \subset \mathfrak{B}(\mathcal{F}) \cap \mathcal{C}(L)$ is evident. For each $F \in \text{co}\,\mathcal{F}$, the support function $h(F, u)$ equals the supremum of continuous linear functions and, therefore, is lower semicontinuous. Furthermore, if $u_n \to u$ as $n \to \infty$ and $u, u_1, u_2, \ldots \in L$, then $h(F, u_n) \to h(F, u)$ for each $F \in \mathcal{C}(L)$. Consider a countable dense set $L' \subset L$ and a compact set K. For each F from $\mathcal{F}^K \cap \mathcal{C}(L)$ there exists a hyperplane which separates F and K, see Hiriart-Urruty and Lemaréchal [258, Cor. 4.1.3]. Because of the continuity of the support function on L, it is possible to find a hyperplane which separates F and K and has a normal from L'. Therefore, $\mathcal{F}^K \cap \mathcal{C}(L)$ is a countable union of sets from $\sigma(L)$. □

By the same arguments as in Theorem 7.8, we obtain the following result.

Proposition 7.11 (Distribution of random convex closed set). *The distribution of a random closed set X with realisations in $\mathcal{C}(L)$ for some non-empty convex set $L \subset \mathbb{S}^{d-1}$ is uniquely determined by the containment functional $C_X(F)$ for $F \in \mathcal{C}(L)$.*

Weak convergence of random convex compact sets

It is well known (see Billingsley [70, p. 15]) that a class of events that determines the distribution is not necessarily a *convergence determining* class. In other words, while the probabilities of a given family of events may determine the distribution uniquely, this does not mean that the pointwise convergence on all continuity events from the same family automatically implies the weak convergence. However, for random convex compact sets, the pointwise convergence of containment functionals does imply the weak convergence.

Theorem 7.12. *A sequence $\{X_n, n \geq 1\}$ of random convex compact sets converges weakly to a random closed set X if*

$$C_{X_n}(K) \to C_X(K) \quad \text{as } n \to \infty \tag{7.6}$$

for every $K \in \text{co}\,\mathcal{K}$ such that $C_X(K) = C_X(\text{Int}\,K)$.

A proof can be produced using support functions of random compact sets, since (7.6) for all $K \in \text{co}\,\mathcal{K}$ implies the convergence of the finite-dimensional distributions of the support functions of X_n. The corresponding tightness condition in the space of continuous functions easily follows from the Lipschitz property of the support functions (see Theorem F.1) and the fact that (7.6) implies $\|X_n\| \xrightarrow{d} \|X\|$ if K is chosen to be an arbitrary ball, see also Proposition 6.13. Furthermore, it is possible to show that it suffices to require (7.6) for K from a suitably chosen separating class of convex compact sets, e.g. from the family of all compact polytopes.

Star-shaped sets

A random closed set X in $\mathbb{R}^d$ is said to be *star-shaped* with respect to a deterministic point $a \in \mathbb{R}^d$ if $c(X - a) \subset (X - a)$ a.s. for all $c \in [0, 1]$. This is equivalent to the requirement that $a \in X$ a.s. and $[a, \xi] \subset X$ a.s. for every selection $\xi \in \mathcal{S}(X)$. Clearly, every convex set is star-shaped. Every star-shaped set corresponds to its radius-vector function $r_X(u) = \sup\{t \geq 0 : a + tu \in X\}$. Since the distribution of a star-shaped set is determined uniquely by the finite-dimensional distributions of its radius-vector function, the containment functional $C_X(F)$ for all star-shaped closed sets F determines uniquely the distribution of X.

A random closed set X is star-shaped with respect to its selection ξ if $X - \xi$ is star-shaped with respect to the origin. Let Y be the set of all selections $\xi \in \mathcal{S}(X)$ such that X is star-shaped with respect to ξ. The set Y is called the *kernel* of X.

Theorem 7.13. *Let X be a random compact set. Then its kernel is a random compact convex set.*

Proof. Without loss of generality we can assume that X is star-shaped with respect to the origin. Otherwise one can consider an appropriate random translation of X. It is easy to see that Y is closed. Furthermore, for any two selections $\xi, \eta \in Y$ and any selection $\zeta \in \mathcal{S}(X)$, the triangle with vertices ξ, η and ζ is contained in X. Hence the set Y is convex.

If Y has non-empty interior (and so is regular closed), it is easy to show that Y is a random closed set. By Theorem 2.6, it suffices to show that $\{y \in Y\}$ is measurable for any y. This event can be represented as the intersection of the events $[y, \xi] \subset X$ for all ξ from a Castaing representation of X.

In the general case, consider the r-envelope X^r. Then the kernel Y_r of X^r contains a neighbourhood of the origin. Indeed, if $\|y\| < r$, then, for any point $x \in X$, the segment $[y, x]$ lies within the Hausdorff distance at most r from $[0, x] \subset X$, whence $[y, x]$ is contained in X^r. Hence Y_r is regular closed and so is a random compact convex set. The proof is finished by observing that Y_r converges almost surely to the kernel of X as $r \downarrow 0$. □

8 Point processes and random measures

8.1 Random sets and point processes

Locally finite measures

A measure μ on the family $\mathfrak{B}$ of Borel sets in $\mathbb{E}$ is called *counting* if it takes only non-negative integer values. A counting measure μ is *locally finite* if μ is finite on bounded subsets of $\mathbb{E}$.

Proposition 8.1 (Support of locally finite measure). *Let $\mathbb{E}$ be a LCHS space. If μ is a locally finite measure on $\mathbb{E}$, then*
  (i) *the support of μ is closed, i.e. $\operatorname{supp}\mu \in \mathcal{F}$;*
  (ii) *for all $G \in \mathcal{G}$, $\operatorname{supp}\mu \cap G \neq \emptyset$ if and only if $\mu(G) > 0$.*

Proof. If $x \notin \operatorname{supp}\mu$, then $x \in G$ for an open set G with $\mu(G) = 0$. Therefore, $G \subset (\operatorname{supp}\mu)^c$, which means that $(\operatorname{supp}\mu)^c$ is open and $\operatorname{supp}\mu$ is closed. Since $\mathbb{E}$ is second countable, $\mu((\operatorname{supp}\mu)^c) = 0$. □

Applied to a counting measure μ, Proposition 8.1 implies that $\operatorname{supp}\mu$ is a locally finite set, i.e. $\operatorname{supp}\mu$ has at most a finite number of points in any compact set.

The family $\mathcal{N}$ of all counting measures can be endowed with a σ-algebra generated by $\{\mu \in \mathcal{N} : \mu(B) = k\}$ for $k = 0, 1, 2, \ldots$ and $B \in \mathfrak{B}$, so that a random counting measure can be defined as a random element N in $\mathcal{N}$. A random counting measure is also called a *point process*. The measurability condition implies that $N(B)$, the number of points in a Borel set B, is a random variable. A point process (or the corresponding counting measure) is called *simple* if $\sup_x N(\{x\}) \leq 1$ a.s. The following important result follows from Proposition 8.1 and the fundamental measurability theorem (Theorem 2.3).

Corollary 8.2 (Counting measures and point processes). *Let $\mathbb{E}$ be LCHS. Then N is a simple point process if and only if $\operatorname{supp} N$ is a locally finite random closed set in $\mathbb{E}$.*

Since the map $F \mapsto \operatorname{card}(F \cap K)$ is measurable on $\mathcal{F}$ for every $K \in \mathcal{K}$, it is easily seen that the family of locally finite sets belongs to $\mathfrak{B}(\mathcal{F})$. Therefore, the event $\{X \text{ is locally finite}\}$ is measurable for every random closed set X.

A point process is said to be *stationary* if its distribution is invariant under translations. This is equivalent to the statement that $\operatorname{supp} N$ is a stationary locally finite random closed set.

Application of the Choquet theorem

Corollary 8.2 yields the following interesting conclusion, which immediately follows from the Choquet theorem. The letter N is used to denote both the random counting measure and the random locally finite set being its support.

Theorem 8.3 (Distribution of a simple point process). *The distribution of a simple point process N in a LCHS space is uniquely determined by the probabilities $\mathbf{P}\{N \cap K = \emptyset\}$ (or $\mathbf{P}\{N(K) = 0\}$) for all $K \in \mathcal{K}$. Alternatively, the distribution of N is determined uniquely by $\mathbf{P}\{N \cap G \neq \emptyset\}$ for all $G \in \mathcal{G}$.*

Let us define by

$$Q_N(K) = \mathbf{P}\{N(K) = 0\}, \quad K \in \mathcal{K},$$

the *avoidance functional* generated by a simple point process N. Since $Q_N(K)$ is the avoidance functional of the random closed set $\operatorname{supp} N$, $T_N(K) = 1 - Q_N(K)$ is a completely alternating upper semicontinuous functional. It is quite natural to ask if it is possible to single out those capacity (or avoidance) functionals which generate distributions of locally finite random sets and so can be interpreted as capacity (or avoidance) functionals corresponding to simple point processes. The following result provides necessary and sufficient conditions for this. It can be equivalently reformulated for the avoidance functional instead of the capacity functional.

Proposition 8.4. *A capacity functional T defines a locally finite random closed set if and only if, for every compact set K and every $\varepsilon > 0$, there exists $k_\varepsilon \geq 1$ such that, whenever $K = \cup_{i=1}^{\infty} B_i$ and the B_i are disjoint, one has*

$$T(K) - \varepsilon \leq -\sum \Delta_{B_{i_k}} \cdots \Delta_{B_{i_1}} T(K \setminus (\cup_{j=1}^{k} B_{i_j})), \tag{8.1}$$

where the sum ranges over all non-empty, finite subcollections $\{B_{i_1}, \ldots, B_{i_k}\} \subset \{B_1, B_2, \ldots\}$ with $k \leq k_\varepsilon$.

Proof. Let X be a random closed set with the capacity functional T. In view of (1.9), condition (8.1) can be reformulated as

$$\mathbf{P}\{X \cap K \neq \emptyset\} \tag{8.2}$$
$$-\sum \mathbf{P}\left\{X \cap B_{i_1} \neq \emptyset, \ldots, X \cap B_{i_k} \neq \emptyset, \ X \cap (K \setminus (\cup_{j=1}^{k} B_{i_j})) = \emptyset\right\} < \varepsilon.$$

The left-hand side of (8.2) is smaller than $\mathbf{P}\{X \cap K \neq \emptyset, \operatorname{card}(X \cap K) > k_\varepsilon\}$. If X is locally finite, then $\mathbf{P}\{\operatorname{card}(X \cap K) > k_\varepsilon\}$ is smaller than ε for sufficiently large k_ε, so that (8.2) is a necessary condition for the local finiteness.

Now assume that (8.1) holds. Let $\{A_i^n, 1 \leq i \leq m_i\}$, $n \geq 1$, be an increasing sequence of finite partitions of $K \in \mathcal{K}$, so that $K = \cup_i A_i^n$ and each set A_i^n, $1 \leq i \leq m_n$, is a union of sets from $\{A_j^{n+1}, 1 \leq j \leq m_{n+1}\}$. Furthermore, let Y_n be the cardinality of $\{i : A_i^n \cap X \neq \emptyset\}$. Then $Y_n \leq Y_{n+1}$ and the number of points in $X \cap K$ is at most $\lim_{n \to \infty} Y_n$. Now (8.2) implies $\mathbf{P}\{Y_n > k_\varepsilon\} < \varepsilon$, so that $\lim_{n \to \infty} Y_n$ is finite almost surely. □

In general, it is difficult and often impossible to verify the assumptions of Proposition 8.4. Consider a particular case of $\mathbb{E} = \mathbb{R}$ and translation-invariant capacity functionals, with the aim of characterising locally finite stationary random closed sets on the line in terms of their capacity functionals. For every $t > 0$ put

$$q(t) = Q_X((0, t)) = 1 - T_X((0, t)) = \mathbf{P}\{X \cap (0, t) = \emptyset\}.$$

Proposition 8.5 (Stationary locally finite random sets on the line). *Let X be a stationary random closed subset of $\mathbb{R}$. If X is locally finite, then*
  (i) $q(0+) = 1$.
On the other hand, X is locally finite if one of the following equivalent conditions is satisfied:
  (ii) *q has a finite right-hand derivative at 0;*

(iii) $\lim_{n\to\infty} 2^n(1 - q(2^{-n})) < \infty$.

Proof. If X is locally finite, then

$$1 - q(0+) = \lim_{t\downarrow 0} \mathbf{P}\{X \cap (0, t) \neq \emptyset\} = 0,$$

so (i) is necessary.

Note that (ii) and (iii) are equivalent because of the monotonicity of q. Let us show that (iii) implies that $X \cap [0, 1]$ is almost surely finite, whence X is necessarily locally finite. For every $n \geq 1$ and $k = 0, \ldots, 2^n - 2$ introduce events

$$A(n, k) = \{X \cap [k2^{-n}, (k+1)2^{-n}) \neq \emptyset, \ X \cap [(k+1)2^{-n}, (k+2)2^{-n}) \neq \emptyset\}$$

and put

$$A(n) = \bigcup_{k=0}^{2^n-2} A(n, k).$$

Then $X \cap [0, 1]$ is almost surely finite if no more than a finite number of events $A_1, A_2, \ldots$ occurs. For this, it suffices to show that

$$\sum_{n\geq 1} \mathbf{P}(A(n)) < \infty. \tag{8.3}$$

By stationarity,

$$\mathbf{P}(A(n, k)) = \mathbf{P}(A(n, 0))$$
$$= 2\mathbf{P}\{X \cap [0, 2^{-n}) \neq \emptyset\} - \mathbf{P}\{X \cap [0, 2^{-n+1}) \neq \emptyset\}$$
$$= 1 - 2\bar{q}(2^{-n}) + \bar{q}(2^{-n+1}) = 2b_n - b_{n-1},$$

where $\bar{q}(t) = \mathbf{P}\{X \cap [0, t) = \emptyset\}$ and $b_n = 1 - \bar{q}(2^{-n})$. Then

$$\sum_{n=1}^{m} \mathbf{P}(A(n, k)) \leq \sum_{n=1}^{m} 2^n \mathbf{P}(A(n, 0)) = \sum_{n=1}^{m} 2^n(2b_n) - \sum_{n=1}^{m} 2^n b_{n-1}$$
$$= 2(2^m b_m - b_0).$$

Note that

$$q(t) \geq \bar{q}(t) \geq q(t) - \mathbf{P}\{0 \notin X\} = q(t) - (1 - q(0+)) = q(t),$$

so that $\bar{q}(t) = q(t)$. Now condition (iii) implies that $2^m b_m = 2^m(1 - q(2^{-m}))$ is bounded, which yields (8.3). □

Ordered coupling and thinning

The following result is a corollary of Theorem 4.42 on the ordered coupling for random sets. It establishes a condition for a point process N' to be a thinning of another point process N. Recall that a *thinning* of N is defined as a point process N' which is a subset of N, see Stoyan, Kendall and Mecke [544, p. 146].

Proposition 8.6 (Thinning of point processes). *Let N' and N be two simple point processes on a LCHS space $\mathbb{E}$. Then N' can be realised as a thinning of N (so that N' is stochastically smaller than N) if and only if*

$$\mathbf{P}\{N' \cap G_1 \neq \emptyset, \ldots, N' \cap G_n \neq \emptyset\} \leq \mathbf{P}\{N \cap G_1 \neq \emptyset, \ldots, N \cap G_n \neq \emptyset\}$$

for every $n \geq 1$ and $G_1, \ldots, G_n \in \mathcal{G}$.

Poisson point process

One particularly important example of a point process is the Poisson point process defined as follows.

Definition 8.7 (Poisson point process). Let Λ be a locally finite measure on a topological space $\mathbb{E}$ with Borel σ-algebra $\mathfrak{B}$. The *Poisson point process* Π_Λ with the *intensity measure* Λ is a random subset of $\mathbb{R}^d$ such that the following properties are satisfied.
(1) For each bounded set B the random variable $\operatorname{card}(\Pi_\Lambda \cap B)$ (number of points in $\Pi_\Lambda \cap B$) has a Poisson distribution with mean $\Lambda(B)$.
(2) Numbers of points of Π_Λ in each of disjoint sets $B_1, \ldots, B_n$ are independent for every $n \geq 2$ and any collection of disjoint Borel sets.

The capacity functional of the locally finite random set corresponding to Π_Λ equals the probability that the Poisson random variable with mean $\Lambda(K)$ does not vanish, whence

$$T_{\Pi_\Lambda}(K) = \mathbf{P}\{\Pi_\Lambda \cap K \neq \emptyset\} = 1 - \exp\{-\Lambda(K)\}. \tag{8.4}$$

If Λ is absolutely continuous with respect to the Lebesgue measure, then the corresponding Radon–Nikodym derivative (or density) λ is called the *intensity function*. If $\mathbb{E} = \mathbb{R}^d$ and Λ is proportional to the Lebesgue measure, then the Poisson point process is said to be *stationary*. It is possible to extend Definition 8.7 for the case when Λ is not locally finite.

Definition 8.8 (Poisson random set). Let F be a closed subset of $\mathbb{E}$. Assume that $\Lambda(K) < \infty$ for any compact set $K \cap F = \emptyset$ and $\Lambda(K) = \infty$ otherwise. A *Poisson random set* Π_Λ with intensity Λ is the union of F and the Poisson point process on F^c with intensity measure Λ.

It should be noted that the capacity functional of the Poisson random set Π_Λ is also given by (8.4).

Example 8.9. Let Λ be a measure on $\mathbb{E} = \mathbb{R}^d$ ($d \geq 2$) with the density $\lambda(x) = \|x\|^{-2}$. Then Λ is infinite in any neighbourhood of the origin, but locally finite on $\mathbb{R}^d \setminus \{0\}$. The corresponding Poisson random set contains the origin almost surely and is a Poisson point process on $\mathbb{R}^d \setminus \{0\}$.

Let $\mathbb{B}$ be a measurable space. A point process in the product space $\mathbb{E} \times \mathbb{B}$ is called a *marked* point process with the second component being the mark and the first component called the location. A point process is called independently marked if the marks at different points are independent.

If N is a general point process, the expectation $\Lambda(K) = \mathbf{E}N(K)$ is called the *intensity measure* of N. The following useful fact, known as the *Campbell theorem*, makes it possible to evaluate expectations of sums defined on point processes.

Theorem 8.10 (Campbell theorem). *If N is a point process with the intensity measure Λ, then*

$$\mathbf{E}\left[\sum_{x \in N} f(x)\right] = \int_{\mathbb{E}} f(x) \Lambda(dx)$$

for each measurable function $f : \mathbb{E} \mapsto \mathbb{R}$.

Weak convergence of point processes

The conditions for the weak convergence of random closed sets can be specified to show the weak convergence of point processes. If N is a simple point process, then the corresponding family of continuity sets is defined as

$$\mathfrak{S}_N = \{B \in \mathfrak{B}_k : N(\partial B) = 0 \text{ a.s.}\}.$$

Alternatively, $\mathfrak{S}_N$ can be defined as the family of all relatively compact Borel sets B such that $N(\operatorname{cl} B) = N(\operatorname{Int} B)$ a.s.

Proposition 8.11 (Continuity sets for point process). *If N is a simple point process, then $\mathfrak{S}_N = \mathfrak{S}_X$, where $X = \operatorname{supp} N$.*

Proof. It suffices to note that

$$\mathbf{P}\{X \cap \operatorname{cl} B \neq \emptyset, \; X \cap \operatorname{Int} B = \emptyset\} = \mathbf{P}\{N(\partial B) = 0\}. \qquad \square$$

Although the distribution of a simple point process is determined by its hitting (or avoidance) probabilities, the pointwise convergence of those probabilities does not suffice to ensure the weak convergence of the point processes. The following result shows that an additional condition ought to be satisfied in order to obtain a simple point process in the limit.

Theorem 8.12 (Weak convergence of point processes). *Let N and $\{N_n, n \geq 1\}$ be point processes in a LCHS space $\mathbb{E}$. Assume that N is simple. Let $\mathcal{A} \subset \mathfrak{B}_k$ be a*

separating class and $\mathcal{A}_0 \subset \mathfrak{S}_N$ be a pre-separating class. Then N_n weakly converges to N if
$$\lim_{n\to\infty} \mathbf{P}\{N_n(A) = 0\} = \mathbf{P}\{N(A) = 0\}, \quad A \in \mathcal{A}, \tag{8.5}$$
and
$$\limsup_{n\to\infty} \mathbf{P}\{N_n(A) > 1\} \leq \mathbf{P}\{N(A) > 1\}, \quad A \in \mathcal{A}_0. \tag{8.6}$$
If $\mathcal{A} \subset \mathfrak{S}_N$, then (8.5) and (8.6) are also necessary for the convergence $N_n \xrightarrow{d} N$.

Proof. By (8.5) and Theorem 6.8, $\operatorname{supp} N_n \xrightarrow{d} \operatorname{supp} N$ as random closed sets. Since both the space of all counting measures and the space of closed sets are Polish with respect to the vague topology (for measures) and the Fell topology (for closed sets) and the map $N \mapsto \operatorname{supp} N$ is measurable, Proposition E.7 implies that we can assume (passing to random elements defined on the same probability space) that
$$\operatorname{supp} N_n \xrightarrow{F} \operatorname{supp} N \quad \text{a.s.} \tag{8.7}$$
First, prove that
$$\limsup_{n\to\infty} \min(N_n(A), 1) \leq N(A) \leq \liminf_{n\to\infty} N_n(A), \quad A \in \mathfrak{S}_N. \tag{8.8}$$
For the first of these inequalities, it suffices to assume that $N(A) = 0$ and so $N(\operatorname{cl} A) = 0$ (since $A \in \mathfrak{S}_N$). Then (8.7) together with the definition of the Fell topology (Appendix B) imply that $\operatorname{supp} N_n \cap A = \emptyset$ for all sufficiently large n, whence $\limsup_{n\to\infty} \min(N_n(A), 1) = 0$. For the second inequality in (8.8), assume that $N(A) = m > 0$. Since $\mathfrak{S}_N$ is a separating class and N is simple, it is possible to choose $A_1, \ldots, A_m \in \mathfrak{S}_N$ such that $N(A_k) = N(\operatorname{Int} A_k) = 1$ for every k. Then $\operatorname{supp} N_n \cap \operatorname{Int} A_k \neq \emptyset$ for all sufficiently large n. Thus, $\liminf N_n(A_k) \geq 1$ and
$$N(A) = m \leq \sum_{k=1}^{m} \liminf_{n\to\infty} N_n(A_k) \leq \liminf_{n\to\infty} \sum_{k=1}^{m} N_n(A_k) = \liminf_{n\to\infty} N_n(A).$$

Let us show that for $A \in \mathcal{A}_0$ it is possible to replace $\min(N_n(A), 1)$ by $N_n(A)$ in the left-hand side of (8.8). Note that, for m and n being non-negative integers,
$$\{m > 1\} \cup \{m < \min(n, 2)\} = \{n > 1\} \cup \{m = 0, n = 1\} \cup \{m > 1 \geq n\},$$
where all unions are disjoint. Substituting $m = N(A)$ and $n = N_n(A)$, (8.8) and (8.6) imply that
$$\lim_{n\to\infty} \mathbf{P}\{N(A) < \min(N_n(A), 2)\} = 0, \quad A \in \mathcal{A}_0. \tag{8.9}$$
For each set $B \subset A \in \mathcal{A}_0$,
$$\{N_n(B) > N(B)\} \subset \{N_n(A) > N(A)\} \cup \{N_n(A \setminus B) < N(A \setminus B)\}$$
$$\subset \{\min(N_n(A), 2) > N(A)\} \cup \{N(A) > 1\}$$
$$\cup \{N_n(A \setminus B) < N(A \setminus B)\}.$$

112   1 Random Closed Sets and Capacity Functionals

Fix any $B \in \mathfrak{S}_N$ and $K \in \mathcal{K}$ such that cl $B \subset$ Int K. Because $\mathcal{A}_0$ is a pre-separating class, it is possible to find $A_1, \ldots, A_m \in \mathcal{A}_0$ with diameters less than a fixed number $\varepsilon > 0$ such that $B \subset (A_1 \cup \cdots \cup A_m) \subset K$. Then (8.8) and (8.9) yield that

$$\limsup_{n \to \infty} \mathbf{P}\{N_n(B) > N(B)\} \leq \mathbf{P}(\cup_{k=1}^m \{N(A_k) > 1\}). \qquad (8.10)$$

Since N is a simple point process, the right-hand side of (8.10) is bounded by $\mathbf{P}\{\alpha < \varepsilon\}$, where α is a positive random variable being the smallest distance between the points of (supp N) $\cap K$. Since $\varepsilon > 0$ is arbitrary,

$$\mathbf{P}\{N_n(A) > N(A)\} \to 0 \quad \text{as } n \to \infty.$$

Combining this with (8.8) implies that $N_n(B)$ converges to $N(B)$ is probability. In particular, the m-tuple $(N_n(A_1), \ldots, N_n(A_m))$ converges in distribution to $(N(A_1), \ldots, N(A_m))$. By Kallenberg [287, Th. 4.2], the point process N_n converges weakly to N. □

8.2 A representation of random sets as point processes

The space $\mathbb{E}$ in the definition of a point process can be a rather general measurable space. Typical examples include the Euclidean space $\mathbb{R}^d$, the space of all compact sets $\mathcal{K}$, the space of all compact convex sets co $\mathcal{K}$, the space of all upper semicontinuous functions, etc.

In particular, a locally finite point process on $\mathcal{K}$ is a countable family of compact sets $K_1, K_2, \ldots$ such that only a finite number of the K_i's hits any given bounded set. This local finiteness property ensures that

$$X = K_1 \cup K_2 \cup \cdots \qquad (8.11)$$

is a closed set, which is also measurable, since $\{X \cap G = \emptyset\} = \cap_i \{K_i \cap G = \emptyset\}$ is measurable for every open G.

The following decomposition theorem states that rather general random closed sets in $\mathbb{E} = \mathbb{R}^d$ can be obtained using (8.11) with $K_1, K_2, \ldots$ being convex compact sets. The random set X is said to belong to the *extended convex ring* $\tilde{\mathcal{R}}$ if $X \cap W$ belongs to the *convex ring* $\mathcal{R}$ for each convex compact set W, i.e. $X \cap W$ is a union of at most a finite number of convex compact sets, see Appendix F.

Theorem 8.13 (Decomposition theorem). *If X is a random closed set in $\mathbb{E} = \mathbb{R}^d$ with values in the extended convex ring, then there exists a point process $N = \{Y_1, Y_2, \ldots\}$ on the family co $\mathcal{K}'$ of non-empty convex compact sets such that*

$$X = Y_1 \cup Y_2 \cup \cdots. \qquad (8.12)$$

If X is stationary, then the point process N can be chosen to be stationary (with respect to translations of sets from co $\mathcal{K}'$).

8 Point processes and random measures   113

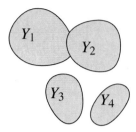

Figure 8.1. Representation (8.12).

Let us start with two auxiliary results. For a non-empty $K \in \mathcal{R}$, let $\mathfrak{n}(K)$ be the minimal number $n \geq 1$ such that K is a union of n convex compact sets. By agreement, put $\mathfrak{n}(\emptyset) = 0$.

Lemma 8.14. *The map* $\mathfrak{n} \colon \mathcal{R} \mapsto \{0, 1, 2, \ldots\}$ *is measurable and* $\mathcal{R}$ *is a union of an at most countable family of closed subsets of* $\mathcal{K}$.

Proof. For each $n \geq 0$ define $\mathcal{R}_n = \{K \in \mathcal{R} : \mathfrak{n}(K) \leq n\}$. Since $\mathcal{R} = \cup_{n \geq 0} \mathcal{R}_n$, it suffices to show that $\mathcal{R}_n$ is a closed subset of $\mathcal{K}$ for every $n \geq 0$. Assume that K_i converges to K in the Fell topology, where $K_i \in \mathcal{R}_n$, $i \geq 1$. Then $K_i = \cup_{j=1}^n K_{ij}$ with $K_{ij} \in \operatorname{co}\mathcal{K}'$. It is easy to see that $\{K_{ij}\}$ are uniformly bounded, so that there exists a subsequence $\{i_k, k \geq 1\}$ such that $K_{i_k j} \to K'_j$ as $k \to \infty$. Hence $K_{i_k} \to \cup_{j=1}^n K'_j \in \mathcal{R}_n$, so that $\mathcal{R}_n$ is closed. □

Let $(\operatorname{co}\mathcal{K})_0^\infty$ be the family of all finite sequences $(K_1, \ldots, K_m)$, $m \geq 0$, of convex compact sets with the standard σ-algebra. Note that $(\operatorname{co}\mathcal{K})^m$ denotes the family of m-tuples of convex compact sets for fixed m.

Lemma 8.15. *There exists a measurable map* $s \colon \mathcal{R} \mapsto (\operatorname{co}\mathcal{K})_0^\infty$ *such that* $s(K) = (K_1, \ldots, K_{\mathfrak{n}(K)})$ *and*

$$\bigcup s(K) = \bigcup_{i=1}^{\mathfrak{n}(K)} K_i = K$$

for each $K \in \mathcal{R}$.

Proof. The families $\mathcal{R}_n = \{K \in \mathcal{R} : \mathfrak{n}(K) = n\}$, $n \geq 0$, are pairwise disjoint. For every fixed n,

$$F(K) = \{(K_1, \ldots, K_n) : \cup_{i=1}^n K_i = K\}$$

is a set-valued map from $\mathcal{R}_n$ into $(\operatorname{co}\mathcal{K})^n$. Consider a closed family $\mathcal{Y} \subset (\operatorname{co}\mathcal{K})^n$. For every $m \geq 1$ let

$$\mathcal{Y}_m = \{(K_1 \cap B_m(0), \ldots, K_n \cap B_m(0)) : (K_1, \ldots, K_n) \in \mathcal{Y}\},$$

where $B_m(0)$ is the ball of radius m centred at the origin. Then

$$\{K \in \mathcal{R}_n : F(K) \cap \mathcal{Y} \neq \emptyset\} = \bigcup_{m=1}^{\infty} \{K \in \mathcal{R}_n : F(K) \cap \mathcal{Y}_m \neq \emptyset\}$$

$$= \bigcup_{m=1}^{\infty} \mathcal{R}_n \cap \{\cup_{i=1}^{n} K_i : (K_1,\ldots,K_n) \in \mathcal{Y}_m\}.$$

Note that $\{\cup_{i=1}^{n} K_i : (K_1,\ldots,K_n) \in \mathcal{Y}_m\}$ is compact in $\mathcal{K}$ and so measurable, which implies the measurability of F. By the fundamental selection theorem (Theorem 2.13), there exists a measurable selection $s\colon \mathcal{R}_n \mapsto (\operatorname{co}\mathcal{K})^n$ such that $s(K) \in F(K)$. The required map on $\mathcal{R}$ can be obtained by combining these measurable selections for $n \geq 0$. □

Proof of Theorem 8.13. Without loss of generality assume that X is almost surely non-empty. Let $C = \{x = (x_1,\ldots,x_d) \in \mathbb{R}^d : |x_i| \leq 1/2,\ 1 \leq i \leq d\}$ be the unit cube in $\mathbb{R}^d$ and let $C_z = C + z$ from z from $\mathbb{Z}^d$ (the integer grid in $\mathbb{R}^d$). Furthermore, let ξ be a random vector uniformly distributed in C. Using the map s from Lemma 8.15 define

$$N(X) = \bigcup_{z \in \mathbb{Z}^d} [s((X \cap (C_z + \xi)) - z) + z].$$

This map is a measurable map from $\bar{\mathcal{R}}$ into the family of locally finite collections of convex compact sets. Thus, $N(X)$ is a point process on $\operatorname{co}\mathcal{K}'$ which satisfies (8.12).

If X is stationary, then, for every $x \in \mathbb{R}^d$, we have

$$N(X) + x = \bigcup_{z \in \mathbb{Z}^d} [s((X \cap (C_z + \xi)) - z) + z + x]$$

$$\stackrel{d}{\sim} \bigcup_{z \in \mathbb{Z}^d} [s(((X - x) \cap (C_z + \xi - x)) - z) + z + x]$$

$$= \bigcup_{z \in \mathbb{Z}^d} [s((X \cap (C_z + \xi)) - z) + z] = N(X),$$

so that $N(X)$ is indeed stationary. □

It is possible to extend the above result to show that the point process N is invariant with respect to any rigid motion for which X is invariant. To prove this, it is possible to use the group of invariant motions to "randomise" the cubic tiling.

Let us associate with each Y_i from (8.12) a point $c(Y_i)$ in such a way that $c(Y_i + x) = Y_i + x$ for all x. For instance, $c(Y_i)$ can be the centre of gravity of Y_i. Then instead of (8.12) X can be represented as

$$X = \bigcup_i (c(Y_i) + Y_i^0) \tag{8.13}$$

with $Y_i^0 = Y_i - c(Y_i)$. This representation is called a *germ-grain model*, where the points $c(Y_i)$ are called germs and the corresponding sets Y_i^0 are grains. A particularly important case of this construction appears if N is a Poisson process on $\operatorname{co}\mathcal{K}$;

then X is called a *Boolean model*, see Molchanov [406] and Stoyan, Kendall and Mecke [544]. In this case the germs form a Poisson point process independently marked by i.i.d. grains.

Similarly to Theorem 8.13, it is possible to show that a general random closed set X can be represented as a union-set for a point process on $\mathcal{K}$. The corresponding point process can be trivially obtained as $Y_z = (X \cap C_z)$ for $z \in \mathbb{Z}^d$.

8.3 Random sets and random measures

Random measures associated with random sets

Similarly to the σ-algebra on the family $\mathcal{N}$ of all counting measures used to define a random point process, the family $\mathbb{M}$ of all locally finite measures on $\mathbb{E}$ can be equipped with a σ-algebra generated by $\{\mu \in \mathbb{M} : \mu(B) > t\}$ for every $B \in \mathfrak{B}$ and $t > 0$. A *random measure* M is an $\mathbb{M}$-valued random element, i.e. $M(B)$ is a random variable for each Borel set B.

The fundamental construction which delivers random measures related to random closed sets can be described as follows. Let X be a random closed set in $\mathbb{E}$ and let μ be a fixed measure on $\mathbb{E}$ such that $\mu(B \cap X)$ is almost surely finite for every bounded Borel set B. Then $M(B) = \mu(B \cap X)$ is a locally finite random measure. This construction can be generalised by taking a measurable random function $\zeta(x)$, $x \in \mathbb{E}$, and letting

$$M(B) = \int_{X \cap B} \zeta(x) d\mu(x).$$

Particularly important examples of random measures associated with random closed sets are related to the Hausdorff measures. If $\mathcal{H}^{d-1}$ is the $(d-1)$-dimensional Hausdorff measure, then $M(B) = \mathcal{H}^{d-1}(X \cap B)$ (if it exists) is called the surface measure generated by X. Further examples are provided by the curvature measures described in Appendix F.

A natural question whether a random measure determines the distribution of a random closed set has been answered positively in Section 8.1 for counting measures and the corresponding locally finite random sets. However, in general, this is not the case. For instance, if $X = \{\xi\}$ is a singleton (or any other random set of zero Lebesgue measure), then the random measure $M(B) = \text{mes}(X \cap B)$ vanishes and does not determine the distribution of X. Note that the *support* of a measure μ is defined to be the set of all $x \in \mathbb{E}$ such that $x \in G$ for an open set G implies $\mu(G) > 0$. The support of μ can be alternatively defined as

$$\operatorname{supp} \mu = \bigcap_{i=1}^{\infty} \operatorname{cl}\{x_j \in \mathbb{Q} : \mu(B_{\varepsilon_i}(x_j)) > 0\},$$

where $\varepsilon_i \downarrow 0$ and $\mathbb{Q}$ is a countable dense set in $\mathbb{E}$. Proposition 8.1 yields the following result.

Proposition 8.16 (Support of a random measure). *For every random measure μ, supp μ is a random closed set whose distribution is uniquely determined by the distribution of μ.*

Intersections of random sets

Let $\{X_n, n \geq 1\}$ be a sequence of i.i.d. random closed sets in a space $\mathbb{E}$ which itself is a second countable compact group G. Assume that all these sets have G-invariant distributions, see Section 4.1. Let ν be the Haar measure on G. It is interesting to find out whether or not the event

$$A_n = \{X_n \cap (\cup_{i=1}^{n-1} X_i) = \emptyset\} \tag{8.14}$$

occurs for infinitely many n, meaning that a new X_n infinitely often fits in the gap left free by the previous sets. By the Hewitt–Savage zero-one law this happens with probability either zero or one, so that $\mathbf{P}(\limsup A_n)$ is either 0 or 1. A necessary condition for

$$\mathbf{P}(\limsup A_n) = 0 \tag{8.15}$$

is $\mathbf{P}\{X_i \cap X_j \neq \emptyset | X_j\} > 0$ for $i \neq j$. A sufficient condition can be formulated using random measures $\{\mu_n, n \geq 1\}$ such that $X_n = \operatorname{supp}\mu_n$, $n \geq 1$. The convolution of μ_1 and inverted μ_2 is defined as

$$(\mu_1 \star \tilde{\mu}_2)(B) = \int_{\mathsf{G}} \mu_1(dx)\mu_2(xB^{-1}), \quad B \in \mathfrak{B}(\mathsf{G}).$$

Proposition 8.17 (Random sets in a group). *Let $\{\mu_n, n \geq 1\}$ be a sequence of i.i.d. random probability measures on G such that $\mu_n(x^{-1}B)$ has the same distribution as $\mu_n(B)$ for all $x \in \mathsf{G}$ and $B \in \mathfrak{B}(\mathsf{G})$. If $\mu_i \star \tilde{\mu}_j$ is almost surely absolutely continuous with respect to ν for $i \neq j$ with a mean square integrable density g, then (8.15) holds for A_n defined by (8.14) with $X_n = \operatorname{supp}\mu_n$, $n \geq 1$.*

Proof. Observe that

$$\mathbf{P}(A_n) = \mathbf{E}\left[\prod_{i=1}^{n-1} \mathbf{P}\{X_i \cap X_n = \emptyset | \mu_n\}\right]$$
$$= \mathbf{E}\left[(\mathbf{P}\{X_1 \cap X_2 = \emptyset | \mu_2\})^{n-1}\right].$$

By the Borel–Cantelli lemma, it suffices to show that

$$\mathbf{E}\left[\sum_{n=1}^{\infty}(\mathbf{P}\{X_1 \cap X_2 = \emptyset | \mu_2\})^{n-1}\right] = \mathbf{E}\left[\mathbf{P}\{X_1 \cap X_2 \neq \emptyset | \mu_2\}^{-1}\right] < \infty.$$

Let B_ε be a ball centred at the identity in any metric on G compatible with the topology. Using the inequality $\mathbf{P}\{\alpha > 0\} \geq (\mathbf{E}\alpha)^2/\mathbf{E}(\alpha^2)$ for a random variable α, we can write

8 Point processes and random measures   117

$$\mathbf{P}\{X_1 \cap X_2 \neq \emptyset \mid \mu_2\} = \lim_{\varepsilon \downarrow 0} \mathbf{P}\left\{\iint 1_{B_\varepsilon}(yz^{-1})\mu_1(dy)\mu_2(dz) > 0 \mid \mu_2\right\}$$

$$\geq \limsup_{\varepsilon \downarrow 0} \frac{(\mathbf{E}\left[(\mu_1 \star \tilde{\mu}_2)(B_\varepsilon) \mid \mu_2\right])^2}{\mathbf{E}\left[(\mu_1 \star \tilde{\mu}_2)(B_\varepsilon)^2 \mid \mu_2\right]}$$

$$= \left(\liminf_{\varepsilon \downarrow 0} \mathbf{E}\left[(\nu(B_\varepsilon))^{-1}(\mu_1 \star \tilde{\mu}_2)(B_\varepsilon))^2 \mid \mu_2\right]\right)^{-1},$$

where we have used the fact that $\mathbf{E}\left[\mu_1 \star \tilde{\mu}\right] = \nu$ for every finite deterministic probability measure μ. Fatou's lemma and Jensen's inequality imply

$$\mathbf{E}\left[\mathbf{P}\{X_1 \cap X_2 \neq \emptyset \mid \mu_2\}^{-1}\right] \leq \liminf_{\varepsilon \downarrow 0} \mathbf{E}\left[(\nu(B_\varepsilon))^{-1}(\mu_1 \star \tilde{\mu}_2)(B_\varepsilon))^2\right]$$

$$= \liminf_{\varepsilon \downarrow 0} \mathbf{E}\left[\left(\frac{1}{\nu(B_\varepsilon)}\int 1_{B_\varepsilon}(y)g(y)\nu(dy)\right)^2\right]$$

$$= \liminf_{\varepsilon \downarrow 0} \mathbf{E}\left[\int \left(\frac{1}{\nu(B_\varepsilon)}\int 1_{B_\varepsilon}(xy)g(y)\nu(dy)\right)^2 \nu(dx)\right]$$

$$\leq \liminf_{\varepsilon \downarrow 0} \mathbf{E}\left[\frac{1}{\nu(B_\varepsilon)}\iint 1_{B_\varepsilon}(xy)g^2(y)\nu(dy)\nu(dx)\right]$$

$$= \mathbf{E}\left[\int g^2(y)\nu(dy)\right] < \infty. \qquad \square$$

8.4 Random capacities

Consider a general *capacity* φ defined on subsets of a LCHS space $\mathbb{E}$. It is only assumed that $\varphi(\emptyset) = 0$, φ is increasing and upper semicontinuous on $\mathcal{K}$. More exactly, such functionals on $\mathcal{K}$ are called topological precapacities (see Appendix E), but we will use the term capacity without risk of ambiguity. The family of capacities is equipped with the *vague topology*, which is generated by $\{\varphi : \varphi(K) < t\}$ and $\{\varphi : \varphi(G) > t\}$ where $K \in \mathcal{K}$, $G \in \mathcal{G}$ and $t > 0$. The Borel σ-algebra generated by the vague topology is the smallest σ-algebra which makes the map $\varphi \mapsto \varphi(K)$ measurable for every $K \in \mathcal{K}$.

A *random capacity* (also denoted by φ) is a random element in the family of all capacities which is measurable with respect to the Borel σ-algebra generated by the vague topology, i.e. $\varphi(K)$ and $\varphi(G)$ are random variables for each compact K and open G. It is also possible to define random capacities that satisfy certain properties, for example, random strongly subadditive capacities, random sup-measures, etc.

A random capacity φ gives rise to the family of continuity sets defined by

$$\mathfrak{S}_\varphi = \{B \in \mathfrak{B}_k : \varphi(\text{Int } B) = \varphi(\text{cl}(B)) \quad \text{a.s.}\}.$$

It is important to know when it is possible to extend a random function defined on a separating class $\mathcal{A}$ to a capacity on Borel subsets of $\mathbb{E}$. A random function φ on $\mathcal{A}$ is a stochastic process indexed by $\mathcal{A}$, defined by means of finite-dimensional distributions $(\varphi(A_1), \ldots, \varphi(A_n))$ where $A_1, \ldots, A_n \in \mathcal{A}$ and $n \geq 1$.

Proposition 8.18 (Extension of random capacity). *Let $\mathbb{E}$ be a LCHS space and let φ be a random process on a separating class $\mathcal{A}$ that consists of relatively compact sets. Assume that $\varphi(\emptyset) = 0$ a.s. and*
(1) $\mathbf{P}\{\varphi(A_1) \leq t, \; \varphi(A_2) \leq t\} = \mathbf{P}\{\varphi(A_2) \leq t\}$ *for all $t > 0$ and $A_1, A_2 \in \mathcal{A}$ such that $A_1 \subset A_2$;*
(2) $\varphi(A_n)$ *weakly converges to* $\varphi(A)$ *if $A_n \downarrow A$ for all $A, A_1, A_2, \ldots \in \mathcal{A}$ with* $\mathrm{cl}(A_{n+1}) \subset \mathrm{Int}\, A_n, n \geq 1$;
(3) $\varphi(A_n)$ *weakly converges to* $\varphi(A)$ *if $A_n \uparrow A$ for all $A, A_1, A_2, \ldots \in \mathcal{A}$ with* $\mathrm{cl}(A_n) \subset \mathrm{Int}\, A_{n+1}, n \geq 1$.
Then there exists a random capacity φ^* such that $\mathcal{A} \subset \mathfrak{S}_{\varphi^*}$ and $\varphi(A) = \varphi^*(A)$ for all $A \in \mathcal{A}$.

Proof. A separating class always contains a countable separating class, so that we can consider a countable separating class $\mathcal{A}_0 \subset \mathcal{A}$. It follows from (1) that φ is increasing on $\mathcal{A}_0$ with probability one. On this set of probability one, it is possible to extend φ to φ^* on $\mathfrak{B}(\mathbb{E})$ by (1.30) and (1.31). Then φ and φ^* coincide on $\mathcal{A}$ and $\mathcal{A} \subset \mathfrak{S}_{\varphi^*}$. □

If $\mathcal{A} \subset \mathcal{K}$, then Proposition 8.18 holds without condition (3). Similarly, if $\mathcal{A} \subset \mathcal{G}$, then (2) can be dropped. If $\mathcal{A} \subset \mathcal{K}$ and $\mathcal{A}$ is closed under finite union, then the properties of φ usually hold for its extension. For example, if φ is completely alternating on $\mathcal{A}$, then so is φ^*. This is easy to see by choosing a countable separating subclass of $\mathcal{A}$, where all those properties hold with probability one.

The convergence of random capacities in distribution is defined in a conventional way using the vague topology.

Proposition 8.19 (Weak convergence of random capacities). *Let $\varphi, \varphi_1, \varphi_2, \ldots$ be random capacities. Then $\varphi_n \xrightarrow{d} \varphi$ if and only if there exists a separating class $\mathcal{A}$ with elements being bounded Borel sets, such that the random vector $(\varphi_n(A), \; A \in \mathcal{A}_0)$ weakly converges to $(\varphi(A), \; A \in \mathcal{A}_0)$ for each finite subfamily $\mathcal{A}_0 \subset \mathfrak{S}_\varphi$.*

Proof. The necessity is a direct implication of Billingsley [70, Th. 5.5]. To prove the sufficiency, note that the map $\varphi \mapsto \varphi^*$ is continuous in the vague topology. Since φ_n converges in distribution to φ as a stochastic process on $\mathcal{A}$, their extensions converge in distribution by the continuous mapping theorem, see Billingsley [70, p. 29]. □

Definition 8.20 (Completely random capacity). A random capacity φ is called *completely random* if $\varphi(K_1), \ldots, \varphi(K_n)$ are independent whenever $K_1, \ldots, K_n$ are disjoint sets from the domain of definition of φ.

The family of capacities can be considered as a lattice. By applying the lattice-theoretic technique, it is possible to show that φ_1 is stochastically smaller than φ_2 (respectively $\varphi_1 \stackrel{d}{\sim} \varphi_2$) if and only if the random vector $(\varphi_1(G_1), \ldots, \varphi_1(G_n))$ is stochastically smaller than (respectively coincides in distribution with) the random vector $(\varphi_2(G_1), \ldots, \varphi_2(G_n))$ for every $n \geq 1$ and each $G_1, \ldots, G_n \in \mathcal{G}$, see Norberg [433].

The concept of a random capacity covers random closed sets (including point processes), random upper semicontinuous functions, random measures and sup-measures. If X is a random set, then

$$\varphi(K) = \mathbf{1}_{X \cap K \neq \emptyset} \tag{8.16}$$

is a random capacity called *indicator* random capacity; every random measure is a random capacity; and the random sup-measure

$$\varphi(K) = f^{\vee}(K) = \sup_{x \in K} f(x)$$

is a random capacity if f is a random upper semicontinuous function.

8.5 Robbin's theorem for random capacities

Carathéodory's extension of a capacity

A random capacity φ can be extended to a measure $\bar{\varphi}$ by *Carathéodory's extension* described in Appendix E. As in Section 5.4, assume that the family $\mathcal{M}$ used to define this extension contains all open balls.

Lemma 8.21 (Carathéodory's extension using a subclass). *Let $\mathcal{M} \subset \mathcal{G}$ consist of open sets and let $\mathcal{B}$ be a subclass of $\mathcal{M}$ which is closed under finite unions. Assume that any set $M \in \mathcal{M}$ is a (possibly uncountable) union of members of $\mathcal{B}$. Then, for each capacity φ, its Carathéodory extensions using $\mathcal{M}$ and $\mathcal{B}$ coincide.*

Proof. It suffices to show that $\bar{\varphi}_\delta^{\mathcal{M}}$ and $\bar{\varphi}_\delta^{\mathcal{B}}$ (defined by (E.1) for the families $\mathcal{M}$ and $\mathcal{B}$ respectively) coincide on the family of all compact sets. Let K be compact; clearly it suffices to consider only finite δ-covers in (E.1). Let $\{M_1^i, \ldots, M_{n_i}^i\}$ be a family of δ-covers from $\mathcal{M}$ such that $\sum \varphi(M_k^i) \to \bar{\varphi}_\delta^{\mathcal{M}}(K)$ as $i \to \infty$. We can find sets $B_k^i \in \mathcal{B}$ such that $B_k^i \subset M_k^i$ and $\{B_1^i, \ldots, B_{n_i}^i\}$ cover K. Thus,

$$\bar{\varphi}_\delta^{\mathcal{B}}(K) \leq \lim \sum \varphi(B_k^i) \leq \lim \sum \varphi(M_k^i) = \bar{\varphi}_\delta^{\mathcal{M}}(K).$$

On the other hand, $\bar{\varphi}_\delta^{\mathcal{B}} \geq \bar{\varphi}_\delta^{\mathcal{M}}$, since $\mathcal{B} \subset \mathcal{M}$. □

For example, Lemma 8.21 applies when $\mathcal{M}$ is the family of all open sets and $\mathcal{B}$ is the class of all finite unions of elements from an open base for the topology.

Proposition 8.22 (Carathéodory's extension as random measure). *Assume that $\mathcal{M}$ is a subfamily of $\mathcal{G}$ that has a countable subclass $\mathcal{B}$ as in Lemma 8.21. If φ is a random capacity, then $\bar{\varphi}$ is a random measure on $\mathbb{R}^d$.*

Proof. Clearly $\mu = \bar{\varphi}$ is a measure almost surely. It suffices to show that the value $\mu(K)$ is a random variable for each compact set K. Since (E.2) involves monotone pointwise limits, it suffices to show that $\bar\varphi_\delta(K)$ is a random variable for each compact set K. For any $a > 0$,

$$\{\bar\varphi_\delta(K) > a\} = \left\{\inf \sum \varphi(M_n) > a\right\} = \bigcap \left\{\sum \varphi(M_n) > a\right\}, \tag{8.17}$$

where the infimum and intersection are taken over the family of all δ-covers of K by sets from $\mathcal{M}$. By Lemma 8.21, $\mathcal{M}$ can be replaced with $\mathcal{B}$. Since K is compact, the intersection in (8.17) is taken over finite subsets of $\mathcal{B}$. Hence $\bar\varphi_\delta(K)$ is a random variable. □

Capacity version of Robbins' theorem

Definition 8.23 (Integrable random capacity). A random capacity φ is called *integrable* if $\mathbf{E}\varphi(K) < \infty$ for each $K \in \mathcal{K}$. For an integrable random capacity φ the functional $(\mathbf{E}\varphi)(K) = \mathbf{E}\varphi(K)$ is a capacity called the *intensity* of φ.

If φ is a random measure, then $\mathbf{E}\varphi$ is the corresponding intensity measure. In particular, if $\varphi = N$ is a counting measure generated by a point process N, then $\mathbf{E}N$ is the intensity measure of N. If X is a random closed set, then

$$\mathbf{E}\mathbf{1}_{X \cap K \neq \emptyset} = \mathbf{P}\{X \cap K \neq \emptyset\} = T_X(K)$$

is the capacity functional of X. Robbins' Theorem 4.21 can now be expressed as follows.

Theorem 8.24 (Robbins' theorem, capacity version). *Let X be a random closed set in $\mathbb{R}^d$ and μ a locally finite measure. Then, for the indicator capacity $\varphi(K) = \mathbf{1}_{X \cap K \neq \emptyset}$, we have*

$$\mathbf{E}\overline{\varphi\mu} = \overline{T_X \mu}.$$

Proof. By Proposition 5.20 the left side is $\mathbf{E}\mu(X \cap \cdot)$. Since T_X is bounded and μ is Radon, Theorem 5.19(i) shows

$$\overline{T_X \mu}(K) = \int_K \mathbf{P}\{x \in X\} \mu(\mathrm{d}x).$$

The two sides are equal by Fubini's theorem. □

General conditions that would guarantee

$$\mathbf{E}\overline{\varphi\upsilon} = \overline{\mathbf{E}\varphi\upsilon} \tag{8.18}$$

for a random capacity φ and a deterministic capacity υ are not known. The following example describes a special case when (8.18) holds although Robbins' theorem is not applicable because the Carathéodory extension $\bar\upsilon$ is not a locally finite measure.

Example 8.25 (Robbins' theorem for counting measures). Let N be a simple point process on $\mathbb{R}^d$. Two principal random capacities associated with N are the random counting measure $N(K)$ and the indicator capacity $\varphi(K) = \mathbf{1}_{X \cap K \neq \emptyset}$ of the associated random closed set $X = \operatorname{supp} N$. Note that Robbins' theorem does not hold for the counting measure as the latter is not locally finite. Let $\mathcal{M} = \mathcal{G}$ be the family of all open sets. If $\upsilon = 1$, then the Carathéodory extension of $\varphi\upsilon$ is the random counting measure N, see Example E.2 and Proposition 5.20. Then $\mathbf{E} N$ is the *intensity measure* λ of N, whereas the expectation $\mathbf{E}\varphi(K)$ of the indicator capacity on the right-hand side of (8.18) is the capacity functional of X. Carathéodory's extension of T_X is the so-called *parametric measure* ν of the point process. Korolyuk's theorem (see Daley and Vere-Jones [117, p. 120]) states that the intensity measure λ and the parametric measure ν coincide if the process is *orderly*. Therefore,

$$\mathbf{E}\overline{\varphi\upsilon} = \lambda = \nu = \overline{\mathbf{E}\varphi\upsilon}$$

showing that (8.18) holds provided N is orderly.

Theorem 8.26 (Campbell theorem, capacity version). *Let N be an orderly point process in $\mathbb{R}^d$ whose intensity measure λ is locally finite. If $f : \mathbb{R}^d \mapsto \mathbb{R}$ is an upper semicontinuous function such that $\int_K f \, d\lambda < \infty$ for all compact sets K, then, for $X = \operatorname{supp} N$ and $\varphi(K) = \mathbf{1}_{X \cap K \neq \emptyset}$,*

$$\mathbf{E}\overline{f^\vee \varphi} = \overline{f^\vee \mathbf{E}\varphi}. \tag{8.19}$$

Proof. Since X is almost surely locally finite, we have

$$\overline{f^\vee \varphi}(A) = \sum_{x_i \in A} f(x_i) \quad \text{a.s.,}$$

where the sum is over all points of N falling in A. Note that this is a random variable by Proposition 8.22. On the right-hand side of (8.19) $\mathbf{E}\varphi = T_X$ and Carathéodory's extension of T_X is the parametric measure ν. Since N is orderly $\nu = \lambda$; since λ is locally finite we may apply Theorem 5.19(ii) to yield

$$\overline{f^\vee \mathbf{E}\varphi}(A) = \int_A f \, d\lambda.$$

The two sides are equal by Theorem 8.10. □

Intrinsic density

Definition 8.27 (Intensity measure and parametric measure). Let φ be a random capacity and υ a deterministic capacity. Then $\mu(K) = \mathbf{E}\overline{\varphi\upsilon}(K)$ is said to be the (υ-weighted) *intensity measure* of φ and $\nu(K) = \overline{\mathbf{E}\varphi\upsilon}(K)$ the (υ-weighted) parametric measure of φ. Both depend implicitly on the family $\mathcal{M}$ used to define Carathéodory's extension.

By applying Fatou's lemma to (E.1) and the monotone convergence theorem to (E.2) it is easily seen that

$$\mu(K) \leq \nu(K) \tag{8.20}$$

for all measurable K, whence $\mu \ll \nu$. In general, the inequality (8.20) cannot be improved to the equality (8.18), so that a counterpart of Robbins' theorem in the form (8.18) does not hold, see Example 8.30. However, a weighted form of Robbins' theorem holds.

Definition 8.28 (Intrinsic density). Assume that the parametric measure ν of φ is locally finite. Then the Radon–Nikodym derivative

$$\iota(x) = \frac{d\mu}{d\nu}(x)$$

is called the (ν-weighted) *intrinsic density* of the random capacity φ (with respect to the class $\mathcal{M}$ used to define Carathéodory's extensions). If φ is the indicator capacity of a random closed set X, then ι is called the intrinsic density of X.

It should be noted that the parametric measure (and hence the intrinsic density) is sensitive to the choice of the family of sets $\mathcal{M}$ used to define Carathéodory's construction. The following result follows from Theorem 5.19(ii).

Theorem 8.29 (Weighted Robbins' formula). *Let φ be a random capacity and let ν be a deterministic capacity. If $\nu = \overline{\mathbf{E}\varphi}\nu$ is locally finite and the intrinsic density ι is upper semicontinuous, then*

$$\mathbf{E}\overline{\varphi}\nu = \overline{\iota^\vee \mathbf{E}\varphi}\nu.$$

A random capacity φ is said to be *stationary* if the finite-dimensional distributions of the random function $\{\varphi(K+a), K \in \mathcal{K}\}$ indexed by compact sets do not depend on $a \in \mathbb{R}^d$. It is easy to see that indicator capacities generated by stationary random sets or sup-measures obtained from stationary upper semicontinuous functions are stationary. If φ is a stationary capacity, then its intensity $\mathbf{E}\varphi$ is translation invariant, i.e. $\mathbf{E}\varphi(K) = \mathbf{E}\varphi(K+a)$ for each $a \in \mathbb{R}^d$. For instance, if φ is the indicator capacity of a stationary random closed set X, then its intrinsic density is constant, i.e. $\iota(x) = a$, whence

$$\mathbf{E}\bar{\nu}(X \cap K) = a\overline{T_X\nu}(K), \quad K \in \mathcal{K}.$$

Clearly, $0 \leq \iota(x) \leq 1$ for ν-almost all x. We have seen that $\iota = 1$ for the cases of a random closed set X (if φ is the indicator capacity and $\bar{\nu}$ is locally finite) and a simple orderly point process (for $\nu = 1$). Below we give examples where the intrinsic density is not unity.

Example 8.30 (Boolean segment process). Let X be a planar Boolean segment process of intensity α (see Molchanov [406] and Stoyan, Kendall and Mecke [544]), i.e. X is defined by (8.13) where the germs form a Poisson point process in $\mathbb{R}^2$ and the grains are isotropic random segments of the constant length l. Then, for a ball B,

$$T(B) = 1 - \exp\{-\alpha(l\operatorname{diam}(B) + \pi(\operatorname{diam}(B))^2/4)\} \sim \lambda l(\operatorname{diam}(B))$$

if the diameter of B is small.

(i) Put $\upsilon(B) = \operatorname{diam}(B)$ so that $\bar{\upsilon} = \mathcal{H}^1$ is the one-dimensional Hausdorff measure. The expected one-dimensional Hausdorff measure of $X \cap K$ is equal to

$$\mu(K) = \mathbf{E}\bar{\upsilon}(X \cap K) = \mathbf{E}\mathcal{H}^1(X \cap K) = \alpha l \operatorname{mes}(K),$$

see Stoyan, Kendall and Mecke [544]. On the other hand,

$$\sum T_X(B_i)\upsilon(B_i) \sim \alpha l \sum (\operatorname{diam}(B_i))^2.$$

Thus,

$$\nu(K) = \overline{T_X\upsilon}(K) = \frac{4}{\pi}\alpha l \operatorname{mes}(K).$$

Therefore, $\iota(x)$ is identically equal to $\pi/4$.

(ii) Let $L \subset \mathbb{R}^2$ be a rectifiable curve. If $\upsilon = 1$, then $\bar{\varphi}(L)$ is the number of intersections of X with L. We have $\mathbf{E}\varphi(L) = \lambda l|L|$, where $|L|$ is the length of L, but $\mathbf{E}\bar{\varphi}(L) = 2\lambda l|L|/\pi$, so that $\iota(x) = 2/\pi$.

Open problem 8.31. Find necessary and sufficient conditions on a random capacity φ (or indicator random capacity) and a set-function υ that would guarantee (8.18), meaning that the υ-weighted intrinsic density is identically equal to one.

Upper bound on Hausdorff dimension

Let φ be the indicator random capacity of X. Since $0 \leq \iota \leq 1$,

$$\mathbf{E}\bar{\upsilon}(X \cap K) \leq \overline{T_X\upsilon}(K).$$

Therefore, $\overline{T_X\upsilon}(K) = 0$ implies that $\bar{\upsilon}(X \cap K) = 0$ a.s. This fact can be used to bound the Hausdorff dimension of $X \cap K$.

Example 8.32. In $\mathbb{E} = \mathbb{R}^d$ choose $\upsilon(K) = \operatorname{diam}(K)^\alpha$ with $\alpha \in [0, d]$ as in Example E.1. Then $\bar{\upsilon} = \mathcal{H}^\alpha$ is the α-dimensional Hausdorff measure. If $\overline{T_X\upsilon}(K) = 0$, then $\mathcal{H}^\alpha(X \cap K) = 0$, i.e. the Hausdorff dimension of $X \cap K$ does not exceed α.

If $K = \mathbb{R}^d$, then $\overline{T_X\upsilon}$ vanishes if $T_X(B_r(x)) \geq cr^\gamma$ for all sufficiently small r uniformly over $x \in \mathbb{R}^d$ with a constant $c > 0$ and $\gamma > d - \alpha$. For instance if $T_X(K) = 1 - \exp\{-C(K)\}$ where $C(K)$ is the Newton capacity of K (see Example 4.1.19) in $\mathbb{R}^d$ with $d \geq 3$, then $T_X(B_r(x)) \sim cr^{d-2}$ for a constant $c > 0$. Therefore, $\overline{T_X\upsilon} = 0$ for $\upsilon(K) = \operatorname{diam}(K)^\alpha$ with $\alpha > 2$. Accordingly, $\dim_H(X) \leq 2$, which corresponds to the fact that X (being the path of a Brownian motion) has the Hausdorff dimension 2.

9 Various interpretations of capacities

9.1 Non-additive measures

Definitions

Assume that $\mathbb{E}$ is a Polish space, unless stated otherwise. A set-function φ defined on an *algebra* $\mathcal{E}$ of subsets of $\mathbb{E}$ is said to be a *non-additive measure* if it is normalised, i.e. $\varphi(\emptyset) = 0$ and $\varphi(\mathbb{E}) = 1$, and monotonic, i.e.

$$\varphi(A) \leq \varphi(B) \quad \text{if } A \subset B.$$

In *game theory*, the elements of $\mathbb{E}$ are called *players*, subsets of $\mathbb{E}$ are *coalitions* and φ acts on coalitions and determines the game. The total value of φ may be different from 1, as is also typical in the theory of fuzzy measures (see Sugeno, Narukawa and Murofushi [552]) which are allowed to take values from the extended half-line.

The *dual* (conjugate) to φ function $\tilde{\varphi}$ is defined as $\tilde{\varphi}(A) = 1 - \varphi(A^c)$. This duality relationship is identical to the relationship between the containment functional and the capacity functional of a random closed set. The non-additive measure φ is said to be *equalised* if $(\tilde{\varphi} + \varphi)/2$ is a measure. For capacity functionals of random closed sets, this property was discussed in Section 5.3.

Definition 9.1 (Symmetric and coherent non-additive measures).
(i) A non-additive measure φ defined on a σ-algebra $\mathcal{E}$ is *weakly symmetric* with respect to a non-atomic probability measure μ on $\mathcal{E}$ if $\mu(A) = \mu(B)$ implies $\varphi(A) = \varphi(B)$ for all $A, B \in \mathcal{E}$.
(ii) $\varphi \colon \mathcal{E} \mapsto [0, 1]$ is *coherent* if there exists a family of probability measures $\mathbb{P}$ on $\mathcal{E}$ such that $\varphi(A) = \sup_{\mu \in \mathbb{P}} \mu(A)$ for every $A \in \mathcal{E}$.

Coherent non-additive measures are also called upper probabilities, see Section 9.3.

Core of non-additive measure

A non-additive measure (game) φ is called *convex* if φ satisfies

$$\varphi(A \cap B) + \varphi(A \cup B) \geq \varphi(A) + \varphi(B)$$

for all $A, B \in \mathcal{E}$, compare with (1.18) which defines a concave capacity. Furthermore, φ is called *decomposable* if

$$\varphi(A) = \sum_{i=1}^{n} \varphi(A \cap H_i)$$

for every $A \in \mathcal{E}$, where $H_1, \ldots, H_n$ is a fixed partition of $\mathbb{E}$.

Definition 9.2 (Feasible measures and the core). A finite-additive measure μ (called a payoff in game theory) is said to be *feasible* if $\mu(\mathbb{E}) \leq \varphi(\mathbb{E})$. The *core* of φ is defined as the set of all feasible μ such that $\mu(A) \geq \varphi(A)$ for all $A \in \mathcal{E}$.

If $\varphi(\mathbb{E}) = 1$, then by a simple duality argument, the core of φ can be defined as the family of all measures μ satisfying $\mu(A) \leq \tilde{\varphi}(A)$ for all $A \in \mathcal{E}$. An important result in game theory states that every convex non-additive measure (game) has a non-empty core, see Shapley [535].

Decomposition of non-additive measures

For every $F \subset \mathbb{E}$, the *unanimity game* on F is defined by

$$u_F(A) = \begin{cases} 1, & F \subset A, \\ 0, & \text{otherwise}, \end{cases} \quad (9.1)$$

which is the containment functional of a deterministic set identically equal to F. Gilboa and Schmeidler [196] showed that if φ is defined on a finite algebra $\mathcal{E}_0$, then

$$\varphi = \sum_{K \in \mathcal{E}_0} \alpha_K^\varphi u_K, \quad (9.2)$$

where α_K^φ, $K \in \mathcal{E}_0$, are uniquely defined coefficients. If φ is the *containment functional* of a random closed set X, then $\alpha_K^\varphi = \mathbf{P}\{X = K\}$ is the probability that X takes a value K. The non-negativity condition on α_K^φ characterises the completely monotone φ. In general, every φ on a finite algebra $\mathcal{E}_0$ may be decomposed as

$$\varphi = \varphi^+ - \varphi^-, \quad (9.3)$$

where φ^+ and φ^- are two completely monotone functionals such that

$$\|\varphi\|_{\mathcal{E}_0} = \|\varphi^+\|_{\mathcal{E}_0} + \|\varphi^-\|_{\mathcal{E}_0},$$

where

$$\|\varphi\|_{\mathcal{E}_0} = \sum_{K \in \mathcal{E}_0} |\alpha_K^\varphi|$$

is called the *composition norm*. In the infinite case, the composition norm is defined as the supremum of $\|\varphi\|_{\mathcal{E}_0}$ over all finite sub-algebras $\mathcal{E}_0 \subset \mathcal{E}$. If φ is additive, then $\|\varphi\|_{\mathcal{E}}$ turns into the *total variation* norm $\|\varphi\|$ of φ. In the infinite case decomposition (9.2) is written as

$$\varphi = \int_{\mathcal{E} \setminus \emptyset} u_K \, d\mu_\varphi(K), \quad (9.4)$$

where μ_φ is a signed finitely additive measure on the algebra in $\mathcal{E}$ generated by $\{A \in \mathcal{E} : A \subset K\}$ for all $K \in \mathcal{E}$. Then $\|\varphi\|_{\mathcal{E}} = \|\mu_\varphi\|$ and φ is completely monotone if and only if μ_φ is non-negative. The Jordan decomposition (9.3) continues to hold in the infinite case provided φ has a finite composition norm and the family of non-additive measures with a finite composition norm forms a Banach space.

Choquet and Sugeno integrals

A non-additive measure can be used to define a *Choquet integral* in the same way as has been done in Section 5.1 for integrals with respect to the capacity functional and the containment functional. Let f be a non-negative function of $\mathbb{E}$ such that $\{f \geq t\} = \{x : f(x) \geq t\} \in \mathcal{E}$ for all $t \geq 0$. The Choquet integral of f is defined by

$$\int f\,d\varphi = \int_0^\infty \varphi(\{f \geq t\})\,dt \,.$$

It is shown by Graf [208] and Schmeidler [516] that a non-additive measure is convex if and only if its core is non-empty and for every measurable function f the Choquet integral equals the infimum of the integrals of f with respect to measures from the core of φ. Decomposition (9.4) leads to

$$\int f\,d\varphi = \int_{\mathcal{E}\setminus\emptyset} (\inf f(K))\mu_\varphi(dK)$$

assuming that μ_φ is σ-additive.

Schmeidler [516] proved that a general non-negative functional $I(f)$ defined on the family of bounded measurable real-valued functions on $\mathbb{E}$ is comonotonic additive (see Definition 5.4) and monotone (i.e. $I(f) \leq I(g)$ if $f \leq g$ pointwise on $\mathbb{E}$) if and only if I is the Choquet integral with respect to a non-additive measure φ given by $\varphi(M) = I(\mathbf{1}_M)$.

It is shown in Sugeno, Narukawa and Murofushi [552] that for every continuous f with a compact support there exists a measure μ such that

$$\int g\,d\varphi = \int g\,d\mu \tag{9.5}$$

for all g which are strongly comonotonic with f. This implies that, for every such f and $\varepsilon > 0$, there exist $a_1, \ldots, a_n \geq 0$ and $x_1, \ldots, x_n \in \mathbb{E}$ such that

$$\left| \int f\,d\varphi - \sum a_i f(x_i) \right| < \varepsilon \,,$$

meaning that the Choquet integral can be approximated by the integral of a step-function. If φ is the capacity functional of a random closed set, the measure μ from (9.5) admits a simple interpretation given in Proposition 5.14.

The *Sugeno integral* of f is defined as

$$(S)\int_A f\,d\varphi = \sup_{r \geq 0} \min(r, \varphi(A \cap \{f \geq r\})) \,. \tag{9.6}$$

As shown by Ralescu and Adams [473], this definition is equivalent to

$$(S)\int_A f\,d\varphi = \sup_{B\in\mathcal{E}, B\subset A} \min(\varphi(B), \inf f(B)).$$

The latter definition extends to the case when the sets $\{f \geq r\}$ do not necessarily belong to $\mathcal{E}$. Then (9.6) holds for an *outer* non-additive measure $\varphi^*(A) = \inf_{B\in\mathcal{E}, B\supset A} \varphi(B)$, see Murofushi [419]. If φ is the capacity functional of a random closed set X, then

$$(S)\int_A f\,dT_X = \sup\{r: \mathbf{P}\{(\sup f(X\cap A)) \geq r\} \geq r\}.$$

Example 9.3 (Integrals of indicator function). Let $f = a\mathbf{1}_A$. Then the Choquet integral of f equals $a\varphi(A)$, while the Sugeno integral is $\min(a, \varphi(A))$.

Another integral introduced in Gerritse [188] in view of applications to the theory of large deviations is

$$\int f\,d\varphi = \sup\{\varphi(A)\inf_{x\in A} f(x): A\in\mathcal{E}\} = \sup_{r\geq 0} r\varphi(A\cap\{f\geq r\}).$$

9.2 Belief functions

Belief and plausibility functions

Belief functions are non-additive measures that satisfy additional monotonicity conditions. In statistics of imprecise data, the belief function replaces the probability that a true value lies in a particular set.

Definition 9.4 (Belief and plausibility functions). A function $\mathrm{Bel}(A)$ for A from an algebra $\mathcal{E}$ with values in $[0, 1]$ is said to be a *belief function* if $\mathrm{Bel}(\emptyset) = 0$, $\mathrm{Bel}(\mathbb{E}) = 1$ and Bel is completely monotone on $\mathcal{E}$. The dual function to Bel is said to be a *plausibility function* and denoted by Pl.

An extension of the belief function to the family of all subsets of $\mathbb{E}$ is defined as

$$\widetilde{\mathrm{Bel}}(A) = \sup\{\mathrm{Bel}(B): B\subset A,\ B\in\mathcal{E}\}.$$

A belief function is called continuous if $\widetilde{\mathrm{Bel}}(A_n) \downarrow \widetilde{\mathrm{Bel}}(A)$ as $A_n \downarrow A$. General belief functions are not assumed to be semicontinuous and also the basic space $\mathbb{E}$ is not necessarily locally compact Hausdorff separable. Although a containment functional of a random closed set is a continuous belief function, a general belief function does not necessarily satisfy the regularity conditions that hold for a containment functional and so cannot be interpreted as $\mathrm{Bel}(A) = \mathbf{P}\{X\subset A\}$ for a random closed set X. If this representation is possible, then X can be interpreted as a region where the true value belongs.

Shafer [534] proved that a general belief function can be represented as

$$\mathrm{Bel}(A) = \mu(r(A)), \tag{9.7}$$

where μ is a finitely additive probability measure on an abstract space Ω and r maps $\mathcal{E}$ into measurable subsets of Ω. In the case of containment functionals of random closed sets, $\Omega = \mathcal{F}$ and $r(A) = \{F \in \mathcal{F} : F \subset A\}$. The map r is called an *allocation of probability*.

The *vacuous* belief function satisfies $\text{Bel}(A) = 1$ for $A = \mathbb{E}$ and is zero otherwise and so represents the state of complete ignorance. This belief function is the containment functional of the deterministic set $X = \mathbb{E}$. The *Boolean* belief function models the information that the true value lies in $F \subset \mathbb{E}$ and is given by $\text{Bel}(A) = 1$ for $F \subset A$ and zero otherwise. This belief function is indeed the *unanimity game* defined in (9.1).

A plausibility function is called *condensable* if its value on every compact set K is equal to the supremum of its values on finite subsets of K. The corresponding belief function is also called condensable. This property is similar to the separability property of random closed sets.

Updating belief functions

The idea of *updating* belief functions is central to statistical reasoning with imprecise probabilities (see Walley [591]) and also to applications in artificial intelligence, economic theory, etc. The *Dempster rule of combination* suggests that, given an event A, the new belief function Bel_A should be

$$\text{Bel}_A(B) = \frac{\text{Bel}((A \cap B) \cup A^c) - \text{Bel}(A^c)}{1 - \text{Bel}(A^c)}. \tag{9.8}$$

Note that the denominator is the plausibility function. Updating belief functions using the Dempster rule can be easily interpreted for containment functionals of random closed sets. The *orthogonal sum* of two containment functionals C_X and C_Y is the containment functional of $X \cap Y$ (with independent X and Y) conditional on the event $\{X \cap Y \neq \emptyset\}$. This reduces to (9.8) if $Y = A$ is a deterministic set. Indeed,

$$\mathbf{P}\{(X \cap A) \subset B \mid X \cap A \neq \emptyset\} = \frac{\mathbf{P}\{(X \cap A) \subset (B \cap A), X \not\subset A^c\}}{T_X(A)}$$

$$= \frac{\mathbf{P}\{X \subset (A \cap B) \cup A^c, X \not\subset A^c\}}{T_X(A)}$$

$$= \frac{C_X((A \cap B) \cup A^c) - C_X(A^c)}{1 - C_X(A^c)}.$$

Therefore, the Dempster rule of combination can be interpreted as taking the conditional containment functional of the intersection of a random closed set with a deterministic set. If the true value belongs to X, given the prior information that the value lies in A, it is natural to believe that it belongs to $X \cap A$.

An adaptation of the classical concept of the conditional probability leads to an alternative *Bayesian* updating rule

$$\text{Bel}_A(B) = \frac{\text{Bel}(A \cap B)}{\text{Bel}(A)}. \tag{9.9}$$

The latter rule can be deduced from decomposition (9.4) applied to $\varphi = \text{Bel}$, then updating μ_φ using the conditional probability formula and then retrieving Bel using the updated μ_φ.

Likelihood-based belief function

If $\mathbb{E}$ represents possible values of the parameters and $L(x)$, $x \in \mathbb{E}$, is a likelihood function for the estimated parameter x (its dependence on the observed sample is suppressed in the notation), then the *likelihood-based* belief function (see Shafer [533] and Wasserman [598]) is defined by its dual (plausibility) function

$$\text{Pl}(A) = \sup_{x \in A} \tilde{L}(x),$$

where

$$\tilde{L}(x) = \frac{L(x)}{\sup\{L(x) : x \in \mathbb{E}\}}$$

is the normalised likelihood function. The so-defined plausibility function is the sup-measure generated by $\tilde{L}$ and the corresponding random closed set is given by

$$X = \{x : \tilde{L}(x) \geq \alpha\}$$

with α uniformly distributed on $[0, 1]$ (see Example 1.15), so that X has realisations being likelihood regions. It is easy to combine the likelihood-based belief function with another belief function. Let Y be a random closed set independent of X and such that $\mathbf{E} \sup L(Y) > 0$. Then the *conditional* capacity functional of the intersection $X \cap Y$ is given by

$$\mathbf{P}\{(X \cap Y) \cap A \neq \emptyset \mid X \cap Y \neq \emptyset\} = \frac{\mathbf{P}\left\{\sup \tilde{L}(Y \cap A) \geq \alpha\right\}}{\mathbf{P}\left\{\sup \tilde{L}(Y) \geq \alpha\right\}}$$
$$= \frac{\mathbf{E} \sup L(Y \cap A)}{\mathbf{E} \sup L(Y)},$$

where $\sup \emptyset = 0$. If $Y = \{\eta\}$ is a singleton, then this rule reduces to the usual Bayesian methodology with the update given by $\mathbf{E}(L(\eta) \mathbf{1}_{\eta \in A})/\mathbf{E} L(\eta)$.

Open problem 9.5. For a containment functional C (or belief function) find a probability measure $\mathbf{P}$ that dominates C and has the maximal entropy. The answer for finite spaces in known, see Jaffray [278].

9.3 Upper and lower probabilities

Definitions

A pair u, v of non-additive measures with total mass 1 is said to be a *lower* and *upper* probability if u and v are *dual*, i.e. $v(A) + u(A^c) = 1$ for every A, v is subadditive and u is superadditive, i.e.

$$u(A \cup B) \geq u(A) + u(B), \quad v(A \cup B) \leq v(A) + v(B),$$

for every two disjoint sets A and B from an algebra $\mathcal{E}$ of sets where u and v are defined. A measure μ dominates u if $\mu(A) \geq u(A)$ for all $A \in \mathcal{E}$. The lower probability is said to be the *lower envelope* if u is the infimum of all measures that dominate it. In this case, v is *coherent*, i.e. equals the upper envelope of the family of measure that it dominates. It is known that there are lower probabilities with no measures that dominate it and not all lower probabilities are lower envelopes, see Papamarcou and Fine [449].

In the following we consider only upper and lower probabilities that are *coherent*, i.e. equal to upper and lower envelopes of a family of measures. Every non-empty family $\mathbb{P}$ of probability measures on $\mathbb{E}$ with a σ-algebra $\mathfrak{F}$ naturally generates two non-additive measures, the upper probability (upper envelope)

$$v(A) = \sup\{\mathbf{P}(A) : \mathbf{P} \in \mathbb{P}\}, \quad A \in \mathfrak{F}, \tag{9.10}$$

and the lower probability (lower envelope)

$$u(A) = \inf\{\mathbf{P}(A) : \mathbf{P} \in \mathbb{P}\}, \quad A \in \mathfrak{F}.$$

The family $\mathbb{P}$ is said to be *m-closed* if the core of u is $\mathbb{P}$ itself, so that $\mathbb{P}$ is exactly the family of measures that dominate u.

The lower probability is monotonic and inherits the properties of the outer continuity and the outer regularity of a probability measure. Every coherent upper probability is σ-subadditive, i.e.

$$v(\cup_{i=1}^{\infty} A_i) \leq \sum_{i=1}^{\infty} v(A_i).$$

Sudadditive capacities as upper probabilities

Theorem 5.13 implies that the capacity functional of a random closed set X is the upper envelope of the family $\mathbb{P}$ that consists of distributions for all selections of X. This fact is, however, more general and holds for every 2-alternating (strongly subadditive) capacity, see (1.18).

Theorem 9.6 (Subadditive capacities as upper envelopes of measures). *Every strongly subadditive capacity equals the upper envelope for the family of measures that it dominates.*

Note that not every coherent upper probability is 2-alternating, see Huber and Strassen [268, Ex. 2]. For instance, let $\mathbb{E} = \{1, 2, 3, 4\}$ have four elements with the two probability distributions given by the vectors $p_1 = (0.5, 0.2, 0.2, 0.1)$ and $p_2 = (0.6, 0.1, 0.1, 0.2)$. For the corresponding upper probability v, one has $v(A) = v(B) = 0.7$ for $A = \{1, 2\}$ and $B = \{1, 3\}$. On the other hand, $v(A \cup B) + v(A \cap B) = 1.5 > v(A) + v(B) = 1.4$, i.e. v is not 2-alternating.

Upper and lower Choquet integrals

The upper and lower probability v and u can be used to construct Choquet integrals that are called *upper* and *lower* expectations. If v and u are respectively 2-alternating and 2-monotone and v is *regular*, i.e.

$$v(B) = \sup\{v(K) : K \subset B, \ K \in \mathcal{K}\}$$
$$= \inf\{v(G) : G \supset B, \ G \in \mathcal{G}\}, \quad B \in \mathcal{B}(\mathcal{E}),$$

then the Choquet integrals can be obtained as infimum and supremum of expectations with respect to the family $\mathbb{P}$ that generates u and v,

$$\int f\,du = \inf\{\int f\,d\mathbf{P} : \mathbf{P} \in \mathbb{P}\},$$
$$\int f\,dv = \sup\{\int f\,d\mathbf{P} : \mathbf{P} \in \mathbb{P}\},$$

see Huber [267, p. 262] for a finite $\mathbb{E}$ and Graf [208, Prop. 2.3] for a general $\mathbb{E}$. If v (respectively u) is completely alternating (respectively monotone), this fact easily follows by identifying all $\mathbf{P} \in \mathbb{P}$ as distributions of selections for the corresponding random closed set. The upper expectation is called *symmetric* if $\int f_1\,du = \int f_2\,du$ whenever $\mu(\{f_1 \geq t\}) = \mu(\{f_2 \geq t\})$ for all t, i.e. f_1 and f_2 are μ-equimeasurable functions. If the upper expectation is symmetric, then u is weakly symmetric. The reverse conclusion is wrong, see Wasserman and Kadane [601, Ex. 4].

Unambiguous events

The interval $[u(A), v(A)]$ characterises the degree of uncertainty associated with the family $\mathbb{P}$ of probability distributions. An event $A \in \mathfrak{F}$ is called *unambiguous* if this interval shrinks to a single point, i.e. $u(A) = v(A)$ or $v(A) + v(A^c) = 1$. The following proposition characterises symmetric coherent upper probabilities.

Proposition 9.7 (Upper probabilities with an unambiguous event). *Let v be an upper probability on a σ-algebra $\mathfrak{F}$ which is symmetric with respect to a non-atomic probability measure μ.*
   (i) *If v is coherent, then there exists a non-trivial unambiguous event if and only if v is a probability measure.*
   (ii) *If v is monotonic and subadditive, then the existence of a non-trivial unambiguous event implies that μ is the unique probability measure dominated by v.*

Proof. Let A be a non-trivial unambiguous event.
   (i) There exists a function $g : [0, 1] \mapsto [0, 1]$ such that $v(D) = g(\mu(D))$ for all $D \in \mathfrak{F}$. Then any set $B \in \mathfrak{F}$ with $\mu(B) = \mu(A)$ is also unambiguous for v, whence $v(B) + v(B^c) = 1$ and $v(B) = \mathbf{P}(B)$ for all $\mathbf{P} \in \mathbb{P}$. Therefore, $\mu(A) = \mu(B)$ implies $\mathbf{P}(A) = \mathbf{P}(B)$ for every $\mathbf{P} \in \mathbb{P}$, whence $\mathbb{P} = \{\mu\}$ and $v = \mu$ identically.

(ii) Since v is symmetric, $v(D) = g(\mu(D))$, $D \in \mathfrak{F}$, for a function $g: [0, 1] \mapsto [0, 1]$ such that $g(0) = 0$ and $g(1) = 1$. We show first that g is non-decreasing. Suppose that $1 > x > y > 0$. Since μ is non-atomic there exist $D, B \in \mathfrak{F}$ such that $\mu(D) = x$, $\mu(B) = y$ and $B \subset D$. Therefore, $v(B) \leq v(D)$, whence $g(x) \geq g(y)$.

Since v is subadditive, $g(x + y) \leq g(x) + g(y)$ and $g(1 - x - y) \leq g(1 - x) + g(1 - y)$ for $x, y \in [0, 1]$ with $x + y \leq 1$. By Wasserman and Kadane [601, Lemma 9], $g(x) \geq x$ for all $x \in [0, 1]$, so μ is dominated by u. The proof finishes similarly to the proof of (i). □

If v is the capacity functional T of a random closed set, then the existence of an unambiguous event A means that $T(A) = C(A)$, i.e. the values of the capacity functional and the containment functional coincide. If $T(K) = g(\mu(K))$ for a certain function g (i.e. T is symmetric), then Proposition 9.7 implies that the corresponding random closed set is necessarily a singleton.

9.4 Capacities in robust statistics

"Contaminated" families of probability measures

In *robust statistics* it is typical to consider a family $\mathbb{P}$ of measures that includes the "true" probability measure **P** and other probability measures interpreted as "contaminations" of **P**. These contaminations can be defined as belonging to a neighbourhood of **P** with respect to some probability metric. The family $\mathbb{P}$ gives rise to the upper probability given by (9.10). Although it is usually very difficult to prove that v is completely alternating, this is easy for the ε-contamination model, which deals with the family $\mathbb{P} = \mathbb{P}_\varepsilon$ of all probability measures that can be represented as the sum of $(1 - \varepsilon)\mathbf{P}$ and $\varepsilon\mathbf{Q}$ for all probability measures **Q**. The corresponding upper probability $v(A) = (1 - \varepsilon)\mathbf{P} + \varepsilon$ is the capacity functional of a random closed set that with probability $(1 - \varepsilon)$ equals to a singleton $\{\xi\}$ with the distribution **P** and otherwise equals the whole space. In many interesting cases v is 2-alternating capacity, which is extremely important in view of applications to robust statistics, see Huber [267].

Neyman–Pearson lemma

In general, it is quite difficult to construct a minimax test that discriminates between the two composite alternatives $\mathbb{P}_0$ and $\mathbb{P}_1$. However, the situation becomes much simpler if $\mathbb{P}_0$ and $\mathbb{P}_1$ are the *cores* of 2-alternating capacities v_0 and v_1 with the corresponding dual (2-monotone) capacities u_0 and u_1. Let A denote a critical region for testing between $\mathbb{P}_0$ and $\mathbb{P}_1$, i.e. $\mathbb{P}_0$ is rejected if $x \in A$ is observed. Then the upper probability of falsely rejecting $\mathbb{P}_0$ is $v_0(A)$ and of falsely accepting $\mathbb{P}_0$ is $v_1(A^c) = 1 - u_1(A)$.

If $\mathbb{P}_0$ is true with prior probability $p = t/(1 + t)$, $0 \leq t \leq \infty$, then the upper *Bayes risk* of the critical region A is by definition

$$\frac{t}{1+t} v_0(A) + \frac{1}{1+t} (1 - u_1(A)).$$

9 Various interpretations of capacities

This risk is a function of $w_t(A) = tv_0(A) - u_1(A)$ that is to be minimised through a suitable choice of A. Note that a similar function appears in the Radon–Nikodym theorem for capacities in order to describe the strong decomposition property of two capacities, see Definition 5.9. It is possible to show that, for each t, there is a critical region A_t minimising $w_t(A)$, and the sets A_t can be chosen to form a decreasing family. Define

$$\pi(x) = \inf\{t : x \notin A_t\}.$$

For instance, if v_0 and v_1 are probability measures (i.e. the alternatives $\mathbb{P}_0$ and $\mathbb{P}_1$ are simple), then π is a version of the Radon–Nikodym derivative dv_1/dv_0 (if v_1 is absolutely continuous with respect to v_0). Interestingly, a similar interpretation holds for pairs of 2-alternating capacities.

Theorem 9.8 (Neyman–Pearson lemma for capacities). *Let v_0 and v_1 be a pair of 2-alternating capacities such that $v_i(A_n) \uparrow v_i(A)$ for all $A_n \uparrow A$ and $v_i(F_n) \downarrow v_i(F)$ for closed $F_n \downarrow F$, $i = 0, 1$. Then there exist probabilities $\mathbf{Q}_0 \in \mathbb{P}_0$ and $\mathbf{Q}_1 \in \mathbb{P}_1$ such that, for all t,*

$$\mathbf{Q}_0(\{\pi > t\}) = v_0(\{\pi > t\}), \quad \mathbf{Q}_1(\{\pi > t\}) = u_1(\{\pi > t\}),$$

and $\pi = d\mathbf{Q}_1/d\mathbf{Q}_0$.

The regularity condition imposed on v_0 and v_1 in Theorem 9.8 guarantees that the families $\mathbb{P}_0$ and $\mathbb{P}_1$ are weakly compact. It should be noted that this is a restrictive condition that for many interesting cases would require the compactness of $\mathbb{E}$.

The pair $\mathbf{Q}_0$ and $\mathbf{Q}_1$ from Theorem 9.8 is called the least favourable pair. Conversely, if v_1 is a probability measure and the conclusion of Theorem 9.8 holds, then v_0 is necessarily 2-alternating. Consider the *Neyman–Pearson test* of level α between $\mathbf{Q}_0$ and $\mathbf{Q}_1$ with the critical function

$$\varphi(x_1, \ldots, x_n) = \begin{cases} 1, & \prod \pi(x_i) > C, \\ \gamma, & \prod \pi(x_i) = C, \\ 0, & \prod \pi(x_i) < C, \end{cases}$$

where γ and C are chosen such that the expectation of φ with respect to $\mathbf{Q}_0$ equals α. Theorem 9.8 implies that, for any sample size n and any level α, this test is also a minimax test between $\mathbb{P}_0$ and $\mathbb{P}_1$ with the same level α and the same minimum power.

Application of belief functions

Consider now an application of belief functions to robust Bayesian inference. Let $\mathbf{P}$ be a fixed prior distribution and let $\mathbf{P}_y$ denote the posterior of $\mathbf{P}$ after observing y. In order to examine the sensitivity of the posterior with respect to the choice of the prior, we can consider a family $\mathbb{P}$ of prior distributions. Then the upper and lower posterior probabilities are given as upper and lower envelopes of the posteriors obtained for all possible priors $\mathbf{P} \in \mathbb{P}$,

$$v_y(A) = \sup_{\mathbf{P} \in \mathbb{P}} \mathbf{P}_y(A), \quad u_y(A) = \inf_{\mathbf{P} \in \mathbb{P}} \mathbf{P}_y(A).$$

These upper and lower envelopes may then be examined to quantify the robustness with respect to the choice of the prior.

Let $L(y|x)$ be the likelihood function of observing y if the unknown parameter is x. Then

$$\mathbf{P}_y(A) = \frac{\int_A L(y|x) \mathbf{P}(dx)}{\int_\mathbb{E} L(y|x) \mathbf{P}(dx)} = \frac{\mathbf{E}(L(y|\xi) \mathbf{1}_{\xi \in A})}{\mathbf{E}(L(y|\xi))},$$

where ξ is a random element with distribution $\mathbf{P}$. If $\mathbb{P}$ is the family of distributions of selections for a random closed set X, then

$$v_y(A) = \frac{\mathbf{E} \sup L(X \cap A)}{\mathbf{E} \sup L(X \cap A) + \mathbf{E} \inf L(X) \mathbf{1}_{X \subset A^c}}, \qquad (9.11)$$

see Wasserman [599, Th. 4.1]. For example, if $\mathbb{P}$ is the ε-contamination family of a prior $\mathbf{P}$, then, for $A \neq \mathbb{E}$,

$$v_y(A) = \frac{(1-\varepsilon) \int_A L(x) \mathbf{P}(dx) + \varepsilon \sup L(A)}{(1-\varepsilon) \int_\mathbb{E} L(x) \mathbf{P}(dx) + \varepsilon \sup L(A)}.$$

Open problem 9.9. Find the order of alternation for the capacity v_y obtained in (9.11) as the Bayesian update of a belief function. In particular, what is the answer for the case of the ε-contamination model?

Notes to Chapter 1

Section 1.1. The definition of a random closed set closely follows Matheron [381, Sec. 2.1]. Sometimes, the σ-algebra $\mathfrak{B}(\mathcal{F})$ is called Matheron's σ-algebra, while we prefer to reconciliate the terminology with the preexisting concepts in the set-valued analysis and call it Effros σ-algebra (also applicable for non-locally compact spaces). The letter $\mathcal{F}$ for the family of closed sets stems from the French word "ferme".

The concept of a random closed set is principally different from the concept of a fuzzy set, see Zadeh [621]. The latter is a function A on $\mathbb{E}$ with values in $[0, 1]$. The value $A(x)$ at $x \in \mathbb{E}$ determines a "degree of membership" of x to the fuzzy set. A random closed set X can be associated with a fuzzy set by considering the coverage function $A(x) = \mathbf{P}\{x \in X\}$. However, taking the coverage function seriously reduces the amount of information available about the distribution of X. Nguyen [427] interpreted the coverage function of a random set as the membership function of a fuzzy set that results in a characterisation of fuzzy sets satisfying some measurability conditions as equivalence classes of random sets, see Goodman and Nguyen [207].

The studies of random sets on the line were initiated by Dellacherie [129, 131] mostly motivated by applications to the general theory of stochastic processes, see also Section 5.2.3.

Various applications of random sets in statistical physics and material science are described by Torquato [564].

Section 1.2. The fact that it is possible to define a measure on the space of closed sets by its values on the families $\mathcal{F}_K$ and the corresponding complete alternation concept go back to Choquet [98]. Later on Matheron [381] formulated and proved this result explicitly for distributions of random closed sets and capacity functionals. At the same time, this fact was realised by Kendall [295] who worked with incidence functions (or hitting processes) and not necessarily closed sets, see Section 7.2. Because of these contributions, the Choquet theorem is sometimes called the *Choquet–Kendall–Matheron* theorem. Similar results in a more general setting of partially ordered topological spaces were obtained by Revuz [486] and Huneycutt [270], see also Talagrand [556].

The definition of a general capacity goes back to Choquet [98], see also Landkof [343] and Matheron [381]. Extensive studies of completely alternating and completely monotone capacities have been initiated by the seminal work by Choquet [98]. There (and in many other places) such capacities appear under the names of alternating of infinite order and monotone of infinite order. They are also sometimes called infinitely alternating (monotone) or totally alternating (monotone). Here we follow the terminology of Berg, Christensen and Ressel [61]. The notation of successive differences follows the pattern used by Norberg [432]. The ∪- and ∩-monotone (alternating) functionals were introduced to cover various concepts that appear when dealing with various functionals related to distributions of random sets and also set functionals in the theory of belief functions and cooperative games. In lattice theory and combinatorial mathematics 2-alternating capacities are called submodular or lower semi-modular.

Proposition 1.14 is a folklore fact that appears, for instance, in King [305].

Maxitive capacity functionals appear under the name of *possibility measures* in the theory of fuzzy sets (Zadeh [622]), where it is assumed that the value of the functional on a countable union of sets coincides with the supremum of the functional's values on the individual sets. Maxitive capacities generated by a likelihood function were considered in Walley and Moral [592]. Proposition 1.16 is due to Nguyen and Nguyen [428], who also gave a different proof of Theorem 1.17.

In the Russian translation of G. Matheron's book [381] the capacity functional of random closed set X is called an accompanying functional of X. This doubly translated term appears in several papers on random closed sets translated from the Russian.

Section 1.3. The presented measure-theoretic proof of the Choquet theorem is taken from Matheron [381]. The original G. Choquet's proof [98] is based on a representation of positive definite functions on cones and is similar to the presented harmonic analysis proof of the Choquet theorem. The harmonic analysis proof is adapted from Berg, Christensen and Ressel [61], where all the missing details can be retrieved. The Choquet theorem appears then as a particular case of a representation of a positive definite function on an idempotent semigroup.

Section 1.4. Separating classes were introduced by Norberg [429] who also used them to characterise the weak convergence of random closed sets. It should be noted that a simple separating class of unions of all closed balls has been already mentioned by Matheron [381]. The uniqueness property in a variant of the Choquet theorem for capacity functionals on open sets was noticed by Fortet and Kambouzia [179] who also considered several relationships with point processes.

136   1 Random Closed Sets and Capacity Functionals

Section 1.5. This section contains mostly well known facts whose origins are difficult to trace. Theorem 1.31 is proved by Crauel [109].

Section 1.6. Various functionals related to random closed sets have been discussed in Matheron [381], although without names used consistently in Section 1.6. Inclusion functionals defined on finite sets (also called point laws) have been studied by Matheron [381] and Engl [162]. In the theory of evidence Shafer [533] for expert systems, capacity functionals appear under the name of plausibility functionals, containment functionals are known as belief functions and inclusion functionals are called commonality functionals, see Thoma [563].

The Möbius inversion originates in combinatorial mathematics, while its application to general set-functions was described by Shafer [533]. More generally, the Möbius inversion can be defined for a not necessarily finite space $\mathbb{E}$ assuming that a random set takes at most a finite number of possible values. The non-negativity property of the Möbius inversion characterises completely monotone capacities, while a modification of the non-negativity condition singles out capacities with a given degree of monotonicity, see Chateauneuf and Jaffray [94].

Rataj [476] considered three-point covariances given by the measure of intersection of X and its two translations. A generalisation of the exponential covariance model is considered by Böhm and Schmidt [75]. The spectral theory for random closed sets has been described by Koch, Ohser and Schladitz [320].

Section 2.1. There is a vast literature on multifunctions and their selections. The main areas of applications are control and optimisation and mathematical economy. The fundamental measurability theorem for multifunctions (Theorem 2.3) is due to Himmelberg [257]. Further variants of this theorem have been surveyed by Wagner [588]. Other proofs of Theorem 2.3 are given by Castaing and Valadier [91, p. 59] and Aubin and Frankowska [30, Th. 8.1.4]. Proposition 2.5 was proved by Papageorgiou [442], while its variant for compact convex multifunctions under weaker restrictions on $\mathbb{E}$ is due to Valadier [569]. Theorem 2.6 was proved by Rockafellar [496] in the Euclidean case; his proof is restated here for a general Polish space $\mathbb{E}$. The predictability properties of the graph of a multifunction as a subset of $\Omega \times \mathbb{E}$ were studied by Ransford [475].

Theorem 2.7 is a synthesis of several results published in the literature. Part (i) is the statement of the Hess theorem [56, Th. 6.5.14] originally proved by Hess [240]. The current proof is a modified original proof by Ch. Hess. Part (ii) is a relatively well known fact, see Castaing and Valadier [91, p. 62]. The separability of $\mathcal{F}$ in the Wijsman topology is shown in Beer [56, Th. 2.5.4]. A number of further results related to part (ii) can be found in Barbati, Beer and Hess [48], in particular, those concerning the case when the Effros σ-algebra is generated by the Attouch–Wets topology. The latter generates the Effros σ-algebra if and only if all bounded subsets of $\mathbb{E}$ are totally bounded (a set F is totally bounded if F can be decomposed into a finite union of sets of diameter not greater than any given $\varepsilon > 0$).

The concept of Hausdorff approximable bounded random closed sets appears in Hiai and Umegaki [255], however without a special name. It was noticed by Hess [240] that each random closed set is Wijsman approximable. Theorem 2.11(ii) follows from a result of Debreu [124], who has shown that $(\mathcal{K}, \rho_H)$ is separable. The two examples of non-approximable random closed sets are due to Hiai and Umegaki [254, 255]. Relationships between the scalar measurability and the Effros measurability were investigated by Barbati and Hess [49].

Section 2.2. Selections (sometimes called selectors, sections or uniformisations) provide a very useful tool for the studies of multifunctions. It is especially important to establish the existence of selections. The studies of selections are summarised in useful surveys by Wagner [588, 589]. Although the fundamental selection theorem (Theorem 2.13) is often associated with the names of Kuratowski and Ryll-Nardzewski [339], similar results were obtained by von Neumann [423], by P.S. Novikov who much earlier used selections to prove implicit functions theorems and by V.A. Rokhlin who formulated the selection theorem and applied it to study dynamical systems.

Selections are discussed in Section 2.1.1 in view of their integrability properties and applications to define an integral of a multifunction (or an expectation of a random closed set). There is a vast literature on selections with special properties, for example, continuous selections (Repoš and Semenov [479]) and differentiable selections (Dentcheva [136]). Generalised Steiner points have been used by Dentcheva [136] to obtain a Castaing representation that consists of the Lipschitz selections.

Theorem 2.16 was proved by Hart and Kohlberg [223] for integrably bounded (see Definition 2.1.11) random compact sets in Euclidean space and further generalised by Artstein and Hart [22] for general random closed sets in $\mathbb{R}^d$. Example 2.15 is taken from Hart and Kohlberg [223]. A direct proof of Theorem 2.20 (based on matching lemma in the spirit of Halmos and Vaughan [218]) is provided by Artstein [19]. A weaker variant of Theorem 2.20 for a measure ν on $\mathcal{K}$ and with K being an arbitrary closed set in (2.4) appeared in Artstein [16]. Another proof involving constructions from nonstandard analysis, relaxing topological assumptions on $\mathbb{E}$ and assuming that the involved measures are Radon is given by Ross [502]. Theorem 2.17 and Proposition 2.18 were proved by Hess [247]. Relationships with Young multimeasures (maps on a probability space with values being collections of measures) have been discussed by Artstein [19].

Proposition 2.21 and Theorem 2.22 are due to Artstein [16]. Selection operators and their applications to set-valued processes were discussed by Gao and Zhang [187].

Section 2.3. Theorem 2.25 is a synthesis of a number of results concerning the measurability of set-theoretic operations with multifunctions. The relevant results can be proved using arguments based on semicontinuity of the relevant maps, as was done by Matheron [381] for random closed sets in locally compact spaces. Many parts of Theorem 2.25 are of folklore nature, while some proofs can be found in Aubin and Frankowska [30] and Hess [240]. Rockafellar [496] discusses in detail the case of multifunctions with values in the Euclidean space. The second part of Theorem 2.26 is Filippov's theorem on implicit functions. Its proof has been adapted from Aubin and Frankowska [30].

Section 2.4. Theorem 2.28 was announced by Hess [238], see also Hess [247, Prop. 2.1]. Its variant in $\mathbb{E} = \mathbb{R}^d$ is given by Salinetti and Wets [512]. The local compactness assumption on $\mathbb{E}$ is difficult to relax in the Choquet theorem. The described counterexample to the Choquet theorem is due to Nguyen and Nguyen [428]. An attempt to prove the Choquet theorem on non-separable spaces was unsuccessful; the main result of Ross [501] is wrong.

Section 2.5. In the general theory of stochastic processes it is quite typical to define random sets as measurable subsets of the product space $\Omega \times \mathbb{E}$, see Dellacherie and Meyer [131, 132]. Countable dense random sets were studied by Kendall [297]. Straka and Štěpán [548] investigated relationships between random Borel sets in [0, 1] and random automorphisms.

Section 3. The lattice theoretic approach to constructions of distributions of random sets has been developed by Norberg [432] who studied measures on continuous posets and, in particular, identified those measures corresponding to distributions of random closed sets. The presentation is adapted from Norberg [432, 434]. General results on continuous posets can be found in a comprehensive monograph by Gierz et al. [192]. A number of further papers concerning probability measures on lattices are collected by Vervaat and Holwerda [573].

Section 4.1. Stationary random closed sets were defined by Matheron [381]. A variant of Proposition 4.4 for avoidance functionals defined on open sets is given by Kendall [295,

Th. 18]. In that case it is possible to prove that every completely monotone functional is lower semicontinuous and to provide an integral representation for the function $q(t) = \mathbf{P}\{X \cap (0, t) = \emptyset\}$, $t > 0$, see Kendall [295, Th. 20]. Kendall [297] proved that all quasi-stationary countable dense sets (which are not closed) satisfy zero-one law, so that every (measurable) event associated with such sets has probability either 0 or 1. General properties of invariant capacities have been investigated by Talagrand [556].

Self-similar random sets on the line have been investigated by Pitman and Yor [459].

Section 4.2. Separable random sets and the corresponding inclusion functionals on finite sets have been studied by Matheron [381] and Engl [162]. Proposition 4.7 is taken from [381]. The capacity functional on the family of finite sets was called a *space law* in Matheron [381], but we avoid using this term.

The definition of **P**-continuous random closed sets is equivalent to Definition 2.5.1 of Matheron [381], but formulated in a different way. Definition 4.13 of an a.s.-continuous random closed set and the relevant results can also be found in Matheron [381], although some parts of the proofs presented here are new, in particular, those related to the separability and stochastic continuity concepts for indicator functions.

Section 4.3. The theory of regenerative events (or regenerative phenomena) was developed by Kingman [308] who generalised the concept of a recurrent event (for discrete time) described by W. Feller. A number of results on regenerative events (or regenerative phenomena) are summarised by Kingman [309]. Mathematical theory of subordinators is presented in depth by Bertoin [65, Ch. III]. The avoidance probabilities for an interval were obtained by Kendall [294]. The current proof follows a modern approach of Bertoin [65]. It is easy to extend (4.11) to calculate avoidance probabilities for a finite union of intervals. Proposition 4.20 is intrinsically related to results on hitting probabilities for stochastic processes with independent increments described by Kesten [298].

Section 4.4. Results similar to the current formulation of Robbins' theorem were apparently first stated by Kolmogorov [321, p. 41], Kolmogorov and Leontovich [322] and Robbins [494, 495], who assumed that the indicator function of X is jointly measurable, whereupon the identity (4.13) is a straightforward application of Fubini's theorem. Further results along this line can be found in Kendall and Moran [296] and Bronowski and Neyman [77]. Matheron [381, Cor. 1, p. 47] wrongly formulated Theorem 4.21 without the local-finiteness assumption on μ. This mistake has been repeated since then several times, for instance by Schmitt and Mattioli [518]. Baddeley and Molchanov [39] discussed possible generalisations of Robbins' theorem, see also Section 8.5.

Section 4.5. The presentation follows the ideas of Peres [453, 454] and Pemantle, Peres and Shapiro [451], see also Peres [455, Sec. 24] for the proofs in the one-dimensional case. Proposition 4.23(i) goes back to Hawkes [225], where also the converse statement is proved. The set $Z(p)$ appears also in the studies of fractal percolation, see Falconer [165, Sec. 15.2].

Section 4.6. Random open sets were defined by Matheron [381]. The formulated propositions are apparently new. A direct proof of Corollary 4.31 is given by Dynkin and Fitzsimmons [153]. Proposition 4.28 appears also in Crauel [109]. It is also possible to define random open sets in a Polish space by replacing $K \in \mathcal{K}$ with $F \in \mathcal{F}$.

Section 4.7. C-additive capacity functionals and their relationships to random convex sets have been thoroughly investigated by Matheron [381, Ch. 4]. The semi-Markov property is discussed in [381, Ch. 5] in relation to the infinite divisibility of random sets with respect to unions, see Section 4.1.2.

Section 4.8. Some parts of Theorem 4.41 go back to the pioneering work by Strassen [550], while its current formulation is taken from Kamae, Krengel and O'Brien [290].

Stochastic orders for random closed sets were introduced by Stoyan and Stoyan [546]. They considered particular cases of comparison of random sets of the same shape and with the same expected volume. The ordering concept was further studied by Norberg [433]. For a non-Hausdorff $\mathbb{E}$, Norberg showed that Theorem 4.42 implies the Strassen theorem, i.e. the equivalence of (1) and (4) in Theorem 4.41. Examples 4.43(i), (ii) are taken from Norberg [433] and Stoyan and Stoyan [546]. It is also possible to prove Theorem 4.42 without referring to Theorem 4.41, but instead using a direct method that involves probability measures on lattices, see Norberg [433]. Representation (4.24) is obtained by Wasserman [599].

If the random sets are almost surely convex, then it is possible to order them using one of the conventional orderings for their support functions considered as a stochastic process on the unit sphere, see Müller and Stoyan [418]. Furthermore, it is possible to order integrable random closed sets using the concepts of the expectation. For example, if $\mathbf{E}$ denotes the selection expectation (see Definition 2.1.1), then we say that X is symmetrically smaller than Y if $\mathbf{E}(X \cup K) \subset \mathbf{E}(Y \cup K)$ for every ball B_r. The properties of this and other orderings based on the selection expectation are studied by Cascos and Molchanov [87].

Section 5.1. The Choquet integral was introduced by Choquet [98] and further has been extensively studied. Denneberg [135] provides an excellent source of information about integration with respect to non-additive measures. Theorem 5.1 goes back to Wasserman [599], but in the current form appeared first in Molchanov [398]. A number of properties of the Choquet integral interpreted from the perspective of game theory can be found in Gilboa and Schmeidler [196]. Applications of the comonotonicity concept in mathematical economics are discussed by Dhaene et al. [138, 139].

Section 5.2. The Radon–Nikodym theorem for capacities is proved by Graf [208] who also analysed related decomposition properties and characterised measures among capacities in terms of the conditional expectation. The Choquet integral and Radon–Nikodym theorem for capacities defined on algebras of sets was also considered by Harding et al. [220]. In particular, the corresponding "local" versions appear if the corresponding algebra is finite. Ghirardato [190] investigated conditions that yield Fubini's theorem for Choquet integrals.

Section 5.3. Theorem 5.13 holds for all strongly subadditive capacities, see Huber and Strassen [268]. A number of related references can be found in Anger and Lembcke [9]. The current proof of Theorem 5.13 is considerably simplified by appealing to the selection theorem, which is not possible for general strongly subadditive capacities as they do not necessarily correspond to distributions of random closed sets. A more general proof appeals to the separation theorem in Banach spaces. Birgé [72] showed that Proposition 5.14 can be extended for all 2-alternating capacities. Proposition 5.15 is due to Chateauneuf and Jaffray [94].

Dichotomous capacities were introduced by Choquet [99]. Further references can be found in Fitzsimmons and Kanda [177]. A characterisation of random closed sets with dichotomous capacity functionals is not known.

The definition of an equalised capacity functional has been adapted from Wolfenson and Fine [611], where they are called coherent. The current name has been chosen to avoid confusion with a number of texts, where a coherent capacity is defined as equal to the upper (lower) envelopes of a family of probability measures that it dominates (dominated by it), see Kadane and Wasserman [285] and Walley [591].

An upper (lower) probability u generates the corresponding Choquet integral that is sometimes called upper (lower) *prevision* and can be used as a risk measure associated with a random variable. Delbaen [127] shows that an upper prevision $\int f \mathrm{d}u$ is the supremum of the expectations taken with respect to finitely additive measures if the upper prevision is subadditive, positively homogeneous and takes negative values on a negative f. In this case the upper prevision is called *coherent* and u is the supremum of a family of finitely-additive measures. This result can be strengthened to show the σ-additivity if $\int f \mathrm{d}u \geq \limsup \int f_n \mathrm{d}u$ if a sequence f_n is uniformly bounded by 1 and converges in probability to f.

Section 5.4. This section is based on results of Baddeley and Molchanov [39]. As shown in Section 8.5 the technique described here is useful for deriving a generalisation of Robbins' formula for random capacities.

Section 5.5. This section presents a new concept of capacity derivative. Definition 5.21 is inspired by the concept of the direct Radon–Nikodym derivative for measures in $\mathbb{R}^d$ introduced by Faro, Navarro and Sancho [166]. A principal difference is that the capacity of a point may be (and often is) different from zero. Note also that the vague limit in Definition 5.21 allows some mass to disappear. Furthermore, the main definition in [166] requires that the limit of $\mu(\{x\} \oplus tL)/t^d$ is finite, while the presented definition allows $\mathrm{d}_L \varphi(K)$ to take infinite values, but not identically for all $L \in \mathcal{K}$. The capacity derivative is similar to the concept of tangent measures, see Mattila [383] and Graf [209].

A variant of the introduced capacity derivative, obtained as the limit of

$$\frac{\varphi(K \cup (x + tL)) - \varphi(K)}{\varphi(tL)}.$$

was considered by Choquet [98] in application to φ being the Newton capacity. This derivative has been used by Fitzsimmons [175] to study hitting probabilities for Markov processes.

Derivatives of capacities are important in the context of data fusion, see Mori [414] for a discussion of application and some theory in the discrete case. Derivatives for a set-function and their applications in the data fusion are discussed by Mahler [371]. The information coming from different sensors is interpreted as random sets that combine the effects of clutter, jamming, measurement noise, false alarms, missed detections, unresolved measurements and target manoeuvres using set-theoretic operations. The concept of density is vital as it leads to applications of the likelihood inference and so yields efficient estimators.

Section 6.1. The weak convergence concept for random closed sets goes back to Lyashenko [367] and Norberg [429]. A parallel development of the theory in view of applications to the epiconvergence (see Section 5.3.1) is due to Salinetti and Wets [512]. Lemma 6.3 is due to Lyashenko [367] and Salinetti and Wets [512]. Theorem 6.8 is presented following Norberg [429]. Theorem 6.14 concerning the convergence of selections is due to King [305]. Proposition 6.11 is taken from Gao and Zhang [187]. Results similar to Theorem 6.15 are often called delta theorems in optimisation, see King [305]. Further results and applications can be found in [136, 305, 342]. Applications to statistical estimation problems are discussed by Dupačová and Wets [149], see also Section 5.3.2. The weak convergence of general lattice-valued random elements is considered by Norberg [434]. Further intricate facts about the weak convergence of general random elements and, in particular, random sets are discussed by Hoffman-Jørgensen [263].

Section 6.2. There is a vast literature concerning the polygonal approximation of convex sets. It is possible to generalise this framework in many ways: change dimension, relax assumptions on the boundary and allow points to be distributed according to certain densities or let them appear only on the boundary. Useful surveys are provided by Schneider [519] and Gruber [213]. Proposition 6.17 was proved by Davis, Mulrow and Resnick [119]. The convergence of random closed sets in probability was studied by Salinetti, Vervaat and Wets [510, 511, 512].

Section 6.3. Probability metrics for random closed sets have been studied by Molchanov [398, 403], although the first definition of distances between upper semicontinuous set functions (the Lévy metric for set functions) goes back to Rachev [469]. A comprehensive account of the theory of probability metrics can be found in Rachev [470]. Applications of probability metric for random sets to some problems in image analysis have been considered by Friel and Molchanov [182].

Section 7.1. Hitting processes (or incidence functions) have been studied by Norberg [429] and Vervaat [572]. Such processes are also called random sup-measures. Theorem 7.1 is taken from Norberg [429].

Section 7.2. The idea of the trapping system representation of a random closed set was used by Kendall [295] to build a foundation for the theory of random sets. This approach is especially important when dealing with general (non-closed) random sets. Section 7.2 is based on Kendall [295], where all details can be found. This approach can be quite naturally put within the framework of the theory of lattices, see Norberg [432]. Related ideas are discussed in Section 2.3.1. Theorem 7.7 is derived by Molchanov [391] by following the measure-theoretic proof of the Choquet theorem.

Section 7.3. The characterisation result for distributions of random convex compact sets is due to Vitale [576]. It was proved independently by Trader [565] and Molchanov [391]. The latter proof is based on Theorem 7.7. Norberg [432] showed that this fact naturally follows from the general results for the lattice of all convex compact sets. Theorem 7.12 follows from the weak convergence criteria for lattice valued random elements, see Norberg [434, Prop. 5.6]. Its direct proof (based on Billingsley [70, Th. 2.2]) can be found in Molchanov [397].

Theorem 7.13 is apparently new.

Section 8.1. An excellent presentation of the theory of point processes is given by Daley and Vere-Jones [117], which can be complemented by Stoyan, Kendall and Mecke [544] and Stoyan and Stoyan [547]. The fact that the avoidance probabilities characterise the distribution of a point process was proved by Mönch [412] and Kallenberg [286]. The direct relationship to the Choquet theorem was noticed by Ripley [492]. Other results relaxing topological assumptions are due to Ripley [490, 491]. Some of these results have been later rediscovered by Baudin [53]. Xia [616] relaxed the assumption on the point process being simple. He showed that if N is a point process orderly outside a single point, then the distribution of N is uniquely characterised by the one-dimensional distributions of $N(A)$ for all bounded Borel sets A. Proposition 8.4 was proved by Kurtz [340] and Ripley [491]. Proposition 8.5 is taken from Ripley [491].

Applications of ordered coupling theorem for random closed sets to thinning of point processes are due to Norberg [433]. Theorem 8.12 is proved by Kallenberg [288]. This result can be extended for the weak convergence for superpositions of null-arrays of point processes (Kallenberg [288]), which is closely related to the limit theorem for unions of random closed sets discussed in Chapter 4.

Although the concept of marking is very fruitful in the theory of point processes, its direct generalisation for random closed sets encounters a problem of ensuring the closedness. For example, if some points of a closed set F have mark 1 and other mark 2, then F is split into two sets F_1 and F_2, so that both F_1 and F_2 cannot be closed if F is connected. This shows that marked random closed sets should be necessarily non-connected. Marked (also called labelled) random closed sets were studied by Molchanov [392] and later on applied in image analysis by Ayala and Simó [36]. Marked random sets generated by excursions of random fields have been investigated by Nott and Wilson [437].

Section 8.2. Theorem 8.13 and the corresponding lemmas were proved by Weil and Wieacker [605, 606].

Section 8.3. The theory of random measures is outlined in Daley and Vere-Jones [117] and Stoyan, Kendall and Mecke [544], while a thorough exposition is provided by the classical book of Kallenberg [287]. Relationships between the distribution of a random measure and its support was studied by Ayala, Ferrandiz and Montes [34] who established a variant of Proposition 8.16 for rectifiable sets and the Hausdorff coverage measures.

Proposition 8.17 is due to Evans and Peres [163], who applied it to derive results concerning eventual intersection for sequences of Lévy processes.

Section 8.4. Random capacities have been studied by Norberg [430] where Propositions 8.18 and 8.19 are proved. Random capacities generalise the concepts of a random measure (see Harris [221] and Kallenberg [287]), random set and random semicontinuous functions. Further extensions of these results for capacities in non-Hausdorff spaces are discussed in Norberg and Vervaat [433, 435].

Section 8.5. Robbins' theorem for random capacities is presented following the paper by Baddeley and Molchanov [39], where also a number of further examples concerning intrinsic densities can be found. An application to the upper bound of the Hausdorff dimension is new.

Section 9.1. Non-additive measures naturally appear in decision-making studies and utility theory in economics, see Gilboa and Schmeidler [196] and Schmeidler [517]. In game theory, a non-additive measure φ is called a game, see Shapley [535] and Delbaen [126], and it is often denoted by v with its dual being u. A non-additive measure φ is sometimes called a fuzzy measure, see Murofushi and Sugeno [420] and Sugeno, Narukawa and Murofushi [552]. Couso, Montes and Gil [106] studied the convergence of Choquet integrals with respect to non-additive measures.

The Jordan decomposition of a non-additive measure (9.3) was described by Gilboa and Schmeidler [196]. The decomposition (9.4) and the corresponding Jordan decomposition for the infinite case is due to Gilboa and Schmeidler [197].

Another definition of the coherent set-functions can be found in de Cooman and Aeyels [104]. Symmetric coherent capacities were studied by Kadane and Wasserman [285].

The integration and calculus with non-additive measures are discussed in detail by Denneberg [135]. A representation theorem for comonotonic additive functionals is due to Schmeidler [516], another variant is described in Sugeno, Narukawa and Murofushi [552]. Sugeno [551] and Ralescu and Adams [473] introduced the integral with respect to non-additive measures and studied its convergence properties.

The weak convergence of non-additive measures was studied by Girotto and Holzer [203]. The weak convergence can be defined by means of convergence of the corresponding integrals, see Wang and Li [593]. Jang and Kwon [280] studied Choquet integrals of set-valued functions defined using selections. The Choquet integral is systematically studied by König [324] under the name of a *horizontal* integral. A generalisation of the Choquet integral (the so-called p-integral) is defined in Gerritse [188] as the Choquet integral of f^p with respect to φ^p subsequently raised to the power $1/p$. Gerritse [188] described an application of these integrals to the theory of large deviations. Laws of large numbers with respect to non-additive probabilities have been considered by Marinacci [377].

A function on $\mathbb{E}$ with values in an abstract set W is called an act with W being a set of possible consequences. A real-valued functional u on W is called an utility function. If a non-additive measure describes uncertainty, then the Choquet integral of $u(f(\cdot))$ with respect to a non-additive measure describes a quality of the decision taken and decisions are made as to maximise the Choquet expected utility. The corresponding order on the family of acts was investigated by Dyckerhoff and Mosler [151]. On the other hand, Schmeidler [517] showed that natural ordering axioms imply that the expected utility is given by the Choquet integral.

Relationships to the updating rules in Bayesian statistics and the theory of evidence are discussed by Gilboa and Schmeidler [195]. The Radon–Nikodym theorem for non-additive measures on finite spaces is proved by Gilboa and Schmeidler [196].

Non-monotonic set-functions were considered by Aumann and Shapley [32].

Section 9.2. The basic text on belief functions is G. Shafer's book *Mathematical Theory of Evidence* [533] that deals mostly with the case of a finite $\mathbb{E}$. The infinite case was considered by Shafer [534] where it is shown that belief functions can be represented as allocations of probability satisfying more specific requirements if the continuity assumption (or more restrictive condensability assumption) holds. Belief functions are often defined on a family of sets that is not necessarily an algebra, but only closed under finite intersections and contains $\mathbb{E}$.

The representation of continuous belief functions given by (9.7) goes back to Choquet [98], see also Huneycutt [270] and Revuz [486]. Applications of belief functions to statistical inference are surveyed by Walley [590, 591].

Relationships with random closed sets have been noticed first by Dempster [133] and further developed by Nguyen [426]. Since belief functions generalise probability measures, it is more typical to define them on subsets of the space Ω of elementary events rather than $\mathbb{E}$. With such an interpretation, the Choquet integral is defined for random variables (functions on Ω) and so extends the concept of expectation of random variables for non-additive measures. Condensable belief functions have been introduced by Nguyen [426].

Gilboa and Lehrer [194] show that every completely monotone game (belief function) defined on a finite algebra can be represented as a linear combination of unanimity games. If the belief function corresponds to a random closed set X, then the coefficient before the unanimity game u_K in the representation is equal to the probability $P(K) = \mathbf{P}\{X = K\}$. The Choquet integral of f with respect to a belief function becomes $\sum_{K \subset \mathbb{E}} P(K) \inf f(K)$ and is equal to the infimum of integrals with respect to measures that dominate it. This fact was obtained by Wasserman [598] and also noticed by a number of other authors, see Gilboa and Schmeidler [196].

The Dempster rule of combination is described by Dempster [134], see also Shafer [533]. An axiomatic approach leading to (9.9) was described by Gilboa and Schmeidler [195, 196]. Using signed measures as distributions of random sets, Kramosil [329] considered the inversion problem for belief functions.

Section 9.3. Upper and lower probabilities generated by a multivalued mapping (random set) as the capacity and containment functionals have been considered by Dempster [133]. Walley [591] provides a comprehensive account of various issues related to upper/lower probabilities and their various applications in statistics. The upper and lower probabilities have been used in the Bayesian decision theory by Wolfenson and Fine [611]. Within this context, upper and lower probabilities can be used to represent indeterminacy when assigning a prior distribution in Bayesian statistics. The interval $[u(A), v(A)]$ is also useful in interval computations where interval probabilities represent uncertainty. Completely monotone lower probabilities (i.e. containment functionals within the random sets framework) and their applications in decision making have been studied by Philippe, Debbs and Jaffray [456]. They also studied the m-closed families of probability measures.

As shown by Grize and Fine [211] and Papamarcou and Fine [450], it is possible to use (undominated) upper and lower probabilities to justify divergence of time averages observed for some stationary time series and unexplainable within the conventional probability framework. Domination properties for lower probabilities were studied by Papamarcou and Fine [449] and Fine [171].

In the capital investment, the interval between the lower and upper expectations (integrals) represents the passive interval, so that no action is required if the current price belongs to it, see Heilpern [230]. The concept of an unambiguous event can be generalised to define unambiguous functions. Families of measures that yield symmetric upper expectations have been characterised by Wasserman and Kadane [601] as those being closed with respect to a sort of a majorisation relation.

The fact that a strongly subadditive capacity coincides with the upper envelope for the family of measures that it dominates goes back to Choquet [98] and Strassen [549]. A full characterisation of upper envelopes in terms of an appropriately strengthened subadditivity property was obtained by Anger and Lembcke [9], see also Adamski [2] and Krätschmer [330]. What we call here upper (lower) envelopes are often called upper (lower) probabilities or coherent upper (lower) probabilities, see Kadane and Wasserman [285]. A characterisation of the supremum/infimum of a family of integrals (in application to asset pricing) is given in Chateauneuf, Kast and Lapied [95] where further references may be found.

Proposition 9.7 is due to Marinacci [378].

Section 9.4. The Neyman–Pearson lemma for 2-alternating capacities was proved by Huber and Strassen [268] and also explained in P.J. Huber's book *Robust Statistics* [267]. Buja [82] showed that it is possible to replace a rather restrictive condition $v(F_n) \downarrow v(F)$ for closed $F_n \downarrow F$ by $v(A_n) \downarrow v(A)$ for Borel $A_n \downarrow A \neq \emptyset$, An essential feature of this change is that it allows the contamination to concentrate in arbitrarily small non-empty sets, such that only the empty set escapes the possibility of hosting contamination.

The Neyman–Pearson lemma for "special capacities" that are superpositions of probability measures and concave functions was obtained by Bednarski [55] and Rieder [489] who elaborated some explicit ways of constructing the Radon–Nikodym derivative instead of just establishing its existence.

In the context of Bayesian inference, the intervals of measures have been used by DeRobertis and Hartigan [137]. An approach to robust Bayesian inference based on families of priors was developed by Berger [63] and Berger and Berliner [62]. It differs from the approach of Walley [591] where the lower probabilities rather than the sets of probabilities are of primary importance. The current presentation follows Wasserman [598, 599]. Similar results in the discrete case concerning envelopes of Bayesian updates of probabilities that dominate a given belief function are discussed by Jaffray [277]. A generalisation of (9.11) for general envelopes of measures and, in particular, for 2-alternating capacities was obtained by Wasserman and Kadane [600].

2
Expectations of Random Sets

1 The selection expectation

The space $\mathcal{F}$ of closed sets (and also the space $\mathcal{K}$ of compact sets) is non-linear, so that conventional concepts of expectations in linear spaces are not directly applicable for random closed (or compact) sets. Sets have different features (that often are difficult to express numerically) and particular definitions of expectations highlight various features important in the chosen context.

To explain that an expectation of a random closed (or compact) set is not straightforward to define, consider a random closed set X which equals $[0, 1]$ with probability $1/2$ and otherwise is $\{0, 1\}$. For another example, let X be a triangle with probability $1/2$ and a disk otherwise. A "reasonable" expectation in either example is not easy to define. Strictly speaking, the definition of the expectation depends on what the objective is, which features of random sets are important to average and which are possible to neglect.

This section deals with the selection expectation (also called the Aumann expectation), which is the best investigated concept of expectation for random sets. Since many results can be naturally formulated for random closed sets in Banach spaces, we assume that $\mathbb{E}$ is a separable Banach space unless stated otherwise. Special features inherent to expectations of random closed sets in $\mathbb{R}^d$ will be highlighted throughout. To avoid unnecessary complications, it is always assumed that all random closed sets are almost surely non-empty.

1.1 Integrable selections

The key idea in the definition of the selection expectation is to represent a random closed set as a family of its integrable selections. The concept of a *selection* of a random closed set was introduced in Definition 1.2.2. While properties of selections discussed in Section 1.2.1 can be formulated without assuming a linear structure on $\mathbb{E}$, now we discuss further features of random selections with the key issue being their *integrability*. We systematically use the *Bochner expectation* (see Vakhaniya,

Tarieladze and Chobanyan [568]) in the space $\mathbb{E}$, so that $\mathbf{E}\xi$ denotes the Bochner expectation of an $\mathbb{E}$-valued random element ξ. If $\mathbb{E} = \mathbb{R}^d$, then the Bochner expectation of the random vector $\xi = (\xi_1, \ldots, \xi_d)$ is the vector $\mathbf{E}\xi = (\mathbf{E}\xi_1, \ldots, \mathbf{E}\xi_d)$ of coordinate expectations.

Fix a complete probability space $(\Omega, \mathfrak{F}, \mathbf{P})$. Let $\mathbf{L}^p = \mathbf{L}^p(\Omega; \mathbb{E})$ denote the space of random elements with values in $\mathbb{E}$ such that the $\mathbf{L}^p$-norm

$$\|\xi\|_p = \mathbf{E}\|\xi\|^p$$

for $p \in [1, \infty)$ or the $\mathbf{L}^\infty$-norm

$$\|\xi\|_\infty = \mathbf{E} \operatorname*{ess\,sup}_{\omega \in \Omega} \xi(\omega)$$

is finite. We treat $\mathbf{L}^p$ as a normed linear space whose strong topology is generated by the norm. The weak topology corresponds to the weak convergence of random variables and gives rise to the concepts of weak closure, weak compactness, etc.

Definition 1.1 (*p-integrable selections*). If X is a random closed set in $\mathbb{E}$, then $\mathcal{S}^p(X)$, $1 \leq p \leq \infty$, denotes the family of all selections of X from $\mathbf{L}^p$, so that

$$\mathcal{S}^p(X) = \mathcal{S}(X) \cap \mathbf{L}^p,$$

where $\mathcal{S}(X)$ denotes the family of all (measurable) selections of X. In particular, $\mathcal{S}^1(X)$ is the family of integrable selections.

The following proposition establishes elementary properties of integrable selections.

Proposition 1.2. *If X is a random closed set in $\mathbb{E}$, then, for any $p \in [1, \infty]$,*
  (i) $\mathcal{S}^p(X)$ *is a closed subset of* $\mathbf{L}^p$;
  (ii) *if* $\mathcal{S}^p(X) \neq \emptyset$, *then there exists a sequence* $\{\xi_n, n \geq 1\} \subset \mathcal{S}^p(X)$ *such that* $X = \operatorname{cl}\{\xi_n, n \geq 1\}$;
  (iii) *if* $\mathcal{S}^p(X) = \mathcal{S}^p(Y) \neq \emptyset$, *then* $X = Y$ *almost surely.*

Proof.
  (i) If $\xi_n \to \xi$ in $\mathbf{L}^p$, then there is a subsequence $\{n(k), k \geq 1\}$ such that $\xi_{n(k)} \to \xi$ a.s., so that $\xi \in \mathcal{S}^p(X)$.
  (ii) By Theorem 1.2.3, there exists a sequence $\{\xi_n, n \geq 1\}$ of (not necessarily integrable) selections such that $X = \operatorname{cl}\{\xi_n, n \geq 1\}$. For a fixed $\xi \in \mathcal{S}^p(X)$ define

$$\xi'_{nm} = \mathbf{1}_{\|\xi_n\| \in [m-1, m)} \xi_n + \mathbf{1}_{\|\xi_n\| \notin [m-1, m)} \xi,$$

which belongs to $\mathcal{S}^p(X)$. Then $X = \operatorname{cl}\{\xi'_{nm}, n, m \geq 1\}$.
  (iii) immediately follows from (ii). □

The following useful lemma says that one can fix a countable family of selections such that every other selection can be approximated using selections from the chosen family.

Lemma 1.3 (Approximation by step-functions). Let $\{\xi_n, n \geq 1\} \subset \mathcal{S}^p(X)$ with $1 \leq p < \infty$ such that $X = \mathrm{cl}\{\xi_n, \, n \geq 1\}$. Then, for each $\xi \in \mathcal{S}^p(X)$ and $\varepsilon > 0$, there exists a finite measurable partition $A_1, \ldots, A_n$ of Ω such that

$$\left\| \xi - \sum_{i=1}^{n} \mathbf{1}_{A_i} \xi_i \right\|_p < \varepsilon.$$

Proof. Without loss of generality assume that $\xi(\omega) \in X(\omega)$ for all ω. Fix a positive random variable α such that $\mathbf{E}\alpha < \varepsilon^p/3$. Then there exists a countable measurable partition $\{B_n, n \geq 1\}$ of Ω such that $\|\xi(\omega) - \xi_i(\omega)\| < \alpha(\omega)$ for all $\omega \in B_i$, $i \geq 1$. Pick n such that

$$\sum_{i=n+1}^{\infty} \mathbf{E}\left(\mathbf{1}_{B_i}\|\xi\|^p\right) < \frac{1}{3}(\varepsilon/2)^p \quad \text{and} \quad \sum_{i=n+1}^{\infty} \mathbf{E}\left(\mathbf{1}_{B_i}\|\xi_1\|^p\right) < \frac{1}{3}(\varepsilon/2)^p.$$

Define

$$A_1 = B_1 \cup (\cup_{i=n+1}^{\infty} B_i) \quad \text{and} \quad A_i = B_i, \ 2 \leq i \leq n.$$

Then

$$\left\| \xi - \sum_{i=1}^{n} \mathbf{1}_{A_i} \xi_i \right\|_p^p = \sum_{i=1}^{n} \mathbf{E}\left(\mathbf{1}_{B_i} \|\xi - \xi_i\|^p\right) + \sum_{i=n+1}^{n} \mathbf{E}\left(\mathbf{1}_{B_i} \|\xi - \xi_1\|^p\right)$$

$$\leq \mathbf{E}\alpha + \sum_{i=n+1}^{n} \mathbf{E}\mathbf{1}_{B_i}(\|\xi\|^p + \|\xi_i\|^p) < \varepsilon^p. \qquad \square$$

The following results establish relationships between families of integrable selections and operations with random closed sets.

Proposition 1.4. Let $X = \mathrm{cl}(X_1 + X_2)$. If both $\mathcal{S}^p(X_1)$ and $\mathcal{S}^p(X_2)$ are non-empty with $1 \leq p < \infty$, then

$$\mathcal{S}^p(X) = \mathrm{cl}(\mathcal{S}^p(X_1) + \mathcal{S}^p(X_2)),$$

the closure and Minkowski sum in the right-hand side taken in $\mathbf{L}^p$.

Proof. By Proposition 1.2(ii), $X_i = \mathrm{cl}\{\xi_{in}, \, n \geq 1\}$ for $\{\xi_{in}, n \geq 1\} \subset \mathcal{S}^p(X_i)$, $i = 1, 2$. Then $X = \mathrm{cl}\{\xi_{1i} + \xi_{2j}, \, i, j \geq 1\}$. By Lemma 1.3, each $\xi \in \mathcal{S}^p(X)$ can be approximated by $\sum_{k=1}^{n} \mathbf{1}_{A_k}(\xi_{1i_k} + \xi_{2j_k})$, whence $\mathcal{S}^p(X) \subset \mathrm{cl}(\mathcal{S}^p(X_1) + \mathcal{S}^p(X_2))$. The reverse inclusion is evident. $\qquad \square$

Proposition 1.5. Let X be a random closed set. If $\mathcal{S}^p(X) \neq \emptyset$ with $1 \leq p < \infty$, then $\mathcal{S}^p(\overline{\mathrm{co}}(X)) = \overline{\mathrm{co}}(\mathcal{S}^p(X))$, the closed convex hull in the right-hand side taken in $\mathbf{L}^p$. Furthermore, $\mathcal{S}^p(X)$ is convex in $\mathbf{L}^p$ if and only if X is almost surely convex.

Proof. Since $\mathcal{S}^p(\overline{\mathrm{co}}(X))$ is a convex closed subset of $\mathbf{L}^p$, $\mathcal{S}^p(\overline{\mathrm{co}}(X)) \supset \overline{\mathrm{co}}(\mathcal{S}^p(X))$. To prove the converse, take $\{\xi_i, i \geq 1\} \subset \mathcal{S}^p(X)$ such that $X = \mathrm{cl}\{\xi_i, i \geq 1\}$. The set Y of all finite convex combinations of $\{\xi_i, i \geq 1\}$ with rational coefficients is a countable subset of $\mathcal{S}^p(\overline{\mathrm{co}}(X))$ and $\overline{\mathrm{co}}(X) = \mathrm{cl}(Y)$. Each $\xi \in \mathcal{S}^p(\overline{\mathrm{co}}(X))$ can be approximated by $\sum \mathbf{1}_{A_k} \eta_k$ with $\{\eta_k, k \geq 1\} \subset Y$, which turns into a convex combination of selections from $\mathcal{S}^p(X)$. Thus, $\xi \in \overline{\mathrm{co}}(\mathcal{S}^p(X))$. The last statement follows from Proposition 1.2(iii). □

The following theorem characterises closed subsets of $\mathbf{L}^p$ that can be represented as $\mathcal{S}^p(X)$ for a random closed set X. A subset $\Xi \subset \mathbf{L}^p$ is called *decomposable* if, for any $\xi_1, \xi_2 \in \Xi$ and $A \in \mathfrak{F}$, we have $\mathbf{1}_A \xi_1 + \mathbf{1}_{\Omega \setminus A} \xi_2 \in \Xi$. This further implies $\sum \mathbf{1}_{A_i} \xi_i \in \Xi$ for all measurable partitions $A_1, \ldots, A_n$ of Ω and $\xi_1, \ldots, \xi_n \in \Xi$.

Theorem 1.6 (Decomposable sets and selections). *Let Ξ be a non-empty closed subset of $\mathbf{L}^p$, $1 \leq p \leq \infty$. Then $\Xi = \mathcal{S}^p(X)$ for a random closed set X if and only if Ξ is decomposable.*

Proof. Clearly, $\mathcal{S}^p(X)$ is closed and decomposable. Let $\Xi \neq \emptyset$ be a closed decomposable subset of $\mathbf{L}^p$ for $1 \leq p < \infty$. Choose a sequence $\{\xi_i, i \geq 1\} \subset \mathbf{L}^p$ which is dense in Ξ for all ω. For each $i \geq 1$, choose $\{\eta_{ij}, j \geq 1\} \subset \Xi$ such that

$$\|\xi_i - \eta_{ij}\|_p \to \alpha_i = \inf_{\eta \in \Xi} \|\xi_i - \eta\|_p \quad \text{as } j \to \infty. \tag{1.1}$$

Define $X = \mathrm{cl}\{\eta_{ij}, i, j \geq 1\}$ with the aim of showing that $\Xi = \mathcal{S}^p(X)$. By Lemma 1.3, for each $\xi \in \mathcal{S}^p(X)$ and $\varepsilon > 0$, there exists a measurable partition $A_1, \ldots, A_n$ of Ω and $\xi'_1, \ldots, \xi'_n \in \{\eta_{ij}, i, j \geq 1\}$ such that

$$\left\| \xi - \sum \mathbf{1}_{A_k} \xi'_k \right\|_p < \varepsilon.$$

Since $\sum \mathbf{1}_{A_k} \xi'_k \in \Xi$, we obtain that $\xi \in \Xi$, so that $\mathcal{S}^p(X) \subset \Xi$.

Suppose that there exists $\xi \in \Xi$ such that

$$\inf_{i,j \geq 1} \|\xi(\omega) - \eta_{ij}(\omega)\|_p \geq \delta > 0$$

for $\omega \in A$ with $\mathbf{P}(A) > 0$. Fix i such that

$$B = A \cap \{\omega : \|\xi(\omega) - \xi_i(\omega)\| < \delta/3\}$$

has a positive probability and let $\eta'_j = \mathbf{1}_B \xi + \mathbf{1}_{\Omega \setminus B} \eta_{ij}$ for $j \geq 1$. Then $\eta'_j \in \Xi$ for $j \geq 1$. Since

$$\|\xi_i - \eta_{ij}\| \geq \|\xi - \eta_{ij}\| - \|\xi - \xi_i\| \geq 2\delta/3, \quad \omega \in B,$$

condition (1.1) implies

$$\|\xi_i - \eta_{ij}\|_p^p - \alpha_i^p \geq \|\xi_i - \eta_{ij}\|_p^p - \|\xi_i - \eta_j'\|_p^p$$
$$= \mathbf{E}\Big[\mathbf{1}_B(\|\xi_i - \eta_{ij}\|_p^p - \|\xi_i - \xi\|_p^p)\Big]$$
$$\geq ((2\delta/3)^p - (\delta/3)^p)\mathbf{P}(B) > 0.$$

Letting $j \to \infty$ leads to a contradiction.

Now consider the case $p = \infty$. Let Ξ' be the closure of Ξ taken in $\mathbf{L}^1$. Since Ξ' is decomposable, $\Xi' = S^1(X)$ for a random closed set X. Let us prove that $\Xi = S^\infty(X)$. For each $\xi \in S^\infty(X)$ choose a sequence $\{\xi_n, n \geq 1\} \subset \Xi$ such that $\|\xi_n - \xi\|_1 \to 0$ and $\xi_n \to \xi$ a.s. If $\alpha > \|\xi\|_\infty$, then $\eta_n = \mathbf{1}_{\xi_n < \alpha}\xi_n + \mathbf{1}_{\xi_n \geq \alpha}\xi_1$ converges to ξ in $\mathbf{L}^\infty$, whence $\xi \in \Xi$. □

The following theorem establishes a necessary and sufficient condition for the existence of integrable selections.

Theorem 1.7 (Existence of integrable selection). *For each $p \in [1, \infty]$, the family $S^p(X)$ is non-empty if and only if the random variable*

$$\alpha = \inf\{\|x\| : x \in X\} = \rho(0, X) \tag{1.2}$$

belongs to $\mathbf{L}^p(\Omega; \mathbb{R})$, i.e. $\mathbf{E}\alpha^p < \infty$ if $p \in [1, \infty)$ or α is essentially bounded if $p = \infty$.

Proof. Since the event $\{\alpha > r\} = \{X \cap B_r(0) = \emptyset\}$ is measurable by Theorem 1.2.3, α is a random variable. Then $Y = \{x : \|x\| = \alpha\}$ is a random closed set. By Theorem 1.2.25, $X \cap Y$ is a random closed set, which is almost surely non-empty by construction. Then there exists a selection ξ of $X \cap Y$ which is also a selection of X. Finally, $\xi \in \mathbf{L}^p$, since $\|\xi\| = \alpha \in \mathbf{L}^p(\Omega; \mathbb{R})$. □

It is often useful to construct selections measurable with respect to the *minimal σ-algebra* $\mathfrak{F}_X$ generated by X. It is easy to see that α defined by (1.2) is $\mathfrak{F}_X$-measurable. Then $Y = \{x : \|x\| = \alpha\}$ is $\mathfrak{F}_X$-measurable, so that $X \cap Y$ possesses at least one $\mathfrak{F}_X$-measurable selection. This leads to the following result.

Proposition 1.8 ($\mathfrak{F}_X$-measurable selections). *Let $p \in [1, \infty]$. If $S^p(X)$ is non-empty, then there exists an $\mathfrak{F}_X$-measurable selection $\xi \in S^p(X)$.*

Two identically distributed random sets may have different families of distributions of random selections. However, it is possible to extract identically distributed selections from identically distributed sets using their minimal generated σ-algebras (or the canonical representations of random sets). Although the following proposition is formulated in the Euclidean space, an appropriate generalisation for infinite-dimensional Banach spaces holds.

Proposition 1.9. *For each closed set $F \subset \mathbb{R}^d$ define F_0 to be the set of all $x \in F$ such that $\|x\| \leq \|y\|$ for all $y \in F$. Let $e(F)$ be the lexicographical minimum of F_0.*

150   2 Expectations of Random Sets

(i) If X a random set in $\mathbb{R}^d$, then $e(X)$ is $\mathfrak{F}_X$-measurable. If $\mathcal{S}^p(X) \neq \emptyset$ for $p \in [1, \infty]$, then $e(X) \in \mathcal{S}^p(X)$.
(ii) Let $\{X_n, n \geq 1\}$ be a sequence of i.i.d. random closed sets in $\mathbb{R}^d$. Then there exists a sequence $\{\xi_n, n \geq 1\}$ of i.i.d. random selections of the corresponding random closed sets. If $\mathcal{S}^p(X_1) \neq \emptyset$, then ξ_n can be chosen from $\mathcal{S}^p(X_n)$, $n \geq 1$.

Proof.
(i) is a direct implication of Proposition 1.8.
(ii) Define $\xi_n(\omega) = e(X_n(\omega))$ and use (i).   □

Representations of random closed sets through the families of their selections are helpful to characterise lower bounds for integral functionals. The probability space $(\Omega, \mathfrak{F}, \mathbf{P})$ is assumed to be complete. Let $\zeta_x(\omega)$ be a function defined for $x \in \mathbb{E}$ and $\omega \in \Omega$ with values in the extended real line $\bar{\mathbb{R}} = [-\infty, \infty]$. Assume that ζ is measurable with respect to $\mathfrak{F} \otimes \mathfrak{B}(\bar{R})$ being the product of $\mathfrak{F}$ and the Borel σ-algebra on $\bar{\mathbb{R}}$. Then ζ is a stochastic process on $\mathbb{E}$ with values in the extended real line. For a random $\mathbb{E}$-valued random element ξ consider $\mathbf{E}\zeta_\xi$, allowing for possible values $\pm\infty$. Note that additive functionals on $\mathbf{L}^p$ can be represented as $\mathbf{E}\zeta_\xi$, see Theorem 5.3.19. The following result can be proved using the Castaing representation of a random closed set, see Hiai and Umegaki [255, Th. 2.2].

Theorem 1.10 (Infimum for integral functionals). *Assume that the function $\zeta_x(\omega)$ defined for x from a separable Banach space $\mathbb{E}$ is upper semicontinuous for every $\omega \in \Omega$ or lower semicontinuous for every $\omega \in \Omega$.*
(i) *If X is a random closed set, then $\inf\{\zeta_x : x \in X\}$ is a random variable.*
(ii) *If, additionally, $\mathcal{S}^p(X) \neq \emptyset$, $\mathbf{E}\zeta_\xi$ is defined for all $\xi \in \mathcal{S}^p(X)$ and $\mathbf{E}\zeta_\eta < \infty$ for at least one $\eta \in \mathcal{S}^p(X)$, then*

$$\inf_{\xi \in \mathcal{S}^p(X)} \mathbf{E}\zeta_\xi = \mathbf{E} \inf_{x \in X} \zeta_x. \tag{1.3}$$

Theorem 1.10(i) strengthens the first assertion of Theorem 1.2.27. For example, if $\zeta_x(\omega) = \|x\|$, then (1.3) yields

$$\inf_{\xi \in \mathcal{S}^1(X)} \mathbf{E}\|\xi\| = \mathbf{E} \inf_{x \in X} \|x\|.$$

1.2 The selection expectation

Integrable random sets

Let X be a random closed set in a separable Banach space $\mathbb{E}$.

Definition 1.11 (Integrable random sets).
(i) A random closed set X is called *integrably bounded* if

$$\|X\| = \sup\{\|x\| : x \in X\}$$

has a finite expectation.

(ii) A random closed set is called *integrable* if $\mathcal{S}^1(X) \neq \emptyset$.

Example 1.1.3(i) demonstrates that $\|X\|$ is a random variable. Clearly, an integrably bounded random closed set X is integrable. Moreover, since $\|\xi\| \leq \|X\|$, all selections of X are integrable. If $\mathbb{E}$ is locally compact, the integrability of $\|X\|$ implies that X is a random compact set, while if $\mathbb{E}$ is infinite-dimensional, then an integrably bounded random closed set is almost surely bounded but not necessarily compact. Furthermore, integrable random closed sets may be unbounded and so are not necessarily integrably bounded.

Definition 1.12 (Selection expectation). The *selection expectation* of X is the closure of the set of all expectations of integrable selections, i.e.

$$\mathbf{E} X = \mathrm{cl}\{\mathbf{E}\xi : \xi \in \mathcal{S}^1(X)\}. \tag{1.4}$$

The selection expectation of X will be sometimes denoted by $\mathbf{E}_A X$, where confusion with other definitions of expectations (introduced later) may occur. It is also often called the *Aumann* expectation. If $A \subset \Omega$ is measurable, define

$$\mathbf{E}(\mathbf{1}_A X) = \mathrm{cl}\{\mathbf{E}(\mathbf{1}_A \xi) : \xi \in \mathcal{S}^1(X)\}.$$

Definition 1.13 (Aumann integral). The *Aumann integral* of X is defined as

$$\mathbf{E}_\mathrm{I} X = \{\mathbf{E}\xi : \xi \in \mathcal{S}^1(X)\}. \tag{1.5}$$

Theorem 1.24 provides some conditions which ensure that $\mathbf{E}_\mathrm{I} X = \mathbf{E} X$, while, in general, $\mathbf{E}_\mathrm{I} X$ is not always closed.

Example 1.14 (Selections of deterministic set). The family of all selections depends on the structure of the underlying probability space. For instance, consider a random set X which equals $\{0, 1\}$ almost surely. Then all selections of X can be obtained as

$$\xi(\omega) = \begin{cases} 0, & \omega \in \Omega_1, \\ 1, & \omega \in \Omega_2, \end{cases}$$

for all measurable partitions $\Omega = \Omega_1 \cup \Omega_2$ of Ω into two disjoint sets. Since $\mathbf{E}\xi = \mathbf{P}(\Omega_2)$, the range of expectations of selections depends on the atomic structure of the underlying probability space. If the probability space has no atoms, then the possible values for $\mathbf{P}(\Omega_2)$ fill in the whole segment $[0, 1]$.

Example 1.14 explains that two identically distributed random sets may have different selection expectations. Let $\Omega' = \{\omega\}$ be a single-point probability space and $\Omega'' = [0, 1]$ with its Borel σ-algebra. Define two random sets on these spaces: $X_1(\omega) = \{0, 1\}$ for $\omega \in \Omega'$, and $X_2(\omega) = \{0, 1\}$ for all $\omega \in \Omega''$. Then X_1 and X_2 have the same distribution (in fact, both sets are deterministic), but $\mathbf{E} X_1 = \{0, 1\}$, while $\mathbf{E} X_2 = [0, 1]$. The main difference between Ω' and Ω'' is that Ω'' is not atomic. Note that Ω' provides the so-called reduced representation of X_1, since the σ-algebra on Ω' coincides with the minimal σ-algebra generated by X_1.

Convexification

The following theorem establishes an important fact that on non-atomic probability spaces the expectation $\mathbf{E}X$ is convex and coincides with the expectation of the convex hull of X. It is instructive to give two formulations and proofs of this result: one for integrably bounded random closed sets in the Euclidean space $\mathbb{E} = \mathbb{R}^d$ and the other for random closed sets in a separable Banach space.

Theorem 1.15 (Convexification in $\mathbb{R}^d$). *Let $\mathbb{E} = \mathbb{R}^d$. If the basic probability space $(\Omega, \mathfrak{F}, \mathbf{P})$ contains no atoms and X is an integrably bounded random compact set, then $\mathbf{E}X$ and $\mathbf{E}_\mathrm{I}X$ are convex, $\mathbf{E}_\mathrm{I}X = \mathbf{E}_\mathrm{I}\mathrm{co}(X)$ and $\mathbf{E}X = \mathbf{E}\,\mathrm{co}(X)$.*

Proof. The proof is based on Lyapunov's theorem for vector measures. Lyapunov's theorem states that if $\mu_1, \ldots, \mu_d$ are finite non-atomic measures on Ω, then the range of the vector measure $\mu = (\mu_1, \ldots, \mu_d)$ is a convex closed set in $\mathbb{R}^d$. Without loss of generality all μ_i can be considered probability measures. In probabilistic terms, Lyapunov's theorem is equivalent to the statement that, for each random variable ξ with $0 \leq \xi \leq 1$ a.s., there exists a measurable set Ω' such that

$$\int_\Omega \xi(\omega)\mu(d\omega) = \mu(\Omega') = (\mu_1(\Omega'), \ldots, \mu_d(\Omega')). \tag{1.6}$$

From (1.6) we get

$$\int_\Omega \xi(\omega)\eta(\omega)\mu(d\omega) = \int_{\Omega'} \eta(\omega)\mu(d\omega) \tag{1.7}$$

first for an indicator random variable η and then for all integrable random variables.

Returning to Theorem 1.15, note that the convexity of $\mathbf{E}X$ and $\mathbf{E}_\mathrm{I}X$ would follow from the equality $\mathbf{E}_\mathrm{I}X = \mathbf{E}_\mathrm{I}\mathrm{co}(X)$. To see this, it suffices to note that the closure of a convex set is convex and to apply the second part of Theorem 1.15 to the random set $\{\xi_1, \xi_2\}$ for two arbitrary integrable selections ξ_1 and ξ_2 of X. Below we will prove that $\mathbf{E}_\mathrm{I}X = \mathbf{E}_\mathrm{I}\mathrm{co}(X)$.

First, prove that every selection ζ of $\mathrm{co}(X)$ can be represented as

$$\zeta = \sum_{i=1}^k \alpha_i \xi_i, \tag{1.8}$$

where $k \leq d+1$, α_i are non-negative random variables such that $\sum \alpha_i = 1$ a.s. and ξ_i are selections of X, $i = 1, \ldots, k$. Define

$$S = \{\alpha = (\alpha_1, \ldots, \alpha_{d+1}) \in \mathbb{R}^{d+1} : \alpha_i \geq 0, \sum \alpha_i = 1\},$$

$$g(\alpha, x_1, \ldots, x_{d+1}) = \sum_{i=1}^{d+1} \alpha_i x_i.$$

Then

$$Y = \{(\alpha, x_1, \ldots, x_{d+1}) \in S \times X \times \cdots \times X : g(\alpha, x_1, \ldots, x_{d+1}) = \zeta\}$$

is a non-empty random closed set. Thus, there exists its selection $(\alpha, \xi_1, \ldots, \xi_d)$ with $\alpha = (\alpha_1, \ldots, \alpha_{d+1})$ which satisfies (1.8).

Second, prove Theorem 1.15 for a finite random set $X = \{\xi_1, \ldots, \xi_m\}$. For this, we use induction with respect to m. The first step (for $m = 1$) is evident. Assume that the statement holds for $m = k - 1$ and prove it for $m = k$. If $Y = \{\xi_2, \ldots, \xi_k\}$, then $\mathrm{co}(X) = \mathrm{co}(Y \cup \{\xi_1\})$, so that each measurable selection of $\mathrm{co}(X)$ can be represented as

$$\xi = \alpha \xi_1 + (1 - \alpha)\eta = \eta + \alpha(\xi_1 - \eta),$$

where η is a selection of $\mathrm{co}(Y)$ and $0 \leq \alpha \leq 1$. If $x \in \mathbf{E}_I \mathrm{co}(X)$, then $x = \mathbf{E}\xi$ for a selection ξ. By (1.7), there exists a measurable set Ω' such that

$$\mathbf{E}\alpha(\xi_1 - \eta) = \mathbf{E}\mathbf{1}_{\Omega'}(\xi_1 - \eta).$$

Thus, $x = \mathbf{E}\mathbf{1}_{\Omega'}\xi_1 + \mathbf{E}\mathbf{1}_{\Omega \setminus \Omega'}\eta$. Note that Y can be considered a random set on the probability space $\Omega \setminus \Omega'$. By the induction assumption, there exists a measurable selection ζ of Y such that $\mathbf{E}\mathbf{1}_{\Omega \setminus \Omega'}\eta = \mathbf{E}\mathbf{1}_{\Omega \setminus \Omega'}\zeta$. Then $\xi' = \xi_1 \mathbf{1}_{\Omega'} + \zeta \mathbf{1}_{\Omega \setminus \Omega'}$ is a measurable selection of X and $\mathbf{E}\xi' = x$.

Finally, we prove that $\mathbf{E}_I X = \mathbf{E}_I \mathrm{co}(X)$ for a general integrably bounded random closed set X. Let $x = \mathbf{E}\zeta$, where $\zeta \in \mathrm{co}(X)$ a.s. By (1.8), $\zeta = \sum_{i=1}^{k} \alpha_i \xi_i$. Consider the set $Y = \{\xi_1, \ldots, \xi_k\}$. Obviously, $\mathbf{E}_I Y \subset \mathbf{E}_I X$ and $\zeta \in \mathrm{co}(Y)$. But $\mathbf{E}_I Y = \mathbf{E}_I \mathrm{co}(Y)$, so that $x \in \mathbf{E}_I \mathrm{co}(Y) \subset \mathbf{E}_I X$. Thus, $\mathbf{E}_I \mathrm{co}(X) \subset \mathbf{E}_I X$. The reverse inclusion is obvious. $\square$

Theorem 1.16 (Convexification in Banach space). *Let $\mathbb{E}$ be a separable Banach space. If $(\Omega, \mathfrak{F}, \mathbf{P})$ has no atoms and X is an integrable random closed set, then $\mathbf{E}X$ is convex.*

Proof. It suffices to show that for any two integrable selections $\xi_1, \xi_2 \in \mathcal{S}^1(X)$, any $\varepsilon > 0$ and $\alpha \in [0, 1]$, there exists $\eta \in \mathcal{S}^1(X)$ such that

$$\|\alpha \mathbf{E}\xi_1 + (1 - \alpha)\mathbf{E}\xi_2 - \mathbf{E}\eta\| < \varepsilon. \tag{1.9}$$

Define an $\mathbb{E} \times \mathbb{E}$-valued measure λ by

$$\lambda(A) = (\mathbf{E}(\mathbf{1}_A \xi_1), \mathbf{E}(\mathbf{1}_A \xi_2)), \quad A \in \mathfrak{F}.$$

A generalisation of Lyapunov's theorem for vector-valued measures [567] implies that the closure of the range of λ is convex in $\mathbb{E} \times \mathbb{E}$. Since $\lambda(\emptyset) = (0, 0)$ and $\lambda(\Omega) = (\mathbf{E}\xi_1, \mathbf{E}\xi_2)$, there exists an $A \in \mathfrak{F}$ such that

$$\|\alpha \mathbf{E}\xi_i - \mathbf{E}(\mathbf{1}_A \xi_i)\| < \varepsilon/2, \quad i = 1, 2.$$

Taking $\eta = \mathbf{1}_A \xi_1 + \mathbf{1}_{\Omega \setminus A} \xi_2$ we get a selection of X which satisfies (1.9). $\square$

It is easy to see that Theorem 1.16 holds if the probability space has atoms, but X takes convex values on the atoms of Ω. The following theorem establishes further properties of the selection expectation. They do not require that the probability space is non-atomic.

Theorem 1.17 (Properties of selection expectation). *Let X and Y be two integrable random closed sets in a Banach space. Then*
(i) $\rho_H(\mathbf{E}X, \mathbf{E}Y) \leq \mathbf{E}\rho_H(X, Y)$, *where ρ_H is the Hausdorff metric (possibly infinite) generated by the norm on $\mathbb{E}$.*
(ii) $\mathbf{E}\overline{\mathrm{co}}\,(X + Y) = \overline{\mathrm{co}}\,(\mathbf{E}X + \mathbf{E}Y)$.
(iii) $\mathbf{E}\overline{\mathrm{co}}\,(X) = \overline{\mathrm{co}}\,(\mathbf{E}X)$. *If $(\Omega, \mathfrak{F}, \mathbf{P})$ contains no atom, then $\mathbf{E}\overline{\mathrm{co}}\,(X) = \mathbf{E}X$.*

Proof. Note that (i) follows from
$$\inf_{\eta \in \mathcal{S}^1(Y)} \|\mathbf{E}\xi - \mathbf{E}\eta\| \leq \inf_{\eta \in \mathcal{S}^1(Y)} \mathbf{E}\|\xi - \eta\| = \mathbf{E}\rho(\xi, Y)$$
$$\leq \mathbf{E}\rho_H(X, Y),$$
where the equality is provided by Theorem 1.10. For (ii) and (iii) similar arguments furnish the proofs. The final part of (iii) refers to Theorem 1.16. □

Theorem 1.17(iii) can be strengthened to $\mathrm{co}(\mathbf{E}_I X) = \mathbf{E}_I \mathrm{co}(X)$, which holds if X is an integrable random closed set in $\mathbb{R}^d$ and $\mathrm{co}(\mathbf{E}_I X)$ does not contain a line, see Artstein and Wets [24].

Properties of selections

The following proposition characterises selections and distributions of random closed sets in terms of their selection expectations. Recall that a bounded random closed set X is said to be Hausdorff approximable if X is an almost sure limit (in the Hausdorff metric) of a sequence of simple random closed sets, see Section 1.2.1.

Proposition 1.18 (Distributions of random sets in terms of their selection expectations). *Let X and Y be Hausdorff approximable integrably bounded random convex closed sets.*
(i) *A random element ξ is an integrable selection of X if and only if $\mathbf{E}(1_A \xi) \in \mathbf{E}(1_A X)$ for all $A \in \mathfrak{F}$. If the dual space $\mathbb{E}^*$ is separable, then the statement holds for each integrably bounded random closed set X without assuming that X is Hausdorff approximable.*
(ii) $X = Y$ *a.s. if and only if* $\mathbf{E}(1_A X) = \mathbf{E}(1_A Y)$ *for all $A \in \mathfrak{F}$.*

Proof. Since X is Hausdorff approximable, it is possible to choose a null-set Ω' (i.e. $\mathbf{P}(\Omega') = 0$) and a sequence $\{F_i, i \geq 1\} \subset \mathrm{co}\,\mathcal{F}$ such that $\{F_i, i \geq 1\}$ is dense (in the Hausdorff metric) in the set $\{X(\omega) : \omega \in \Omega \setminus \Omega'\}$ of values of X. For each i, let $\{y_{ij}, j \geq 1\}$ be a countable dense subset of $\mathbb{E} \setminus F_i$. For $i, j \geq 1$, by the separation theorem (see Dunford and Schwartz [148, Th. V.1.12]), there exists an element $u_{ij} \in \mathbb{E}^*$ with $\|u_{ij}\| = 1$ such that

$$\langle y_{ij}, u_{ij}\rangle \geq \sup_{y\in F_i} \langle y, u_{ij}\rangle.$$

Let $U=\{u_{ij}, i, j \geq 1\}$. Then for each $\omega \in \Omega \setminus \Omega'$ we have $x \in X(\omega)$ if and only if

$$\langle x, u\rangle \leq \sup_{y\in X(\omega)} \langle y, u\rangle, \quad u \in U.$$

Assume $\xi \notin \mathcal{S}^1(X)$. Then there exists a set $A \in \mathfrak{F}$ of positive probability such that

$$\langle \xi(\omega), u\rangle > \sup_{y\in X(\omega)} \langle y, u\rangle, \quad \omega \in A,$$

for some $u \in U$. By Theorem 1.10,

$$\langle \mathbf{E}\mathbf{1}_A\xi, u\rangle = \mathbf{E}\left(\mathbf{1}_A\langle \xi, u\rangle\right) > \mathbf{E}(\mathbf{1}_A \sup_{y\in X(\omega)} \langle y, u\rangle)$$

$$= \sup_{\eta\in\mathcal{S}^1(X)} \mathbf{E}(\mathbf{1}_A\langle \eta, u\rangle) = \sup_{x\in\mathbf{E}(\mathbf{1}_AX)} \langle x, u\rangle,$$

whence $\mathbf{E}(\mathbf{1}_A\xi) \notin \mathbf{E}(\mathbf{1}_AX)$. If $\mathbb{E}^*$ is separable, then the above proof is applicable with U being a countable sense subset of $\mathbb{E}^*$. The second statement easily follows from (i). □

The following two theorems deal with the *weak compactness* of the set of integrable selections of an integrably bounded random closed set.

Theorem 1.19 (Weak compactness of integrable selections). *Let X be a bounded random closed set in a reflexive Banach separable space $\mathbb{E}$. Then the following conditions are equivalent.*
  (i) *X is an integrably bounded random convex closed set.*
  (ii) *$\mathcal{S}^1(X)$ is a non-empty bounded convex subset of $\mathbf{L}^1$.*
  (iii) *$\mathcal{S}^1(X)$ is non-empty weakly compact convex in $\mathbf{L}^1$.*

Proof. Conditions (i) and (ii) are equivalent by Proposition 1.5; (ii) trivially follows from (iii). The implication (i)⇒(iii) follows from the known fact (cf. Dunford and Schwartz [148, Th. IV.8.9]) saying that $\Xi \subset \mathbf{L}^1$ is relatively weakly compact if Ξ is bounded and the countable additivity of the integrals $\int_A \|f(\omega)\|\mathbf{P}(d\omega)$ is uniform with respect to $f \in \Xi$. □

It is shown in Papageorgiou [442] that $\mathcal{S}^1(X)$ is a non-empty weakly compact convex set if X is an integrably bounded weakly compact convex random set. The following theorem is proved by Byrne [85].

Theorem 1.20 (Weak compactness for families of selections). *Let X be a Hausdorff approximable integrably bounded random set with almost surely convex weakly compact values. If $\{X_n, n \geq 1\}$ is a sequence of simple random sets with convex weakly compact values such that $X_n \to X$ a.s., then $\mathcal{S}^1(X) \cup (\cup_{n\geq 1}\mathcal{S}^1(X_n))$ is weakly compact in $\mathbf{L}^1$.*

Debreu expectation

According to the *Rådström embedding theorem* (see Rådström [472]) co $\mathcal{K}$ can be embedded as a convex cone in a real Banach space $\mathbb{Y}$ in such a way that the embedding is isometric; addition in $\mathbb{Y}$ induces addition in co $\mathcal{K}$; multiplication by nonnegative real numbers in $\mathbb{Y}$ induces the corresponding operation in co $\mathcal{K}$. Then an integrably bounded random convex compact set X corresponds to a random element in $\mathbb{Y}$ which is Bochner integrable. Its Bochner expectation (in the space $\mathbb{Y}$) can be identified with an element of co $\mathcal{K}$ which is called the Debreu expectation of X and denoted by $\mathbf{E}_\mathrm{B} X$ (the subscript "B" stands for the Bochner expectation in $\mathbb{Y}$ used in this construction). A similar construction is applicable if X is a bounded random convex closed set in a reflexive space $\mathbb{E}$.

If X is Hausdorff approximable, then this construction corresponds to the expectation defined as a limit of expectations of simple random sets X_n such that $X_n \to X$ a.s. If X_n is a simple random convex closed set which takes values $F_1, \ldots, F_k$ with probabilities $p_1, \ldots, p_k$, then the Debreu expectation of X_n is given by the weighted Minkowski sum

$$\mathbf{E}_\mathrm{B} X_n = \mathrm{cl}(p_1 F_1 + \cdots + p_k F_k),$$

and $\mathbf{E}_\mathrm{B} X$ is the limit of $\mathbf{E}_\mathrm{B} X_n$ in the Hausdorff metric.

Theorem 1.21 (Debreu and selection expectations). *Let X be an integrably bounded random closed set. The Debreu expectation of X coincides with the selection expectation of X if one of the following assumptions holds:*
(i) $X \in \mathrm{co}\,\mathcal{K}$ a.s.;
(ii) $\mathbb{E}$ is reflexive, X is Hausdorff approximable and a.s. convex;
(iii) X is Hausdorff approximable and a.s. weakly compact convex.

Proof.
(i) Since the image of co $\mathcal{K}$ is a closed convex cone in $\mathbb{Y}$, $\mathbf{E}_\mathrm{B} X \in \mathrm{co}\,\mathcal{K}$. If X is simple, then by writing down all selections of X, it is easily seen that $\mathbf{E} X = \mathbf{E}_\mathrm{B} X$. Note that each random convex compact set X is Hausdorff approximable, see Theorem 1.2.11. Take a sequence $\{X_n, n \geq 1\}$ of simple random convex compact sets such that $\mathbf{E} \rho_\mathrm{H}(X_n, X) \to 0$ as $n \to \infty$. By Theorem 1.17(i), $\rho_\mathrm{H}(\mathbf{E} X_n, \mathbf{E} X) \to 0$. The definition of $\mathbf{E}_\mathrm{B} X$ implies $\rho_\mathrm{H}(\mathbf{E}_\mathrm{B} X_n, \mathbf{E}_\mathrm{B} X) \to 0$, whence $\mathbf{E} X = \mathbf{E}_\mathrm{B} X$. The same arguments are applicable to prove (ii).
(iii) Let $z \in \mathbf{E}_\mathrm{B} X$. Then there are $z_n \in \mathbf{E}_\mathrm{B} X_n = \mathbf{E} X_n$, $n \geq 1$, such that $z_n \to z$. Note that $z_n = \mathbf{E} \xi_n$ with $\xi_n \in \mathcal{S}^1(X_n)$. By Theorem 1.20, we may assume that ξ_n converges weakly to $\xi \in \mathbf{L}^1$, whence $\mathbf{E} \xi_n \to \mathbf{E} \xi$. It is easy to see that ξ is a selection of X, whence $\mathbf{E}_\mathrm{B} X \subset \mathbf{E} X$. In the other direction, for each $\xi \in \mathcal{S}^1(X)$ there is a sequence $\{\xi_n, n \geq 1\}$ such that $\xi_n \in \mathcal{S}^1(X_n)$ and $\xi_n \to \xi$ a.s., which implies $\mathbf{E} \xi \in \mathbf{E}_\mathrm{B} X$ and $\mathbf{E} X \subset \mathbf{E}_\mathrm{B} X$. □

Theorem 1.21(i) implies that $\mathbf{E} X \in \mathrm{co}\,\mathcal{K}$ if X is an integrably bounded random convex compact set.

Selection expectation and support function

An embedding of the family of bounded convex closed sets into a Banach space provided by the *Hörmander embedding theorem* (see Hörmander [264]) is realised by support functions of random sets, see (F.1). Each random convex closed set X corresponds uniquely to its support function $h(X, u)$ which is a random function on the dual space $\mathbb{E}^*$. Note that $h(X, u) \leq \|X\| \|u\|$ for all $u \in \mathbb{E}^*$. If $\mathbf{E}\|X\| < \infty$, then $\mathbf{E}h(X, u)$ is finite for all $u \in \mathbb{E}^*$. Support functions are characterised by their *sublinear* property (homogeneity and subadditivity) which is kept after taking expectation. Therefore, $\mathbf{E}h(X, u)$, $u \in \mathbb{E}^*$, is the support function of a convex set. As the following theorem states, this fact leads to an equivalent definition of the selection expectation if X is convex or if the probability space is non-atomic.

Theorem 1.22 (Selection expectation and support functions). *If X is an integrably bounded random set and $X \in \operatorname{co} \mathcal{F}$ a.s. or the probability space $(\Omega, \mathfrak{F}, \mathbf{P})$ is non-atomic, then the selection expectation of X is the unique convex closed set $\mathbf{E}X$ satisfying*

$$\mathbf{E}h(X, u) = h(\mathbf{E}X, u) \tag{1.10}$$

for all u from the dual space $\mathbb{E}^*$.

Proof. In both cases $\mathbf{E}X$ is convex, so it suffices to consider only the case of a convex X and show that $\mathbf{E}X$ satisfies (1.10). The proof can be provided by applying Theorem 1.10 to $\zeta_x = -\langle x, u \rangle$. An alternative proof can be carried over as follows.

For each point $x \in \mathbf{E}X$, there exist selections $\xi_n \in \mathcal{S}^1(X)$, $n \geq 1$, such that $\mathbf{E}\xi_n \to x$ as $n \to \infty$. Thus,

$$h(\{x\}, u) = \lim_{n \to \infty} \langle \mathbf{E}\xi_n, u \rangle = \lim_{n \to \infty} \mathbf{E} \langle \xi_n, u \rangle \leq \mathbf{E}h(X, u), \quad u \in \mathbb{E}^*.$$

Hence $h(\mathbf{E}X, u) \leq \mathbf{E}h(X, u)$ for all $u \in \mathbb{E}^*$.

For each $u \in \mathbb{E}^*$ define

$$Y_\varepsilon = \{x \in \mathbb{E} : \langle x, u \rangle \geq h(X, u) - \varepsilon\}, \quad \varepsilon > 0.$$

Then $Y_\varepsilon \cap X$ is a non-empty random closed set, which possesses a selection ξ_ε. Since $h(\xi_\varepsilon, u) \geq h(X, u) - \varepsilon$, passing to expectations yields

$$\mathbf{E}h(X, u) - \varepsilon \leq h(\{\mathbf{E}\xi_\varepsilon\}, u) \leq h(\mathbf{E}X, u).$$

Letting $\varepsilon \downarrow 0$ finishes the proof. □

Aumann integral

The *Aumann integral* $\mathbf{E}_\mathrm{I} X$ is defined in (1.5) similarly to Definition 1.12, but without taking closure of the set of expectations of integrable selections. The following examples show that the set $\mathbf{E}_\mathrm{I} X$ is not always closed, so that $\mathbf{E}_\mathrm{I} X$ may constitute a proper subset of $\mathbf{E}X$.

Example 1.23 (Random sets with non-closed Aumann integrals).
(i) Let $\Omega = [0, 1]$ with the Lebesgue measure. Consider a random closed set $X(\omega) = \{0, \mathbf{1}_{[0,\omega]}(t)\}$ in the space of square integrable functions on $[0, 1]$. Then

$$\mathbf{E}_{\mathrm{I}} X = \left\{ \int_0^1 \eta(\omega) \mathbf{1}_{[0,\omega]}(t) \mathrm{d}\omega : \eta(\omega) = 0, 1 \right\} = \left\{ \int_t^1 \mathbf{1}_A(\omega) \mathrm{d}\omega : A \in \mathfrak{B}([0, 1]) \right\},$$

while

$$\mathbf{E} X = \mathbf{E} \overline{\mathrm{co}}\,(X) \supset \mathbf{E}_{\mathrm{I}} \overline{\mathrm{co}}\,(X) = \left\{ \int_t^1 \alpha(\omega) \mathrm{d}\omega : 0 \leq \alpha(\omega) \leq 1 \right\}, \qquad (1.11)$$

see Theorem 1.17(iii). Then $x(t) = (1-t)/2$ appears in the right-hand side of (1.11) if $\alpha(\omega) = 1/2$, while it cannot be obtained as $\int_t^1 \mathbf{1}_A(\omega) \mathrm{d}\omega = \mu(A \cap [t, 1])$ for all Borel sets A. Thus, x belongs to $\mathbf{E} X$, but $x \notin \mathbf{E}_{\mathrm{I}} X$.

(ii) Let $\mathbb{E}$ be a non-reflexive separable Banach space. Then there exist two disjoint sets $F_1, F_2 \in \mathrm{co}\,\mathcal{F}$ which cannot be separated by a hyperplane, so that $\inf\{\|x_1 - x_2\| : x_1 \in F_1, x_2 \in F_2\} = 0$. Let X be equal to F_1 with probability $1/2$ and to $\tilde{F}_2 = \{-x : x \in F_2\}$ otherwise. Then $\mathbf{E}_{\mathrm{I}} X = \{(x_1 - x_2)/2 : x_1 \in F_1, x_2 \in F_2\}$. Since F_1 and F_2 are disjoint, $0 \notin \mathbf{E}_{\mathrm{I}} X$, but $0 \in \mathbf{E} X = \mathrm{cl}(\mathbf{E}_{\mathrm{I}} X)$.

In the following we describe several particular cases when $\mathbf{E}_{\mathrm{I}} X$ is closed and, therefore, coincides with $\mathbf{E} X$. The Banach space $\mathbb{E}$ is said to have the *Radon–Nikodym property* if for each finite measure space $(\Omega, \mathfrak{F}, \mu)$ and each $\mathbb{E}$-valued measure λ on $\mathfrak{F}$ which is of bounded variation and absolutely continuous with respect to μ, there exists an integrable function f such that $\lambda(A) = \int_A f \mathrm{d}\mu$ for all $A \in \mathfrak{F}$. It is known that reflexive spaces have the Radon–Nikodym property.

Theorem 1.24 (Closedness of Aumann integral). *Let X be an integrably bounded random closed set. Then $\mathbf{E}_{\mathrm{I}} X$ is closed if one of the following conditions is satisfied.*
  (i) $\mathbb{E}$ *is a finite-dimensional space.*
  (ii) $\mathbb{E}$ *has the Radon–Nikodym property and $X \in \mathrm{co}\,\mathcal{K}$ a.s.*
  (iii) $\mathbb{E}$ *is reflexive and $X \in \mathrm{co}\,\mathcal{F}$ a.s.*
  (iv) X *is Hausdorff approximable with a.s. bounded weakly compact convex values.*

Proof.
(i) follows from Fatou's lemma in finite-dimensional spaces. Let $X_n = X$ for all $n \geq 1$. Theorem 1.37 implies

$$\mathbf{E}_{\mathrm{I}} X = \mathbf{E}_{\mathrm{I}} \limsup X_n \supset \limsup \mathbf{E}_{\mathrm{I}} X_n = \mathrm{cl}(\mathbf{E}_{\mathrm{I}} X) = \mathbf{E} X\,,$$

and hence $\mathbf{E}_{\mathrm{I}} X$ is closed.
(ii) Let $\{\xi_n, n \geq 1\} \subset \mathcal{S}^1(X)$ and $\|\mathbf{E}\xi_n - x\| \to 0$ for some $x \in \mathbb{E}$. Consider a countable algebra $\mathfrak{F}_0 \subset \mathfrak{F}$ such that X and all random elements $\{\xi_n, n \geq 1\}$ are measurable with respect to the minimal σ-field $\sigma(\mathfrak{F}_0)$ generated by $\mathfrak{F}_0$. For each

$A \in \mathcal{F}_0$, $\mathbf{E}(\mathbf{1}_A \xi_n) \in \mathbf{E}(\mathbf{1}_A X) = \mathbf{E}_\mathbf{B}(\mathbf{1}_A(X))$, the latter being a convex compact set. Thus, $\{\mathbf{E}(\mathbf{1}_A \xi_n), n \geq 1\}$ has a convergent subsequence. By the diagonal method, there exists a sequence $\eta_k = \xi_{n(k)}$, $k \geq 1$, such that $\mathbf{E}(\mathbf{1}_A \eta_n) \to \lambda(A)$ for every $A \in \mathfrak{F}_0$. Since $\|\eta_n\| \leq \|X\|$ a.s., it follows from [148, Lemma IV.8.8, Th. IV.10.6] that the limit $\lambda(A)$ exists for all $A \in \sigma(\mathfrak{F})$ and λ is an $\mathbb{E}$-valued measure which is of bounded variation and absolutely continuous with respect to $\mathbf{P}$. By the Radon–Nikodym property, there exists a $\sigma(\mathfrak{F}_0)$-measurable function $\eta \in \mathbf{L}^1$ such that $\lambda(A) = \int_A \eta d\mathbf{P} = \mathbf{E}(\mathbf{1}_A \eta)$ for all $A \in \sigma(\mathfrak{F}_0)$. Since $\mathbf{E}(\mathbf{1}_A \eta) \in \mathbf{E}(\mathbf{1}_A X)$, Proposition 1.18(i) yields $\eta \in \mathcal{S}^1(X)$, so that $x = \mathbf{E}\eta \in \mathbf{E}_\mathbf{I} X$.

(iii) follows from Theorem 1.19 and the weak continuity of the mapping $\xi \mapsto \mathbf{E}\xi$ from $\mathbf{L}^1$ to $\mathbb{E}$.

(iv) is an immediate corollary of Theorem 1.20. □

For further results along the same line, note that Fatou's lemma for random closed sets in Banach spaces (see Theorem 1.42) implies that $\mathbf{E}_\mathbf{I} X$ is closed in weak topology. If X is an integrably bounded weakly compact convex random set, then $\mathbf{E}_\mathbf{I} X$ is weakly compact and convex, see Klee [317]. Yannelis [619] showed that $\mathbf{E}_\mathbf{I} X = \mathbf{E} X$ if X is a random convex closed set such that $X \subset Y$ a.s. with Y being an integrably bounded weakly compact convex random set.

Open problem 1.25. Find conditions that guarantee $\mathbf{E} X = \mathbf{E}_\mathbf{I} X$ in a general Banach space $\mathbb{E}$ for possibly non-compact X.

Selection expectation in $\mathbb{R}^d$

Let us summarise several facts inherent to the case when X is an integrably bounded random compact set in the Euclidean space.

Theorem 1.26 (Selection expectation for random sets in $\mathbb{R}^d$). *If X is an integrably bounded random compact set in $\mathbb{R}^d$, then*

$$\mathbf{E} X = \{\mathbf{E}\xi : \xi \in \mathcal{S}^1(X)\}$$

is a compact set in $\mathbb{R}^d$. If the basic probability space is non-atomic, then $\mathbf{E} X$ is convex and coincides with $\mathbf{E}\operatorname{co}(X)$, and

$$\mathbf{E} h(X, u) = h(\mathbf{E} X, u), \quad u \in \mathbb{S}^{d-1}, \tag{1.12}$$

which identifies $\mathbf{E} X$ uniquely from the family of convex compact sets. If X is almost surely convex, then $\mathbf{E} X = \mathbf{E}_\mathbf{B} X$.

The *mean width* of a convex compact set is defined by (F.5). If X is a random convex compact set in $\mathbb{R}^d$, then (1.12) yields

$$\mathbf{E}\mathsf{b}(X) = \frac{2}{\omega_d} \int_{\mathbb{S}^{d-1}} (\mathbf{E} h(X, u)) \mathcal{H}^{d-1}(du) = \mathsf{b}(\mathbf{E} X).$$

The same relationship holds for the first intrinsic volume $V_1(X)$ related to the mean width by (F.8). Since in the plane the perimeter $U(X)$ is $2V_1(X)$, we obtain the following result.

Proposition 1.27 (Expected perimeter). *If X is an integrably bounded random compact set in the plane, then the perimeter of $\mathbf{E}X$ equals the expected perimeter of X, i.e. $\mathbf{E}U(X) = U(\mathbf{E}X)$. In a general dimension, the mean width of $\mathbf{E}X$ equals the expected mean width of X.*

1.3 Applications to characterisation of distributions

A famous characterisation result for *order statistics* due to Hoeffding [260] states that if $\alpha_1, \alpha_2, \ldots$ are i.i.d. integrable random variables, then the distribution of α_1 is uniquely determined by the sequence $\mathbf{E} \max(\alpha_1, \ldots, \alpha_n)$, $n \geq 1$. Below we describe a generalisation for random elements in separable Banach spaces.

Theorem 1.28 (A characterisation of multivariate distributions). *Let $\xi, \xi_1, \xi_2, \ldots$ be i.i.d. Bochner integrable random elements in a separable Banach space. Then the distribution of ξ is uniquely determined by the nested (increasing) sequence of convex compact sets $\mathbf{E} \operatorname{co}\{\xi_1, \ldots, \xi_n\}$, $n \geq 1$.*

Proof. Put $X_n = \operatorname{co}\{\xi_1, \ldots, \xi_n\}$. Then $\|X_n\| \leq \max(\|\xi_1\|, \ldots, \|\xi_n\|) \leq \sum \|\xi_i\|$, so that $\mathbf{E}\|X_n\| < \infty$ for all $n \geq 1$, which means that the X_n's are integrably bounded. Since $X_n \subset X_{n+1}$, $\{\mathbf{E}X_n, n \geq 1\}$ is an increasing (nested) sequence of convex compact sets, see Theorem 1.21(i). For each $u \in \mathbb{E}^*$,

$$h(\mathbf{E}X_n, u) = \mathbf{E}\max\{\langle \xi_1, u\rangle, \ldots, \langle \xi_n, u\rangle\}.$$

By the one-dimensional Hoeffding theorem, the sequence $\{\mathbf{E}X_n, n \geq 1\}$ uniquely determines the distribution of $\langle \xi, u\rangle$, which is the probability that ξ is contained in a half-space. These probabilities determine uniquely the distribution of ξ. □

Example 1.29 (Nested sequences of selection expectations).
(i) If ξ is a Gaussian random element in a Hilbert space $\mathbb{E}$, then

$$\mathbf{E}X_n = \gamma_n\{x \in \mathbb{E} : \langle x, u\rangle \leq \sqrt{\operatorname{Var}\langle \xi, u\rangle} \quad \text{for all } u \in \mathbb{E}^*\},$$

where $\gamma_n = \mathbf{E} \max(\alpha_1, \ldots, \alpha_n)$ with $\alpha_1, \ldots, \alpha_n$ being i.i.d. standard normal random variables. For instance, if ξ is a centred Gaussian vector in $\mathbb{R}^d$ with covariance matrix A, then $\mathbf{E}X_n = \gamma_n\{Ap : p \in \mathbb{R}^d, \langle p, Ap\rangle \leq 1\}$.
(ii) If $\mathbb{E}$ is the space of continuous functions on $[0, 1]$ and ξ is the Wiener process, then

$$\mathbf{E}X_n = \gamma_n \left\{ f \in \mathbb{E} : f(t) = \int_0^t y(s)ds, \int_0^1 y^2(s)ds \leq 1 \right\}.$$

The set $\mathbf{E}X_n/\gamma_n$ appears in connection to the Wiener process as the unit ball of a Hilbert space associated with an underlying measure and also in Strassen's law of the iterated logarithm.

Now consider a result of another kind, which concerns the case when the selection expectation of X is degenerated.

Proposition 1.30. *Assume that $\mathbb{E}^*$ is separable.*
  (i) *If X is integrable and $\mathbf{E}X = \{x\}$ for some $x \in \mathbb{E}$, then X is a singleton almost surely.*
  (ii) *If X and Y are weakly compact convex random sets, $X \subset Y$ a.s. and $\mathbf{E}X = \mathbf{E}Y$, then $X = Y$ a.s.*

Proof.
  (i) Take $\xi \in \mathcal{S}^1(X)$ and define $Y = X - \xi$. Then $\mathbf{E}Y = \{0\}$ and $0 \in Y$ almost surely. By Theorem 1.17(iii), $\mathbf{E}\overline{\operatorname{co}}(Y) = \{0\}$. Theorem 1.22 implies that $\mathbf{E}h(Y, u) = 0$ for all $u \in \mathbb{E}^*$. Since $h(Y, u)$ is non-negative almost surely, this implies $h(Y, u) = 0$ a.s., so that $Y = \{0\}$ a.s.
  (ii) is proved similarly. □

1.4 Variants of the selection expectation

Reduced selection expectation

The convexifying property of the selection expectation is largely determined by the richness of the σ-algebra $\mathfrak{F}$ on the space of elementary events Ω. Let $\mathcal{S}_{\mathfrak{H}}^1(X)$ be the family of integrable selections of X which are measurable with respect to a sub-σ-algebra $\mathfrak{H}$ of $\mathfrak{F}$. Define

$$\mathbf{E}^{\mathfrak{H}} X = \operatorname{cl}\{\mathbf{E}\xi : \xi \in \mathcal{S}_{\mathfrak{H}}^1(X)\}. \tag{1.13}$$

For example, if $\mathfrak{H} = \{\emptyset, \Omega\}$ is the trivial σ-algebra, then $\mathbf{E}^{\mathfrak{H}} X = \{x : x \in X \text{ a.s.}\}$ is the set of fixed points of X. This shows that $\mathcal{S}_{\mathfrak{H}}^1(X)$ can be empty even if X is integrably bounded.

A canonical sub-σ-algebra $\mathfrak{F}_X$ of $\mathfrak{F}$ is generated by the random closed set X itself. Considering $\mathbf{E}^{\mathfrak{F}_X} X$ reduces the convexifying effect of the selection expectation if X takes a finite or a countable number of possible values. This is equivalent to redefining X as a random closed set $X^{\natural}$ on the probability space being the space of sets itself. If X is an integrable random closed set, then $\mathcal{S}_{\mathfrak{F}_X}^1(X)$ is not empty by Proposition 1.8.

Definition 1.31 (Reduced selection expectation). *Let X be an integrable random closed set. The reduced selection expectation of X is defined by $\mathbf{E} X^{\natural} = \mathbf{E}^{\mathfrak{F}_X} X$.*

Since $\mathcal{S}_{\mathfrak{F}_X}^1(X) \subset \mathcal{S}^1(X)$,

$$\mathbf{E} X^{\natural} \subset \mathbf{E} X. \tag{1.14}$$

Therefore, $\mathbf{E} X^{\natural}$ is the intersection of $\mathbf{E} Y$ for all random closed sets Y sharing the distribution with X. If X is a simple random closed set, which takes a finite number of values $F_1, \ldots, F_n$ with the corresponding probabilities $p_1, \ldots, p_n$, then $\mathfrak{F}_X$ consists of a finite number of events and

$$\mathbf{E}X^\sharp = \mathrm{cl}(p_1 F_1 + p_2 F_2 + \cdots + p_n F_n), \tag{1.15}$$

while if the basic probability space is non-atomic, then

$$\mathbf{E}X = \overline{\mathrm{co}}\,(p_1 F_1 + p_2 F_2 + \cdots + p_n F_n).$$

Note that (1.15) also holds if X takes a countable family of values with the finite sum replaced by the sum of series. The inclusion in (1.14) can be strict if X takes non-convex values on at least one atom of Ω.

Theorem 1.32. *Let $\mathbb{E}$ be a separable Banach space.*
(i) *For each integrable random closed set X, $\overline{\mathrm{co}}\,\mathbf{E}X = \overline{\mathrm{co}}\,\mathbf{E}X^\sharp$.*
(ii) *Let X and Y be identically distributed integrable random closed sets. For each integrable $\mathfrak{F}_X$-measurable selection ξ of X there exists an integrable $\mathfrak{F}_Y$-measurable selection η of Y such that ξ and η are identically distributed.*
(iii) *If X and Y are identically distributed and integrable, then $\mathbf{E}X^\sharp = \mathbf{E}Y^\sharp$.*
(iv) *If $\mathcal{S}^1_{\mathfrak{F}_X}(X) = \mathcal{S}^1_{\mathfrak{F}_Y}(Y) \neq \emptyset$, then X and Y are identically distributed.*

Proof.
(i) Since $\overline{\mathrm{co}}\,X$ is $\mathfrak{F}_X$-measurable (see Section 1.2.1) we have (using the notation of Section 1.6)

$$\mathcal{S}_{\mathfrak{F}_X}(\overline{\mathrm{co}}\,X) = \{\mathbf{E}(\xi|\mathfrak{F}_X) : \xi \in \mathcal{S}^1(\overline{\mathrm{co}}\,X)\}.$$

Proposition 1.5 implies $\mathcal{S}^1(\overline{\mathrm{co}}\,X) = \overline{\mathrm{co}}\,\mathcal{S}^1(X)$ and $\mathcal{S}^1_{\mathfrak{F}_X}(\overline{\mathrm{co}}\,X) = \overline{\mathrm{co}}\,\mathcal{S}^1_{\mathfrak{F}_X}(X)$. Hence

$$\overline{\mathrm{co}}\,\mathbf{E}X = \mathbf{E}\overline{\mathrm{co}}\,X = \mathrm{cl}\{\mathbf{E}(\mathbf{E}\xi|\mathfrak{F}_X)) : \xi \in \mathcal{S}^1(\overline{\mathrm{co}}\,X)\}$$
$$= \mathrm{cl}\{\mathbf{E}(\xi) : \xi \in \mathcal{S}^1_{\mathfrak{F}_X}(\overline{\mathrm{co}}\,X)\} = \overline{\mathrm{co}}\,\mathbf{E}X^\sharp.$$

(ii) Since $\mathbb{E}$ is separable and ξ is $\mathfrak{F}_X$ measurable, there exists a $(\mathfrak{B}(\mathcal{F}), \mathfrak{B}(\mathbb{E}))$-measurable function $\Phi \colon \mathcal{F} \mapsto \mathbb{E}$ satisfying $\xi(\omega) = \Phi(X(\omega))$ for every $\omega \in \Omega$. If $\eta = \Phi(Y(\omega))$, then η and ξ are identically distributed. Furthermore,

$$\mathbf{E}\|\eta\| = \mathbf{E}\|\Phi(Y)\| = \mathbf{E}\|\Phi(X)\| = \mathbf{E}\|\xi\| < \infty.$$

Because the function $(x, F) \mapsto \rho(x, F) = \inf\{\|x - y\| : y \in F\}$ is $\mathfrak{B}(\mathbb{E}) \otimes \mathfrak{B}(\mathcal{F})$-measurable, $\rho(\xi, X)$ and $\rho(\eta, Y)$ are identically distributed, whence $\rho(\eta, Y) = 0$ almost surely. Thus, η is a selection of Y.
(iii) immediately follows from (ii).
(iv) Assume that $\mathbf{P}\{X \cap G \neq \emptyset\} > \mathbf{P}\{Y \cap G \neq \emptyset\}$ for an open set $G \in \mathcal{G}$. Following the proof of Theorem 1.2.28 is is possible to see that G may be assumed to be bounded. Consider a Castaing representation $\{\xi_n, n \geq 1\}$ of X whose members are $\mathfrak{F}_X$-measurable. Define events $A_n = \{\xi_n \in G\}, n \geq 1$, and further events $B_1 = A_1$, $B_n = A_n \setminus (\cup_{j<n} A_j), n \geq 2$. Fix $\xi_0 \in \mathcal{S}^1_{\mathfrak{F}_X}(X)$ and let

$$\xi = \sum_{n \geq 1} \mathbf{1}_{B_n} \xi_n + \mathbf{1}_A \xi_0,$$

where A is the complement to $\cup_{n\geq 1} A_n$. Then ξ is an $\mathfrak{F}_X$-measurable integrable selection of X and $\xi \in G$ whenever $X \cap G \neq \emptyset$. For any $\eta \in \mathcal{S}^1_{\mathfrak{F}_Y}(Y)$ we then have

$$\mathbf{P}\{\eta \in G\} \leq \mathbf{P}\{Y \cap G \neq \emptyset\} < \mathbf{P}\{X \cap G \neq \emptyset\} = \mathbf{P}\{\xi \in G\},$$

which shows that no such η shares the distribution with ξ. □

Translative expectation

Random translations of X affect the shape of the reduced selection expectation. For example, if $X = \{0, 1\}$ is a deterministic subset of the real line and $Y = X + \xi$ for a random variable ξ uniformly distributed on $[-1, 1]$, then $\mathbf{E}X^\natural = \{0, 1\}$, whereas $\mathbf{E}Y^\natural = [0, 1] \neq \mathbf{E}X^\natural + \mathbf{E}\xi = \mathbf{E}X^\natural$. This happens because adding ξ makes the relevant canonical σ-algebra $\mathfrak{F}_Y$ non-atomic. A possible way to eliminate such dependence is to consider the "smallest" possible expectation for all translations of X.

Definition 1.33 (Translative expectation). For an integrable random compact set X in $\mathbb{R}^d$, its translative expectation is defined by

$$\mathbf{E}_T A = \bigcap_{\mathbf{E}\|\xi\|<\infty} \left(\mathbf{E}(X - \xi)^\natural + \mathbf{E}\xi \right), \quad (1.16)$$

where the intersection is taken over all integrable random vectors ξ.

The following result shows that the intersection in the right-hand side of (1.16) can be effectively computed using the special choice of translation given by its Steiner point, $\mathbf{s}(X)$, see Appendix F. This is not possible in infinite-dimensional spaces because an analogue of the Steiner point does not exist there, see Giné and Hahn [200].

Theorem 1.34. *Let X be an integrable random compact set in $\mathbb{R}^d$. Then*

$$\mathbf{E}_T X = \mathbf{E}(X - \mathbf{s}(X))^\natural + \mathbf{E}\mathbf{s}(X).$$

Proof. First, note that $\mathbf{E}\mathbf{s}(X) = \mathbf{s}(\mathbf{E}X)$ by definition of the Steiner point (F.6). If X and X' are two identically distributed integrable random sets, then $\mathbf{s}(X)$ and $\mathbf{s}(X')$ are identically distributed, whence $\mathbf{s}(\mathbf{E}X) = \mathbf{s}(\mathbf{E}X')$. Therefore, $\mathbf{s}(\mathbf{E}X^\natural) = \mathbf{s}(\mathbf{E}X)$.
Define $Y = X - \mathbf{s}(X)$ and $\eta = \mathbf{s}(X) - \xi$. It suffices to show that

$$\mathbf{E}Y^\natural + \mathbf{E}\eta \subset \mathbf{E}(Y + \eta)^\natural \quad (1.17)$$

for each random compact set Y with $\mathbf{s}(Y) = 0$ a.s. and any integrable random vector η. If Y has no atoms, then the selection expectation of Y and $(Y + \eta)$ and their canonical representations coincide with the selection expectations of the corresponding convex hulls, whence (1.17) holds with the exact equality.
If $Y = K$ with probability 1 and η is atomic (takes values $z_1, z_2, \ldots$ with probabilities $p_1, p_2, \ldots$), then (1.17) follows from the fact that

$$K \subset p_1 K + p_2 K + \cdots.$$

For the non-atomic part of η one can write

$$\mathbf{E}(Y + \eta)^\natural = \mathbf{E}\,\mathrm{co}(Y + \eta) = \mathbf{E}_A Y + \mathbf{E}\eta,$$

whence (1.17) holds. The general cases of Y having an atomic or mixed distribution are handled similarly. □

Multivalued measures

The selection expectation of an integrably bounded random closed set X gives rise to a *multivalued* (set-valued) measure on $\mathfrak{F}$. For each $A \in \mathfrak{F}$ put

$$M(A) = \mathbf{E}(\mathbf{1}_A X).$$

Then M is a multivalued measure on Ω, which means that $M(\emptyset) = \{0\}$ and, for each disjoint sequence of sets $\{A_n, n \geq 1\} \subset \mathfrak{F}$,

$$\mathrm{cl}\, M(\cup_{n \geq 1} A_n) = \mathrm{cl} \sum_{n \geq 1} M(A_n),$$

where the latter sum is understood as the set of $x = \sum_{n=1}^{\infty} x_n$ given by the absolutely convergent sum of $x_n \in M(A_n), n \geq 1$. The *variation* of M is defined as

$$|M|(A) = \sup \sum_{i=1}^{n} \|M(A_i)\|,$$

where the supremum is taken over all measurable partitions $A_1, \ldots, A_n$ of $A \in \mathfrak{F}$. We say that M has a bounded variation if $|M|(\Omega) < \infty$. The multivalued measure M is called *absolutely continuous* with respect to $\mathbf{P}$ if $\mathbf{P}(A) = 0$ for any $A \in \mathfrak{F}$ implies $M(A) = \{0\}$.

Theorem 1.35 (Convexity and representation of multivalued measures). *Let $M: \mathfrak{F} \mapsto \mathcal{F}$ be a multivalued measure on the space $\mathbb{E}$ having the Radon–Nikodym property.*
  (i) *If $|M|$ is non-atomic, then $M(A)$ is convex for every $A \in \mathfrak{F}$.*
  (ii) *If M is absolutely continuous with respect to $\mathbf{P}$, then there exists an integrably bounded random closed set X such that $M(A) = \mathbf{E}(\mathbf{1}_A X)$ for all $A \in \mathfrak{F}$.*

Proof. For (i) we refer to Godet-Thobie [205]. To prove (ii), consider all measures μ such that $\mu(A) \in M(A)$ for all $A \in \mathfrak{F}$ (then μ is called a *selection* of M). Then show that $\mu(A) = \mathbf{E}(\mathbf{1}_A \xi)$, $A \in \mathfrak{F}$, for some $\xi \in \mathbf{L}^1$, whence the set of the corresponding ξ's is decomposable, so that the statement follows from Theorem 1.7. □

An *integral* of a real-valued function f with respect to a multivalued measure M is defined as the set of integrals of f with respect to all measures μ being selections of M. For multivalued measures with convex compact values, Hiai [254] and Luu [361] provide necessary and sufficient conditions for the existence of the multivalued Radon–Nikodym derivative.

Proposition 1.36. *Let $\mathbb{E}^*$ be separable. If M is a multivalued measure of bounded variation with weakly compact values, which is absolutely continuous with respect to $\mathbf{P}$, then there is an integrably bounded random closed set X such that*

$$\int \alpha(\omega) M(d\omega) = \mathbf{E}(\alpha X)$$

for every integrable real-valued random variable α.

1.5 Convergence of the selection expectations

Fatou's lemma for bounded sets in $\mathbb{R}^d$

Many results concerning the convergence of selection expectations deal with various generalisations of Fatou's lemma for random closed sets. First, consider *Fatou's lemma* for integrably bounded random closed sets in $\mathbb{R}^d$. It is traditionally formulated for the Aumann integral $\mathbf{E}_I X$ since the effect of taking closure is quite intricate in this setting.

Theorem 1.37 (Fatou's lemma in $\mathbb{R}^d$). *Let $\{X_n, n \geq 1\}$ be a sequence of random compact sets in $\mathbb{R}^d$ such that $\alpha = \sup_{n \geq 1} \|X_n\|$ is integrable. Then*

$$\limsup \mathbf{E}_I X_n \subset \mathbf{E}_I \limsup X_n \,. \tag{1.18}$$

Proof. Assume first that the probability space is non-atomic. If $x \in \limsup \mathbf{E}_I X_n$, then x is a limit point of a sequence $\{\mathbf{E}\xi_n, n \geq 1\}$, where $\xi_n \in \mathcal{S}^1(X_n), n \geq 1$. Since $\|\mathbf{E}\xi_n\| \leq \mathbf{E}\|\xi_n\| \leq \mathbf{E}\alpha$, without loss of generality, assume that $\mathbf{E}\xi_n \to x$ as $n \to \infty$. Because the norms $\|\mathbf{E}\xi_n\|$ are all bounded by an integrable random variable α, there is a subsequence of $\{\xi_n, n \geq 1\}$ with a weak limit, see Dunford and Schwartz [148, Th. IV.8.9]. Therefore, we can assume that ξ_n converges weakly to an $\mathbb{E}$-valued random element ξ.

From [148, Cor. V.3.14] it follows that there is a sequence $\{\eta_n, n \geq 1\}$ of convex combinations of $\xi_n, \xi_{n+1}, \dots$ such that η_n converges to ξ in $\mathbf{L}^1(\Omega; \mathbb{R}^d)$, so that $\|\eta_n - \xi\|_1 \to 0$ as $n \to \infty$. Since there exists a subsequence of $\{\eta_n, n \geq 1\}$ that converges to ξ almost surely, it is possible to assume that $\eta_n \to \xi$ a.s. as $n \to \infty$.

Since $\xi_n \in \mathbb{R}^d, n \geq 1$, it is possible to represent η_n as a finite convex combination of $\xi_n, \xi_{n+1}, \dots$, i.e.

$$\eta_n = \sum_{j=0}^{d} \theta_{jn} \zeta_{jn} \,,$$

where θ_{in}, $0 \le j \le d$, are non-negative and sum to 1 and ζ_{jn}, $0 \le j \le d$, are chosen from among $\xi_n, \xi_{n+1}, \ldots$ By passing to subsequences, it is possible to assume that

$$\xi = \lim_{n \to \infty} \eta_n = \sum_{j=0}^{d} \theta_j \zeta_j \quad \text{a.s.},$$

where θ_j, $0 \le j \le d$, are non-negative and sum to 1, and ζ_j, $0 \le j \le d$, are the limiting points of $\{\xi_n, n \ge 1\}$. If Y is the set of the limiting points of $\{\xi_n, n \ge 1\}$, then $\xi \in \text{co}(Y)$. Thus,

$$x = \mathbf{E}\xi \in \mathbf{E}_I \text{co}(Y) \subset \mathbf{E}_I \text{co}(\limsup X_n) = \mathbf{E}_I \limsup X_n,$$

since the probability space is non-atomic, see Theorem 1.15.

If the probability space is purely atomic, then (in the above notation) the weak convergence of ξ_n implies that $\xi_n \to \xi$ almost surely, so that ξ is a limiting point of $\{\xi_n, n \ge 1\}$. A general probability space Ω can be decomposed into its purely atomic part Ω' and the non-atomic part Ω''. Then

$$\limsup \mathbf{E}_I X_n = \limsup(\mathbf{E}_I(\mathbf{1}_{\Omega'} X_n) + \mathbf{E}_I(\mathbf{1}_{\Omega''} X_n))$$
$$\subset \limsup \mathbf{E}_I(\mathbf{1}_{\Omega'} X_n) + \limsup \mathbf{E}_I(\mathbf{1}_{\Omega''} X_n),$$

i.e. the proof follows from the results in the non-atomic and purely atomic cases. □

It was shown in the proof of Theorem 1.24(i) that Theorem 1.37 implies that the Aumann integral $\mathbf{E}_I X$ equals the selection expectation $\mathbf{E}X$ for any integrably bounded random compact set X in $\mathbb{R}^d$. Formulated for the selection expectation instead of the Aumann integral, (1.18) becomes a weaker statement:

$$\limsup \mathbf{E} X_n \subset \mathbf{E} \limsup X_n. \tag{1.19}$$

It is possible to prove a complementary result to Theorem 1.37 which establishes the reverse inclusion for lower limits and is useful to derive a dominated convergence theorem. In view of Theorem 1.24(i), it is formulated for the selection expectation instead of the Aumann integral, since both coincide for integrably bounded random closed sets in $\mathbb{R}^d$.

Theorem 1.38 (Dominated convergence for selection expectations). *Let X_n, $n \ge 1$, be random compact sets in $\mathbb{R}^d$ such that $\alpha = \sup_{n \ge 1} \|X_n\|$ is integrable. Then*

$$\mathbf{E} \liminf X_n \subset \liminf \mathbf{E} X_n \subset \limsup \mathbf{E} X_n \subset \mathbf{E} \limsup X_n. \tag{1.20}$$

If, additionally, $X_n \to X$ almost surely in the Hausdorff metric, then

$$\rho_H(\mathbf{E} X_n, \mathbf{E} X) \to 0 \quad \text{as } n \to \infty. \tag{1.21}$$

Proof. Let us show that

$$\mathbf{E}_\mathrm{I} \liminf X_n \subset \liminf \mathbf{E}_\mathrm{I} X_n. \tag{1.22}$$

If $x \in \mathbf{E}_\mathrm{I} \liminf X_n$, then $x = \mathbf{E}\xi$ for ξ being an integrable selection of $\liminf X_n$. Define a Borel random subset of $\mathbb{R}^d \times \mathbb{R}^d \times \cdots$ as

$$Z = \{(x_1, x_2, \ldots) : x_1 \in X_1, x_2 \in X_2, \ldots, \lim x_n = \xi\}.$$

Then the statement $\xi \in \liminf X_n$ a.s. is equivalent to $Z \neq \emptyset$ a.s. A *selection theorem* of von Neumann [423] implies that Z has a measurable selection being a sequence $(\xi_1, \xi_2, \ldots)$ such that $\xi_n \in X_n$ a.s. and $\xi_n \to \xi$ for all $\omega \in \Omega$. Since $\|\xi_n\| \leq \alpha$, the dominated convergence theorem yields $\mathbf{E}\xi_n \to \mathbf{E}\xi = x$, whence $x \in \liminf \mathbf{E}_\mathrm{I} X_n$. Since $\liminf \mathbf{E}_\mathrm{I} X_n = \liminf \mathbf{E} X_n$, (1.22) together with (1.19) imply (1.20). Finally, (1.21) follows from (1.20) if $\liminf X_n = \limsup X_n$. □

Fatou's lemma for unbounded random sets

It is possible to generalise Theorem 1.37 in two directions: by relaxing conditions on the boundedness and uniform integrability of $\{X_n, n \geq 1\}$ or by generalising it to infinite-dimensional spaces. The first generalisation causes appearance of an additional additive term in the right-hand side of (1.18). Infinite dimensional variants of Fatou's lemma generally involve taking a closure in the right-hand side of (1.18); the corresponding results are often called *approximate* Fatou's lemmas.

Let us formulate a general Fatou's lemma which holds for unbounded random closed sets. Its proof (and the proof of Theorem 1.42 below) are rather technical and can be found in Balder and Hess [45]. Even the formulations involve several further concepts from convex analysis. If F is a non-empty closed convex set, then $\mathrm{As}(F)$ denotes the asymptotic (or recession) cone of F which is the largest convex cone $\mathbb{C}$ satisfying $x_0 + \mathbb{C} \subset F$ for some $x_0 \in F$. Then

$$\mathrm{As}(F) = \bigcap_{t>0} t(F - x_0)$$

does not depend on $x_0 \in F$. If $\mathbb{C}$ is a cone in $\mathbb{E}^*$, then its polar cone $\mathbb{C}^\circ$ is the set of all $x \in \mathbb{R}^d$ such that $\langle x, u \rangle \leq 0$ for all $u \in \mathbb{C}$. Furthermore, $\check{\mathbb{C}}^\circ = \{-x : x \in \mathbb{C}^\circ\}$ is the central symmetric cone to $\mathbb{C}^\circ$.

Theorem 1.39 (Fatou's lemma: finite-dimensional). *Let $\{X_n, n \geq 1\}$ be a sequence of random closed sets in $\mathbb{R}^d$ such that*

$$X_n \subset Y_n + \alpha_n L, \quad n \geq 1,$$

where $\{Y_n, n \geq 1\}$ are random compact sets with $\sup_{n \geq 1} \mathbf{E}\|Y_n\| < \infty$, $\{\alpha_n, n \geq 1\}$ is a uniformly integrable sequence of random variables and L is a deterministic closed set such that $\overline{\mathrm{co}}(L)$ does not contain any line. Let $\mathbb{C}$ be a convex cone which consists of all $u \in \mathbb{R}^d$ such that $\max(0, h(Y_n, -u))$ is uniformly integrable. Then

$$\limsup \mathbf{E}_\mathrm{I} X_n \subset \mathbf{E}_\mathrm{I}(\limsup X_n) + \mathrm{As}(L + \check{\mathbb{C}}^\circ). \tag{1.23}$$

If $\{\|Y_n\|, n \geq 1\}$ is a uniformly integrable sequence, then $\mathbb{C}^\circ = \{0\}$ and

$$\limsup \mathbf{E}_I X_n \subset \mathbf{E}_I \limsup X_n + \mathrm{As}(L).$$

If $\sup_{n \geq 1} \|X_n\|$ is integrable, then $L = \{0\}$ and (1.23) turns into (1.18).

Example 1.40 (Fatou's lemma applies). Let $\Omega = [0, 1]$ with the Lebesgue measure, $\mathbb{E} = \mathbb{R}$, and let $X_n(\omega) = L = [0, \infty)$ if $\omega \in [0, 1/n]$ and $X_n(\omega) = \{0\}$ otherwise. Then $\limsup X_n = \{0\}$, the conditions of Theorem 1.39 hold for $Y_n = \{0\}$, $\alpha = 1$, $\mathbb{C}^\circ = \{0\}$ and

$$\limsup \mathbf{E}_I X_n = L \subset \{0\} + \mathrm{As}(L + \{0\}) = L$$

in agreement with Theorem 1.39.

Example 1.41 (Fatou's lemma does not apply). Let $\mathbb{E} = \mathbb{R}$ and let X_n be either $\{n\}$ or $\{-n\}$ with equal probabilities. Then $X_n \subset Y_n + L$, where $L = \mathbb{R}$ in order to ensure the boundedness of $\mathbf{E}\|Y_n\|$, $n \geq 1$. Such L contains the whole line and does not satisfy Theorem 1.39. Then $\mathbf{E}_I X_n = \{0\}$ for all n, so that the left-hand side of (1.23) is $\{0\}$. However $\limsup X_n$ is empty almost surely, so that the right-hand side of (1.23) is empty, which shows that Fatou's lemma does not hold in this case.

Approximate Fatou's lemma: infinite-dimensional case

To formulate the result in infinite-dimensional spaces, we need several further concepts. A set $F \subset \mathbb{E}$ is called weakly ball-compact if its intersection with every closed ball if weakly compact. If $\mathbb{E}$ is reflexive, then all weakly closed sets are automatically weakly ball-compact. If $\{F_n, n \geq 1\}$ is a sequence of subsets of $\mathbb{E}$, then its weak sequential upper limit w-seq-$\limsup F_n$ is the set of all $x \in \mathbb{E}$ such that x is a weak limit of x_{n_k} where $x_{n_k} \in F_{n_k}$, $k \geq 1$. The weak non-sequential upper limit is the intersection of the weak closures of $\cup_{k \geq n} F_k$ for all $n \geq 1$.

Theorem 1.42 (Fatou's lemma: infinite-dimensional). *Let $\{X_n, n \geq 1\}$ be a sequence of random closed subsets of a separable Banach space $\mathbb{E}$ such that*

$$X_n \subset Y_n + \alpha_n L, \quad n \geq 1,$$

where $\{Y_n, n \geq 1\}$ are random weakly compact sets with $\sup_{n \geq 1} \mathbf{E}\|Y_n\| < \infty$, $\{\alpha_n, n \geq 1\}$ is a uniformly integrable sequence of random variables and L is a deterministic subset of $\mathbb{E}$ such that $\overline{\mathrm{co}}(L)$ is locally weakly compact and does not contain any line. If $\mathbb{E}$ is not reflexive, assume additionally that $\cup_{n \geq 1} X_n \subset Z$ a.s., where Z is a weakly closed weakly ball-compact set. Let $\mathbb{C}$ be a convex cone which consists of all $u \in \mathbb{E}^*$ such that $\max(0, h(Y_n, -u))$ is uniformly integrable. If $X = $ w-$\limsup X_n$, then

$$\text{w-seq-}\limsup \mathbf{E}_I X_n \subset \mathbf{E}_I(\mathbf{1}_{\Omega^{\mathrm{pa}}} X) + \mathrm{cl}\, \mathbf{E}_I(\mathbf{1}_{\Omega^{\mathrm{na}}} X) + \mathrm{As}(L + \check{\mathbb{C}}^\circ), \quad (1.24)$$

where the closure in the right-hand side is taken in the strong topology, Ω^{pa} is the purely atomic part of Ω and Ω^{na} is the non-atomic part of Ω.

The following theorem (which is an extension of the dominated convergence theorem to infinite-dimensional spaces) provides useful sufficient conditions for the convergence of selection expectations. Its proof can be found in Hiai [253, Th. 2.8]. Note that part (ii) follows from Theorem 1.42.

Theorem 1.43 (Dominated convergence theorem in Banach spaces). *Let $X_n, n \geq 1$ be integrably bounded random closed sets in a separable Banach space $\mathbb{E}$.*
(i) *If the sequence $\{\rho(0, X_n), n \geq 1\}$ is uniformly integrable and $X = $ s-lim inf X_n is an integrable random closed set, then*
$$\mathbf{E}X \subset \text{s-lim inf } \mathbf{E}X_n .$$
(ii) *Assume that $\mathbb{E}$ is reflexive and $\{\|X_n\|, n \geq 1\}$ is uniformly integrable. If $X = $ w-lim sup X_n is integrable, then*
$$\text{w-lim sup } \mathbf{E}X_n \subset \mathbf{E}X .$$

If, additionally, X_n converges to X almost surely in the Mosco topology, then $\mathbf{E}X_n$ converges to $\mathbf{E}X$ in the Mosco topology as $n \to \infty$.
(iii) *If $\{\|X_n\|, n \geq 1\}$ is uniformly integrable and $\rho_\mathrm{H}(X_n, X) \to 0$ in probability, then $\rho_\mathrm{H}(\mathbf{E}X_n, \mathbf{E}X) \to 0$ as $n \to \infty$.*

Monotone and weak convergence

The monotone convergence theorem holds under rather general assumptions.

Theorem 1.44 (Monotone convergence theorem). *Assume that $\{X_n, n \geq 1\}$ is a non-decreasing sequence of random closed sets and X_1 is integrable. If $X = \mathrm{cl}(\bigcup_{n \geq 1} X_n)$, then*
$$\mathbf{E}X = \mathrm{cl}(\bigcup_{n \geq 1} \mathbf{E}X_n) .$$

Proof. For each $\xi \in \mathcal{S}^1(X)$, Theorem 1.10 applied to $\zeta_x = \|\xi - x\|$ yields
$$\inf_{\eta \in \mathcal{S}^1(X_n)} \|\xi - \eta\|_1 = \mathbf{E}\rho(\xi, X_n) .$$
The right-hand side tends to zero, since $\rho(\xi, X_1)$ is integrable and $\rho(\xi, X_n) \downarrow 0$ a.s. Therefore, $\mathcal{S}^1(X) = \mathrm{cl}(\bigcup_n \mathcal{S}^1(X_n))$, which finishes the proof. □

The weak convergence of a sequence of random compact sets implies the convergence of their expectations.

Theorem 1.45 (Expectations of weakly convergent sequences). *Let $\{X_n, n \geq 1\}$ and X be random convex compact sets in $\mathbb{R}^d$ such that $\alpha = \sup_{n \geq 1} \|X_n\|$ is integrable. If X_n weakly converges to X as $n \to \infty$, then $\mathbf{E}X_n$ converges to $\mathbf{E}X$ in the Hausdorff metric and the Lebesgue measure of $\mathbf{E}X_n$ converges to the Lebesgue measure of $\mathbf{E}X$ as $n \to \infty$. The statement holds if X_n and X are not necessarily convex and the probability space is non-atomic.*

Proof. The dominated convergence theorem yields $\mathbf{E}h(X_n, u) \to \mathbf{E}h(X, u)$ as $n \to \infty$ for each $u \in \mathbb{S}^{d-1}$. Assume that $\{u_n, n \geq 1\} \subset \mathbb{S}^{d-1}$ and $u_n \to u_0$ as $n \to \infty$. By subadditivity of support functions,

$$\mathbf{E}h(X_n, u_0) - \mathbf{E}h(X_n, u_0 - u_n) - \mathbf{E}h(X, u_0) \leq \mathbf{E}h(X_n, u_n) - \mathbf{E}h(X, u_0)$$
$$\leq \mathbf{E}h(X_n, u_0) + \mathbf{E}h(X_n, u_n - u_0) - \mathbf{E}h(X, u_0).$$

Clearly,
$$|\mathbf{E}h(X, u_0 - u_n)| \leq \mathbf{E}\alpha \|u_0 - u_n\| \to 0 \quad \text{as } n \to \infty.$$

Therefore,
$$\sup_{u \in \mathbb{S}^{d-1}} |\mathbf{E}h(X_n, u) - \mathbf{E}h(X, u)| \to 0 \quad \text{as } n \to \infty,$$

whence $\mathbf{E}X_n$ converges to $\mathbf{E}X$ in the Hausdorff metric. The convergence of measures immediately follows from convexity. The statement for non-convex random closed sets follows from Theorem 1.15. □

It is easy to see that Theorem 1.45 holds if the sequence $\{\|X_n\|, n \geq 1\}$ is uniformly integrable. If $\{\|X_n\|, n \geq 1\}$ are not necessarily bounded by an integrable random variable a truncation argument yields $\mathbf{E}X \subset \liminf \mathbf{E}X_n$.

1.6 Conditional expectation

Existence

Let $\mathfrak{H}$ be a sub-σ-algebra of $\mathfrak{F}$. By $\mathbf{L}_{\mathfrak{H}}^1$ we denote the family of $\eta \in \mathbf{L}^1 = \mathbf{L}^1(\Omega; \mathbb{E})$ which are measurable with respect to $\mathfrak{H}$. The conditional expectation, $\mathbf{E}(\xi|\mathfrak{H})$, of an integrable random element ξ is $\eta \in \mathbf{L}_{\mathfrak{H}}^1$ such that $\mathbf{E}\mathbf{1}_A \eta = \mathbf{E}\mathbf{1}_A \xi$ for every $A \in \mathfrak{H}$. If $\mathbb{E}$ is a Banach space, then the conditional expectation exists and is unique up to a.s. equivalence for each $\xi \in \mathbf{L}^1$ and σ-algebra $\mathfrak{H} \subset \mathfrak{F}$.

The following theorem defines the *conditional expectation* for an integrable random closed set and at the same time establishes its existence and uniqueness. Recall that $\mathcal{S}_{\mathfrak{H}}^1(X)$ denotes the family of $\mathfrak{H}$-measurable integrable selections of X.

Theorem 1.46 (Existence of the conditional expectation). *Let X be an integrable random closed set. For each σ-algebra $\mathfrak{H} \subset \mathfrak{F}$ there exists a unique integrable $\mathfrak{H}$-measurable random closed set Y (denoted by $Y = \mathbf{E}(X|\mathfrak{H})$ and called the* conditional selection expectation *of X) such that*

$$\mathcal{S}_{\mathfrak{H}}^1(Y) = \mathrm{cl}\{\mathbf{E}(\xi|\mathfrak{H}) : \xi \in \mathcal{S}^1(X)\}, \qquad (1.25)$$

where the closure is taken with respect to the norm in $\mathbf{L}_{\mathfrak{H}}^1$. If X is integrably bounded, so is Y.

Proof. The set $\{\mathbf{E}(\xi|\mathfrak{H}) : \xi \in \mathcal{S}^1(X)\}$ is decomposable in $\mathbf{L}_{\mathfrak{H}}^1$, whence its closure is also decomposable. By Theorem 1.6, there exists a unique random closed set Y satisfying (1.25). If X is integrably bounded, then the set in the right-hand side of (1.25) is bounded in $\mathbf{L}_{\mathfrak{H}}^1$, whence Y is also integrably bounded. □

Properties of conditional expectation

It should be noted that the conditions (ii) or (iii) of Theorem 1.24 imply that $\{\mathbf{E}(\xi|\mathfrak{H}) : \xi \in \mathcal{S}^1(X)\}$ is a closed set if $\mathfrak{H}$ is countably generated, see Li and Ogura [352]. Many properties of the conditional expectation are easily recognisable counterparts of the properties for the (unconditional) selection expectation.

Theorem 1.47 (Properties of conditional expectation). *Let X and Y be integrable random closed sets and let $\mathfrak{H}$ be a sub-σ-algebra of $\mathfrak{F}$. Then*
(i) $\mathbf{E}(\overline{\mathrm{co}}\,(X)|\mathfrak{H}) = \overline{\mathrm{co}}\,(\mathbf{E}(X|\mathfrak{H}))$ *a.s.*
(ii) $\mathbf{E}(\mathrm{cl}(X+Y)|\mathfrak{H}) = \mathrm{cl}(\mathbf{E}(X|\mathfrak{H}) + \mathbf{E}(Y|\mathfrak{H}))$ *a.s.*
(iii) If αX is integrable for a $\mathfrak{H}$-measurable random variable α, then

$$\mathbf{E}(\alpha X|\mathfrak{H}) = \alpha \mathbf{E}(X|\mathfrak{H}) \quad \text{a.s.}$$

(iv) For every $u \in \mathbb{E}^*$,

$$h(\mathbf{E}(X|\mathfrak{H}), u) = \mathbf{E}(h(X, u)|\mathfrak{H}) \quad \text{a.s.}$$

(v) If both X and Y are integrably bounded, then

$$\mathbf{E}\rho_{\mathrm{H}}(\mathbf{E}(X|\mathfrak{H}), \mathbf{E}(Y|\mathfrak{H})) \le \mathbf{E}\rho_{\mathrm{H}}(X, Y).$$

(vi) Assume that $\mathbb{E}^*$ is separable. If $\mathbf{E}(X|\mathfrak{H})$ is a singleton almost surely, so is X.

Proof. Denote $X' = \mathbf{E}(X|\mathfrak{H})$ and $Y' = \mathbf{E}(Y|\mathfrak{H})$.
(i) follows from

$$\begin{aligned}\mathcal{S}^1_\mathfrak{H}(\mathbf{E}(\overline{\mathrm{co}}\,(X)|\mathfrak{H})) &= \mathrm{cl}\{\mathbf{E}(\xi|\mathfrak{H}) : \xi \in \mathcal{S}^1(\overline{\mathrm{co}}\,(X))\} \\ &= \overline{\mathrm{co}}\,\{\mathbf{E}(\xi|\mathfrak{H}) : \xi \in \mathcal{S}^1(X)\} \\ &= \overline{\mathrm{co}}\,(\mathcal{S}^1_\mathfrak{H}(X')) = \mathcal{S}^1_\mathfrak{H}(\overline{\mathrm{co}}\,(X')).\end{aligned}$$

Similar arguments lead to (ii).
(iii) We have

$$\begin{aligned}\mathcal{S}^1_\mathfrak{H}(\mathbf{E}(\alpha X|\mathfrak{H})) &= \mathrm{cl}\{\mathbf{E}(\eta|\mathfrak{H}) : \eta \in \mathcal{S}^1(\alpha X)\} \\ &= \mathrm{cl}\{\mathbf{E}(\alpha \xi|\mathfrak{H}) : \xi \in \mathcal{S}^1(\alpha X),\ \alpha\xi \in \mathbf{L}^1\} \\ &= \mathrm{cl}\{\alpha \mathbf{E}(\xi|\mathfrak{H}) : \xi \in \mathcal{S}^1(\alpha X),\ \alpha\xi \in \mathbf{L}^1\} \subset \mathcal{S}^1_\mathfrak{H}(\alpha X').\end{aligned}$$

Now prove the reverse inclusion. Note that

$$\mathcal{S}^1_\mathfrak{H}(\alpha X') = \mathrm{cl}\{\alpha\xi : \xi \in \mathcal{S}^1_\mathfrak{H}(X'),\ \alpha\xi \in \mathbf{L}^1\}.$$

For each $\xi \in \mathcal{S}^1_\mathfrak{H}(X')$ with $\alpha\xi \in \mathbf{L}^1$ choose a sequence $\{\xi_n, n \ge 1\} \subset \mathcal{S}^1(X)$ such that $\|\mathbf{E}(\xi_n|\mathfrak{H}) - \xi\|_1 \to 0$ and define $\eta_{nk} = \alpha(\mathbf{1}_{B_k}\xi_n + \mathbf{1}_{\Omega\setminus B_k}\xi_0) \in \mathcal{S}^1(\alpha X)$, where $B_k = \{\omega : \|\xi\| \le k\} \in \mathfrak{H}$ and ξ_0 is an integrable selection of X such that $\alpha\xi_0$ is also integrable. Since

$$\mathbf{E}(\eta_{nk}|\mathfrak{H}) = \alpha(\mathbf{1}_{B_k}\mathbf{E}(\xi_n|\mathfrak{H}) + \mathbf{1}_{\Omega\setminus B_k}\mathbf{E}(\xi_0|\mathfrak{H})),$$

the conditional expectation $\mathbf{E}(\eta_{nk}|\mathfrak{H})$ tends to $\alpha\xi$ in $\mathbf{L}^1$-norm, whence $\alpha\xi$ belongs to $\mathcal{S}^1_{\mathfrak{H}}(\mathbf{E}(\alpha X|\mathfrak{H}))$.

(iv) For every $A \in \mathfrak{H}$, by applying Theorem 1.10 twice we get

$$\begin{aligned}\mathbf{E}(\mathbf{1}_A h(X', u)) &= \sup_{\xi' \in \mathcal{S}^1_{\mathfrak{H}}(X')} \mathbf{E}(\mathbf{1}_A \langle \xi', u \rangle) \\ &= \sup_{\xi \in \mathcal{S}^1(X)} \mathbf{E}(\mathbf{1}_A \langle \mathbf{E}(\xi|\mathfrak{H}), u \rangle) \\ &= \sup_{\xi \in \mathcal{S}^1(X)} \mathbf{E}(\mathbf{1}_A \langle \xi, u \rangle) = \mathbf{E}(\mathbf{1}_A h(X, u)),\end{aligned}$$

so that $h(X', u) = \mathbf{E}(h(X, u)|\mathfrak{H})$ a.s.

(v) Write the Hausdorff distance between X' and Y' as

$$\mathbf{E}\rho_{\mathrm{H}}(X', Y') = \mathbf{E}(\mathbf{1}_A \sup_{x \in X'} \rho(x, Y')) + \mathbf{E}(\mathbf{1}_{\Omega\setminus A} \sup_{y \in Y'} \rho(y, X'))$$

with

$$A = \{\omega : \sup_{x \in X'} \rho(x, Y') \geq \sup_{y \in Y'} \rho(y, X')\}.$$

The proof is finished by applying Theorem 1.10 several times, see Hiai and Umegaki [255, Th. 5.2(1)].

(vi) Because of (ii), assume without loss of generality that $0 \in X$ a.s. and $\mathbf{E}(X|\mathfrak{H}) = \{0\}$. Furthermore, (iv) implies

$$0 = h(\mathbf{E}(X|\mathfrak{H}), u) = \mathbf{E}(h(X, u)|\mathfrak{H}) \quad \text{a.s.}$$

Taking the expectations of both sides yields $\mathbf{E}h(X, u) = 0$. Since $0 \in X$ a.s., $h(X, u) = 0$ a.s. for all $u \in \mathbb{E}^*$. □

The following properties are derived in Hiai and Umegaki [255] for integrably bounded random convex closed sets. As shown by Hiai [254], easy modifications of the arguments prove the identical statements for integrable random convex closed sets as formulated below.

Theorem 1.48. *Let X be a random convex closed set.*

(i) *If α is a non-negative integrable random variable and X is $\mathfrak{H}$-measurable with non-empty $\mathcal{S}^1_{\mathfrak{H}}(\alpha X)$, then $\mathbf{E}(\alpha X|\mathfrak{H}) = \mathbf{E}(\alpha|\mathfrak{H})X$ a.s. In particular, $\mathbf{E}(X|\mathfrak{H}) = X$ a.s. if $\mathcal{S}^1_{\mathfrak{H}}(X)$ is non-empty.*

(ii) *If $\mathfrak{H}' \subset \mathfrak{H} \subset \mathfrak{F}$ and X is $\mathfrak{H}$-measurable with non-empty $\mathcal{S}^1_{\mathfrak{H}}(X)$, then $\mathbf{E}(X|\mathfrak{H}')$ on the probability space $(\Omega, \mathfrak{F}, \mathbf{P})$ is a.s. identical to the conditional expectation of X relative to $\mathfrak{H}'$ taken on the probability space $(\Omega, \mathfrak{H}, \mathbf{P})$.*

(iii) *If $\mathfrak{H}' \subset \mathfrak{H} \subset \mathfrak{F}$ and X is integrable, then $\mathbf{E}(\mathbf{E}(X|\mathfrak{H})|\mathfrak{H}') = \mathbf{E}(X|\mathfrak{H}')$ a.s.*

If α in (i) is allowed to be negative, then the result no longer holds for a general X. Indeed, if $\mathbf{E}\alpha = 0$, then $\mathbf{E}(\alpha|\mathfrak{H})X = \{0\}$, whence the only X being a singleton ensures $\mathbf{E}(\alpha X|\mathfrak{H}) = \{0\}$, see Theorem 1.47(vi).

The following results justify the introduced concept of the conditional expectation. The proofs can be found in Hiai and Umegaki [255].

Theorem 1.49.
(i) *If X is an integrable random closed set, then*
$$\mathbf{E}^{\mathfrak{H}}(\mathbf{1}_A \mathbf{E}(X|\mathfrak{H})) = \mathbf{E}(\mathbf{1}_A X), \quad A \in \mathfrak{H}. \tag{1.26}$$

(ii) *If X is an integrable random convex closed set, then*
$$\mathbf{E}(\mathbf{1}_A \mathbf{E}(X|\mathfrak{H})) = \mathbf{E}(\mathbf{1}_A X), \quad A \in \mathfrak{H}. \tag{1.27}$$

(iii) *If $\mathbb{E}^*$ is separable and X is an integrably bounded random convex closed set, then $\mathbf{E}(X|\mathfrak{H})$ is uniquely determined as an integrably bounded random convex closed set satisfying either (1.26) or (1.27). If $\mathbb{Q}^*$ is a countable dense set in $\mathbb{E}^*$, then*
$$\mathbf{E}(X|\mathfrak{H}) = \bigcap_{u \in \mathbb{Q}^*} \{x \in \mathbb{E} : \langle x, u \rangle \le \mathbf{E}(h(X, u)|\mathfrak{H})\} \quad \text{a.s.}$$

(iv) *If X is an integrably bounded random convex compact set, then $\mathbf{E}(X|\mathfrak{H})$ is almost surely convex compact,*
$$\mathbf{E}_{\mathrm{I}}^{\mathfrak{H}}(\mathbf{1}_A \mathbf{E}(X|\mathfrak{H})) = \mathbf{E}_{\mathrm{I}}(\mathbf{1}_A \mathbf{E}(X|\mathfrak{H})) = \mathbf{E}_{\mathrm{I}}(\mathbf{1}_A X), \quad A \in \mathfrak{H}, \tag{1.28}$$
and
$$\mathcal{S}_{\mathfrak{H}}^1(\mathbf{E}(X|\mathfrak{H})) = \{\mathbf{E}(\xi|\mathfrak{H}) : \xi \in \mathcal{S}^1(X)\}. \tag{1.29}$$

(v) *If $\mathbb{E}$ is reflexive, then (1.28) and (1.29) hold for every integrably bounded random convex closed set X.*

An event $A \in \mathfrak{H}$ is called a $\mathfrak{H}$-atom if for each $A' \in \mathfrak{F}$ with $A' \subset A$, there exists a $B \in \mathfrak{H}$ satisfying $\mathbf{P}((A \cap B) \triangle A') = 0$. It has been shown by Valadier [571] that if the probability space has no $\mathfrak{H}$-atom, then $\mathbf{E}(\overline{\mathrm{co}}\,(X)|\mathfrak{H}) = \mathbf{E}(X|\mathfrak{H})$ a.s. for every integrable random closed set X.

Convergence of conditional expectations

It is possible to provide a whole spectrum of results on convergence of conditional expectations of random closed sets, which are more or less exact counterparts of the results from Section 1.5. The monotone convergence theorem keeps its exact formulation for conditional expectations. Here are several other results taken from Hiai [253].

Theorem 1.50 (Convergence of conditional expectations).
(i) *Assume that* $\sup_{n\geq 1} \rho(0, X_n)$ *is integrable. If* $X = \text{s-lim inf } X_n$ *a.s. is integrable, then* $\mathbf{E}(X|\mathfrak{H}) \subset \text{s-lim inf } \mathbf{E}(X_n|\mathfrak{H})$ *a.s.*
(ii) *Assume that* $\mathbb{E}$ *is reflexive and* $\sup_{n\geq 1} \|X_n\|$ *is integrable. If* $X = \text{w-lim sup } X_n$ *a.s., then*
$$\text{w-lim sup } \mathbf{E}(X_n|\mathfrak{H}) \subset \mathbf{E}(\overline{\text{co}}(X)|\mathfrak{H}) \quad \text{a.s.}$$
If the probability space contains no $\mathfrak{H}$-atom or X is almost surely convex, then the Mosco convergence $X_n \to X$ a.s. implies that $\mathbf{E}(X_n|\mathfrak{H})$ almost surely converges in the Mosco topology to $\mathbf{E}(X|\mathfrak{H})$ a.s. as $n \to \infty$.
(iii) *If* $\sup_{n\geq 1} \|X_n\|$ *is integrable and* $\rho_{\mathbf{H}}(X_n, X) \to 0$ *a.s. (respectively in probability), then* $\rho_{\mathbf{H}}(\mathbf{E}(X_n|\mathfrak{H}), \mathbf{E}(X|\mathfrak{H})) \to 0$ *a.s. (respectively in probability).*

2 Further definitions of expectations

2.1 Linearisation approach

Since the space $\mathcal{F}$ is not linear, conventional tools suitable for defining an expectation of a random element in a linear space are not applicable for general random closed sets. A common approach to handle similar situations is to *linearise* $\mathcal{F}$ using a map (or maps) from $\mathcal{F}$ to a linear space, where it is easy to define an expectation.

In a general situation, a random closed set X is associated with a random element ξ_X taking values in a Banach space $\mathbb{Y}$. This is done by mapping $\mathcal{F}$ into $\mathbb{Y}$, so that ξ_X becomes the image of X under this map. Then the expectation of ξ_X is defined in $\mathbb{Y}$ with the aim to map it back into $\mathcal{F}$. If $\mathbf{E}\xi_X$ has a unique *inverse* image, i.e.

$$\mathbf{E}\xi_X = \xi_F \tag{2.1}$$

for some $F \in \mathcal{F}$, then F is said to be the *expectation* of X. For example, the selection expectation in $\mathbb{R}^d$ can be defined using embedding of convex compact sets in the space of continuous functions on the unit sphere, so that $\xi_X(\cdot) = h(X, \cdot)$ is the support function of X. By Theorem 1.22, the expected support function is the support function of the selection expectation of X.

However, it is quite typical that the possible values for ξ_F for $F \in \mathcal{F}$ do not form a convex set in $\mathbb{Y}$, so that $\mathbf{E}\xi_X$ may not be representable as ξ_F for any $F \in \mathcal{F}$. In this case the aim is to find a closed set F such that ξ_F "mimics" $\mathbf{E}\xi_X$ in some sense or is "sufficiently near" to $\mathbf{E}\xi_X$. For this, it is necessary to equip $\mathbb{Y}$ with a metric or pseudometric $\mathfrak{d}$ which assesses the discrepancy between $\mathbf{E}\xi_X$ and ξ_F for possible "candidates" F. The *pseudometric* $\mathfrak{d}$ satisfies the triangle inequality and is symmetric in its arguments, whereas $\mathfrak{d}(f, g) = 0$ for $f, g \in \mathbb{Y}$ does not necessarily imply $f = g$. The *expectation* of X is defined by

$$\mathbf{E}X = \operatorname{argmin}_{F\in\mathcal{Z}} \mathfrak{d}(\mathbf{E}\xi_X, \xi_F), \tag{2.2}$$

i.e. $\mathbf{E}X$ is an element of $\mathcal{Z}$ such that ξ_F is the closest to $\mathbf{E}\xi_X$. Here $\mathcal{Z}$ is a subfamily of $\mathcal{F}$ which consists of possible candidates for the expectation. A proper choice of $\mathcal{Z}$ is important, since it is difficult to solve minimisation problems over the whole family of closed sets. In general, several sets $F \in \mathcal{Z}$ may minimise $\mathfrak{d}(\mathbf{E}\xi_X, \xi_F)$. However, it is possible to avoid this ambiguity by imposing extra conditions on $\mathbf{E}X$, for example, assuming that $\mathbf{E}X$ is convex or regular closed. The linearisation approach can be illustrated by the following diagram

$$
\begin{array}{ccc}
X & \longrightarrow & \xi_X \in \mathbb{Y} \\
& & \downarrow \\
\mathbf{E}X & \underset{(2.1)\text{ or }(2.2)}{\longleftarrow} & \mathbf{E}\xi_X .
\end{array}
\qquad (2.3)
$$

In many examples, the space $\mathbb{Y}$ is a space of functions defined on a parameter space U and $\xi_X(u)$, $u \in U$, is a function on U. Then $\mathfrak{d}$ can be either the uniform ($\mathbf{L}^\infty$) or $\mathbf{L}^p$ metric (if U is equipped with a measure μ). In the first case $\mathbf{E}X$ is the set $F \in \mathcal{Z}$ that minimises

$$\|\mathbf{E}\xi_X(u) - \xi_F(u)\|_\infty = \sup_{u \in U} |\mathbf{E}\xi_X(u) - \xi_F(u)|,$$

while the choice of the $\mathbf{L}^p$-metric leads to the minimisation of

$$\|\mathbf{E}\xi_X(u) - \xi_F(u)\|_p = \left(\int_U (\mathbf{E}\xi_X(u) - \xi_F(u))^p \mu(du) \right)^{1/p}.$$

Clearly, various definitions of expectations utilise different features of the realisations of X. The situation can be explained by the following lucid example. Imagine that X is a "cat" with probability 1/2 and a "dog" otherwise. Clearly, it is pointless to average them, there is no known animal that might serve as their average. However, the question becomes sensible if we aim to average several features of a "cat" and a "dog" (weight, tail length, etc.) and then find an existing animal with the features matching the obtained averages as exactly as possible. The values $\xi_X(u)$ for various u represent those features or measurements that are being matched when defining an expectation.

Therefore, the expectation of X can be determined by the following ingredients:
(1) the Banach space $\mathbb{Y}$;
(2) the map $\xi_F : \mathcal{F} \mapsto \mathbb{Y}$;
(3) the metric $\mathfrak{d}$ on $\mathbb{Y}$;
(4) the family of closed sets $\mathcal{Z}$ providing candidates for $\mathbf{E}X$.

A generic notation $\mathbf{E}X$ for an expectation of a random set X will be equipped with different subscripts in order to designate various particular expectations, for instance, the selection expectation is denoted by $\mathbf{E}_A X$. From the point of view of many applications it suffices to assume that X is a random compact set in $\mathbb{R}^d$, although several definitions of expectation are also applicable for unbounded random closed sets and random closed sets in a Polish space $\mathbb{E}$.

2.2 The Vorob'ev expectation

Indicator and coverage functions

Let $\mathbb{E}$ be a space with a σ-finite measure μ. Put $U = \mathbb{E}$ and define

$$\xi_X(u) = \mathbf{1}_X(u) = \begin{cases} 1, & u \in X, \\ 0, & \text{otherwise}, \end{cases}$$

to be the *indicator function* of X. If $\mu(X)$ is finite almost surely, then X is associated with a random element $\xi_X(\cdot)$ in the space $\mathbb{Y}$ of square integrable functions on $\mathbb{E}$.

The expectation of the indicator function

$$\mathbf{E}\xi_X(u) = \mathbf{E}\mathbf{1}_X(u) = \mathbf{P}\{u \in X\} = p_X(u)$$

is called the *coverage function* of X, see Section 1.1.6. Unless X is deterministic or p_X vanishes everywhere, the coverage function is not an indicator function itself. Therefore, there is no set F which satisfies (2.1). The approach outlined in (2.3) suggests finding a closed set F such that its indicator function mimics the coverage function as exactly as possible. Natural candidates for F are determined by the coverage function itself as excursion (thresholded) sets of $p_X(u)$

$$\{p_X \geq t\} = \{u \in \mathbb{E}: \ p_X(u) \geq t\}, \quad t \in (0, 1], \tag{2.4}$$

called the *t-th quantile* of X, see Figure 2.1. By Proposition 1.1.34, the coverage function p_X is upper semicontinuous, so that all excursion sets are closed and $\{p_X \geq t\}$ is left-continuous as a function of t.

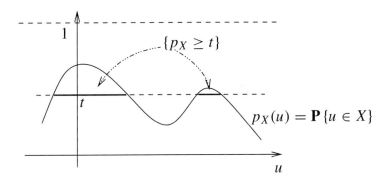

Figure 2.1. Excursion set of the coverage function.

Assume that $\mathbf{E}\mu(X) < \infty$. By Robbins' theorem (see Section 1.4.4),

$$\int_{\mathbb{E}} p_X(u) du = \mathbf{E}\mu(X) < \infty,$$

whence $\mu(\{p_X \geq t\})$ is finite for all $t \in (0, 1]$.

Definition 2.1 (Vorob'ev expectation). The *Vorob'ev expectation*, $\mathbf{E_V} X$, is defined as the set $\{p_X \geq t\}$ for $t \in [0, 1]$ which is determined from the equation $\mathbf{E}\mu(X) = \mu(\{p_X \geq t\})$ if this equation has a solution, or, in general, from the condition

$$\mu(\{p_X \geq s\}) \leq \mathbf{E}\mu(X) \leq \mu(\{p_X \geq t\})$$

for all $s > t$.

In other words, $\mathbf{E_V} X$ is a quantile of X (or the excursion set of the coverage function) such that its Lebesgue measure is the closest to $\mathbf{E}\mu(X)$.

Example 2.2. Let $X = B_\xi(0)$ be a random disk in $\mathbb{E} = \mathbb{R}^2$ centred at the origin with the radius ξ. If μ in Definition 2.1 is the Lebesgue measure, then $\mathbf{E_V} X = B_r(0)$ with $r = (\mathbf{E}\xi^2)^{1/2}$.

It should be noted that the Vorob'ev expectation treats all random sets with almost sure vanishing coverage functions as non-essential, so that the Vorob'ev expectation of X equals the Vorob'ev expectation of $X \cup Y$ if $\mathbf{P}\{x \in Y\} = 0$ for μ-almost all x. If $\mathbb{E}$ is the Euclidean space $\mathbb{R}^d$ and μ is the Lebesgue measure in $\mathbb{R}^d$, then all singletons or curve pieces with absolutely continuous distributions are not taken into account when calculating the expectation.

Vorob'ev expectation as minimiser

The following result shows that the Vorob'ev expectation minimises the expected measure of the symmetric difference.

Theorem 2.3 (Vorob'ev expectation and minimisation problem). *For each measurable set M with $\mu(M) = \mathbf{E}\mu(X)$,*

$$\mathbf{E}\mu(X \triangle \mathbf{E_V} X) \leq \mathbf{E}\mu(X \triangle M).$$

Proof. Denote $F = \mathbf{E_V} X = \{p_X \geq t\}$. Then

$$\mathbf{E}\mu(X \triangle M) - \mathbf{E}\mu(X \triangle F) = 2[\mu(X \cap (F \setminus M)) - \mu(X \cap (M \setminus F))]$$
$$= 2 \int_{F \setminus M} p_X(u) \mu(du) - 2 \int_{M \setminus F} p_X(u) \mu(du)$$
$$\geq 2[t\mu(F \setminus M) - t\mu(M \setminus F)]$$
$$= 2t[\mu(F \setminus M) - \mu(M \setminus F)]$$
$$= 2t[\mu(F) - \mu(M)] \geq 0,$$

because $\mu(M) = \mathbf{E}\mu(X) \leq \mu(F)$. □

Definition 2.1 appears as a particular case of (2.2) with the family $\mathcal{Z}$ that consists of all excursion sets given by (2.4) and a pseudometric

$$\mathfrak{d}(f,g) = \left| \int_{\mathbb{E}} (f(u) - g(u))\mu(du) \right|.$$

If $\mathfrak{d}$ is the uniform metric, then the solution of (2.2) is the set $\{p_X \geq 1/2\} = \{x \in \mathbb{E} : p_X(u) \geq 1/2\}$ called the *Vorob'ev median* of X. The following proposition establishes a property of the *Vorob'ev median*, which is similar to the classical property of the median which minimises the first absolute central moment.

Proposition 2.4. *For every measurable set M with $\mu(M) < \infty$,*

$$\mathbf{E}\mu(X \triangle \{p_X \geq 1/2\}) \leq \mathbf{E}\mu(X \triangle M). \tag{2.5}$$

Proof. Denote $F = \{p_X \geq 1/2\}$. By Robbins' theorem,

$$\mathbf{E}\mu(X \triangle M) = \int_{\mathbb{E}} p_X(u)\mu(du) - \int_M p_X(u)\mu(du) + \int_M (1 - p_X(u))\mu(du)$$

$$= \int_{\mathbb{E}} p_X(u)\mu(du) + \int_{M \cap F} (1 - 2p_X(u))\mu(du)$$

$$+ \int_{M \setminus F} (1 - 2p_X(u))\mu(du)$$

$$\geq \int_{\mathbb{E}} p_X(u)\mu(du) + \int_F (1 - 2p_X(u))\mu(du) = \mathbf{E}\mu(X \triangle F). \qquad \square$$

It is possible to generalise the Vorob'ev expectation by choosing other distances. For instance, if $\mathfrak{d}$ is either $\mathbf{L}^1$ or $\mathbf{L}^2$ metrics, then (2.2) yields other expectations given by $\{p_X \geq t\}$ with t chosen to minimise

$$\|p_X(u) - \mathbf{1}_{\{p_X \geq t\}}(u)\|_1 = \mathbf{E}\mu(X) + \mu(\{p_X \geq t\}) - 2\int_{\{p_X \geq t\}} p_X(u)du$$

or

$$\|p_X(u) - \mathbf{1}_{\{p_X \geq t\}}(u)\|_2 = \int_{\mathbb{E}} p_X^2(u)du + \mu(\{p_X \geq t\}) - 2\int_{\{p_X \geq t\}} p_X(u)du.$$

2.3 Distance average

Distance functions

Let $\mathbb{E}$ be a metric space with metric ρ. For each set $F \subset \mathbb{E}$ all points in $\mathbb{E}$ can be classified according to their positions with respect to F. For example, each point can be assigned its distance to F. However, this is not the only possible way.

2 Further definitions of expectations

Definition 2.5 (Distance function). Let $\mathcal{F}'$ be the space of all non-empty closed sets. A function $\mathbf{d}\colon \mathbb{E} \times \mathcal{F}' \mapsto \mathbb{R}$ is said to be a *distance function* if it is lower semicontinuous with respect to its first argument, measurable with respect to the second and satisfies the following two conditions:

(D1) If $F_1 \subset F_2$, then $\mathbf{d}(x, F_1) \geq \mathbf{d}(x, F_2)$ for all $x \in \mathbb{E}$ (monotonicity);
(D2) $F = \{x : \mathbf{d}(x, F) \leq 0\}$ for every $F \in \mathcal{F}'$ (consistency).

Example 2.6 (Various distance functions).

(i) The *metric* distance function $\mathbf{d}(x, F)$ is equal to the distance between $x \in \mathbb{E}$ and $F \in \mathcal{F}$ in the metric ρ, that is

$$\mathbf{d}(x, F) = \rho(x, F) = \inf\{\rho(x, y) : y \in F\}, \quad x \in \mathbb{E}.$$

(ii) The *square* distance function is defined as $\mathbf{d}(x, F) = \rho^2(x, F)$.
(iii) The *signed* distance function is given by

$$\mathbf{d}(x, F) = \begin{cases} \rho(x, F), & x \notin F, \\ -\rho(x, F^c), & x \in F. \end{cases}$$

If F has empty interior, then the signed distance function is equal to the metric distance function. A rationale for using the signed distance function is that it treats the set symmetrically with respect to exchanging black and white.

(iv) The *indicator* distance function is defined by letting $\mathbf{d}(x, F)$ be the indicator of the complement F^c, i.e. $\mathbf{d}(x, F) = \mathbf{1}_{F^c}(x)$. Formally, this is a particular case of the ρ-distance function, taking ρ to be the discrete metric.

The map $F \mapsto \mathbf{d}(\cdot, F)$ linearises $\mathcal{F}'$ by embedding it into the space $\mathbb{Y}$ of functions on $\mathbb{E}$. Let $\mathfrak{d}$ be a pseudometric on $\mathbb{Y}$. For example, if $\mathfrak{d}$ is the uniform distance, then the uniform distance between two distance functions equals the Hausdorff distance between the corresponding closed sets. Other metrics can be defined using $\mathbf{L}^p$ metrics on the family of distance functions, see Definition C.12.

If the space $\mathbb{E}$ is not compact, then some "bounded" or "restricted" versions of these $\mathbf{L}^p$ metrics are needed, for example,

$$\mathfrak{d}(f, g) = \left(\int_W (f(x) - g(x))^p dx \right)^{1/p},$$

where W is a compact set (window). This metric induces a pseudometric on $\mathcal{F}$ as

$$\Delta_W^p(F_1, F_2) = \left(\int_W |\mathbf{d}(x, F_1) - \mathbf{d}(x, F_2)|^p dx \right)^{1/p}. \quad (2.6)$$

These pseudometrics are convenient in image analysis, since they are less sensitive to small transformations of sets than the Hausdorff distance.

In general,
$$\mathfrak{d}_W(f, g) = \mathfrak{d}(\mathbf{1}_W f, \mathbf{1}_W g)$$
denotes the restricted version of $\mathfrak{d}$. Assume that $\mathfrak{d}_W(f, g) \leq \mathfrak{d}_{W_1}(f, g)$ if $W \subset W_1$, which automatically holds for Δ_W^p metrics. We also write $\mathfrak{d}(F, G)$ instead of $\mathfrak{d}(\mathbf{d}(\cdot, F), \mathbf{d}(\cdot, G))$ and $\mathfrak{d}(F, g)$ instead of $\mathfrak{d}(\mathbf{d}(\cdot, F), g(\cdot))$. In most cases $\mathfrak{d}$ is either the uniform metric or $\mathbf{L}^p$ metric. It is useful to put $\mathbf{d}(x, \emptyset) = \infty$ and $\mathfrak{d}(\emptyset, \emptyset) = 0$.

Mean distance function and distance average

Assume that $\mathbf{d}(x, X)$ is integrable for all $x \in \mathbb{E}$ and define the *mean* distance function
$$\bar{\mathbf{d}}(x) = \mathbf{E}\mathbf{d}(x, X) . \tag{2.7}$$

Proposition 2.7. *If* $\mathbf{d}$ *is a non-negative distance function, then* $\bar{\mathbf{d}}(x) = \mathbf{d}(x, F)$ *for some* $F \in \mathcal{F}$ *if and only if* X *is deterministic, i.e.* $X = F$ *almost surely.*

Proof. Sufficiency is evident. To prove *necessity*, suppose that
$$\bar{\mathbf{d}}(x) = \mathbf{E}\mathbf{d}(x, X) = \mathbf{d}(x, F) \quad \text{for all } x \in \mathbb{E}. \tag{2.8}$$
By (D2), $F = \{x : \bar{\mathbf{d}}(x) = 0\}$. Since the distance function is non-negative, $\mathbf{d}(x, X) = 0$ a.s. for all $x \in F$. Thus, $X \supset F$ a.s. By (D1), $\mathbf{d}(x, X) \leq \mathbf{d}(x, F)$. Finally, by (2.8), $\mathbf{d}(x, X) = \mathbf{d}(x, F)$ a.s. for all x. □

For the signed distance function (and other non-positive distance functions) the conclusion of Proposition 2.7 is not true.

Since, in general, $\bar{\mathbf{d}}$ is not a distance function itself, the minimisation problem (2.2) ought to be solved. Fix a closed set W (window) and define an increasing family of sets
$$\{\bar{\mathbf{d}} \leq t\} = \{x \in W : \bar{\mathbf{d}}(x) \leq t\}, \quad t \in \mathbb{R} .$$
by introducing a moving (upper) threshold for the mean distance function $\bar{\mathbf{d}}(x)$. The lower semicontinuity of $\bar{\mathbf{d}}$ follows from Fatou's lemma and, in turn, yields the closedness of $\{\bar{\mathbf{d}} \leq t\}$. The family $\mathcal{Z} = \{\{\bar{\mathbf{d}} \leq t\} : t \in \mathbb{R}\}$ provides candidates for the expectation of X.

Definition 2.8 (Distance average). *Let* $\bar{t}$ *be the value of* $t \in \mathbb{R}$ *which minimises the* $\mathfrak{d}_W$*-distance* $\mathfrak{d}_W(\{\bar{\mathbf{d}} \leq t\}, \bar{\mathbf{d}})$ *between the distance function of* $\{\bar{\mathbf{d}} \leq t\}$ *and the mean distance function of* X. *If* $\mathfrak{d}_W(\{\bar{\mathbf{d}} \leq t\}, \bar{\mathbf{d}})$ *achieves its minimum at several points, then* $\bar{t}$ *is chosen to be their infimum. The set*
$$\mathbf{E}_{DA} X = \mathbf{E}_{DA, \mathfrak{d}_W} X = \{\bar{\mathbf{d}} \leq \bar{t}\}$$
is said to be the distance average of X.

Note that $\partial_W(\{\bar{\mathbf{d}} \leq t\}, \bar{\mathbf{d}})$ attains its minimum, since $\{\bar{\mathbf{d}} \leq t\}$ is a right-continuous function of t. Mostly we omit the subscripts ∂ and W, but always remember that the distance average depends on the choice of the metric ∂ and the window W. The definition of the distance average does not use the linear structure of the underlying space $\mathbb{E}$. Thus, it is applicable for random sets in curved spaces (e.g. on the sphere).

Example 2.9 (Deterministic set). If X is a deterministic compact subset of W, then $\mathbf{E}_{DA} X = X$.

Example 2.10 (Random singleton). Let $X = \{\xi\}$ be a random singleton on the line.

(i) Assume that $\xi = 1$ with probability $1/2$ and $\xi = 0$ otherwise. Then, for the metric distance function d,

$$\bar{\mathbf{d}}(x) = \frac{1}{2}|x-1| + \frac{1}{2}|x|.$$

If ∂_W is either the Hausdorff metric or $\mathbf{L}^p$ metric with $W \supset [0,1]$, then $\bar{t} = 1/2$ and $\mathbf{E}_{DA} X = [0,1]$. The square distance function yields $\bar{\mathbf{d}}(x) = x^2 - x + 1/2$, so that $\mathbf{E}_{DA} X = \{1/2\}$ with $\bar{t} = 1/4$.

(ii) If ξ is uniformly distributed in $[0, a]$, then $\mathbf{E}_{DA} X = \{a/2\}$ for the metric distance function and ∂ being the Hausdorff metric. However, in general, the distance average of a random singleton may contain several points.

Example 2.11 (Segment and its boundary). Let $\mathbb{E} = \mathbb{R}$. Suppose $X = \{0, 1\}$ with probability $1/2$ and $X = [0, 1]$ otherwise. Then

$$\bar{\mathbf{d}}(x) = \begin{cases} -x, & x < 0, \\ x/2, & 0 \leq x < 1/2, \\ 1/2 - x/2, & 1/2 \leq x < 1, \\ x - 1, & x \geq 1. \end{cases}$$

If ∂ is the uniform metric (with $W = \mathbb{R}$), then $\bar{t} = 1/12$, and

$$\mathbf{E}_{DA} X = [-1/12, 1/6] \cup [5/6, 13/12].$$

Example 2.12 (Two-point set). Let $X = \{0, \xi\}$ be a two-point random set, where ξ is uniformly distributed in the unit interval $[0, 1]$ on the x-axis on the plane $\mathbb{R}^2$. Then, for $v = (x, 0) \in \mathbb{R}^2$ and the Euclidean distance function, we get

$$\bar{\mathbf{d}}(v) = \begin{cases} -x, & x < 0, \\ x - x^2, & 0 \leq x \leq 1/2, \\ x^2 - x + 1/2, & 1/2 < x \leq 1, \\ x - 1/2, & x > 1. \end{cases}$$

If ∂_W is the uniform metric with sufficiently large W, then $\bar{t} > 0$. Conversely, for each $t > 0$ the set $\{\bar{\mathbf{d}} \leq t\}$ contains a certain neighbourhood of the origin. Hence

$\mathbf{E}_{DA} X$ is not a subset of $[0, 1] \times \{0\}$, although $X \subset [0, 1] \times \{0\}$ almost surely. Therefore, we conclude that the property $X \subset K_0$ a.s. for non-random convex compact set K_0 does not yield $\mathbf{E}_{DA} X \subset K_0$.

It is easily seen that $\mathbf{E}_{DA} X$ always contains the set of minimum points for the mean distance function $\bar{\mathbf{d}}$. For instance, if X is a singleton, then $\mathbf{E}_{DA} X$ contains the set of points which minimise the expectation $\mathbf{E}d(x, \{\xi\})$. Therefore, for the metric distance function, $\mathbf{E}_{DA} X$ contains the set of *spatial medians* of ξ, see Small [538].

Proposition 2.13. *If $\mathbf{d}$ is the square distance function on a Hilbert space $\mathbb{E}$ and $X = \{\xi\}$ is a random point with integrable norm, then $\mathbf{E}_{DA} X = \{\mathbf{E}\xi\}$.*

Proof. In a Hilbert space, the expectation $a = \mathbf{E}\xi$ minimises the expected square distance function $\bar{\mathbf{d}}(x) = \mathbf{E}[\rho(x, \xi)^2]$. Therefore, $\mathbf{E}\xi \in \mathbf{E}_{DA} X$, since each non-empty set $\{\bar{\mathbf{d}} \le t\}$ contains the minimum point of $\bar{\mathbf{d}}(\cdot)$. By the monotonicity property (D1),

$$\rho(x, \mathbf{E}_{DA} X)^2 \le \rho(x, \mathbf{E}\xi)^2 \le \mathbf{E}[\rho(x, \xi)^2] = \bar{\mathbf{d}}(x)$$

for all x. Thus, $\partial_W(\{\bar{\mathbf{d}} \le t\}, \bar{\mathbf{d}}) \ge \partial_W(\{\mathbf{E}\xi\}, \bar{\mathbf{d}})$, whence $\mathbf{E}_{DA} X = \{\bar{\mathbf{d}} \le t\} = \{\mathbf{E}\xi\}$ for $\bar{t} = \mathbf{E}[(\xi - \mathbf{E}\xi)^2]$. □

It should be noted that the distance average, like many morphological operators (see Heijmans [228]) is non-linear, that is, the average of the union or the Minkowski (element-wise) sum of two random sets does not coincide with the union or Minkowski sum of the corresponding averages. Moreover, the distance average is not associative in general and $\mathbf{E}_{DA}(cX)$ is not necessarily equal to $c\mathbf{E}_{DA} X$ for $c > 0$.

Open problem 2.14. When is $\mathbf{E}_{DA}\{\xi\}$ equal to the conventional expectation of a random element ξ in a Banach space $\mathbb{E}$?

2.4 Radius-vector expectation

Let $\mathbb{E} = \mathbb{R}^d$ be the Euclidean space. A compact set K is called *star-shaped* with respect to the origin 0 if, for each $x \in X$, the segment $[0, x]$ with end-points 0 and x is contained in K. The *radius-vector function* of K is defined by

$$r_K(u) = \sup\{t : tu \in K, \ t \ge 0\}$$

for u from the unit sphere $\mathbb{S}^{d-1}$, see Figure 2.2.

The representation of star-shaped compact sets by means of their radius-vector functions establishes a correspondence $X \mapsto r_X(\cdot) = \xi_X(\cdot)$ between star-shaped random compact sets and random functions on the unit sphere and so can be used to define an expectation of X. The expected values $\mathbf{E}r_X(u)$, $u \in \mathbb{S}^{d-1}$, define a function which is the radius-vector function of a deterministic star-shaped set K called the *radius-vector expectation* of X. The major shortcomings are the necessity to work with star-shaped sets and the non-linearity with respect to translations of the sets, since the radius-vector function depends non-linearly on the location of the origin within the set, whereas its natural location is difficult to identify in many applications.

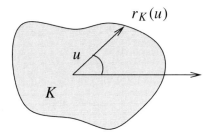

Figure 2.2. The radius-vector function of a star-shaped set.

3 Expectations on lattices and in metric spaces

3.1 Evaluations and expectations on lattices

Let ℓ be a *sup-generating* family in a lattice $\mathbb{L}$, i.e. assume that every element in $\mathbb{L}$ can be written as a supremum $\vee D$ for some $D \subset \ell$. A mapping $u \colon \ell \mapsto \bar{\mathbb{R}}$ is called an *evaluation* if $x \leq \vee D$ implies $u(x) \leq \vee\{u(y) : y \in D\}$ for all $x \in \ell$ and $D \subset \ell$. Define the mappings $\delta_u \colon \mathbb{L} \mapsto \bar{\mathbb{R}}$ and $\varepsilon_u \colon \bar{\mathbb{R}} \mapsto \mathbb{L}$ as

$$\delta_u(F) = \sup\{u(x) : x \in \ell, \, x \leq F\}, \quad F \in \mathbb{L},$$
$$\varepsilon_u(y) = \vee\{x \in \ell : u(x) \leq y\}, \quad y \in \bar{\mathbb{R}}.$$

The pair $(\varepsilon_u, \delta_u)$ forms an *adjunction* between $\bar{\mathbb{R}}$ and $\mathbb{L}$, which means that $\delta_u(F) \leq y$ if and only if $F \leq \varepsilon_u(y)$, see Heijmans [228].

Let U be a family of evaluations. For each $F \in \mathbb{L}$ define

$$\mathrm{cl}(F; U) = \bigwedge_{u \in U} \varepsilon_u(\delta_u(X)),$$

which is called the *U-closure* of F. If the lattice $\mathbb{L}$ contains the objects under study, then the evaluations represent the available information and may be regarded as measurements. Then $\mathrm{cl}(F; U)$ is an element of $\mathbb{L}$ which is retrievable from the measurements of F. The family U is called *unbiased* if $\mathrm{cl}(\{x\}; U) = x$ for all $x \in \ell$. This is the case if for any two elements $x, y \in \ell$ with $x \not\leq y$ there exists an evaluation $u \in U$ such that $u(x) \not\leq u(y)$.

Let X be an $\mathbb{L}$-valued random element. Assume that each $u \in U$ is measurable with respect to the σ-algebra $\sigma(\mathbb{L})$ generated by the Scott topology on $\mathbb{L}$. Then $\delta_u(X)$ is a random variable for each $u \in U$ and X is said to be *U-integrable* if $\delta_u(X)$ is integrable for each $u \in U$. The *U-expectation* of X is defined as

$$\mathbf{E}_U X = \bigwedge_{u \in U} \varepsilon_u(\mathbf{E}\delta_u(X)). \tag{3.1}$$

If X is deterministic, then $\mathbf{E}_U X = \mathrm{cl}(X; U)$ is the U-closure of X. In general, $\mathbf{E}_U X$ is U-closed and

$$\mathbf{E}_U X = \mathbf{E}_U \mathrm{cl}(X; U). \tag{3.2}$$

Example 3.1 (*U*-expectation of random closed sets). Let $\mathbb{L} = \mathcal{F}$ be the lattice of closed sets in $\mathbb{E}$ ordered by inclusion. The sup-generating family ℓ is the family of all singletons, i.e. ℓ can be identified with $\mathbb{E}$. A function $u \colon \mathbb{E} \mapsto \bar{\mathbb{R}}$ is an evaluation if $x \in \mathrm{cl}(D)$ implies $u(x) \leq u^\vee(D) = \sup\{u(y) : y \in D\}$ for all $x \in \mathbb{E}$ and $D \subset \mathbb{E}$. Then $\delta_u(F) = u^\vee(F)$ and $\varepsilon_u(y) = \mathrm{cl}(\{x \in \mathbb{E} : u(x) \leq y\})$. The *U*-expectation of a random closed set X is given by

$$\mathbf{E}_U X = \bigcap_{u \in U} \varepsilon_u(\mathbf{E}\delta_u(X)) = \bigcap_{u \in U} \{x : u(x) \leq \mathbf{E} u^\vee(X)\}. \quad (3.3)$$

For instance, the selection expectation is obtained if U is the family of all linear functions on $\mathbb{R}^d$. Then $u^\vee(X) = h(X, u)$ is the support function, the *U*-closure is the convex hull, (3.3) becomes the definition of the selection expectation using the support functions and (3.2) corresponds to Theorem 1.26.

3.2 Fréchet expectation

It is well known that if $\mathbb{E}$ is a Hilbert space, then the expectation $\mathbf{E}\xi$ of a square integrable $\mathbb{E}$-valued random element ξ can be defined as the (necessarily unique) element $a \in \mathbb{E}$ which minimises $\mathbf{E}\|\xi - a\|^2$. This definition can be extended to general metric spaces, although then the uniqueness property might be lost.

Definition 3.2 (Fréchet expectation). Let ξ be a random element in the space $\mathbb{E}$ with metric (or pseudometric) ρ such that $\mathbf{E}\rho(\xi, x_0)^2 < \infty$ for some $x_0 \in \mathbb{E}$. The *Fréchet expectation* of ξ (denoted by $\mathbf{E}_F\xi$) is the set of all $a \in \mathbb{E}$ such that $x = a$ minimises $\mathbf{E}\rho(\xi, x)^2$ over $x \in \mathbb{E}$. The value of $\mathbf{E}\rho(\xi, a)^2$ is called the *Fréchet variance* of ξ.

This general definition is applicable to the family $\mathcal{K}'$ of non-empty compact subsets of $\mathbb{E}$ equipped with the Hausdorff metric (if $\mathbb{E}$ is infinite-dimensional, it is possible to consider the family of non-empty bounded closed sets instead of $\mathcal{K}'$). Let X be a random set with values in $\mathcal{K}'$ such that $\mathbf{E}\rho_\mathrm{H}(X, L)^2 < \infty$ for some $L \in \mathcal{K}'$. The *Fréchet expectation*, $\mathbf{E}_F X$, of X is a family of sets which consists of all sets $K \in \mathcal{K}'$ providing a global minimum for $\mathbf{E}\rho_\mathrm{H}(X, K)^2$.

Example 3.3. If $X \subset \mathbb{R}^1$ takes values $[0, 1]$ and $\{0, 1\}$ with probabilities $1/2$, then its Fréchet expectation is unique and equals $[0, 0.4] \cup [0.6, 1]$.

This approach is very general and can be used if the Hausdorff metric ρ_H is replaced by another metric on $\mathcal{K}'$, for example, by one of the Δ-metrics from (2.6) or Definition C.12. Unfortunately, in most practical cases it is not possible to solve the basic minimisation problem, since the parameter space $\mathcal{K}'$ is too rich.

The Fréchet expectation can be defined for a *sample* in a metric space, so that $a \in \mathbb{E}$ is called a *mean* of $x_1, \ldots, x_n \in \mathbb{E}$ if

$$\sum_{i=1}^n \rho(x_i, a)^2 = \inf_{b \in \mathbb{E}} \sum_{i=1}^n \rho(x_i, b)^2.$$

The set of all means of $x_1, \ldots, x_n$ is denoted by $M(x_1, \ldots, x_n)$. The following result provides a strong law of large numbers for the Fréchet expectation. It is formulated for general metric spaces, and so holds for the space of compact sets equipped with the Hausdorff metric.

Theorem 3.4 (Law of large numbers for Fréchet expectation). *Let $\xi, \xi_1, \xi_2, \ldots$ be i.i.d. random elements in a separable space $\mathbb{E}$ with a finite pseudometric ρ. If $\mathbf{E}\rho(\xi, x_0)^2 < \infty$ for some $x_0 \in \mathbb{E}$, then*

$$M = \bigcap_{k=1}^{\infty} \mathrm{cl}(\bigcup_{n=k}^{\infty} M(\xi_1, \ldots, \xi_n)) \subset \mathbf{E}_F\xi \quad \text{a.s.}$$

Proof. Let $\mathbb{Q}$ be a countable dense subset of $\mathbb{E}$. By the strong law of large numbers, the event

$$A = \left\{ \lim_{n\to\infty} \frac{1}{n} \sum_{i=1}^{n} \rho(\xi_i, x)^m = \mathbf{E}\rho(\xi_i, x)^m \text{ for all } x \in \mathbb{Q}, \, m = 1, 2 \right\}$$

occurs with probability 1. Consider an arbitrary point $x \in \mathbb{E}$. There exists a sequence $\{x_n, n \geq 1\} \subset \mathbb{Q}$ such that $x_n \to x$. Then

$$\left| \frac{1}{n} \sum_{i=1}^{n} [\rho(\xi_i, x_k)^2 - \rho(\xi_i, x)^2] \right| \leq \rho(x_k, x)^2 + 2\rho(x_k, x) \frac{1}{n} \sum_{i=1}^{n} \rho(\xi_i, x_k).$$

The continuity of $\mathbf{E}\rho(\xi, \cdot)$ and $\mathbf{E}\rho(\xi, \cdot)^2$ implies

$$\lim_{n\to\infty} \frac{1}{n} \sum_{i=1}^{n} \rho(\xi_i, x)^2 = \mathbf{E}\rho(\xi, x)^2 \quad \text{a.s.}$$

for all $x \in \mathbb{E}$.

Fix an elementary event $\omega \in A$ and any point $z \in M$. For every $k \geq 1$ pick

$$z_k \in \cup_{n=k}^{\infty} M(\xi_1(\omega), \ldots, \xi_n(\omega))$$

with $\rho(z_k, z) \leq 1/k$ and define

$$n_k = \min\{n \geq 1 : z_k \in M(\xi_1(\omega), \ldots, \xi_n(\omega))\}.$$

From

$$\left(\frac{1}{n_k} \sum_{i=1}^{n_k} \rho(\xi_i, z)^2 \right)^{1/2} \leq \left(\frac{1}{n_k} \sum_{i=1}^{n_k} \rho(\xi_i, z_k)^2 \right)^{1/2} + \left(\frac{1}{n_k} \sum_{i=1}^{n_k} \rho(z_k, z)^2 \right)^{1/2}$$

$$\leq \left(\frac{1}{n_k} \sum_{i=1}^{n_k} \rho(\xi_i, z_k)^2 \right)^{1/2} + \frac{1}{k}$$

it follows that

$$(\mathbf{E}\rho(\xi,z)^2)^{1/2} \leq \liminf_{k\to\infty} \left(\frac{1}{n_k}\sum_{i=1}^{n_k}\rho(\xi_i,z_k)^2\right)^{1/2}.$$

Since $z_k \in M(\xi_1, \ldots, \xi_n)$,

$$\mathbf{E}\rho(\xi,z)^2 \leq \mathbf{E}\rho(\xi,x)^2$$

for every $x \in \mathbb{E}$, whence $z \in \mathbf{E}_F\xi$. □

3.3 Expectations of Doss and Herer

Doss expectation in metric spaces

The following definition is applicable for random elements in a metric space $(\mathbb{E}, \rho)$.

Definition 3.5 (Doss expectation). Let ξ be a random element in $\mathbb{E}$ such that $\mathbf{E}\rho(\xi, x)$ is finite for some $x \in \mathbb{E}$. The *Doss expectation* of $\xi \in \mathbb{E}$ is the set $\mathbf{E}_D\xi$ of all points $a \in \mathbb{E}$ such that $\rho(a, x) \leq \mathbf{E}\rho(\xi, x)$ for all $x \in \mathbb{E}$.

The triangle inequality implies that $\mathbf{E}\rho(\xi, x)$ is finite for all $x \in \mathbb{E}$ if it is finite for at least one x. In comparison with the Fréchet expectation, the Doss expectation does not require the existence of the second moment of $\rho(\xi, x)$. The following proposition shows that if $\mathbb{E}$ is a Banach space, then the Doss expectation becomes the conventional expectation of ξ.

Theorem 3.6 (Doss and Bochner expectations). *If $\mathbb{E}$ is the Banach space with the metric $\rho(x, y) = \|x - y\|$ generated by the corresponding norm, then each Bochner integrable random element $\xi \in \mathbb{E}$ satisfies $\mathbf{E}_D\xi = \{\mathbf{E}\xi\}$, where $\mathbf{E}\xi$ is the Bochner expectation of ξ.*

Proof. We have to show that $\mathbf{E}\xi$ is the unique element of $\mathbb{E}$ satisfying

$$\|\mathbf{E}\xi - x\| \leq \mathbf{E}\|\xi - x\|$$

for all $x \in \mathbb{E}$. If $u \in \mathbf{E}_D\xi$ and $a \in \mathbb{E}$, then $u + a \in \mathbf{E}_D(\xi + a)$.

Let ξ and η be two Bochner integrable random elements in $\mathbb{E}$ with distributions $\mathbf{P}_\xi$ and $\mathbf{P}_\eta$ respectively. If $a \in \mathbf{E}_D\xi$ and $b \in \mathbf{E}_D\eta$, then $a + b \in \mathbf{E}_D(\xi + \eta)$, since, for all $x \in \mathbb{E}$,

$$\mathbf{E}\|\xi + \eta - x\| = \int_{\mathbb{E}}\left(\int_{\mathbb{E}}\|v - (x-u)\|\mathbf{P}_\eta(dv)\right)\mathbf{P}_\xi(du)$$

$$\geq \int_{\mathbb{E}}\|(x-u) - b\|\mathbf{P}_\xi(du) \geq \|a + b - x\|.$$

Let $\{\xi_n, n \geq 1\}$ be a sequence of i.i.d. random elements having the same distribution as ξ. Without loss of generality assume that $\mathbf{E}\xi = 0$. Let $a_i \in \mathbf{E}_D\xi_i$, $i \geq 1$. Then

$$\frac{1}{n}\sum_{i=1}^{n} a_i \in \mathbf{E}_D\left(\frac{1}{n}\sum_{i=1}^{n}\xi_i\right),$$

that is, for all $x \in \mathbb{E}$,

$$\left\|\frac{1}{n}\sum_{i=1}^{n} a_i - x\right\| \leq \mathbf{E}\left\|\frac{1}{n}\sum_{i=1}^{n}\xi_i - x\right\|.$$

The right-hand side converges to $\|x\|$ by the strong law of large numbers in $\mathbb{E}$. Therefore,

$$\limsup_{n\to\infty}\left\|\frac{1}{n}\sum_{i=1}^{n} a_i - x\right\| \leq \|x\|$$

for all $x \in \mathbb{E}$. If we assume that $\mathbf{E}_D\xi$ contains a point $a \neq 0$, then we come to a contradiction by putting $a_i = a$, $i \geq 1$. □

Doss expectation for random sets and Herer expectation

It is possible to specialise the Doss expectation for non-empty bounded random closed sets, since the family $\mathcal{F}'_b$ of non-empty bounded closed sets forms a metric space under the Hausdorff metric. Let X be an almost surely non-empty bounded random closed set such that $\rho_H(X, x_0)$ has a finite expectation for some $x_0 \in \mathbb{E}$ (so that X is integrably bounded). Then $\mathbf{E}\rho_H(X, \{x\})$ is finite for all $x \in \mathbb{E}$ and, moreover, $\mathbf{E}\rho_H(X, L) < \infty$ for all $L \in \mathcal{F}'_b$. By Definition 3.5, the Doss expectation of X is given by

$$\mathbf{E}_D X = \{K \in \mathcal{F}'_b : \rho_H(K, L) \leq \mathbf{E}\rho_H(X, L) \text{ for all } L \in \mathcal{F}'_b\}.$$

Thus, $\mathbf{E}_D X$ is a family of sets. It is possible to come up with a modification of this expectation which yields a single set.

Definition 3.7 (Herer expectation). Let X be an integrable random closed set. The Herer expectation of X is defined as

$$\mathbf{E}_H X = \{a \in \mathbb{E} : \rho(a, x) \leq \mathbf{E}\rho_H(X, \{x\}) \text{ for all } x \in \mathbb{E}\}. \quad (3.4)$$

Example 3.8. Let $\mathbb{E} = \mathbb{R}$ and let X take values $\{0\}$ and $\{0, 2\}$ with equal probabilities. Then $\mathbf{E}_D X$ is the family of all compact sets K such that $\{0, 1\} \subset K \subset [0, 1]$ and $\mathbf{E}_H X = [0, 1]$.

It is easy to show that $\mathbf{E}_H X$ is the union of all sets K from $\mathbf{E}_D X$. Indeed,

$$\mathbf{E}_D X \subset \{K \in \mathcal{F}'_b : \rho_H(K, \{x\}) \leq \mathbf{E}\rho_H(X, \{x\}) \text{ for all } x \in \mathbb{E}\},$$

while the union of all K from the right-hand side equals the Herer expectation of X.

The Herer expectation appears as a particular case of the expectation on lattices defined by (3.1), if $\mathbb{L}$ is the lattice of closed bounded subsets of $\mathbb{E}$ and $u(x) = \rho(x, u)$ with $u \in U = \mathbb{R}^d$. If $X = \{\xi\}$ is a singleton, then (3.4) turns into the definition of the Doss expectation of ξ. Theorem 3.6 implies $\mathbf{E}_H\{\xi\} = \mathbf{E}\xi$, so that the Herer expectation of a singleton in a Banach space coincides with its Bochner expectation.

It is possible to rewrite (3.4) as

$$\mathbf{E}_H X = \bigcap_{x \in \mathbb{E}} B_{\mathbf{E}\rho_H(X, \{x\})}(x). \tag{3.5}$$

For example, if $\mathbb{E} = \mathbb{R}^d$, then $\mathbf{E}_H X$ is the intersection of closed balls and so is a convex set. The following proposition immediately follows from (3.5).

Proposition 3.9 (Monotonicity of Herer expectation). *If $X \subset Y$ and Y is an integrable random closed set, then $\mathbf{E}_H X \subset \mathbf{E}_H Y$.*

If applied to all selections $\xi \in X$, Proposition 3.9 yields the following relationship between the selection and Herer expectations.

Theorem 3.10 (Herer and selection expectations).
(i) *If $\mathbb{E}$ is a Banach space and X is an integrable random closed set, then $\mathbf{E}_A X \subset \mathbf{E}_H X$. In particular, $\mathbf{E}_H X$ is not empty.*
(ii) *Let X be an integrably bounded random compact set in a Hilbert space $\mathbb{E}$. Assume that X is either almost surely convex or the probability space is non-atomic. Then $\mathbf{E}_A X = \mathbf{E}_H X$.*

Proof.
(i) For each selection $\xi \in X$ we have $\mathbf{E}_H\{\xi\} \subset \mathbf{E}_H X$ by Proposition 3.9. Since $\mathbf{E}_H\{\xi\} = \{\mathbf{E}\xi\}$, the Herer expectation $\mathbf{E}_H X$ contains expectations of all selections of X, whence the statement of the theorem immediately follows.

(ii) In view of (i), it suffices to show that the Herer expectation is contained in the selection expectation assuming that $\mathbf{E}_A X$ is convex. By the Hahn–Banach theorem, for every $y \notin \mathbf{E}_A X$ and $\varepsilon > 0$, there exists an $u \in \mathbb{R}^d$ such that

$$h(\mathbf{E}_A X, u) \leq \langle y, u \rangle - \varepsilon.$$

For every $x \in \mathbb{E}$, referring to Theorem 1.10, we have

$$\mathbf{E}\rho_H(X, \{x\})^2 - \|x - y\|^2 = \mathbf{E}\left(\sup_{z \in X} \|x - z\|^2 - \|x - y\|^2\right)$$

$$= \sup_{\xi \in \mathcal{S}(X)} \mathbf{E}\left(\|x - \xi\|^2 - \|x - y\|^2\right)$$

$$\leq \sup_{\xi \in \mathcal{S}(X)} \mathbf{E}\|y - \xi\|^2 + 2 \sup_{\xi \in \mathcal{S}(X)} \mathbf{E}\langle y - \xi, x - y \rangle.$$

In particular, for $x_c = y - cu$ the above inequality implies that

$$\mathbf{E}\rho_H(X, \{x_c\})^2 - \|x_c - y\|^2 \leq \sup_{\xi \in \mathcal{S}(X)} \mathbf{E}\|y - \xi\|^2 + 2 \sup_{\xi \in \mathcal{S}(X)} \mathbf{E}\langle y - \xi, -cu\rangle$$
$$= \mathbf{E}\rho_H(X, \{y\})^2 - 2c(\langle y, u\rangle - \sup_{\xi \in \mathcal{S}(X)} \langle \mathbf{E}\xi, u\rangle)$$
$$= \mathbf{E}\rho_H(X, \{y\})^2 - 2c(\langle y, u\rangle - h(\mathbf{E}_A X, u))$$
$$\leq \mathbf{E}\rho_H(X, \{y\})^2 - 2c\varepsilon.$$

For $c > \mathbf{E}\rho_H(X, \{y\})^2/(2\varepsilon)$ the right-hand side becomes negative, meaning that

$$\mathbf{E}\rho_H(X, \{x_c\}) < \|x_c - y\|,$$

whence y does not belong to the Herer expectation of X. □

Open problem 3.11. Find minimal conditions on a general Banach space $\mathbb{E}$ that would guarantee that Theorem 3.10(ii) holds.

Doss convexity

It should be noted that the Herer expectation can be defined in general metric spaces, where the selection expectation makes no sense. However, the Herer expectation can be empty, as the following example shows.

Example 3.12 (Empty Herer expectation). Let $\mathbb{E} = \mathbb{S}^1$ be the unit circle in $\mathbb{R}^2$ with the geodesic distance as the metric. Consider a random singleton $X = \{\xi\}$ with ξ taking the values $(0, 1)$ and $(0, -1)$ with equal probabilities. Then both the Doss expectation of ξ and the Herer expectation of X are empty. Indeed, $\mathbf{E}\rho(\xi, x) = \pi/2$ and $\mathbf{E}_D\xi = \cap_{x \in \mathbb{E}}\{y : \rho(y, x) \leq \pi/2\}$ is empty.

Incidentally, in this case the distance average with the metric distance function yields $\mathbf{E}_{DA} X = \mathbb{S}^1$. The distance average with the square distance function yields $\mathbf{E}_{DA} X = \{(1, 0), (-1, 0)\}$. The latter coincides with the Fréchet expectation of X.

Definition 3.13 (Doss convexity). A metric space $(\mathbb{E}, \rho)$ is called *convex* in the sense of Doss if for any two elements $x_1, x_2 \in \mathbb{E}$ there exists an element $a \in \mathbb{E}$ such that

$$\rho(x, a) \leq \frac{1}{2}(\rho(x, x_1) + \rho(x, x_2)) \tag{3.6}$$

for all $x \in \mathbb{E}$.

It is easy to see that (3.6) implies that the Doss expectation exists for all random elements which take two values with equal probabilities. The following theorem proved by Herer [234] shows that the convexity in the sense of Doss guarantees that the Herer expectation is not empty. Clearly, a Banach space is convex in the sense of Doss.

Theorem 3.14 (Herer expectation in Doss-convex space). *Assume that all bounded closed sets in $\mathbb{E}$ are compact. Any integrable random closed set X has a non-empty Herer expectation if and only if $\mathbb{E}$ is convex in the sense of Doss.*

3.4 Properties of expectations

It is possible to formulate several basic properties of a "reasonable" expectation $\mathbf{E}X$ of a random closed set X. The first group of properties is related to inclusion relationships.

(A1) If X is deterministic, then $\mathbf{E}X = X$.
(A2) If $K \subset X$ a.s., where K is deterministic, then $K \subset \mathbf{E}X$.
(A3) If $X \subset W$ a.s. for a deterministic set W, then $\mathbf{E}X \subset W$.
(A4) If $X \subset Y$ a.s., then $\mathbf{E}X \subset \mathbf{E}Y$.

Clearly, (A2) and (A3) imply (A1), while (A1) and (A4) yield both (A2) and (A3).

The second group consists of the properties related to invariance with respect to some transformations.

(B1) If X and gX coincide in distribution for every g from a certain group G, then $g\mathbf{E}X = \mathbf{E}X$ for each $g \in \mathsf{G}$.
(B2) Translation-equivariance: $\mathbf{E}(X + x) = \mathbf{E}X + x$ for all translations x (if $\mathbb{E} = \mathbb{R}^d$).
(B3) Homogeneity: $\mathbf{E}(cX) = c\mathbf{E}(X)$ for each $c \in \mathbb{R}$.

The third group of properties relates expectations of sets and "usual" expectations of random variables and vectors.

(C1) If $X = \{\xi\}$ is a random singleton, then $\mathbf{E}X = \{\mathbf{E}\xi\}$.
(C2) If $X = B_\eta(\xi)$ is a ball of random radius η and centre ξ, then $\mathbf{E}X = B_{\mathbf{E}\eta}(\mathbf{E}\xi)$.
(C3) If $X = \text{co}(\xi_1, \ldots, \xi_n)$ is the convex hull of a finite number of random elements, then $\mathbf{E}X = \text{co}(\mathbf{E}\xi_1, \ldots, \mathbf{E}\xi_n)$.
(C4) If $X = \{\xi_1, \ldots, \xi_n\}$, then $\mathbf{E}X = \{\mathbf{E}\xi_1, \ldots, \mathbf{E}\xi_n\}$.

Some of these natural properties are non-compatible and have far-reaching consequences. For example, (A4) and (C1) imply that $\mathbf{E}\xi \in \mathbf{E}X$ for each selection $\xi \in X$, so that $\mathbf{E}X$ contains the selection expectation of X. For instance, the Doss expectation satisfies (C1) and (A4), whence it contains the selection expectation. However the convexity of both selection and Doss expectations severely restricts possible applications, e.g. in image analysis, where most images are non-convex. The distance average seems to be the most versatile expectation, which can have a wide range of possible applications in image analysis, see Lewis, Owens and Baddeley [350].

Notes to Chapter 2

Section 1.1. Most of results of Section 1.1 are taken from Hiai and Umegaki [255], who considered a more general case of multivalued functions defined on a measurable space with a not necessarily finite measure. It should be noted that the results for the selection (Aumann) expectation are often formulated for integrals of multivalued functions.

The results concerning $\mathfrak{F}_X$-measurable selections are due to Hart and Kohlberg [223] for the case of random compact sets. They have been generalised by Artstein and Hart [22] for general random closed sets. Some of them have been reformulated here for random closed sets in Banach spaces.

A converse statement to Proposition 1.2(iii) is wrong, as Example 1.2.15 shows. Theorem 1.2.16 concludes that the weak closures of the families of all selections coincide for two identically distributed random closed sets on non-atomic probability spaces. The same result holds for the families of integrable selections provided that the random closed sets are integrable, i.e. possess at least one integrable selection. This has been proved by Hart and Kohlberg [223] for random compact sets, while a simple truncation argument extends this statement for any two integrable random closed sets assuming that $\mathbb{E}$ is locally compact, see Artstein [16] and Artstein and Hart [22].

Theorem 1.10 goes back to Hiai and Umegaki [255], but is reformulated here in a probabilistic language for stochastic processes and random sets.

Section 1.2. The selection expectation stems from the theory of multivalued functions. It is sometimes called the Kudo–Aumann integral, acknowledging the contribution by Kudo [336]. Aumann [31] defined an integral of a multivalued function in $\mathbb{R}^d$ as a family of all integrals of its selections, while its random sets meaning was discovered by Artstein and Vitale [23]. Another integral of a set-valued function was defined by Debreu [124] as a limit of sums for simple functions. Both definitions are equivalent, as shown by Byrne [85]. A construction of the expectation for random convex sets in the Euclidean space using approximations by simple random sets is described by Hess [246]. Artstein and Burns [20] introduced a set-valued integral of a closed-valued multifunction using Riemann sums, which can be used to define an expectation if the probability space is [0, 1]. Then it coincides with the Debreu approach for random compact sets in $\mathbb{R}^d$. The Aumann and Bochner integrals for multivalued functions can be defined with respect to any finitely-additive measure, see Martellotti and Sambucini [379].

Theorem 1.15 goes back to Richter [488]. The current proof is adapted from Ioffe and Tihomirov [271], where also various relationships with optimisation are discussed. Theorem 1.16 was proved by Hiai and Umegaki [255]. Relationships with the so-called bang-bang principle in optimisation are discussed by Artstein [15]. A proof of Lyapunov's theorem for set-valued measures can be found in Ioffe and Tihomirov [271]. A generalisation of Lyapunov's theorem for vector valued measures is given by Cerf and Mariconda [93].

Example 1.23(i) is given in Giné and Hahn [198] and goes back to Z. Artstein and R.A. Vitale. Example 1.23(ii) is taken from Li and Ogura [352]. Theorem 1.24 is a synthesis of results from Hiai and Umegaki [255] and Byrne [85]. These results have been generalised for conditional expectations by Li and Ogura [352]. It should be noted that necessary and sufficient conditions for $\mathbf{E}X = \mathbf{E}_\mathbb{I} X$ in a general separable Banach space $\mathbb{E}$ are not known so far. Theorem 1.22 is a folklore fact that also appears in Papageorgiou [442].

Stich [542] extended the Aumann integral for non-measurable maps X from $(\Omega, \mathfrak{F}, \mathbf{P})$ into the family of closed subsets of the real line. Khan and Sun [300] proved a number of results on integrals of set-valued functions with a countable family of values giving rather general statements about their convexity and the closedness. These results can be immediately interpreted as concerning selection expectations of random closed sets with at most a countable family of values. Pucci and Vitiliaro [460] considered the Aumann integral of not necessarily closed set-valued functions in a separable reflexive Banach space.

The concept of a Pettis integral is essentially weaker than the Bochner integral in infinite-dimensional spaces. Then the Aumann–Pettis expectation (or integral) is defined as the set of all Pettis integrals for Pettis integrable selections. El Amri and Hess [160] study the Pettis integrability for random closed sets with possibly unbounded values. Hess [248] provides a useful survey on set-valued integration and set-valued probability theory.

Section 1.3. Theorem 1.28 was proved by Vitale [579] and subsequently generalised for random convex sets by Molchanov [400].

Section 1.4. Although the idea of a canonical representation of a random set goes back to Matheron [381], its use within the context of expectations of random sets was realised by Vitale [582]. The current presentation incorporates some ideas from Hiai [253], in particular, this relates to Theorem 1.32. Theorem 1.32(ii),(iv) appears also in Hess [247]. The translative expectation was defined by Vitale [584] with the aim to tighten the Brunn–Minkowski inequality for random sets, see Theorem 3.1.11. Another proof of Theorem 1.34 is given in Vitale [584].

Multivalued measures were studied by Artstein [14], Costé [105], Debreu and Schmeidler [125], Godet-Thobie [205], Hiai [250] and Luu [361]. The current presentation follows Hiai [254]. Proposition 1.36 is proved by Papageorgiou [442].

Section 1.5. The first variant of Fatou's lemma for multivalued functions is due to Aumann [31], see Theorem 1.37 for the non-atomic case. Subsequent improvements have been done by Schmeidler [515] (where another proof of Theorem 1.37 can be found), Hildenbrand and Mertens [256], Hiai [253], Khan and Majumdar [299], Yannelis [619] and Balder [43, 44]. The most general case so far is treated by Balder and Hess [45]. It should be noted that the formulations of [45] are more general than Theorems 1.39 and 1.42 by imposing less restrictive conditions on measurability of multivalued functions. Examples 1.40 and 1.41 are taken from Hess [241] and Balder and Hess [45].

Theorems 1.43 and 1.44 are due to Hiai [253]. It is shown by Jacobs [276] that $\mathbf{E}(X \cap B_r(0))$ converges to $\mathbf{E}(X)$ as $r \uparrow \infty$ if X has a bounded selection.

The convergence of expectations for a weakly convergent sequence of random sets was considered by Artstein and Wets [24] (in a different setup involving integrals of a fixed multifunction with respect to a weakly convergent sequence of probability measures) and Molchanov [398], where Theorem 1.45 was proved. Further results on the convergence of conditional expectations are due to Castaing, Ezzaki and Hess [90].

Section 1.6. The presentation of conditional expectations of random closed sets follows Hiai [254] and Hiai and Umegaki [255], where all proofs can be found. Conditional expectations of random sets are widely investigated in the literature, mostly because of applications to the studies of multivalued martingales, see Section 5.1.1. Other results on conditional expectations are due to Dynkin and Evstigneev [152], Neveu [425], Sainte-Beuve [508], Valadier [570] and van Cutsem [114]. The convergence results for conditional expectations are due to Hiai [253] and Hess [241]. The latter paper deals with unbounded random closed sets exploiting the truncation argument to prove the results.

Similar to a definition of an essential supremum, it is possible to define an essential convex closure For a family of random closed sets, Wang [595] considered the properties of conditional expectation in this framework and applied them to prove an optional sampling theorem and the convergence results for set-valued martingales.

Section 2.1. The general linearisation approach to defining expectation in the space $\mathcal{F}$ was described in Molchanov [405].

Section 2.2. The Vorob'ev expectation was originally defined by Vorob'ev [586] for discrete random sets (in a finite space $\mathbb{E}$) with measures replaced by cardinalities of the corresponding sets. This concept was reformulated for the general case by Stoyan and Stoyan [547]. Further generalisations of the Vorob'ev expectation are suggested by Molchanov [408]. Note that the coverage function is called the membership function in the theory of fuzzy sets and the thresholded sets are called α-cuts, see Li and Lee [351].

Section 2.3. The distance average was introduced by Baddeley and Molchanov [40] who also discuss applications to image analysis and properties of the empirical estimator of the distance average.

The metric distance function is a well known *binary distance transform* from the image processing literature. Distance functions are widely used in image analysis and can be computed very efficiently, see Rosenfeld and Pfalz [500]. Applications of the signed distance function to shape analysis were considered by Delfour and Zolésio [128]. It should be noted that the distance average includes the Vorob'ev expectation as a particular case for the indicator distance function defined in Example 2.6(iv). A characterisation theorem for random closed sets by their distance functions (called processes of first contact variables) is obtained by Ayala and Montes [35].

The Δ-metric (2.6) was defined by Baddeley [38], who showed that it has a number of advantages over the Hausdorff metric when assessing dissimilarities between binary images.

The mean distance function was studied by Molchanov and Terán [411] and applied to the thresholding problem in image analysis by Friel and Molchanov [183].

Section 2.4. The radius-vector expectation is described in Stoyan and Stoyan [547]. Radius-vector functions are very popular in the engineering literature, where it is usual to apply Fourier methods for shape description, see Beddow and Mellow [54] and Réti and Czinege [485]. An application of the radius-vector expectation to averaging and modelling of images is discussed by Hobolth and Vedel Jensen [259], Mancham and Molchanov [375] and Stoyan and Stoyan [547]. The common idea is to perturb the template given by the radius-vector expectation using a stochastic process on the unit circle (or unit sphere).

Section 3.1. The presentation follows Heijmans and Molchanov [229] who elaborate this idea for maps between two lattices and discuss a number of further concepts including convolution operation on lattices and the corresponding strong law of large numbers. Section 3.1 describes the simplest variant, where the second lattice is the extended real line. A related concept appears in Molchanov [398, Ch. 2].

An expectation for a lattice-valued random element can be defined using approximations by simple functions. A systematic study of this definition including the related concepts of conditional expectation, martingales and the convergence is published by Jonasson [282].

Section 3.2. The Fréchet expectation on metric spaces is due to Fréchet [181]. The strong law of large numbers for the Fréchet expectation (Theorem 3.4) was proved by Ziezold [627]. Its variant for the centroids of order $k \geq 1$ (the minimisers of $\mathbf{E}\rho(\xi, x)^k$) is due to Sverdrup-Thygeson [553].

Section 3.3. The Doss expectation was introduced by Doss [144], who proved Theorem 3.6 for $\mathbb{E}$ being the real line. The general variant of Theorem 3.6 is due to Bru, Heinich and Lootgieter [81]. Definition 3.7 is due to Herer [234]. T. Okon (unpublished report) studied relationships between the Doss and Herer expectations. Theorem 3.10(ii) is due to Ch. Hess, see Aubin [28, Th. 3.10.3]. The Herer expectation described here is often also called the Doss expectation, see Aubin [28].

Further results concerning the Doss expectation in metric spaces of negative curvature (which are convex in the sense of Doss) can be found in Herer [235] and Raynaud de Fitte [173]. The corresponding definition of a conditional expectation and a martingale is explored by Doss [145] and Herer [235, 236]. The Doss definition can be applied to define an average of several points in a metric space, which leads to classification of metric spaces, see Gähler and Murphy [185] and Pick [458]. The ideas similar to Doss and Fréchet expectations are used to define barycentres for probability measures on manifolds, see Emery and Mokobodzki [161] and Picard [457]. General ergodic theorems and laws of large numbers for the Herer expectation in a metric space are obtained by Raynaud de Fitte [173].

Section 3.4. The properties of expectations are presented here following Molchanov [405]. Many concepts of the expectation can be extended to define the corresponding conditional expectations.

While the described expectations are derived from the probability distributions of random sets, in practice they are estimated as averages if a sample of independent identically distributed realisations is given. Such a sample of sets $X_1, \ldots, X_n$ can be interpreted as a random closed set X which takes the enlisted values with equal probabilities $1/n$. This allows us to reformulate all expectations for samples of sets. In statistical language this approach means substituting the empirical distribution instead of the theoretical distribution of X. The corresponding "naive" estimators are unbiased for the case of the selection and radius-vector expectations and asymptotically unbiased in other cases. The strong law of large numbers for the selection expectation is treated in detail in Section 3.1.2.

It is possible to define the *variance* for general random compact sets using the Fréchet variance being the minimum value of $\mathbf{E}\rho_H(X, K)^2$ for $K \in \mathcal{K}$. If X is a random convex compact set, then its variance can be identified with the covariance operator of the support function $h(X, \cdot)$. Kruse [334] defined a variance of a random closed set as a set of variances of all its square integrable selections. This definition leads to a set-valued variance, which is however far too complicated to evaluate even for random intervals on the real line.

3
Minkowski Addition

1 Strong law of large numbers for random sets

1.1 Minkowski sums of deterministic sets

Shapley–Folkman–Starr theorem

Minkowski addition is a natural operation for sets in linear spaces, see Appendix A. If M and L are two subsets of a linear space $\mathbb{E}$, then their Minkowski (or elementwise) sum is defined as
$$M \oplus L = \{x + y : x \in M, y \in L\}.$$
We systematically write $M + L$ instead of $M \oplus L$.

This chapter deals with laws of large numbers and limit theorems for Minkowski sums of random sets. The relevant methods are closely related to probabilities in Banach spaces, since the Minkowski addition of convex sets can be identified with the conventional arithmetic addition of their support functions. Therefore, a number of results for Minkowski sums of random sets can be obtained using well-known results for sums of random functions. Many results hold for random closed sets in a general Banach space $\mathbb{E}$, which is the typical setting in this chapter. The space $\mathbb{E}$ is often assumed to be separable and we also specify results for random closed sets in the Euclidean space $\mathbb{R}^d$.

Minkowski addition of sets has a "convexifying" property, which means that the sum is "more convex" than the summands. This property is formalised by the following important result. Recall that $\|K\|$ is the norm of set K, see (A.3) and ρ_H is the Hausdorff metric.

Theorem 1.1 (Shapley–Folkman–Starr). *Let $K_1, \ldots, K_n$ be compact subsets of $\mathbb{R}^d$ for $n \geq 1$. Then*
$$\rho_H(K_1 + \cdots + K_n, \mathrm{co}(K_1 + \cdots + K_n)) \leq \sqrt{d} \max_{1 \leq i \leq n} \|K_i\|. \tag{1.1}$$

This theorem holds for sets in any finite-dimensional linear space $\mathbb{E}$. A stronger variant of Theorem 1.1 is Theorem C.13 proved in Appendix C. An important feature of (1.1) is that the upper bound does not depend on n if the norms $\|K_i\|$, $1 \leq i \leq n$, are bounded by a constant c.

Corollary 1.2 (Sums of identical sets). *If K is a compact subset of $\mathbb{R}^d$, then*

$$\rho_H(K^{(n)}, \operatorname{co}(K^{(n)})) \leq \sqrt{d}\|K\|, \tag{1.2}$$

where $K^{(n)} = K + \cdots + K$ is the Minkowski sum of n identical summands.

It follows from (1.2) that

$$\rho_H(n^{-1}K^{(n)}, n^{-1}\operatorname{co}(K^{(n)})) \leq \frac{\sqrt{d}}{n}\|K\|$$

with the right-hand side converging to 0 as $n \to \infty$. Therefore, $n^{-1}K^{(n)}$ converges to a convex set as the number of summands n tends to infinity. A deterministic compact set K is said to be *infinitely divisible for Minkowski summation* (or *M-infinitely divisible*) if, for each $n \geq 2$, there exists a convex set L_n such that K is equal to the Minkowski sum of n identical sets equal to L_n.

Theorem 1.3 (M-infinitely divisible deterministic sets). *A compact set K is M-infinitely divisible if and only if K is convex.*

Proof. Sufficiency is obvious with $L_n = n^{-1}K$. For the proof of *necessity*, note that Corollary 1.2 implies $\rho_H(K, \operatorname{co}(K)) \leq \sqrt{d}\|L_n\|$. The statement now follows from the fact that $\|L_n\| \leq n^{-1}\|K\|$. □

It is interesting to note that every continuous linear map g from $\mathcal{K}(\mathbb{R}^d)$ into a linear space satisfies $g(K) = g(\operatorname{co} K)$. Indeed,

$$g(K) = \frac{1}{n}\sum g(K) = g(n^{-1}(K + \cdots + K)) \to g(\operatorname{co} K).$$

Therefore, it is not possible to embed the family of compact sets isomorphically into a linear space.

Convexification in Banach spaces

The explicit constant in the right-hand side of (1.1) makes no sense if the space $\mathbb{E}$ is infinite-dimensional. The following result replaces Theorem 1.1 in the infinite-dimensional setting.

Theorem 1.4 (Convexification of deterministic sums). *Let $\{K_n, n \geq 1\}$ be a sequence of compact sets in Banach space $\mathbb{E}$ and let K_0 be compact and convex. If*

$$\rho_H(n^{-1}(\operatorname{co}(K_1) + \cdots + \operatorname{co}(K_n)), K_0) \to 0 \quad \text{as } n \to \infty, \tag{1.3}$$

then

$$\rho_H(n^{-1}(K_1 + \cdots + K_n), K_0) \to 0 \quad \text{as } n \to \infty. \tag{1.4}$$

Proof. Let e be an exposed point in K_0, i.e. there exists a linear functional f such that $f(e) > f(x)$ for $x \in K_0$, $x \neq e$. For each i, let $x_i \in K_i$ be a maximiser of f on K_i, i.e. $f(x_i) \geq f(x)$ for every $x \in K_i$.

Then $y_n = n^{-1}(x_1 + \cdots + x_n)$ is a maximiser of f on $n^{-1}(\mathrm{co}(K_1) + \cdots + \mathrm{co}(K_n))$. By (1.3), each limiting point of y_n lies in K_0, whence

$$y_n \to e \quad \text{as } n \to \infty. \tag{1.5}$$

Let $0 \leq \lambda \leq 1$ and let $p(n) = [\lambda n]$ be the integer part of λn. Choose $m(n)$ such that $1 \leq m(n) \leq n - p(n)$. Then

$$n^{-1}(x_{m(n)} + \cdots + x_{m(n)+p(n)}) \to \lambda e \quad \text{as } n \to \infty.$$

Now consider a finite number of exposed points e^j and the corresponding maximisers x_i^j, $1 \leq j \leq k$, $i \geq 1$. Let $m_j(n) = [(\lambda_1 + \cdots + \lambda_j)n]$, where $\lambda_j \in [0, 1]$, and $\lambda_1 + \cdots + \lambda_k = 1$. Put

$$z_{i,n} = x_i^j \quad \text{if} \quad m_j(n) < j \leq m_{j+1}(n).$$

Then

$$n^{-1}(z_{1,n} + \cdots + z_{n,n}) \to \lambda_1 e^1 + \cdots + \lambda_k e^k. \tag{1.6}$$

If $w_n \in n^{-1}(K_1 + \cdots + K_n)$, then every limiting point of $\{w_n, n \geq 1\}$ belongs to K_0. The exposed points are these limits by (1.5) and convex combinations of them by (1.6). Since convex combinations of the exposed points are dense in K_0 (see Köthe [328, p. 337]), we obtain (1.4). $\square$

In $\mathbb{R}^d$ the assertion of Theorem 1.4 can be derived from the Shapley–Folkman–Starr theorem. Denote $S_n = K_1 + \cdots + K_n$ and note that $\mathrm{co}(S) = \mathrm{co}(K_1) + \cdots + \mathrm{co}(K_n)$. The triangle inequality yields

$$\rho_{\mathrm{H}}(n^{-1} S_n, K_0) \leq \rho_{\mathrm{H}}(n^{-1} S_n, n^{-1} \mathrm{co}(S_n)) + \rho_{\mathrm{H}}(n^{-1} \mathrm{co}(S_n), K_0).$$

The second summand converges to zero by (1.3). By (1.3), $\|\mathrm{co}(K_n)\| = o(n)$, whence

$$\max_{1 \leq i \leq n} \|K_i\|/n \to 0 \quad \text{as } n \to \infty.$$

By (1.1),

$$\rho_{\mathrm{H}}(n^{-1} S_n, n^{-1} \mathrm{co}(S_n)) \leq \sqrt{d} \max_{1 \leq i \leq n} \|K_i\|/n \to 0 \quad \text{as } n \to \infty.$$

A direct generalisation of Theorem 1.1 for compact sets in a Banach space $\mathbb{E}$ is possible under additional geometric assumptions on $\mathbb{E}$. The space $\mathbb{E}$ is said to be of *type p* if there exists a constant $C > 0$ such that

$$\mathbf{E}(\|\sum_{i=1}^n \xi_i\|^p) \leq C \sum_{i=1}^n \mathbf{E}(\|\xi_i\|^p)$$

for each $n \geq 2$ and independent $\mathbb{E}$-valued elements $\xi_1, \ldots, \xi_n$ with mean zero. Every Hilbert space is of type 2, the spaces $\mathbf{L}^p$ with $1 < p < \infty$ are of type $\min(p, 2)$, while the space $\mathbf{C}([0, 1])$ of continuous functions with the uniform metric and $\mathbf{L}^\infty$ are not of type p for any $p > 1$.

Theorem 1.5. *Let $\mathbb{E}$ be a Banach space of type $p > 1$. Then there exists a constant $C > 0$ such that, for all $K_1, \ldots, K_n \in \mathcal{K}$,*

$$\rho_{\mathrm{H}}(K_1 + \cdots + K_n, \mathrm{co}(K_1 + \cdots + K_n)) \leq C^{1/p} \left(\sum_{i=1}^{n} \|K_i\| \right)^{1/p}. \tag{1.7}$$

Proof. See Puri and Ralescu [464] for the proof of a stronger inequality with the norms of K_i in (1.7) replaced by the *inner radius* of K_i given by (C.6), which is one of the non-convexity measures. The particular case of a Hilbert space $\mathbb{E}$ was considered by Cassels [88]. □

1.2 Strong law of large numbers

Euclidean case

A useful tool suitable to derive the strong law of large numbers (SLLN) for random convex sets is based on their representation as elements of functional spaces. For the time being, assume that $\mathbb{E} = \mathbb{R}^d$. A set K from the family co $\mathcal{K}$ of convex compact sets in $\mathbb{R}^d$ gives rise to its support function

$$h(K, u) = \sup\{\langle x, u \rangle : x \in K\}, \quad u \in \mathbb{R}^d.$$

In the following we usually consider the support function as a function defined on the unit ball B_1 or the unit sphere $\mathbb{S}^{d-1}$. The support function is Lipschitz on $\mathbb{R}^d$, see Theorem F.1. The properties

$$h(K_1 + K_2, u) = h(K_1, u) + h(K_2, u)$$

and

$$h(cK, u) = ch(K, u)$$

make it possible to convert the Minkowski sums of convex sets into the arithmetic sums of the corresponding support functions. Furthermore,

$$\rho_{\mathrm{H}}(K_1, K_2) = \sup_{\|u\|=1} |h(K_1, u) - h(K_2, u)| \tag{1.8}$$

$$= \sup_{\|u\| \leq 1} |h(K_1, u) - h(K_2, u)|.$$

and, in particular,

$$\|K\| = \|\mathrm{co}(K)\| = \rho_{\mathrm{H}}(K, \{0\}) = \sup_{\|u\|=1} |h(K, u)|.$$

Thus, the support function provides an isometric embedding of the family co $\mathcal{K}$ of convex compact subsets of $\mathbb{E}$ into the Banach space $C(B_1)$ (or $C(\mathbb{S}^{d-1})$) of continuous functions on B_1 (or $\mathbb{S}^{d-1}$) with the uniform norm.

The general approach to derive strong laws of large numbers and limit theorems for random compact sets X_n, $n \geq 1$, consists of two steps:

Step 1: Reduce consideration to the case of random convex compact sets.
Step 2: Derive results for random convex sets by invoking the corresponding results for probabilities in Banach spaces and applying them to the sequence $h(X_n, \cdot)$ of the support functions of X_n, $n \geq 1$.

Recall that the representation of random sets through their support functions can be used to define the selection (or the Aumann) expectation, see Theorem 2.1.22. Therefore, it is natural that the selection expectation appears in the strong law of large numbers for random sets with respect to Minkowski addition. In its simplest form this law of large numbers establishes the almost sure convergence with respect to the Hausdorff metric of normalised Minkowski sums of i.i.d. random closed sets to the selection expectation of a summand. The mere existence of a sequence of independent random elements implies that the underlying probability space is non-atomic, so that the selection expectation is convex. In this chapter $\mathbf{E}X$ always means the selection expectation. Recall that a random compact set X is called integrably bounded if $\|X\|$ is integrable, see Definition 2.1.11(i).

Theorem 1.6 (SLLN for random sets in $\mathbb{R}^d$). *Let $X, X_1, X_2, \ldots$ be i.i.d. integrably bounded random compact sets in $\mathbb{R}^d$ and let $S_n = X_1 + \cdots + X_n$, $n \geq 1$. Then*

$$\rho_H(n^{-1}S_n, \mathbf{E}X) \to 0 \quad \text{a.s. as } n \to \infty. \tag{1.9}$$

Proof. If X is almost surely convex, then

$$h(n^{-1}S_n, u) = \frac{1}{n}\sum_{i=1}^{n} h(X_i, u) \to \mathbf{E}h(X, u) = h(\mathbf{E}X, u)$$

by a strong law of large numbers in a Banach space (see Mourier [416]) specialised for the space $C(B_1)$ of continuous functions on B_1 with the uniform metric. By (1.8) the uniform metric on $C(B_1)$ corresponds to the Hausdorff metric on co $\mathcal{K}$, whence (1.9) holds. A not necessarily convex X can be replaced with $\mathrm{co}(X)$, so that an application of Theorem 1.1 or Theorem 1.4 finishes the proof. The integrable boundedness of X implies $\mathbf{E}\rho_H(n^{-1}S_n, \mathbf{E}X) \to 0$ as $n \to \infty$. □

SLLN in Banach space

Now consider the case when $\mathbb{E}$ is a general Banach space. The Rådström embedding theorem (see Rådström [472]) provides a normed linear space $\mathbb{Y}$ and a linear isometry g between co $\mathcal{K}$ and $\mathbb{Y}$. However, this space $\mathbb{Y}$ is not complete in general. The Hörmander embedding theorem says more specifically that g maps co $\mathcal{K}$ into the space $C_b(B_1^*)$ of bounded continuous functions on the unit ball B_1^* in the dual space $\mathbb{E}^*$. This embedding is realised by the support function

$$h(K, u) = \sup\{\langle x, u\rangle : x \in K\}, \quad u \in B_1^*. \tag{1.10}$$

Applying the strong law of large numbers in the space $C_b(B_1^*)$ yields a law of large numbers for integrably bounded random convex compact sets in a Banach space. Finally, the result can be extended to non-convex compact sets by applying Theorem 1.4.

Theorem 1.7 (SLLN for random sets in Banach space). *Let $\{X_n, n \geq 1\}$ be a sequence of i.i.d. Hausdorff approximable integrably bounded random sets in a Banach space $\mathbb{E}$. Then $n^{-1}(X_1 + \cdots + X_n)$ converges almost surely in the Hausdorff metric to a compact convex set being the selection expectation of $\mathrm{co}(X_1)$.*

Theorem 1.2.11(ii) implies that a random compact set is Hausdorff approximable, whence Theorem 1.7 holds for a sequence of integrably bounded random compact sets in $\mathbb{E}$. Theorem 1.7 also holds for a general (possibly non-separable) Banach space $\mathbb{E}$, see Artstein and Hansen [21]. The limit of the normalised sums in Theorem 1.7 is the expectation of the convex hull of X_1, since in a general Banach space, $\mathbf{E}X$ may be a proper subset of $\mathbf{E}\,\mathrm{co}(X)$, see Example 2.1.23(i).

Open problem 1.8. Prove the strong law of large numbers for *operator-normalised* sums of random closed sets, e.g. for the sequences $A_n(X_1 + \cdots + X_n)$ or $A_{n1}X_1 + \cdots + A_{nn}X_n$ with $A_n, A_{n1}, \ldots, A_{nn}, n \geq 1$, being linear operators. The corresponding results for operator normalised sums of random vectors can be found, e.g. in Buldygin and Solntsev [83].

Blaschke sums

Let $S_{d-1}(K, \cdot)$ be the *area measure* of a convex compact set $K \subset \mathbb{R}^d$, see Appendix F. The area measure is a measure on the unit sphere that depends on K and can be extended for sets K from the convex ring $\mathcal{R}$. If X is a random set with values in $\mathcal{R}$, then the *Blaschke expectation* of X is defined (up to a translation) by

$$S_{d-1}(\mathbf{E}_{\mathrm{BL}} X, \cdot) = \mathbf{E} S_{d-1}(X, \cdot).$$

The *Blaschke sum* of X_1 and X_2 from $\mathcal{R}$ is a convex set with the surface area measure equal to the sum of the surface area measures of X_1 and X_2.

Using this embedding of sets in the space of measures on the unit sphere it is possible to derive the strong law of large numbers for normalised Blaschke sums. Note that we do not need to resort to convexification arguments, since the Blaschke sum of two sets from the convex ring is always convex by definition.

1.3 Applications of the strong law of large numbers

The Brunn–Minkowski theorem

The classical Brunn–Minkowski inequality (see Schneider [520, p. 309]) states that

$$V^{1/d}(\lambda K + (1-\lambda)L) \geq \lambda V^{1/d}(K) + (1-\lambda)V^{1/d}(L), \tag{1.11}$$

where $V = \text{mes}$ is the d-dimensional Lebesgue measure (volume) in $\mathbb{R}^d$, $\lambda \in [0,1]$ and K, L are arbitrary compact sets (in fact, their measurability suffices).

Theorem 1.9 (Brunn–Minkowski inequality for random sets). *If X is an integrably bounded random compact set in $\mathbb{R}^d$, then*

$$V^{1/d}(\mathbf{E}X) \geq \mathbf{E}V^{1/d}(X). \tag{1.12}$$

Proof. We prove the stronger statement

$$V^{1/d}(\mathbf{E}X^\natural) \geq \mathbf{E}V^{1/d}(X), \tag{1.13}$$

where $X^\natural$ is the *reduced representation* of X, see Section 2.1.4. Indeed, $\mathbf{E}X^\natural$ can be a proper subset of $\mathbf{E}X$, while $\mathbf{E}V^{1/d}(X) = \mathbf{E}V^{1/d}(X^\natural)$.

Consider first the non-atomic case when $\mathbf{E}X = \mathbf{E}X^\natural$. Let $X, X_1, X_2, \ldots$ be i.i.d. random sets. Then (1.11) can be iterated to get

$$V^{1/d}(n^{-1}(X_1 + \cdots + X_n)) \geq n^{-1} \sum_{i=1}^{n} V^{1/d}(X_i). \tag{1.14}$$

Since $\mathbf{E}\|X\| < \infty$ and $V^{1/d}(X) \leq c\|X\|$ for some $c > 0$ (depending on the dimension of the space), the Kolmogorov strong law of large numbers implies that the right-hand side of (1.14) converges to $\mathbf{E}V^{1/d}(X) < \infty$ almost surely. Theorem 1.6 yields

$$V^{1/d}(\mathbf{E}\,\text{co}(X)) \geq \mathbf{E}V^{1/d}(X),$$

since V is continuous on $\text{co}\,\mathcal{K}$. Now (1.12) follows from the fact that the selection expectation of X coincides with the selection expectation of $\text{co}(X)$, see Theorem 2.1.15.

In the pure atomic case, assume that X takes values $K_1, K_2, \ldots$ with probabilities $p_1, p_2, \ldots$. Then (1.11) implies

$$V^{1/d}(p_1 K_1 + \cdots + p_n K_n) \geq \sum_{i=1}^{n} p_i V^{1/d}(K_i) \tag{1.15}$$

for each $n \geq 2$. The right-hand side converges as $n \to \infty$ to the associated infinite sum. For the left-hand side, one can refer to (2.1.15) and the inclusion

$$\mathbf{E}X^\natural = p_1 K_1 + p_2 K_2 + \cdots \supset p_1 K_1 + \cdots + p_n K_n.$$

In the general case, $X^\natural$ can be considered as a mixture of two random sets, X_1 with the probability distribution attaching the masses $p_1, p_2, \ldots$ to the compact sets $K_1, K_2, \ldots$ and X_2 distributed on $\mathcal{K} \setminus \{K_1, K_2, \ldots\}$. If θ is the sum of all p_i's, then

$$\mathbf{E}X = \theta \mathbf{E}X_1 + (1-\theta)\mathbf{E}X_2.$$

By (1.11) and the two cases discussed above,
$$V^{1/d}(\mathbf{E}X) \geq \theta V^{1/d}(\mathbf{E}X_1) + (1-\theta)V^{1/d}(\mathbf{E}X_2)$$
$$\geq \theta \mathbf{E}V^{1/d}(X_1) + (1-\theta)\mathbf{E}V^{1/d}(X_2) \geq \mathbf{E}V^{1/d}(X).\quad\square$$

Note that (1.12) implies the classical Brunn–Minkowski inequality (1.11) for X taking two values K and L with probabilities λ and $(1-\lambda)$ respectively.

Proposition 1.10. *Let $K \in \text{co}\,\mathcal{K}$ and let X be an integrably bounded random compact set such that $X \cap K \neq \emptyset$ a.s. Then*
$$V^{1/d}((\mathbf{E}X) \cap K) \geq \mathbf{E}V^{1/d}(X \cap K). \tag{1.16}$$

Proof. Note that $(\mathbf{E}X) \cap K \supset \mathbf{E}(X \cap K)$ because every selection ξ of $X \cap K$ is a selection of both X and K and $\mathbf{E}\xi \in K$ by the convexity of K. Then apply (1.12) to the random set $(X \cap K)$. $\quad\square$

The Brunn–Minkowski inequality for random compact sets can be sharpened further by noticing that translations of X are immaterial for the inequality. The translative expectation from Section 2.1.4 can be used to formulate the following variant of (1.12).

Theorem 1.11. *Let X be a random compact set such that $\mathbf{E}\|X - \xi\| < \infty$ for a random vector ξ. Then*
$$V^{1/d}(\mathbf{E}(X - \xi)^{\natural}) \geq \mathbf{E}V^{1/d}(X). \tag{1.17}$$

If such ξ exists, it is possible to sharpen the inequality by taking the Steiner point $s(X)$ as ξ. The equality in (1.17) holds if X is a random translate of a fixed convex set.

Proof. First, (1.17) easily follows from (1.13) since translations leave the volume invariant. Replacement of ξ by $s(X)$ makes the inequality tighter because of Theorem 2.1.34. If $X = K + \xi$, then $(X - s(X))^{\natural} = K - s(K)$ and (1.17) turns into an equality. $\quad\square$

Translations of unimodal functions

The conventional Brunn–Minkowski inequality has been applied by Anderson [8] to study translates of multivariate density functions.

Theorem 1.12 (Anderson inequality). *If K is a symmetric convex set in $\mathbb{R}^d$ and $f: \mathbb{R}^d \mapsto \mathbb{R}$ is a non-negative symmetric unimodal function (so that $\{x \in \mathbb{R}^d : f(x) \geq u\}$ is convex for each u), then*
$$\int_K f(x + \lambda y)\,dx \geq \int_K f(x + y)\,dx \tag{1.18}$$
for all $y \in \mathbb{R}^d$ and $\lambda \in [0, 1]$.

1 Strong law of large numbers for random sets

Proof. It suffices to consider a centrally symmetric $K \in \text{co}\,\mathcal{K}$ and $f(x) = \mathbf{1}_L(x)$ with symmetric $L \in \text{co}\,\mathcal{K}$. Then (1.18) is equivalent to

$$V(K \cap (L - \lambda y)) \geq V(K \cap (L - y)).$$

Consider random compact set $X = \eta(L - y)$, where η takes values 1 and -1 with probabilities $p \in [0.5, 1]$ and $(1 - p)$ respectively. By symmetry of K,

$$\mathbf{E}V(K \cap X) = pV(K \cap (L - y)) + (1 - p)V(-K \cap (L - y))$$
$$= V(K \cap (L - y)).$$

Since L is also symmetric,

$$\mathbf{E}X = \mathbf{E}\eta(L - y) = \mathbf{E}[\eta L - \eta y] = \mathbf{E}[L - \eta y].$$

Note that $\mathbf{E}(L - \eta y)^{\natural} = L - \lambda y$ for $\lambda = 2p - 1$. Now (1.18) follows from (1.13). □

Isoperimetric inequality

Theorem 1.9 yields the *isoperimetric inequality* in $\mathbb{R}^2$, which states that the disk has the maximal area among all sets with the given perimeter. Clearly, it suffices to prove this for convex sets. Let $K \in \text{co}\,\mathcal{K}$ and let $\mathbf{w}$ be an *isotropic random rotation*, i.e. rotation to an angle uniformly distributed over $[0, 2\pi]$. Then $X = \mathbf{w}K$ is an isotropic random set. Its selection expectation is a disk B whose perimeter is equal to the expected perimeter of X (or the perimeter of K), see Proposition 2.1.27. Finally, (1.12) for $d = 2$ yields

$$(\text{mes}_2(B))^{1/2} \geq \mathbf{E}(\text{mes}_2(X))^{1/2} = (\text{mes}_2(K))^{1/2},$$

where $\text{mes}_2(K)$ is the area of K. Thus the area of B does not exceed the area of K.

Random convex hulls

The Brunn–Minkowski inequality may be used to bound the tail of the distribution for the volume of random convex hulls. Let X_n be the convex hull of i.i.d. random points $\xi_1, \ldots, \xi_n$ in $\mathbb{R}^d$ with a radially symmetric distribution. The selection expectation of X_n can be found from

$$h(\mathbf{E}X_n, u) = \mathbf{E} \max_{1 \leq i \leq n} \langle \xi_i, u \rangle.$$

By radial symmetry, this support function is a constant denoted by a_n. Then $\mathbf{E}X_n = B_{a_n}$ is the ball of radius a_n. By (1.12),

$$\mathbf{E}V^{1/d}(X_n) \leq V^{1/d}(B_{a_n}) = a_n \varkappa_d^{1/d},$$

where $\varkappa_d$ is the volume of the unit ball in $\mathbb{R}^d$. By Markov's inequality, for $t > 0$,

$$\mathbf{P}\{V(X_n) \geq t\} = \mathbf{P}\left\{V^{1/d}(X_n) \geq t^{1/d}\right\}$$
$$\leq \frac{\mathbf{E}V^{1/d}(X_n)}{t^{1/d}} \leq a_n \left(\frac{\varkappa_d}{t}\right)^{1/d}. \qquad (1.19)$$

Isotropic rotations and rounding of sets

By applying isotropic random rotations $\mathbf{w}_1, \mathbf{w}_2, \ldots$ to a fixed compact set K and taking their Minkowski averages it is possible to "round" the set K. Indeed,

$$n^{-1}(\mathbf{w}_1 K + \cdots + \mathbf{w}_n K)$$

converges to $\mathbf{E}(\mathbf{w}_1 K)$ which is a ball. Thus, K can be *rounded* by almost all sequences of independent rotations. Repeating the same argument for a countable dense collection of compact sets in $\mathcal{K}$ shows that almost every sequence of rotations rounds every K.

Zonoids

Convex bodies that appear as finite sums of linear segments are called *zonotopes*; limits of zonotopes in the Hausdorff metrics are called *zonoids*. The following theorem provides a probabilistic interpretation of zonoids by applying the strong law of large numbers for sums of random segments.

Theorem 1.13 (Representation of zonoids). *A random convex closed set Z is a zonoid if and only if there is $x \in \mathbb{R}^d$ and random vector ξ with integrable norm such that $Z = x + \mathbf{E}\{0, \xi\}$.*

Proof. Necessity. Let $Y = L_1 + \cdots + L_n$ be a zonotope given by a sum of linear segments. This sum can be rewritten as

$$Y = p_1 \tilde{L}_1 + \cdots + p_n \tilde{L}_n,$$

where $p_i = \|L_i\| / \sum \|L_i\|$ and $\tilde{L}_i = p_i^{-1} L_i$. If $e_1, \ldots, e_d$ is the standard basis in $\mathbb{R}^d$, then

$$\sum_i \|L_i\| \leq \sum_i \sum_k [h(L_i, e_k) + h(L_i, -e_k)]$$
$$= \sum_k \sum_i [h(L_i, e_k) + h(L_i, -e_k)]$$
$$\leq \sum_k [h(Y, e_k) + h(Y, -e_k)] \leq 2d \|Y\|.$$

Hence $\|\tilde{L}_i\| \leq 2d \|Y\|$. One concludes that Y is the selection expectation of the segment $[\zeta, \eta]$ with the end-points ζ and η having the discrete distributions with uniformly bounded supports.

A general zonoid, Z, is a limit of $Z_n = \mathbf{E}[\zeta_n, \eta_n]$, where all η_n's and ζ_n's are uniformly bounded. Without loss of generality, the joint distribution of ζ_n and η_n converges to a distribution of a pair (ζ, η), so that

$$h(Z_n, u) = \mathbf{E} \max(\langle \zeta_n, u \rangle, \langle \eta_n, u \rangle) \to \mathbf{E} \max(\langle \zeta, u \rangle, \langle \eta, u \rangle) \quad \text{as } n \to \infty$$
$$= \langle \mathbf{E}\zeta, u \rangle + \mathbf{E}h(\{0, \eta - \zeta\}, u),$$

which yields the required representation with $\xi = \eta - \zeta$.
Sufficiency is easily seen by approximating the distribution of ξ using random vectors having a finite number of values. □

Definition 1.14 (Zonoid and lift zonoid of a random vector). Let ξ be a random vector in $\mathbb{R}^d$. Its *zonoid*, Z_ξ, is the selection expectation of $X = \{0, \xi\}$. The *lift zonoid*, $\tilde{Z}_\xi$, of ξ is the selection expectation of the segment in $\mathbb{R}^{d+1}$ with the endpoints being the origin and the $(d+1)$-dimensional vector $(1, \xi)$.

Although different distributions on $\mathbb{R}^d$ may share the same zonoid, the lift zonoid uniquely characterises the corresponding multivariate distribution.

Theorem 1.15 (Characterisation of distribution by lift zonoid). *The lift zonoid $\tilde{Z}_\xi$ characterises uniquely the probability distribution of ξ in $\mathbb{R}^d$.*

Proof. The support function of $\tilde{Z}_\xi$ is given by

$$h(\tilde{Z}_\xi, (t, u)) = \mathbf{E}(\max(0, (t + \langle \xi, u \rangle))), \quad (1.20)$$

where $u \in \mathbb{R}^d$ and $t \in \mathbb{R}$. It is easy to see that the right-hand side of (1.20) as a function of t determines uniquely the distribution of $\langle \xi, u \rangle$ and thereupon the distribution of ξ itself. □

Theorem 1.13 can be generalised for a more general subfamily $\mathcal{M}$ of co $\mathcal{K}$ instead of the family of segments. Assume that $cK \in \mathcal{M}$ for each $K \in \mathcal{M}$ and $c \in (0, 1]$. By $\widetilde{\mathcal{M}}$ denote the family of all convex bodies which appear as limits (in the Hausdorff metric) of sums of elements of $\mathcal{M}$. For instance, $\widetilde{\mathcal{M}}$ is the family of zonoids if $\mathcal{M}$ consists of all segments. The following proposition can be proved similarly to Theorem 1.13.

Proposition 1.16. *For each $K \in \widetilde{\mathcal{M}}$ there exists a random convex closed set X such that $X \in \mathcal{M}$ almost surely and $K = \mathbf{E}X$.*

Random determinants

Now formulate results on determinants for a *random matrix* whose columns are given by i.i.d. realisations of a random vector ξ.

Theorem 1.17. *If ξ is a random vector in $\mathbb{R}^d$ with $\mathbf{E}\|\xi\| < \infty$, then*

$$\mathbf{E}|\det M_\xi| = d!V(\mathbf{E}\{0, \xi\}),$$

where M_ξ denotes a $d \times d$ matrix whose columns are i.i.d. copies of ξ.

Proof. Consider a sequence $\{\xi_n, n \geq 1\}$ of i.i.d. copies of ξ. For each $n \geq 1$ define the zonotope

$$Z_n = n^{-1}([0, \xi_1] + \cdots + [0, \xi_n]).$$

The strong law of large numbers implies $V(Z_n) \to V(\mathbf{E}\{0,\xi\})$ a.s. as $n \to \infty$. It is possible (see Shephard [536]) to decompose the volume of Z_n as follows:

$$V(Z_n) = \frac{1}{n^d} \sum_{i_1 < i_1 < \cdots < i_d} |\det M(\xi_{i_1}, \ldots, \xi_{i_d})|,$$

where $M(\xi_{i_1}, \ldots, \xi_{i_d})$ is the matrix with the columns $\xi_{i_1}, \ldots, \xi_{i_d}$. It follows from the theory of U-statistics (see Serfling [531]) that $V(Z_n) \to (d!)^{-1}\mathbf{E}|\det M_\xi|$ almost surely, which completes the proof. □

1.4 Non-identically distributed summands

As in the i.i.d. case, the principal technique for non-identically distributed summands relies on the corresponding laws of large numbers in Banach spaces. However, now either geometric conditions on Banach spaces and/or a kind of uniform integrability condition for random sets are needed. As usual, the Banach space $\mathbb{E}$ is assumed to be separable.

Weighted sums

The simplest case of non-identically distributed summands appears while considering weighted sums of random compact sets. The weak law of large numbers for weighted sums of random elements in type p Banach spaces (see Adler, Rosalsky and Taylor [3]) yields the following result for weighted sums of random compact sets.

Theorem 1.18 (Weighted sums of random compact sets). *Let $\{X_n, n \geq 1\}$ be a sequence of i.i.d. random compact sets in a separable Banach space $\mathbb{E}$. Let a_n and b_n, $n \geq 1$, be positive constants such that b_n/a_n is an increasing sequence, and*

$$\frac{b_n}{na_n} \to \infty, \quad \sum_{i=1}^n a_i = \mathcal{O}(na_n), \quad \sum_{i=1}^n \frac{b_i}{i^2 a_i} = \mathcal{O}(b_n(\sum_{i=1}^n a_i)^{-1}).$$

If $n\mathbf{P}\{\|X_1\| > b_n/a_n\} \to 0$, then

$$\rho_H\left(b_n^{-1}\sum_{i=1}^n a_i X_i, b_n^{-1}\sum_{i=1}^n a_i \mathbf{E}(\mathrm{co}(X_1)\mathbf{1}_{\|X_1\| \leq a_n^{-1}b_n})\right) \to 0$$

in probability as $n \to \infty$.

The identical distribution condition in Theorem 1.18 cannot be easily dropped, as is shown in Taylor [558, Ex. 4.1.1] for the "classical" case when X_i's are singletons.

Conditions of uniform integrability type

Further laws of large numbers can be proved under some conditions on distributions of the summands. The following conditions stems from Daffer and Taylor [115]. A sequence of random sets $\{X_n, n \geq 1\}$ is said to be *tight* if, for each $\varepsilon > 0$, there exists a compact in the Hausdorff metric set $\mathcal{D}_\varepsilon \subset \mathcal{K}$, such that $\mathbf{P}\{X_n \notin \mathcal{D}_\varepsilon\} < \varepsilon$ for all $n \geq 1$. A sequence of i.i.d. random sets is tight, while the converse is not true. Furthermore, $\{X_n, n \geq 1\}$ is *compactly uniform integrable* if

$$\mathbf{E}\|X_n \mathbf{1}_{X_n \notin \mathcal{D}_\varepsilon}\| < \varepsilon.$$

The compactly uniform integrability implies both the tightness and the uniform integrability of the sequence $\{\|X_n\|, n \geq 1\}$, but not conversely.

Theorem 1.19. *Let $\{X_n, n \geq 1\}$ be a compactly uniform integrable sequence of independent random compact sets in a separable Banach space. If*

$$\sum_{n=1}^{\infty} n^{-p} \mathbf{E}\|X_n\|^p < \infty \tag{1.21}$$

for some $1 \leq p \leq 2$, then

$$\rho_H(n^{-1}\sum_{i=1}^{n} X_i, n^{-1}\sum_{i=1}^{n} \mathbf{E}\,\mathrm{co}(X_i)) \to 0 \quad \text{a.s. as } n \to \infty.$$

Proof. Let $\varepsilon > 0$ be given. Fix a finite ε-net $K_1, \ldots, K_m \in \mathcal{D}_\varepsilon$, so that each $K \in \mathcal{D}_\varepsilon$ can be approximated by at least one of K_i with $\rho_H(K, K_i) \leq \varepsilon$. Define random sets $Y'_n, n \geq 1$, as being equal to the set K_j with the minimum possible j such that $\rho_H(X_n, K_j)$ is minimal among $K_1, \ldots, K_m$. Let $Y_n = Y'_n \mathbf{1}_{X \in \mathcal{D}_\varepsilon}$, so that $Y_n = Y'_n$ if $X \in \mathcal{D}_\varepsilon$ and $Y_n = \{0\}$ otherwise. Then, for each n,

$$\rho_H(n^{-1}\sum_{k=1}^{n} X_k, n^{-1}\sum_{k=1}^{n} \mathbf{E}\,\mathrm{co}(X_k))$$

$$\leq \rho_H(n^{-1}\sum_{k=1}^{n} X_k, n^{-1}\sum_{k=1}^{n} X_k \mathbf{1}_{X_k \in \mathcal{D}_\varepsilon}) \tag{I}$$

$$+ \rho_H(n^{-1}\sum_{k=1}^{n} X_k \mathbf{1}_{X_k \in \mathcal{D}_\varepsilon}, n^{-1}\sum_{k=1}^{n} Y_k) \tag{II}$$

$$+ \rho_H(n^{-1}\sum_{k=1}^{n} Y_k, n^{-1}\sum_{k=1}^{n} \mathbf{E}\,\mathrm{co}(Y_k)) \tag{III}$$

$$+ \rho_H(n^{-1}\sum_{k=1}^{n} \mathbf{E}\,\mathrm{co}(Y_k), n^{-1}\sum_{k=1}^{n} \mathbf{E}(\mathrm{co}(X_k) \mathbf{1}_{X_k \in \mathcal{D}_\varepsilon})) \tag{IV}$$

$$+ \rho_H(n^{-1}\sum_{k=1}^{n} \mathbf{E}(\mathrm{co}(X_k) \mathbf{1}_{X_k \in \mathcal{D}_\varepsilon}), n^{-1}\sum_{k=1}^{n} \mathbf{E}\,\mathrm{co}(X_k)). \tag{V}$$

Since $\|X_k \mathbf{1}_{X_k \in \mathcal{D}_\varepsilon}\|$, $k \geq 1$, is a sequence of independent random variables and because of (1.21), term (I) is bounded from above by

$$\limsup_{n \to \infty} n^{-1} \sum_{k=1}^{n} \rho_H(X_k, X_k \mathbf{1}_{X_k \in \mathcal{D}_\varepsilon}) = \limsup_{n \to \infty} n^{-1} \sum_{k=1}^{n} \|X_k \mathbf{1}_{X_k \notin \mathcal{D}_\varepsilon}\|$$

$$\leq \limsup_{n \to \infty} n^{-1} \sum_{k=1}^{n} \mathbf{E}\|X_k \mathbf{1}_{X_k \notin \mathcal{D}_\varepsilon}\| < \varepsilon \text{ a.s.} \quad (1.22)$$

Term (II) is bounded by ε, since $\rho_H(X_k \mathbf{1}_{X_k \in \mathcal{D}_\varepsilon}, Y_k) < \varepsilon$ for each k. By Theorem 1.4,

$$\rho_H(n^{-1} \sum_{k=1}^{n} Y_k, n^{-1} \sum_{k=1}^{n} \mathrm{co}(Y_k))$$

$$\leq \sum_{j=1}^{m} \rho_H(n^{-1} \sum_{k=1}^{n} K_j \mathbf{1}_{Y_k = K_j}, n^{-1} \sum_{k=1}^{n} \mathrm{co}(K_j) \mathbf{1}_{Y_k = K_j}) \to 0.$$

The strong law of large numbers in Banach spaces (Daffer and Taylor [115]) yields:

$$\rho_H(n^{-1} \sum_{k=1}^{n} \mathrm{co}(Y_k), n^{-1} \sum_{k=1}^{n} \mathbf{E}\, \mathrm{co}(Y_k)) \to 0 \quad \text{a.s. as } n \to \infty.$$

Thus, term (III) converges to 0 almost surely. Term (IV) can be made arbitrarily small, since $\rho_H(\mathbf{E}X, \mathbf{E}Y) \leq \mathbf{E}\rho_H(X, Y)$ for any two integrably bounded convex compact sets. The proof is finished by noticing that the reasons that led to (1.22) are also applicable for term (V). $\square$

Note that (1.21) holds if

$$\sup_{n \geq 1} \mathbf{E}\|X_n\|^p < \infty \quad (1.23)$$

for some $p > 1$. If $\mathbb{E} = \mathbb{R}^d$, (1.23) implies the compactly uniform integrability, since

$$\mathbf{E}\|X_n \mathbf{1}_{\|X_n\| > c}\| \leq (\mathbf{E}\|X_n\|^p)^{1/p} (\mathbf{P}\{\|X_n\| > c\})^{(p-1)/p}$$

$$\leq (\mathbf{E}\|X_n\|^p)^{1/p} (c^{-1} \mathbf{E}\|X_n\|)^{(p-1)}.$$

It is also possible to consider weighted Minkowski sums of the form

$$a_{n1} X_1 + a_{n2} X_2 + \cdots + a_{nn} X_n,$$

where $\{a_{ni} : 1 \leq i \leq n, n \geq 1\}$ is a triangular array of non-negative constants such that $\sum_{i=1}^{n} a_{ni} \leq 1$ for all $n \geq 1$. The following result can be proved similarly to Theorem 1.19, see Taylor and Inoue [559].

Theorem 1.20. *Let $\{X_n, n \geq 1\}$ be a compactly uniform integrable sequence of independent compact sets in a separable Banach space. Let ξ be a random variable such that $\mathbf{P}\{\|X_n\| > t\} \leq \mathbf{P}\{\xi > t\}$ for all n and $t > 0$. If $\mathbf{E}\xi^{1+1/\gamma} < \infty$ for some $\gamma > 0$ and $\max_{1 \leq i \leq n} a_{ni} = \mathcal{O}(n^{-\gamma})$, then*

$$\rho_H(a_{n1}X_1 + \cdots + a_{nn}X_n, a_{n1}\mathbf{E}X_1 + \cdots + a_{nn}\mathbf{E}X_n) \to 0 \quad \text{a.s. as } n \to \infty.$$

1.5 Non-compact summands

Consider a sequence $\{X_n, n \geq 1\}$ of i.i.d. integrable (but possibly unbounded and non-compact) random closed sets in a Banach space $\mathbb{E}$. Since Minkowski sums of non-compact closed sets are not necessarily closed, define

$$S_n = \mathrm{cl}(X_1 + \cdots + X_n), \quad n \geq 1.$$

Euclidean case

The strong law of large numbers for random closed sets in $\mathbb{R}^d$ is formulated with respect to the Painlevé–Kuratowski convergence of closed sets, see Definition B.5.

Theorem 1.21 (SLLN for integrable random sets in $\mathbb{R}^d$). *Let $X, X_1, X_2, \ldots$ be a sequence of i.i.d. integrable random convex closed sets in $\mathbb{R}^d$. Then $n^{-1}S_n$ converges in the Painlevé–Kuratowski sense to the closed convex hull of $\mathbf{E}X$.*

Proof. First, prove that $\limsup n^{-1}S_n \subset C = \overline{\mathrm{co}}\,(\mathbf{E}X)$ almost surely. Since C is convex and closed, this inclusion is equivalent to the inequality $h(n^{-1}S_n, u) \leq h(C, u)$ for all $u \in \mathbb{R}^d$. It suffices to check this inequality for a countable set of those u where $h(C, u) < \infty$. The strong law of large numbers for random variables implies

$$h(\limsup n^{-1}S_n, u) \leq \limsup h(n^{-1}S_n, u)$$

$$= \limsup n^{-1}\sum_{i=1}^{n} h(X_i, u) = h(C, u)$$

for every such u, whence $\limsup n^{-1}S_n \subset C$.

It remains to prove that $\liminf n^{-1}S_n \supset C$ almost surely. Assume that $0 \in X_n$ almost surely for each $n \geq 1$. Otherwise, we can replace X_n with $X_n - e(X_n)$, where $e(X_i)$ is defined in Lemma 2.1.9 and note that the sequence $\{e(X_n), n \geq 1\}$ satisfies the strong law of large numbers. Let us show that almost surely every point in C is a limit of a sequence of points in $n^{-1}S_n$. It suffices to show this for a dense family of points $y \in C$ having representation $y = \sum_{j=1}^{d+1} c_j e_j$, where $c_1, \ldots, c_{d+1}$ are non-negative rational numbers that sum up to 1 and each $e_j = \mathbf{E}\xi_j$ is the expectation of a bounded $\mathfrak{F}_X$-measurable selection ξ_j of X. This choice is possible since the expectations of bounded selections are dense in $\mathbf{E}^{\tilde{\mathfrak{F}}_X} X$ (which is the reduced selection expectation from Section 2.1.4) and the convex hull of the latter is dense in C. The proof is finished by applying the strong law of large numbers to the integrably bounded random compact (and finite) set $\{\xi_1, \ldots, \xi_{d+1}\}$. $\square$

210   3 Minkowski Addition

A converse to Theorem 1.21 can be formulated as follows.

Theorem 1.22. *Let $X, X_1, X_2, \ldots$ be i.i.d. random closed sets in $\mathbb{R}^d$ such that $X \subset F$ a.s. for a closed set F that does not contain any whole line. If X is not integrable, then $n^{-1} S_n$ converges in the Painlevé–Kuratowski sense to the empty set.*

SLLN in Mosco and Wijsman topologies

While Theorem 1.21 provides the full generalisation of the strong law of large numbers for random closed sets in $\mathbb{E} = \mathbb{R}^d$, the situation is more complicated if $\mathbb{E}$ is a general Banach space. Then many further variants of the strong law of large numbers are possible, according to a number of meaningful concepts of convergence for closed sets in Banach spaces. The simplest result follows directly from the strong law of large numbers for support functions if $\{X_n, n \geq 1\}$ is a family of bounded random convex closed sets (which are not necessarily compact if $\mathbb{E}$ is infinite-dimensional). Note a difference with Theorem 1.7 which deals with random compact sets.

Theorem 1.23. *If $\{X_n, n \geq 1\}$ is a sequence of i.i.d. integrably bounded random convex closed sets in a separable Banach space $\mathbb{E}$, then $n^{-1} S_n$ converges to $\mathbf{E} X_1$ in the Hausdorff metric, where $\mathbf{E} X_1$ is the convex set that satisfies $\mathbf{E} h(X_1, u) = h(\mathbf{E} X_1, u)$.*

Further conditions are required to ensure the convexification in the limit (guaranteed by Theorem 1.4 in the case of compact summands). The following theorem establishes the strong law of large numbers with respect to the Mosco convergence defined in Appendix B.

Theorem 1.24 (SLLN for Mosco convergence). *Let $\{X_n, n \geq 1\}$ be a sequence of i.i.d. non-deterministic integrable random closed sets in a separable Banach space $\mathbb{E}$. Then $n^{-1} S_n$ Mosco-converges to $\overline{\mathrm{co}}\,(\mathbf{E} X_1)$ as $n \to \infty$.*

Proof. Let $C = \overline{\mathrm{co}}\,(\mathbf{E} X_1)$ and $Y_n = n^{-1} S_n$. For any $x \in C$ and $\varepsilon > 0$, by Theorem 2.1.32(i) we can choose $\xi_j \in \mathcal{S}^1_{\mathfrak{F} X_j}(X_j)$, $1 \leq j \leq m$, such that

$$\left\| m^{-1} \sum_{j=1}^{m} x_j - x \right\| < \varepsilon,$$

where $x_j = \mathbf{E} \xi_j$, $1 \leq j \leq m$. By Theorem 2.1.32(ii), there exists a sequence $\{\xi_n, n \geq 1\}$ of selections $\xi_n \in \mathcal{S}^1_{\mathfrak{F} X_n}(X_n)$ such that $\xi_{(k-1)m+j}$, $k \geq 1$, are identically distributed for each $j = 1, \ldots, m$. If $n = (k-1)m + l$ with $1 \leq l \leq m$, then

$$\left\|\frac{1}{n}\sum_{i=1}^{n}\xi_i - \frac{1}{m}\sum_{j=1}^{m}x_j\right\|$$

$$= \left\|\frac{1}{n}\sum_{j=1}^{m}\sum_{i=1}^{k}\xi_{(i-1)m+j} - \frac{1}{n}\sum_{j=l+1}^{m}\xi_{(k-1)m+j} - \frac{1}{m}\sum_{j=1}^{m}x_j\right\|$$

$$\leq \frac{k}{n}\sum_{j=1}^{m}\left\|\frac{1}{k}\sum_{i=1}^{k}\xi_{(i-1)m+j} - x_j\right\|$$

$$+ \frac{k}{n}\sum_{j=1}^{m}\frac{1}{k}\|\xi_{(k-1)m+j}\| + \left(\frac{k}{n} - \frac{1}{m}\right)\left\|\sum_{j=1}^{m}x_j\right\|.$$

Since $\xi_{(k-1)m+j}, k \geq 1$, are i.i.d. integrable random variables for $j = 1, \ldots, m$,

$$\left\|\frac{1}{k}\sum_{i=1}^{k}\xi_{(i-1)m+j} - x_j\right\| \to 0 \quad \text{a.s. as } k \to \infty,$$

whence $k^{-1}\|\xi_{(k-1)m+j}\| \to 0$ a.s. as $k \to \infty$. Therefore,

$$\left\|\frac{1}{n}\sum_{i=1}^{n}\xi_i - \frac{1}{m}\sum_{j=1}^{m}x_j\right\| \to 0 \quad \text{a.s. as } n \to \infty.$$

Since $n^{-1}\sum_{i=1}^{n}\xi_i \in Y_n$ a.s., we have $m^{-1}\sum_{j=1}^{m}x_j \in \liminf Y_n$ a.s. Thus, $C \subset$ s-$\liminf Y_n$ a.s.

Let $\{x_j, j \geq 1\}$ be a dense sequence in $\mathbb{E} \setminus C$. By the separation theorem, there exists a sequence $\{u_j, j \geq 1\}$ of linear continuous functionals with unit norms in $\mathbb{E}^*$ such that

$$\langle x_j, u_j\rangle - \rho(x_j, C) \geq h(C, u_j), \quad j \geq 1.$$

Then $x \in C$ if and only if $\langle x, u_j\rangle \leq h(C, u_j)$ for all $j \geq 1$. Because the function $h(F, u_j)$ is $\mathfrak{B}(\mathcal{F})$-measurable as a function of $F \in \mathcal{F}$ and $\mathbf{E}h(X_1, u_j) = h(C, u_j) < \infty$ for all $j \geq 1$, it follows that $\{h(X_n, u_j), n \geq 1\}$ is a sequence of i.i.d. integrable random variables for each $j \geq 1$. By the strong law of large numbers applied for each j, $h(Y_n, u_j) \to h(C, u_j)$ a.s. as $n \to \infty$ for all $j \geq 1$ simultaneously. Except an event of probability zero, $x \in$ w-$\limsup Y_n$ implies that $x_k \to x$ weakly for some $x_k \in Y_{n_k}$ and hence

$$\langle x, u_j\rangle = \lim_{k\to\infty}\langle x_k, u_j\rangle \leq \lim_{k\to\infty} h(Y_{n_k}, u_j) = h(C, u_j), \quad j \geq 1,$$

which implies $x \in C$. Thus, w-$\limsup Y_n \subset C$ a.s. □

Since the Mosco convergence coincides with the Painlevé–Kuratowski convergence if $\mathbb{E} = \mathbb{R}^d$, Theorem 1.21 follows from Theorem 1.24. If the (possibly unbounded) summands are not identically distributed, then it is possible to derive a

212   3 Minkowski Addition

strong law of large numbers assuming that $\mathbb{E}$ is a Banach space of type $p \in (1, 2]$ and imposing a moment condition on the sequence $\{X_n, n \geq 1\}$, see Hiai [253, Th. 3.3].

Theorem 1.25 (SLLN in Wijsman topology). *If $\{X_n, n \geq 1\}$ is a sequence of i.i.d. integrable random closed sets in a separable Banach space $\mathbb{E}$, then $n^{-1}S_n$ converges almost surely in the Wijsman topology to $\overline{\mathrm{co}}\,(\mathbf{E}X_1)$, that is $\rho(x, n^{-1}S_n) \to \rho(x, \overline{\mathrm{co}}\,(\mathbf{E}X_1))$ almost surely simultaneously for all $x \in \mathbb{E}$.*

Proof. Denote $C = \overline{\mathrm{co}}\,(\mathbf{E}X_1)$. Choose a countable subset Q^* of the unit ball B_1^* in the dual space $\mathbb{E}^*$ such that

$$\rho(x, C) = \sup_{u \in Q^*} [\langle x, u \rangle - h(u, C)] \quad \text{for all } x \in \mathbb{E}.$$

Indeed, if $y \in C$ is a point nearest to x, then $u \in B_1^*$ can be chosen to be the normal of the support hyperplane to C that passes through y. Passing to a countable family of u is possible because $\mathbb{E}$ is separable.

Note that $\{h(u, X_n), n \geq 1\}$ is a sequence of i.i.d. random variables and $\rho(0, X_i) = \sup_{u \in B_1^*}(-h(u, X_i))$. By the strong law of large numbers, $h(u, S_n) \to h(u, C)$ a.s. as $n \to \infty$ for every $u \in B_1^*$. Hence this convergence holds almost surely simultaneously for all $u \in Q^*$. Then

$$\liminf_{n \to \infty} \rho(x, S_n) \geq \liminf_{n \to \infty} \rho(x, \overline{\mathrm{co}}\,(S_n))$$

$$\geq \sup_{u \in Q^*} [\langle x, u \rangle - \limsup_{n \to \infty} h(u, \overline{\mathrm{co}}\,(S_n))]$$

$$= \sup_{u \in Q^*} [\langle x, u \rangle - \limsup_{n \to \infty} h(u, S_n)]$$

$$\geq \sup_{u \in Q^*} [\langle x, u \rangle - h(u, C)] = \rho(x, C).$$

Let C' be the set of all rational convex combinations of members of a countable dense subset of C. For every $x \in \mathbb{E}$ and $m \geq 1$, there exists $y \in C'$ such that

$$\rho(x, y) \leq \rho(x, C) + 1/m.$$

Using rather delicate arguments (see Hess [247, Prop. 3.3]) it is possible to show that there exists a sequence of selections $\xi_n \in \mathcal{S}^1_{\mathfrak{F}_{X_n}}(X_n)$ such that $n^{-1}(\xi_1 + \cdots + \xi_n) \to y$ a.s. as $n \to \infty$. The corresponding negligible event depends on x and m. Taking the countable union of these negligible events we arrive at

$$\limsup_{n \to \infty} \rho(x, S_n) \leq \rho(x, y) \leq \rho(x, C) + 1/m$$

that holds almost surely simultaneously for all x from a countable dense set in $\mathbb{E}$ and $m \geq 1$. Letting m go to infinity and using the Lipschitz property of the distance function we conclude that

$$\limsup_{n\to\infty} \rho(x, S_n) \leq \rho(x, C)$$

almost surely for all $x \in \mathbb{E}$ simultaneously. □

If $\mathbb{E}$ is finite-dimensional then the Wijsman topology coincides with the Fell topology which is equivalent to the Painlevé–Kuratowski convergence. If $\mathbb{E}$ is reflexive, then the Wijsman topology is weaker than the Mosco topology and Theorem 1.24 implies Theorem 1.25.

SLLN in the Hausdorff metric

To prove a strong law of large numbers in the Hausdorff metric on the family of bounded closed sets in a Banach space, one should restrict the consideration to random closed sets with special values. Let $\mathcal{F}_{co}$ be the family of all closed sets F that satisfy

$$\rho_H(n^{-1}(F^{(n)}, \overline{co}\, F) \to 0 \quad \text{as } n \to \infty,$$

where $F^{(n)}$ is the sum of n identical summands equal to F. By Theorem 1.3, $\mathcal{K} \subset \mathcal{F}_{co}$. It should be noted that $\mathcal{F}_{co}$ is a proper subclass of all bounded closed sets.

Example 1.26. Let $\mathbb{E} = \ell^1$ be the space of all summable sequences and let $e_i = (0,\ldots,0,1,0,\ldots)$ with 1 on the ith place, $i \geq 1$. Then $F = \{e_i, i \geq 1\}$ is bounded, but $F \notin \mathcal{F}_{co}$. To see this, it is easy to check that $F_n = n^{-1}(F + \cdots + F)$ is not a Cauchy sequence with respect to the Hausdorff metric. Take $x_m = (e_1 + \cdots + e_m)/m$ and $n < m/2$. Then

$$\inf_{y \in F_n} \|x_m - y\| \geq \frac{m-n}{m} > \frac{1}{2}.$$

The following theorem was proved by Uemura [566] similarly to Theorem 1.19 (also in the case of non-identically distributed random sets).

Theorem 1.27. *Let $\{X_n, n \geq 1\}$ be a sequence of i.i.d. random closed sets with values in $\mathcal{F}_{co}$. Assume that*
1. $\mathbf{E}\|X_1\|^p < \infty$ *for some* $p \in (1,2]$;
2. *for any $\varepsilon > 0$ there exists a compact in the Hausdorff metric set $\mathcal{D}_\varepsilon \subset \mathcal{F}_{co}$ such that $\mathbf{P}\{X_1 \notin \mathcal{D}_\varepsilon\} < \varepsilon$.*

Then $\rho_H(n^{-1} S_n, \mathbf{E}(\overline{co}\, X_1)) \to 0$ a.s. as $n \to \infty$.

Probably the most important unbounded random closed sets appear from stochastic processes as epigraphs of random functions. An application of the strong law of large numbers in this context is described in Section 5.3.5.

2 The central limit theorem

2.1 A central limit theorem for Minkowski averages

Euclidean case

By the strong law of large numbers, the Minkowski averages

$$\bar{X}_n = n^{-1}(X_1 + \cdots + X_n)$$

converge to the selection expectation $\mathbf{E}X$ under rather mild assumptions. In the classical probability theory the speed of this convergence is assessed by taking the difference between $\bar{X}_n$ and $\mathbf{E}X$ and normalising it with a growing sequence. However, Minkowski addition is not invertible, so that it is not possible to subtract $\mathbf{E}X$ from $\bar{X}_n$. Nor it is possible to circumvent the problem by considering random sets with zero expectation, since all integrable random compact sets with zero expectation are singletons.

It is possible to avoid subtraction by considering the normalised Hausdorff distance $\sqrt{n}\rho_{\mathrm{H}}(\bar{X}_n, \mathbf{E}X)$ between the Minkowski averaged and the selection expectation. The central limit theorem for Minkowski sums of random compact sets yields the weak convergence of this Hausdorff distance to a random variable equal to the maximum of a Gaussian random function on the unit ball B_1 in $\mathbb{R}^d$. Similar to the strong law of large numbers, a central limit theorem (CLT) for random convex compact sets in $\mathbb{R}^d$ follows from the corresponding results in the Banach space $\mathsf{C}(B_1)$ of continuous functions on the unit ball.

Let X be a random compact set which is *square integrable*, i.e. $\mathbf{E}\|X\|^2 < \infty$. Define its *covariance function* $\Gamma_X : \mathsf{C}(B_1) \times \mathsf{C}(B_1) \mapsto \mathbb{R}$ as

$$\Gamma_X(u,v) = \mathbf{E}\big[h(X,u)h(X,v)\big] - \mathbf{E}h(X,u)\mathbf{E}h(X,v), \qquad (2.1)$$

i.e. $\Gamma_X(\cdot,\cdot)$ is the covariance of the support function $h(X,\cdot)$ considered to be a random element in $\mathsf{C}(B_1)$. Note that $\Gamma_X = \Gamma_{\mathrm{co}(X)}$.

Theorem 2.1 (CLT for random sets in $\mathbb{R}^d$). *Let $X, X_1, X_2, \ldots$ be i.i.d. square integrable random sets. Then*

$$\sqrt{n}\rho_{\mathrm{H}}(n^{-1}(X_1 + \cdots + X_n), \mathbf{E}X) \to \sup_{u \in B_1} \|\zeta(u)\|, \qquad (2.2)$$

where $\{\zeta(u), u \in B_1\}$ is a centred Gaussian random function in $\mathsf{C}(B_1)$ with the covariance $\mathbf{E}[\zeta(u)\zeta(v)] = \Gamma_X(u,v)$.

Proof. First, note that $\mathrm{co}(X), \mathrm{co}(X_1), \mathrm{co}(X_2), \ldots$ is a sequence of i.i.d. random convex compact sets and $\mathbf{E}\|\mathrm{co}(X)\|^2 = \mathbf{E}\|X\|^2 < \infty$. Thus,

$$\mathbf{E}\|h(\mathrm{co}(X), \cdot)\|_\infty^2 < \infty,$$

where $\|f\|_\infty = \sup\{f(u) : u \in B_1\}$ is the uniform norm of $f \in \mathsf{C}(B_1)$. Then

$$\sqrt{n}\rho_{\mathrm{H}}(n^{-1}(\mathrm{co}(X)_1 + \cdots + \mathrm{co}(X)_n), \mathbf{E}X_1)$$

$$= \sqrt{n}\left\|\frac{1}{n}\sum_{i=1}^n h(\mathrm{co}(X_i), \cdot) - \mathbf{E}h(\mathrm{co}(X), \cdot)\right\|_\infty. \qquad (2.3)$$

The key argument of the proof is an application of the central limit theorem in $\mathsf{C}(B_1)$, see Araujo and Giné [12, Cor. 7.17] and Jain and Markus [279]. For this, we have to check the *entropy condition*

$$\int_0^1 H^{1/2}(\alpha)\,d\alpha < \infty, \tag{2.4}$$

where $H(\alpha) = \log N(\alpha)$ for the metric entropy $N(\alpha)$ of the unit ball, i.e. the smallest number of balls of radius α covering B_1. By replacing B_1 with a cube we can conclude that $N(\alpha) \leq c_d \alpha^{-d}$ for some dimension-dependent constant c_d, whence integral (2.4) is finite.

By Theorem F.1, the centred support function is Lipschitz, i.e.

$$|h(\mathrm{co}(X), u) - h(\mathbf{E}X, u) - h(\mathrm{co}(X), v) + h(\mathbf{E}X, v)|$$
$$\leq |h(\mathrm{co}(X), u) - h(\mathrm{co}(X), v)| + |h(\mathbf{E}X, u) - h(\mathbf{E}X, v)|$$
$$\leq \Big(\|h(\mathrm{co}(X), \cdot)\|_\infty + \|h(\mathbf{E}X, \cdot)\|_\infty\Big)\|u - v\|.$$

The central limit theorem in $\mathbf{C}(B_1)$ implies that

$$n^{-1/2} \sum_{i=1}^n \Big(h(\mathrm{co}(X_i), \cdot) - h(\mathbf{E}X, \cdot)\Big)$$

converges weakly in the space $\mathbf{C}(B_1)$ to the Gaussian random function ζ with the covariance Γ_X. This weak convergence implies the convergence in distribution of the maximum for the corresponding random functions and (2.3) yields (2.2). This finishes the proof for random convex sets.

It remains to show that it is possible to replace X_n, $n \geq 1$, by their convex hulls without changing the limiting distribution. Note that $\mathrm{co}(\bar{X}_n) = n^{-1}(\mathrm{co}(X_1) + \cdots + \mathrm{co}(X_n))$. The triangle inequality together with Theorem 1.1 yields that

$$|\sqrt{n}\rho_\mathrm{H}(\bar{X}_n, \mathbf{E}X_1) - \sqrt{n}\rho_\mathrm{H}(\mathrm{co}(\bar{X}_n), \mathbf{E}X_1)| \leq \sqrt{n}\rho_\mathrm{H}(\bar{X}_n, \mathrm{co}(\bar{X}_n))$$
$$= n^{-1/2}\rho_\mathrm{H}(X_1 + \cdots + X_n, \mathrm{co}(X_1) + \cdots + \mathrm{co}(X_n))$$
$$\leq cn^{-1/2} \max_{1 \leq i \leq n} \|X_i\|$$

for a constant c. Since $\|X_1\|, \|X_2\|, \ldots$ are i.i.d. random variables with a finite second moment, the right-hand side converges in distribution to zero as $n \to \infty$. Indeed, if $M_n = n^{-1/2}\max(\|X_1\|, \ldots, \|X_n\|)$, then

$$\mathbf{P}\{M_n < x\} = \mathbf{P}\Big\{\|X\| < n^{1/2}x\Big\}^n.$$

Now $\mathbf{E}\|X\|^2 < \infty$ implies that

$$\mathbf{P}\Big\{\|X\| > n^{1/2}x\Big\} \leq n^{-1}x^{-2}\mathbf{E}\Big[\mathbf{1}_{\|X\|>n^{1/2}x}\|X\|^2\Big] \to 0 \quad \text{as } n \to \infty,$$

whence $\limsup \mathbf{P}\{M_n < x\} = 1$. $\square$

Since
$$\rho_H(K, L) = \sup_{u \in B_1} |h(K, u) - h(L, u)| = \sup_{u \in \mathbb{S}^{d-1}} |h(K, u) - h(L, u)|$$

for all $K, L \in \mathcal{K}$, it is easy to show that the suprema of ζ over the unit ball B_1 and over the unit sphere $\mathbb{S}^{d-1}$ are almost surely equal. Therefore, the Gaussian random function ζ can be alternatively defined on the unit sphere.

Proposition 2.2 (Random sets with ζ being a support function). *If the limiting random field $\zeta(u)$, $u \in \mathbb{S}^{d-1}$, in Theorem 2.1 is almost surely a support function of a random compact set, then the summands are distributed as $X = \eta + K$, where η is a random vector in $\mathbb{R}^d$ and K is a deterministic convex compact set.*

Proof. If ζ is the support function of a random convex compact set Z, then $\mathbf{E}\zeta(u) = \mathbf{E}h(Z, u) = 0$ for all u, whence $Z = \{\xi\}$ for a centred Gaussian random vector ξ, see Proposition 2.1.30(i). Hence $\Gamma_X(u, v)$ is a bilinear form determined by the covariance matrix of ξ. Fix any $u \in \mathbb{S}^{d-1}$. Then both $\alpha_1 = h(X, u)$ and $\alpha_2 = h(X, -u)$ have identical variances and the correlation coefficient between α_1 and α_2 is -1. Hence $\alpha_1 + \alpha_2$ is constant, i.e. X has a deterministic width in the direction u. The same argument applies for a countable set of u. Therefore, $X = \eta + K$ for a deterministic set K. The random vector η, however, does not necessarily have the Gaussian distribution, cf. Theorem 2.11. □

Open problem 2.3. Find a geometrical interpretation of the Gaussian function ζ that appears in Theorem 2.1. Can it be interpreted geometrically as determined by a certain random set Z?

Open problem 2.4. Formulate the central limit theorem for Minkowski sums of random sets using asymptotic expansions on the same probability space as suggested by Zolotarev [629] for the classical central limit theorem. The idea is to avoid subtraction of sets by decomposing the average $\bar{X}_n$ as

$$\bar{X}_n = \mathbf{E}X + n^{-1/2}Z + \cdots,$$

where all relevant random sets are defined on the same probability space.

Non-square integrable summands

The following theorem gives other sufficient conditions for the central limit theorem without assuming that X is square integrable.

Theorem 2.5. *Let a random compact set X in $\mathbb{R}^d$ satisfy the following conditions:*
(1) $\mathbf{E}\left[\|X\|^2 \mathbf{1}_{\|X\| \leq t}\right]$ *is a slowly varying function;*
(2) $\Gamma_X(u, u) = \lim_{t \to \infty} \mathbf{E}\left[(h(X, u))^2 \mathbf{1}_{|h(X,u)| \leq t}\right] \Big/ \mathbf{E}\left[\|X\|^2 \mathbf{1}_{\|X\| \leq t}\right].$

Then $na_n^{-1}\rho_H(\bar{X}_n, EX)$ converges in distribution to $\|\zeta\|_\infty$, where

$$a_n = \sup\{t : t^{-2}\mathbf{E}(\|X\|^2 \mathbf{1}_{\|X\|\leq t}) \geq n^{-1}\}$$

and ζ is a centred Gaussian element in $C(B_1)$ with covariance (2.1).

Similar ideas yield a limit theorem with non-Gaussian limits. Here is a theorem from Giné, Hahn and Zinn [202] which provides the condition for the convergence to a *p-stable* law with $1 < p < 2$.

Theorem 2.6 (Convergence to stable laws). *Let X be a random convex compact set in $\mathbb{R}^d$ and, for some $p \in (1, 2)$,*
(1) *the function $t^p \mathbf{P}\{\|X\| > t\}$ is slowly varying;*
(2) *there exists a finite measure μ on $\mathcal{K}_1 = \{K \in \text{co}\,\mathcal{K} : \|K\| = 1\}$ such that*

$$\lim_{t\to\infty} \frac{\mathbf{P}\{X/\|X\| \in \mathcal{D}, \|X\| \geq t\}}{\mathbf{P}\{\|X\| \geq t\}} = \frac{\mu(\mathcal{D})}{\mu(\mathcal{K}_1)}$$

for every μ-continuity set $\mathcal{D} \subset \mathcal{K}_1$.
If $a_n = \sup\{t : n\mathbf{P}\{\|X\| > t\} \geq \mu(\mathcal{K}_1)/p\}$, $n \geq 1$, then $na_n^{-1}\rho_H(\bar{X}_n, EX)$ converges in distribution to the maximum of a p-stable random function on B_1.

CLT in Banach space

Similar to the finite-dimensional case, each convex compact set in a separable Banach space $\mathbb{E}$ corresponds to its support function which is a continuous functional on the unit ball B_1^* in the dual space $\mathbb{E}^*$, see (1.10). Although this unit ball B_1^* is not compact and also is not separable in the strong topology on $\mathbb{E}^*$, this unit ball is compact in the weak*-topology on $\mathbb{E}^*$, i.e. the space (B_1^*, w^*) is compact. This topology is metrised by

$$\rho_*(u, v) = \sum_{n\geq 1} 2^{-n} |\langle x_n, u\rangle - \langle x_n, v\rangle|, \quad u, v \in B_1^*,$$

where $\{x_n, n \geq 1\}$ is a countable dense set in the unit ball B_1 in $\mathbb{E}$. Since the support function of a finite set is weak*-continuous, the continuity in the general case is derived by approximation. Thus, $h(X, u)$ belongs to the space $C(B_1^*)$ of weak* continuous functions on B_1^*.

For $K \in \mathcal{K}$, define a variant of its support function

$$\bar{h}(K, u) = \sup_{x\in K} |\langle x, u\rangle| = \max(h(K, u), h(K, -u)).$$

Consider a function $g : \mathbb{R}_+ \mapsto \mathbb{R}_+$, which is non-decreasing continuous sublinear (i.e. $g(r+s) \leq g(r) + g(s)$) and satisfies $g(0) = 0$. Let L be a fixed compact convex centrally symmetric set in $\mathbb{E}$. A semi-metric on B_1^* can be defined as

$$\rho_{g,L}(u, v) = g(\bar{h}(L, u - v)).$$

For $K \in \mathrm{co}\,\mathcal{K}$, let $\|K\|_{g,L}$ be the smallest $t \geq 0$ such that, for any $u, v \in B_1^*$, one has $\bar{h}(K, u - v) \leq t\rho_{g,L}(u, v)$. Furthermore, $H(B_1^*, \rho_{g,L}, \varepsilon)$ denotes the $\rho_{g,L}(u,v)$-metric entropy of B_1^*, i.e. the logarithm of the minimal number of balls of radius ε in metric $\rho_{g,L}$ that cover B_1^*.

Theorem 2.7. *Let $\{X_n, n \geq 1\}$ be a sequence of i.i.d. random convex compact sets in a separable Banach space $\mathbb{E}$ such that $\mathbf{E}\|X_1\|_{g,L}^2 < \infty$ and*

$$\int_0^1 H^{1/2}(B_1^*, \rho_{g,L}, \varepsilon)\mathrm{d}\varepsilon < \infty. \tag{2.5}$$

Then $\sqrt{n}\rho_{\mathrm{H}}(\bar{X}_n, \mathbf{E}X_1)$ converges in distribution to $\|\zeta\|_\infty$, where ζ is a centred Gaussian random element in $\mathbf{C}(B_1^*)$ with the covariance given by (2.1) for $u, v \in B_1^*$.

Proof. For each $K \in \mathrm{co}\,\mathcal{K}$ with $\|K\|_{g,L} < \infty$ and $u, v \in B_1^*$,

$$|h(K, u) - h(K, v)| \leq \bar{h}(K, u - v) \leq = \|K\|_{g,L}\,\rho_{g,L}(u, v)\,.$$

Thus,
$$|h(X_1, u) - h(X_1, v)| \leq \|X_1\|_{g,L}\,\rho_{g,L}(u, v)\,.$$

This condition together with the metric entropy condition (equivalent to (2.5)) are known to imply the central limit theorem in the Banach space of weak*-continuous functions on B_1^*, see Jain and Markus [279]. This finishes the proof because of the isometry between this Banach space and the family of convex compact sets. □

It is more complicated to prove the central limit theorem for non-convex random sets in Banach spaces, than for random sets in $\mathbb{R}^d$. This is explained by the lack of an analogue of the Shapley–Folkman–Starr theorem in infinite-dimensional spaces, while quantitative results like Theorem 1.4 are no longer useful to show that in a central limit theorem non-convex sets can be replaced by their convex hulls.

2.2 Gaussian random sets

Lipschitz functionals

Gaussian random functions on $\mathbb{S}^{d-1}$ (or on the unit ball B_1) appear naturally in the limit theorem for Minkowski sums of random compact sets in $\mathbb{R}^d$, since the normalised Hausdorff distance between $\bar{X}_n = n^{-1}(X_1 + \cdots + X_n)$ and $\mathbf{E}X_1$ converges in distribution to the maximum of a Gaussian random function on B_1 or $\mathbb{S}^{d-1}$. It is well known that Gaussian random elements in linear spaces can be defined through Gaussian distributions of all linear continuous functionals. A similar approach can be applied to the (non-linear) space $\mathcal{K}$ of compact sets.

Let $\mathrm{Lip}^+(\mathcal{K}, \mathbb{R})$ denote the family of functionals $g: \mathcal{K} \mapsto \mathbb{R}$ which satisfy the following conditions.

(i) *g* is *positively linear*, i.e. for all $a, b \geq 0$ and $K, L \in \mathcal{K}$,

$$g(aK + bL) = ag(K) + bg(L). \tag{2.6}$$

(ii) *g* is *Lipschitz* with respect to the Hausdorff metric.

Definition 2.8 (Gaussian random sets). A random compact set X in $\mathbb{R}^d$ is said to be *Gaussian*, if $g(X)$ is a Gaussian random variable for each $g \in \text{Lip}^+(\mathcal{K}, \mathbb{R})$.

It is possible to consider random convex sets whose support functions are Gaussian, i.e. have *Gaussian finite-dimensional distributions*. The following result establishes the equivalence of this property of support functions and Definition 2.8.

Proposition 2.9 (Gaussian support functions). *A random convex compact set X is Gaussian if and only if its support function $h(X, u)$ is a Gaussian random function on $\mathbb{S}^{d-1}$, i.e. $h(X, u_1), \ldots, h(X, u_m)$ are jointly Gaussian for all $u_1, \ldots, u_m \in \mathbb{S}^{d-1}$ and $m \geq 1$.*

Proof. If f is a linear functional on the space $\mathsf{C}(\mathbb{S}^{d-1})$, then $f(h(X, \cdot)) = g(X)$ for some $g \in \text{Lip}^+(\mathcal{K}, \mathbb{R})$. Thus, $f(h(X, \cdot))$ is a Gaussian random variable, whence $h(X, \cdot)$ is a Gaussian random element in $\mathsf{C}(\mathbb{S}^{d-1})$.

In the other direction, assume that $h(X, \cdot)$ is a Gaussian random function. Each functional $g \in \text{Lip}^+(\mathcal{K}, \mathbb{R})$ can be regarded as a functional of $h(X, \cdot) \in \mathsf{C}(\mathbb{S}^{d-1})$. It is possible to extend g onto the linear subspace of $\mathsf{C}(\mathbb{S}^{d-1})$ obtained as the linear span of all support functions. The Lipschitz property implies the continuity of the extended version of g. By the Hahn–Banach theorem, g is extendable to a bounded linear functional on $\mathsf{C}(\mathbb{S}^{d-1})$, which is Gaussian by Definition 2.8. □

Since the support function does not make a distinction between X and its convex hull, we obtain the following corollary.

Corollary 2.10. *A random compact set X is Gaussian if and only if its convex hull $\text{co}(X)$ is Gaussian.*

Proof. For the alternative proof, note that $g(K) = g(\text{co}(K))$ for any $g \in \text{Lip}^+(\mathcal{K}, \mathbb{R})$ and (possibly non-convex) $K \in \mathcal{K}$. The Lipschitz property of g yields its continuity, whence

$$g(K) = g(n^{-1}K + \cdots + n^{-1}K) \to g(\text{co}(K)) \quad \text{as } n \to \infty. \qquad \square$$

Characterisation theorem

The following important result characterises Gaussian random compact sets as those having degenerate shapes. More exactly, each Gaussian random set is a translation of a deterministic convex compact set by a Gaussian random vector. The main idea of the proof is to show that the support function can be made positive and then argue that a non-negative Gaussian random variable is degenerate.

Theorem 2.11 (Characterisation of Gaussian sets). *A random compact set X in $\mathbb{R}^d$ is Gaussian if and only if*

$$\mathrm{co}(X) = K + \xi \quad \text{a.s.}$$

for a deterministic non-empty convex compact set $K \subset \mathbb{R}^d$ and a Gaussian random vector ξ.

Proof. Sufficiency. Note that $g(X) = g(K) + g(\{\xi\})$ is Gaussian for any $g \in \mathrm{Lip}^+(\mathcal{K}, \mathbb{R})$, since $g(K)$ is deterministic and $g(\{\xi\})$ is a linear bounded functional of ξ.

Necessity. With each $K \in \mathrm{co}\,\mathcal{K}$ it is possible to associate its Steiner point $\mathbf{s}(K)$ defined by (F.6). If $K \in \mathcal{K}$ is non-convex, set $\mathbf{s}(K) = \mathbf{s}(\mathrm{co}(K))$. All coordinates of $\mathbf{s}(K)$ are functionals from $\mathrm{Lip}^+(\mathcal{K}, \mathbb{R})$. By the assumption, $\mathbf{s}(X)$ is a Gaussian random vector. It is known that $\mathbf{s}(K) \in K$ for each K from $\mathrm{co}\,\mathcal{K}$. Then $X_0 = X - \mathbf{s}(X)$ is a translate of X such that $h(X_0, u) \geq 0$ for all $u \in \mathbb{S}^{d-1}$. Hence $h(X_0, u)$ is Gaussian and non-negative almost surely, which is possible only for a degenerate distribution, i.e. for $h(X_0, u)$ being a constant. This holds for a countable set of u and, by continuity, for all $u \in \mathbb{S}^{d-1}$. Thus, $X_0 = X - \mathbf{s}(X) = K$ is a deterministic set, whence $X = K + \xi$ with $\xi = \mathbf{s}(X)$. □

A similar characterisation result holds for *square integrable* convex compact random sets in a separable Banach space $\mathbb{E}$. For this, define

$$g(u) = h(X, u) - h(\mathbf{E}X, u), \quad u \in \mathbb{E}^*,$$

then prove that g is weak*-continuous on $\mathbb{E}^*$, so that $g(u) = \langle \xi, u \rangle$ for some $\xi \in \mathbb{E}$. Then it is possible to show that ξ is measurable and Gaussian.

Open problem 2.12. Suggest an alternative and "natural" definition of Gaussian random sets that yields sets with variable shapes.

2.3 Stable random compact sets

The following definition is similar to the classical definition of a stable random element with values in a Banach space.

Definition 2.13 (p-stable random sets). A random convex compact set X in a Banach space $\mathbb{E}$ is called *p-stable*, $0 < p \leq 2$, if, for X_1, X_2 i.i.d. with the same distribution as X and for all $\alpha, \beta \geq 0$, there exist sets $C, D \in \mathrm{co}\,\mathcal{K}$ such that

$$\alpha X_1 + \beta X_2 + C \stackrel{d}{\sim} (\alpha^p + \beta^p)^{1/p} X + D. \tag{2.7}$$

X is said to be *strictly p-stable* if (2.7) holds with $C = D = \{0\}$.

It should be noted that D may be chosen to be $\{0\}$ for $0 < p < 1$, and C may be set to $\{0\}$ for $1 \leq p \leq 2$. By considering a deterministic set X, it is easy to see that it is not possible to get rid of both C and D simultaneously for the whole range of p, so there are p-stable sets that are not strictly p-stable. Let $\text{Lip}^+(\mathcal{K},\mathbb{R}^d)$ denote the family of $\mathbb{R}^d$-valued Lipschitz functions on $\mathcal{K}$, which are *positively linear* as in (2.6).

Theorem 2.14. *The following statements are equivalent for a random convex compact set X in a separable Banach space $\mathbb{E}$.*
(i) X is p-stable.
(ii) The support function of X is a p-stable random element in $C(B_1^*)$.
(iii) $\varphi(X)$ is p-stable for all $\varphi \in \text{Lip}^+(\mathcal{K},\mathbb{R}^2)$.
(iv) $\varphi(X)$ is p-stable for all $\varphi \in \text{Lip}^+(\mathcal{K},\mathbb{R}^d)$ and all $d \geq 1$.

Since the Gaussian distribution is p-stable with $p = 2$, Theorem 2.14 shows that Gaussian random sets appear as a particular case of p-stable sets.

Theorem 2.15 (Characterisation of p-stable sets). *If X is a p-stable random convex compact set with $1 \leq p \leq 2$, then $X = K + \xi$, where $K \in \text{co}\,\mathcal{K}$ and ξ is a p-stable random element in $\mathbb{E}$. If $0 < p < 1$, then*

$$X = K + \int_{\mathcal{K}_1} L \mathbf{M}(\mathrm{d}L),$$

where the integral is understood as the limit of Minkowski sums and $\mathbf{M}$ is a completely random p-stable measure on $\mathcal{K}_1 = \{L \in \text{co}\,\mathcal{K} : \|L\| = 1\}$.

The complete randomness of $\mathbf{M}$ in Theorem 2.15 means that the values of $\mathbf{M}$ on disjoint sets are independent and its p-stability signifies that $\mathbf{M}(\mathcal{D})$ coincides in distribution with $\mu(\mathcal{D})^{1/p}\theta$ for $\mathcal{D} \in \mathfrak{B}(\mathcal{K}_1)$, where μ is a measure on $\mathcal{K}_1$ called the *spectral measure* and θ is a p-stable random variable. The proof of Theorem 2.15 is based on an interpretation of $h(X, u)$ as a stable random vector in the Banach space $C(B_1^*)$ using the characterisation of p-stable elements in Banach spaces from Araujo and Giné [12]. The degenerate representation for $1 \leq p \leq 2$ is obtained by showing that the corresponding spectral measure is supported by linear functions, which implies that K is deterministic. Therefore, p-stable sets with $p \in [1, 2]$ are degenerate similarly to Gaussian random sets characterised in Theorem 2.11. However, non-trivial p-stable sets may appear if $p < 1$.

2.4 Minkowski infinitely divisible random compact sets

Compound Poisson law and Lévy measure

Following the representation of random convex compact sets by their support functions, it is possible to invoke results on infinitely divisible random elements in Banach spaces in order to characterise random sets which are infinitely divisible with respect to Minkowski addition. The key idea is to use the fact that infinitely divisible laws in Banach spaces have a *Lévy–Khinchin representation*, i.e. they are convolutions of point masses, Gaussian laws and limits of compound Poisson laws.

Definition 2.16 (M-infinite divisibility). A random convex compact set X is called *M-infinitely divisible*, if, for each $n \geq 1$, there exist i.i.d. random convex compact sets $Z_{n1}, \ldots, Z_{nn}$ such that

$$Z_{n1} + \cdots + Z_{nn} \stackrel{d}{\sim} X. \tag{2.8}$$

A finite measure Λ on $\operatorname{co} \mathcal{K}$ determines a finite *Poisson process* Π_Λ whose "points" are convex compact sets, see Definition 1.8.7.

Definition 2.17 (Compound Poisson sets). If $\Pi_\Lambda = \{X_1, \ldots, X_N\}$ is a finite Poisson process on $\operatorname{co} \mathcal{K}$ with the intensity measure Λ, then the random convex compact set $Z = X_1 + \cdots + X_N$ is called *compound Poisson* with measure Λ (notation $Z \in \operatorname{Pois}_+(\Lambda)$). Note that Z is empty if at least one summand is empty or $N = 0$.

The following definition extends the above concept to σ-finite measures.

Definition 2.18 (Lévy measure). A σ-finite measure Λ on $\operatorname{co} \mathcal{K}$ is a *Lévy measure*, if there exist finite measures Λ_n on $\operatorname{co} \mathcal{K}$ and convex compact sets K_n such that $\Lambda_n \uparrow \Lambda$ and the sequence of random closed sets $\{K_n + Z_n\}$ with $Z_n \in \operatorname{Pois}_+(\Lambda_n)$ converges weakly to a random closed set Z whose distribution is denoted by $\operatorname{Pois}_+(\Lambda)$.

It follows from the general results of Araujo and Giné [12], that a σ-finite measure Λ on $\operatorname{co} \mathcal{K}$ is a Lévy measure if and only if

$$\int_{\operatorname{co} \mathcal{K}} \min(1, \|L\|) \Lambda(dL) < \infty. \tag{2.9}$$

Characterisation of infinite divisibility

First, we provide a characterisation for M-infinitely divisible random compact sets in $\mathbb{R}^d$ containing the origin almost surely.

Theorem 2.19. *A random convex compact set X in $\mathbb{R}^d$ containing the origin almost surely is M-infinitely divisible if and only if there exist a deterministic convex compact set K containing the origin and a σ-finite measure Λ on the family of convex compact sets containing the origin which satisfies (2.9) such that X coincides in distribution with $K + Z$ for $Z \in \operatorname{Pois}_+(\Lambda)$.*

Proof. Sufficiency is easy, since $K = n^{-1}K + \cdots + n^{-1}K$ and $Z \in \operatorname{Pois}_+(\Lambda)$ can be represented as a sum of i.i.d. random sets $Z_{n1}, \ldots, Z_{nn}$ from $\operatorname{Pois}_+(n^{-1}\Lambda)$.
Necessity follows from the Lévy–Khinchin representation of infinitely divisible laws in Banach spaces. The fact that $0 \in X$ almost surely implies that the support functions are almost surely non-negative. Therefore, all one-dimensional projections of the distribution law in $C(B_1)$ are supported by half-lines. Since Lévy measures supported by half-lines must integrate $\min(1, \|\cdot\|)$, a bound based on the dimension d yields (2.9). □

If the condition that $0 \in X$ is dropped, then one has to translate X, for example, using the Steiner point $\mathbf{s}(X)$ as in Section 2.2.

Theorem 2.20. *A random convex compact set X in $\mathbb{R}^d$ is M-infinitely divisible if and only if there exist a deterministic set $K \in \text{co}\,\mathcal{K}$, a centred Gaussian random vector $\xi \in \mathbb{R}^d$ and a σ-finite measure Λ on $\text{co}\,\mathcal{K}$ satisfying*

$$\int_{\text{co}\,\mathcal{K}} \min(1, \|\mathbf{s}(L)\|^2) \Lambda(\mathrm{d}L) < \infty,$$

$$\int_{\text{co}\,\mathcal{K}} \min(1, \|L - \mathbf{s}(L)\|) \Lambda(\mathrm{d}L) < \infty,$$

such that $X = \xi + K + Z$, where Z is the weak limit of the sequence of random convex compact sets

$$Z_n = \int_{n^{-1} < \|L\| \le 1} \mathbf{s}(L) \Lambda(\mathrm{d}L) + Z'_n, \quad n \ge 1,$$

with $Z'_n \in \text{Pois}_+(\Lambda_n)$ and Λ_n being the restriction of Λ onto the family of convex compact sets with the norm greater than n^{-1}.

While the M-infinite divisibility of X implies that $\mathbf{s}(X)$ is an infinitely divisible random vector and $X - \mathbf{s}(X)$ is an M-infinitely divisible random set which contains the origin, the converse is not true.

Example 2.21. Let $\mathbb{E} = \mathbb{R}$ and let $X = [\eta_1 - \eta_2, \eta_1 + \eta_2]$, where η_1, η_2 are non-negative infinitely divisible random variables such that their sum is not infinitely divisible. Then $\mathbf{s}(X) = \eta_1$ is infinitely divisible and $X - \mathbf{s}(X) = [-\eta_2, \eta_2]$ is M-infinitely divisible, while X is not M-infinitely divisible. Indeed, if X satisfies (2.8), then $Z_{ni} = [\eta'_{ni}, \eta''_{ni}]$, $1 \le i \le n$, are i.i.d random segments such that $\eta_1 + \eta_2 = \eta''_{n1} + \cdots + \eta''_{nn}$, which contradicts the choice of $\eta_1 + \eta_2$ as not being infinitely divisible.

Open problem 2.22. Characterise M-infinitely divisible random closed (compact) sets in a separable Banach space. Some results in this direction are given in Giné and Hahn [200], where it is explained that the main obstacle is the non-existence of a generalised Steiner functional for convex compact subsets of a general Banach space.

3 Further results related to Minkowski sums

3.1 Law of iterated logarithm

The *law of iterated logarithm* for a square integrable random compact set X in $\mathbb{R}^d$ states

$$\limsup_{n\to\infty} \frac{\sqrt{n}}{\sqrt{2\log\log n}} \rho_H(\bar{X}_n, \mathbf{E}\,\mathrm{co}(X)) \leq \sqrt{\mathbf{E}\|X\|^2}, \tag{3.1}$$

where $\bar{X}_n$ denotes the Minkowski average of i.i.d. random sets $X_1,\ldots,X_n$ having the same distribution as X.

The proof is based on the convexification argument and passing to support functions. Denote $a_n = \sqrt{2n\log\log n}$. By Theorem 1.1,

$$\rho_H(a_n^{-1}\sum_{i=1}^n X_i, a_n^{-1}\sum_{i=1}^n \mathrm{co}(X_i)) \leq (2n\log\log n)^{-1/2}\sqrt{d}\max_{1\leq i\leq n}\|X_i\|,$$

which converges almost surely to zero. So we may assume in (3.1) that $X_1,\ldots,X_n$ are almost surely convex and pass to their support functions. Since the support functions satisfy the central limit theorem and have integrable norms, they also satisfy the compact law of iterated logarithm. The bound in the right-hand side of (3.1) corresponds to the fact that the cluster set of $\sum(h(X_i,\cdot) - \mathbf{E}h(X_i,\cdot))/a_n$ is a bounded set contained in the ball $\{u \in B_1^* : \|u\| \leq \sqrt{\mathbf{E}\|X\|^2}\}$.

Similarly, the conditions of Theorem 2.7 imply that the corresponding random sets satisfy the law of iterated logarithm in the Banach space $\mathbb{E}$.

3.2 Three series theorem

The classical Kolmogorov three series theorem can be generalised for random compact sets. Recall that this theorem characterises convergent series of independent random variables in terms of the convergence of three other series built from means, variances and truncation probabilities. Consider here only the case of random compact sets in the Euclidean space. For a random compact set X, define

$$\mathrm{Var}_A X = \mathbf{E}(\rho_H(X, \mathbf{E}X))^2.$$

Note that the smallest value of $\mathbf{E}(\rho_H(X, K)^2)$ over $K \in \mathcal{K}$ is called the *Fréchet variance* of X, see Definition 2.3.2. Therefore, $\mathrm{Var}_A X$ is not smaller than the Fréchet variance of X. Below we discuss the almost sure convergence of the series

$$\sum_{n=1}^\infty X_n = X_1 + X_2 + \cdots \tag{3.2}$$

for i.i.d. random compact sets $X_1, X_2,\ldots$ which share the same distribution with X. If $0 \in X$ a.s., then (3.2) converges if and only if the sum of the norms $\sum \|X_n\|$ converges. For each $c > 0$, introduce a *truncated* variant of X, defined as

$$X(c) = \begin{cases} X, & \|X\| \leq c, \\ \{0\}, & \text{otherwise.} \end{cases}$$

3 Further results related to Minkowski sums

Theorem 3.1 (Three series theorem). *Let $\{X_n, n \geq 1\}$ be a sequence of independent random compact sets. Then $\sum X_n$ converges almost surely if and only if the following three series*

$$\sum \mathbf{P}\{\|X_n\| > c\}, \tag{3.3}$$

$$\sum \mathbf{E} X_n(c), \tag{3.4}$$

$$\sum \mathrm{Var}_A X_n(c) \tag{3.5}$$

converge for some $c > 0$.

Proof. First, assume that $0 \in X_n$ almost surely for all $n \geq 1$. Denote $\xi_n = \|X_n\|$ and $\xi_n(c) = \|X_n(c)\|$. To prove sufficiency, note that $\sum \mathbf{P}\{\|\xi_n\| > c\}$ converges. The convergence of (3.4) implies the convergence of $\sum \mathbf{E}\|X_n(c)\|$ and thereupon of both $\sum \mathbf{E}\xi_n(c)$ and $\sum (\mathbf{E}\xi_n(c))^2$. Since

$$\|X_n(c)\|^2 \leq 2\rho_{\mathrm{H}}(X_n(c), \mathbf{E} X_n(c))^2 + 2\rho_{\mathrm{H}}(\mathbf{E} X_n(c), \{0\})^2,$$

the convergence of (3.5) implies that $\sum \mathbf{E}\xi_n(c)^2$ converges. Therefore, the conditions of Kolmogorov's three series theorem are fulfilled for the sequence $\{\xi_n, n \geq 1\}$, so that $\sum \xi_n$ converges, which implies the convergence of $\sum X_n$. Necessity is proved similarly.

Now prove *sufficiency* for the general case. Let $X_n = Y_n + \eta_n$, where Y_n contains the origin almost surely and η_n is a random vector. The particular choice of this decomposition is not important. Then (3.3) implies that both $\sum \mathbf{P}\{\|Y_n\| > c\}$ and $\sum \mathbf{P}\{\|\eta_n\| > c\}$ converge for some $c > 0$. Set $\eta_n(c) = \eta_n$ if $\|\eta_n\| \leq c$ and $\eta_n(c) = 0$ otherwise. The convergence of (3.4) implies that $\sum \mathbf{E}\eta_n(c)$ converges, so that $\sum \mathbf{E} Y_n(c)$ converges, whence both $\sum \|\mathbf{E} Y_n(c)\|$ and $\sum \|\mathbf{E} Y_n(c)\|^2$ also converge. Taking (3.5) into account we deduce that $\sum \mathbf{E}(\|Y_n(c)\|^2)$ converges, whence both $\sum \mathrm{Var}_A\{\eta_n(c)\}$ and $\sum \mathrm{Var}_A Y_n(c)$ converge. Thus, the convergence of (3.2) follows from the first part of the proof (for Y_n) and Kolmogorov's three series theorem for the random vectors η_n, $n \geq 1$.

Necessity in the general case can be proved by arguing that the convergence of (3.2) implies that both $\sum Y_n$ and $\sum \eta_n$ converge with subsequent application of the three series theorem and noticing that

$$\mathrm{Var}_A(X_n) \leq 2(\mathrm{Var}_A(Y_n) + \mathrm{Var}_A(\{\eta_n\})). \qquad \square$$

Using a similar (but simpler) proof one can deduce from the "two series" theorem in classical probability theory the following result.

Proposition 3.2. *If $\{X_n, n \geq 1\}$ is a sequence of independent random compact sets such that both $\sum \mathbf{E} X_n$ and $\sum \mathrm{Var}_A X_n$ converge, then $\sum X_n$ a.s. converges.*

3.3 Komlós theorem

A sequence of random variables $\{\xi_n, n \geq 1\}$ is said to *K-converge* to a random variable ξ_0 if

$$\frac{1}{m} \sum_{k=1}^{m} \xi_{n(k)} \to \xi_0 \quad \text{a.s. as } m \to \infty$$

for every subsequence $\{\xi_{n(k)}, k \geq 1\}$. This concept can be generalised for sequences of random closed sets using any of the possible definitions of convergence for sets, see Appendix B.

The theorem of J. Komlós [323] states that if $\{\xi_n, n \geq 1\}$ is a sequence of integrable random variables with $\sup_{n \geq 1} \mathbf{E}|\xi_n| < \infty$, then there exists a random variable ξ_0 such that ξ_n K-converges to ξ_0.

Theorem 3.3 (Komlós theorem for random sets). *Let $\{X_n, n \geq 1\}$ be a sequence of integrably bounded random convex compact sets in a separable Banach space such that $\sup_{n \geq 1} \mathbf{E}\|X_n\| < \infty$ and $\overline{\mathrm{co}}\,(\cup_{n \geq 1} X_n)$ has a compact intersection with every ball. Then there exists a subsequence $\{X_{n(k)}, k \geq 1\}$ of $\{X_n, n \geq 1\}$ and an integrably bounded random convex compact set X_0 such that $X_{n(k)}$ K-converges to X_0 in the Hausdorff metric and $X_0 \subset \overline{\mathrm{co}}\,(\text{s-lim}\sup X_{n(k)})$ a.s.*

3.4 Renewal theorems for random convex compact sets

Multivariate renewal theorem

The *elementary renewal theorem* for i.i.d. non-negative random variables $\xi_1, \ldots, \xi_n$ states that $H(t)/t$ converges to $1/\mathbf{E}\xi_1$ as $t \to \infty$ for the *renewal function*

$$H(t) = 1 + \sum_{n=0}^{\infty} \mathbf{P}\{\xi_1 + \cdots + \xi_n \leq t\}. \tag{3.6}$$

We will make use of the following theorem (proved in Molchanov, Omey and Kozarovitzky [410]) which is a *multivariate* analogue of the elementary renewal theorem. Let $\xi_n = (\xi_{n1}, \ldots, \xi_{nd})$, $n \geq 1$, be a sequence of i.i.d. random vectors with the cumulative distribution function $F(x_1, \ldots, x_d) = \mathbf{P}\{\xi_{11} \leq x_1, \ldots, \xi_{1d} \leq x_d\}$. Put $S_n = \sum_{i=1}^{n} \xi_i$, $S_0 = 0$ and define the renewal function by

$$H(x_1, \ldots, x_d) = \sum_{n=0}^{\infty} \mathbf{P}\{S_{n1} \leq x_1, \ldots, S_{nd} \leq x_d\}.$$

Theorem 3.4 (Multivariate renewal theorem). *Assume that all $\mathbf{E}\xi_{1i} = \mu_i$, $1 \leq i \leq d$, are finite and that $\max(\mu_i)$ is strictly positive. Furthermore, let $\mathbf{E}(\xi_{1i}^-)^2 < \infty$, $1 \leq i \leq d$, where ξ_{1i}^- is the negative part of ξ_{1i}. Then, for all finite positive $x_1, \ldots, x_d$,*

$$\lim_{t \to \infty} \frac{1}{t} H(tx_1, \ldots, tx_d) = \min_{1 \leq i \leq d,\, \mu_i > 0} \frac{x_i}{\mu_i}. \tag{3.7}$$

Containment renewal function for random sets

Let $S_n = X_1 + \cdots + X_n$, $n \geq 1$, be partial sums of i.i.d. random compact sets $X, X_1, X_2, \ldots$ in $\mathbb{R}^d$ ($S_0 = \{0\}$ is the origin). For a closed set K define the *containment renewal function*

$$H(K) = \sum_{n=0}^{\infty} \mathbf{P}\{S_n \subset K\}.$$

If $d = 1$, $K = [0,1]$ and $X = \{\xi\}$ for a non-negative random variable ξ, then $H(tK)$ becomes the renewal function from (3.6). In contrast to the strong law of large numbers, Banach space variants of renewal theorems are not known, which makes it impossible to obtain the result simply by reformulating the problem for the support functions. For $K \in \mathcal{K}$ define

$$S_K^+ = \{u \in \mathbb{S}^{d-1} : h(K, u) > 0\}.$$

Theorem 3.5 (Elementary renewal theorem for random sets). *Assume that X is an integrably bounded random compact set and $\mathbf{E}\rho(0, \operatorname{co}(X))^2 < \infty$, where $\rho(0, \operatorname{co}(X))$ is the distance between $\operatorname{co}(X)$ and the origin. Then*

$$\lim_{t \to \infty} \frac{H(tK)}{t} = \inf_{u \in S_{\mathbf{E}X}^+} \frac{h(K, u)}{h(\mathbf{E}X, u)} \tag{3.8}$$

for each convex compact set K such that $0 \in \operatorname{Int} K$.

Proof. First, note that $S_n \subset tK$ if and only if the support function of S_n is not greater than the support function of tK, that is

$$h(S_n, u) = \sum_{i=1}^{n} h(X_i, u) \leq th(K, u), \quad u \in \mathbb{S}^{d-1}.$$

Note also that

$$\sup_{u \in \mathbb{S}^{d-1}} h(X, u)^- < \rho(0, \operatorname{co}(X)),$$

where $h(X, u)^-$ is the negative part of $h(X, u)$. Choose an ε-net $u_1, \ldots, u_m$ on the unit sphere $\mathbb{S}^{d-1}$. Then

$$K'_\varepsilon = \bigcap_{j=1}^{m} \mathbb{H}_{u_j}^-(s'_j) \subset K \subset K''_\varepsilon = \bigcap_{j=1}^{m} \mathbb{H}_{u_j}^-(s''_j)$$

for suitable positive reals s'_j, s''_j, $1 \leq j \leq m$, where $\mathbb{H}_u^-(s) = \{y \in \mathbb{R}^d : \langle y, u \rangle \leq s\}$. Thus, K can be approximated by polyhedra K'_ε and K''_ε with facets orthogonal to $u_1, \ldots, u_m$, so that

$$\rho_H(K'_\varepsilon, K''_\varepsilon) \to 0 \quad \text{as } \varepsilon \to 0. \tag{3.9}$$

By Theorem 3.4,

$$\frac{H(tK)}{t} \geq t^{-1} \sum_{n=0}^{\infty} \mathbf{P}\left\{\sum_{i=1}^{n} h(X_i, u_j) \leq ts'_j; \ 1 \leq j \leq m\right\}$$

$$\to \min_{1 \leq j \leq m;\ u_j \in S^+_{\mathbf{E}X}} \frac{s'_j}{\mathbf{E}h(X, u_j)} \quad \text{as } t \to \infty$$

$$\geq \min_{1 \leq j \leq m;\ u_j \in S^+_{\mathbf{E}X}} \frac{h(K, u_j) - \rho_{\mathbf{H}}(K'_\varepsilon, K)}{\mathbf{E}h(X, u_j)}.$$

A similar bound from above yields

$$\min_{1 \leq j \leq m;\ u_j \in S^+_{\mathbf{E}X}} \frac{h(K, u_j) - \rho_{\mathbf{H}}(K'_\varepsilon, K)}{\mathbf{E}h(X, u_j)} \leq \lim_{t \to \infty} \frac{H(tK)}{t}$$

$$\leq \min_{1 \leq j \leq m;\ u_j \in S^+_{\mathbf{E}X}} \frac{h(K, u_j) + \rho_{\mathbf{H}}(K''_\varepsilon, K)}{\mathbf{E}h(X, u_j)}.$$

The continuity of the support function and (3.9) finish the proof. □

It is easy to see that $H(tK)/t = (\|\mathbf{E}X\|)^{-1}$ as $t \to \infty$ if K is the unit ball B_1.

Example 3.6. If X is *isotropic* and $0 \in \text{Int } K$, then

$$\lim_{t \to \infty} H(tK)/t = 2(\mathbf{E}b(X))^{-1} \sup\{r : B_r(0) \subset K\},$$

where $b(X)$ is the mean width of X defined by (F.5). If X is the ball $B_\xi(\eta)$ with rotation invariant distribution of its centre η, then $\mathbf{E}b(X) = 2\mathbf{E}\xi$ and

$$\lim_{t \to \infty} H(tK)/t = (\mathbf{E}\xi)^{-1} \sup\{r : B_r(0) \subset K\}.$$

Example 3.7. Let $X = \{\xi\}$ be a random singleton. If $\mathbf{E}\|\xi\|^2 < \infty$, then Theorem 3.5 yields

$$\lim_{t \to \infty} \frac{H(tK)}{t} = \sup\{r : r\mathbf{E}\xi \in K\} = \frac{1}{g(K, \mathbf{E}\xi)},$$

where $g(K, x) = \inf\{r \geq 0 : x \in rK\}$ is the *gauge function* of K, see Schneider [520, p. 43].

In the following the assumption $0 \in \text{Int } K$ is dropped. The proofs are similar to the proof of Theorem 3.5 and can be found in Molchanov, Omey and Kozarovitzky [410].

Theorem 3.8. Assume that X is integrably bounded and $\mathbf{E}\rho(0, \text{co}(X))^2 < \infty$. Let K be a convex compact set.

(i) If $0 \notin K$, then

$$\lim_{t \to \infty} \frac{H(tK)}{t} = \alpha_K - \min(\alpha_K, \beta_K),$$

where α_K (respectively β_K) is the infimum of $h(K, u)/h(\mathbf{E}X, u)$ taken over all $u \in \mathbb{S}^{d-1}$ such that $h(K, u) > 0$ and $h(\mathbf{E}X, u) > 0$ (respectively $h(K, u) < 0$ and $h(\mathbf{E}X, u) < 0$). Here $\inf \emptyset = \infty$ and $\infty - \infty = 0$.

(ii) Let the origin be a boundary point of K. If X is a.s. a subset of the minimum cone containing K, then

$$\lim_{t\to\infty} \frac{H(tK)}{t} = \inf_{u\in S^+_{\mathbf{E}X}} \frac{h(K,u)}{h(\mathbf{E}X,u)}. \qquad (3.10)$$

The same result holds if $S^+_{\mathbf{E}X} \subset S^+_K$ and $0 \notin \mathbf{E}X$. If $S^+_{\mathbf{E}X} \not\subset S^+_K$, then the limit in (3.10) is zero.

Open problem 3.9. Provide a geometric interpretation of the limit of $H(tK)/t$ from Theorems 3.5 and 3.8.

Further renewal functions

It is also possible to consider the *inclusion* renewal function defined as

$$J(K) = \sum_{n=1}^{\infty} \mathbf{P}\{K \subset S_n\}, \quad K \in \text{co}\,\mathcal{K}.$$

If $0 \in K$, then $J(tK)$ decreases, whence either $J(K) = \infty$ or $J(tK)/t$ converges to zero as $t \to \infty$. Thus we have to consider only the case $0 \notin K$.

Theorem 3.10. *Let $0 \notin K$. Under the conditions and in the notation of Theorem 3.8(i)*

$$\lim_{t\to\infty} \frac{J(tK)}{t} = \beta_K - \min(\alpha_K, \beta_K),$$

if $0 \notin \mathbf{E}X$. If $0 \in \text{Int}\,\mathbf{E}X$, then $J(K)$ is infinite. If 0 belongs to the boundary of $\mathbf{E}X$, then $J(tK)/t$ converges to zero if $h(K,u) \neq 0$ and $h(\mathbf{E}X,u) = 0$ for some u. Otherwise $J(K)$ is infinite.

It is interesting also to consider the *hitting* renewal function

$$U(K) = \sum_{n=1}^{\infty} \mathbf{P}\{S_n \cap K \neq \emptyset\}.$$

If X is a singleton, then U coincides with the containment renewal function H, while Theorem 3.10 can be applied if K is a singleton. If $X = [\xi, \eta]$ is a convex subset of the line ($d = 1$), then a renewal theorem for the hitting function follows from Theorem 3.4. It is easily seen that

$$\mathbf{P}\{X \cap [a,b] \neq \emptyset\} = \mathbf{P}\{\xi \leq a, \eta \geq a\} + \mathbf{P}\{\xi \in (a,b]\}.$$

If $a, b > 0$ and $\mathbf{E}\xi, \mathbf{E}\eta > 0$, then

$$\lim_{t\to\infty} U(t[a,b])/t = b/\mathbf{E}\xi - \min(b/\mathbf{E}\eta, a/\mathbf{E}\xi).$$

Similar results can be obtained for other a and b. It is possible to derive further results for K being a finite union of disjoint segments.

Open problem 3.11. Find the limit of $U(tK)/t$ as $t \to \infty$ for a general compact set K, where U is the hitting renewal function.

3.5 Ergodic theorems

The pointwise ergodic theorem for families of random sets goes back to the *subadditive ergodic theorem* of Kingman [312] (further generalised to the Banach valued case by Ghoussoub and Steele [191]) and M. Abid's extension of Birkhoff's pointwise ergodic theorem for super-stationary and subadditive processes, see [1]. Since the general pointwise ergodic theorem does not hold in the Banach space of continuous functions, this calls for a specific generalisation of the ergodic theorem for random closed sets.

Consider a *triangular array* $\mathbf{X} = \{X_{m,n}, \ m, n = 0, 1, 2, \ldots, \ m < n\}$ of random convex compact sets in a separable Banach space $\mathbb{E}$. A sequence $\{X_n, n \geq 1\}$ of random compact sets is said to be *superstationary* if $\mathbf{E}f(X_1, X_2, \ldots) \geq \mathbf{E}f(X_2, X_3, \ldots)$ for all bounded Borel coordinatewise increasing functions f.

Definition 3.12 (Subadditive superstationary arrays). A triangular array $\mathbf{X}$ is called

(i) *subadditive* if $X_{0,n} \subset X_{0,m} + X_{m,n}$ a.s. for all $0 < m < n$;
(ii) *superstationary* if $\{X_{(m-1)k, mk}, \ m \geq 1\}$ is a superstationary sequence for each $k \geq 1$ and, for each $m \geq 0$, the joint distributions of $\{X_{m,m+n}, \ n \geq 1\}$ dominate those of $\{X_{m+1,m+n+1}, \ n \geq 1\}$, i.e. $\mathbf{E}f(X_{m,m+1}, X_{m,m+2}, \ldots) \geq \mathbf{E}f(X_{m+1,m+2}, X_{m+1,m+3}, \ldots)$ for all coordinatewise increasing bounded Borel functions f.

Theorem 3.13 (Ergodic theorem). *Let $\mathbf{X}$ be a subadditive, superstationary family of random convex compact sets such that $\mathbf{E}\|X_{0,1}\| < \infty$. Then there exists a random convex compact set X_∞ such that $\rho_H(n^{-1}X_{0,n}, X_\infty) \to 0$ as $n \to \infty$.*

Proof. Since $X_{0,n} \subset \sum_{i=1}^{n} X_{i-1,i}$ and $\{X_{n-1,n}, n \geq 1\}$ is a superstationary sequence, Lemma 3.14 is applicable, whence $\mathrm{cl}(\cup_{n=1}^\infty n^{-1} X_{0,n})$ is compact. By Proposition 1.6.18, it suffices to show that $d_H(n^{-1} X_{0,n}, K)$ (given by (C.2)) converges for all $K \in \mathrm{co}\,\mathcal{K}$. Since $d_H(\cdot, K)$ is an increasing function, the random variables $\{\rho_H(X_{m,n}, (n-m)K)\}$ form a superstationary subadditive family. Indeed, for $0 < m < n$ and $x \in X_{0,n}$, there exist $x_1 \in X_{0,m}$ and $x_2 \in X_{m,n}$ such that $x = x_1 + x_2$. For any $y_1, y_2 \in K$, $y = n^{-1}(my_1 + (n-m)y_2) \in K$ by convexity, and $\|x - ny\| \leq \|x_1 - my_1\| + \|x_2 - (n-m)y_2\|$, whence

$$d_H(X_{0,n}, K) \leq d_H(X_{0,m}, mK) + d_H(X_{m,n}, (n-m)K).$$

Furthermore, $d_H(X_{0,n}, nK) \leq \|X_{0,n}\| + \|nK\|$ and $\|X_{0,n}\| \leq \sum_{i=1}^{n} \|X_{i-1,i}\|$ a.s. Hence

$$\mathbf{E}d_H(X_{0,n}, nK) \leq n(\mathbf{E}\|X_{0,1}\| + \|K\|) < \infty.$$

By the subadditive ergodic theorem for random variables (see Liggett [356]), the random variable $d_H(n^{-1} X_{0,n}, K)$ converges a.s., whence the result follows. □

Lemma 3.14. *Let $\{X_n, n \geq 1\}$ be a superstationary sequence of random compact sets such that X_1 is integrably bounded. Then $\mathrm{cl}(\cup_{n=1}^\infty n^{-1} \sum_{i=1}^n X_i)$ is compact almost surely.*

3 Further results related to Minkowski sums

Proof. Let $\mathbb{Q} = \{x_k, k \geq 0\}$ be a countable dense set in $\mathbb{E}$, with $x_0 = 0$. Define $V_k = \mathrm{co}\{x_0, \ldots, x_k\}$ and $d_k(K) = d_H(K, V_k)$ for $k \geq 1$ and $V \in \mathcal{K}$. For each k, $\{d_k(X_n), n \geq 1\}$ is a superstationary sequence of random variables and $d_k(X_1) \leq \|X_1\|$. Since X_1 is almost surely compact, $d_k(X_1) \to 0$ a.s. as $k \to \infty$, whence $\mathbf{E} d_k(X_1) \to 0$ as $k \to \infty$. Given $\varepsilon > 0$, choose k such that $\mathbf{E} d_k(X_1) \leq \varepsilon^2/4$. By the superstationary ergodic theorem, there exists a random variable ξ such that $n^{-1} \sum_{i=1}^{n} d_k(X_i) \to \xi$ a.s. as $n \to \infty$ and $\mathbf{P}\{\xi > \varepsilon/2\} \leq \varepsilon/2$. By Egoroff's theorem using the uniform convergence on the subset of measure at least $1 - \varepsilon/2$, one obtains that

$$\mathbf{P}\left\{\sup_{n \geq N} n^{-1} \sum_{i=1}^{n} d_k(X_i) > \varepsilon\right\} \leq \varepsilon$$

for some number N. Since V_k is convex,

$$\mathbf{P}\left\{\sup_{n \geq N} d_k(n^{-1} \sum_{i=1}^{n} d_k(X_i)) > \varepsilon\right\} \leq \varepsilon.$$

For $\varepsilon \geq 0$, define an event

$$A_\varepsilon = \bigcup_{k,N} \left\{ \bigcup_{n=N}^{\infty} n^{-1} \sum_{i=1}^{n} X_i \subset V_k^\varepsilon \right\}.$$

Then $\mathbf{P}(A_\varepsilon) \geq 1 - \varepsilon$ and $\mathbf{P}(\cap_{\varepsilon > 0} A_\varepsilon) = 1$. Given $\varepsilon > 0$, one can choose k and N such that

$$\bigcup_{n=N}^{\infty} \left(n^{-1} \sum_{i=1}^{n} X_i \right) \subset V_k^\varepsilon$$

almost surely. Note that V_k^ε is contained in a finite number of balls of radius ε. Since $\cup_{n=1}^{N-1} n^{-1} \sum_{i=1}^{n} X_i$ is contained in a finite number of balls of radius ε, the set $\cup_{n=1}^{\infty} n^{-1} \sum_{i=1}^{n} X_i$ is totally bounded, whence the result follows. $\square$

It is easy to see that the limit in Theorem 3.13 is deterministic if the sequences $\{d_H(X_{mn,(m+1)n}, nK), m \geq 1\}$ are ergodic for every $n \geq 1$ and $K \in \mathrm{co}\mathcal{K}$. An important example of a subadditive family $\mathbf{X} = \{X_{m,n}\}$ appears if $X_{m,n} = Y_{n-m}$, where $\{Y_k, k \geq 1\}$ is a subadditive sequence of random convex compact sets, i.e.

$$Y_{m+n} \subset Y_m + Y_n \tag{3.11}$$

for all $m, n \geq 0$. Clearly, the partial sums $Y_n = X_1 + \cdots + X_n$ of a sequence $\{X_n, n \geq 1\}$ of i.i.d. random compact sets satisfy (3.11). Theorem 3.13 implies the following result.

Corollary 3.15. *If $\{Y_n, n \geq 1\}$ is a subadditive family of integrably bounded random convex compact sets, then $n^{-1} Y_n$ converges in the Hausdorff metric.*

Theorem 3.16 (Mean ergodic theorem). *Let $\mathbf{X}$ and X_∞ be as in Theorem 3.13. If $\mathbf{E}\|X_{0,1}\|^p < \infty$ for $p \geq 1$, then*

$$\mathbf{E} \rho_H (n^{-1} X_{0,n}, X_\infty)^p \to 0 \quad \text{as } n \to \infty.$$

3.6 Large deviation estimates

It is possible to give a *large deviation* type estimate for the Hausdorff distance between the Minkowski averages of random sets and the averages of their selection expectations.

Theorem 3.17 (Large deviation for Minkowski averages). Let $X_1, \ldots, X_n$ be independent random convex compact sets in $\mathbb{R}^d$ such that $\|X_i\| \leq b$ a.s. for a constant b and all $i = 1, \ldots, n$. Define β to be the essential supremum of $\rho_H(X_i, \mathbf{E}X_i)$ for all $i = 1, \ldots, n$. Let $\rho < 1$ be fixed. Then, for every $\varepsilon > 0$,

$$\mathbf{P}\{\rho_H(\bar{X}_n, \overline{\mathbf{E}X}_n) > \varepsilon\} \leq c \exp\{-\rho \varepsilon^2 n/(4b^2)\}, \tag{3.12}$$

where $\bar{X}_n = n^{-1}(X_1 + \cdots + X_n)$ and $\overline{\mathbf{E}X}_n = n^{-1}(\mathbf{E}X_1 + \cdots + \mathbf{E}X_n)$. The constant c depends on ε/b and ρ.

Proof. The proof is based on the inequality

$$\mathbf{P}\{n^{-1}|\xi_1 + \cdots + \xi_n| > \varepsilon\} \leq 2\exp\{-\varepsilon^2 n/4\} \tag{3.13}$$

valid for independent centred random variables $\xi_1, \ldots, \xi_n$ with absolute values not exceeding 1. For each u with norm 1, $|h(X_i, u) - h(\mathbf{E}X_i, u)|$, $i \geq 1$, are independent and bounded by b almost surely. Therefore, (3.13) yields

$$\mathbf{P}\{|h(\bar{X}_n, u) - h(\overline{\mathbf{E}X}_n, u)| > \varepsilon \rho^{1/2}\} \leq 2\exp\{-\rho \varepsilon^2 n/(4b^2)\}.$$

Let $u_1, \ldots, u_l$ be unit vectors, which form a 2δ-net on the unit sphere with $2\delta = \varepsilon(1 - \rho^{1/2})(2b)^{-1}$. With this choice, $|h(\bar{X}_n, u) - h(\overline{\mathbf{E}X}_n, u)| > \varepsilon$ for any u implies $|h(\bar{X}_n, u_i) - h(\overline{\mathbf{E}X}_n, u_i)| > \varepsilon \rho^{1/2}$ for at least one u_i from the chosen 2δ-net. Therefore, (3.12) holds with $c = 2l$ which depends on ε/b and ρ. □

If the random sets are not necessarily convex, then (3.12) holds with ε under the probability sign replaced by $\varepsilon\theta$ for any $\theta > 1$ provided that n is large enough in order to ensure that the Shapley–Folkman–Starr theorem yields an effective bound for the Hausdorff distance. The following is a general large deviation theorem for Minkowski sums of i.i.d. random compact sets.

Theorem 3.18 (Large deviation for random sets). Let $\{X_n, n \geq 1\}$ be a sequence of i.i.d. random compact sets in a separable Banach space $\mathbb{E}$ such that $\mathbf{E}e^{\alpha\|X_1\|} < \infty$ for all $\alpha > 0$. For a signed measure λ on the unit ball B_1^* in the dual space $\mathbb{E}^*$ define

$$\Phi(\lambda) = \log \mathbf{E}\left[\exp\left\{\int_{B_1^*} h(X_1, u)\lambda(du)\right\}\right]$$

and for a convex compact set K put

$$\Phi^*(K) = \sup_\lambda \left(\int_{B_1^*} h(K, u)\lambda(du) - \Phi(\lambda) \right),$$

where the supremum is taken over all signed measures λ on B_1^*. Set $\Phi^*(K) = \infty$ if K is a non-convex compact set. Then the Minkowski averages $\bar{X}_n$, $n \geq 1$, satisfy a large deviations principle with rate function Φ^*, i.e. for each measurable family $\mathcal{M}$ of compact sets

$$-\inf_{K \in \text{Int}\,\mathcal{M}} \Phi^*(K) \leq \liminf_{n \to \infty} \frac{1}{n} \log \mathbf{P}\{\bar{X}_n \in \mathcal{M}\}$$
$$\leq \limsup_{n \to \infty} \frac{1}{n} \log \mathbf{P}\{\bar{X}_n \in \mathcal{M}\} \leq - \inf_{K \in \text{cl}(\mathcal{M})} \Phi^*(K),$$

where the closure and the interior of $\mathcal{M}$ are taken in the topology generated by the Hausdorff metric.

Open problem 3.19. Derive an analogue of Theorem 3.18 for the case of non-compact or unbounded summands (as in Section 1.5).

3.7 Convergence of functionals

The strong law of large numbers implies that all *continuous* in the Hausdorff metric functionals of $\bar{X}_n$ converge almost surely to their values on $\mathbf{E}X$, where X has the distribution common to all summands $\{X_n, n \geq 1\}$.

The classical law of large numbers for random variables immediately yields the almost sure convergence of *mean widths* (perimeters in the planar case), since they can be easily represented as integrals of the support function and as such commute with the selection expectation, i.e.

$$\mathsf{b}(\bar{X}_n) = \frac{1}{n} \sum_{i=1}^n \mathsf{b}(X_i) \to \mathbf{E}\mathsf{b}(X) = \mathsf{b}(\mathbf{E}X) \quad \text{a.s. as } n \to \infty.$$

The corresponding central limit theorem follows from the classical results for sums of random variables.

The convergence of volumes is more intricate. The key idea is to use the expansion (F.9) for the Minkowski sums of convex compact sets using the *mixed volumes* $\mathsf{V}(K_{i_1}, \ldots, K_{i_d})$. Then

$$\mathsf{V}(\bar{X}_n) = \frac{\binom{n}{d}}{n! n^d} U_n + \mathcal{O}(n^{-1}),$$

where the sum of mixed volumes

$$U_n = \frac{1}{\binom{n}{d}} \sum_{1 \leq i_1 < i_2 < \cdots < i_d \leq n} \mathsf{V}(X_{i_1}, \ldots, X_{i_d}) \qquad (3.14)$$

can be regarded as an *U-statistic*. It follows from the theory of U-statistics (see Serfling [531]) that $\mathbf{EV}(\bar{X}_n)$ converges to $\mathbf{EV}(X_1,\ldots,X_d)$ and

$$\mathbf{V}(\bar{X}_n) \to \mathbf{EV}(X_1,\ldots,X_d) \quad \text{a.s. as } n \to \infty.$$

This implies, by the way, that

$$\mathbf{V}(\mathbf{E}X) = \mathbf{EV}(X_1,\ldots,X_d).$$

Applying the theory of the U-statistic to the deviation $\Delta_n = \mathbf{V}(\bar{X}_n) - \mathbf{V}(\mathbf{E}X_1)$ we obtain

$$\Delta_n = \mathbf{V}(\bar{X}_n) - \mathbf{EV}(\bar{X}_n) + \mathcal{O}(n^{-1}).$$

Then $\sqrt{n}\Delta_n$ converges in distribution to the centred normal random variance with the variance

$$\sigma^2 = \mathrm{Var}[\mathbf{E}(\mathbf{V}(X_1,\ldots,X_n)|X_1)]$$

provided σ^2 is positive. If $\sigma^2 = 0$, then applying the results of Rubin and Vitale [506] it is possible to deduce that $n\Delta_n$ converges in distribution to a random variable given by

$$\sum_{j=1}^{\infty} \lambda_j(Z_j^2 - 1) + \binom{d}{2}[\mathbf{EV}(X_1,\ldots,X_d) - \mathbf{V}(\mathbf{E}X_1)],$$

where $\{Z_n, n \geq 1\}$ are i.i.d. standard Gaussian random variables.

Open problem 3.20. Investigate asymptotic distributions of $V_j(\bar{X}_n)$ where V_j is the jth intrinsic volume, see Appendix F.

3.8 Convergence of random broken lines

Let ξ be a random vector distributed on the right half-plane of $\mathbb{R}^2$. Consider a sequence $\xi_1, \xi_2, \ldots, \xi_n$ of i.i.d. copies of ξ. For each $i = 1,\ldots,n$, let θ_i be the angle between ξ_i and the abscissa axis. Order the vectors according to the growing angles θ_i to get a sequence $\xi'_1,\ldots,\xi'_n$. If several vectors share the same angles, they can be rearranged in an arbitrary manner. The successive sums $0, \xi'_1, \xi'_1+\xi'_2,\ldots,\xi'_1+\cdots+\xi'_n$ form the vertices of a *broken line* C_n which is called a *convex rearrangement* of the sequence $\xi_1,\ldots,\xi_n$, see Figure 3.1.

Assume that $\mathbf{E}\|\xi\| < \infty$ and denote by F the cumulative distribution function of θ, which is concentrated on $[-\pi/2, \pi/2]$, by the assumption.

Theorem 3.21. *The normalised convex rearrangements $n^{-1}C_n$ converge in the Hausdorff metric to the convex curve $C = \{(x(t), y(t)) : t \in [-\pi/2, \pi/2]\}$ with*

$$x(t) = \int_{-\pi/2}^{t} a(s)\cos(s)\mathrm{d}F(s), \quad y(t) = \int_{-\pi/2}^{t} a(s)\sin(s)\mathrm{d}F(s),$$

where $a(t) = \mathbf{E}(\|\xi\| \mid \theta = t)$.

3 Further results related to Minkowski sums

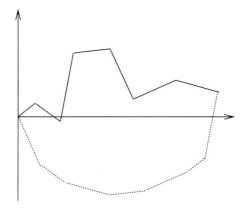

Figure 3.1. Convex rearrangement.

Proof. Let $(x_n(t), y_n(t))$ be the coordinates of the vertex of $n^{-1}C_n$ such that $\theta_i < t$ for all $i < n$ and $\theta_i \geq t$ for all $i \geq n$. Then

$$x_n(t) = n^{-1} \sum_{i=1}^{n} \|\xi_i\| \cos(\theta_i) \mathbf{1}_{[-\pi/2, t)}(\theta_i).$$

This is a sum of i.i.d random variables. The strong law of large numbers yields the desired representation of the curve C.

Notice that C_n is the "bottom half" of the Minkowski sum $[0, \xi_1] + \cdots + [0, \xi_n]$. By the strong law of large numbers the normalised sum converges almost surely in the Hausdorff metric to a convex set with the bottom half being the curve C. □

It is possible to extend Theorem 3.21 to the case when ξ is distributed on the whole plane. Then the limiting set is closed if and only if $\mathbf{E}\xi = 0$.

3.9 An application to allocation problem

The strong law of large numbers can be applied to optimisation problems concerning *optimal allocations* of resources. A general problem of this kind can be formulated as

$$J(x) \to \max \quad \text{subject to } x \in Y, \qquad (3.15)$$

$$x = \frac{1}{n}(x_1 + \cdots + x_n), \quad x_i \in X_i, \ 1 \leq i \leq n,$$

where $X_1, \ldots, X_n$ are independent identically distributed random closed sets (or multifunctions on Ω in the usual optimisation terms) and Y is a random compact set. For simplicity, assume that all random sets are subsets of the Euclidean space $\mathbb{R}^d$.

The applied nature of this problem can be explained as follows. There are n firms, and each receives its share of the total resources determined by the vector

$nq = (nq_1, \ldots, nq_d)$. If firm number i receives amount f_i, it produces $u_i(f_i)$ where $f_i \in F_i$ with F_i being the constraint set (usually $F_i = [0, \infty)^d$). The objective function is $n^{-1} \sum u_i(f_i)$ and the global constraint is $f = n^{-1}(f_1 + \cdots + f_n) \leq q$ coordinatewise. This problem can be reformulated as a particular case of (3.15) for $x = (f, \alpha) \in \mathbb{R}^d \times \mathbb{R}$ with the objective function $J(x) = \alpha$ under the constraints $Y = \{(f, \alpha) : f \leq q\}$ and $X_i = \{(f, \alpha) : f \in F_i, \alpha = u_i(f)\}$.

For every n, the problem (3.15) is solved by specifying a *programme*, or a sequence of n measurable functions $\xi_{1n}(\omega), \ldots, \xi_{nn}(\omega)$ which give the optimal allocations among the n firms for each ω. Because of the constraints,

$$\zeta_n = n^{-1}(\xi_{1n} + \cdots + \xi_{nn}) \in Y(\omega) \tag{3.16}$$

for each ω. Such a programme is called *exact*. In many cases instead of the exact condition (3.16) one defines a *relaxed* programme satisfying $\rho(\zeta_n(\omega), Y(\omega)) \to 0$ as $n \to \infty$ for almost all ω. The *stationary* programmes are determined by the constraints only, so that $\xi_{in}(\omega) = \xi_i(\omega)$. The following theorem follows from the strong law of large numbers for (possibly unbounded) random closed sets, see Theorem 1.21. The proofs can be found in Artstein [18] together with further results which deal with the existence of stationary solutions.

Theorem 3.22. *Assume that X_1 is integrable. A relaxed programme exists if and only if $Y_0 = Y \cap \overline{\mathrm{co}}\,(\mathbf{E}X_1) \neq \emptyset$ almost surely. If $Y_0 \neq \emptyset$ a.s., then the following statements hold.*

(i) For every relaxed programme,

$$v = \sup\{J(x) : x \in Y_0\} \geq \limsup J(\zeta_n) \text{ a.s.}$$

(ii) There exists a relaxed programme with

$$v = \limsup J(\zeta_n) \text{ a.s.} \tag{3.17}$$

(iii) If $\{x \in \mathrm{Int}\, Y_0 : J(x) = v\}$ is non-empty almost surely, then there exists an exact programme which satisfies (3.17).

3.10 Infinite divisibility in positive convex cones

Let $\mathbb{L}$ be a LCHS space, which is also assumed to be a *positive convex cone* with respect to an operation $+$. The latter means that $\mathbb{L}$ is closed under addition and multiplication by non-negative scalars and the sum of two elements is zero if and only if both elements are zeros. A character on $\mathbb{L}$ is a bounded continuous real- or complex-valued function χ on $\mathbb{L}$ such that

$$\chi(x + y) = \chi(x)\chi(y), \quad \chi(0) = 1, \quad |\chi(y)| \leq 1$$

for every $x, y \in \mathbb{L}$. The family of all characters is denoted by $\hat{\mathbb{L}}$. If μ is a finite measure on $\mathbb{L}$, its *Laplace transform* is a function on $\hat{\mathbb{L}}$ given by

3 Further results related to Minkowski sums 237

$$\hat{\mu}(\chi) = \int_{\mathbb{L}} \chi(x)\mu(dx).$$

As in the classical situation, the measure μ is completely specified by its Laplace transform. Moreover, it suffices to consider $\hat{\mu}(\chi)$ for χ from a subset $\hat{\mathbb{L}}$, e.g. on the subset given by $\{e^{-f} : f \in \mathbb{F}\} \cup \{1\}$, where $\mathbb{F}$ is a class of continuous, positively homogeneous, additive functionals $f: \mathbb{L} \mapsto \mathbb{R}_+$ such that $\{x : f(x) = 0\} = 0$. Assume that for every $x, y \in \mathbb{L}$ with $x \neq y$ there exists an $f \in \mathbb{F}$ such that $f(x) \neq f(y)$.

An $\mathbb{L}$-valued random element ξ is said to be infinitely divisible if, for every $n \geq 2$, ξ coincides in distribution with the sum of i.i.d. $\mathbb{L}$-valued random elements $\xi_1, \ldots, \xi_n$.

Theorem 3.23. *A probability measure $\mathbf{P}$ on $\mathbb{L}$ is infinitely divisible if and only if its Laplace transform can be uniquely represented as*

$$\hat{\mathbf{P}}(e^{-f}) = \exp\left\{-f(x_0) + \int_{\mathbb{L}} (e^{-f(x)} - 1)\frac{1 + g(x)}{g(x)}\mu(dx)\right\},$$

where f runs through $\mathbb{F}$, x_0 is a fixed point in $\mathbb{L}$, g is a fixed functional from $\mathbb{F}$ and μ is a finite measure on $\mathbb{L}$ with $\mu(\{0\}) = 0$.

Following the classical terminology for infinite divisible random variables, the measure μ is called the Lévy–Khinchin measure of $\mathbf{P}$. Theorem 3.23 yields the well-known representation of infinitely divisible non-negative random variables. In this case $\mathbb{L} = \mathbb{R}_+$ and $\mathbb{F} = \{sx : s > 0\}$. Theorem 3.23 is related to the studies of general semigroups, see Appendix G and Berg, Christensen and Ressel [61, Th. 4.3.20, 3.2.3].

If $\mathbb{L}$ is the family of all all convex compact sets in $\mathbb{R}^d$ which contain the origin with Minkowski addition as the additive operation, then Theorem 3.23 yields the characterisation of M-infinite divisible sets, see Theorem 2.19 with the Lévy measure given by $(1 + g(x))g(x)^{-1}\mu(dx)$.

Notes to Chapter 3

Section 1.1. The Shapley–Folkman–Starr theorem is relevant in many areas, see Arrow and Hahn [13] for further references. Its proof in Appendix C is reproduced from the book by Arrow and Hahn [13, pp. 396–398], who also discuss numerous applications of this result to the mathematical economy. Further results in this direction are due to Artstein [15] and

Cassels [88]. The Banach space generalisation was proved by Puri and Ralescu [464]. Theorem 1.3 goes back to Matheron [381], although his original proof did not refer to the Shapley–Folkman–Starr theorem.

Section 1.2. Although the concept of the Aumann integral of a multivalued function goes back to the early 1960s, see Richter [488], Aumann [31] and Debreu [124], the relationship between this concept of expectation and the first strong law of large numbers for Minkowski sums of random sets was discovered in 1975 by Artstein and Vitale [23]. Their paper gave rise to a large number of further studies which successfully applied methods from the theory of Banach space valued random elements to investigate Minkowski sums of random compact sets. An independent approach was developed by Lyashenko [364].

Further results concerning strong laws of large numbers for random sets are due to Hess [237], Giné, Hahn and Zinn [202] (for separable spaces) and Puri and Ralescu [462, 464]. The latter derived the convexifying part of the strong law of large numbers from the generalised Shapley–Folkman–Starr theorem (see Theorem 1.5). Finally, Artstein and Hansen [21] established the strong law of large numbers without both convexity and separability assumptions by proving the convexifying Theorem 1.4. Theorem 1.7 is proved in [21]. The latter theorem and the strong law of large numbers for random compact sets in Banach spaces have been obtained independently by Hiai [252].

It usually suffices to assume that the sequence of random closed sets is pairwise independent in order to ensure that the strong law of large numbers holds.

Blaschke addition of random sets is important in the context of statistical estimation for Boolean models of random sets, see Weil [604]. It is also possible to use the surface area measure representation instead of the support function in order to prove the strong law of large numbers for random convex compact sets.

Section 1.3. Various applications of the strong law of large numbers have been explored by Vitale [581, 582, 584, 585]. Theorem 1.12 is proved by Anderson [8]. Rounding of compact sets is discussed in detail by Vitale [585].

The studies of zonoids and zonotopes in the literature on convex geometry are surveyed in Schneider and Weil [521] and Goodey and Weil [206]. Theorems 1.13 and 1.17 are due to Vitale [583], where further results including bounds on absolute determinants and comparisons of the expected absolute determinants can be found. Proposition 1.16 is apparently new. Applications of zonoids and lift zonoids to statistics and characterisation of multivariate probability distributions have been developed by Koshevoi and Mosler [327, 415].

It is possible to prove the Brunn–Minkowski inequality for random sets by applying its classical variant first to simple random sets (with a finite number of values) and then using approximations, see Hess [246]. This yields, in particular, a variant of the Brunn–Minkowski inequality for conditional expectations.

Section 1.4. A number of results on the strong law of large numbers for non-identically distributed summands are due to R.L. Taylor and his collaborators. Section 1.4 is adapted from the survey by Taylor and Inoue [561] where further references can be found. Theorem 1.19 was proved by Taylor and Inoue [559].

Section 1.5. The strong law of large numbers for random closed (not necessarily compact) sets in $\mathbb{R}^d$ (Theorem 1.21) is due to Artstein and Hart [22]. Theorem 1.22 is proved by Hess [239] for random closed sets in Banach spaces. Theorem 1.24 is due to Hiai [253].

Further generalisations for random closed sets in Banach spaces (including random sets with non-identical distributions and pairwise independent) are due to Hiai [252, 253] and

Hess [243]. The latter paper deals with the strong law of large numbers in the slice topology, which is stronger than the Mosco topology in some cases. Theorem 1.25 is proved by Hess [247]. The strong law of large numbers in the Hausdorff metric is due to Uemura [566]. The latter paper deals also with non-identically distributed random sets using the approach similar to Taylor and Inoue [560]. Castaing and Ezzaki [89] showed how to derive a strong law of large numbers for random convex closed sets using the Mosco convergence of reversed martingales.

Note that the results related to the strong law of large numbers for Minkowski sums of unbounded random sets will appear later in Sections 5.3.5 and 5.3.6 where similar addition schemes are considered for random upper semicontinuous functions.

Section 2.1. The first central limit theorem for random sets with a finite number of values has been proved by Cressie [110] using explicit calculations of probabilities. A further variant is due to Lyashenko [364]. Theorem 2.1 is the most general central limit theorem for sums of i.i.d. square integrable random compact sets in the Euclidean space. It was obtained by Weil [602] followed by more general studies of random compact sets in Banach spaces by Giné, Hahn and Zinn [202] where Theorem 2.5 is proved and Puri and Ralescu [464]. Theorem 2.7 is proved in Giné, Hahn and Zinn [202], where further examples for particular choices of g can be found.

A number of results on limit theorems for random sets are summarised by Li, Ogura and Kreinovich [354].

Section 2.2. The concept of a Gaussian random set was suggested by Lyashenko [366] who defined such sets as those having Gaussian support functions. Using a rather technical proof he showed that Gaussian random sets have a degenerate shape. Section 2.2 follows Vitale [577] who suggested another definition of a Gaussian random set (equivalent to one given by N.N. Lyashenko by Proposition 2.9) and came up with a simple proof of the characterisation theorem (see Theorem 2.11). These characterisation results show that Gaussian random sets are not very interesting from the point of view of modelling, since these sets have degenerated shapes.

Gaussian random sets in Banach spaces have been characterised by Puri, Ralescu and Ralescu [467]. A generalisation for random upper semicontinuous functions is given by Puriand Ralescu [463].

Section 2.3. The concept of a Gaussian random set was substantially generalised by Giné and Hahn [198] who proved Theorems 2.14 and 2.15 and characterised Gaussian random sets as a special case of their general characterisation theorem for p-stable random sets. Further results (including a characterisation of Gaussian set-valued processes) can be found in Meaya [386] and Davydov, Paulauskas and Račkauskas [120].

Section 2.4. Studies of infinite divisible random sets with respect to Minkowski addition were initiated by Mase [380] who proved Theorem 2.19 without resorting to the technique of probabilities in Banach spaces. This theorem was proved using the Lévy–Khinchin theorem in Banach spaces by Giné and Hahn [199] who also generalised it for random sets which do not necessarily contain the origin. Theorem 2.20 is proved in Giné and Hahn [199]. A partial generalisation for random convex compact sets in Banach spaces has been reported in Giné and Hahn [200]. Intrinsic volumes for M-infinitely divisible random sets have been studied by Mase [380].

Section 3.1. The law of iterated logarithm for random compact sets goes back to Giné, Hahn and Zinn [202].

Section 3.2. The three series theorem for random sets goes back to Lyashenko [364], where further results for random convex sets based on their support function representation are derived. Wang [594] proved a set-valued Wald's formula. Three series theorem for increasing set-valued processes (interpreted as thresholds of a random upper semicontinuous function) is proved by Feng [170].

Section 3.3. The Komlós theorem for random sets is proved by Balder and Hess [46]. Further results are obtained by Krupa [332]. Hess and Ziat [249] extended the Komlós theorem for Pettis integrable random sets.

Section 3.4. The elementary renewal theorem for random convex sets was proved by Molchanov, Omey and Kozarovitzky [410]. This theorem generalises the elementary renewal theorem in finite-dimensional spaces (Bickel and Yahav [69]). Further results along this line for counting processes generated by Minkowski sums of random fuzzy sets (see Section 5.3.6) are obtained by Dozzi, Merzbach and Schmidt [147].

Section 3.5. The studies of the ergodic theorem for families of random compact sets were initiated by Hess [237] and Schürger [528]. A systematic approach to ergodic theorems for random sets was developed by Krupa [332], who considered also the case of unbounded random sets in Banach spaces. Section 3.5 follows Hansen and Hulse [219]. Theorem 3.13 entails a strong law of large numbers for pairwise independent identically distributed integrably bounded random compact sets.

Section 3.6. Large deviation estimates for sums of random sets have been reported by Artstein [17], who also gave an explicit bound for the constant c in (3.12). Theorem 3.18 was proved by Cerf [92].

Section 3.7. The results on the convergence of volumes of Minkowski sums were obtained by Vitale [580] who suggested using the theory of U-statistics (symmetric statistics) to derive a limit theorem for the volume.

Section 3.8. The convergence of random broken lines was studied by Davydov and Vershik [122], Davydov and Thilly [121] and Vilkov [575], where a number of further results, for example, concerning the smoothness of the limiting curve can be found. Applications to Lorenz curves when observations are stationary and either short-range or long-range dependent are discussed by Davydov and Zitikis [123].

Section 3.9. Applications to allocation problems have been worked out by Artstein [18] and Artstein and Hart [22].

Section 3.10. Infinitely divisible random elements in locally compact positive convex cones have been characterised by Jonasson [281].

4
Unions of Random Sets

1 Union-infinite-divisibility and union-stability

1.1 Extreme values: a reminder

Unions of random sets generalise the concept of extremes for random variables. The classical theory of extreme values deals with maximum or minimum of i.i.d. random variables $\xi_1, \ldots, \xi_n$, for instance the distribution of $\max(\xi_1, \ldots, \xi_n)$ is of interest, see Galambos [186], Leadbetter, Lindgren and Rootzen [346]. It is easy to see that every random variable ξ is *max-infinitely divisible*, i.e. $\xi \stackrel{d}{\sim} \max(\xi_{1n}, \ldots, \xi_{nn})$ for i.i.d. random variables $\xi_{1n}, \ldots, \xi_{nn}$.

A random variable ξ is said to be *max-stable* if, for all $n > 1$, there exist $a_n > 0$ and $b_n \in \mathbb{R}$ such that
$$a_n \xi + b_n \stackrel{d}{\sim} \max(\xi_1, \ldots, \xi_n),$$
where $\xi_1, \ldots, \xi_n$ are i.i.d. copies of ξ. It is well known (see Galambos [186]) that, up to an affine transform, any non-degenerate max-stable random variable can be identified as having a distribution from the parametric family
$$F_\gamma(x) = \exp\left\{-(1+\gamma x)^{-1/\gamma}\right\}, \quad \gamma x \geq -1, \gamma \in \mathbb{R}. \tag{1.1}$$

Besides, if $\gamma > 0$ (type I), then $F_\gamma(x) = 0$ for $x \leq -1/\gamma$, if $\gamma < 0$ (type II) then $F_\gamma(x) = 1$ for $x \geq -1/\gamma$, and if $\gamma = 0$ (type III), then $(1 + \gamma x)^{-1/\gamma}$ is an abuse of language for e^{-x}. These cumulative distribution functions can also be written as

$$F_\alpha(x) = \begin{cases} 0, & x \leq 0, \\ \exp\{-x^{-\alpha}\}, & x > 0, \end{cases} \quad \alpha > 0 \text{ (type I)}; \tag{1.2}$$

$$F_\alpha(x) = \begin{cases} \exp\{-(-x)^{-\alpha}\}, & x < 0, \\ 1, & x \geq 0, \end{cases} \quad \alpha < 0 \text{ (type II)}; \tag{1.3}$$

$$F(x) = \exp\{-e^{-x}\}, \quad x \in \mathbb{R}, \quad \text{(type III)}, \tag{1.4}$$

where $\alpha = 1/\gamma$ if $\gamma \neq 0$.

If $\zeta_n = \max(\xi_1, \ldots, \xi_n)$ is the maximum of i.i.d. random variables, then the asymptotic properties of the normalised sequence $(a_n \zeta_n + b_n)$ are determined by the regular variation property of the cumulative distribution function or the probability density of ξ_1, see Galambos [186] and Seneta [530]. See Appendix H for the definition of regularly varying functions. The limiting distribution of $(a_n \zeta_n + b_n)$ (if it is non-degenerate) is max-stable and belongs to one of the three types given by (1.1) (up to an affine transform of x).

Similar concepts can be formulated for random vectors (although not all vectors are max-infinitely divisible) and random elements in partially ordered spaces, semigroups or lattices.

1.2 Infinite divisibility for unions

Definition

It is easy to see that the maximum of random variables can be easily "translated" into operations with random closed sets. For instance, if $X_i = (-\infty, \xi_i]$, $1 \leq i \leq n$, are random half-lines, then $X_1 \cup \cdots \cup X_n$ is the half-line bounded by $\max(\xi_1, \ldots, \xi_n)$. Example 1.1.5(ii) shows that the cumulative distribution function of a random variable ξ is closely related to the capacity functional of $X = (-\infty, \xi]$. In contrast to the studies of Minkowski sums which rely on representation of random sets through their support functions, the main tool suitable to work with unions of random sets is the capacity functional of random sets. Since we rely on the Choquet theorem and extensively use the concept of a capacity functional, it is assumed throughout this chapter that $\mathbb{E}$ is a locally compact Hausdorff second countable (LCHS) space (unless another type of space is specifically mentioned). Quite often, $\mathbb{E}$ is assumed to be the Euclidean space $\mathbb{R}^d$.

It is easy to express the capacity functional of the union of independent random closed sets X and Y as

$$T_{X \cup Y}(K) = T_X(K) + T_Y(K) - T_X(K) T_Y(K).$$

If $X_1, \ldots, X_n$ are i.i.d. random closed sets with the common capacity functional T, then the capacity functional of $Z_n = X_1 \cup \cdots \cup X_n$ is given by

$$T_{Z_n}(K) = 1 - (1 - T(K))^n. \tag{1.5}$$

Definition 1.1 (Union-infinite-divisibility). A random closed set X is said to be *infinitely divisible for unions* (or union-infinitely-divisible) if, for every $n \geq 1$, there exist i.i.d. random closed sets $X_{n1}, \ldots, X_{nn}$ such that

$$X \stackrel{d}{\sim} X_{n1} \cup \cdots \cup X_{nn}.$$

Fixed points of random closed set

Clearly, each deterministic set is infinitely divisible in the sense of Definition 1.1. To exclude this trivial case it is useful to remove from consideration the deterministic part of a random closed set X.

Definition 1.2 (Fixed point). A point $x \in \mathbb{E}$ is said to be a *fixed point* of X if

$$\mathbf{P}\{x \in X\} = T(\{x\}) = 1,$$

where T is the capacity functional of X. The set of all fixed points is denoted by F_X.

The random closed set X is said to be *non-trivial* if $\mathbf{P}\{X = F_X\} < 1$, i.e. X does not coincide almost surely with the set of its fixed points. This excludes both the case of a deterministic X and the set X which is empty with probability 1.

Proposition 1.3. *Let X be a random closed set in a LCHS space. Then*
(i) *F_X is a closed set;*
(ii) *$\mathbf{P}\{F_X \subset X\} = 1$.*

Proof.
(i) Note that $F_X = \{x : p(x) \geq 1\}$, where $p_X(x) = \mathbf{P}\{x \in X\}$ is the coverage function of X. Then F_X is closed, since p_X is an upper semicontinuous function by Proposition 1.1.34.
(ii) There exists a countable set D such that $F_X = \mathrm{cl}(F_X \cap D)$. For example, if $\mathbb{E}$ is a metric space, then D can be constructed as the set of points $x \in (F_X \cap B_{1/n}(q))$, where q belongs to a countable dense set in $\mathbb{E}$ and $n \geq 1$. Then $\mathbf{P}\{D \subset X\} = 1$, so that (ii) follows from the fact that X is closed. □

Example 1.4. Let $X = (-\infty, \xi]$ be a random subset of $\mathbb{R}^1$. If the random variable ξ is positive almost surely, then F_X contains $(-\infty, 0]$.

If $\xi \stackrel{d}{\sim} \max(\xi_1, \ldots, \xi_n)$ for some $n \geq 2$ with i.i.d. random variables $\xi_1, \ldots, \xi_n$ having the same distribution as ξ, then ξ is deterministic almost surely. The following proposition generalises this fact for random closed sets.

Proposition 1.5. *Let X be a random closed set in a LCHS space $\mathbb{E}$. If, for some $n \geq 2$, either*

$$X \stackrel{d}{\sim} X_1 \cap \cdots \cap X_n \tag{1.6}$$

or

$$X \stackrel{d}{\sim} X_1 \cup \cdots \cup X_n \tag{1.7}$$

with $X_1, \ldots, X_n$ being i.i.d. copies of X, then $X = F_X$ almost surely.

244   4 Unions of Random Sets

Proof. By (1.5) and (1.7), we obtain

$$1 - T_X(K) = (1 - T_X(K))^n.$$

Therefore, $T_X(K)$ equals either 0 or 1, whence X is a deterministic set. If (1.6) holds, then $T_X(K) \leq T_X(K)^n$ for all $K \in \mathcal{K}$, whence $T_X(K)$ is either 0 or 1. □

Characterisation theorem

Clearly, $T(K) = 1$ if K has a non-empty intersection with the set F_X of fixed points. To exclude such sets K, introduce the family

$$\mathcal{K}^X = \{K \in \mathcal{K} : K \cap F_X = \emptyset\}$$

of compact sets that miss F_X. Having replaced $\mathbb{E}$ by the space $\mathbb{E}\setminus F_X$, we can consider only random sets without fixed points, as was done by Matheron [381]. The following theorem provides a slight modification of G. Matheron's result, which can either be proved directly as it done below or derived by the instrumentality of the harmonic analysis on semigroups (see Berg, Christensen and Ressel [61, Prop. 4.6.10]) or the theory of lattices (see Norberg [432, Th. 6.2]).

Theorem 1.6 (Characterisation of infinite divisibility for unions). *A random closed set X in a LCHS space $\mathbb{E}$ is infinitely divisible for unions if and only if its capacity functional is represented as*

$$T(K) = 1 - \exp\{-\Psi(K)\}, \qquad (1.8)$$

where $\Psi(K)$ is a completely alternating non-negative upper semicontinuous capacity such that $\Psi(\emptyset) = 0$ and $\Psi(K)$ is finite for each $K \in \mathcal{K}^X$.

Lemma 1.7. *Let $q(s) = \mathbb{E}s^\zeta$ be the probability generating function of a non-negative discrete random variable ζ. Then $T = 1 - q(1 - T')$ is a capacity functional, if T' is a capacity functional. In particular, $T = 1 - \exp\{-\lambda T'\}$ is a capacity functional if T' is.*

Proof. Let T' be the capacity functional of a random closed set X'. Put $X = X'_1 \cup \cdots \cup X'_\zeta$ for i.i.d. realisations of X' which are also independent of ζ. Then

$$1 - T_X(K) = \mathbf{P}\{X \cap K = \emptyset\}$$

$$= \sum_{n=0}^\infty \mathbf{P}\{\zeta = n\} (1 - T'(K))^n = q(1 - T'(K)).$$

Thus, $T(K) = 1 - q(1 - T'(K))$ is a capacity functional of X. In particular, if ζ is a Poisson random variable with mean λ, then $q(s) = e^{-\lambda(1-s)}$, whence $T(K) = 1 - \exp\{-\lambda T'(K)\}$ is a capacity functional. □

1 Union-infinite-divisibility and union-stability   245

Lemma 1.7 can be alternatively derived from Proposition 1.1.11 that refers to the results of harmonic analysis on semigroups.

Lemma 1.8. *If X is infinitely divisible for unions, then its capacity functional satisfies $T(K) < 1$ for all $K \in \mathcal{K}^X$.*

Proof. Assume that $T(K) = 1$ for some $K \in \mathcal{K}^X$. By Zorn's lemma, there exists a minimal set $K_0 \subset K$ such that $T(K_0) = 1$. If we show that K_0 is a singleton, this would prove the lemma. Assume that K_0 contains at least two points. Then $K_0 = K_1 \cup K_2$ with both K_1 and K_2 strictly included in K_0. This implies $T(K_1) < 1$, $T(K_2) < 1$ and also

$$\mathbf{P}\{X \cap K_1 = \emptyset, \ X \cap K_2 = \emptyset\}$$
$$= \mathbf{P}\{X \cap K_1 = \emptyset\} + \mathbf{P}\{X \cap K_2 = \emptyset\} - \mathbf{P}\{X \cap (K_1 \cup K_2) = \emptyset\}$$
$$= \mathbf{P}\{X \cap K_1 = \emptyset\} + \mathbf{P}\{X \cap K_2 = \emptyset\} \,.$$

The random closed set X_{n1} from Definition 1.1 has the same fixed points as X, whence

$$\mathbf{P}\{X_{n1} \cap K_1 = \emptyset, \ X_{n1} \cap K_2 = \emptyset\}$$
$$= (\mathbf{P}\{X \cap K_1 = \emptyset\})^{1/n} + (\mathbf{P}\{X \cap K_2 = \emptyset\})^{1/n} \,.$$

The right-hand side can be made strictly larger than 1 for sufficiently large n. The obtained contradiction confirms that K_0 is a singleton. □

Proof of Theorem 1.6. Necessity. If X is infinitely divisible for unions, then

$$T = 1 - (1 - T_n)^n \,,$$

with T_n being a capacity functional. By Lemma 1.8, $T(K) < 1$ for each $K \in \mathcal{K}^X$, whence

$$\Psi(K) = -\log(1 - T(K)) < \infty \,.$$

Clearly, Ψ is upper semicontinuous on $\mathcal{K}^X$. Furthermore,

$$nT_n(K) = n(1 - (1 - T(K))^{1/n}) \to -\log(1 - T(K)) = \Psi(K) \quad \text{as } n \to \infty,$$

so that Ψ is a pointwise limit of completely alternating capacities nT_n, whence Ψ is completely alternating.

Sufficiency. Let us show that $T(K)$ given by (1.8) is a capacity functional of a random closed set. For this, it suffices to prove that T is completely alternating. Let $\{K_n, n \geq 1\}$ be a sequence of compact sets such that $K_n \uparrow (\mathbb{E} \setminus F_X)$. By assumption, $\Psi(K_n) < \infty$. Then $T'_n(K) = \Psi(K \cap K_n)/\Psi(K_n)$ is a capacity functional. For $\lambda = \Psi(K_n)$, Lemma 1.7 implies that

$$T_n(K) = 1 - \exp\{-\Psi(K \cap K_n)\}$$

is a capacity functional of a random closed set X_n such that $X_n \subset K_n$ a.s. Put $T(K) = T_n(K)$ where n is such that $K \subset K_n$. Then T is a capacity functional of a random closed set $\tilde{X}$, such that $\tilde{X} \subset (\mathbb{E} \setminus F_X)$ a.s. and

$$T_{\tilde{X}}(K) = \mathbf{P}\left\{\tilde{X} \cap K \neq \emptyset\right\} = 1 - \exp\{-\Psi(K)\}, \quad K \in \mathcal{K}^X.$$

Finally, $X = F_X \cup \tilde{X}$ is infinitely divisible for unions, since

$$1 - (1 - T(K))^{1/n} = 1 - \exp\{-n^{-1}\Psi(K)\}$$

is a capacity functional for each $n \geq 1$. □

Representation by Poisson process

It is known from the theory of multivariate extremes and pointwise extremes of upper semicontinuous functions (see Norberg [431, 481]) that a max-infinitely divisible random element can be obtained as the maximum of points of a *Poisson point process* whose intensity measure is called the *Lévy measure* of the corresponding max-infinitely divisible law. Let us show that every union-infinitely-divisible random set can be obtained as a union of sets that constitute a Poisson point process on $\mathcal{F}$. This statement can be compared with Theorem 1.8.13 saying that each random closed set in the extended convex ring can be represented as the union of sets that appear as a (not necessarily Poisson) point process on co $\mathcal{K}'$ and the representation of M-infinite divisible random compact sets as the Minkowski sum of a Poisson process on co $\mathcal{K}$, see Section 3.2.4.

Assume first that $F_X = \emptyset$. It follows from the general results on probability measures on lattices (see Corollary 1.3.18) that Ψ generates a locally finite measure Λ on $\mathcal{F}$ such that

$$\Lambda(\{F \in \mathcal{F} : F \cap K \neq \emptyset\}) = \Lambda(\mathcal{F}_K) = \Psi(K) \qquad (1.9)$$

for all $K \in \mathcal{K}$. Note that $\mathcal{F}^K = \{F \in \mathcal{F} : F \cap K = \emptyset\}$ is open in the Fell topology.

The locally finite measure Λ on $\mathcal{F}$ can be interpreted as the *intensity measure* of a Poisson point process $\Pi_\Lambda = \{F_1, F_2, \dots\}$ on $\mathcal{F}$, so that the number of the F_is in any measurable $\mathcal{D} \subset \mathcal{F}$ is Poisson distributed with mean $\Lambda(\mathcal{D})$ and these numbers are independent for disjoint sets $\mathcal{D}_1, \dots, \mathcal{D}_k \subset \mathcal{F}$, see Definition 1.8.7. Note that the union of all sets from Π_Λ is closed since the intensity measure Λ is locally finite. The distribution of the random closed set $X = F_1 \cup F_2 \cup \cdots$ is denoted by $\mathrm{Pois}_\cup(\Lambda)$, cf. Definition 3.2.17. It is easy to see that (1.8) follows from this representation of X, since

$$\begin{aligned} T(K) &= 1 - \mathbf{P}\{X \cap K = \emptyset\} \\ &= 1 - \mathbf{P}\{\text{no points of } \Pi_\Lambda(\mathcal{F}) \text{ in } \mathcal{F}_K\} \\ &= 1 - e^{-\Lambda(\mathcal{F}_K)} \\ &= 1 - e^{-\Psi(K)}. \end{aligned}$$

If $F_X \neq \emptyset$, then the intensity measure Λ is concentrated on the family of closed sets $F \in \mathcal{F}$ such that $F \supset F_X$ and $X = F_X \cup F_1 \cup F_2 \cup \cdots$.

As explained in Section 1.4.7, C-additive capacities correspond to random convex sets. Similar to Theorem 1.4.35, it is easy to prove that if $\Psi(K) = -\log(1 - T(K))$ is a C-additive capacity, then the measure Λ defined by (1.9) is supported by the family co$\mathcal{F}$ of convex closed sets, whence X is the union of a Poisson process of convex closed sets. The corresponding random closed set X is *semi-Markov*, see Definition 1.4.38.

1.3 Union-stable random sets

Definitions

In the following (unless otherwise stated) we assume that $\mathbb{E}$ is the Euclidean space $\mathbb{R}^d$, noticing that all results also hold in every finite-dimensional linear space. First, define union-stable sets that form a subclass of random closed sets infinitely divisible for unions.

Definition 1.9 (Union-stability). A random closed set X is said to be *union-stable* if, for every $n \geq 2$, there exists $a_n > 0$ such that

$$a_n X \stackrel{d}{\sim} X_1 \cup \cdots \cup X_n, \tag{1.10}$$

where $X_1, \ldots, X_n$ are independent random closed sets distributed as X.

Proposition 1.10 (Alternative definitions of union-stability). *A random closed set X is union-stable if and only if one of the following conditions holds for i.i.d. random closed sets X_1, X_2 having the same distribution as X.*
  (i) *For each $s > 0$, there exists $t > 0$ such that*

$$X_1 \cup s X_2 \stackrel{d}{\sim} t X. \tag{1.11}$$

  (ii) *There exists $\gamma \neq 0$ such that for all $s, t > 0$*

$$t^\gamma X_1 \cup s^\gamma X_2 \stackrel{d}{\sim} (t+s)^\gamma X.$$

Proof.
  (i) Note that (1.11) implies (1.10) by successive applications. Conversely, (1.10) implies (1.11) first for s and t being positive rational numbers and then for general s and t by approximation.
  (ii) easily follows from the characterisation of union-stable random closed sets in Theorem 1.12. Note that $\gamma = -1/\alpha$ for α that appears in (1.12). □

Characterisation of distributions for union-stable sets

The main difficulty in the characterisation of union-stable random sets is caused by the possible self-similarity of random sets. If a random variable ξ coincides in distribution with $c\xi$, then $\xi = 0$ a.s. However, there are non-trivial random closed sets (called *self-similar*) that satisfy $X \stackrel{d}{\sim} tX$ for all $t > 0$, see Example 1.4.5. If X does not contain fixed points, then X cannot be self-similar, which immediately simplifies the arguments used to characterise union-stable sets. Unlike the infinite divisibility for unions, possible fixed points cannot be simply eliminated by considering the space $\mathbb{R}^d \setminus F_X$, since $\mathbb{R}^d \setminus F_X$ is not necessarily invariant with respect to multiplication by positive numbers. However, even in $\mathbb{R}^1$ there are simple examples of union-stable sets that do have fixed points.

Example 1.11 (Union-stable random set with fixed points). Let $X = (-\infty, \xi]$ be a random subset of $\mathbb{R}^1$. Then X is union-stable if and only if ξ is a max-stable random variable. For max-stable laws of type I, ξ is positive almost surely, whence the corresponding set of fixed points $F_X = (-\infty, 0]$ is not empty.

Union-stable sets without fixed points have been characterised by Matheron [381] whose result is covered by Theorem 1.12 for the case $\alpha > 0$.

Theorem 1.12 (Characterisation of union-stable random sets). *A non-trivial random closed set X is union-stable if and only if its capacity functional T_X is given by (1.8), where $\Psi(K)$ is a completely alternating upper semicontinuous capacity, $\Psi(\emptyset) = 0$ and there exists $\alpha \neq 0$ such that*

$$\Psi(sK) = s^\alpha \Psi(K), \tag{1.12}$$

$\Psi(K) < \infty$ *and*

$$sF_X = F_X \tag{1.13}$$

for all $s > 0$ and $K \in \mathcal{K}^X$.

Proof. The proof of necessity falls into several steps. The key idea is to show that a union-stable set is not self-similar.

Step I. Let $T(aK) = T(a_1 K)$ for all $K \in \mathcal{K}$ and some $a, a_1 > 0$. Prove that $a = a_1$. It suffices to consider the case $a_1 = 1$, $a < 1$. Then $T(K) = T(a^n K)$ for all $n \geq 1$ and $K \in \mathcal{K}$. Hence $T(K) \leq T(B_\varepsilon(0))$ for each $\varepsilon > 0$. The semicontinuity of T implies $T(K) \leq T(\{0\})$ for each $K \in \mathcal{K}$. Thus, $T(\{0\}) \geq T(\mathbb{R}^d) > 0$, since X is non-empty with positive probability. It follows from (1.10) that

$$T(\{0\}) = 1 - (1 - T(\{0\}))^n,$$

whence $T(\{0\}) = 1$, i.e. $0 \in F_X$. If X has no fixed points, then the first step has been proved by contradiction. If F_X is non-void, then (1.10) yields

1 Union-infinite-divisibility and union-stability   249

$$T(K) = 1 - (1 - T(a_n K))^n, \quad n \geq 1, \ K \in \mathcal{K}, \tag{1.14}$$

for some $a_n > 0$. Since $a \neq 1$,

$$a_n = a^{m(n)} \delta_n, \quad n \geq 1,$$

for integer $m(n)$ and δ_n belonging to $(a, 1]$. Then, for each compact set K and $n \geq 1$,

$$T(K) = 1 - \left(1 - T(a_n K/a^{m(n)})\right)^n = 1 - (1 - T(\delta_n K))^n. \tag{1.15}$$

Without loss of generality assume $\delta_n \to \delta \in [a, 1]$ as $n \to \infty$. Take any $K \in \mathcal{K}$ such that $K \cap F_X = \emptyset$. Then $K^\varepsilon \cap F_X = \emptyset$ for some $\varepsilon > 0$, so that $T(K^\varepsilon) < 1$, where $K^\varepsilon = \{x : \rho(x, K) \leq \varepsilon\}$ is the ε-envelope of K. Therefore, $T(\delta_n K^\varepsilon) \to 0$ as $n \to \infty$. Then $T(\delta K) = 0$ for each K which misses F_X. It is easy to derive from (1.15) that $\delta_n F_X = F_X$, whence $\delta F_X = F_X$, meaning that $K \cap F_X = \emptyset$ if and only if $\delta K \cap F_X = \emptyset$. Therefore, $T(K) = 0$ for each K that misses F_X, whence $X \cap (\mathbb{R}^d \setminus F_X)$ is empty almost surely. Thus, $X = F_X$ almost surely, contrary to the condition of Theorem 1.12. The obtained contradiction shows that $a = 1$.

Step II. Since a union-stable random closed set is infinitely divisible for unions, its capacity functional is given by (1.8). It follows from (1.10) that $n\Psi(a_n K) = \Psi(K)$ and $a_n F_X = F_X$ for each $n \geq 1$ and $K \in \mathcal{K}^X$. For any positive rational number $s = m/n \in \mathbb{Q}_+$ put $a(s) = a_m/a_n$. It is easy to show that $a(s)$ does not depend on the representation of s. Then, for any s from $\mathbb{Q}_+$, $a(s)F_X = F_X$ and

$$s\Psi(a(s)K) = \Psi(K), \quad K \in \mathcal{K}^X. \tag{1.16}$$

It follows from (1.16) that $\Psi(a(s)a(s_1)K) = \Psi(a(ss_1)K)$ for each $s, s_1 \in \mathbb{Q}_+$. The first step of the proof and (1.8) yield

$$a(ss_1) = a(s)a(s_1). \tag{1.17}$$

Step III. It follows from (1.8) and (1.16) that $T(a(s_n)K) \to T(K)$ as $s_n \to 1$ for each $K \in \mathcal{K}$. Without loss of generality assume that the sequence $\{a(s_n), n \geq 1\}$ has a limit (which is allowed to be infinite). Let this limit be finite and equal to $a > 0$. Then, for any $\varepsilon > 0$ and sufficiently large n,

$$T(a(s_n)K) \leq T(aK^\varepsilon).$$

Hence $T(K) \leq T(aK)$. Similarly, $T(K/a) \geq T(K)$. Thus, $T(aK) = T(K)$ for each $K \in \mathcal{K}$, whence $a = 1$.

Since $\Psi(a(s_n)K) = \Psi(K)/s_n$ and $\Psi(K/a(s_n)) = s_n \Psi(K)$, it suffices to assume that either $a(s_n) \to 0$ or $a(s_n) \to \infty$ as $n \to \infty$. Choose an integer $m > 1$. Let $a_m > 1$. Suppose that $a(s_n) \to \infty$ as $n \to \infty$. Then, for all sufficiently large n,

$$a(s_n) = (a_m)^{k(n)} \delta_n,$$

where $1 \leq \delta_n < a_m$ and $k(n)$ is a positive integer. It follows from (1.14) and (1.17) that $(a_m)^{k(n)} = a_{m^{k(n)}}$. Hence

$$T(\delta_n K) = 1 - \left[1 - T\left((a_m)^{k(n)}\delta_n K\right)\right]^{m^{k(n)}}$$
$$= 1 - [1 - T(a(s_n)K)]^{m^{k(n)}}.$$

If $T(K) > 0$, then $T(\delta_n K) \to 1$, since $T(a(s_n)K) \to T(K) > 0$. Without loss of generality assume that $\delta_n \to \delta$ as $n \to \infty$. The semicontinuity of T implies $T(\delta K) = 1$. Hence $\delta K \cap F_X \neq \emptyset$ whenever $T(K) > 0$. It is easy to show that $\delta_n F_X = F_X$ for all $n \geq 1$, whence $\delta F_X = F_X$. Thus, $K \cap F_X \neq \emptyset$ whenever $T(K) > 0$, so that $X = F_X$ almost surely.

It is obvious that $a_m \neq 1$. If $a_m < 1$, then suppose $a(s_n) \to 0$ as $n \to \infty$ and arrive at a contradiction using the same arguments as above. Thus, $a(s_n) \to 1 = a(1)$ as $s_n \to 1$, i.e. the function $a(s)$ is continuous on $\mathbb{Q}_+$ at $s = 1$.

Step IV. If $s_n \to s \in \mathbb{Q}_+$ as $n \to \infty$, then $a(s_n) = a(s)a(s_n/s) \to a(s)$ as $n \to \infty$, since $a(s_n/s) \to 1$. Thus, $a(s)$ is continuous on $\mathbb{Q}_+$.

Extend $a(s)$ onto the positive half-line by continuity, i.e. for any positive s denote $a(s) = \lim a(s_n)$, where $s_n \to s$ as $n \to \infty$, $s_n \in \mathbb{Q}_+$. Then the function $a(s)$ is continuous on $\mathbb{R}_+$ and $a(ss_1) = a(s)a(s_1)$ for each s, $s_1 > 0$. Thus, $a(s) = s^\gamma$ for a real number γ. If $\gamma = 0$, then $s\Psi(K) = \Psi(K)$, i.e. $X = F_X$ almost surely. Hence $\gamma \neq 0$, so that (1.12) and (1.13) hold with $\alpha = -1/\gamma$.

Sufficiency. The capacity functional of $X_1 \cup \cdots \cup X_n$ is equal to

$$T_n(K) = 1 - \exp\{-n\Psi(K)\}.$$

The capacity functional of $a_n X$ is given by

$$T'_n(K) = 1 - \exp\{-\Psi(K/a_n)\}.$$

If $a_n = n^{-1/\alpha}$, then $T_n = T'_n$ on $\mathcal{K}$. Now (1.10) follows from the Choquet theorem. □

Corollary 1.13.
(i) *If a union-stable random set X has no fixed points, then $\alpha > 0$ in (1.12).*
(ii) *If $\alpha < 0$, then F_X is non-empty and $0 \in F_X$.*
(iii) *A stationary non-trivial union-stable random closed set X has a positive parameter α in (1.12).*
(iv) *If X is union-stable with parameter α, then in (1.10) one has $a_n = n^{1/\alpha}$, $n \geq 1$.*

Proof.
(i) If $F_X = \emptyset$, then $\Psi(B_r(0)) < \infty$ for each $r > 0$. The homogeneity property implies that $\Psi(B_r(0)) = r^\alpha \Psi(B_1(0))$. Thus, $\alpha > 0$.
(ii) Let $\alpha < 0$. Then $\Psi(sK) \to \infty$ as $s \downarrow 0$, whence $T(\{0\}) = 1$ by the semicontinuity of the capacity functional.
(iii) If α is negative, then X has a fixed point, whence $X = \mathbb{R}^d$.
(iv) Follows from the proof of sufficiency in Theorem 1.12. □

1 Union-infinite-divisibility and union-stability 251

Corollary 1.14. *A random closed set X with the closed inverse*

$$X^* = \{x\|x\|^{-2} : x \in X, \, x \neq 0\} \tag{1.18}$$

is union-stable with parameter $\alpha \neq 0$ if and only if X^* is union-stable with parameter $(-\alpha)$.

Proof. Evidently,

$$\mathbf{P}\{X^* \cap K \neq \emptyset\} = \mathbf{P}\{X \cap K^* \neq \emptyset\} = 1 - \exp\{-\Psi(K^*)\}.$$

By (1.12), $\Psi(sK^*) = s^\alpha \Psi(K^*)$, whence $\Psi((sK)^*) = s^{-\alpha}\Psi(K^*)$. □

Examples of union-stable random sets

Example 1.15 (Max-stable laws of types I and II). The random set $X = (-\infty, \xi] \subset \mathbb{R}^1$ is union-stable if and only if ξ is max-stable with parameter $\gamma = \alpha^{-1} \neq 0$. If X is union-stable, then (1.13) implies that the set of fixed points of X is either empty or is $(-\infty, 0]$. Furthermore, (1.12) yields

$$T(\{x\}) = \begin{cases} 1 - \exp\{-c_1 x^\alpha\}, & x > 0, \\ 1 - \exp\{-c_2(-x)^\alpha\}, & x < 0, \end{cases}$$

for some non-negative constants c_1 and c_2. Since $\mathbf{P}\{\xi < x\} = 1 - T(\{x\})$ is monotone, we immediately obtain $c_1 = 0$ if $\alpha > 0$ and $c_2 = 0$ if $\alpha < 0$. The first case corresponds to the law of type II, see (1.3), and the second one to the law of type I, see (1.2). Therefore, Theorem 1.12 yields the representation of cumulative distribution functions of stable laws of types I and II.

Example 1.16 (Poisson point process in $\mathbb{R}^d$). Let $X = \Pi_\Lambda$ be a Poisson point process in $\mathbb{R}^d$ with intensity measure Λ, see Definition 1.8.7. The capacity functional of X is equal to

$$T(K) = \mathbf{P}\{\Pi_\Lambda \cap K \neq \emptyset\} = 1 - \exp\{-\Lambda(K)\}.$$

Each Borel measure is an upper semicontinuous completely alternating capacity, so that Theorem 1.6 implies that Π_Λ is infinitely divisible for unions. Then X is union-stable if and only if Λ is homogeneous, i.e., for some $\alpha \neq 0$,

$$\Lambda(sK) = s^\alpha \Lambda(K), \quad K \in \mathcal{K}, s \geq 0. \tag{1.19}$$

If Λ is absolutely continuous with respect to the Lebesgue measure. Then X is union-stable if and only if the density λ is *homogeneous*, i.e.

$$\lambda(su) = s^{\alpha-d}\lambda(u), \quad s > 0, \, u \in \mathbb{R}^d,$$

for a real α. If $\alpha < d$, then Λ is not locally finite at the origin, and the origin is a fixed point of X. For $\alpha = d$, Π_Λ becomes the stationary Poisson point process.

Example 1.17 (Poisson-rescaled sets). Let Π_Λ be a Poisson point process on $\mathbb{R}_+$ with intensity measure Λ. For a deterministic compact set Z define

$$X = \bigcup_{x_i \in \Pi_\Lambda} x_i^\gamma Z,$$

where $\gamma > 0$. Then

$$T_X(K) = 1 - \exp\{-\Lambda(K/Z)\},$$

where $K/Z = \{t \geq 0 : (t^\gamma Z) \cap K \neq \emptyset\}$. If Λ is homogeneous of degree α, then X is union-stable with parameter α/γ. Indeed,

$$\Lambda((cK)/Z) = \Lambda(\{t : (t^\gamma Z) \cap cK \neq \emptyset\})$$
$$= \Lambda(c^{1/\gamma}\{t : (t^\gamma Z) \cap K \neq \emptyset\}) = c^{\alpha/\gamma} \Lambda(K/Z).$$

Example 1.18 (Maxitive union-stable random sets). If $f: \mathbb{R}^d \mapsto [0, \infty]$ is an upper semicontinuous function, then

$$\Psi(K) = \sup_{x \in K} f(x)$$

is a *maxitive* capacity, see Example 1.1.15. The capacity functional (1.8) corresponds to the random closed set $X = \{x : f(x) \geq \xi\}$, where ξ is a random variable exponentially distributed with parameter 1. The random closed set X is union-stable if and only if f is homogeneous, i.e. $f(sx) = s^\alpha f(x)$ for each $s > 0$ and $x \in \mathbb{R}^d$. Note that $F_X = \{x : f(x) = \infty\}$.

Example 1.19 (Riesz capacities). Let C be the capacity defined by (E.7) for kernel $k(x, y)$. Assume that C is an upper semicontinuous completely alternating capacity. By Theorem 1.6, the functional $T(K) = 1 - \exp\{-C(K)\}$ is the capacity functional of a random closed set X. Then X is union-stable if and only if $k(sx, sy) = s^{-\alpha} k(x, y)$ for some $\alpha \neq 0$, all $s > 0$ and $x, y \in \mathbb{R}^d$. If X is stationary and isotropic, then $k(x, y)$ depends on $\|x - y\|$ only. Thus, for some $a > 0$ and $q \in [0, \infty]$,

$$k(x, y) = \begin{cases} a\|x - y\|^{-\alpha}, & x \neq y, \\ q, & x = y. \end{cases}$$

If $q = 0$, then $C(\{x\}) = \infty$, so that $X = \mathbb{R}^d$ a.s. by the stationarity. If $0 < q < \infty$, then $C(\{x\}) = 1/q$ and $C(\{x, y\}) = 2(q + k(x, y))$. If $\alpha < 0$, then

$$\lim_{y \to x} C(\{x, y\}) = 2/q > C(\{x\}),$$

i.e. C is not upper semicontinuous. If $\alpha > 0$, then $C(\{x, y\}) < C(\{x\})$ for sufficiently small $\|x - y\|$, i.e. C is not increasing. Thus, $q = \infty$ and $k(x, y) = a\|x - y\|^{-\alpha}$ equals (up to a constant factor) the *Riesz* kernel $k_{d,\gamma}\|x - y\|^{\gamma - d}$ with $\gamma = d - \alpha$. Then C is the Riesz capacity, see Appendix E.

If $\alpha = d - 2$ and $d \geq 3$, then C is the *Newton* capacity up to a proportionality constant and the corresponding random closed set X can be constructed as follows.

1 Union-infinite-divisibility and union-stability

Let μ be the equilibrium probability measure on $B_r(0) = \{x : \|x\| \leq r\}$ with respect to the kernel $k(x, y) = \|x - y\|^{2-d}$, i.e.

$$U_\mu^k(x) = \int_K k(x, y)\mu(dy) = \frac{1}{C(B_r(0))}, \quad x \in B_r(0),$$

and let N be the Poisson random variable of mean $C(B_r(0))$. At time $t = 0$ we launch N mutually independent and independent of N random Wiener processes $w_i(t)$, $1 \leq i \leq N$, with the initial distribution μ. The path of each process until it leaves $B_r(0)$ is denoted by X_i. Let $X = X_1 \cup \cdots \cup X_N$ be the union of these paths. The capacity functional of X is equal to

$$T(K) = 1 - \exp\{-C(B_r(0))T_{X_1}(K)\},$$

where $T_{X_1}(K)$ is the capacity functional of X_1. Let $\mathbf{P}_x$ be the distribution of the process $w_1(\cdot)$ which starts from x, $\tau_G = \inf\{t : w_1(t) \in G\}$ for some open set $G \subset B_r(0)$. Since $\mathbf{P}_x\{\tau_G < \infty\}$ is the potential of the equilibrium measure μ_G on G (see Itô and McKean [272, p. 352]) we obtain

$$T_{X_1}(G) = \int_{B_r} \mathbf{P}_x\{\tau_G < \infty\}\mu(dx)$$

$$= \int_{B_r}\left[\int_K k(x, y)\mu_G(dy)\right]\mu(dx)$$

$$= \int_G \frac{\mu_G(dy)}{C(B_r(0))} = \frac{C(G)}{C(B_r(0))}, \quad G \subset B_r(0).$$

By approximation, the same formula holds with G replaced by a compact set K, whence $T(K) = 1 - e^{-C(K)}$. The above construction defines a random closed subset of the ball $B_r(0)$ or, equivalently, the probability measure on the family of compact subsets of $B_r(0)$. Theorem E.3 establishes the existence of a random closed set X in $\mathbb{R}^d$ such that $X \cap B_r(0)$ has the above specified distribution.

Open problem 1.20. Characterise stationary isotropic union-stable random closed sets. Note that the Riesz capacities from Example 1.19 may not exhaust all examples of their capacity functionals.

1.4 Other normalisations

Different scaling factors

It is possible to generalise the union-stability definition by using non-equal normalising constants along different axes in $\mathbb{R}^d$. A particular case appears if the various axes are being rescaled according to various powers of the same constant. Consider a transformation in $\mathbb{R}^d$ given by

$$a^{\mathbf{g}} \circ x = (a^{\gamma_1} x_1, \ldots, a^{\gamma_d} x_d), \qquad (1.20)$$

where $\mathbf{g} = (\gamma_1, \ldots, \gamma_d) \in (0, \infty)^d$ and $x = (x_1, \ldots, x_d)$. For every $x \in \mathbb{R}^d$ define

$$f_{\mathbf{g}}(x) = (\operatorname{sign}(x_1)|x_1|^{\gamma_1}, \ldots, \operatorname{sign}(x_d)|x_d|^{\gamma_d}).$$

The normalisation (1.20) can be applied to a random closed set and used in the left-hand side of (1.10) to define the corresponding stable random closed sets. The union-stable random sets from Definition 1.9 appear if $\mathbf{g} = (1, \ldots, 1)$.

Proposition 1.21. *A random closed set X is stable with respect to the normalisation (1.20) if and only if there exists a union-stable random closed set Y such that $X = f_{\mathbf{g}}(Y)$.*

Proof. Since $f_{\mathbf{g}}$ is a continuous bijection, $X = f_{\mathbf{g}}(Y)$ for a random closed set Y. Then

$$a_n^{\mathbf{g}} \circ X \overset{d}{\sim} X_1 \cup \cdots \cup X_n$$

implies that $f_{\mathbf{g}}(a_n \circ Y)$ coincides in distribution with $f_{\mathbf{g}}(Y_1 \cup \cdots \cup Y_n)$, whence Y is union-stable. □

Proposition 1.21 shows that an example of stable random sets with respect to the normalisation (1.20) can be constructed by applying the function $f_{\mathbf{g}}$ to union-stable random closed sets. In particular, Example 1.17 is applicable with $x_i^{\gamma} Z$ replaced by $x_i^{\mathbf{g}} \circ Z$ and $\{x_i, i \geq 1\}$ that form a Poisson point process in $[0, \infty)^d$.

Affine normalisation

Possible generalisations of union-stable random sets are based on analogues of (1.10) using normalisations other than the pure multiplicative one.

Definition 1.22 (Affine union-stable random sets). A random closed set X in $\mathbb{R}^d$ is said to be *affine union-stable* if, for every $n \geq 2$ and $X_1, \ldots, X_n$ being independent copies of X, there exist $a_n > 0$ and $b_n \in \mathbb{R}^d$ such that

$$a_n X + b_n \overset{d}{\sim} (X_1 \cup \cdots \cup X_n). \qquad (1.21)$$

The main difficulty in the way of characterising affine union-stable random closed sets is caused by the lack of a convergence of types theorem. Let $\mathbb{M}$ be a family of probability measures and let $\mathbb{A}$ be a topological group of transformations. Any $A \in \mathbb{A}$ acts on $\mu \in \mathbb{M}$ in the standard way $A\mu(F) = \mu(A^{-1}F)$. The pair $(\mathbb{A}, \mathbb{M})$ fulfils the *convergence of types* condition if for any sequence $\{\mu_n, n \geq 1\} \subset \mathbb{M}$ and $\{A_n, n \geq 1\} \subset \mathbb{A}$ such that μ_n converges weakly to $\mu \in \mathbb{M}$ and $A_n \mu_n$ converges weakly to $\nu \in \mathbb{M}$, the sequence $\{A_n, n \geq 1\}$ is relatively compact in $\mathbb{A}$, see Hazod [227]. This fact is known also under the name of the *Khinchin lemma* (see Leadbetter, Lindgren and Rootzen [346]) that, in its classical formulation, applies to affine transformations of random variables.

1 Union-infinite-divisibility and union-stability   255

Unfortunately, the convergence of types theorem does not hold even for $\mathbb{A}$ being the group of homothetical transformations used to define union-stable random sets. For instance, if μ is the distribution of a self-similar random closed set, then $A\mu = \mu$ for every homothetical transformation A, so that one can put $\mu_n = \mu$ and obtain $A_n\mu_n \to \mu$ for every sequence of homotheties. If we allow all affine transformations (as in Definition 1.22), then we should take into account the fact that the distribution of a random closed set X may coincide with the distribution of $X + u$ for some (and possibly all) $u \in \mathbb{R}^d$. The convergence $T(a_n K + b_n) \to T(K)$ as $n \to \infty$ for each $K \in \mathcal{K}$ does not automatically imply the boundedness of the sequences $\{a_n, n \geq 1\}$ and $\{\|b_n\|, n \geq 1\}$.

The characterisation theorem of union-stable sets relies on the fact that a union-stable random set cannot be self-similar. This also implies that the family of distributions for union-stable random closed sets together with the family of homothetical transformations satisfy the convergence of types condition. However, an affine union-stable set X may satisfy $X + b \stackrel{d}{\sim} X$ for $b \neq 0$, so that the convergence of types fails. Indeed, this is the case if X is a stationary Poisson point process from Example 1.16.

Since we have to pay special attention to the case when the distribution of X is invariant with respect to translations, define the set

$$H_X = \left\{ u \in \mathbb{R}^d : X \stackrel{d}{\sim} X + u \right\}$$

of all translations which leave the distribution of X invariant. If X is stationary, then $H_X = \mathbb{R}^d$, while if X is compact, then $H_X = \{0\}$. It is easy to see that H_X is centrally symmetric with respect to the origin and is closed for sums.

Theorem 1.23 (Reduction to union-stable sets). *Assume that H_X is a linear subspace of $\mathbb{R}^d$. If X is an affine union-stable random closed set and a_m from (1.21) is not equal to 1 for some $m \geq 2$, then there exists $b \in \mathbb{R}^d$ such that $X + b$ is union-stable.*

Proof. For the chosen m and any $b \in \mathbb{R}^d$ we have

$$\bigcup_{i=1}^{m}(X_i + b) \stackrel{d}{\sim} a_m X + b_m + b = a_m(X + b) + (b_m + b - a_m b).$$

Denote $b = b_m/(1 - a_m)$. The random closed set $Y = X + b$ is affine union-stable with the same multiplicative constants a_n as X and with the additive constants $b'_n = b_n + b(1 - a_n), n \geq 1$. Note that $b'_m = 0$.

Let $\{Y_n, n \geq 1\}$ and $\{Y'_n, n \geq 1\}$ be two independent sequences of i.i.d. copies of Y. Then, for each $k \geq 2$,

$$\bigcup_{i=1}^{mk} Y_i \stackrel{d}{\sim} \bigcup_{j=1}^{k} a_m Y'_j = a_m \bigcup_{j=1}^{k} Y'_j \stackrel{d}{\sim} a_m(a_k Y + b'_k).$$

Furthermore,

$$\bigcup_{i=1}^{mk} Y_i \stackrel{d}{\sim} \bigcup_{j=1}^{m}(a_k Y'_j + b'_k) = a_k \bigcup_{j=1}^{k} Y'_j + b'_k \stackrel{d}{\sim} a_k a_m Y + b'_k.$$

Therefore,

$$a_m a_k Y + a_m b'_k \stackrel{d}{\sim} a_k a_m Y + b'_k.$$

Note that $H_Y = H_X$, whence

$$\frac{a_m - 1}{a_m a_k} b'_k = u_k \in H_X.$$

By the assumption on H_X, we get

$$a_k Y + b'_k = a_k(Y + b'_k a_k^{-1}) \stackrel{d}{\sim} a_k Y,$$

whence $a_k Y \stackrel{d}{\sim} Y_1 \cup \cdots \cup Y_k$ for each $k \geq 2$. □

Additive normalisation

If at least one of the multiplicative constants a_n, $n \geq 2$, in Definition 1.22 is not 1, then it is possible to reduce the consideration to the case of union-stable random sets. Now assume that $a_n = 1$ for all $n \geq 2$.

Definition 1.24 (Additive union-stable random sets). A random set X is said to be *additive union-stable* if, for every $n \geq 2$, there exists $b_n \in \mathbb{R}^d$ such that

$$X + b_n \stackrel{d}{\sim} (X_1 \cup \cdots \cup X_n).$$

The following definition is helpful to establish the convergence of types theorem for the pure additive normalisation.

Definition 1.25 (Homogeneity at infinity). A non-stationary random closed set X is said to be *homogeneous at infinity* if H_X is a linear subspace of $\mathbb{R}^d$ and, for each sequence $\{b_n, n \geq 1\} \subset \mathbb{R}^d$,

$$\lim_{n \to \infty} T(K + b_n) = \lim_{n \to \infty} T(K + u + b_n), \quad K \in \mathcal{K}, \ u \in \mathbb{R}^d, \tag{1.22}$$

given that at least one limit exists and $\rho(b_n, H_X) \to \infty$ as $n \to \infty$.

Each random compact set is homogeneous at infinity.

Lemma 1.26. *If X is homogeneous at infinity, then X satisfies the convergence of types theorem, i.e. for each sequence $\{b_n, n \geq 1\} \subset \mathbb{R}^d$, the convergence $T(K + b_n) \to T(K)$ as $n \to \infty$ for $K \in \mathcal{K}$ implies $\sup\{\rho(b_n, H_X) : n \geq 1\} < \infty$.*

Proof. Suppose that $\rho(b_n, H_X) \to \infty$ as $n \to \infty$. It follows from (1.22) that the limit of $T(K + u + b_n)$ exists and is equal to $T(K)$. Furthermore, $T(K + u + b_n) \to T(K + u)$ as $n \to \infty$. Thus $T(K) = T(K + u)$ for each u, i.e. $H_X = \mathbb{R}^d$. The obtained contradiction shows that the sequence $\{\rho(b_n, H_X), n \geq 1\}$ is bounded. □

Open problem 1.27. Which other conditions (apart from the homogeneity at infinity) imply that the convergence of types theorem holds for the additive normalising scheme?

Theorem 1.28 (Characterisation of additive union-stable random sets). *A random closed set X is additive union-stable if (and only if in the case X is homogeneous at infinity and H_X is a linear space) its capacity functional is given by (1.8), where $\Psi(\emptyset) = 0$ and, for some v orthogonal to H_X,*

$$\Psi(K + vs) = e^{-s}\Psi(K), \quad F_X + vs = F_X \qquad (1.23)$$

whatever $K \in \mathcal{K}^X$ and $s \in \mathbb{R}$ may be.

Proof. Sufficiency can be obtained from (1.23) and (1.21) for $b_n = v \log n$.
Necessity. Since X is infinitely divisible for unions, (1.8) holds. It is easy to prove the existence of a function $b(s)$, $s \in \mathbb{Q}_+$, with values in the orthogonal complement to H_X such that

$$s\Psi(b(s) + K) = \Psi(K), \quad b(s) + F_X = F_X, \qquad (1.24)$$

for all $s \in \mathbb{R}$ and $K \in \mathcal{K}^X$. As in the proof of Theorem 1.6, we can show that $b(ss_1) = b(s) + b(s_1)$ for all positive rational numbers s, s_1. It follows from (1.24) that $T(b(s_n) + K) \to T(K)$ for $K \in \mathcal{K}$, if $s_n \to 1$ and $n \to \infty$. By Lemma 1.26, the norms $\|b(s_n)\| = \rho(b_n, H_X)$, $n \geq 1$, are bounded. Without loss of generality assume that $b(s_n) \to b$ as $n \to \infty$. It is easy to show that $T(K + b) = T(K)$ for each $K \in \mathcal{K}$, whence $b = 0$. Thus, $b(s)$ is continuous at $s = 1$ and, therefore, may be continuously extended onto the positive half-line. Hence $b(s) = v \log s$ for some v orthogonal to H_X. □

Example 1.29 (Additive union-stable sets).
(i) The random set $X = (-\infty, \xi]$ is additive union-stable if and only if ξ is a max-stable random variable of type III, see (1.4).
(ii) The Poisson process Π_Λ from Example 1.16 is additive union-stable if there exists $v \in \mathbb{R}^d \setminus \{0\}$ such that $\lambda(u + vs) = e^{-s}\lambda(u)$ for all $u \in \mathbb{R}^d$ and $s \in \mathbb{R}$.

Further generalisations

It is possible to generalise Definition 1.22 by replacing the a_n in the left-hand side of (1.21) with an invertible linear operator A_n. Then it is possible to use the same reasoning as in Theorems 1.23 and 1.12. For this, X is assumed to be compact and $(I - A_m)$ should be invertible for some m.

A far reaching generalisation of the stability concept for random closed sets can be introduced as follows. Let $\mathcal{M}_1$ and $\mathcal{M}_2$ be two families of compact sets. A random closed set X is called $(\mathcal{M}_1, \mathcal{M}_2)$-union-stable if, for each $n \geq 2$,

$$A_n X + L_n \stackrel{d}{\sim} (X_1 \cup \cdots \cup X_n) + K_n \qquad (1.25)$$

258   4 Unions of Random Sets

for an invertible linear operator A_n, $L_n \in \mathcal{M}_1$ and $K_n \in \mathcal{M}_2$. The problem is getting considerably more complicated if K_n and L_n from (1.25) are not singletons, for example, if one of them is allowed to be a ball. A peculiar situation that might appear in this context is that the stability condition

$$X + L_n \stackrel{d}{\sim} X_1 \cup \cdots \cup X_n, \quad n \geq 1,$$

does not imply anymore that X is infinitely divisible for unions.

It is possible also to consider random closed sets which are infinitely divisible or stable with respect to intersection. However, in many cases this can be reduced to consideration of unions of their complements.

1.5 Infinite divisibility of lattice-valued random elements

The infinite divisibility is a general concept which can be defined for random elements in cones, semigroups or lattices. Here we concentrate on the latter case with the aim to characterise infinite divisible random elements on continuous posets. The notation of Section 1.3.1 is used without further comments.

Consider a random element ξ in a continuous semi-lattice $\mathbb{L}$. Assume that $\mathbb{L}$ has a top and a second countable Scott topology. We say that ξ (or its distribution $\mathbf{P}$) is infinitely divisible if, for every $n \geq 2$, there exist i.i.d. $\mathbb{L}$-valued random elements $\xi_1, \ldots, \xi_n$ such that

$$\xi \stackrel{d}{\sim} \bigwedge_{1 \leq i \leq n} \xi_i. \qquad (1.26)$$

Recall that $\mathcal{L}$ denotes the collection Ofilt($\mathbb{L}$) of Scott open filters on $\mathbb{L}$. Since each $F \in \mathcal{L}$ is an upper set, (1.26) implies, for every $n \geq 2$,

$$\mathbf{P}(F) = \mathbf{P}_n(F)^n, \quad F \in \mathcal{L}, \qquad (1.27)$$

where $\mathbf{P}_n$ is a probability measure on the family $\sigma(\mathbb{L})$ of Borel sets in $\mathbb{L}$. In the following an infinite-divisible distribution $\mathbf{P}$ on $\mathbb{L}$ will be associated with a measure ν on $\mathcal{L}$ and the latter, in turn, with a completely $\cap$-alternating function Ψ on $\mathcal{L}$, see Definition 1.1.9(i). This programme will be first carried over in case $\mathbf{P}(F) > 0$ for each non-empty $F \in \mathcal{L}$ (the case of no fixed points) and then the strict positivity assumption will be dropped.

The lattice $\mathcal{L}$ can be considered an idempotent semigroup $(\mathcal{L}, \cap)$, so that (1.27) implies that $\mathbf{P}$ is an infinitely divisible function on $\mathcal{L}$. By duality, the semicharacters on $\mathcal{L}$ can be identified with elements of $\mathbb{L}$. Therefore, (G.3) implies the following result.

Proposition 1.30. *Let $\mathbb{L}$ be a continuous semi-lattice with a top $\mathbb{I}$ and a second countable Scott topology. Then the formula*

$$\mathbf{P}(F) = \exp\{-\nu(\mathbb{L} \setminus F)\}, \quad F \in \mathcal{L}, \qquad (1.28)$$

defines a bijection between the family of infinitely divisible probability measures **P** on $\mathbb{L}$ satisfying $\mathbf{P}(F) > 0$ for each non-empty $F \in \mathcal{L}$ and the family of measures ν on $\mathbb{L}$ which concentrate their mass on $\mathbb{L} \setminus \{\mathbb{I}\}$ and satisfy

$$\nu(\mathbb{L} \setminus F) < \infty, \quad F \in \mathcal{L}. \tag{1.29}$$

By a *Lévy–Khinchin measure* on $\mathbb{L}$ we understand a measure ν on $\mathbb{L}$ which may appear in the right-hand side of (1.28), so that ν concentrates its mass on $\mathbb{L} \setminus \{\mathbb{I}\}$ and satisfies (1.29). It is possible to show that (1.29) is equivalent to one of the following conditions:
(1) $\nu(\mathbb{L} \setminus F) < \infty$ for all non-empty $F \in \mathrm{Scott}(\mathbb{L})$;
(2) $\nu(\mathbb{L} \setminus (\uparrow x)) < \infty$ for all $x \in \mathbb{L}$ such that $x \ll \mathbb{I}$.

Theorem 1.31. *Let $\mathbb{L}$ be a continuous semi-lattice with a top and a second countable Scott topology. Then*

$$\Psi(F) = \nu(\mathbb{L} \setminus F), \quad F \in \mathcal{L},$$

defines a bijection between the family of Lévy–Khinchin measures ν on $\mathbb{L}$ and the family of completely $\cap$-alternating functionals $\Psi : \mathcal{L} \mapsto \mathbb{R}_+$ satisfying $\Psi(\mathbb{L}) = 0$ and $\Psi(F_n) \to \Psi(F)$ as $F_n \uparrow F$ for $F, F_1, F_2, \ldots \in \mathcal{L}$.

Proof. Necessity follows from $\Psi(\mathbb{L}) = \nu(\mathbb{L} \setminus \mathbb{L}) = 0$, the continuity of ν and the fact that $\nabla_{F_n} \cdots \nabla_{F_1} \Psi(F) = \nu(F \setminus \cup_i F_i)$.
Sufficiency. For each $H \in \mathcal{L}$ the function $\Psi(F \cap H) - \Psi(F)$, $F \in \mathcal{L}$, satisfies the condition of Corollary 1.3.18. Therefore, there is a locally finite measure ν_H on $\mathbb{L}$ satisfying

$$\nu_H(F) = \Psi(F \cap H) - \Psi(F)$$

for all $F \in \mathcal{L}$. Since $\nu_H(\mathbb{L}) = \Psi(H) - \Psi(\mathbb{L}) = \Psi(H)$,

$$\nu_H(\mathbb{L} \setminus F) = \Psi(H) - \Psi(F \cap H) + \Psi(F), \quad F \in \mathcal{L}.$$

Define a measure ν

$$\nu(B) = \sup_{H \in \mathcal{L}} \nu_H(B), \quad B \in \sigma(\mathbb{L}).$$

If $F \in \mathcal{L}$, then

$$\nu(\mathbb{L} \setminus F) = \Psi(F) + \sup_{H \in \mathcal{L}} (\Psi(H) - \Psi(F \cap H)) = \Psi(F).$$

Furthermore, each $H \in \mathcal{L}$ contains the top $\mathbb{I}$, whence

$$\nu(\{\mathbb{I}\}) = \sup_{H \in \mathcal{L}} \nu_H(\{\mathbb{I}\}) = 0. \qquad \square$$

We proceed to generalise Proposition 1.30 for the general case which would allow $\mathbf{P}(F) = 0$ for non-empty $F \in \mathcal{L}$. Define the support of a random element ξ as

$$\mathbb{L}_\xi = \bigcap \{\mathbb{L} \setminus F : F \in \mathcal{L}, \mathbf{P}(F) = 0\}.$$

Hence $\mathbf{P}\{\xi \in \mathbb{L}_\xi\} = 1$. Since $\mathbb{L}_\xi$ is a continuous semi-lattice itself, it is possible to define a family $\mathcal{L}_\xi = \mathrm{OFilt}(\mathbb{L}_\xi)$ if all open filters in $\mathbb{L}_\xi$. Then

$$\mathcal{L}_\xi = \{F \cap \mathbb{L}_\xi : F \in \mathcal{L}, \ F \cap \mathbb{L}_\xi \neq \emptyset\}$$

and $F \cap \mathbb{L}_\xi \neq \emptyset$ for $F \in \mathcal{L}$ if and only if $\mathbf{P}\{\xi \in F\} > 0$. The following proposition characterises the supports of infinitely divisible random elements and can be regarded as an abstract variant of Lemma 1.8.

Proposition 1.32. *Let $\mathbb{L}$ be a continuous semi-lattice with a second countable Scott topology and let ξ be an infinitely divisible random variable in $\mathbb{L}$. Let*

$$x = \vee\{y \in \mathbb{L} : \mathbf{P}\{y \leq \xi\} > 0\}. \tag{1.30}$$

Then $\mathbb{L}_\xi = \downarrow x = \{y \in \mathbb{L} : y \leq x\}$ and so $\mathbf{P}\{\xi \leq x\} = 1$.

Proof. Let $F_1, F_2 \in \mathcal{L}$ and assume that $\mathbf{P}(F_1 \cap F_2) = 0$. Using (1.27), we conclude that

$$1 = \mathbf{P}(F_1^c \cup F_2^c) = \mathbf{P}_n(F_1^c \cup F_2^c) \leq \mathbf{P}_n(F_1^c) + \mathbf{P}_n(F_2^c).$$

Therefore,

$$n \leq n(1 - \mathbf{P}(F_1)^{1/n}) + n(1 - \mathbf{P}(F_2)^{1/n}).$$

By letting $n \to \infty$ we deduce that $\mathbf{P}(F_1) = 0$ or $\mathbf{P}(F_2) = 0$. This shows that $\mathcal{L}_\xi$ is a semi-lattice. A general argument from the theory of lattices implies that its dual $\mathbb{L}_\xi$ has a top z. If $y \ll z$, then we may choose $F \in \mathcal{L}$ with $z \in F \subset \uparrow y$ and $F \cap \mathbb{L}_\xi \neq \emptyset$. Then $\mathbf{P}\{y \leq \xi\} > 0$, so that $y \leq x$ and $z \leq x$.

Assume that $\mathbf{P}\{y \leq \xi\} > 0$ for some $y \ll x$. Then $\mathbf{P}\{\xi \in F\} > 0$ for some $F \in \mathcal{L}$ and $F \subset \uparrow y$. This yields $F \cap \mathbb{L}_\xi \neq \emptyset$, i.e. $x \in F$ and $x \leq z$. Thus, the top of $\mathbb{L}_\xi$ is given by (1.30). □

A general infinitely divisible random element in $\mathbb{L}$ can be characterised by reducing the consideration to the lattice $\mathbb{L}_\xi$. The following result is a generalisation of Proposition 1.30.

Theorem 1.33. *Let $\mathbb{L}$ be a continuous semi-lattice with a top and a second countable Scott topology. The formulae*

$$x = \vee\{y \in \mathbb{L} : \mathbf{P}(\uparrow y) > 0\},$$
$$\mathbf{P}(F) = \exp\{-\nu(\downarrow x \setminus F)\}, \quad F \in \mathcal{L},$$

define a bijection between the set of all infinitely divisible probability measures $\mathbf{P}$ on $\mathbb{L}$ and the set of all pairs (x, ν), where $x \in \mathbb{L}$ and ν is a Lévy–Khinchin measure on $\downarrow x$.

The following result generalises Theorem 1.31 and exploits the duality between $\mathbb{L}$ and $\mathcal{L}$.

Theorem 1.34. *Let* **P** *be a probability measure on* $\mathbb{L}$. *Define*

$$\Psi(F) = -\log \mathbf{P}(F), \quad F \in \mathcal{L}, \tag{1.31}$$

and $\mathcal{L}_\Psi = \{F \in \mathcal{L} : \Psi(F) < \infty\}$. *Then* **P** *is infinitely divisible if and only if* $\mathcal{L}_\Psi$ *is a semi-lattice and* $\Psi : \mathcal{L} \mapsto [0, \infty]$ *is completely* $\cap$-*alternating on* $\mathcal{L}_\Psi$. *If also* $\Psi(\mathbb{L}) = 0$ *and* $\Psi(F_n) \to \Psi(F)$ *as* $F_n \uparrow F$ *for* $F, F_1, F_2, \ldots \in \mathcal{L}$, *then there exists a unique infinitely divisible probability measure* **P** *on* $\mathbb{L}$ *satisfying* (1.31).

Note that Ψ extends to a unique locally finite measure Λ on $\sigma(\mathbb{L})$. The measure μ determines a *Poisson point process* $\Pi_\Lambda = \{\eta_1, \eta_2, \ldots\}$ on $\mathbb{L}$, so that the numbers of points in each $F \in \sigma(\mathbb{L})$ is Poisson and the numbers of points in disjoint sets are independent. Then the random element ξ with distribution **P** can be obtained as

$$\xi = x \wedge \bigwedge_{\eta \in \Pi_\Lambda} \eta$$

for x defined by (1.30).

Example 1.35 (Infinite divisibility for unions). Let X be a random closed set in a second countable sober space $\mathbb{E}$ (which is the case if $\mathbb{E}$ is LCHS). Let $\mathcal{F}$ be a lattice of all closed subsets of $\mathbb{E}$ with the reversed inclusion, so that $F_1 \vee F_2 = F_1 \cap F_2$ and $F_1 \wedge F_2 = F_1 \cup F_2$. The infinite divisibility concept in the lattice $\mathcal{F}$ turns into the infinite divisibility concept for unions of random closed sets. Furthermore, (1.30) defines the set

$$H = \cap \{F \in \mathcal{F} : \mathbf{P}\{F \supset X\} > 0\}.$$

If X is infinitely divisible for unions, Proposition 1.32 yields

$$\mathbf{P}\{X \leq H\} = \mathbf{P}\{H \subset X\} = 1,$$

whence H coincides with the set F_X of fixed points of X. Assume that $F_X \neq \mathbb{E}$. It was shown in Section 1.3.4 that $\text{OFilt}(\mathcal{F})$ consists of the families $\mathcal{F}^K$ for $K \in \mathcal{K}$. Theorem 1.34 implies that

$$\Psi(K) = -\log \mathbf{P}\left\{X \in \mathcal{F}^K\right\} = -\log(1 - T_X(K))$$

is a completely alternating capacity on $\mathcal{K}^X$. Conversely, for every pair (H, Ψ), where $H \in \mathcal{F}$, $H \neq \mathbb{E}$ and Ψ is a completely alternating capacity on $\mathcal{K}^H$ with $\Psi(\emptyset) = 0$, there exists an infinitely divisible random closed set X with the capacity functional

$$T_X(K) = \begin{cases} 1 - \exp\{-\Psi(K)\}, & K \cap H = \emptyset, \\ 1, & K \cap H \neq \emptyset, \end{cases} \quad K \in \mathcal{K}. \tag{1.32}$$

Theorem 1.34 says that, for each upper semicontinuous and completely alternating capacity $\Psi : \mathcal{K} \mapsto [0, \infty]$ such that the family $\{K \in \mathcal{K} : \Psi(K) < \infty\}$ is closed under finite non-empty unions, there exists an infinitely divisible random closed set X with the capacity functional given by (1.32).

A random element X in $\mathbb{L}$ is called *self-decomposable* if, for each $t > 0$, there exists a random element X_t such that

$$X \stackrel{d}{\sim} A^t X \vee X_t,$$

where A^t belongs to a predetermined group $\mathbb{A} = \{A^t, t \in \mathbb{R}\}$ of bijections $\mathbb{L} \mapsto \mathbb{L}$ and $A^t X$ and X_t are independent. The bijections are assumed to be order preserving and anti-extensive for $t > 0$, i.e. $A^t x \leq x$ for all $x \in \mathbb{L}$ and $t > 0$. The self-decomposable random elements form a class sandwiched between infinitely divisible and stable laws. However, this concept is not meaningful for random closed sets without some restrictions on the families of their realisations. For instance, let $\mathbb{L}$ be the family $\mathcal{F}$ of all closed subsets of a T_1 space $\mathbb{E}$ (so that all singletons are closed and belong to $\mathcal{F}$). Then the conditions on the transformations A^t imply that A^t leave singletons invariant and so are identical transformations. It is possible to get a non-trivial situation if $\mathbb{E}$ is not a T_1 space (and so has some non-closed singletons) or if $\mathbb{L}$ is a proper subfamily of $\mathcal{F}$, for example, if $\mathbb{L}$ consists of all epigraphs of upper semicontinuous functions, see Section 5.3.1.

2 Weak convergence of normalised unions

2.1 Sufficient conditions

Similar to the classical theory of extreme values, limit theorems for scaled unions of random sets can be formulated using regular variation conditions imposed on the capacity functionals, see Appendix H for information on regularly varying functions. In this section we consider limit theorems for scaled unions of random sets, where union-stable sets appear as weak limits. Define

$$Z_n = X_1 \cup \cdots \cup X_n, \quad n \geq 1,$$

for independent identically distributed random closed sets $X, X_1, X_2, \ldots$ with the common capacity functional T. We study the weak convergence of $a_n^{-1} Z_n$ where $\{a_n, n \geq 1\}$ is a sequence of normalising constants. Naturally, while dealing with unions of random sets we use methods similar to the theory of extreme values. The function $1 - T(xK)$, $x > 0$, plays in our consideration the same role as the cumulative distribution function in limit theorems for extremes, but this function is no longer monotone and may not converge to 1 as $x \to \infty$.

We first consider the case $a_n \to \infty$ as $n \to \infty$. Then the limiting random set Z has the origin as a fixed point, so that Corollary 1.13 yields $\alpha < 0$ for the parameter α of Z. For each compact set K, define

$$a_n(K) = \sup \{x > 0 : T(xK) \geq 1/n\}, \tag{2.1}$$

where $\sup \emptyset = 0$. Introduce the function $v_K(x) = T(xK)$ for $x \geq 0$ and the family $\mathcal{V}$ of compact sets by

$$\mathcal{V} = \left\{ K \in \mathcal{K} : \liminf_{x \to \infty} T(xK) = 0 \right\}.$$

2 Weak convergence of normalised unions

Theorem 2.1 (Weak convergence of scaled unions). *Assume that for each K from $\mathcal{V}$ there exists a (possibly infinite) limit of $a_n(K)/a_n$ and $v_K(x)$ is a regularly varying function with a negative exponent α (i.e. $v_K \in RV_\alpha$). Then $a_n^{-1} Z_n$ weakly converges to the union-stable random closed set Z with the capacity functional*

$$T_Z(K) = 1 - \exp\{-\Psi_Z(K)\}, \qquad (2.2)$$

where

$$\Psi_Z(K) = \begin{cases} \lim(a_n(K)/a_n)^{-\alpha}, & K \in \mathcal{V}, \\ \infty, & \text{otherwise}. \end{cases} \qquad (2.3)$$

The following lemma can be derived from the representation (H.3) of slowly varying functions, see Molchanov [398, Lemma 1.2].

Lemma 2.2. *Let $f \in RV_\alpha$ with $\alpha < 0$ and let $g(x)$ be a non-negative function such that $xg(x) \to \infty$ as $x \to \infty$ and $g(x)$ has a (possibly infinite) limit as $x \to \infty$. Then*

$$\lim_{x \to \infty} \frac{f(xg(x))}{f(x)} = \lim_{x \to \infty} g(x)^\alpha.$$

Proof of Theorem 2.1. If T is the capacity functional of the random set X_1, then $a_n^{-1} Z_n$ has the capacity functional T_n given by

$$T_n(K) = 1 - (1 - T(a_n K))^n. \qquad (2.4)$$

Step I. If $K \notin \mathcal{V}$, then $T_n(K) \to 1 = T_Z(K)$ as $n \to \infty$, i.e. (2.3) holds. Further suppose that $K \in \mathcal{V}$. It follows from (2.4) that

$$T_Z(K) = \lim_{n \to \infty} T_n(K) = 1 - \exp\{-\Psi_Z(K)\}$$

if $nT(a_n K) \to \Psi_Z(K)$ as $n \to \infty$. Note that $\Psi_Z(K)$ can be infinite, which is the case if K hits the set of fixed points of Z.

Step II. Suppose that

$$\limsup_{x \to \infty} T(xK) \geq \varepsilon > 0.$$

Then $a_n(K) = \infty$ for all $n \geq n_0$, i.e. $a_n(K)/a_n = \infty$ for $n \geq n_0$. Let $\lambda > 1$ be specified. Then, for each $n \geq n_0$, there exists $\lambda_n > \lambda$ such that $T(a_n \lambda_n K) \geq 1/n$. Hence

$$\lim_{n \to \infty} nT(a_n K) = \lim_{n \to \infty} n \frac{T(a_n \lambda_n K) v_K(a_n)}{v_K(a_n \lambda_n)}$$
$$\geq \liminf_{n \to \infty} \lambda_n^{-\alpha} \geq \lambda^{-\alpha}.$$

Letting λ go to infinity yields $nT(a_n K) \to \infty$ as $n \to \infty$. Hence $T_Z(K) = 1$, i.e. (2.3) holds with $\Psi_Z(K) = \infty$.

Step III. It remains to consider the case $v_K(x) \to 0$ as $x \to \infty$. Then $a_n(K) < \infty$ for all $n \geq 1$. If the sequence $\{a_n(K), n \geq 1\}$ is bounded, then $T(xK) = 0$ for all sufficiently large x, so that $T_n(K) \to 0$ as $n \to \infty$ and (2.3) holds, since

$$\Psi_Z(K) = \lim_{n \to \infty} (a_n(K)/a_n)^{-\alpha} = 0.$$

Assume that $v_K(x) \to 0$ as $x \to \infty$ and $a_n(K) \to \infty$ as $n \to \infty$. Lemma 2.2 yields

$$\lim_{n \to \infty} \frac{v_K(a_n)}{v_K(a_n(K))} = \lim_{n \to \infty} (a_n(K)/a_n)^{-\alpha}. \qquad (2.5)$$

Let us prove that $nT(a_n(K)K) \to 1$ as $n \to \infty$. For arbitrary $n \geq 1$, choose $\{x_m, m \geq 1\}$ such that

$$a_n(K) - 1/m \leq x_m \leq a_n(K)$$

and $T(x_m K) \geq 1/n$ for all $m \geq 1$. The upper semicontinuity of T implies that

$$1/n \leq \lim_{m \to \infty} T(x_m K) \leq T(a_n(K)K).$$

Thus $nT(a_n(K)K) \geq 1$ for all $n \geq 1$. It follows from (2.1) that $T(a_n(K)\lambda K) \leq 1/n$ for $\lambda > 1$. Since v_K is regularly varying,

$$1 \leq \lim_{n \to \infty} nT(a_n(K)K) = \lim_{n \to \infty} nT(a_n(K)\lambda K) \frac{v_K(a_n(K))}{v_K(\lambda a_n(K))} \leq \lambda^{-\alpha}.$$

Letting λ go to 1 yields $nT(a_n(K)K) \to 1$ as $n \to \infty$. From (2.5) we get

$$\lim_{n \to \infty} nT(a_n K) = \lim_{n \to \infty} nT(a_n(K)K) \frac{v_K(a_n)}{v_K(a_n(K))}$$
$$= \lim_{n \to \infty} (a_n(K)/a_n)^{-\alpha}.$$

Thus (2.3) holds for each compact set K. It is easy to verify that, for each $K \in \mathcal{V}$ and $s > 0$, the set sK belongs to $\mathcal{V}$ and also

$$\Psi_Z(sK) = \lim_{n \to \infty} (a_n(sK)/a_n)^{-\alpha} = s^\alpha \Psi_Z(K).$$

Therefore, Z is union-stable with parameter α. The limiting random set Z in Theorem 2.1 has the origin as a fixed point, since $a_n(B_r(0)) = \infty$ for each $r > 0$. □

We have proved the pointwise convergence of the capacity functional of $a_n^{-1} Z_n$ to the capacity functional of Z on *all* compact sets and not only on those compact sets which are continuous for the limiting capacity functional T_Z. As in Section 1.6.1, it is easy to see that $a_n^{-1} Z_n$ converges weakly if the conditions of Theorem 2.1 hold for the family $\mathcal{V} \cap \mathcal{A}$ instead of $\mathcal{V}$ for any separating class $\mathcal{A} \subset \mathcal{K}$.

If the normalising factor a_n converges to a positive constant a, then $a_n^{-1} X_n$ converges almost surely in the Fell topology to a deterministic set M equal to the complement of

$$\bigcup \left(\{ aB_r(x) : T(B_r(x)) = 0, r > 0, x \in \mathbb{R}^d \} \right).$$

The case $a_n \to 0$ as $n \to \infty$ can be reduced to Theorem 2.1 by using the inverse transform defined in (1.18).

Different scaling factors

Consider a more general normalisation scheme for unions of random sets which allows for different normalising factors along different axes. For $y = (y_1, \ldots, y_d)$ from $\text{Int}\,\mathbb{R}_+^d = (0, \infty)^d$ and $K \subset \mathbb{R}^d$ define

$$y \circ K = \{(y_1 x_1, \ldots, y_d x_d) : (x_1, \ldots, x_d) \in K\}. \tag{2.6}$$

Let $Z_n = X_1 \cup \cdots \cup X_n$ for i.i.d. random closed sets $X_1, X_2, \ldots$ with the common capacity functional T. Consider a sequence $\{a_n = (a_{n1}, \ldots, a_{nd}), n \geq 1\}$. Define

$$q_n(K) = \sup\{t \geq 0 : T((ta_n) \circ K) \geq n^{-1}\},$$

$$\mathcal{V} = \{K \in \mathcal{K} : \liminf_{t \to \infty} T(ty \circ K) = 0 \text{ for every } y \in (0, \infty)^d\}.$$

Note that $q_n(K)$ is the analogue of $a_n(K)/a_n$ from Theorem 2.1. The following limit theorem for $a_n^{-1} \circ Z_n$ and can be proved similarly to Theorem 2.1.

Theorem 2.3. *Assume that, for every $K \in \mathcal{V}$, $q_n(K)$ has a (possibly infinite) limit as $n \to \infty$. Let $T(y \circ K)$ (as a function of y) belong to $\mathrm{RV}_\alpha^u(\mathbb{R}_+^d)$ with $\alpha < 0$. If $\lim_{n \to \infty} a_n \|a_n\|^{-1}$ exists and belongs to $\text{Int}\,\mathbb{R}_+^d$, then $a_n^{-1} \circ Z_n$ converges weakly to a random closed set Z with the capacity functional*

$$T_Z(K) = \begin{cases} 1 - \exp\{-\lim(q_n(K))^\alpha\}, & K \in \mathcal{V}, \\ 1, & \text{otherwise}. \end{cases}$$

The limiting random set Z is stable in the following sense. For each $n \geq 1$ and i.i.d. copies $Z_1, \ldots, Z_n$ of Z there exists $a_n \in (0, \infty)^d$ such that

$$a_n \circ Z \overset{d}{\sim} Z_1 \cup \cdots \cup Z_n.$$

Open problem 2.4. Prove analogues of limit theorems for more general normalisation schemes than purely multiplicative (e.g. $a_n^{-1} Z_n + b_n$ or $A_n Z_n$, where $A_n, n \geq 1$, are linear operators). Similar to the characterisation problem, the main obstacle here is the lack of an analogue of the convergence of types results for random sets.

2.2 Necessary conditions

It is well known that regular variation conditions are both sufficient and necessary in limit theorems for extremes of random variables, see Galambos [186, Th. 2.4.3]. For unions of random sets, the situation is different to some extent, since the pointwise convergence of $T_n(x K)$ for all positive x no longer implies the uniform convergence (recall that $T_n(x K)$ is not necessarily monotone as a function of x).

Theorem 2.5 (Necessary condition for weak convergence of scaled unions). *Let the capacity functional T_n of $a_n^{-1} Z_n$ converge uniformly on $\mathcal{K}$ to the capacity functional T_Z of a union-stable random closed set Z with parameter α. Consider a compact set K such that $T_Z(K) < 1$. If $\alpha < 0$, then $K \in \mathcal{V}$ and the function $v_K(x)$ is regularly varying with exponent α.*

Proof. Let T be the capacity functional of X_1. Denote $F(x) = 1 - T(xK)$ for $x > 0$. Since $T_Z(K) < 1$, $\Psi_Z(K) = -\log(1 - T_Z(K))$ is finite. It follows from (2.4) and the condition of the theorem that

$$F^n(a_n x) \to \exp\{-\Psi_Z(xK)\} = \exp\{-x^\alpha \Psi_Z(K)\} = \tilde{F}(x) \quad \text{as } n \to \infty \qquad (2.7)$$

uniformly for $x > 0$. If $\alpha < 0$, then $\tilde{F}(x)$ is the distribution function of a max-stable random variable of type I, see (1.2). Denote

$$F_*(x) = \inf_{x < t} F(t), \quad F^*(x) = \sup_{0 < t \leq x} F(t), \quad x > 0.$$

The uniform convergence in (2.7) yields

$$(F^*(a_n x))^n \to \tilde{F}(x), \quad (F_*(a_n x))^n \to \tilde{F}(x) \quad \text{as } n \to \infty \qquad (2.8)$$

for each positive x. The functions F^* and F_* are right-continuous. Since $T_Z(K) < 1$, $\Psi(K)$ is finite, whence $F^*(\infty) = F_*(\infty) = 1$, i.e. F^* and F_* are distribution functions and K belongs to $\mathcal{V}$. From (2.8) and the necessary conditions in the limit theorem for the maximum of random variables (see Galambos [186, Th. 2.4.3]) we derive that the functions $1 - F_*(x)$ and $1 - F^*(x)$ are regularly varying with exponent α. Evidently,

$$1 - F^*(x) \leq 1 - F(x) \leq 1 - F_*(x). \qquad (2.9)$$

For some $s > 1$, let $n(k) = [s^k]$ be the integer part of s^k, $k \geq 1$. Then, for all sufficiently large t, there exists k such that $a_{n(k)} \leq t < a_{n(k+1)}$ and also

$$F_*(a_{n(k)}) \leq F_*(t) \leq F_*(a_{n(k+1)}),$$
$$F^*(a_{n(k)}) \leq F^*(t) \leq F^*(a_{n(k+1)}).$$

Hence

$$\frac{\log F_*(a_{n(k+1)})}{\log F^*(a_{n(k)})} \leq \frac{\log F_*(t)}{\log F^*(t)} \leq \frac{\log F_*(a_{n(k)})}{\log F^*(a_{n(k+1)})}.$$

It follows from (2.8) that

$$\frac{\log F_*(t)}{\log F^*(t)} \to 1 \quad \text{as } t \to \infty,$$

whence

$$\lim_{x \to \infty} \frac{1 - F_*(x)}{1 - F^*(x)} = 1.$$

Now (2.9) and the regular variation of F^* and F_* imply that $v_K(x) = 1 - F(x)$ is regularly varying. □

The following example shows that the capacity functional T_n in Theorem 2.5 may converge pointwisely but not uniformly.

2 Weak convergence of normalised unions

Example 2.6. Let $Y \subset (-\infty, 0]$ be a union-stable random closed set with $\alpha = -1$. Put $M = \cup_{k\geq 2} M_k$ for a sequence $M_2, M_3, \ldots$ of independent random closed sets such that $M_k = [k, k+1/2]$, with probability k^{-3}, $k \geq 2$ and $M_k = \emptyset$ otherwise. Let $\{X_n, n \geq 1\}$ be i.i.d. copies of the random closed set $X = Y \cup M$. Put $a_n = n$ and $Z_n = X_1 \cup \cdots \cup X_n$. The capacity functional of $n^{-1} Z_n$ is

$$T_n(K) = 1 - (1 - T_Y(nK))^n (1 - T_M(nK))^n$$
$$= 1 - (1 - T_Y(K))(1 - T_M(nK))^n.$$

Let $K \subset [a, b]$ for $a > 0$. Then

$$1 - T_M(nK) \geq \left(1 - [na]^{-3}\right)^{[n(b-a)]}.$$

Hence $(1 - T_M(nK))^n \to 1$ as $n \to \infty$, whence $T_n(K) \to T_Y(K)$ as $n \to \infty$, i.e. $n^{-1} Z_n \xrightarrow{d} Y$ as $n \to \infty$. However, the corresponding capacity functionals do not converge uniformly on the family $\{xK, x > 0\}$ even for $K = \{1\}$. Indeed, $T_Y(\{x\}) = 0$ for all $x > 0$ and also

$$\sup_{x>0} |T_n(\{x\}) - T_Y(\{x\})| \geq T_n(\{x_n\})$$
$$= 1 - \left(1 - [nx_n]^{-3}\right)^n \to 1 - e^{-1} \quad \text{as } n \to \infty$$

for $x_n = [n^{1/3}] n^{-1}$, $n \geq 1$. Note that the function $T_X(xK)$ is not regularly varying in this example.

It is an open problem to derive necessary and sufficient conditions for the weak convergence of scaled unions of random closed sets. It follows from Example 2.6 that the regular variation condition for the function $v_K(x)$ is too restrictive. But it cannot be weakened, since for the random set $X = (-\infty, \xi]$ the corresponding regular variation condition is necessary and sufficient. It should be noted that Example 2.6 is quite artificial, since, in fact, the limiting random set is degenerated on $[0, \infty)$.

Open problem 2.7. Find necessary and sufficient conditions for the pointwise convergence of the capacity functions for normalised unions of random closed sets. Is it possible to construct an example of a random set X such that $T(xK)$ is regularly varying as a function of x, but the capacity functional of $a_n^{-1} Z_n$ does not converge uniformly to T_Z?

Let us show that the weak convergence of normalised unions together with the regular variation property for the sequence of normalising constants imply that the limiting random closed set is union-stable.

Theorem 2.8 (Union-stability of weak limits). Let $\{X_n, n \geq 1\}$ be a sequence of i.i.d. random closed sets such that $a_n^{-1} Z_n = a_n^{-1}(X_1 \cup \cdots \cup X_n)$ converges in distribution to a non-trivial random closed set Z, where either $a_n \to 0$ or $a_n \to \infty$. If there exists $\alpha \neq 0$ such that

$$\lim_{n\to\infty} \frac{a_{[nt]}}{a_n} = t^{-1/\alpha} \tag{2.10}$$

for every $t \in \mathbb{R}$, then Z is union-stable with parameter α.

Proof. Note that

$$\Psi_Z(sK) = -\lim_{n\to\infty} \log \mathbf{P}\left\{a_n^{-1} Z_n \cap sK = \emptyset\right\} = -\lim_{n\to\infty} \log(1 - T_{X_1}(a_n sK))^n.$$

Without loss of generality K can be assumed regular closed. Then

$$a_{[ns^{-\alpha}]} K^{-\delta_n} \subset a_{ns} K \subset a_{[ns^{-\alpha}]} K^{\delta_n}$$

with $\delta_n \downarrow 0$, where K^δ is the δ-envelope of K and $K^{-\delta}$ is the inner envelope of K. Each $K \in \mathfrak{S}_Z$ (the continuity set for Z) can be approximated by $K^{-\delta_n}$ and K^{δ_n}, whence

$$\Psi_Z(sK) = -\lim_{n\to\infty} \log(1 - T_X(a_{[ns^{-\alpha}]} K))^n$$
$$= -\frac{n}{[ns^{-\alpha}]} \lim_{n\to\infty} \log(1 - T_X(a_{[ns^{-\alpha}]} K))^{[ns^{-\alpha}]} = s^\alpha \Psi_Z(K).$$

By the semicontinuity of Ψ, this equality holds for all $K \in \mathcal{K}^Z$. □

If $X_n = \{\xi_n\}$, $n \geq 1$, are random singletons, then the weak convergence of $a_n^{-1} Z_n$ already implies (2.10), so that the limit is union-stable. To see this, it suffices to apply the necessary conditions for extremes of random variables (see Galambos [186, Sec. 2.4]) to the sequence of random variables $\alpha_n = \inf\{s > 0 : \xi_n \in sF\}$ (or $\alpha_n = \sup\{s > 0 : \xi_n \in sF\}$), where $F = (-\infty, x_1] \times \cdots \times (-\infty, x_d]$. This idea can be extended for random closed sets using their support functions.

Theorem 2.9 (Regular valuation of normalising constants). *Let $X, X_1, X_2, \ldots$ be a sequence of i.i.d. random closed sets such that $a_n^{-1} Z_n$ weakly converges to a non-trivial random closed set Z. Assume that one of the following conditions holds:*
  (i) $a_n \to \infty$ and $\max(h(X, u), h(Z, u)) < \infty$ a.s. for some $u \neq 0$.
  (ii) $a_n \to 0$ and $T_X(\{0\}) = T_Z(\{0\}) = 0$.
Then (2.10) holds and Z is union-stable.

Proof.
  (i) $\alpha_n = h(X_n, u)$ is a sequence of i.i.d. almost surely finite random variables. Furthermore, $h(a_n^{-1} Z_n, u) = a_n^{-1} \max(\alpha_1, \ldots, \alpha_n)$. By Galambos [186, Th. 2.4.3], (2.10) holds, so that Z is union-stable by Theorem 2.8.
  (ii) Let $\alpha_n = \rho(X_n, 0)$. Then $\rho(a_n^{-1} Z_n, 0) = a_n^{-1} \min(\alpha_1, \ldots, \alpha_n)$ and the result follows from Galambos [186, Th. 2.4.4]. □

Open problem 2.10. Find general conditions under which $a_n^{-1} Z_n \xrightarrow{d} Z$ for a non-trivial random closed set Z implies $a_{[nt]}/a_n \to t^\alpha$ as $n \to \infty$.

2.3 Scheme of series for unions of random closed sets

Let $\{X_{nj}, n \geq 1, j \in J_n\}$ be a family of random closed sets, where J_n is a finite set for each $n \geq 1$. For instance, a triangular array corresponds to $J_n = \{1, \ldots, n\}$. Note that J_n is allowed to be infinite if $\cup_{j \in J_n} X_{nj}$ is closed almost surely.

Definition 2.11 (Null-array). Say that $\{X_{nj}\}$ is a *null-array* if, for each $n \geq 1$, the random closed sets X_{nj}, $j \in J_n$, are independent and

$$\sup_{j \in J_n} T_{X_{nj}}(K) \to 0 \quad \text{as } n \to \infty$$

for all $K \in \mathcal{K}$.

Theorem 2.12 (Limit theorem in the scheme of series). Let $\{X_{nj}\}$ be a null-array of random closed sets. Put $X_n = \cup_{j \in J_n} X_{nj}$, $n \geq 1$. If X_n converges in distribution to a random closed set Z, then Z is infinitely divisible for unions and there exists a (possibly infinite) limit

$$\lim_{n \to \infty} \sum_{j \in J_n} T_{X_{nj}}(K) = -\log(1 - T_Z(K))$$

for all $K \in \mathfrak{S}_Z$. Conversely, if there exists a capacity Ψ_Z on $\mathcal{K}$ such that

$$\Psi_Z(\text{Int } K) \leq \liminf_{n \to \infty} \sum_{j \in J_n} T_{X_{nj}}(K) \leq \limsup_{n \to \infty} \sum_{j \in J_n} T_{X_{nj}}(K) \leq \Psi_Z(K)$$

for all $K \in \mathcal{K}$ (allowing infinite values for Ψ_Z and the limits), then X_n converges in distribution to a random closed set Z with the capacity functional $T_Z(K) = 1 - \exp\{-\Psi_Z(K)\}$.

Proof. We will use the inequalities

$$x \leq -\log(1 - x) \leq c_1 x$$

and

$$0 \leq -x - \log(1 - x) \leq c_1 x^2/2$$

valid for all $x \in [0, c]$ for some $c \in (0, 1)$ with $c_1 = 1/(1 - c)$. Since $\{X_{nj}\}$ is a null-array, assume that $T_{X_{nj}}(K) \leq c$ for some $c \in (0, 1)$. Then

$$\sum_{j \in J_n} T_{X_{nj}}(K) \leq \sum_{j \in J_n} -\log(1 - T_{X_{nj}}(K))$$
$$= -\log(1 - T_{X_n}(K)) \leq c_1 \sum_{j \in J_n} T_{X_{nj}}(K).$$

Furthermore,

$$0 \le -\sum_{j \in J_n} T_{X_{nj}}(K) - \log(1 - T_{X_n}(K)) \le \frac{c_1}{2} \sum_{j \in J_n} T_{X_{nj}}(K)^2$$
$$\le \frac{c_1}{2} \sum_{j \in J_n} T_{X_{nj}}(K) \sup_{j \in J_n} T_{X_{nj}}(K).$$

Therefore, $\sum_j T_{X_{nj}}(K) \to a$ is equivalent to $T_{X_n}(K) \to 1 - e^{-a}$. The proof is finished by referring to Theorem 1.6.5. □

It is possible to strengthen Theorem 2.12 by replacing the family $\mathcal{K}$ with a separating class of compact sets. For non-identically distributed summands the limiting distribution corresponds to a union-infinitely-divisible random closed set, which is not necessarily union-stable.

Limit theorems for unions of random sets can be formulated within the framework of general lattice-valued elements, see Norberg [434]. This formulation avoids the concepts of regular variation and normalisation by scaling and instead formulates the limit theorem in the scheme of series similarly to Theorem 2.12.

It should be pointed out that all results on the pointwise convergence of capacity functionals remain true for unions of random sets in Banach spaces. However, they do not imply the weak convergence, since distributions of random closed sets in infinite-dimensional Banach spaces are no longer determined by the corresponding capacity functionals.

3 Convergence with probability 1

3.1 Regularly varying capacities

Definition

We keep the setting of Section 2.1 and find conditions which ensure that $a_n^{-1} Z_n$ converges almost surely as $n \to \infty$ to a deterministic limit. The almost sure convergence of random sets in a LCHS space is defined in the Fell topology. The special case of the convergence to a deterministic limit was characterised in Proposition 1.6.17. To apply this proposition, fix a pre-separating class $\mathcal{A} \subset \mathcal{K}$ that satisfies the following assumption.

Assumption 3.1. A pre-separating class $\mathcal{A}$ consists of regularly closed sets and $cK \in \mathcal{A}$ for each $c > 0$ and $K \in \mathcal{A}$. Denote $\mathcal{A}' = \{\text{Int } K : K \in \mathcal{A}\}$.

Assume that $R: \mathcal{A} \mapsto [0, \infty]$ is a non-increasing lower semicontinuous capacity without any restrictions on signs of higher differences inherent to alternating capacities.

Definition 3.2 (Regularly varying capacity). The capacity R is said to be *regularly varying* on $\mathcal{A}$ with the limiting capacity Θ and exponent β (notation $R \in \text{RV}_{\beta,g,\Theta}(\mathcal{A})$) if

3 Convergence with probability 1   271

$$\lim_{t\to\infty} \frac{R(tK)}{g(t)} = \Theta(K) \tag{3.1}$$

for each $K \in \mathcal{A}$, where $g: (0, \infty) \mapsto (0, \infty)$ is a regularly varying function of exponent β and the limiting capacity $\Theta(K)$ is allowed to take zero or infinite values but is not identically equal to zero or infinity.

Limiting capacity

It is easily seen that Θ is a non-increasing functional on $\mathcal{A}$ and $\Theta(cK) = c^\beta \Theta(K)$ for each $c > 0$ and $K \in \mathcal{A}$. In the following we mostly consider the case of $R(K)$ given by $-\log T(K)$ for a capacity functional T.

Lemma 3.3. *Let T be the capacity functional of a random closed set X and let*

$$R(K) = -\log T(K)$$

belong to $\mathrm{RV}_{\beta,g,\Theta}(\mathcal{A})$ *with a positive β. Then, for any K_1 and K_2 from $\mathcal{A}$, the limit (3.1) exists for the set $K = K_1 \cup K_2$ and*

$$\Theta(K) = \min(\Theta(K_1), \Theta(K_2)). \tag{3.2}$$

Proof. It is evident that

$$\limsup_{t\to\infty} \frac{R(tK)}{g(t)} \leq \lim_{t\to\infty} \frac{R(tK_i)}{g(t)} \leq \min(\Theta(K_1), \Theta(K_2)), \quad i = 1, 2.$$

If either $\Theta(K_1)$ or $\Theta(K_2)$ vanishes then (3.2) is evident. Let both $\Theta(K_1)$ and $\Theta(K_2)$ be finite and strictly positive. Then, for any $\varepsilon > 0$ and sufficiently large t,

$$T(tK_i) \leq \exp\{-g(t)\Theta(K_i)(1-\varepsilon)\}, \quad i = 1, 2. \tag{3.3}$$

The subadditivity of T yields

$$\liminf_{t\to\infty} \frac{R(tK)}{g(t)} \geq \liminf_{t\to\infty} -\frac{\log(T(tK_1) + T(tK_2))}{g(t)}$$

$$\geq \liminf_{t\to\infty} -\frac{\log\left(2\exp\{-g(t)\min(\Theta(K_1), \Theta(K_2))(1-\varepsilon)\}\right)}{g(t)}$$

$$= \min(\Theta(K_1), \Theta(K_2))(1-\varepsilon).$$

Hence (3.2) holds.

If $\Theta(K_1) = \infty$, then (3.3) is replaced with $T(tK_1) \leq \exp\{-g(t)c\}$, which holds for any $c > 0$ and sufficiently large t. Then, for $c > \Theta(K_2)$,

$$\liminf_{t\to\infty} \frac{R(tK)}{g(t)} \geq \liminf_{t\to\infty} -\frac{\log\left(\exp\{-g(t)c\} + \exp\{-g(t)\Theta(K_2)(1-\varepsilon)\}\right)}{g(t)}$$

$$\geq \liminf_{t\to\infty} -\frac{\log\left(2\exp\{-g(t)\Theta(K_2)(1-\varepsilon)\}\right)}{g(t)}$$

$$= \Theta(K_2)(1-\varepsilon)$$

$$= \min(\Theta(K_1), \Theta(K_2))(1-\varepsilon). \qquad \square$$

272   4 Unions of Random Sets

It follows from Lemma 3.3 that Θ is a *minitive* capacity, compare with maxitive capacities introduced in (1.1.21). It is tempting to deduce that $\Theta(K)$ is equal to the infimum of $\Theta(\{x\})$ for $x \in K$. However, this is wrong in general, since $\mathcal{A}$ does not necessarily contain singletons and $\Theta(\{x\})$ may be infinite for all x.

Definition 3.4 (Strictly increasing functional). The functional Θ is said to be strictly decreasing on $\mathcal{A}$ if $\Theta(K_1) > \Theta(K)$ for every $K, K_1 \in \mathcal{A}$ such that $K_1 \subset \text{Int } K$ and $\Theta(K) < \infty$.

The limiting function Θ determines the following closed set in $\mathbb{R}^d$

$$Z(\Theta; \mathcal{A}) = \mathbb{R}^d \setminus \bigcup \{\text{Int } K : K \in \mathcal{A}, \ \Theta(K) > 1\}.$$

The inequality $\Theta(sK) = s^\beta \Theta(K) > \Theta(K)$ for $s > 1$ yields $sZ(\Theta; \mathcal{A}) \subset Z(\Theta; \mathcal{A})$ for all $s \geq 1$ if Θ is the limiting capacity in (3.1).

3.2 Almost sure convergence of scaled unions

Unions of random sets

The following theorem provides a sufficient condition for the almost sure convergence of scaled normalised unions of i.i.d. random closed sets $X, X_1, X_2, \ldots$. For a compact set K denote

$$\hat{K} = \cup \{sK : s \geq 1\}.$$

It is evident that $\hat{K}$ is closed and $s\hat{K} \subset \hat{K}$ for all $s \geq 1$.

Theorem 3.5 (Almost sure convergence in the Fell topology). *Let X be a random closed set with the capacity functional T. Define the capacity R with possibly infinite values by $R(K) = -\log T(K)$ for K from a pre-separating class $\mathcal{A}$ satisfying Assumption 3.1. Assume that $R \in \text{RV}_{\beta,g,\Theta}(\mathcal{A})$ with $\beta > 0$ and a strictly monotone capacity Θ on $\mathcal{A}$ such that*

$$\lim_{t \to \infty} \frac{R(t\hat{K})}{g(t)} = \Theta(\hat{K}) = \Theta(K) \tag{3.4}$$

for each $K \in \mathcal{A}$. Since $\beta > 0$, we can define a_n to satisfy $g(a_n) \sim \log n$ as $n \to \infty$. Then $a_n^{-1} Z_n = a_n^{-1}(X_1 \cup \cdots \cup X_n)$ converges in the Fell topology (or in the Painlevé–Kuratowski sense) to $Z(\Theta; \mathcal{A})$.

Proof. Let K belong to $\mathcal{A}$ and miss $Z(\Theta; \mathcal{A})$. Then K is covered by the sets $\text{Int } L$ for $L \in \mathcal{A}$ with $\Theta(L) > 1$. By compactness, K is covered by a finite collection $\text{Int } L_1, \ldots, \text{Int } L_m$ of such sets. It follows from (3.1) and the choice of a_n that

$$\lim_{n \to \infty} \frac{R(a_n L_i)}{\log n} = \Theta(L_i).$$

Lemma 3.3 yields

3 Convergence with probability 1   273

$$\lim_{n\to\infty} \frac{R(a_n K)}{\log n} = \Theta(K) \geq \min_{1\leq i\leq m} \Theta(L_i) = a > 1.$$

It follows from (3.4) that

$$\lim_{n\to\infty} \frac{R(a_n \hat{K})}{\log n} = \Theta(\hat{K}) = \Theta(K) > a.$$

Pick $\zeta > 0$ such that $a - \zeta > 1$. Then

$$T(a_n \hat{K}) \leq n^{-(a-\zeta)} \tag{3.5}$$

for all sufficiently large n. Note that

$$\mathbf{P}\left\{a_n^{-1} Z_n \cap \hat{K} \neq \emptyset \text{ i.o.}\right\} = \mathbf{P}\left\{X_{i_n} \cap a_n \hat{K} \neq \emptyset \text{ i.o.}\right\},$$

where i.o. abbreviates infinitely often and $1 \leq i_n \leq n$ for $n \geq 1$. It is easy to show that the sequence $\{i_n, n \geq 1\}$ is unbounded. Since $a_n \hat{K} \subset a_{n+1} \hat{K}$ for $n \geq 1$,

$$\mathbf{P}\left\{a_n^{-1} Z_n \cap \hat{K} \neq \emptyset \text{ i.o.}\right\} = \mathbf{P}\left\{(X_1 \cup \cdots \cup X_n) \cap a_n \hat{K} \neq \emptyset \text{ i.o.}\right\}$$
$$= \mathbf{P}\left\{X_{i_n} \cap a_n \hat{K} \neq \emptyset \text{ i.o.}\right\}$$
$$\leq \mathbf{P}\left\{X_n \cap a_n \hat{K} \neq \emptyset \text{ i.o.}\right\}.$$

The right-hand side vanishes because of (3.5) and the Borel–Cantelli lemma, since

$$\sum_{n=1}^{\infty} \mathbf{P}\left\{X_n \cap a_n \hat{K} \neq \emptyset\right\} = \sum_{n=1}^{\infty} T(a_n \hat{K}) \leq \sum_{n=1}^{\infty} n^{-(a-\zeta)} < \infty.$$

If $x \in (G \cap Z(\Theta; \mathcal{A}))$ with $G \in \mathcal{A}'$, then

$$G \not\subset \bigcup \{\text{Int } K : \Theta(K) > 1, K \in \mathcal{A}\}.$$

Choose an open neighbourhood $U(x) \subset G$ and pick K and K_1 from $\mathcal{A}$ such that

$$U(x) \subset K_1 \subset \text{Int } K \subset K \subset G.$$

If $\Theta(K) \geq 1$, then $\Theta(K_1) > 1$, since Θ is strictly monotone. Hence $x \in \text{Int } K_1$ and $\Theta(K_1) > 1$, so that $x \notin Z(\Theta; \mathcal{A})$. Thus, $\Theta(K) = a < 1$ and $K \cap Z(\Theta; \mathcal{A}) \neq \emptyset$. Clearly,

$$\mathbf{P}\left\{a_n^{-1} Z_n \cap G = \emptyset \text{ i.o.}\right\} \leq \mathbf{P}\left\{a_n^{-1} Z_n \cap K = \emptyset \text{ i.o.}\right\}.$$

Pick $\zeta > 0$ such that $a + \zeta < 1$. Then $T(a_n K) \geq n^{-(a+\zeta)}$ for all sufficiently large n, whence

$$\mathbf{P}\{(X_1 \cup \cdots \cup X_n) \cap a_n K = \emptyset\} = (1 - T_X(a_n K))^n$$
$$\leq \exp\{-n T_X(a_n K)\}$$
$$\leq \exp\{-n^{1-(a+\zeta)}\}.$$

Since $\delta = 1 - (a + \zeta) > 0$,

$$\sum_{n=1}^{\infty} \mathbf{P}\left\{a_n^{-1} Z_n \cap K = \emptyset\right\} \leq \sum_{n=1}^{\infty} \exp\{-n^\delta\} < \infty.$$

Hence $\mathbf{P}\{a_n^{-1} Z_n \cap G = \emptyset \text{ i.o.}\} = 0$. An application of Proposition 1.6.17 concludes the proof. □

Theorem 3.5 can be formulated for $a_n \to 0$ if the regular variation condition (3.1) is appropriately modified with g being regularly varying at zero. The following result concerns the convergence in the Hausdorff metric as $a_n \to \infty$.

Corollary 3.6. *Let X be a random compact set which satisfies the conditions of Theorem 3.5. Assume that the limit in (3.1) exists and is different from 0 and infinity for a convex compact set K_0 such that $0 \in \text{Int } K_0$ and $K_0^c \in \mathcal{A}'$. Then $\rho_H(a_n^{-1} Z_n, Z(\Theta; \mathcal{A})) \to 0$ a.s. as $n \to \infty$.*

Proof. We have to check that $\sup_{n \geq 1} \|a_n^{-1} Z_n\|$ is a.s. bounded, see also Davis, Mulrow and Resnick [119]. It suffices to show that

$$\mathbf{P}\left\{\sup_{n \geq 1} \inf\{t > 0: a_n^{-1} Z_n \subset t K_0\} < \infty\right\}$$

$$= \mathbf{P}\left\{\sup_{n \geq 1} a_n^{-1} \inf\{t > 0: X_1 \cup \cdots \cup X_n \subset t K_0\} < \infty\right\} = 1.$$

If $\zeta_n = \inf\{t > 0: X_1 \cup \cdots \cup X_n \subset t K_0\}$ for $n \geq 1$, then $\zeta_n = \max(\eta_1, \ldots, \eta_n)$ for i.i.d. random variables $\eta_1, \ldots, \eta_n$ with the common distribution $\mathbf{P}\{\eta_1 > y\} = T(y K_0^c)$. Hence $-\log(\mathbf{P}\{\eta_1 > y\})$ is a regularly varying function. This suffices for the almost sure stability of $\sup_{n \geq 1} a_n^{-1} \zeta_n$, see Resnick and Tomkins [484]. □

Application to random singletons

Let $X = \{\xi\}$ be a random singleton which almost surely has all non-negative coordinates. If $\mathcal{A}$ is the class of all bounded parallelepipeds, then $\Theta(K)$ from (3.4) for each parallelepiped K depends on the lower-left vertex of K only. Define $r(x) = -\log \mathbf{P}\{\xi \leq x\}$, where the inequality is understood coordinatewisely. Theorem 3.5 yields the following result.

Theorem 3.7 (A.s. convergence of scaled multivariate samples). *Assume that $r \in \text{RV}^u_{\beta, g, \theta}(\mathbb{R}^d_+)$ with a coordinatewisely strictly increasing function θ. If $g(a_n) \sim \log n$, then $a_n^{-1}\{\xi_1, \ldots, \xi_n\}$ converges almost surely as $n \to \infty$ in the Hausdorff metric to the compact set $K = \{x \in \mathbb{R}^d_+ : \theta(x) \leq 1\}$.*

It should be noted that the lack of preferable directions in Theorem 3.5 (in contrast to Theorem 3.7 where ξ is distributed within $\mathbb{R}_+^d$) makes it possible to apply it for random samples in all quadrants of $\mathbb{R}^d$ without any changes. For this, it is useful to choose $\mathcal{A}$ to be the family of sets $\{ux : u \in S, a \leq x \leq b\}$ for S running through the family of regular closed (in the induced topology) subsets of $\mathbb{S}^{d-1}$. This choice of $\mathcal{A}$ is similar to the formulation of the problem in polar coordinates efficiently used in Kinoshita and Resnick [313] to deduce a necessary and sufficient condition for the almost sure convergence for scaled samples of random vectors, see Section 3.3.

Example 3.8. Let $\xi = (\xi_1, \xi_2)$ be a random vector in $\mathbb{R}_+^2$ with i.i.d. marginals such that $-\log(1 - F_1) \in RV_\alpha$ with $\alpha > 0$, where F_1 is the cumulative distribution function of ξ_1. Then Theorem 3.7 is applicable with $K = \{(x_1, x_2) \in \mathbb{R}_+^2 : (x_1^\alpha + x_2^\alpha) \leq 2\}$.

Example 3.9. Define $X = (-\infty, \xi_1] \times \cdots \times (-\infty, \xi_d]$, where $\xi = (\xi_1, \ldots, \xi_d)$ has the cumulative distribution function F. If $[a, b] = [a_1, b_1] \times \cdots \times [a_d, b_d]$, then $\mathbf{P}\{A \cap t[a, b] \neq \emptyset\} = \mathbf{P}\{\xi \geq ta\}$. The regular variation property of the function $-\log \mathbf{P}\{\xi \geq ta\}$ for all $a \in \mathbb{R}_+^d$ with the strictly monotone limiting function θ ensures the almost sure convergence of the normalised unions to the deterministic limit $\{y \in \mathbb{R}^d : y \leq x, x \in \mathbb{R}_+^d, \theta(x) \leq 1\}$.

3.3 Stability and relative stability of unions

Let $\{X_n, n \geq 1\}$ be a sequence of i.i.d. random compact sets. Below we will give a necessary and sufficient condition for the relative stability of the sequence $Z_n = X_1 \cup \cdots \cup X_n$, $n \geq 1$, and characterise possible limits of $a_n^{-1} Z_n$. Because the random sets are compact, it is appropriate to consider the convergence in the Hausdorff metric.

Definition 3.10 (Stable and relatively stable sequences).
(i) A sequence $\{Z_n, n \geq 1\}$ of random closed sets in $\mathbb{R}^d$ is called (almost surely) *relatively stable* if there exists a sequence $\{a_n, n \geq 1\}$ of normalising multiplicative constants such that $a_n^{-1} Z_n$ has a non-trivial deterministic Painlevé–Kuratowski limit (not equal to $\emptyset$ or $\mathbb{R}^d$).
(ii) A sequence $\{Z_n, n \geq 1\}$ of random compact sets in $\mathbb{R}^d$ is called (almost surely) *stable* if there exists a sequence $\{K_n, n \geq 1\}$ of deterministic compact sets such that $\rho_H(Z_n, K_n) \to 0$ a.s. as $n \to \infty$.

Definition 3.10(i) is a generalisation of the stability concept for random variables. The sequence of random variables $\{\zeta_n, n \geq 1\}$ is called *relatively stable* if there exists a sequence $a_n < 0$, $n \geq 1$, such that $a_n^{-1} \zeta_n$ converges almost surely as $n \to \infty$ with a deterministic limit in $(0, \infty)$.

The following proposition says that the relative stability of the unions implies that the norms of random sets are relatively stable and that possible limits (in the Hausdorff metric) of $a_n^{-1} Z_n$ are always star-shaped.

Proposition 3.11 (Relative stability of unions). *Let $\{X_n, n \geq 1\}$ be i.i.d. random compact sets in $\mathbb{R}^d$ and let $Z_n = X_1 \cup \cdots \cup X_n$, $n \geq 1$. If there exist $a_n \to \infty$ and a* non-empty deterministic compact set Z such that

$$\rho_H(a_n^{-1} Z_n, Z) \to 0 \quad \text{a.s. as } n \to \infty, \tag{3.6}$$

then the sequence $\zeta_n = \max(\|X_1\|, \ldots, \|X_n\|)$, $n \geq 1$, is relatively stable and Z is star-shaped.

Proof. Since $Z_n \subset Z^\varepsilon$ and $Z \subset Z_n^\varepsilon$ for each $\varepsilon > 0$ and sufficiently large n,

$$\|Z\| - \varepsilon \leq a_n^{-1} \zeta_n \leq \|Z\| + \varepsilon,$$

whence $\{\zeta_n, n \geq 1\}$ is relatively stable.

A result on the almost sure stability of a sequence of random variables $\{\zeta_n, n \geq 1\}$ (see Kinoshita and Resnick [313, Prop. 3.1]) implies that, for given $t \in (0, 1)$, there is a subsequence $\{a_{k(n)}, n \geq 1\}$ such that $k(n) \leq n$ for all sufficiently large n, $k(n) \to \infty$ as $n \to \infty$ and $a_{k(n)}/a_n \to t$ as $n \to \infty$. Since $Z_{k(n)} \subset Z_n$, (3.6) yields $a_n^{-1} Z_{k(n)} \subset Z^\varepsilon$ for all $\varepsilon > 0$ and sufficiently large n. Therefore,

$$Z^\varepsilon \supset a_n^{-1} Z_{k(n)} = a_{k(n)}^{-1} Z_{k(n)} \frac{a_{k(n)}}{a_n}.$$

The right-hand side converges to tZ in the Hausdorff metric, whence $Z^\varepsilon \supset tZ$ for all $\varepsilon > 0$. Therefore, $Z \supset tZ$ for all $t \in (0, 1)$, meaning that Z is star-shaped. □

The compactness assumption in Proposition 3.11 is indispensable. To produce a counterexample, consider a relatively stable sequence $\{\xi_n, n \geq 1\}$ of non-negative random variables such that $a_n^{-1} \min(\xi_1, \ldots, \xi_n)$ converges almost surely to a constant $c > 0$. Then $a_n^{-1} \cup_{n \geq 1} [\xi_n, \infty)$ converges to $[c, \infty)$ with the limit not being star-shaped.

The following result can be proved similarly to Kinoshita and Resnick [313, Th. 4.6]. The main idea of the proof is to replace the star-shaped sets by their radius-vector functions and use the fact that compact star-shaped sets converge in the Hausdorff metric if and only if their radius-vector functions converge sup-vaguely, see Vervaat [572]. For each $S \subset \mathbb{S}^{d-1}$ define

$$\mathbb{C}_S(t) = \{sx : s > t, \, x \in S\}.$$

Write $\mathbb{C}_S$ instead of $\mathbb{C}_S(0-) = \{sx : s \geq 0, \, x \in S\}$.

Theorem 3.12 (Star-shaped limits). *Let $\{X_n, n \geq 1\}$ be a sequence of i.i.d. random compact sets. Then there exists a sequence $\{a_n, n \geq 1\}$ of non-negative constants, $a_n \to \infty$ as $n \to \infty$, such that $a_n^{-1} Z_n$ a.s. converges to a star-shaped random compact set Z as $n \to \infty$, if and only if the following two conditions hold.*
(i) *$\{\zeta_n = \max(\|X_1\|, \ldots, \|X_n\|), n \geq 1\}$ is relatively stable.*
(ii) *If $\|Z \cap \mathbb{C}_S\| = \|Z \cap \mathbb{C}_{\mathrm{Int}\, S}\|$ for a regular closed set $S \subset \mathbb{S}^{d-1}$, then $\mathbf{P}\{X_1 \cap \mathbb{C}_S = \emptyset\} = 0$ implies $Z \cap \mathbb{C}_S = \emptyset$ and otherwise*

$$\lim_{s \to \infty} \frac{\mathbf{P}\{X_1 \cap \mathbb{C}_S(ts) \neq \emptyset\}}{\mathbf{P}\{\|X_1\| > s\}} = \begin{cases} 0, & t > \|Z \cap \mathbb{C}_S\|, \\ \infty, & t < \|Z \cap \mathbb{C}_S\|. \end{cases}$$

Open problem 3.13. Characterise the almost sure relative stability property for sequences of random closed (not necessarily compact) sets.

3.4 Functionals of unions

Consider now limit theorems for functionals of unions of i.i.d. random closed sets $X_1, X_2, \ldots$, e.g. for $\mu(Z_n^c)$, where μ is the probability measure on $\mathbb{E}$ and $Z_n = X_1 \cup \cdots \cup X_n$. However, this is a particular case of the following more general situation.

Let Y be a random closed set in $\mathbb{E}$ with the avoidance functional $Q_Y(F) = \mathbf{P}\{Y \cap F = \emptyset\}$. For example,

$$\mu(Z_n^c) = Q_Y(Z_n)$$

if Y is a singleton distributed according to μ. The value of the avoidance functional and the value of the capacity functional on a random closed set are random variables themselves. Let

$$F(t) = \mathbf{P}\{T_X(Y) \leq t\}$$

be the cumulative distribution function of $T_X(Y)$. Then

$$\mathbf{P}\{Z_n \cap Y = \emptyset\} = \mathbf{E}(1 - T_X(Y))^n = \int_{[0,1]} (1-t)^n \, dF(t).$$

Therefore,

$$\frac{\log((1-1/n)^n F(1/n))}{\log n} \leq \frac{\mathbf{P}\{Z_n \cap Y = \emptyset\}}{\log n} \leq \frac{\log(F(\varepsilon_n) + (1-\varepsilon_n)^n)}{\log n},$$

where $\varepsilon_n = (\log n)^2/n$. By an argument similar to the one used in the proof of Lemma 3.3, one shows that

$$-\limsup_{t \to 0} \frac{\log F(t)}{\log t} \leq \liminf_{n \to \infty} \frac{\mathbf{P}\{Z_n \cap Y = \emptyset\}}{\log n}$$

$$\leq \limsup_{n \to \infty} \frac{\mathbf{P}\{Z_n \cap Y = \emptyset\}}{\log n}$$

$$\leq -\liminf_{t \to 0} \frac{\log F(t)}{\log t}.$$

Theorem 3.14 (Asymptotics for expected avoidance functional). *Assume that $F(0) = 0$ and F is absolutely continuous in a neighbourhood of zero with density $\psi(t) = ct^\alpha + o(t^\alpha)$, where c is a positive constant and $\alpha > -1$. Then*

$$\mathbf{E} Q_Y(Z_n) = c\Gamma(\alpha+1)n^{-(\alpha+1)} + o(n^{-(\alpha+1)}).$$

Proof. For arbitrary $a > 0$, $c(1-a)t^\alpha \leq \varphi(t) \leq c(1+a)t^\alpha$ for $t \in (0, \delta]$ and some $\delta > 0$. Then

$$n^{\alpha+1}\mathbf{P}\{Z_n \cap Y = \emptyset\} = n^{\alpha+1}\int_0^\delta (1-t)^n \varphi(t)dt + n^{\alpha+1}\int_\delta^1 (1-t)^n dF(t).$$

The second summand converges exponentially to 0, while the first one lies between $c(1-a)\int_0^\delta (1-t)^n t^\alpha dt$ and $c(1+a)\int_0^\delta (1-t)^n t^\alpha dt$. The proof is finished by observing that

$$\lim_{n\to\infty} n^{\alpha+1}\int_0^\delta (1-t)^n t^\alpha dt = \Gamma(\alpha+1). \qquad \square$$

The following results concern the higher moments of $Q_Y(Z_n)$. The proofs are based on replacing Y with $Y_1 \cup \cdots \cup Y_k$ for i.i.d. random closed sets $Y_1, \ldots, Y_k$ distributed as Y. Denote

$$T_X(F_1, F_2) = \mathbf{P}\{X \cap F_1 \neq \emptyset, \ X \cap F_2 \neq \emptyset\} = -\Delta_{F_1}\Delta_{F_2} T_X(\emptyset).$$

Theorem 3.15 (Higher-order asymptotics for avoidance functional). *If there exists a (possibly infinite) limit*

$$\gamma = \lim_{t\to 0} \frac{\log F(t)}{\log t}.$$

Then

$$\lim_{n\to\infty} \frac{\mathbf{E}(Q_Y(Z_n)^k)}{\log n} = -k\gamma.$$

If there also exists an $\varepsilon > 0$ such that

$$\mathbf{E}\left(T_X(Y_1, Y_2)\mathbf{1}_{T_X(Y_1)\leq t}\mathbf{1}_{T_X(Y_2)\leq t}\right) = \mathcal{O}(t^{1+\varepsilon} F^2(t))$$

for independent Y_1 and Y_2 sharing the distribution with Y, then under the conditions of Theorem 3.14

$$\mathbf{E}(Q_Y(Z_n)^k) = (c\Gamma(\alpha+1))^k n^{-k(\alpha+1)} + o(n^{-k(\alpha+1)}),$$

and $n^{\alpha+1}Q_Y(Z_n)$ converges in probability to $c\Gamma(\alpha+1)$.

4 Convex hulls

4.1 Infinite divisibility with respect to convex hulls

Consider convex hulls of unions

$$Y_n = \overline{\mathrm{co}}\,(X_1 \cup \cdots \cup X_n)$$

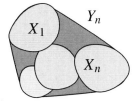

Figure 4.1. Convex hull of random compact sets.

for random closed sets in the Euclidean space $\mathbb{R}^d$. It is essential to take the closure of the convex hull, since the convex hull of a closed set is not always closed. If $X_1, \ldots, X_n$ are almost surely compact, then the closure can be omitted.

Convergence results for convex hulls of random compact sets typically follow from the corresponding results for unions of random sets, since the map $K \mapsto \operatorname{co}(K)$ is continuous on $\mathcal{K}$ in the myopic topology. In many other cases taking convex hulls allows us to derive results not available for unions of random sets.

Definition 4.1 (Infinite divisibility for convex hulls). A random convex closed set X is said to be *infinitely divisible for convex hulls* if, for each $n \geq 2$, there exist i.i.d. random closed sets $X_1, \ldots, X_n$ such that

$$X \stackrel{d}{\sim} \overline{\operatorname{co}}(X_1 \cup \cdots \cup X_n). \tag{4.1}$$

When studying convex hulls of random closed sets, it is natural to work with the containment functional $C_X(F) = \mathbf{P}\{X \subset F\}$ for $F \in \operatorname{co}\mathcal{F}$. However, as Example 1.7.9 shows, the containment functional defined on $\operatorname{co}\mathcal{F}$ does not determine uniquely the distribution of a (non-compact) random convex closed set. Because of this reason, we often restrict attention to the case of random convex compact sets. The family $\operatorname{co}\mathcal{K}$ of convex compact sets is an idempotent semigroup with the semigroup operation being the convex hull of the union. For each $L \in \operatorname{co}\mathcal{K}$, the function $\chi(K) = \mathbf{1}_{K \subset L}$, $K \in \operatorname{co}\mathcal{K}$, is a semicharacter, since

$$\mathbf{1}_{\operatorname{co}(K_1 \cup K_2) \subset L} = \mathbf{1}_{K_1 \subset L} \mathbf{1}_{K_2 \subset L}$$

for all $K_1, K_2 \in \operatorname{co}\mathcal{K}$. In terms of the containment functionals, (4.1) implies

$$C_X(L) = \mathbf{P}\{X \subset L\} = (\mathbf{P}\{X_1 \subset L\})^n = C_{X_1}(L)^n, \quad L \in \operatorname{co}\mathcal{K}, \tag{4.2}$$

i.e. $\mathbf{E}\chi(X) = (\mathbf{E}\chi(X_1))^n$. The results of Appendix G yield a representation of the containment functionals of random compact convex sets which are infinitely divisible for convex hulls. However, this requires a careful handling of the situation when the containment functional vanishes. Note that (4.2) makes it possible to characterise the containment functional of X in the same manner as was done in Section 1.2 for the union-infinitely-divisibility using entirely elementary tools. However, it is more instructive to derive this characterisation from the general approach worked out in Section 1.5 for lattice-valued random elements.

Consider the family co $\mathcal{K}$ as a partially ordered set with the reverse inclusion and the top element being the empty set. Then $K \vee L = K \cap L$ and $K \wedge L = \mathrm{co}(K \cup L)$. It is not difficult to show that $K \ll L$ is equivalent to $L \subset \mathrm{Int}\, K$, see also Gierz et al. [192, p. 50]. The Scott topology on co $\mathcal{K}$ is generated by $\{K \in \mathrm{co}\,\mathcal{K} : K \subset G\}$ for $G \in \mathcal{G}$. Since co $\mathcal{K}$ contains a countable separating subset (for example, convex hulls of finite unions of balls with rational centres and radii), the Scott topology is second countable.

Let X be an infinitely divisible random element in co $\mathcal{K}$. Define
$$H = \vee\{K \in \mathrm{co}\,\mathcal{K} : \mathbf{P}\{K \leq X\} > 0\} = \cap\{K \in \mathrm{co}\,\mathcal{K} : \mathbf{P}\{X \subset K\} > 0\}.$$

Note that H is convex as the intersection of convex compact sets. Proposition 1.32 implies that $\mathbf{P}\{X \leq H\} = \mathbf{P}\{H \subset X\} = 1$. Therefore, H coincides with the set F_X of fixed points for X. Consider the family
$$\mathrm{co}\,\mathcal{K}_X = \{K \in \mathrm{co}\,\mathcal{K} : F_X \subset K\}$$
as a sub-lattice of co $\mathcal{K}$ with the top being F_X.

Theorem 4.2 (Infinite divisibility for convex hulls). *A random convex compact set X is infinitely divisible for convex hulls if and only if its containment functional is given by*
$$C(K) = \exp\{-\Lambda(\{L \in \mathrm{co}\,\mathcal{K}_X : L \not\subset K\})\}, \quad K \in \mathrm{co}\,\mathcal{K}_X, \tag{4.3}$$
where Λ is a measure concentrated on $\mathrm{co}\,\mathcal{K}_X \setminus \{F_X\}$ such that
$$\Lambda(\{L \in \mathrm{co}\,\mathcal{K}_X : L \not\subset \mathrm{Int}\, K\}) < \infty, \quad K \in \mathrm{co}\,\mathcal{K}_X. \tag{4.4}$$
If $\Pi_\Lambda = \{K_1, K_2, \dots\}$ is a Poisson point process on $\mathrm{co}\,\mathcal{K}_X$ with the intensity measure Λ, then
$$X \stackrel{\mathrm{d}}{\sim} \overline{\mathrm{co}}\,(F_X \cup K_1 \cup K_2 \cup \cdots). \tag{4.5}$$
If X is non-empty almost surely, then $F_X \neq \emptyset$.

Proof. Representation (4.3) and the fact that Λ is locally finite follow from Proposition 1.30 applied to the semi-lattice $\mathrm{co}\,\mathcal{K}_X$. Since Λ is locally finite, it defines a Poisson point process on $\mathrm{co}\,\mathcal{K}_X$, such that only a finite number of "points" (actually sets) from the process are not contained in any $K \in \mathrm{co}\,\mathcal{K}_X$ such that $\mathrm{Int}\, K \supset F_X$. Therefore, the set in the right-hand side of (4.5) is compact almost surely. For each $K \in \mathrm{co}\,\mathcal{K}_X$, we have
$$\mathbf{P}\{\overline{\mathrm{co}}\,(F_X \cup K_1 \cup K_2 \cup \cdots) \subset K\} = \exp\{-\Lambda(\{L \in \mathrm{co}\,\mathcal{K}_H : L \not\subset K\})\}.$$

Thus, (4.5) follows from the fact that distributions of random convex compact sets are uniquely determined by their containment functionals.

If $F_X = \emptyset$, then (4.5) shows that X is empty with a positive probability unless the total mass of Λ is infinite. For two disjoint regular closed sets $K_1, K_2 \in \mathrm{co}\,\mathcal{K}$, one has $\Lambda(\{L \in \mathrm{co}\,\mathcal{K} : L \not\subset \mathrm{Int}\, K_i\}) < \infty$ for $i = 1, 2$, contrary to the fact that the total mass of Λ is infinite, since any $L \in \mathrm{co}\,\mathcal{K}$ satisfies $L \not\subset K_1$ or $L \not\subset K_2$. □

The following corollary follows from (4.3) and Theorem 1.31.

Corollary 4.3. *A random convex compact set X is infinitely divisible for convex hulls if and only if its containment functional is given by*

$$C_X(K) = \exp\{-\Psi(K)\}, \quad K \in \operatorname{co} \mathcal{K}_X, \tag{4.6}$$

where Ψ is a completely $\cap$-alternating non-negative functional on $\operatorname{co} \mathcal{K}_X$ such that $\Psi(K) < \infty$ for $K \neq F_X$ and $\Psi(K_n) \uparrow \Psi(K)$ if $K_n \uparrow K$ with $K, K_1, K_2, \ldots$ from $\operatorname{co} \mathcal{K}_X$.

Open problem 4.4. Characterise random closed (not necessarily compact) sets that are infinitely divisible for convex hulls.

4.2 Convex-stable sets

Distributions and containment functionals

A natural stability concept associated with the convex hull operation can be formulated as follows.

Definition 4.5 (Convex-stable random sets). *A random convex closed set X is said to be convex-stable if, for every $n \geq 2$ and independent copies $X_1, \ldots, X_n$ of X,*

$$a_n X \stackrel{d}{\sim} \overline{\operatorname{co}}\,(X_1 \cup \cdots \cup X_n) + K_n \tag{4.7}$$

for some $a_n > 0$ and $K_n \in \mathcal{K}$, where the sum in the right-hand side is understood as the Minkowski sum of sets. If $K_n = \{b_n\}$, $n \geq 1$, i.e. K_n is a singleton for all $n \geq 1$, then X is said to be strictly *convex-stable.*

It should be noted that, for every union-stable random closed set X, its closed convex hull $\overline{\operatorname{co}}\,(X)$ is strictly convex-stable. Convex hulls of sets correspond to a pointwise maximum of their support functions. Therefore, (4.7) yields

$$a_n h(X, u) \stackrel{d}{\sim} \max\{h(X_1, u), \ldots, h(X_n, u)\} + h(K_n, u), \quad u \in \mathbb{S}^{d-1}. \tag{4.8}$$

If X is compact and $h(X, u)$ has a non-degenerate distribution for each $u \in \mathbb{S}^{d-1}$, then $h(X, \cdot)$ is a random max-stable sample continuous process. Define L_X as the set of all $u \in \mathbb{S}^{d-1}$ such that $h(X, u)$ is almost surely finite, see (1.7.4).

Proposition 4.6. *If X is a convex-stable random closed set, then*

$$\operatorname{dom} h(X, \cdot) = \{u \in \mathbb{S}^{d-1} : h(X, u) < \infty\}$$

is a deterministic set equal to L_X.

Proof. It can be shown that $\operatorname{dom} h(X, \cdot)$ is a random closed subset of $\mathbb{S}^{d-1}$, while (4.7) yields

$$\operatorname{dom} h(X, \cdot) \stackrel{d}{\sim} \operatorname{dom} h(\operatorname{co}(X_1, \ldots, X_n) + K_n, \cdot) = \bigcap_{i=1}^{n} \operatorname{dom} h(X_i, \cdot).$$

Proposition 1.5 implies that $\operatorname{dom} h(X, \cdot) = L_X$ a.s. □

Corollary 4.7. *The distribution of a convex-stable random closed set X is uniquely determined by its containment functional $C_X(F) = \mathbf{P}\{X \subset F\}$ for $F \in \operatorname{co} \mathcal{F}$.*

Proof. The functional $C_X(F)$ determines uniquely the finite-dimensional distributions of $h(X, u)$, $u \in L_X$, and, therefore, the distribution of X. Note that the compactness of X is not required, cf. Example 1.7.9 and Theorem 1.7.8. □

Characterisation in terms of containment functionals

All realisations of a convex-stable random closed set X belong to the family

$$\mathcal{C}_X = \mathcal{C}(L_X) \cap \{F \in \operatorname{co} \mathcal{F} : F_X \subset F\},$$

where $\mathcal{C}(L_X)$ is defined in (1.7.5) as the family of sets $F \in \operatorname{co} \mathcal{F}$ such that $h(F, u) < \infty$ if and only if $u \in L_X$ and F_X is the set of fixed points of X. Then $C_X(F) = 0$ if a convex set F does not belong to $\mathcal{C}_X$.

Theorem 4.8 (Characterisation of strictly convex-stable sets). *A random (non-deterministic) convex closed set X is strictly convex-stable if and only if*

$$C(F) = \exp\{-\Psi(F)\}, \quad F \in \mathcal{C}_X, \tag{4.9}$$

where Ψ is a completely $\cap$-alternating non-negative functional such that $\Psi(F) < \infty$ if $F \neq F_X$, $\Psi(F_n) \uparrow \Psi(F)$ if $F_n \uparrow F$ for $F, F_1, F_2, \ldots \in \mathcal{C}_X$ and there exists $\alpha \neq 0$ such that $\Psi(sF) = s^\alpha \Psi(F)$ and $sF_X = F_X$ for all $F \in \mathcal{C}_X$ and $s > 0$. Then (4.7) holds with $a_n = n^{-1/\alpha}$, $n \geq 1$.

Proof. Representation (4.9) does not follow immediately from (4.6), since Theorem 4.2 and Corollary 4.3 do not cover the case of non-compact random closed sets. However, (4.9) follows from the general results on infinitely divisible random elements in lattices applied to the lattice $\mathcal{C}_X$ with the reverse inclusion and the top being F_X. This general result is applicable because of Proposition 4.6 and Corollary 4.7. The homogeneity property of Ψ can be proved similarly to Theorem 1.12. Its proof is even simpler, since $cX \stackrel{d}{\sim} X$ for a convex-stable random closed set X necessarily implies that either $c = 1$ or X is a deterministic cone. □

Note that representation (4.9) holds also for convex-stable sets which are not necessarily strictly convex-stable. The following theorem deals with general convex-stable sets. Recall that $\check{H}$ is the centrally symmetric set to H and $\ominus$ denotes the Minkowski subtraction, see Appendix A.

Theorem 4.9 (Characterisation of convex-stable sets). *A non-trivial random convex set X is convex-stable in one and only one of the following three cases.*

(i) *There exist a strictly convex-stable random set Y with $\alpha < 0$ and a compact convex set H such that $X = Y + H$. Then $K_n = (n^{-1/\alpha} - 1)H$ and $a_n = n^{-1/\alpha}$ in (4.7). If, additionally, X is compact and $h(X, u)$ has a non-degenerate distribution for each $u \in \mathbb{S}^{d-1}$, then $0 \in \text{Int}\, Y$ almost surely.*

(ii) *There exists a convex compact set H such that $Y = X + H$ is a strictly convex-stable random set with $\alpha > 0$. Then $a_n = n^{-1/\alpha}$ and $K_n = (1 - n^{-1/\alpha})H$, $n \geq 1$.*

(iii) *There exists a convex compact set H such that X satisfies (4.7) with $a_n = 1$ and $K_n = H \log n$ for all $n \geq 1$. This happens if and only if its containment functional is given by (4.9) where Ψ is a completely $\cap$-alternating non-negative functional on $\mathcal{C}_X$ such that $\Psi(F) < \infty$ if $F \neq F_X$, $\Psi(F_n) \uparrow \Psi(F)$ if $F_n \uparrow F$ with $F, F_1, F_2, \ldots \in \mathcal{C}_X$ and*

$$\Psi(F) = s\Psi(F \ominus \check{H} \log s), \quad F_X \ominus \check{H} \log s = F_X,$$

for all $s > 0$.

Proof. (i) and (ii) can be proved similarly to Theorem 1.23. The part of case (i) concerning compact convex-stable random sets with non-degenerate support functions is contained in Giné, Hahn and Vatan [201]. Statement (iii) follows from the results of [201], which make it possible to identify K_n as $H \log n$ using their support function representation and (4.8). Then

$$X \stackrel{d}{\sim} \overline{\text{co}}(X_1 \cup \cdots \cup X_n) + H \log n$$

and (4.9) imply $n\Psi(F \ominus \check{H} \log n) = \Psi(F)$. The proof can be finished similarly to the proof of Theorem 1.28 by reducing the problem to functional equations. □

Note that if X is a compact convex-stable random set, then only the case (i) of Theorem 4.9 is possible with $\alpha < 0$.

Example 4.10. Let $\xi = (\xi_1, \ldots, \xi_d)$ be a random vector in $\mathbb{R}^d$. Define

$$X = \overline{\text{co}}\{e_i x_i : 1 \leq i \leq d, x_i \leq \xi_i\}, \tag{4.10}$$

where $e_1, \ldots, e_d$ is the basis in $\mathbb{R}^d$, see Figure 4.2. Then X is strictly convex-stable if and only if ξ is a max-stable random vector with respect to coordinate-wise maximum, see Balkema and Resnick [47] and Galambos [186]. Evidently, $C(\downarrow x) = \mathbf{P}\{X \subset (\downarrow x)\} = F_\xi(x)$, where F_ξ is the cumulative distribution function of ξ and $(\downarrow x) = (-\infty, x_1] \times \cdots \times (-\infty, x_d]$ for $x = (x_1, \ldots, x_d)$. Thus, X is strictly convex-stable if and only if $F_\xi(x) = \exp\{-\psi(x)\}$, where $\psi(x) \leq \psi(y) \leq 0$ if $x \leq y$ coordinatewisely, ψ is right-continuous, satisfies (3.7) and there exists $v \in \mathbb{R}^d$ such that either $\psi(x) = s\psi(x + v \log s)$ or $\psi(s(x+v)) = s^\alpha \psi(x+v)$ for each $s > 0$ and $x \in \mathbb{R}^d$. This is a known representation of max-stable random vectors.

284   4 Unions of Random Sets

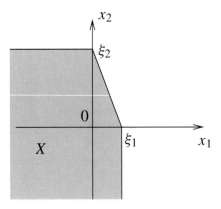

Figure 4.2. A random set X generated by a random vector ξ in $\mathbb{R}^2$.

Example 4.11. If Π_Λ is a Poisson point process from Example 1.16 with $\alpha < 0$, then the random closed set $X = \overline{\text{co}}\,(\Pi_\Lambda)$ is strictly convex-stable and $C_X(F) = \exp\{-\Lambda(F^c)\}$ for any convex closed set F such that $0 \in \text{Int}\,F$.

Open problem 4.12. Characterise distributions of random vectors that give rise to not necessarily strictly stable random closed sets by (4.10).

4.3 Convergence of normalised convex hulls

Weak convergence

Results concerning the weak convergence of convex hulls are simpler than those concerning general unions of random sets, since it is possible to handle convex sets by means of their support functions. Let $\{X_n, n \geq 1\}$ be i.i.d. random closed sets, possibly non-convex and unbounded, with

$$C(F) = \mathbf{P}\{\overline{\text{co}}\,(X_1) \subset F\}$$

being the containment functional of $\overline{\text{co}}\,(X_1)$. Define $Y_n = \overline{\text{co}}\,(X_1 \cup \cdots \cup X_n)$. Evidently, for each convex F,

$$C_n(F) = \mathbf{P}\left\{a_n^{-1} Y_n \subset F\right\} = (C(a_n F))^n\,.$$

Define $L = L_{X_1}$ to be the set of all $u \in \mathbb{S}^{d-1}$ such that $h(X_1, u)$ is almost surely finite. Then the weak limit of $a_n^{-1} Y_n$ almost surely belongs to the family $\mathcal{C}(L)$ introduced in (1.7.5).

Let $a_n \to \infty$ as $n \to \infty$. Similar to Section 2.1, define for any convex F

$$a_n(F) = \sup\{x : C(xF) \leq 1 - 1/n\}\,,$$

where $\sup \emptyset = 0$, and introduce a subfamily of $\mathcal{C}(L)$ by

$$\mathcal{V}_c = \{F \in \mathcal{C}(L) : \limsup_{x \to \infty} C(xF) = 1\}.$$

The following theorem resembles Theorem 2.1 although its proof is simpler, since the limit theorem for convex hulls can be reduced directly to the limit theorem for the maximum of random variables.

Theorem 4.13 (Weak convergence of convex hulls). *Let $F \in \mathcal{V}_c$. Suppose that $a_n(F)/a_n \to q(F)$ as $n \to \infty$, where $q(F)$ is allowed to be infinite. Then*

$$\lim_{n \to \infty} C_n(F) = \exp\{-q(F)^{-\alpha}\}, \quad (4.11)$$

if (and only if in the case $0 < q(F) < \infty$) the function $v_F(x) = 1 - C(xF)$, $x \geq 0$, is regularly varying with exponent $\alpha < 0$. If $F \notin \mathcal{V}_c$ then $C_n(F) \to 0$ as $n \to \infty$.

If the above conditions hold for each $F \in \mathcal{V}_c$, then $a_n^{-1} Y_n$ converges weakly to a strictly convex-stable random closed set Z with the containment functional given by

$$C_Z(F) = \begin{cases} \exp\{-q(F)^\alpha\}, & F \in \mathcal{V}_c, \\ 0, & \text{otherwise}. \end{cases}$$

Proof. If $0 \notin F$, then $C_n(F) \to 0$, so assume $0 \in F$ and define

$$\xi(X_i) = \inf\{s \geq 0 : X_i \subset sF\}.$$

The cumulative distribution function of the random variable $\xi(X_i)$ is evaluated as

$$\mathbf{P}\{\xi(X_i) \leq x\} = \mathbf{P}\{X_i \subset xF\} = C(xF).$$

Hence $a_n(F) = \sup\{x : F_\xi(x) \leq 1 - 1/n\}$, and F belongs to $\mathcal{V}_c$ if and only if $\mathbf{P}\{\xi(X_i) \leq x\} \to 1$ as $x \to \infty$. Moreover,

$$\zeta_n = \xi(Y_n) = \max\{\xi(X_i), 1 \leq i \leq n\}.$$

The classical limit theorem for extremes (Galambos [186, Th. 2.1.1]) yields that

$$\mathbf{P}\{\zeta_n < a_n(F)x\} \to \exp\{-x^\alpha\} \quad \text{as } n \to \infty.$$

Then (4.11) holds, since

$$C_n(a_n F) = \mathbf{P}\{\zeta_n < a_n\} = \mathbf{P}\left\{\zeta_n < a_n(F) \frac{a_n}{a_n(F)}\right\}.$$

If $0 < q(F) < \infty$, then the necessity follows from the corresponding theorem for extremes of random variables, see Galambos [186, Th. 2.4.3].

For each $F \in \mathcal{V}_c$ and $s > 0$, the set sF belongs to $\mathcal{V}_c$ and $q(sF)^{-\alpha} = s^\alpha q(F)$, so that the limiting distribution corresponds to a strictly convex-stable set with parameter $\alpha < 0$. The weak convergence follows from the results of Section 1.6.1. □

Limit theorems for convex hulls of non-compact random sets do not follow immediately from the corresponding results for unions, since the map $F \mapsto \overline{\mathrm{co}}(F)$ is not continuous in the Fell topology, but only lower semicontinuous. In contrast, the map $K \mapsto \mathrm{co}(K)$ is continuous in the Hausdorff metric, so that the weak convergence of unions in the myopic topology implies the weak convergence of their convex hulls.

Convergence almost surely

The following theorem follows from Resnick and Tomkins [484].

Theorem 4.14. *Let $X, X_1, X_2, \ldots$ be i.i.d. random compact sets. Assume that for every $u \in \mathbb{S}^{d-1}$ and $x > 0$ there exists a finite limit*

$$\lim_{t \to \infty} \frac{-\log \mathbf{P}\{h(X, ux) \geq t\}}{g(t)} = \varphi(ux).$$

Then $a_n^{-1} \operatorname{co}(X_1 \cup \cdots \cup X_n)$ a.s. converges in the Hausdorff metric to

$$\{ux : u \in \mathbb{S}^{d-1}, x \geq 0, \varphi(ux) \geq 1\}.$$

5 Unions and convex hulls of random functions

5.1 Random points

Consider convergence of random samples in $\mathbb{R}^d$ and their convex hulls. In this case $X_n = \{\xi_n\}$, $n \geq 1$, are i.i.d. random singletons and $Z_n = \{\xi_1, \ldots, \xi_n\}$, $n \geq 1$. The following well-known result states that the scaled sample converges weakly to a Poisson point process if ξ_1 has a regular varying distribution.

Theorem 5.1. *Let $X_1 = \{\xi_1\}$, where ξ_1 is a random vector with a regularly varying positive density $f \in \mathrm{RV}^{\mathrm{u}}_{\alpha-d}(\mathbb{R}^d)$ with $\alpha < 0$. Assume that f can be decomposed as $f = \varphi L$ for a slowly varying function L and a homogeneous function φ. Define*

$$a_n = \sup\{x : x^\alpha L(xe) \geq 1/n\}$$

for some $e \in \mathbb{R}^d \setminus \{0\}$. Then $a_n^{-1} Z_n = a_n^{-1}\{\xi_1, \ldots, \xi_n\}$ converges weakly to the Poisson random set $Z = \Pi_\Lambda$ with the capacity functional $T_Z(K) = 1 - \exp\{-\Lambda(K)\}$ and the intensity measure

$$\Lambda(K) = \int_K \varphi(u) du, \quad K \in \mathcal{K}. \tag{5.1}$$

Proof. The proof is based on application of Theorem 2.1. Since $\alpha < 0$, $a_n \to \infty$ as $n \to \infty$. If $0 \in K$, then $T(xK)$ does not converge to 0 as $x \to \infty$, so that $T_Z(K) = 1$. If $0 \notin K$, then

$$v_K(x) = \mathbf{P}\{\xi \in xK\} = x^\alpha \int_K \varphi(u) L(xu) du.$$

It follows from (H.8) that $v_K(x) \to 0$ as $x \to \infty$ and $v_K \in \mathrm{RV}_\alpha$. Note that $v_K(x) \sim x^\alpha L(xe) \Lambda(K)$. It follows from the inversion theorem for regularly varying functions (see Appendix H) and Theorem 2.1 that $\Psi(K) = \Lambda(K)$ (if $0 \in \text{Int } K$ we assume $\Lambda(K) = \infty$). The random closed set Z is compact almost surely, since

$$\mathbf{P}\{Z \subset B_r(0)\} = \exp\left\{-\int_{\|u\|>r} \varphi(u)du\right\} \to 1 \quad \text{as } r \uparrow \infty,$$

and φ is integrable outside a neighbourhood of the origin. □

The limiting Poisson point process in Theorem 5.1 with the intensity measure Λ given by (5.1) is not locally finite with the origin being its accumulation point, since the function φ is not integrable near the origin.

If Z is the limiting random set in Theorem 5.1, then

$$\mathbf{P}\{h(Z, v) < x\} = \mathbf{P}\{Z \subset xH_v^c\} = \exp\{-x^\alpha a(v)\},$$

where

$$a(v) = \int_{\langle u,v\rangle \geq 1} \varphi(u)du$$

$$= \int_{S_v^+} du \int_{\langle u,v\rangle^{-1}}^{\infty} \varphi(yu)y^{d-1}dy = -\alpha^{-1}\int_{S_v^+} \varphi(u)\langle u, v\rangle^{-\alpha}du,$$

where

$$S_v^+ = \{u \in \mathbb{S}^{d-1} : \langle u, v\rangle \geq 0\}.$$

Elementary calculations yield the support function of the selection expectation of Z:

$$h(\mathbf{E}Z, v) = \mathbf{E}h(Z, v) = a(v)^{-1/\alpha}\Gamma(1/\alpha + 1)$$

$$= \Gamma(1 + 1/\alpha)\left[-\alpha^{-1}\int_{S_v^+} \varphi(u)\langle u, v\rangle^{-\alpha}du\right]^{-1/\alpha}.$$

where Γ is the gamma-function.

Below we outline an analogue of the previous results for convergence to union-stable and convex-stable random sets with $\alpha > 0$.

Theorem 5.2. *Let ξ be distributed with the density g in a convex cone $\mathbb{C} \subset \mathbb{R}^d$. Suppose that the function $f(u) = g(u\|u\|^{-2})$, $u \neq 0$, belongs to the class $\text{RV}_{d-\alpha}^u(\mathbb{C})$ with $\alpha > 0$ and $f = L\varphi$. For $e \in \mathbb{C} \setminus \{0\}$ put $a_n = \inf\{t > 0 : t^\alpha L(t^{-1}e) \geq n^{-1}\}$. If $\xi_1, \ldots, \xi_n$ are i.i.d. copies of ξ, then $a_n^{-1}\{\xi_1, \ldots, \xi_n\}$ weakly converges to Z as $n \to \infty$, where*

$$T_Z(K) = 1 - \exp\left\{-\int_{\mathbb{C} \cap K} \varphi(u\|u\|^{-2})du\right\}, \quad K \in \mathcal{K}.$$

If $\mathbb{C}$ does not contain any line, then $a_n^{-1}\,\mathrm{co}\{\xi_1,\ldots,\xi_n\}$ weakly converges to a strictly convex-stable random set $\mathrm{co}(Z)$ with the containment functional

$$\mathbf{P}\{\mathrm{co}\,Z \subset F\} = \exp\left\{-\int_{\mathbb{C}\setminus F} \varphi(u\|u\|^{-2})du\right\}, \quad F \in \mathrm{co}\,\mathcal{F}.$$

5.2 Multivalued mappings

Weak convergence

Let $\{X_n, n \geq 1\}$ be i.i.d. copies of the random closed set $X = M(\xi)$, where M is a multifunction continuous in the Hausdorff metric and ξ is a random vector in $\mathbb{R}^m$ with a regularly varying density. The function $M: \mathbb{R}^m \mapsto \mathcal{K}$ is called *homogeneous* if $M(su) = s^\eta M(u)$ for all $s > 0$ and $u \in \mathbb{R}^m$.

Theorem 5.3. *Let ξ be a random vector distributed in a cone $\mathbb{C} \subset \mathbb{R}^m$ with a positive density $f \in \mathrm{RV}^{\mathrm{u}}_{\alpha-m}(\mathbb{C})$, $\alpha < 0$ and let M be a homogeneous multifunction of order $\eta > 0$. Denote*

$$M^-(K) = \{u \in \mathbb{C}: M(u) \cap K \neq \emptyset\},$$
$$a_n = \sup\{x^\eta : x^\alpha L(xe) \geq 1/n\},$$

where e is a fixed point in $\mathbb{R}^m \setminus \{0\}$ and $f = \varphi L$ for a slowly varying function L and homogeneous φ. Then $a_n^{-1} Z_n = a_n^{-1}(X_1 \cup \cdots \cup X_n)$ converges weakly to the union-stable set Z with

$$T_Z(K) = \begin{cases} 1 - \exp\left\{-\int_{M^-(K)} \varphi(u)du\right\}, & 0 \notin K, \\ 1, & \text{otherwise.} \end{cases}$$

If $\alpha/\eta < -1$, then Z is integrably bounded and

$$h(\mathbf{E}Z, v) = \Gamma(1 + \eta/\alpha)\left[-\frac{\eta}{\alpha}\int_{\{u:\, h(M(u),v) \geq 1\}} \varphi(u)du\right]^{-\eta/\alpha}, \quad v \in \mathbb{S}^{d-1}.$$

Proof. Note that

$$a_n^{-1}(X_1 \cup \cdots \cup X_n) = M(a_n^{-1/\eta}\{\xi_1,\ldots,\xi_n\})$$

and refer to Theorem 5.1. The evaluation of the support function of $\mathbf{E}Z$ is straightforward. □

Example 5.4. Let $m = d = 1$ and let $X = \xi M$ be a random subset of $\mathbb{R}$, where ξ is a random variable having the Cauchy distribution and M is a deterministic compact set missing the origin. Then $n^{-1}(\xi_1 M \cup \cdots \cup \xi_n M)$ converges weakly to a union-stable random set Z with $\alpha = -1$ and the capacity functional

$$T_Z(K) = 1 - \exp\left\{-\int_{K/M} u^{-2} du\right\},$$

where $K/M = \{x/y : x \in K, y \in M\}$.

Theorem 5.3 makes it possible to obtain limit theorems for unions and convex hulls of *random balls* ($m = d+1$, $M(u_1, \ldots, u_{d+1})$ is the ball in $\mathbb{R}^d$ of radius u_{d+1} centred at $(u_1, \ldots, u_d)$) or *random triangles* ($m = 3d$ and $M(u_1, \ldots, u_{3d})$ is the triangle with the vertices $(u_1, \ldots, u_d), (u_{d+1}, \ldots, u_{2d}), (u_{2d+1}, \ldots, u_{3d})$). In these cases $M(su) = sM(u)$ for all $u \in \mathbb{R}^m$ and $s > 0$, whence $\eta = 1$. The following result follows from Theorem 5.3 for $M(u_1, \ldots, u_{d+1}) = (u_1, \ldots, u_d) + u_{d+1}K$, where K is a compact set.

Proposition 5.5. *Let $(\xi_1, \ldots, \xi_d, \zeta)$ be a random vector in $\mathbb{R}^d \times [0, \infty)$ with the regularly varying density $f(u; y) \in \mathrm{RV}_{\alpha-d-1}(\mathbb{R}^d \times [0, \infty))$, $\alpha < 0$, and let $X = (\xi_1, \ldots, \xi_d) + \zeta K$, where $\xi = (\xi_1, \ldots, \xi_d)$ and K is a compact set in $\mathbb{R}^d$ containing the origin. Furthermore, let $f(u; y) = \varphi(u; y)L(u; y)$ for $u \in \mathbb{R}^d$ and $y > 0$, where φ is homogeneous and L is slowly varying. Define*

$$a_n = \sup\{x : x^\alpha L(xe; xt) \geq 1/n\}$$

for some $(e; t) \in (\mathbb{R}^d \setminus \{0\}) \times (0, \infty)$. Then $a_n^{-1}Z_n$ converges weakly to a union-stable random set Z with the capacity functional

$$T_Z(K) = 1 - \exp\left\{-\int_0^\infty dy \int_{K+y\check{K}} \varphi(u; y) du\right\}.$$

If $\alpha < -1$, then $\mathbf{E}Z$ exists and has the support function

$$h(\mathbf{E}Z, v) = \Gamma(1+1/\alpha)\left[-\alpha^{-1} \int_0^\infty dy \int_{S_v^+} \varphi(w; y)(\langle w, v\rangle + yh(K, v))^{-\alpha} dw\right]^{-1/\alpha}.$$

Example 5.6. Let $d = 2$, $\alpha = -2$ and let ξ have a circular symmetric distribution with

$$f(u; y) = 2\pi^{-1}(1 + (\|u\|y)^4)^{-1}.$$

Then

$$h(\mathbf{E}Z, v) = \frac{\pi}{2^{5/4}}\left(\frac{\pi}{2} + 2^{3/2}h(K, v) + h(K, v)^2\right)^{1/2}.$$

Almost sure convergence

Let $h \in \mathrm{RV}_{\alpha,\varphi}^{\mathrm{u}}(\mathbb{C})$ be a regularly varying function on a cone $\mathbb{C} \subset \mathbb{R}^m$. Fix a vector $e \in \mathbb{C} \setminus \{0\}$ in the definition of regular variation, see (H.6). Assume that $\alpha > 0$ and φ does not vanish on $\mathbb{S}^{m-1} \cap \mathbb{C}$.

Let ξ be a random vector in $\mathbb{C}$ with the probability density function $\exp\{-h(x)\}$ and let $X_1, X_2, \ldots$ be independent copies of the random compact set $X = M(\xi)$. For simplicity, assume that $M(u)$ hits any open cone $\mathbb{G} \subset \mathbb{R}^d$ for the set of $u \in \mathbb{C}$ with a positive Lebesgue measure. Otherwise the range of possible values of X is a cone $\mathbb{G}'$, so that all formulations can be amended by replacing $\mathbb{R}^d$ with $\mathbb{G}'$ and $\mathbb{S}^{d-1}$ with $\mathbb{S}^{d-1} \cap \mathbb{G}'$. For each $S \subset \mathbb{S}^{d-1}$ denote

$$\mathbb{C}_S = \{xv : x \geq 0, v \in S\},$$
$$A_S = \{u \in \mathbb{S}^{m-1} \cap \mathbb{C} : M(u) \cap \mathbb{C}_S \neq \emptyset\},$$
$$q_S(u) = \|M(u) \cap \mathbb{C}_S\|, \quad u \in A_S.$$

For $S = \{v\}$, the corresponding left-hand sides are written as $\mathbb{C}_v$, A_v and $q_v(u)$ respectively. A rather technical proof of the following theorem relies on Corollary 3.6, see Molchanov [398, 402].

Theorem 5.7. *Let $X = M(\xi)$ for ξ defined above and a homogeneous multifunction M of order η. Assume that, for each regular closed (in the induced topology) set $S \subset \mathbb{S}^{d-1}$, $w \in A_S$, $\delta > 0$ and $u \in M(w) \cap \mathbb{C}_S$, there exists $v \in A_S$ such that $M(v) \cap \mathrm{Int}\,\mathbb{C}_S \cap B_\delta(u) \neq \emptyset$. If $h(a_n^{1/\eta} e) \sim \log n$ as $n \to \infty$, then $Y_n = a_n^{-1}(X_1 \cup \cdots \cup X_n)$ converges almost surely in the Hausdorff metric to*

$$Z = \{vx : v \in \mathbb{S}^{d-1}, 0 \leq x \leq f(v)\},$$

where $f(v) = \sup\{q_v(u)\varphi(u)^{-\eta/\alpha} : u \in A_v\}$.

In particular, if $\varphi(u) = c$ for all $u \in \mathbb{S}^{d-1}$, then

$$Z = \left\{vx : v \in \mathbb{S}^{d-1}, 0 \leq x \leq c^{-\eta/\alpha} \sup\{q_v(u) : u \in A_v\}\right\}. \tag{5.2}$$

The condition of Theorem 5.7 is satisfied for $M(u) = \{u\}$. Then $\eta = 1$, $X = \{\xi\}$, $A_v = \{v\}$, $q_v(u) = \|u\|$ and

$$Z = \{vx : v \in \mathbb{S}^{d-1}, x \geq 0, \varphi(vx) \leq 1\}.$$

The condition on M in Theorem 5.7 holds for a function M defined as

$$M(x_1, \ldots, x_{dn+l}) = \mathrm{co}\left\{(x_1, \ldots, x_d), \ldots, (x_{(n-1)d+1}, \ldots, x_{nd})\right\}$$
$$+ x_{dn+1} M_1 + \cdots + x_{dn+l} M_l,$$

where $M_1, \ldots, M_l$ are compact subsets of $\mathbb{R}^d$. This representation covers many important examples of random sets, for instance, random balls if $n = 1$, $l = 1$ and $M_1 = B_1(0)$.

5 Unions and convex hulls of random functions   291

Example 5.8 (Random rotation). Let M_0 be a non-random convex subset of $\mathbb{R}^2$ and let $\mathbf{w}_\beta$ denote a rotation (say clockwise) to the angle β. Denote $M(tu) = t^\eta \mathbf{w}_\beta M_0$ for $t > 0$ and $u = (\cos\beta, \sin\beta)$. Then the conditions of Theorem 5.7 are satisfied and $q_v(u) = q(\mathbf{w}_\beta^{-1} v)$, where $q(u) = \sup\{r : ru \in M_0\}$ for $u \in \mathbb{S}^1$. Similarly,

$$A_v = \{u = (\cos\beta, \sin\beta) \in \mathbb{S}^1 : \mathbf{w}_\beta^{-1} v \in S_0\},$$

where $S_0 = \{u\|u\|^{-1} : u \in M_0 \setminus \{0\}\}$. The limiting set in Theorem 5.7 is given by

$$Z = \left\{ xv : v \in \mathbb{S}^1, 0 \le x \le \sup_{u=(\cos\beta, \sin\beta) \in S_0} \left(\frac{q(u)}{\varphi(\mathbf{w}_\beta^{-1} v)}\right)^{\eta/\alpha} \right\}.$$

In the isotropic case $\varphi(u) = c$ for all $u \in \mathbb{S}^{m-1}$, whence $Z = B_r(0)$ with $r = c^{-\eta/\alpha}\|M_0\|$.

Example 5.9 (Random balls). Let $X_1 = B_\xi(\zeta)$ be a random ball in $\mathbb{R}^d$. Then $X_1 = M(\xi, \zeta)$, where $M(u) = B_{u_0}(u_1, \ldots, u_d)$ for $u = (u_0, u_1, \ldots, u_d) \in \mathbb{R}^m$, $u_0 \ge 0$, $m = d+1$. The function M satisfies the conditions of Theorem 5.7 with $\eta = 1$. For any $v \in \mathbb{S}^{d-1}$ and $u = (u_0, u_1, \ldots, u_d) \in \mathbb{S}^{m-1}$ we get

$$q_v(u) = \sup\{r : rv \in B_{u_0}(u_1, \ldots, u_d)\}.$$

In general the evaluation of $f(v)$ in Theorem 5.7 is very complicated. If $\varphi(u) = c$ is a constant on the unit sphere, then (5.2) holds. Since $q_v(u)$ attains its maximum for $(u_1, \ldots, u_d) = tv$ and $u_0^2 + u_1^2 + \cdots + u_d^2 = 1$,

$$\sup\{q_v(u) : u \in A_v\} = \sup\{t + u_0 : t^2 + u_0^2 = 1, t, e_0 \ge 0\} = \sqrt{2}.$$

Thus, the scaled unions converge to $B_r(0)$ with $r = c^{-1/\alpha}\sqrt{2}$.

Example 5.10 (Random triangles). Let $m = 6$, $d = 2$ and let $M(u)$ for $u = (u_1, \ldots, u_6)$ be the triangle with the vertices (u_1, u_2), (u_3, u_4) and (u_5, u_6). Then

$$Z = \left\{ vx : v \in \mathbb{S}^{d-1}, 0 \le x \le \sup_{u \in A_v} \frac{q_v(u)}{\varphi(u)^{\eta/\alpha}} \right\},$$

where

$$A_v = \{u = (u_1, \ldots, u_6) \in \mathbb{S}^{m-1} : M(u) \cap S_v \ne \emptyset\}.$$

The function $q_v(u) = q_v(u_1, \ldots, u_6)$ attains its minimum for $(u_{2i-1}, u_{2i}) = t_i v$, $i = 1, 2, 3$, i.e. for the degenerated triangle $M(u)$. In the isotropic case $\varphi(u) = c$ and

$$f(v) = \sup\{q_v(u)\varphi(u)^{-1/\alpha} : u \in A_v\}$$
$$= c^{-1/\alpha} \sup\{\max(t_1, t_2, t_3) : t_1^2 + t_2^2 + t_3^2 = 1\} = 1.$$

Hence Y_n almost surely converges to $Z = B_r(0)$ with $r = c^{-1/\alpha}$.

Regularly varying functions of random vectors

Results of Section 5.2 can be extended for unions of random sets defined as regularly varying multivalued functions of random vectors. Let ξ be a random point in a cone $\mathbb{C} \subset \mathbb{R}^m$ having the probability density function $f \in \mathrm{RV}^{\mathrm{u}}_{\alpha-m}(\mathbb{C})$ with $\alpha < 0$, so that $f = \varphi L$, where φ is homogeneous and L is a slowly varying function on $\mathbb{C}$. Consider independent copies $\{X_n, n \geq 1\}$ of a random compact set $X = M(\xi)$, where M is a multivalued function from $\mathrm{RV}^{\mathrm{u}}_{\alpha,g,\Phi}(\mathbb{C}; \mathcal{K})$ with $\eta > 0$. Define $a_n^{-1} Z_n = a_n^{-1}(X_1 \cup \cdots \cup X_n)$ for

$$a_n = \sup\{g(s) : s^\alpha L(se) \geq 1/n\}, \quad n \geq 1,$$

with a fixed $e \in \mathbb{C}' = \mathbb{C} \setminus \{0\}$.

The following result can be proved using the inversion theorem for multivalued regularly varying functions, see Theorem H.5.

Theorem 5.11. *Assume that $M \in \mathrm{RV}^{\mathrm{u}}_{\alpha,g,\Phi}(\mathbb{C}; \mathcal{K})$ and, for every $u_0 \in \mathbb{C}'$, positive r and K from a separating class $\mathcal{A} \subset \mathcal{K}$, the conditions $\Phi(u_0) \cap K \neq \emptyset$ and $\Phi(u_0) \cap \mathrm{Int}\, K = \emptyset$ yield the existence of $u_1, u_2 \in B_r(u_0)$ such that $\Phi(u_1) \cap K = \emptyset$ and $\Phi(u_2) \cap \mathrm{Int}\, K \neq \emptyset$. If M satisfies (H.17), then $a_n^{-1} Z_n$ converges weakly to the random closed set Z with the capacity functional given by*

$$T_Z(K) = \begin{cases} 1 - \exp\left\{-\int_{\Phi^-(K)} \varphi(u)\,du\right\}, & 0 \notin K, \\ 1, & \text{otherwise}, \end{cases}$$

where Φ^- is the inverse function to Φ, see (H.16).

Intersections of random half-spaces

For $u \in \mathbb{R}^d$ define the corresponding half-space $H(u)$ as

$$H(u) = \{x \in \mathbb{R}^d : \langle x, u \rangle \leq \|u\|^2\}.$$

Consider normalised intersections of random half-spaces defined as

$$a_n^{-1} Z_n = a_n^{-1}(H(\xi_1) \cap \cdots \cap H(\xi_n)),$$

where $a_n \to 0$ as $n \to \infty$ and $\xi, \xi_1, \xi_2, \ldots$ are i.i.d. random vectors in $\mathbb{R}^d$. Suppose that ξ has a density f which is regularly varying at zero with exponent $\alpha - d$ for $\alpha > 0$. This means that the function $\tilde{f}(u) = f(u\|u\|^{-2})$ belongs to the class $\mathrm{RV}^{\mathrm{u}}_{d-\alpha}(\mathbb{R}^d)$, see (H.6).

Let Y_n be the closure of $\mathbb{R}^d \setminus (a_n^{-1} Z_n)$. Then $a_n^{-1} Z_n$ converges in distribution to a random convex closed set Z if and only if Y_n converges in distribution to $Y = \mathrm{cl}(\mathbb{R}^d \setminus Z)$. Furthermore,

$$\mathbf{P}\{Y \cap K = \emptyset\} = \mathbf{P}\{K \subset \mathrm{Int}\, Z\} \tag{5.3}$$

for all $K \in \mathcal{K}$. The random closed set Y_n is the normalised union of the complements to the half-spaces $H(\xi_i)$, that is

$$Y_n = a_n^{-1}(M(\xi_1) \cup \cdots \cup M(\xi_n)),$$

where $M(u) = \{x \in \mathbb{R}^d : \langle x, u \rangle \geq \|u\|^2\}$ for $u \in \mathbb{R}^d$. Theorem 5.3 is applicable for the random set $X = M(\xi)$. Evidently,

$$M^-(K) = \{u : K \cap M(u) \neq \emptyset\}$$
$$= \{yv : v \in S_K^+, \ y \geq 0, \ h(K, v) \geq y\},$$

where $S_K^+ = \{v \in \mathbb{S}^{d-1} : h(K, v) > 0\}$. Then

$$\mathbf{P}\{X \cap tK \neq \emptyset\} = \mathbf{P}\{M(\xi) \cap tK \neq \emptyset\}$$

$$= t^d \int_{S_K^+} dv \int_0^{h(K,v)} y_1^{d-1} f(y_1 v t) dy_1$$

$$\sim t^\alpha \alpha^{-1} L(te) \int_{S_K^+} h(K, v)^\alpha \varphi(v) dv \quad \text{as } t \to 0,$$

where $e \in \mathbb{R}^d \setminus \{0\}$, L is slowly varying at zero and φ is a homogeneous function, such that $f = \varphi L$. If $a_n = \inf\{t \geq 0 : t^\alpha L(te) \geq n^{-1}\}$, then $Y_n \xrightarrow{d} Y$ with

$$T_Y(K) = 1 - \exp\left\{-\alpha^{-1} \int_{S_K^+} h(K, v)^\alpha \varphi(v) dv\right\}.$$

From (5.3) and using the continuity of φ we get

$$\mathbf{P}\{K \subset Z\} = \mathbf{P}\{K \subset \text{Int } Z\} = \exp\left\{-\alpha^{-1} \int_{S_K^+} h(K, v)^\alpha \varphi(v) dv\right\}.$$

For instance,

$$\mathbf{P}\{B_r(0) \subset Z\} = \exp\left\{-r^\alpha \alpha^{-1} \int_{\mathbb{S}^{d-1}} \varphi(v) dv\right\}.$$

6 Probability metrics method

6.1 Inequalities between metrics

The probability metrics method and its applications to limit theorems were elaborated by Zolotarev [630], see also Rachev [470]. The probability metrics method makes it possible to prove limit theorems for the most convenient metric. Afterwards, the speed of convergence in other metrics can be assessed using inequalities between the metrics.

Example 6.1 (Proof of central limit theorem). To explain why probability metrics are important, we will show how to prove the central limit theorem using probability metrics. Suppose that the metric $\mathfrak{m}(\xi, \eta)$ depends only on the marginal distributions of the random variables ξ and η and

(1) for some $\gamma > 2$, $\mathfrak{m}(c\xi, c\eta) = |c|^\gamma \mathfrak{m}(\xi, \eta)$, $c \in \mathbb{R}$, for all random variables ξ and η with finite second moments;

(2) $\mathfrak{m}(\xi + \zeta, \eta + \zeta) \leq \mathfrak{m}(\xi, \eta)$ if ζ is independent of both ξ and η.

Examples of such metrics can be found in Zolotarev [630] and Rachev [470]. Consider a random variable ξ such that $\mathfrak{m}(\xi, \zeta) < \infty$ for a centred Gaussian random variable ζ. Note that

$$\mathfrak{m}(\xi_1 + \xi_2, \eta_1 + \eta_2) \leq \mathfrak{m}(\xi_1 + \xi_2, \eta_1 + \xi_2) + \mathfrak{m}(\eta_1 + \xi_2, \eta_1 + \eta_2)$$
$$\leq \mathfrak{m}(\xi_1, \eta_1) + \mathfrak{m}(\xi_2, \eta_2).$$

Since $\zeta = n^{-1/2}(\zeta_1 + \cdots + \zeta_n)$ for $\zeta_1, \ldots, \zeta_n$ being i.i.d. copies of ζ,

$$\mathfrak{m}(n^{-1/2}(\xi_1 + \cdots + \xi_n), \zeta) = \mathfrak{m}(n^{-1/2}(\xi_1 + \cdots + \xi_n), n^{-1/2}(\zeta_1 + \cdots + \zeta_n))$$
$$= n^{-\gamma/2} \mathfrak{m}(\xi_1 + \cdots + \xi_n, \zeta_1 + \cdots + \zeta_n)$$
$$\leq n^{1-\gamma/2} \mathfrak{m}(\xi_1, \zeta) \to 0 \quad \text{as } n \to \infty.$$

By the way, the above expression provides an estimate for the speed of the convergence with respect to the chosen metric $\mathfrak{m}$.

Basic concepts of probability metrics for random closed sets are discussed in Section 1.6.3, in particular, the uniform metric $\mathfrak{u}(X, Y; \mathcal{M})$ and the Lévy metric $\mathfrak{L}(X, Y; \mathcal{M})$ for two random closed sets X and Y, where $\mathcal{M}$ is a subfamily of $\mathcal{K}$.

It is often desirable to obtain bounds for the speed of convergence in the uniform metric by relating it to other metrics that are better designed for the particular summation scheme. Recall that the classical inequality between uniform and Lévy metrics involves concentration functions of random variables, see Hengartner and Theodorescu [233]. The *concentration function* of a random closed set X is defined as

$$Q(\varepsilon, X; \mathcal{M}) = \sup\{T_X(K^\varepsilon) - T_X(K) : K \in \mathcal{M}\}, \quad \varepsilon > 0.$$

Evidently, $Q(\varepsilon, X; \mathcal{M})$ coincides with the uniform distance between the distributions of X and X^ε, i.e. $Q(\varepsilon, X; \mathcal{M}) = \mathfrak{u}(X, X^\varepsilon; \mathcal{M})$. The following theorem provides an inequality between the uniform and Lévy metrics.

Theorem 6.2 (Uniform and Lévy metrics). *If* $L = \mathfrak{L}(X, Y; \mathcal{M})$, *then*

$$L \leq \mathfrak{u}(X, Y; \mathcal{M}) \leq L + \min\{Q(L, X; \mathcal{M}), Q(L, Y; \mathcal{M})\}.$$

Proof. It suffices to refer to the obvious inequalities $L \leq u(X, Y; \mathcal{M})$ and

$$T_X(K) - T_Y(K) = T_X(K) - T_Y(K^\varepsilon) + T_Y(K^\varepsilon) - T_Y(K)$$
$$\leq \mathcal{L}(X, Y; \mathcal{M}) + Q(\varepsilon, Y; \mathcal{M})$$

for $\varepsilon \leq \mathcal{L}(X, Y; \mathcal{M})$. □

Example 6.3 (Half-lines). Let $X = (-\infty, \xi]$ and $Y = (-\infty, \eta]$ be random subsets of $\mathbb{R}^1$. Assume that $\{\inf K : K \in \mathcal{M}\} = \mathbb{R}^1$, i.e. the class $\mathcal{M}$ is sufficiently rich. Then $u(X, Y; \mathcal{M})$ coincides with the uniform distance between distribution functions of ξ and η and $\mathcal{L}(X, Y; \mathcal{M})$ equals the Lévy distance between them. The concentration functions of X and Y are equal to the concentration functions of the random variables ξ and η, see Hengartner and Theodorescu [233].

Example 6.4 (Distance between Poisson processes). Let X and Y be the Poisson point processes in $\mathbb{R}^d$ with intensity measures Λ_X and Λ_Y respectively. Then

$$|T_X(K) - T_Y(K)| \leq |\Lambda_X(K) - \Lambda_Y(K)|,$$

so that $u(X, Y; \mathcal{K})$ is bounded by the total variation distance between Λ_X and Λ_Y. If X and Y are stationary and have intensities λ_X and λ_Y, then $u(X, Y; \mathcal{K}) = g(\lambda_X/\lambda_Y)$, where

$$g(x) = \left| x^{1/(1-x)} - x^{x/(1-x)} \right|, \quad 0 < x < 1. \tag{6.1}$$

Example 6.5 (Distance between union-stable sets). If X and Y are union-stable with the same parameter α then

$$u(X, Y; \mathcal{M}_K) = \sup \left\{ |\exp\{-s^\alpha \Psi_X(K)\} - \exp\{-s^\alpha \Psi_Y(K)\}| : s \geq 0 \right\},$$

where $\mathcal{M}_K = \{sK : s \geq 0\}$ is the family of scale transformations of $K \in \mathcal{K}$. Without loss of generality assume that $\Psi_X(K) < \Psi_Y(K)$. Then

$$u(X, Y; \mathcal{M}_K) = g\left(\Psi_X(K)/\Psi_Y(K)\right),$$

where g is given by (6.1). Similarly, $Q(\varepsilon, X; \mathcal{M}_K) = g(\Psi_X(K)/\Psi_X(K^\varepsilon))$. If the random closed sets X and Y are stationary and union-stable and $\mathcal{M}_0$ is the family of all balls in $\mathbb{R}^d$, then

$$u(X, Y; \mathcal{M}_0) = u(X, Y; \mathcal{M}_{B_1(0)}) = g\left(\frac{\Psi_X(B_1(0))}{\Psi_Y(B_1(0))}\right),$$

$$Q(\varepsilon, X; \mathcal{M}_0) = g((1+\varepsilon)^{-|\alpha|}). \tag{6.2}$$

6.2 Ideal metrics and their applications

Definition

It was shown in Zolotarev [628, 630] that the so-called ideal metrics are particularly useful in the studies of limit theorems for random variables.

4 Unions of Random Sets

Definition 6.6 (Ideal metrics for random sets). A probability metric m is said to be:
  (i) *homogeneous* of degree γ if $\mathsf{m}(cX, cY) = |c|^\gamma \mathsf{m}(X, Y)$ for all $c \neq 0$;
  (ii) *regular* with respect to unions if $\mathsf{m}(X \cup Z, Y \cup Z) \leq \mathsf{m}(X, Y)$ for each random closed set Z independent of X and Y (regular metrics with respect to the Minkowski addition are defined similarly);
  (iii) *ideal* if m is homogeneous and regular.

Ideal metrics of zero degree

Unless otherwise stated we assume that with every set K the family $\mathcal{M}$ contains cK for all $c > 0$ and K^r for all $r > 0$. The typical example is the family $\mathcal{M}_0 = \{B_r(x) : r \geq 0, \; x \in \mathbb{R}^d\}$ of all closed balls. Then the uniform metric $\mathsf{u}(X, Y; \mathcal{M})$ is ideal of zero degree. Indeed, $\mathsf{u}(cX, cY; \mathcal{M}) = \mathsf{u}(X, Y; \mathcal{M}/c) = \mathsf{u}(X, Y; \mathcal{M})$ and

$$\mathsf{u}(X \cup Z, Y \cup Z; \mathcal{M}) = \sup\{|T_{X \cup Z}(K) - T_{Y \cup Z}(K)| : K \in \mathcal{M}\}$$
$$= \sup\{|T_X(K) + T_Z(K) - T_X(K)T_Z(K) - T_Y(K) - T_Z(K)$$
$$+ T_Z(K)T_Y(K)| : K \in \mathcal{M}\}$$
$$= \sup\{|(T_X(K) - T_Y(K))|(1 - T_Z(K)) : K \in \mathcal{M}\}$$
$$\leq \mathsf{u}(X, Y; \mathcal{M}).$$

The metric $\mathsf{u}(X, Y; \mathcal{K})$ is regular with respect to the Minkowski addition, since

$$\mathsf{u}(X + Z, Y + Z) = \sup\{|\mathbf{E}[T_X(K + \check{Z}) - T_Y(K + \check{Z}) | Z]| : K \in \mathcal{K}\} \leq \mathsf{u}(X, Y).$$

The Lévy metric $\mathcal{L}(X, Y; \mathcal{M})$ is regular with respect to unions and with respect to the Minkowski addition in the case $\mathcal{M} = \mathcal{K}$.

Ideal metric of a positive degree

Although u and $\mathcal{L}$ are ideal metrics of zero degree, an ideal metric of a positive degree γ is desirable to prove limit theorems as explained in Example 6.1. Such a metric can be constructed by generalising the uniform metric u. Put

$$\mathsf{u}_\upsilon(X, Y; \mathcal{M}) = \sup\{\upsilon(K)|T_X(K) - T_Y(K)| : K \in \mathcal{M}\},$$

where $\upsilon \colon \mathcal{K} \mapsto [0, \infty)$ is non-negative increasing and homogeneous of degree $\gamma > 0$, i.e. $\upsilon(sK) = s^\gamma \upsilon(K)$ for all $s > 0$ and $K \in \mathcal{K}$. Possible choices are $\upsilon(K) = (\mathrm{mes}(K))^{\gamma/d}$ or $\upsilon(K) = (C(K))^\gamma$, where mes is the Lebesgue measure and C is the Newton capacity. It is easy to see that u_υ is an ideal metric of degree γ with respect to unions.

Assume that $\upsilon(K^\delta) \to \upsilon(K^\varepsilon)$ as $\delta \to \varepsilon > 0$ for each $K \in \mathcal{K}$ and denote

$$\mathcal{U}_\delta(K_0) = \{K \in \mathcal{K} : K_0 \subset K \subset K_0^\delta\}$$

for $K_0 \in \mathcal{K}$ and $\delta > 0$. It is well known that the Lévy distance between random variables is equal to the side of the maximal square inscribed between the graphs of their cumulative distribution functions. Proposition 6.7 generalised this property for the Lévy distance between capacity functionals. The family $\mathcal{U}_\delta(K_0)$ plays the role of the square inscribed between the graphs of capacity functionals.

Proposition 6.7. *The value $\mathfrak{L}(X, Y; \mathcal{M})$ is equal to the supremum $\hat{L}$ of all positive δ such that for a compact set $K_0 \in \mathcal{M}$ with $|T_X(K_1) - T_Y(K_2)| \geq \delta$ for all $K_1, K_2 \in \mathcal{U}_\delta(K_0) \cap \mathcal{M}$.*

Proof. Let $\hat{L} < \delta$. Fix some $K \in \mathcal{M}$. Then $|T_X(K_1) - T_Y(K_2)| \leq \delta$ for some K_1, K_2 from $\mathcal{U}_\delta(K) \cap \mathcal{M}$. Hence

$$T_X(K) \leq T_X(K_1) \leq T_Y(K_2) + \delta \leq T_Y(K^\delta) + \delta.$$

Similarly, $T_Y(K) \leq T_X(K^\delta) + \delta$. Thus, $\mathfrak{L}(X, Y; \mathcal{M}) \leq \delta$ for all $\delta > \hat{L}$.

Let $\mathfrak{L}(X, Y; \mathcal{M}) < \delta$. For sufficiently small $\varepsilon > 0$ there exists a compact set K_0 such that $T_X(K_0) \geq T_Y(K_0^{\delta-\varepsilon}) + \delta - \varepsilon$ or $T_Y(K_0) \geq T_X(K_0^{\delta-\varepsilon}) + \delta - \varepsilon$. Then, for each K from $\mathcal{U}_{\delta-\varepsilon}(K_0)$,

$$T_X(K) \geq T_X(K_0) \geq T_Y(K_0^{\delta-\varepsilon}) + \delta - \varepsilon \geq T_Y(K) + \delta - \varepsilon$$

or

$$T_Y(K) \geq T_X(K) + \delta - \varepsilon.$$

Hence $\hat{L} \leq \delta - \varepsilon$ and, therefore, $\hat{L} \leq \mathfrak{L}(X, Y; \mathcal{M})$. □

The following result provides an inequality between $\mathfrak{u}_\upsilon$ and the Lévy metric $\mathfrak{L}$.

Theorem 6.8 (Weighted uniform and Lévy metrics). *If $L = \mathfrak{L}(X, Y; \mathcal{M})$, then*

$$\mathfrak{u}_\upsilon(X, Y; \mathcal{M}) \geq L^{1+\gamma} \inf_{x \in \mathbb{R}^d} \upsilon(B_1(x)).$$

Proof. If $\delta < L$, then Proposition 6.7 yields $|T_X(K_0^\delta) - T_Y(K_0^\delta)| \geq \delta$ for some $K_0 \in \mathcal{M}$. Hence

$$\mathfrak{u}_\upsilon(X, Y; \mathcal{M}) \geq \upsilon(K_0^\delta)|T_X(K_0^\delta) - T_Y(K_0^\delta)| \geq \delta \upsilon(K_0^\delta).$$

The homogeneity property of υ yields

$$\mathfrak{u}_\upsilon(X, Y; \mathcal{M}) \geq L \inf_{K \in \mathcal{M}} \upsilon(K^L)$$
$$\geq L \inf_{x \in \mathbb{R}^d} \upsilon(B_L(x))$$
$$= L^{1+\gamma} \inf_{x \in \mathbb{R}^d} \upsilon(B_1(x)). \quad \square$$

298   4 Unions of Random Sets

If the functional υ is translation invariant, then

$$\mathfrak{u}_\upsilon(X, Y; \mathcal{M}) \geq \mathfrak{L}(X, Y; \mathcal{M})^{1+\gamma} \upsilon(B_1(0)).$$

Theorem 6.2 yields the following inequality between $\mathfrak{u}_\upsilon$ and $\mathfrak{u}$:

$$\mathfrak{u}(X, Y; \mathcal{M}) \leq q + \min(Q(q, X; \mathcal{M}), Q(q, Y; \mathcal{M})), \quad (6.3)$$

where

$$q = (\mathfrak{u}_\upsilon(X, Y; \mathcal{M})/\upsilon(B_1(0)))^{1/(1+\gamma)}.$$

For example, (6.2) yields $\mathfrak{u}(X, Y; \mathcal{M}) \leq q + h((1+q)^{-\alpha})$ if X is stationary and union-stable with $\alpha > 0$.

Speed of convergence for normalised unions

Let $\{X_n, n \geq 1\}$ be i.i.d. copies of a random closed set X. If

$$Y_n = n^{1/\alpha}(X_1 \cup \cdots \cup X_n)$$

weakly converges to a non-trivial random set Z, then Z is union-stable with parameter α. To cover the cases of α being either positive or negative, we use the "inverse" set Z^* defined by (1.18).

Theorem 6.9. *Let Z be a union-stable random set with a parameter α and let υ be an increasing functional which is homogeneous of degree γ.*
(i) *If $\gamma > -\alpha > 0$ and $\mathfrak{u}_\upsilon(X_1, Z; \mathcal{M}) < \infty$, then*

$$\mathfrak{u}_\upsilon(Y_n, Z; \mathcal{M}) \leq n^{1+\gamma/\alpha} \mathfrak{u}_\upsilon(X_1, Z; \mathcal{M}). \quad (6.4)$$

(ii) *If $\gamma > \alpha > 0$ and $\mathfrak{u}_\upsilon(X_1^*, Z^*; \mathcal{M}) < \infty$, then*

$$\mathfrak{u}_\upsilon(Y_n, Z; \mathcal{M}) \leq n^{1-\gamma/\alpha} \mathfrak{u}_\upsilon(X_1^*, Z^*; \mathcal{M}).$$

Proof.
(i) The proof follows the scheme described in Example 6.1. Since Z is union-stable,

$$Z \stackrel{d}{\sim} n^{1/\alpha}(Z_1 \cup \cdots \cup Z_n),$$

for i.i.d. random closed sets $Z_1, \ldots, Z_n$ having the same distribution as Z. Then (6.4) follows from the homogeneity and regularity properties of m. Indeed,

$$\mathfrak{u}_\upsilon(Y_n, Z; \mathcal{M}) = \mathfrak{u}_\upsilon\left(n^{1/\alpha}(X_1 \cup \cdots \cup X_n), n^{1/\alpha}(Z_1 \cup \cdots \cup Z_n); \mathcal{M}\right)$$

$$\leq n^{\gamma/\alpha} \sum_{k=1}^{n} \mathfrak{u}_\upsilon(X_k, Z_k; \mathcal{M})$$

$$= n^{1+\gamma/\alpha} \mathfrak{u}_\upsilon(X_1, Z; \mathcal{M}).$$

(ii) Since $(cF)^* = c^{-1}F^*$, we have $Y_n^* = n^{-1/\alpha}(X_1^* \cup \cdots \cup X_n^*)$, so that (ii) immediately follows from (6.4). □

Theorem 6.9 and inequality (6.3) yield that

$$\mathfrak{u}(Y_n, Z; \mathcal{M}) \leq q + Q(q, Z; \mathcal{M}),$$

where

$$q = \left(n^{1+\gamma/\alpha} \mathfrak{u}_\upsilon(A_1, Z; \mathcal{M})/\upsilon(B_1(0))\right)^{1/(1+\gamma)}.$$

Notes to Chapter 4

Section 1.1. There is a vast literature on extreme values of random variables and extensions to the multivariate case, see, e.g. monographs by Galambos [186], Leadbetter, Lindgren and Rootzen [346], Resnick [481] and Reiss [477].

The studies of max-stable vectors in $\mathbb{R}^d$ with respect to coordinatewise maximum were initiated by Balkema and Resnick [47] and de Haan and Resnick [216]. These ideas were extended for pointwise maximum of random functions in [201, 215, 216, 431] and for a rather general case of random elements in a lattice in Gerritse [189] and Norberg [432, 434]. Schlather [514] considered models for stationary max-stable random fields.

Section 1.2. The concept of infinite divisibility for unions is due to Matheron [381] whose proof of Theorem 1.6 is given here for a general case that allows for possible fixed points. Matheron also characterised union-infinitely-divisible semi-Markov random closed sets.

Section 1.3. The first to study the stability of random closed sets with respect to unions was Matheron [381]. He considered the simplest case of union-stable sets without fixed points. General union-stable random closed sets were characterised by Molchanov [395], where Theorem 1.12 was proved.

Following Example 1.19, it is shown in Molchanov [398] that a random closed set with the capacity functional derived from the Riesz capacity can be obtained by considering paths of stable processes with exponent $d - \alpha$. Recall that a homogeneous stochastic process $\zeta(t)$ with independent increment is said to be stable if its increment has a stable distribution, so that the characteristic function of process $\zeta(t) - \zeta(0)$ is $\exp\{-c|t|^{d-\alpha}\}$ for $d - 2 < \alpha \leq d$.

Section 1.4. The concepts of stability and infinite divisibility for random sets were discussed from a very general point of view by Trader [565]. However D.A. Trader's approach evaded some difficulties. For instance, the characterisation problem was merely reduced to some functional equations. Theorem 1.23 was proved in Trader [565] for $m = 2$ implicitly assuming that X is not invariant under any non-trivial translation. The current proof rectifies this deficiency. There are close connections between max-stable stochastic processes and union-stable random sets, since the hypograph of any max-stable process is a union-stable random set (cf. Section 5.3.3). Operator-stable distributions and the corresponding convergence of types theorems in linear spaces are considered by Jurek and Mason [284].

Proposition 1.21 is new. Different normalisations along different axes were used by Penrose [452] to define *semi-min-stable* processes. These processes have epigraphs that satisfy Proposition 1.21.

Section 1.5. This section refers to the results of Norberg [432], which go back to the studies of infinitely divisible functions on semigroups, see Appendix G. However, the lattice framework allows us to incorporate very naturally topological considerations which are missed if one exclusively follows the semigroup approach. Self-decomposable random elements in a lattice and self-decomposable random vectors with respect to coordinatewise maximum have been studied by Gerritse [189].

Section 2.1. A limit theorem for normalised unions of random closed sets was proved by Molchanov [398, 399]. It is quite natural that the regular variation condition plays the key role in limit theorems for unions of random sets. Appendix H surveys some results on numerical regularly varying functions.

Random closed sets with capacity functionals related to derivatives of capacities appear as conditional limiting distributions for unions with $a_n \to 0$, see Section 1.5.5.

Theorem 2.3 is taken from Molchanov [398]. Pancheva [440] and Zolotarev [630] considered a very general normalising scheme for the maximum of random variables and showed that the use of non-linear normalisations led to unification of max-stable and self-decomposable laws. The main difficulty in generalising this idea for random sets lies in the solving of some functional equations in the space of closed sets.

Section 2.2. Theorem 2.5 and the corresponding example are taken from Molchanov [398]. Theorems 2.8 and 2.9 are apparently new.

Section 2.3. The results of Norberg [429] concern the scheme of series, where the regular variation of the capacity functional appears as an implicit condition. A generalisation for lattice-valued random elements is described in Norberg [434].

Superpositions of point processes can be naturally interpreted as unions of random closed sets. Along the same lines, Nagel and Weiss [421] consider limits of rescaled superpositions of stationary planar tessellations. The limits are given by tessellations formed by a stationary Poisson line process.

Section 3.1. The concept of regularly varying capacities in Definition 3.2 was introduced by Molchanov [402]. It can be extended to general functions defined on a subclass of $\mathcal{K}$ that is closed under homotheties.

Section 3.2. Results on the almost sure convergence of scaled unions are due to Molchanov [402], see also Molchanov [398, Ch. 5]. The almost sure convergence for random samples was investigated by Eddy [154] and Eddy and Gale [156] for the case of Gaussian samples. Theorem 3.7 is due to Davis, Mulrow and Resnick [119], while its proof based on application of Theorem 3.5 can be found in Molchanov [398, Th. 3.3]. Example 3.8 replicates the result of Fisher [172]. Further examples for the almost sure limits of random samples can be found in Davis, Mulrow and Resnick [119] and Kinoshita and Resnick [313].

Section 3.3. It is possible to mention a number of studies concerned with almost sure limits for a random sample (or its convex hull) in $\mathbb{R}^d$ as the sample size increases. This problem was finally solved by Kinoshita and Resnick [313] who found necessary and sufficient conditions for the convergence of scaled samples of random vectors. Almost sure stability of random samples was studied by McBeth and Resnick [384]. Section 3.3 uses the technique of Kinoshita and Resnick [313] to formulate (apparently new) necessary and sufficient conditions for the stability of unions of random compact sets.

Section 3.4. The presentation follows Schreiber [526, 525]. Theorem 3.15 is proved in Schreiber [526]. It is also possible to use these results in order to prove a limit theorem for the mean width of a random polyhedron formed by a random sample in a ball. Schreiber [525] presented refinements of these results that can be obtained if the avoidance functional $Q_Y(Z_n)$ is replaced by $\mu(Z_n^c)$ for a probability measure μ. A central limit theorem for the measures of the complement of Z_n is proved in Schreiber [525].

Yurachkivsky [620] applied martingale methods in order to derive a central limit theorem for the measure of the union of random sets.

Section 4.1. The concepts of infinite divisibility and stability of random closed sets with respect to convex hulls go back to Trader [565], who obtained the first characterisation result in this direction. D.A. Trader's proof was based on semigroup techniques and did not give an explicit representation of the containment functional, but just the fact that infinitely divisible sets appear as convex hulls of a Poisson point process on the space of compact sets. The infinite divisibility property for convex hulls is characterised by Norberg [432] and Giné, Hahn and Vatan [201]. T. Norberg's proof is essentially lattice theoretic (as outlined in Section 4.1), while Giné, Hahn and Vatan obtained their characterisation through representations of continuous max-infinitely-divisible functions on a metric space.

Section 4.2. Definition 4.5 of the stability of random convex compact sets with respect to convex hulls appeared first in Giné, Hahn and Vatan [201] where such a set X is additionally assumed to be compact with support function having a non-degenerate distribution. Some unnatural restrictions have been removed by Molchanov [397, 398]. The current formulations of the results of Section 4.2 are new, although they can be traced back to their remote origins in Molchanov [397]. Assuming the compactness and a non-degenerate distribution of the support functions at each direction, it is possible to represent the functional ψ as an integral, which provides a spectral representation of distributions for convex-stable random sets. The details can be found in Giné, Hahn and Vatan [201].

Section 4.3. The weak convergence of convex hulls was considered in Molchanov [397]. The presented results provide a slight extension of [397] which dealt mostly with random convex compact sets.

Almost sure and weak convergence results for convex hulls of random closed sets can be deduced from pointwise limit theorems for the maximum of support functions, the latter extensively studied by de Haan and Resnick [215, 216] and Norberg [431].

Section 5.1. The presentation follows Molchanov [398], where further results can be found. Theorem 5.1 goes back to Resnick [480], while its proof based on application of Theorem 2.1 can be found in Molchanov [398, 399] (in a slightly different form). Some statistical applications of the estimates for tail probabilities of volumes of convex hulls are discussed in Molchanov [398, Sec. 8.2]. Results on the almost sure convergence of scaled random samples can be obtained from Theorem 3.12, see further examples in Davis, Mulrow and Resnick [119] and Kinoshita and Resnick [313]. The convergence of distributions of scaled samples may be also used as an alternative definition of regularly varying distributions, see Resnick [481].

Another direction of research concerns limiting behaviour of convex hulls of random samples. This vast programme of research was initiated by Rényi and Sulanke [478], Efron [158] and Carnal [86]. The major direction of work was the development of asymptotic formulae for geometric quantities (number of vertices, perimeter, area, etc.) of the convex hull as the sample size increases. For more recent results, surveys and further references see Schneider [519] and Groeneboom [212].

The fact that asymptotic behaviour of convex hulls is related to the limit theorems for the corresponding samples considered as random closed sets has not been explicitly used in the above mentioned studies. Brozius and de Haan [79, 80] and Davis, Mulrow and Resnick [118] showed how to deduce some asymptotic formulae for the geometric characteristics of the convex hull from the weak convergence of the corresponding random convex sets and how to deduce the convergence of moments. The convex hull of a sample can be used to estimate the support of distribution (Moore [413] and Ripley and Rasson [493]), as an order statistic for a multivariate sample (Barnett [50] and Eddy [155]) and as multivariate quantiles. The convex hull can be used to assess circular symmetry of the sample and to trim the sample. Stam [540] studied expectation and the variance of the volume of unions of random sets obtained as unions of random translations of a deterministic set.

Section 5.2. Limit theorems for unions of homogeneous multivalued functions were considered in Molchanov [398]. Theorem 5.7 on the almost sure convergence is proved in Molchanov [398, 402]. Proposition 5.5 is proved in Molchanov [398, Th. 4.8]. Applied to random singletons, Theorem 5.7 turns into Theorem 6.3 from Davis, Mulrow and Resnick [119] without a strict monotonicity condition imposed on φ in [119].

Multivalued regular varying functions of random vectors are discussed in Molchanov [398, 401]. Intersections of random half-spaces using the approach based on unions of random sets were studied by Molchanov [398, Sec. 8.5]. They can be interpreted as feasible sets in linear programming problems with random constraints.

Section 6. The probability metric method in application to limit theorems for random variables was developed by Zolotarev [628, 630]. A comprehensive monograph by Rachev [470] covers many aspects of probability metrics. Properties of concentration functions for random closed sets are considered in Molchanov [398, Sec. 7.2]. Applications to unions of random sets are due to Molchanov [398, 403], where further results along the lines described in Section 6 can be found.

5
Random Sets and Random Functions

1 Random multivalued functions

A random set is a multivalued measurable function defined on a probability space. If this multivalued function depends on the second argument (time), then random processes of sets (or random multivalued functions) appear. Important examples are provided by growth processes, multivalued martingales and solutions of stochastic differential inclusions. We consider often the case of the discrete time, i.e. sequences of random closed sets.

1.1 Multivalued martingales

Definition and main properties

The concept of a multivalued (or set-valued) martingale relies on the definition of the conditional selection expectation, see Section 2.1.6. Let $\{\mathfrak{F}_n, \ n \geq 1\}$ be a *filtration* on $(\Omega, \mathfrak{F}, \mathbf{P})$, i.e. family of complete sub-σ-algebras of $\mathfrak{F}$ such that $\mathfrak{F}_m \subset \mathfrak{F}_n$ if $m \leq n$.

An adapted family $\{\xi_n, n \geq 1\}$ of $\mathbb{E}$-valued integrable random elements is called a *martingale* if $\mathbf{E}(\xi_{n+1}|\mathfrak{F}_n) = \xi_n$ for every $n \geq 1$, see Chatterji [96] and Vakhaniya, Tarieladze and Chobanyan [568]. When discussing martingales the filtration is usually mentioned along with the relevant random elements. However, we often omit the filtration if it is $\mathfrak{F}_n, n \geq 1$.

Consider a sequence of random closed sets $\{X_n, n \geq 1\}$ such that X_n is Effros measurable with respect to $\mathfrak{F}_n$, i.e. $\{X_n \cap G \neq \emptyset\} \in \mathfrak{F}_n$ for every open G. Then the set-valued process $\{X_n, n \geq 1\}$ is called *adapted*. A sequence $\{X_n, n \geq 1\}$ is adapted with respect to the minimal filtration $\mathfrak{F}_n = \sigma(X_m, m \leq n)$ generated by X_m for $m \leq n$.

Definition 1.1 (Multivalued martingales). An adapted sequence $(X_n, \mathfrak{F}_n), n \geq 1$, of multivalued integrable random convex closed sets is called a multivalued (or set-valued)

(i) *martingale* if $\mathbf{E}(X_{n+1}|\mathfrak{F}_n) = X_n$ a.s. for all $n \geq 1$;
(ii) *supermartingale* if $\mathbf{E}(X_{n+1}|\mathfrak{F}_n) \subset X_n$ a.s. for all $n \geq 1$;
(iii) *submartingale* if $\mathbf{E}(X_{n+1}|\mathfrak{F}_n) \supset X_n$ a.s. for all $n \geq 1$.

All random closed sets in Definition 1.1 are almost surely convex. In the classical case of single-valued processes with real values, submartingales and supermartingales are closely related, i.e. $-\xi$ is a submartingale if and only if ξ is a supermartingale. This is not the case for set-valued random processes, so that multivalued submartingales and supermartingales require separate treatments. In the single-valued case Definition 1.1 complies with the definition of a martingale in a Banach space and makes no difference between single-valued martingales, submartingales or supermartingales.

Example 1.2. Let ξ_n be a real-valued martingale. Then $X_n = \mathrm{co}(0, \xi_n)$ (the segment with end-points zero and ξ_n) is a multivalued martingale if ξ_n is almost surely positive for all n or almost surely negative. If ξ_n is a single-valued martingale in $\mathbb{R}^d$, then $X_n = \mathrm{co}(0, \xi_n)$ is a multivalued submartingale.

A multivalued martingale is called *integrably bounded* if $\mathbf{E}\|X_n\| < \infty$ (i.e. X_n is integrably bounded) for every $n \geq 1$. Multivalued martingales give rise to a number of martingale-like sequences with values in the real line.

Theorem 1.3 (Numerical martingales generated by multivalued ones).
(i) If X_n is a multivalued submartingale such that X_n is bounded for every $n \geq 1$, then $\|X_n\|$ is a submartingale.
(ii) If $x \in \mathbb{E}$ and X_n is a multivalued supermartingale, then $\rho(x, X_n)$ is a submartingale.
(iii) If X_n is an integrably bounded multivalued martingale (respectively submartingale, supermartingale), then their support functions $h(X_n, u)$, $n \geq 1$, form a martingale (respectively submartingale, supermartingale) for every linear functional u.

Proof.
(i) By the definition of conditional expectation,

$$\|X_n\| \leq \|\mathbf{E}(X_{n+1}|\mathfrak{F}_n)\| = \sup\{\|\mathbf{E}(\xi|\mathfrak{F}_n)\| : \xi \in \mathcal{S}^1_{\mathfrak{F}_{n+1}}(X_{n+1})\}$$
$$\leq \mathbf{E}(\sup\{\|\xi\| : \xi \in \mathcal{S}^1_{\mathfrak{F}_{n+1}}(X_{n+1})\}|\mathfrak{F}_n)$$
$$= \mathbf{E}(\|X_{n+1}\||\mathfrak{F}_n).$$

(ii) It suffices to show that $\rho(x, \mathbf{E}(X|\mathfrak{H})) \leq \mathbf{E}(\rho(x, X)|\mathfrak{H})$ for any sub-σ-algebra $\mathfrak{H} \subset \mathfrak{F}$ and any integrable random closed set X. By the fundamental selection theorem, for each $\delta > 0$ there exists an integrable selection $\eta \in \mathcal{S}^1(X)$ satisfying

$$\|x - \eta\| \leq \rho(x, X) + \delta \quad \text{a.s.}$$

Taking conditional expectations and applying Jensen's inequality yields

$$\|x - \mathbf{E}(\eta|\mathfrak{H})\| = \|\mathbf{E}(x - \eta|\mathfrak{H})\| \leq \mathbf{E}(\rho(x, F)|\mathfrak{H}) + \delta.$$

Since $\mathbf{E}(\eta|\mathfrak{H}) \in \mathbf{E}(X|\mathfrak{H})$,

$$\rho(x, \mathbf{E}(X|\mathfrak{H})) \leq \mathbf{E}(\rho(x, F)|\mathfrak{H}) + \delta,$$

which immediately yields the required statement.
(iii) follows from Theorem 2.1.47(iv). □

The following result easily follows from the definition of the conditional expectation.

Proposition 1.4. *If X_n and Y_n are multivalued submartingales (respectively supermartingales), then $X_n \cup Y_n$ (respectively $X_n \cap Y_n$) is a multivalued submartingale (respectively supermartingale).*

Convergence of multivalued (super-) martingales

The embedding technique (see Theorem 2.1.21) allows us to replace the selection expectation with the Bochner integral in a linear space where all convex sets can be embedded. This approach works if $\{X_n, n \geq 1\}$ is a sequence of either random compact sets or integrably bounded Hausdorff approximable random convex closed sets in a reflexive space. Then it is possible to apply the martingale convergence theorem in Banach spaces (see Chatterji [96]) to deduce the following result.

Theorem 1.5 (Convergence of conditional expectations). *Let $X_n = \mathbf{E}(X|\mathfrak{F}_n)$, $n \geq 1$, where X is an integrably bounded random convex compact set (if $\mathbb{E}$ is reflexive, the compactness can be replaced by the Hausdorff approximability assumption). Then $\mathbf{E}\rho_H(X_n, X_\infty) \to 0$ and $\rho_H(X_n, X_\infty) \to 0$ a.s., where $X_\infty = \mathbf{E}(X|\mathfrak{F}_\infty)$ and $\mathfrak{F}_\infty$ is the σ-algebra generated by $\cup_{n=1}^\infty \mathfrak{F}_n$.*

The following result concerns general integrably bounded multivalued martingales.

Theorem 1.6 (Convergence of multivalued martingales). *Assume that $\mathbb{E}$ has the Radon–Nikodym property and that $\mathbb{E}^*$ is separable. Let $(X_n, \mathfrak{F}_n)$ be an integrably bounded multivalued martingale. If $\{\|X_n\|, n \geq 1\}$ is uniformly integrable, then there exists a unique integrably bounded random convex closed set X_∞ such that $X_n = \mathbf{E}(X_\infty|\mathfrak{F}_n)$ for all $n \geq 1$.*

Proof. Without loss of generality assume that $\mathfrak{F} = \mathfrak{F}_\infty$. Let $\varXi$ be the family of all integrable $\mathbb{E}$-valued random elements ξ such that $\mathbf{E}(\xi|\mathfrak{F}_n) \in \mathcal{S}_{\mathfrak{F}_n}^1(X_n)$ for every $n \geq 1$. It is easy to see that $\varXi$ is a convex closed subset of $\mathbf{L}^1 = \mathbf{L}^1(\varOmega; \mathbb{E})$. Let us show that $\varXi$ is bounded and decomposable.

Let $\xi, \eta \in \varXi$ and $A \in \mathfrak{F}$. Denote $\xi_n = \mathbf{E}(\xi|\mathfrak{F}_n)$ and $\eta_n = \mathbf{E}(\eta|\mathfrak{F}_n)$, $n \geq 1$. The convergence theorem for real-valued martingales implies that $\xi_n \to \xi$ and $\eta_n \to \eta$ a.s. Then

$$\zeta_n = \mathbf{E}(\mathbf{1}_A|\mathfrak{F}_n)\xi_n + \mathbf{E}(\mathbf{1}_{\Omega\setminus A}|\mathfrak{F}_n)\eta_n$$

is an integrable $\mathfrak{F}_n$-measurable selection of X_n, so that $\{\|\zeta_n\|, n \geq 1\}$ is uniformly integrable. Since $\zeta_n \to \zeta = \mathbf{1}_A \xi + \mathbf{1}_{\Omega\setminus A}\eta$ a.s., it follows that $\mathbf{E}\|\zeta_n - \zeta\| \to 0$. If $m \geq n$, then $\mathbf{E}(\zeta_m|\mathfrak{F}_n) \in \mathcal{S}^1_{\mathfrak{F}_n}(X_n)$ and

$$\mathbf{E}\|\mathbf{E}(\zeta|\mathfrak{F}_n) - \mathbf{E}(\zeta_m|\mathfrak{F}_n)\| \leq \mathbf{E}\|\zeta - \zeta_m\|.$$

Letting m go to infinity yields $\mathbf{E}(\zeta|\mathfrak{F}_n) \in \mathcal{S}^1_{\mathfrak{F}_n}(X_n)$, so that $\zeta \in \varXi$, which means that $\varXi$ is decomposable. For every $\xi \in \varXi$,

$$\mathbf{E}\|\xi\| = \lim_{n\to\infty} \mathbf{E}\|\mathbf{E}(\xi|\mathfrak{F}_n)\| \leq \sup_{n\geq 1} \mathbf{E}\|X_n\|,$$

so that $\varXi$ is bounded.

Let us show that for every $n \geq 1$, $\xi \in \mathcal{S}^1_{\mathfrak{F}_n}(X_n)$ and $\varepsilon > 0$ there exists a $\eta \in \varXi$ such that

$$\|\mathbf{E}(\mathbf{1}_A \xi) - \mathbf{E}(\mathbf{1}_A \eta)\| \leq \varepsilon, \quad A \in \mathfrak{F}_n. \tag{1.1}$$

Without loss of generality put $n = 1$. Since $X_j = \mathbf{E}(X_{j+1}|\mathfrak{F}_j)$, $j \geq 1$, we can choose a sequence $\{\xi_i, i \geq 1\}$ with $\xi_1 = \xi$ such that $\xi_j \in \mathcal{S}^1_{\mathfrak{F}_j}(X_j)$ and

$$\mathbf{E}\|\xi_j - \mathbf{E}(\xi_{j+1}|\mathfrak{F}_j)\| < 2^{-j}\varepsilon, \quad j \geq 1.$$

If $m > j \geq k$, then

$$\|\mathbf{E}(\mathbf{1}_A \xi_j) - \mathbf{E}(\mathbf{1}_A \xi_m)\| = \left\|\sum_{i=j}^{m-1} \mathbf{E}(\mathbf{1}_A(\xi_i - \mathbf{E}(\xi_{i+1}|\mathfrak{F}_i)))\right\|$$

$$\leq \sum_{i=j}^{m-1} \mathbf{E}\|\xi_i - \mathbf{E}(\xi_{i+1}|\mathfrak{F}_i)\| < 2^{-j+1}\varepsilon \tag{1.2}$$

for all $A \in \mathfrak{F}_k$. Therefore,

$$\lambda(A) = \lim_{m\to\infty} \mathbf{E}(\mathbf{1}_A \xi_m)$$

exists for every $A \in \bigcup_{k=1}^{\infty} \mathfrak{F}_k$. Because of the uniform integrability, the limit exists for all $A \in \mathfrak{F}$. The Radon–Nikodym property implies that there exists $\eta \in \mathbf{L}^1$ such that $\lambda(A) = \mathbf{E}(\mathbf{1}_A \eta)$ for all $A \in \mathfrak{F}$. Since $\mathbf{E}(\xi_m|\mathfrak{F}_j) \in \mathcal{S}^1_{\mathfrak{F}_j}(X_j)$ for $m \geq j$,

$$\mathbf{E}(\mathbf{1}_A \mathbf{E}(\eta|\mathfrak{F}_j)) = \mathbf{E}(\mathbf{1}_A \eta) = \lim_{m\to\infty} \mathbf{E}(\mathbf{1}_A \xi_m)$$

$$= \lim_{m\to\infty} \mathbf{E}(\mathbf{1}_A \mathbf{E}(\xi_m|\mathfrak{F}_j)) \in \mathbf{E}^{\mathfrak{F}_j}(\mathbf{1}_A X_j)$$

for all $A \in \mathfrak{F}_j$ and $j \geq 1$, see (2.1.13). By Proposition 2.1.18(i), $\mathbf{E}(\eta|\mathfrak{F}_j) \in \mathcal{S}^1_{\mathfrak{F}_j}(X_j)$ for all $j \geq 1$, so that $\eta \in \varXi$. Letting $m \to \infty$ and $j = k = 1$ in (1.2) yields (1.1). This also shows that $\varXi$ is not empty.

By Theorem 2.1.6 there exists a random closed set X_∞ such that $\Xi = \mathcal{S}^1(X_\infty)$. Furthermore, X_∞ is convex by Proposition 2.1.5 and integrably bounded by Theorem 2.1.19. By (1.1),

$$\begin{aligned}\mathbf{E}^{\mathfrak{F}_n}(X_n) &= \mathrm{cl}\{\mathbf{E}(\mathbf{1}_A \xi) : \xi \in \mathcal{S}^1_{\mathfrak{F}_n}(X_n)\} \\ &= \mathrm{cl}\{\mathbf{E}(\mathbf{1}_A \eta) : \eta \in \Xi\} \\ &= \mathbf{E}(\mathbf{1}_A X_\infty), \quad A \in \mathfrak{F}_n,\, n \geq 1.\end{aligned}$$

By Theorem 2.1.49(iii), $X_n = \mathbf{E}(X_\infty | \mathfrak{F}_n)$ for $n \geq 1$.

It remains to confirm the uniqueness of the limiting random closed set. Let Y be an integrably bounded random closed set such that $X_n = \mathbf{E}(Y|\mathfrak{F}_n)$ for $n \geq 1$. Then $\mathbf{E}(\mathbf{1}_A Y) = \mathbf{E}(\mathbf{1}_A X_\infty)$ for all $A \in \cup_{n=1}^\infty \mathfrak{F}_n$ and, by an approximation argument, for all $A \in \mathfrak{F}$. Proposition 2.1.18(i) implies that $Y = X_\infty$ a.s. □

Combining Theorem 1.6 with Theorem 1.5 and noticing that in the Euclidean space every random convex closed set is Hausdorff approximable, we obtain that any multivalued martingale in $\mathbb{R}^d$ with uniformly integrable norms converges. The following result establishes the convergence for multivalued supermartingales.

Theorem 1.7 (Multivalued supermartingales in $\mathbb{R}^d$). *Consider a multivalued supermartingale $(X_n, \mathfrak{F}_n)$ in $\mathbb{R}^d$, such that $\sup_{n \geq 1} \mathbf{E}\|X_n\| < \infty$. Then there exists an integrably bounded random convex compact set X_∞ such that $\rho_\mathrm{H}(X_n, X_\infty) \to 0$ a.s. If $\{\|X_n\|, n \geq 1\}$ is uniformly integrable, then also $\mathbf{E}\rho_\mathrm{H}(X_n, X_\infty) \to 0$ as $n \to \infty$.*

Proof. For every $u \in \mathbb{R}^d$ the support function $h(X_n, u)$, $n \geq 1$, is a supermartingale. The supermartingale convergence theorem implies that $h(X_n, u)$ $\mathbf{L}^1$-converges to an integrable random variable $\alpha(u)$. Thus, for all $A \in \mathfrak{F}$,

$$f_n(A, u) = \mathbf{E}(\mathbf{1}_A h(X_n, u)) \to f(A, u) = \mathbf{E}(\mathbf{1}_A \alpha(u)),$$

whence $|f(A, u)| \leq c = \sup_{n \geq 1} \mathbf{E}\|X_n\|$. It is easy to see that $f(A, u)$ is a support function for every u. By the Radon–Nikodym theorem for multivalued measures (Theorem 2.1.35(ii)), $f(A, u) = h(\mathbf{E}(\mathbf{1}_A X_\infty), u)$. Therefore, $h(X_n, u) \to h(X_\infty, u)$ a.s. for all u from a countable dense subset of $\mathbb{R}^d$. The proof is finished by using the fact that the pointwise convergence of support functions to a continuous limit implies the Painlevé–Kuratowski convergence of the corresponding sets, which in turn yields the convergence in the Hausdorff metric provided that the limit is compact and convex, see Papageorgiou [444, Th. 2.2]. □

The Banach space variant of Theorem 1.7 holds if the uniform integrability assumption is replaced by a stronger one that requires that $\{X_n, n \geq 1\}$ is *uniformly integrably bounded*, i.e. $\sup\{\|X_n\| : n \geq 1\}$ is integrable.

Theorem 1.8 (Mosco convergence of multivalued maringales). *Assume that $\mathbb{E}$ has the Radon–Nikodym property and $\mathbb{E}^*$ is separable. If $(X_n, \mathfrak{F}_n), n \geq 1$, is a uniformly integrably bounded multivalued martingale, then X_n a.s. converges in the Mosco sense as $n \to \infty$ to an integrably bounded random convex closed set X_∞.*

Proof. Theorem 1.6 yields the existence of X_∞ such that $\mathbf{E}(X_\infty|\mathfrak{F}_n) = X_n$. For every integrable selection $\xi \in \mathcal{S}^1(X_\infty)$, the sequence $\xi_n = \mathbf{E}(\xi|\mathfrak{F}_n)$ is a single-valued martingale in $\mathbb{E}$. By the convergence theorem for $\mathbb{E}$-valued martingales, $\xi_n \to \xi$ a.s. in the strong topology on $\mathbb{E}$. Thus $X_\infty \subset \text{s-lim inf } X_n$ a.s.

It will be shown in Theorem 1.12(ii) below that $X_n = \text{cl}\{\xi_n^{(k)}, k \geq 1\}$ for every $n \geq 1$, where $\{\xi_n^{(k)}, n \geq 1\}$ is a martingale for every $k \geq 1$. Since $h(X_n, u) = \sup\{\langle \xi_n^{(k)}, u \rangle : k \geq 1\}$, using the fact that $\langle \xi_n^{(k)}, u \rangle$ is a real-valued martingale, it is possible to deduce that

$$\sup_{k \geq 1} \langle \xi_n^{(k)}, u \rangle \to \sup_{k \geq 1} \langle \xi^{(k)}, u \rangle \quad \text{a.s. as } n \to \infty,$$

where $\{\xi^{(k)}, k \geq 1\}$ are integrable selections of X_∞. The exceptional sets in the above almost sure convergence can be combined using the separability of $\mathbb{E}^*$, which implies

$$\limsup h(X_n, u) \leq h(X_\infty, u) \quad \text{a.s.,} \quad u \in \mathbb{E}^*.$$

Therefore, w-$\limsup X_n \subset X_\infty$ a.s., so that $X_n \xrightarrow{M} X_\infty$ a.s. □

The following result provides a rather general convergence theorem for possibly unbounded supermartingales in Banach spaces.

Theorem 1.9 (Convergence of supermartingales in Banach space). *Let $\mathbb{E}$ be a separable Banach space and let $(X_n, \mathfrak{F}_n)$, $n \geq 1$, be a multivalued supermartingale such that $\sup_{n \geq 1} \mathbf{E}\rho(0, X_n) < \infty$ and $X_n \subset Y$ a.s. for every $n \geq 1$, where Y is a random closed set such that the intersection of Y with an arbitrary closed ball is almost surely weakly compact. Then X_n converges in the Mosco sense to an integrable random closed set X_∞ such that X_∞ has almost surely weakly compact intersection with every ball. If, additionally, $\{\rho(0, X_n), n \geq 1\}$ is uniformly integrable, then $\mathbf{E}(X_\infty|\mathfrak{F}_n) \subset X_n$ a.s. for every $n \geq 1$.*

Proof. Let us outline the proof which is based on truncation arguments. All missing details can be retrieved from Hess [242]. By Theorem 1.3(ii), $\alpha_n^k = \rho(0, X_n) + k$, $n \geq 1$, is a positive submartingale for every fixed k. By Krickeberg's decomposition theorem, $\alpha_n^k = \beta_n^k - s_n^k$, where β_n^k is a positive integrable martingale and s_n^k is a positive integrable supermartingale. Then $Y_n^k = X_n \cap B_{\beta_n^k}(0)$ is a multivalued supermartingale, so that $Y_n^k \subset Y \cap B_{w_k}(0)$, where w_k is an integrable random variable which bounds the a.s. convergent martingale β_n^k. By the assumption on Y, the random closed set Y_n^k is a.s. weakly compact. By Theorem 1.3(iii), $h(Y_n^k, u)$ forms a supermartingale. Using separability arguments and the boundedness of Y_n^k together with Theorem F.1, it is possible to show that $h(Y_n^k, u) \to h(Y_\infty^k, u)$ a.s. as $n \to \infty$ for every $k \geq 1$. The latter implies the Mosco convergence of Y_n^k to Y_∞^k a.s. Proposition B.10 establishes the weak convergence of X_n to $X_\infty = \cup_{k \geq 1} Y_\infty^k$. The inclusion $\mathbf{E}(X_\infty|\mathfrak{F}_n) \subset X_n$ follows from Theorem 2.1.50. □

Corollary 1.10 (Convergence of supermartingales with possibly unbounded values in $\mathbb{R}^d$). If $(X_n, \mathfrak{F}_n)$, $n \geq 1$, is a multivalued supermartingale in $\mathbb{R}^d$ such that $\sup_{n \geq 1} \mathbf{E}\rho(0, X_n) < \infty$, then there exists an integrable random closed set X_∞ such that X_n converges to X a.s. in the Fell topology as $n \to \infty$.

Martingale selections

The following concept relates multivalued and single-valued martingales.

Definition 1.11 (Martingale selection). An $\mathbb{E}$-valued martingale $(\xi_n, \mathfrak{F}_n)$, $n \geq 1$, is said to be a *martingale selection* of $\{X_n, n \geq 1\}$ if $\xi_n \in \mathcal{S}^1(X_n)$ for every $n \geq 1$. The family of all martingale selections is denoted by $\mathrm{MS}(X_n, n \geq 1)$.

The concept of the projective limit from Appendix A makes it possible to establish the existence of martingale selections and to provide a Castaing representation of X_n that consists of martingale selections. Note that $\mathrm{proj}_k(\{\xi_n, n \geq 1\}) = \xi_k$ denotes a projection of a sequence $\{\xi_n, n \geq 1\}$ onto its kth coordinate. For example, $\mathrm{proj}_k(\mathrm{MS}(X_n, n \geq 1))$ is the family of all martingale selections at time k.

Theorem 1.12 (Existence of martingale selections). Let $\mathbb{E}$ be a separable Banach space.
(i) Any multivalued martingale admits at least one martingale selection.
(ii) For every $k \geq 1$, $\mathrm{proj}_k(\mathrm{MS}(X_n, n \geq 1))$ is dense in $\mathcal{S}^1_{\mathfrak{F}_k}(X_k)$.
(iii) There exists a countable subset D of $\mathrm{MS}(X_n, n \geq 1)$ such that $\mathrm{proj}_k(D)$ is a Castaing representation of X_k for any $k \geq 1$.

Proof.
(i) For $m \leq n$ define $u_{m,n}(f) = \mathbf{E}(f|\mathfrak{F}_m)$ for $f \in \mathcal{S}^1_{\mathfrak{F}_n}(X_n)$. The sequence $\{\mathcal{S}^1_{\mathfrak{F}_n}(X_n), n \geq 1\}$ together with the maps $u_{m,n}$ is a projective system of non-empty complete subsets of $\mathbf{L}^1$. By the definition of the multivalued conditional expectation,

$$u_{n,n+1}(\mathcal{S}^1_{\mathfrak{F}_{n+1}}(X_{n+1})) = \{\mathbf{E}(\xi|\mathfrak{F}_n) : \xi \in \mathcal{S}^1_{\mathfrak{F}_{n+1}}(X_{n+1})\}$$

is dense in $\mathcal{S}^1_{\mathfrak{F}_n}(X_n)$. Theorem A.4 implies that this projective system has a non-empty projective limit. Any member $\xi = \{\xi_n, n \geq 1\}$ of the projective limit satisfies

$$\mathrm{proj}_m(\xi) = u_{m,n}(\mathrm{proj}_n(\xi)), \quad m \leq n,$$

or, equivalently, $\xi_m = \mathbf{E}(\xi_n|\mathfrak{F}_n)$, so that $(\xi_n, \mathfrak{F}_n)$ is a required martingale selection.
(ii) is an immediate consequence of Theorem A.4.
(iii) By Proposition 2.1.2(ii), X_n has a Castaing representation $\{\xi_n^{(k)}, k \geq 1\}$ which consists of integrable selections for all $n \geq 1$. Since $\mathrm{proj}_n(\mathrm{MS}(X_n, n \geq 1))$ is dense in $\mathcal{S}^1_{\mathfrak{F}_n}(X_n)$, for every $k \geq 1$ there exists a sequence

$$\{\eta_n^{(kj)}, j \geq 1\} \subset \mathrm{MS}(X_n, n \geq 1)$$

such that $\eta_n^{(kj)} \to \xi_n(k)$ in $\mathbf{L}^1$ as $j \to \infty$. Therefore, $\eta_n^{(kj(i))} \to \xi_n(k)$ a.s. for a subsequence $\{j(i),\ i \geq 1\}$. By taking the unions of the exceptional sets where the convergence fails for all n and k, it is readily seen that $\{\eta_n^{(kj)},\ k, j \geq 1\}$ is dense in X_n a.s., and so provides a Castaing representation that consists of martingale selections. □

Using a similar idea it is possible to show that every weakly compact submartingale X_n admits a martingale selection if $X_n \subset Y$ for a weakly compact convex random set Y, see Papageorgiou [442, Th. 2.2]. In particular, this immediately implies that $\liminf X_n \neq \emptyset$ a.s.

The technique based on projective limits yields that an integrably bounded multivalued supermartingale admits at least one martingale selection if X_n has weakly compact values, see Hess [242, Prop. 3.6]. Indeed, then $\mathcal{S}^1_{\mathfrak{F}_n}(X_n)$ is weakly compact in $\mathbf{L}^1$, which allows us to refer to Proposition A.3. This result holds also for supermartingales with unbounded values.

Theorem 1.13 (Martingale selections of possibly unbounded supermartingales). Let X_n, $n \geq 1$, be a multivalued supermartingale in a reflexive space $\mathbb{E}$.
(i) If $\{\mathbf{E}\rho(0, X_n), n \geq 1\}$ is bounded, then there exists a martingale selection ξ_n, $n \geq 1$, such that $\{\mathbf{E}\|\xi_n\|, n \geq 1\}$ is bounded.
(ii) If, additionally, $\{\rho(0, X_n),\ n \geq 1\}$ is a uniformly integrable sequence of random variables, then there exists a martingale selection, which is uniformly integrable in $\mathbf{L}^1$.

Proof. By Theorem 1.3(ii), $\alpha_n = \rho(0, X_n) + 1$ is a positive integrable submartingale verifying $\sup_{n \geq 1} \mathbf{E}\alpha_n < \infty$. By Krickeberg's decomposition theorem, it is possible to decompose α_n as $\alpha_n = \beta_n - s_n$, where β_n is a positive integrable martingale and s_n is a positive integrable supermartingale. It is readily seen that $B_{\beta_n}(0)$ is a multivalued martingale. It follows from Theorem 1.2.25(iv) that $Y_n = X_n \cap B_{\beta_n}(0)$, $n \geq 1$, is a sequence of random closed sets, which are non-empty by the choice of β_n. Proposition 1.4(ii) implies that $(Y_n, \mathfrak{F}_n)$ is a multivalued supermartingale. Since $\mathbb{E}$ is reflexive, Y_n has weakly compact values. Then Y_n has at least one martingale selection $\{\xi_n, n \geq 1\}$ which satisfies

$$\sup_{n \geq 1} \mathbf{E}\|\xi_n\| \leq \sup_{n \geq 1} \mathbf{E}\beta_n \leq \sup_{n \geq 1} \mathbf{E}\alpha_n < \infty,$$

proving (i). The uniform integrability of $\rho(0, X_n)$ implies that $\|\xi_n\|$ is uniformly integrable. □

It is shown in Luu [362] that, under some conditions, the closure of a set of martingale selections is a multivalued martingale. Theorem 1.14 follows from a representation theorem for multivalued amarts (see Definition 1.16) proved by Luu [363].

Theorem 1.14 (Castaing representation of multivalued martingale). *An adapted sequence $\{X_n, n \geq 1\}$ of integrably bounded random convex closed sets in a separable Banach space is a multivalued martingale if and only if the following two conditions hold.*
(1) *There exists a family $\{\xi_n^{(k)}, n \geq 1\}, k \geq 1$, of martingale selections such that $X_n = \text{cl}\{\xi_n^{(k)}, k \geq 1\}$.*
(2) *If $\eta_n \in S(X_n)$, $n \geq 1$, and $\|\eta_n - \tilde{\eta}_n\| \to 0$ a.s. for a martingale $\tilde{\eta}_n$, then $\{\tilde{\eta}_n, n \geq 1\}$ is a martingale selection of $\{X_n, n \geq 1\}$.*

Optional sampling

A random variable τ with values in $\{1, 2, \ldots\}$ is said to be a *stopping time* if $\{\tau = n\} \in \mathfrak{F}_n$ for every $n \geq 1$. The associated stopping σ-algebra $\mathfrak{F}_\tau$ is the family of all $A \in \mathfrak{F}$ such that $A \cap \{\tau = n\} \in \mathfrak{F}_n$ for every $n \geq 1$.

Theorem 1.15 (Optional sampling theorem). *Let $(X_n, \mathfrak{F}_n)$, $n \geq 1$, be an integrably bounded multivalued martingale. If $\{\tau_m, m \geq 1\}$ is an increasing sequence of stopping times, $\hat{X}_m = X_{\tau_m}$ and $\hat{\mathfrak{F}}_m = \mathfrak{F}_{\tau_m}$, then $(\hat{X}_m, \hat{\mathfrak{F}}_m)$, $m \geq 1$, is a multivalued martingale.*

Proof. Let $\{\xi_n^{(k)}, n \geq 1\}$ be a family of martingale selections such that $X_n = \text{cl}\{\xi_n^{(k)}, k \geq 1\}$ for every $n \geq 1$. By the optional sampling theorem for single-valued martingales, $\hat{\xi}_m = \xi_{\tau_m}$ is a martingale, so that the proof is finished by referring to Theorem 1.14. □

Theorem 1.15 is equivalent to the fact that $\mathbf{E}(X_\sigma | \mathfrak{F}_\tau) = X_\tau$ for every two stopping times σ and τ such that $\sigma \geq \tau$ almost surely. The optional sampling theorem implies $\mathbf{E}(X_\tau) = \mathbf{E}(X_0)$ for every stopping time τ, which is *Wald's identity* for multivalued martingales.

Multivalued supermartingales can be applied to decision-making problems. If i is an action and j is its outcome, $I_{ij}^{(n)}$ is an a priori range for the score that a decision-maker associates with the pair (i, j) that appears in the step n. If $\hat{p}^{(n)}$ is an estimate on the nth step of the conditional probability of outcome j given that the course of action i was chosen, then the expected score is estimated by

$$\hat{X}_i^{(n)} = \sum_j I_{ij}^{(n)} \hat{p}^{(n)},$$

which is the Minkowski sum of sets, for instance, line segments if the score ranges are given by intervals. If the decision-maker has less uncertainty as n increases, then $I_{ij}^{(n+1)} \subset I_{ij}^{(n)}$. A supermartingale property of $\hat{X}_i^{(n)}$ makes it possible to prove the convergence of $\hat{X}_i^{(n)}$ to a random closed set X_i, see de Korvin and Kleyle [325].

312   5 Random Sets and Random Functions

Martingale-like set-valued processes

It is possible to obtain meaningful results for set-valued processes which satisfy a weaker form of the martingale property.

Definition 1.16. An adapted sequence $\{(X_n, \mathfrak{F}_n), n \geq 1\}$ of integrably bounded random weakly compact convex sets is
(i) a *mil* (martingale in the limit) if for every $\varepsilon > 0$, there exists $m \geq 1$ such that

$$\mathbf{P}\left\{\sup_{m \leq k \leq n} \rho_H(X_k, \mathbf{E}(X_n|\mathfrak{F}_k)) > \varepsilon\right\} < \varepsilon, \quad n \geq m;$$

(ii) a multivalued *quasi-martingale* if

$$\sum_{n=1}^{\infty} \mathbf{E}\rho_H(X_n, \mathbf{E}(X_{n+1}|\mathfrak{F}_n)) < \infty;$$

(iii) a multivalued *amart* if $\mathbf{E}X_{\tau_n}$ converges in ρ_H for a sequence of stopping times $\{\tau_n, n \geq 1\}$ increasing to infinity;
(iv) a multivalued *subpramart* (respectively *superpramart*) if for any $\varepsilon > 0$ there exists a stopping time σ_0 such that $\mathbf{P}\{X_\sigma \subset (\mathbf{E}(X_\tau|\mathfrak{F}_\sigma))^\varepsilon\} \geq 1-\varepsilon$ (respectively $\mathbf{P}\{\mathbf{E}(X_\tau|\mathfrak{F}_\sigma) \subset X_\sigma^\varepsilon\} \geq 1-\varepsilon$) for any two stopping times σ and τ such that $\sigma_0 \leq \sigma \leq \tau$, where F^ε denotes the ε-envelope of $F \subset \mathbb{E}$;
(v) a multivalued *pramart* if it is both sub- and superpramart, i.e. if the Hausdorff distance between X_σ and $\mathbf{E}(X_\tau|\mathfrak{F}_\sigma)$ converges to zero in probability uniformly over the family of all stopping times $\tau \geq \sigma \geq \sigma_0$ as $\sigma_0 \uparrow \infty$.

Every submartingale is a subpramart, every supermartingale is a superpramart and every martingale is a pramart. Furthermore, every pramart is a mil and every quasi-martingale is a pramart, see Papageorgiou [448, Prop. 13].

1.2 Set-valued random processes

A function $X: \Omega \times \mathbb{T} \mapsto \mathcal{F}$ is called a *set-valued process* on $\mathbb{T}$, where the parameter set $\mathbb{T}$ is usually the half-line, the whole line, the set of non-negative integers, or the set $\mathbb{E}$ itself. It is assumed that X_t is a random closed set for every $t \in \mathbb{T}$. Many particular definitions of special set-valued processes (Markov, stationary) are applicable for stochastic processes in general state spaces and can be reformulated for set-valued processes without difficulties. For some other families of stochastic process (with independent increments, diffusion processes) it is quite difficult to define their set-valued analogues.

A question specific to the studies of set-valued processes concerns the existence of a single-valued process $\xi_t, t \in \mathbb{T}$, such that ξ_t is a selection of X_t for every $t \in \mathbb{T}$ and ξ satisfies a particular property, for example, is Markov or stationary, etc.

Set-valued Markov processes

A set-valued process X_t, $t \geq 0$, is said to be *Markov* if it is a Markov process in the state space $\mathcal{F}$ with its Effros σ-algebra. The sub-σ-algebra of $\mathfrak{F}$ generated by X_s, $s \leq t$, is denoted by $\mathfrak{F}_t$. Set-valued Markov processes that are related to the extremal processes will be considered in Section 3.3.

Proposition 1.17 (Markov selections). *If X_t is a set-valued Markov process, then there exists a family of $\mathbb{E}$-valued stochastic processes $\xi_t^{(n)}$, $n \geq 1$, such that $X_t = \mathrm{cl}\{\xi_t^{(n)}, n \geq 1\}$ for all $t \geq 0$ and $(\xi_t^{(1)}, \xi_t^{(2)}, \ldots)$, $t \geq 0$, is a Markov process in the space $\mathbb{E}^{\{1,2,\ldots\}}$.*

Proof. Define $\xi_t^{(n)} = \mathfrak{f}_n(X_t)$, where $\mathfrak{f}_n$, $n \geq 1$, is a sequence of selection operators introduced in Proposition 1.2.23. Since the sequence $(\xi_t^{(1)}, \xi_t^{(2)}, \ldots)$ generates the same σ-algebras as X_t, it also satisfies the Markov property. □

Stationary processes

A set-valued process X_t, $t \in \mathbb{R}$ is said to be *strictly stationary* if for every $t_1, \ldots, t_n \in \mathbb{R}$ and $s \in \mathbb{R}$ the joint distribution of $(X_{t_1}, \ldots, X_{t_n})$ coincides with the joint distribution of $(X_{t_1+s}, \ldots, X_{t_n+s})$. This concept is a specialisation of a general concept of strictly stationary stochastic processes in an abstract measurable space. A process ξ_t, $t \in \mathbb{R}$, is called a *stationary selection* of X_t if ξ_t is a stationary process and $\xi_t \in X_t$ a.s. for every t. Applying Proposition 1.2.23 and following the proof of Proposition 1.17 it is easy to deduce the following result.

Proposition 1.18 (Stationary selections). *A set-valued process X_t, $t \in \mathbb{R}$, is strictly stationary if and only if there exists a sequence $\{\xi_t^{(k)}, k \geq 1\}$ of single-valued strictly stationary processes such that $X_t = \mathrm{cl}\{\xi_t^{(k)}, k \geq 1\}$ for every $t \in \mathbb{R}$ and $(\xi_t^{(1)}, \ldots, \xi_t^{(n)})$ is an $\mathbb{E}^n$-valued stationary process for every $n \geq 1$.*

Ergodic theorems for strictly stationary set-valued processes follow from the ergodic theorem for superstationary subadditive families of random closed sets considered in Section 3.3.5.

A process X_t, $t \in \mathbb{R}$, is said to be the first-order stationary if the distribution of X_t does not depend on t; the process is second-order stationary if the distribution of (X_{t_1+s}, X_{t_2+s}) does not depend on s, etc.

Open problem 1.19. Investigate filtering problems for second-order stationary set-valued processes.

Increasing set-valued processes

A set-valued process X_t, $t \geq 0$, is said to be *increasing* if $X_t \subset X_s$ a.s. for $t \leq s$. Examples of such processes are readily provided by taking successive *Minkowski*

314   5 Random Sets and Random Functions

sums of random closed sets. For example, consider a sequence $\{X_n, n \geq 1\}$ of i.i.d. integrable random compact sets containing the origin in $\mathbb{E} = \mathbb{R}^d$ and define

$$S_n(t) = \sum_{j \leq nt} X_j, \quad n \geq 1, \quad 0 \leq t \leq 1.$$

Then $S_n(t)$ is a piecewise constant increasing set-valued process. The strong law of large numbers for Minkowski sums (Theorem 3.1.6) implies that

$$\rho_H(n^{-1} S_n(t), t\mathbf{E}X_1) \to 0 \quad \text{a.s. as } n \to \infty$$

uniformly in $t \in [0, 1]$, where $\mathbf{E}X_1$ is the selection expectation of X_1.

Another family of increasing set-valued processes is obtained by taking *unions* of random closed sets $X_1, X_2, \ldots$ as

$$Z_n(t) = \bigcup_{j \leq nt} X_j. \tag{1.3}$$

The properties of $Z_n(t)$ as $n \to \infty$ follow from the limit theorems for unions of random closed sets, see Chapter 4. It is also possible to prove a large deviation principle for the sequence of multifunctions Z_n, $n \geq 1$, assuming that the random closed sets $X_1, X_2, \ldots$ form a Markov chain, see Schreiber [524].

Example 1.20 (Large deviation principle for convex hulls). Let $\xi_1, \xi_2, \ldots$ be i.i.d. random points uniformly distributed in a unit ball B in $\mathbb{R}^d$. Define $Z_n(t) = \text{co}\{\xi_1, \ldots, \xi_n\}$. Then Z_n satisfies the *large deviation principle* in the family $\mathcal{U}$ of increasing set-valued functions on $[0, 1]$ with the metric generated by the Hausdorff metric between their graphs, i.e. for each open set $G \subset \mathcal{U}$

$$\liminf_{n\to\infty} \frac{1}{n} \log \mathbf{P}\{Z_n \in G\} \geq -\inf_{U \in G} I(U)$$

and for each closed $F \subset \mathcal{U}$

$$\limsup_{n\to\infty} \frac{1}{n} \log \mathbf{P}\{Z_n \in F\} \leq -\inf_{U \in F} I(U).$$

The *rate function* I is given by

$$I(U) = \log \varkappa_d - \int_0^1 \log(\text{mes}(U(t))) \, dt.$$

The proof relies on a representation of convex hulls by pointwise maxima of the corresponding support functions, see (4.4.8) and Theorem 3.34.

Random differential inclusions

Deterministic *differential inclusions* extend the concept of a differential equation. For example, the differential equation $dx/dt = f(t, x)$ is a particular case of the differential inclusion

$$\frac{dx}{dt} \in F(t, x),$$

where $F(t, x)$ is a multifunction for $t, x \in \mathbb{R}$. This concept is particularly useful if the right-hand side $f(t, x)$ of the differential equation is discontinuous. Then $F(t, x)$ equals $\{f(t, x)\}$ at all continuity points while at a discontinuity point (t_0, x_0) of f, the set $F(t_0, x_0)$ is the convex hull of all partial limits of $f(t, x)$ as $(t, x) \to (t_0, x_0)$.

Random differential inclusions give rise to set-valued random processes. Let $[a, b]$ be a bounded closed interval in $\mathbb{R}_+$. Consider a random element ξ_0 in a separable Banach space $\mathbb{E}$ and a multifunction F defined on $\Omega \times [a, b] \times \mathbb{E}$ with values being non-empty closed sets in $\mathbb{E}$. The random multivalued Cauchy problem can be formulated as follows

$$\frac{\partial}{\partial t}\xi(\omega, t) \in F(\omega, t, \xi), \quad \xi(\omega, 0) = \xi_0(\omega). \quad (1.4)$$

By a random solution of (1.4) we understand a stochastic process $\xi(\omega, t)$ with almost surely differentiable sample paths.

Theorem 1.21 (Existence of solution for random differential inclusion). *The random differential inclusion (1.4) admits a solution if the following conditions hold.*
(1) *F has compact convex values.*
(2) *For all $x \in \mathbb{E}$, $(\omega, t) \mapsto F(\omega, t, x)$ is measurable.*
(3) *For all $(\omega, t) \in \Omega \times [a, b]$, $x \mapsto F(\omega, t, x)$ is continuous in the Hausdorff metric.*
(4) *For almost all ω and all $x \in \mathbb{E}$,*

$$\|F(\omega, t, x)\| \leq a(\omega, t) + b(\omega, t)\|x\|,$$

where a and b are jointly measurable and integrable with respect to t for all ω.
(5) *For all bounded $B \subset \mathbb{E}$,*

$$\mathrm{nc}(F(\omega, t, B)) \leq \varphi(\omega, t)\,\mathrm{nc}(B) \quad \text{a.s.},$$

where $\mathrm{nc}(B)$ is the Kuratowski measure of non-compactness of $B \subset \mathbb{E}$, φ is jointly measurable and $\int_a^b \varphi(\omega, t)dt < 1/2$ for all ω.

Set-valued stochastic integrals

Let X_t, $t \geq 0$, be an $\mathfrak{F}_t$-adapted set-valued random process in $\mathbb{E} = \mathbb{R}^d$ such that

$$\int_0^t \|X_s\|^2 ds < \infty \quad \text{a.s.}$$

for every $t > 0$. The fundamental selection theorem implies that there exists an adapted stochastic process ξ_t, $t \geq 0$, such that ξ_t is a measurable selection of X_t for every $t \geq 0$ and $\int_0^t \|\xi_s\|^2 ds < \infty$ for every $t > 0$. The process ξ_t is called an *adapted selection* of X_t.

A *set-valued stochastic integral* is defined by taking integrals of all adapted selections. For example, if w_t is a Wiener process in $\mathbb{R}$, then $\int_0^t X_s dw_s$ is the set of $\int_0^t \xi_s dw_s$ for all adapted selections ξ_t of X_t. It is easy to see that the integral of a convex process is convex. Although $\mathbf{E}[\int_0^t \xi_s dw_s] = 0$ for every adapted selection ξ_t, the selection expectation $\mathbf{E}[\int_0^t X_s dw_s]$ of the set-valued stochastic integral is not zero. The following result follows from the fact that $\int_0^t \xi_s dw_s$ is a martingale.

Proposition 1.22 (Submartingale property of the stochastic integral). *If X_t is almost surely convex for every $t \geq 0$, then $\int_0^t X_s dw_s$ is a multivalued submartingale.*

Let $F(t, x)$ and $H(t, x)$ be multifunctions defined for $t \geq 0$ and $x \in \mathbb{R}^d$ such that their norms are square integrable over $t \in [0, s]$ for every $s > 0$ and $x \in \mathbb{R}^d$. The corresponding *stochastic differential inclusion* can be written as

$$\zeta_t \in \zeta_0 + \mathrm{cl}_{\mathbf{L}^2}\left(\int_0^t H(s, \zeta_s)ds + \int_0^t F(s, \zeta_s)dw_s\right), \tag{1.5}$$

where the closure in the right-hand side is taken in the space of square-integrable $\mathfrak{F}_t$-measurable functions and the addition under the closure is in the Minkowski sense. An alternative (however less precise) notation is $d\zeta_t \in H(t, \zeta_t)dt + F(t, \zeta_t)dw_t$. The closure Z_t of the family of all ζ_t satisfying (1.5) is called the set-valued solution of (1.5). The existence theorem for solutions of stochastic differential equations can be formulated for the set-valued case if the functions F and H admits (say, Lipschitz) selections, so that the corresponding single-valued stochastic differential equation has a solution. For instance, if both F and H admit Lipschitz selections and $\rho(0, F(t, x))^2 + \rho(0, H(t, x))^2 \leq c(1 + \|x\|^2)$ for some constant c, then the solution exists by Ito's existence theorem. The existence theorems for *Lipschitz selections* is considered by Dentcheva [136].

It is quite difficult to find all single-valued solutions of set-valued stochastic differential inclusions, since a set-valued function usually possesses a large family of measurable selections.

Open problem 1.23. Suggest a way to calculate efficiently the stochastic integral of set-valued processes (e.g. $\int_0^t X_s dw_s$ for $X_t = \zeta_t M$ with a deterministic convex $M \subset \mathbb{R}^d$ and a single-valued stochastic process ζ_t) and solve stochastic differential inclusions (e.g. $d\zeta_t \in F\zeta_t dw_t$, where F is a deterministic interval on the real line).

Open problem 1.24. Define a set-valued analogue of the Wiener process and the corresponding stochastic integral.

Stochastic control processes

Set-valued random functions appear naturally in *stochastic control* problems. Consider the following controlled stochastic process

$$\xi(t, \omega, u), \quad u \in U(t),$$

where u denotes the control and $U(t)$ is the set of admissible controls at time t. Then all possible values of the controlled process form a *set-valued* process $X(t, \omega) = \{\xi(t, \omega, u) : u \in U(t)\}$. This often helps to establish the existence of optimal or approximately optimal control strategies if the objective function does not directly depend on the values of controls. For example, the minimisation of $\sup_t |\xi(t, \omega, u)|$ for the controlled process reduces to the minimisation of $\sup_t \|X(t, \omega)\|$, i.e. the controlled optimisation is reduced to optimisation without control but for a set-valued process.

Set-valued shot-noise processes

Let $N = \{x_i, i \geq 1\}$ be a stationary point process in $\mathbb{R}^d$ and let $\{\beta_i, i \geq 1\}$ be a sequence of i.i.d. random variables. If $f : \mathbb{R}^d \times \mathbb{R} \mapsto \mathbb{R}$ is a measurable function which vanishes outside a compact set, then the random function

$$v(x) = \sum_{i \geq 1} f(x - x_i, \beta_i) \tag{1.6}$$

is called a *shot-noise process*. The *response function* $f(x - x_i, \beta_i)$ is interpreted as the effect at $x \in \mathbb{R}^d$ caused by an event which is characterised by the random position x_i at which the event occurs and by the random mark β_i giving additional information about the event, for example, the event's "magnitude". Then $v(x)$ is the total effect observed at x. Two multivalued generalisations of the shot-noise process are presented below.

Definition 1.25 (Set-valued shot-noise processes). Let $F : \mathbb{R}^d \times \mathbb{R} \mapsto \mathcal{K}(\mathbb{R}^m)$ be a multivalued response function with values being compact subsets of $\mathbb{R}^m$ and let $N_\lambda = \{x_i, i \geq 1\}$ be a stationary point process in $\mathbb{R}^d$ with intensity λ. Consider a sequence $\{\beta_i, i \geq 1\}$ of i.i.d. random variables.
(i) The set-valued stochastic process

$$\Xi_\lambda(x) = \sum_{i \geq 1} F(x - x_i, \beta_i), \quad x \in \mathbb{R}^d, \tag{1.7}$$

(with the summation being the Minkowski sum) is called the *Minkowski shot-noise process*.
(ii) The set-valued stochastic process

$$\Xi_\lambda(x) = \mathrm{cl}\left(\bigcup_{i \geq 1} F(x - x_i, \beta_i)\right), \quad x \in \mathbb{R}^d,$$

is called the *union* shot-noise process.

The sum in (1.7) is well defined if $\mathbf{E}\|F(x,\beta)\|$ is integrable over $\mathbb{R}^d$, where β is a random element that shares the distribution with the β_is. One is mainly interested in the limiting behaviour of the suitably normalised stochastic process $\Xi_\lambda(t)$ when the intensity λ of the underlying point process N tends either to ∞ ("high density case") or to 0 ("low density case").

Consider first the Minkowski shot-noise process. By passing to support functions, (1.7) yields a family of single-valued shot-noise processes

$$v_\lambda^u(x) = h(\Xi_\lambda(x), u) = \sum_{i \geq 1} h(F(x - x_i, \beta_i), u), \quad x \in \mathbb{R}^d, \ u \in \mathbb{S}^{d-1}.$$

By the Campbell theorem (see Theorem 1.8.10),

$$\mathbf{E}v_\lambda^u(x) = \int_{\mathbb{R}^d} \mathbf{E}(h(F(y, \beta), u)) dy,$$

noticing that the left-hand side yields the support function of the selection expectation of $\Xi_\lambda(x)$. By stationarity, it suffices to let $x = 0$. A limit theorem for $v_\lambda^u(0)$ for any single u follows from a limit theorem for a single-valued shot-noise process proved by Heinrich and Schmidt [232]. A limit theorem for finite-dimensional distributions for several values of u can be obtained using the Cramér–Wold device. Indeed, for every $u_1, \ldots, u_k \in \mathbb{S}^{d-1}$ and $a_1, \ldots, a_k \in \mathbb{R}$,

$$\tilde{v}_\lambda(x) = \sum_{j=1}^k a_j v_\lambda^{u_j}(x) = \sum_{i \geq 1} \tilde{f}(x - x_i, \beta_i),$$

where $\tilde{f}(x, \beta) = \sum_{j=1}^k a_j h(F(x, \beta), u_j)$. Define

$$\tilde{\sigma}^2 = \int_{\mathbb{R}^d} \mathbf{E}(\tilde{f}(x, \beta))^2 dx = \sum_{ij=1}^k a_i a_j \sigma^{u_i u_j},$$

where

$$\sigma^{uv} = \int_{\mathbb{R}^d} \mathbf{E}[h(F(x, \beta), u) h(F(x, \beta), v)] dx. \tag{1.8}$$

If N_λ is a Poisson point process, then $\tilde{\sigma}^2 < \infty$ implies that $\lambda^{-1/2}(\tilde{v}_\lambda(0) - \mathbf{E}\tilde{v}_\lambda(0))$ converges weakly as $\lambda \to \infty$ to the normal distribution with the variance $\tilde{\sigma}^2$. Thus, the finite-dimensional distributions of $\lambda^{-1/2}(v_\lambda^u(0) - \mathbf{E}v_\lambda^u(0))$ converge to finite-dimensional distributions of a centred Gaussian process on $\mathbb{S}^{d-1}$ with the covariance given by (1.8). This readily implies the weak convergence in the space of continuous functions on $\mathbb{S}^{d-1}$ arguing similarly to the proof of Theorem 3.2.1.

Theorem 1.26 (Weak convergence for Minkowski shot-noise). *Let $\mathbf{E}\|F(x, \beta)\|^2 < \infty$ for all $x \in \mathbb{R}^d$ and let $\int_{\mathbb{R}^d} \mathbf{E}\|F(x, \beta)\|^2 dx < \infty$. If Ξ_λ is the Minkowski shot-noise process generated by a Poisson point process, then*

$$\lambda^{-1/2}(h(\Xi_\lambda(0), u) - \mathbf{E}h(\Xi_\lambda(0), u)), \quad u \in \mathbb{S}^{d-1},$$

converges weakly as $\lambda \to \infty$ to a centred Gaussian process ζ on $\mathbb{S}^{d-1}$ with the covariance (1.8). If $F(x, \beta)$ is convex almost surely for all x, then

$$\lambda^{-1/2} \rho_H(\Xi_\lambda(0), \mathbf{E}\Xi_\lambda(0)) \xrightarrow{d} \sup_{u \in \mathbb{S}^{d-1}} |\zeta(u)|.$$

Limit theorems for union shot-noise processes rely on regular variation properties of the response function F. We give here only a basic result that provides a sufficient condition for the weak convergence of the suitably scaled random closed set $\Xi_{\lambda_t}(0)$ as $\lambda_t \to \infty$ or $\lambda_t \to 0$ in the case N_λ is Poisson and F is homogeneous. As in (H.13),

$$F^-(K, \beta) = \{u \in \mathbb{R}^d : F(u, \beta) \cap K \neq \emptyset\}, \quad K \in \mathcal{K},$$

denotes the inverse function to F.

Theorem 1.27 (Weak convergence of union shot-noise). *Assume that*
(1) *F is homogeneous, i.e. $F(su, \beta) = s^\alpha F(u, \beta)$ for all $s > 0$ and $u \in \mathbb{R}^d$;*
(2) *$F(u, \beta)$ is continuous in the Hausdorff metric with respect to β for all $u \in \mathbb{R}^d$;*
(3) *$\mathrm{mes}(F^-(K, \beta) + B_\delta(0))$ is integrable for each $\delta > 0$ and each compact set K missing the origin.*

If Ξ_λ is the union shot-noise process generated by the stationary Poisson point process of intensity λ, then $\lambda_t^{1/(d\alpha)} \Xi_{\lambda_t}(0)$ converges weakly as $\lambda_t \to \infty$ or $\lambda_t \to 0$ to

$$\Xi = \bigcup_{x_i \in \Pi_1} F(x_i, \beta),$$

where Π_1 is the Poisson point process in $\mathbb{R}^d$ of intensity 1.

1.3 Random functions with stochastic domains

In the theory of Markov processes it is quite usual to consider processes defined until a random time moment. This concept can be extended by restoring the symmetry and adding a random birth time, so that a process "lives" on a random interval on the line. The basic arguments are quite general and can be extended to the non-Markovian and multidimensional cases.

Finite-dimensional distributions

Let Y be a *random open set* in $\mathbb{R}^d$, see Section 1.4.6. Suppose that ζ_x, $x \in \mathbb{R}^d$, are random elements in a measurable space $(\mathbb{E}, \mathfrak{E})$ defined on a probability space $(\Omega, \mathfrak{F}, \mathbf{P})$.

Definition 1.28 (Stochastic process with random domain). A family of random elements ζ_x, $x \in \mathbb{R}^d$, is said to be a *stochastic process with random open domain* Y if $\{\omega : x \in Y, \zeta_x \in B\} \in \mathfrak{F}$ for each $x \in \mathbb{R}^d$ and $B \in \mathfrak{E}$.

Note that it suffices to define $\zeta_x(\omega)$ only on the graph of X, i.e. for $x \in X(\omega)$. The finite-dimensional distributions of ζ are defined by

$$I(L; B_L) = \mathbf{P}\{L \subset Y, \ \zeta_{x_i} \in B_i, \ 1 \leq i \leq n\},$$

where $L = \{x_1, \ldots, x_n\}$ is a finite set (i.e. $L \in \mathfrak{I}$, see Section 1.4.2) and $B_L = (B_1, \ldots, B_n)$ for $B_1, \ldots, B_n \in \mathfrak{E}$. Clearly, $I_Y(L) = I(L; \mathbb{E}_L)$ is the inclusion functional of Y, where $\mathbb{E}_L = (\mathbb{E}, \ldots, \mathbb{E})$. For $L' \supset L$ define $I(L'; B_L) = I(L'; B_L \times \mathbb{E}_{L' \setminus L})$.

Theorem 1.29 (Distribution of stochastic process with random open domain).
(i) A functional $I(L; B_L)$ determines finite-dimensional distributions of a stochastic process with a random open domain if and only if $\varphi(L) = 1 - I(L' \cup L; B_{L'})$ is a capacity functional on $L \in \mathfrak{I}$ for every $L' \in \mathfrak{I}$ and $B_{L'} \in \mathfrak{E} \times \cdots \times \mathfrak{E}$.
(ii) A functional $I(L; B_L)$ determines finite-dimensional distributions of a stochastic process on a random open convex domain Y if and only if $I(L' \cup L; B_{L'})$ is a completely monotone functional on $L \in \mathfrak{I}$,

$$I(L \cup L'; B_L \times \mathbb{E}_{L'}) = I(L; B_L)$$

for all $L, L' \in \mathfrak{I}$ satisfying $\operatorname{co}(L \cup L') = \operatorname{co}(L)$ and

$$I(L \cup L_n; B_L \times B_{L_n}) \to I(L; B_L)$$

if $\operatorname{co}(L \cup L_n) \downarrow \operatorname{co}(L)$ as $n \to \infty$.

Proof.
(i) follows from a general inverse limit theorem proved in Dynkin and Fitzsimmons [153, Th. 3.1].
(ii) can be proved similarly to Corollary 1.4.31. □

Multivalued operators with stochastic domain

If a set-valued random process $Z_x(\omega)$ is defined for x from the space $\mathbb{E}$ itself, i.e. Z_x is a random closed set in $\mathbb{E}$ for each $x \in \mathbb{E}$, then it is usually called a *random multivalued operator*. It is essential to know when a random multivalued operator has a *fixed point*, i.e. a random element ξ satisfying $\xi \in Z_\xi$ almost surely. The fixed point x_0 can be regarded as an equilibrium point of the discrete set-valued dynamical system $x_{n+1} \in Z_{x_n}$. The deterministic case is treated by the following famous theorem, see Aubin and Frankowska [30, Th. 3.2.3].

Theorem 1.30 (Kakutani fixed point theorem). *Let Z_x be a function defined for x from a Banach space $\mathbb{E}$ with values being non-empty closed convex subsets of a convex compact set $K \subset \mathbb{E}$. If the support function $h(Z_x, u)$ is upper semicontinuous for each $u \in \mathbb{E}^*$, then Z has a fixed point x_0 which satisfies $x_0 \in K \cap Z_{x_0}$.*

Let X be a random closed set. A multifunction $Z\colon \operatorname{Graph}(X) \mapsto \mathcal{F}\setminus\{\emptyset\}$ is said to be a multivalued operator with *stochastic domain* if

$$\{\omega\colon Z_x(\omega)\cap G \neq \emptyset,\ x\in X(\omega)\} \in \mathfrak{F}$$

for all $x\in \mathbb{E}$ and $G\in\mathcal{G}$. The following result provides the basic fixed point theorem for random operators with stochastic domains.

Theorem 1.31 (Stochastic fixed point theorem). *Let $Z_x(\omega)$ be a random operator with stochastic domain X being a separable almost surely non-empty random closed set. If $Z_x(\omega)$ is Wijsman-continuous with respect to x and, for almost all $\omega\in\Omega$, there exists $x\in X(\omega)$ such that $x\in Z_x(\omega)$, then there exists a measurable selection $\xi\in\mathcal{S}(X)$ (called a random fixed point) such that $\xi(\omega) = Z_{\xi(\omega)}(\omega)$ almost surely.*

Proof. The proof relies on applying the fundamental selection theorem to the multifunction
$$Y(\omega) = \{x\in X\colon x\in \tilde{Z}_x(\omega)\},$$
where $\tilde{Z}_x(\omega)\} = Z_x(\omega)$ if $x\in X(\omega)$ and $\tilde{Z}_x(\omega) = C$ otherwise, where C is an arbitrary non-empty closed set, see Papageorgiou [445, Th. 3.1]. □

Theorem 1.31 can be applied to prove the existence of a random fixed point for a Wijsman-continuous random operator with stochastic domain F such that $F(\omega, X(\omega)) \subset X(\omega)$ almost surely and F is bounded and condensing so that $\operatorname{nc}(F(\omega,A)) < \operatorname{nc}(A)$ for every $A\subset\mathbb{E}$ with a positive Kuratowski measure of non-compactness.

Allocation problems

Random functions with stochastic domains can be used to formulate random *allocation problems* in mathematical economy. Assume that $\mathbb{C}$ is a closed convex cone in $\mathbb{E} = \mathbb{R}^d$ which defines a partial order so that $x \leq y$ if and only if $y - x \in \mathbb{C}$. Let the probability space $(\Omega, \mathfrak{F}, \mathbf{P})$ describe random events that influence economic processes. With every $\omega\in\Omega$ we associate a realisation $X(\omega)$ of a random convex closed set X which describes all feasible consumption plans.

Definition 1.32 (Utility function). A *utility function* $u\colon \Omega\times\mathbb{R}^d \mapsto \mathbb{R}$ is a real-valued stochastic process such that
 (i) $u(\omega,\cdot)$ is continuous and convex for all $\omega\in\Omega$;
 (ii) $u(\cdot,\cdot)$ is jointly measurable;
 (iii) $|u(\omega,x)| \leq \alpha(\omega)$ with $\mathbf{E}\alpha < \infty$;
 (iv) $u(\omega,\cdot)$ is a.s. monotone increasing, so that $x\leq y$ and $x\neq y$ imply $u(\omega,x) < u(\omega,y)$.

An *allocation* ξ (i.e. an integrable selection of X) is said to be *X-efficient* if $\xi \notin (X + \operatorname{Int}\mathbb{C})$. Furthermore, an allocation ξ is called (u,X)-*optimal* if $u(\omega,\xi) \leq u(\omega,\eta)$ a.s. for every selection $\eta\in\mathcal{S}^1(X)$. The linear utility function is determined by a system of prices p, which is an element of $\mathbf{L}^\infty(\Omega; \mathbb{R}^d)$. An allocation ξ is said to be (p, X)-efficient if $\mathbf{E}(\xi p) \leq \mathbf{E}(\eta p)$ for all $\eta \in \mathcal{S}^1(X)$.

It is proved by Papageorgiou [441] that an (u, X)-optimal allocation is X-efficient. Furthermore, if $\operatorname{Int} X \neq \emptyset$ a.s. and ξ is X-efficient, then ξ is (p, X)-efficient for a certain price system p and $\langle \xi, p \rangle = \inf \langle X, p \rangle$ a.s.

2 Levels and excursion sets of random functions

2.1 Excursions of random fields

Important examples of random sets appear as level sets of random functions. For instance, if $\zeta_x, x \in \mathbb{R}^d$, is an almost surely continuous random function, then

$$\{\zeta = a\} = \{x \in \mathbb{R}^d : \zeta_x = a\}$$

is a random closed set called the *level set* of ζ, where a takes a value from the state space of ζ_x. If ζ is not necessarily continuous (but jointly measurable in x and ω), then $\{\zeta = a\}$ is a random Borel set, see Section 1.2.5. Indeed, for every Borel B, $\{\{\zeta = a\} \cap B \neq \emptyset\}$ is the projection on Ω of $\{(x, \omega) : \zeta_x(\omega) = a, x \in B\}$, which is measurable by the projection theorem, see Theorem E.5.

If the state space is $\mathbb{R}$ (and so is totally ordered), then $\{\zeta = a\}$ is the boundary of the *excursion* (upper level) set

$$\{\zeta \geq a\} = \{x \in \mathbb{R}^d : \zeta_x \geq a\}.$$

The excursion set is closed for every a if and only if ζ is almost surely upper semi-continuous.

Smooth random functions on $\mathbb{R}$

Consider the case when both the state space and the parameter space are $\mathbb{R}$. The following result is a famous theorem concerning level sets of differentiable random functions.

Theorem 2.1 (Bulinskaya's theorem). *Let $\zeta_t, t \in \mathbb{R}$, be a stochastic process with almost surely continuously differentiable sample paths. Suppose that ζ_t admits a density for every t and these probability density functions are bounded uniformly in t. Then*

$$\mathbf{P}\{\zeta_t = a \text{ and } \zeta'_t = 0 \text{ for some } t \in [0, 1]\} = 0$$

and $\{\zeta = a\} \cap [0, 1]$ is finite almost surely.

If ζ_t is a stationary random function, then $\{\zeta = a\}$ is a stationary random closed set, which is also locally finite if the realisations of ζ_t are smooth. Assume that ζ_t is a stationary Gaussian process normalised to have zero mean and unit variance. Its covariance function is

$$r(s) = \mathbf{E}(\zeta_t \zeta_{t+s}).$$

The second derivative $\lambda_2 = -r''(0)$ (if it exists) is called the second spectral moment. Under conditions of Theorem 2.1, the number of points $t \in [0, 1]$ with $\zeta_t = a$ is twice the number of upcrossings (intersections of a level moving upward) of the level a.

Theorem 2.2 (Rice's formula). *If λ_2 is finite, then the mean number of upcrossings of the level a in the unit time interval is finite and given by*

$$\frac{1}{2\pi} \lambda_2^{1/2} \exp\left\{-a^2/2\right\}.$$

Non-differentiable processes

If ζ_t has non-differentiable trajectories, then $\{\zeta = a\}$ may have a fractal nature. For instance, if w_t is the Wiener process, then $X = \{t \geq 0 : w_t = 0\}$ is a fractal set whose *Hausdorff dimension* $\dim_H X$ equals $1/2$ almost surely. The capacity functional of X on a segment $[t, t+s]$ is given by

$$T_X([t, t+s]) = \mathbf{P}\{X \cap [t, t+s] \neq \emptyset\} = 1 - \frac{2}{\pi} \arcsin\sqrt{\frac{t}{t+s}}.$$

Further results of this kind can be found in Itô and McKean [272], Lévy [349] and Peres [455].

Random fields

Consider a stationary *Gaussian random field* ζ_x parametrised by $x \in \mathbb{R}^d$. The following results aim to determine the expectation of the *Euler–Poincaré characteristic* $\chi(Z_t)$ (see Appendix F) of the random closed set $Z_t = \{x \in W : \zeta_x \geq t\}$, where W is a compact subset of $\mathbb{R}^d$ with the boundary being a regular $(d-1)$-dimensional manifold and with at most a finite number of connected components.

Assuming that the realisations of ζ are sufficiently regular (in the Gaussian case it suffices to assume that the third derivatives of ζ exist and have finite variances), Worsley [612, Th. 1] proved that, for all x,

$$\chi(Z_t) = \sum_{x \in W} \mathbf{1}_{\zeta_x \geq t} \mathbf{1}_{\dot{\zeta}_x = 0} \operatorname{sign}(\det(-\ddot{\zeta}_x))$$

$$+ \sum_{x \in \partial W} \mathbf{1}_{\zeta_x \geq t} \mathbf{1}_{\dot{\zeta}_x^\natural = 0} \mathbf{1}_{\dot{\zeta}_x^\top < 0} \operatorname{sign}(\det(-\ddot{\zeta}_x^\natural - \dot{\zeta}_x^\top M_x)), \quad (2.1)$$

where $\dot{\zeta}$ (respectively $\ddot{\zeta}$) is the gradient (respectively the Hessian matrix) of ζ, $\dot{\zeta}^\top$ is the directional derivative of ζ in the direction of the inside normal of ∂W, $\dot{\zeta}^\natural$ and $\ddot{\zeta}^\natural$ are the $(d-1)$-dimensional gradient and the $(d-1) \times (d-1)$ Hessian of ζ in the tangent plane to ∂W and M_x is the $(d-1) \times (d-1)$ inside curvature matrix of

∂W. The reasons that led to (2.1) are essentially deterministic and its proof is based on Morse's theorem from differential topology, which is applicable because of the regularity conditions imposed on ζ.

By taking expectations it is possible to deduce the following result valid under the same assumptions as (2.1)

$$\mathbf{E}\chi(Z_t) = \int_W \mathbf{E}\left[\mathbf{1}_{\zeta_x \geq t} \det(-\ddot{\zeta}_x)|\dot{\zeta}_x = 0\right]\theta_0(x)dx$$

$$+ \int_{\partial W} \mathbf{E}\left[\mathbf{1}_{\zeta_x \geq t}\mathbf{1}_{\dot{\zeta}_x^\top <0} \det(-\ddot{\zeta}_x^\sharp - \dot{\zeta}_x^\top M_x)|\dot{\zeta}_x^\sharp = 0\right]\theta_0^\sharp(x)dx,$$

where $\theta_0(x)$ and $\theta_0^\sharp(x)$ are the densities (evaluated at zero) of $\dot{\zeta}_x$ and $\dot{\zeta}_x^\sharp$ respectively. If ζ is stationary, then θ_0 does not depend on x whence the first term in the right-hand side can be written as $\text{mes}(W)\delta(\zeta,t)$ with

$$\delta(\zeta,t) = \mathbf{E}\left[\mathbf{1}_{\zeta \geq t} \det(-\ddot{\zeta}_x)|\dot{\zeta}_x = 0\right]\theta_0$$

being the rate (or intensity) of the Euler–Poincaré characteristic of Z_t.

The result becomes much simpler if ζ is isotropic. For instance, if $d = 2$ and $\mathbf{E}\zeta_x$ is identically zero, then

$$\mathbf{E}\chi(Z_t) = \text{mes}(W)(2\pi)^{-3/2}\det(\Gamma)^{1/2}\sigma^{-3}t\exp\{-u^2/(2\sigma^2)\}, \qquad (2.2)$$

where Γ is the covariance matrix of $\dot{\zeta}$ and $\sigma^2 = \mathbf{E}\zeta_0^2$.

Open problem 2.3. Characterise classes of random processes by intrinsic properties of their level sets. For instance, if X is a stationary random closed set, then there is a stationary stochastic process with level set X. Which conditions on $X \subset \mathbb{R}^d$ must be imposed in order to ensure the existence of a Gaussian (continuous) random process ζ_x such that $X = \{x : \zeta_x = a\}$?

Hitting times

Let ζ_t be a continuous Markov process in $\mathbb{E}$. For every $t \geq 0$, let

$$X_t = \{\zeta_s : 0 \leq s \leq t\}$$

be the image of ζ until time t. The first hitting time of a set K is defined as

$$\tau_K = \inf\{t \geq 0 : \zeta_t \in K\} = \inf\{t \geq 0 : X_t \cap K \neq \emptyset\}.$$

Note that $\mathbf{P}^x$ (respectively $\mathbf{E}^x$) designates the probability (respectively expectation) taken with respect to the distribution of the process that starts at x. Then, for every constant $q > 0$,

$$T^x(K) = \mathbf{E}^x(e^{-q\tau_K}) \qquad (2.3)$$

is a capacity functional of a random closed set $X_\tau = \{\zeta_t : 0 \leq t \leq \tau\}$ which is the image of ζ (starting from x) until the exponentially distributed time moment τ with mean q^{-1} and independent of ζ. Indeed,

$$\mathbf{P}^x\{X_\tau \cap K \neq \emptyset\} = \mathbf{P}^x\{\tau_K \leq \tau\} = \mathbf{E}^x(e^{-q\tau_K}).$$

The integral of $T^x(K)$ over all x

$$C^q(K) = q \int_{\mathbb{R}^d} \mathbf{E}^x(e^{-q\tau_K}) dx$$

is called the q-capacity of K. If ζ is a transient process and K is a compact set, then the 0-capacity of K is defined as the limit of $C^q(K)$ as $q \downarrow 0$.

The following result provides another interpretation of the capacity functional T from (2.3). Note that the set $X_t + K$ can be viewed as a sausage of shape K drawn around the path of the stochastic process ζ.

Proposition 2.4 (Sausages and capacities). *Let K be a compact set. Then*

$$\int_{\mathbb{R}^d} T^x(K) dx = \int_0^\infty \mathbf{E}\, \mathrm{mes}(X_t + \check{K}) e^{-qt} dt.$$

Proof. Note that the expectation $\mathbf{E}^x\, \mathrm{mes}(X_t + \check{K})$ does not depend on x by the translation-invariance of the Lebesgue measure. By Robbins' theorem,

$$\mathbf{E}\, \mathrm{mes}(X_t + \check{K}) = \int_{\mathbb{R}^d} \mathbf{P}^0\{-x \in X_t + \check{K}\} dx = \int_{\mathbb{R}^d} \mathbf{P}^x\{X_t \cap K \neq \emptyset\} dx.$$

Therefore,

$$\int_0^\infty \mathbf{E}\, \mathrm{mes}(X_t + \check{K}) e^{-qt} dt = \int_0^\infty \int_{\mathbb{R}^d} \mathbf{P}^x\{X_t \cap K \neq \emptyset\} e^{-qt} dx dt$$

$$= \int_{\mathbb{R}^d} \mathbf{E}^x(e^{-q\tau_K}) dx. \qquad \square$$

2.2 Random subsets of the positive half-line and filtrations

A random closed set X in $\mathbb{R}_+ = [0, \infty)$ naturally gives rise to a filtration $\mathfrak{F}_t$, $t \geq 0$, where $\mathfrak{F}_t$ is the completion of the minimal complete σ-algebra generated by the set $X_t = X \cap [0, t]$. The filtration $\mathfrak{F}_t$ is called the *natural filtration* of X. Recall that the natural filtration of a stochastic process ζ_t, $t \geq 0$, is formed by the minimal complete σ-algebras $\mathfrak{F}_t$ generated by ζ_s for $s \leq t$.

By the fundamental measurability theorem, the graph of the random closed set X belongs to the product σ-algebra $\mathfrak{B}([0,\infty)) \otimes \mathfrak{F}$. By the construction of the natural filtration, the graph of $X \cap [0, t]$ is measurable with respect to $\mathfrak{B}([0, t]) \otimes \mathfrak{F}$ for each $t \geq 0$, i.e. X is *progressively measurable*, see Dellacherie [131, III-O8]. Therefore, the indicator of a random closed set in $\mathbb{R}_+$ is progressively measurable with respect to the natural filtration.

Important stochastic processes associated with X are *forward recurrence process* (or residual lifetime process)

$$x_t^+ = \inf(X \cap (t, \infty)) - t$$

(with the convention $\inf \emptyset = \infty$) and the *backward recurrence process* (or age process)

$$x_t^- = t - \sup(X \cap [0, t]),$$

see Figure 2.1.

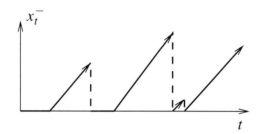

Figure 2.1. A realisation of the age process.

The both recurrence processes x_t^+ and x_t^- are right-continuous. Furthermore, $X = \{t : x_t^- = 0\}$. The zero set of x_t^+ is not X, but the set $X^\flat$ that consists of the isolated points from X or those points that are limits of a decreasing sequence from X. Denote also

$$z_t^+ = \inf(X \cap (t, \infty)).$$

Lemma 2.5. *The natural filtration of X coincides with the natural filtration generated by the process x_t^-.*

Proof. The process x_t^- is measurable with respect to $\mathfrak{F}_t$, since $\{x_t^- > a\} = \{X_t \cap [t-a, t] = \emptyset\} \in \mathfrak{F}_t$ for every a. If x_t^- is adapted with respect to a complete filtration, then this process is progressive measurable. The measurability of X_t then follows from the projection theorem, see Theorem E.5. □

The stopping time τ is a random variable with values in $[0, \infty]$ such that $\{\tau \leq t\} \in \mathfrak{F}_t$ for each t. The corresponding stopping σ-algebra $\mathfrak{F}_\tau$ is the family of all $A \in \mathfrak{F}$ such that $A \cap \{\tau \leq t\} \in \mathfrak{F}_t$ for all $t \geq 0$. The following result provides another interpretation of the stopping σ-algebra.

2 Levels and excursion sets of random functions 327

Theorem 2.6. Let ζ_t be a right-continuous stochastic process in a locally compact space $\mathbb{E}$ with the natural filtration $\mathfrak{F}_t$. If τ is a stopping time, then $\mathfrak{F}_\tau$ coincides with the minimal complete σ-algebra generated by the process

$$\tilde{\zeta}_t = \begin{cases} \zeta_t, & t \leq \tau, \\ \delta, & t > \tau, \end{cases} \quad t \geq 0,$$

where δ is a point that does not belong to $\mathbb{E}$.

Proof. Without loss of generality assume that $\tau < \infty$ almost surely. Let $\tilde{\mathfrak{F}}_t$ be the (not necessarily complete) σ-algebra generated by ζ_s for $s \leq t$. Denote by $\mathfrak{F}_{\tau-}$ (respectively $\tilde{\mathfrak{F}}_{\tau-}$) the σ-algebra generated by the sets $A \cap \{\tau > t\}$ for $A \in \mathfrak{F}_t$ (respectively $A \in \tilde{\mathfrak{F}}_t$) and $t \geq 0$. Define

$$\check{\zeta}_t = \begin{cases} \zeta_t, & t < \tau, \\ \delta, & t \geq \tau, \end{cases} \quad t \geq 0.$$

Then the σ-algebra $\sigma(\check{\zeta}_t, t \geq 0)$ generated by the process $\check{\zeta}_t$ includes $\tilde{\mathfrak{F}}_{\tau-}$.

For a σ-algebra $\tilde{\mathfrak{F}}$, its **P**-completion is denoted by $\tilde{\mathfrak{F}}^\mathbf{P}$. The completion theorem for σ-algebras (see Gihman and Skorohod [193, p. 125]) implies that $\mathfrak{F}_{\tau-} \subset \tilde{\mathfrak{F}}_{\tau-}^\mathbf{P}$. For $n \geq 1$, define

$$\zeta_t^{(n)} = \begin{cases} \zeta_t, & t < \tau + n^{-1}, \\ \delta, & t \geq \tau + n^{-1}, \end{cases} \quad t \geq 0.$$

Using the above arguments and Dellacherie [131, III-T35], one obtains

$$\mathfrak{F}_\tau = \bigcap_{n \geq 1} \mathfrak{F}_{\tau+1/n} \subset \bigcap_{n \geq 1} \tilde{\mathfrak{F}}_{\tau+1/n-}^\mathbf{P} \subset \bigcap_{n \geq 1} \sigma(\zeta_t^{(n)}, t \geq 0)^\mathbf{P}$$

$$\subset \left(\bigcap_{n \geq 1} \sigma(\zeta_t^{(n)}, t \geq 0) \right)^\mathbf{P}.$$

Since $\sigma(\zeta_t^{(n)}, t \geq 0)$ is generated by $\sigma(\tilde{\zeta}_t, t \geq 0)$ together with $\sigma(\zeta_{s+\tau}, 0 < s < n^{-1})$,

$$\bigcap_{n \geq 1} \sigma(\zeta_t^{(n)}, t \geq 0) = \sigma(\tilde{\zeta}_t, t \geq 0),$$

whence $\mathfrak{F}_\tau \subset \sigma(\tilde{\zeta}_t, t \geq 0)^\mathbf{P}$.

The process $\tilde{\zeta}$ is $\mathfrak{F}_\tau$-measurable. Indeed, for all $t, s \geq 0$ and Borel set B,

$$\{\tilde{\zeta}_s \in B\} \cap \{\tau \leq t\} = \{\zeta_s \in B\} \cap \{s \leq \tau\} \cap \{\tau \leq t\} \in \mathfrak{F}_t$$

and $\{\tilde{\zeta} = \delta\} \cap \{\tau \leq t\} = \{\tau < s\} \cap \{\tau \leq t\} \in \mathfrak{F}_t$. Since $\mathfrak{F}_\tau$ is complete, $\sigma(\tilde{\zeta}_t, t \geq 0)^\mathbf{P} \subset \mathfrak{F}_\tau$. □

Theorem 2.6 together with Lemma 2.5 yield the following result.

Proposition 2.7. *Let X be a random closed set in $\mathbb{R}_+$ with the natural filtration $\mathfrak{F}_t$. If τ is a stopping time, such that $\tau \in X^b \cup \{\infty\}$ a.s., then $\mathfrak{F}_\tau$ coincides with the minimal complete σ-algebra $\sigma(X_\tau)^{\mathbf{P}}$ generated by $X_\tau = X \cap [0, \tau]$.*

Proof. By Lemma 2.5, $\mathfrak{F}_t = \sigma(x_s^-, s \leq t)^{\mathbf{P}}$. Consider the process

$$\tilde{x}_t = \begin{cases} x_t^-, & t \leq \tau, \\ \delta, & t > \tau, \end{cases} \quad t \geq 0.$$

For all $t \geq 0$, $\{\tilde{x}_t = \delta\} = \{X_\tau \cap (t, \infty) = \emptyset\} \in \mathfrak{F}(X_\tau)^{\mathbf{P}}$. For every $a \geq 0$,

$$\{\tilde{x}_t \geq a\} = \{[t - a, t] \cap X_\tau = \emptyset\} \cap \{\tilde{x}_t \neq \delta\} \in \mathfrak{F}(X_\tau)^{\mathbf{P}}.$$

Therefore, X_τ is measurable with respect to the completion of the σ-algebra generated by the process $\tilde{x}_t$, $t \geq 0$. The proof is finished by applying Theorem 2.6. □

For a pair of stopping times τ_1 and τ_2 such that $\tau_1 \leq \tau_2$ a.s. define the random interval $Z = [\tau_1, \tau_2)$. An important σ-algebra on $\mathbb{R}_+ \times \Omega$ (called the optional σ-algebra) is generated by the graphs of such random intervals Z, i.e. sets $\{(t, \omega) : t \in Z(\omega)\}$. If the graph of X is measurable with respect to the optional σ-algebra, then X is said to be the *optional* random set. It follows from Dellacherie [131, IV-T10] that if X is optional, then for every $\varepsilon > 0$ there exists a stopping time τ such that $\tau \in X$ almost surely on $\{\tau < \infty\}$ and $\mathbf{P}\{X \neq \emptyset\} \leq \mathbf{P}\{\tau < \infty\} + \varepsilon$.

It is known that every closed set F in $\mathbb{R}$ can be decomposed into the union of at most a countable set and a perfect set F'. The following result proved by Dellacherie [131, Ch. VI] also holds for not necessarily closed sets. It provides a sort of the Castaing representation for optional random closed sets. Note that a selection of X provided by the fundamental selection theorem is not necessarily a stopping time.

Theorem 2.8 (Decomposition of optional random closed sets). *Let X be an optional random closed set in $\mathbb{R}_+$.*
  (i) *X can be decomposed as $X' \cup \{\tau_i, i \geq 1\}$, where X' is a random perfect set and $\tau_i, i \geq 1$, are stopping times.*
  (ii) *If X is almost surely finite or countable, then X can be represented as $\{\tau_i, i \geq 1\}$ for stopping times $\tau_i, i \geq 1$.*

If ξ_t is a non-decreasing process, then $t \geq 0$ is called a point of increase if $\xi_{t+\varepsilon} - \xi_t > 0$ or $\xi_t - \xi_{t-\varepsilon} > 0$ for all $\varepsilon > 0$. The set of all points of increase is a random closed set, which is also perfect if ξ is almost surely continuous, see Dellacherie [131, VI-T35].

Theorem 2.9 (Increasing process supported by a random perfect set). *Every random perfect set X is the set of points of increase for a continuous and bounded non-decreasing stochastic process ξ.*

Proof. Since $[0, \infty]$ is homeomorphic to $[0.5, 1]$, it is possible to assume that $X \subset [0.5, 1]$ a.s. Then $X = X_1 \cup X_2$ for two random closed sets X_1 and X_2, where

2 Levels and excursion sets of random functions

$X_1 = \operatorname{cl}(\operatorname{Int} X)$ is regular closed and $X_2 = \operatorname{cl}(X \setminus X_1)$ has an empty interior. A bounded continuous non-decreasing process corresponding to X_1 is given by

$$\xi_t = \int_0^t e^{-t} \mathbf{1}_{s \in X_1} ds .$$

The increasing process corresponding to X_2 can be constructed by an induction argument, see Dellacherie [131, VI-T37]. □

If $\zeta_t = \inf\{s : \xi_s > t\}$ is the right-continuous inverse to ξ from Theorem 2.9, then X can be represented as the image of ζ.

2.3 Level sets of strong Markov processes

Below we deal with random closed subsets of the positive half-line that appear as level sets of strong Markov processes. Recall that the strong Markov property extends the usual Markov property for random moments being stopping times. Assuming the strong Markov property leads to a full characterisation of level sets that are often called strong Markov random sets or regenerative sets. The latter term emphasises their close relationships to the regenerative events considered in Section 1.4.3.

Let ξ_t, $t \geq 0$, be a right-continuous real-valued *strong Markov process* with $\xi_0 = 0$. Although its *level set* $\{\xi = 0\}$ is not necessarily closed, it is right-closed, i.e. $\{\xi = 0\}$ contains limits of every converging decreasing sequences of its points. The closure of the level set $\{\xi = 0\}$ will be denoted by X. If ξ_t has a finite state space, then X consists of segments of independent exponentially distributed lengths separated by independent random intervals. This set is a particular example of the alternating renewal process.

The *coverage function* $p_X(t) = \mathbf{P}\{t \in X\}$ is called the *p-function* of X. If $p_X(t) \to 1$ as $t \to 0$, then X and its *p*-function are said to be *standard*. In this case $p_X(t)$ has nice analytic properties and $p_X(t)$, $t \geq 0$, determines the distribution of X, see Section 1.4.3. If $p_X(t) \not\to 1$ as $t \to 0$ (e.g. if p_X vanishes identically), then the *p*-function is no longer useful to analyse probabilistic properties of X. One should use a different technique based on filtrations and stopping times.

Strong Markov random sets

Let X be a random subset of $[0, \infty)$. Recall that $X^\flat$ denotes the set of all isolated or right-limit points of X. For $t \geq 0$, define

$$\theta_t(X) = X \cap [t, \infty) - t .$$

Definition 2.10 (Strong Markov set). A random closed set X in $[0, \infty)$ with the natural filtration $\mathfrak{F}_t$, $t \geq 0$, is said to be *homogeneous strong Markov random set* if $0 \in X$ a.s. and, for every $\mathfrak{F}_t$-stopping time τ such than $\tau \in X^\flat$ a.s. on $\{\tau < \infty\}$,
(i) $\theta_\tau(X)$ and $X \cap [0, \tau]$ are conditionally independent given $\{\tau < \infty\}$ and

(ii) conditional distribution of $\theta_\tau(X)$ given $\{\tau < \infty\}$ coincides with the distribution of X.

The conditions of Definition 2.10 are intrinsic, i.e. they are formulated as properties of X only. It is possible to weaken slightly the condition and define a regenerative set as follows. Let $\mathfrak{H}_t$ be the σ-algebra generated by z_t^+ and $X \cap [0, t]$.

Definition 2.11 (Regenerative set). A random closed set X in $[0, \infty)$ is said to be *regenerative* if there exists a random closed set X_0 such that, for each $t \in [0, \infty)$, the conditional distribution of $\theta_{z_t^+}(X)$ given $\mathfrak{H}_t$ coincides with the distribution of X_0 on $\{z_t^+ < \infty\}$.

The random variable z_t^+ is an $\mathfrak{F}_t$-stopping time and $z_t^+ \in X^b$ a.s., whence the strong Markov property implies the regenerative one. Since the requirement $0 \in X$ is dropped in Definition 2.11, it may be used to define stationary regenerative sets, see Fitzsimmons and Taksar [178]. The following result shows that a regenerative property imposed in Definition 2.11 is not much weaker than the property required in Definition 2.10(i).

Proposition 2.12. *Let $\mathfrak{F}_t$ be the natural filtration of a regenerative set X. For each $\mathfrak{F}_t$-stopping time τ such that $\tau \in X^b$ a.s. on $\{\tau < \infty\}$, the conditional distribution of $\theta_\tau(X)$ given $\mathfrak{F}_\tau$ coincides with the distribution of X_0 on $\{\tau < \infty\}$.*

It has been shown by Krylov and Yushkevitch [335] and Hoffman-Jørgensen [261] that the backward recurrence process x_t^- generated by X is strong Markov if and only if X is a strong Markov random set. Noticing that $X = \{t \geq 0 : x_t^- = 0\}$, this immediately leads to the following intrinsic characterisation of level sets of strong Markov processes.

Theorem 2.13 (Level sets of strong Markov processes). *A random closed set X in $[0, \infty)$ is strong Markov if and only if there exists a right-continuous real-valued strong Markov process ξ_t such that $X = \mathrm{cl}\{t : \xi_t = 0\}$ and $\xi_0 = 0$ almost surely.*

Open problem 2.14. Find an intrinsic characterisation of level sets for Markov processes without assuming the strong Markov property.

Subordinators and local time

The Markov property also holds for the forward recurrence process x_t^+. Denote its transition probabilities by $P_t(x, B) = \mathbf{P}\{x_{t+s}^+ \in B \mid x_s^+ = x\}$ for B being a Borel subset of $[0, \infty)$. The semi-linear structure of trajectories of x_t^+ implies $P_t(x, B) = P_{t-x}(B)$ for $t \geq x$, where $P_t(B) = P_t(0, B)$, $t > 0$. The Chapman–Kolmogorov equation for x_t^+ can be written as

$$P_{s+t}(B) = \int_{[0,t]} P_s(\mathrm{d}y) P_{t-y}(B) + \int_{[t,\infty)} P_s(\mathrm{d}y) \mathbf{1}_B(y - t).$$

From this, it is possible to show that

$$\int_0^\infty \int_0^\infty e^{-\alpha t - \theta y} P_t(dy) dt = \frac{\Phi(\theta) - \Phi(\alpha)}{(\theta - \alpha)\Phi(\alpha)},$$

for $\alpha > 0$, $\theta \geq 0$, $\theta \neq \alpha$, where

$$\Phi(\theta) = \varepsilon\theta + \int_{(0,\infty]} (1 - e^{-\theta x}) \mu(dx) \tag{2.4}$$

with $\varepsilon \geq 0$ and a measure μ on $(0, \infty]$ such that $\int (1 - e^{-x}) \mu(dx) < \infty$. The function Φ is called the *Laplace exponent* of X, while μ is the *Lévy measure*. The Laplace exponent also appears in (1.4.10) with $\varepsilon = 1$.

Theorem 2.9 implies that there exists an increasing process ξ_t called the *local time* such that X constitutes its points of increase. If X is a regenerative set, it is possible to define ξ in such a way that its right-continuous inverse ζ is a non-decreasing process with independent increments called a *subordinator*. Its *cumulant* is given by (2.4), i.e.

$$\mathbf{E} e^{-\theta \zeta_s} = e^{-s\Phi(\theta)}, \quad s \geq 0.$$

If $\varepsilon > 0$ in (2.4), then X is *standard*. We say that X is *light* if the drift coefficient ε vanishes. Kesten [298] showed that $\varepsilon = 0$ implies $p_X(x) = \mathbf{P}\{x \in X\} = 0$ for all $x \neq 0$, whence the Lebesgue measure of X vanishes almost surely. If $\varepsilon = 0$ and μ is finite, then X is a renewal process. It is easy to see (Kingman [310] and Fitzsimmons, Fristedt and Maisonneuve [176]) that X is discrete a.s. or perfect a.s. according to whether both $\varepsilon = 0$ and $\mu(0, \infty] < \infty$ or not. Furthermore, X has empty interior a.s. if $\mu(0, \infty] = \infty$, while X is a union of disjoint closed non-degenerate intervals if $\varepsilon > 0$ and $\mu(0, \infty] < \infty$. A non-negative value $\mu(\{\infty\})$ implies that X is almost surely bounded.

It is possible to define the local time constructively as

$$\xi_t = \lim_{\delta \to 0} \frac{\operatorname{mes}(X_\delta(t))}{l(\delta)}$$

where

$$X_\delta(t) = \bigcup_{s \in X \cap [0,t]} (s, s+\delta),$$

$$l(\delta) = \int_{(0,\infty]} \min(x, \delta) \mu(dx).$$

Example 2.15 (Zero set of the Wiener process). Let $X = \{t : w_t = 0\}$ be the zero set for the Wiener process. Then $\Phi(\theta) = \theta^{1/2}$, $\mu(dx) = cx^{-3/2}dx$ and $l(\delta) = c\delta^{1/2}$ for a constant $c > 0$ and the Hausdorff dimension of X is $1/2$.

Example 2.16 (Stable subordinator). The *stable* subordinator arises when $\varepsilon = 0$ and there exists $\alpha \in (0, 1)$ and $c > 0$ such that the measure μ is absolutely continuous with density $c\alpha x^{-(1+\alpha)}/\Gamma(1-\alpha)$ for $x > 0$, whence

332   5 Random Sets and Random Functions

$$\mu((x, \infty)) = cx^{-\alpha}/\Gamma(1-\alpha)$$

and $\Phi(\theta) = c\theta^\alpha$. Then X is said to be a *stable* strong Markov random set. It is shown by Hawkes [226, Lemma 1] that stable strong Markov random sets are characterised by the property that

$$\mathbf{P}\{X \cap (a, b] \neq \emptyset\} = \mathbf{P}\{X \cap (ta, tb] \neq \emptyset\}$$

whenever $0 < a < b$ and $t > 0$. In this case X is also *self-similar*, i.e. $X \stackrel{d}{\sim} cX$ for every $c > 0$.

The following theorem provides a variant of Proposition 1.4.20 for strong Markov random closed sets.

Theorem 2.17 (Hitting probability for the range of subordinator). *Let ζ_t be a drift-free subordinator (i.e. $\varepsilon = 0$ in (2.4)) having continuous distributions. If $0 < a < b$, then*

$$\mathbf{P}\{\zeta_t \in (a, b] \text{ for some } t\} = \int_a^b H(b-s)dU(s),$$

where $H(x) = \mu((0, \infty])$ is the tail of the Lévy measure and

$$U(A) = \mathbf{E}\int_0^\infty \mathbf{1}_A(\zeta_t)dt$$

is the occupation measure of ζ. In particular, if ζ is a stable subordinator of index α, then

$$\mathbf{P}\{\zeta_t \in (a, b] \text{ for some } t\} = \frac{\sin \pi \alpha}{\pi} \int_0^{1-a/b} t^{-\alpha}(1-t)^{\alpha-1}dt.$$

Proof. Note that $\zeta_t \in (a, b]$ for some t if and only if $\zeta_{\eta_a} \leq b$ for $\eta_a = \inf\{t : \zeta_t > a\}$. Furthermore,

$$\mathbf{P}\{\zeta_{\eta_a} > b\} = \int_0^\infty \mathbf{P}\{\eta_a \in dt, \zeta_{\eta_a} > b\}$$

$$= \int_0^\infty \int_0^a \mathbf{P}\{\zeta_t \in ds, \text{ jump exceeding } b - s \text{ in } dt\}$$

$$= \int_0^\infty \int_0^a \mathbf{P}\{\zeta_t \in ds\} H(b-s)dt$$

$$= \int_0^a H(b-s)dU(s).$$

By using Laplace transforms one can show that $\int_0^b H(b-s)dU(s) = 1$ for all b, see Kesten [298, pp. 117-118]. Then

$$\mathbf{P}\{\zeta_{\eta_a} \in (a,b]\} = 1 - \mathbf{P}\{\zeta_{\eta_a} > b\} = \int_a^b H(b-s)dU(s).\qquad\square$$

Weak convergence and embedding

The regenerative property is kept under taking weak limits, meaning that if a sequence of regenerative sets converges weakly, then its limit is also regenerative.

Theorem 2.18 (Weak convergence of strong Markov sets). *A sequence of strong Markov random sets $\{X_n, n \geq 1\}$ with Laplace exponents $\{\Phi_n(\theta), n \geq 1\}$ converges weakly to a (necessarily strong Markov) random closed set X with the Laplace exponent Φ if and only if Φ_n converges pointwisely toward Φ.*

If X_1 and X_2 are independent strong Markov sets, then their intersection $X_1 \cap X_2$ is strong Markov too. If both X_1 and X_2 are standard, then it is easy to find the distribution of their intersection using its p-function, since

$$p_{X_1 \cap X_2}(t) = \mathbf{P}\{t \in (X_1 \cap X_2)\} = p_{X_1}(t)p_{X_2}(t).$$

This argument is no longer applicable if the p-function of either X_1 or X_2 vanishes.

Theorem 2.19 (Intersection of strong Markov sets). *Let X_1 and X_2 be independent strong Markov random closed sets. Assume that their occupation measures U_1 and U_2 have the densities u_1 and u_2. Then*
(i) $X_1 \cap X_2 \neq \emptyset$ a.s. if and only if $\mathrm{cap}_{u_1}(X_2) > 0$, where cap_{u_1} is the capacity defined for the kernel u_1, see Appendix E.
(ii) If $X_1 \cap X_2 \neq \emptyset$ a.s., then $\int_0^1 u_1(t)U_2(dt)$ is finite. If u_1 is monotone, then the converse holds.
(iii) If u_1 exists and is continuous and monotone, then $X_1 \cap X_2 \neq \emptyset$ a.s. if and only if $u_1(t)U_2(dt)$ defines a locally finite measure, in which case this measure is proportional to the occupation measure of $X_1 \cap X_2$.

Example 2.20 (Intersection of stable sets). If X_1 and X_2 are stable strong Markov random sets with parameters α_1 and α_2, then $u_1 u_2$ is proportional to $t^{(\alpha_1+\alpha_2-1)-1}$ which is locally integrable if and only if $\alpha_1 + \alpha_2 > 1$. Hence X_1 and X_2 have a nontrivial intersection if and only if $\alpha_1 + \alpha_2 > 1$, in which case the intersection coincides in distribution with the stable strong Markov random set of index $\alpha_1 + \alpha_2 - 1$. For instance if X_1 and X_2 are zero sets of two independent Wiener processes, then $\alpha_1 = \alpha_2 = 1/2$ whence $X_1 \cap X_2$ is empty a.s.

Note that $X_1 \cap X_2$ is a strong Markov set that can be embedded or coupled as a subset of either X_1 and X_2, see Section 1.4.8. We say that a strong Markov random set X can be *regeneratively embedded* into a strong Markov random set Y if it is possible to realise X and Y on a same probability space such that $X \subset Y$ a.s.

Theorem 2.21 (Embedding of strong Markov sets). *Let X and Y be strong Markov sets with Laplace exponents Φ_X and Φ_Y. Then X is regeneratively embedded into Y if and only if Φ_X/Φ_Y is a completely monotone function.*

Example 2.22. If X is stable strong Markov set with parameter α, then Φ_X/Φ_Y is the Laplace transform of the fractional derivative of order α of the renewal measure U_Y for Y. Then X can be embedded into Y if and only if the α-fractional derivative of the renewal measure of Y is a Radon measure on $[0, \infty)$.

Open problem 2.23. Provide an intrinsic characterisation of all random closed sets on the positive half-line that may appear as level sets of a diffusion process, see Itô and McKean [272, p. 217]. The corresponding result for quasi-diffusions can be found in Knight [319]. The same question may be posed for zero sets of Lévy processes.

Open problem 2.24. Characterise all regenerative sets that are infinitely divisible for intersections, see also Fristedt [184].

2.4 Set-valued stopping times and set-indexed martingales

Set-indexed filtration

Examples of set-indexed stochastic processes are provided by counting measures related to point processes, general random measures and random capacities. For such processes it is possible to explore the natural partial order on the family of sets and introduce the concepts of progressive measurability, predictability and martingale properties. The starting point is a set-indexed filtration on a probability space $(\Omega, \mathfrak{F}, \mathbf{P})$. Assume that $\mathbb{E}$ is LCHS.

Definition 2.25 (Set-indexed filtration). *A family of complete σ-algebras $\mathfrak{F}_K$, $K \in \mathcal{K}$, is a set-indexed filtration if is*
  (i) *monotone, i.e. $\mathfrak{F}_{K_1} \subset \mathfrak{F}_{K_2}$ whenever $K_1 \subset K_2$;*
  (ii) *continuous from above, i.e. $\mathfrak{F}_K = \cap_{n=1}^\infty \mathfrak{F}_{K_n}$ if $K_n \downarrow K$.*

Without loss of generality assume that $\mathfrak{F}$ is the minimal σ-algebra that contains all $\mathfrak{F}_K$ for $K \in \mathcal{K}$.

Example 2.26 (Set-indexed filtrations).
  (i) $\mathfrak{F}_K = \sigma(\zeta_x, x \in K)$ is the minimal σ-algebra generated by ζ_x, $x \in K$, where ζ_x, $x \in \mathbb{E}$, is a random field on $\mathbb{E}$.
  (ii) $\mathfrak{F}_K = \sigma(X \cap K)$ is the minimal σ-algebra generated by $X \cap K$ if X is a random closed set in $\mathbb{E}$.
  (iii) $\mathfrak{F}_K$ generated by $\varphi(L)$ for $L \subset K$, $L \in \mathcal{K}$, where φ is a random capacity.

A set-indexed process ζ_K is said to be $\mathfrak{F}_K$-*adapted* if ζ_K is $\mathfrak{F}_K$-measurable for each $K \in \mathcal{K}$. Applied to the hitting process of a random closed set X (see Section 1.7.1), this means that X is $\mathfrak{F}_K$-adapted if $\{X \cap K \neq \emptyset\} \in \mathfrak{F}_K$ for each $K \in \mathcal{K}$.

Stopping set

Definition 2.27 (Stopping set). A random compact set Z is called a *stopping set* if $\{Z \subset K\} \in \mathfrak{F}_K$ for every $K \in \mathcal{K}$. The *stopping σ-algebra* $\mathfrak{F}_Z$ is the family of all $A \in \mathfrak{F}$ such that $A \cap \{Z \subset K\} \in \mathfrak{F}_K$ for every $K \in \mathcal{K}$.

Many examples of stopping sets are related to point processes. If a filtration is generated by a point process N with an infinite number of points, then the smallest ball centred at a given point containing a fixed number k of points of the process is a stopping set. Further examples are related to Delaunay triangulation and Voronoi tessellation generated by N, see Zuyev [631]. Below we discuss several measurability issues related to stopping sets.

Proposition 2.28 (Measurability with respect to the stopping σ-algebra). *If X is an adapted random closed set and Z is a stopping set, then $X \cap Z$ is $\mathfrak{F}_Z$-measurable. For instance, Z is $\mathfrak{F}_Z$-measurable.*

Proof. For any $K \in \mathcal{K}$ put $Z_K = X$ if $Z \subset K$ and $Z_K = K$ otherwise. Since

$$\{Z_K \subset L\} = \{Z_K \subset (L \cap K)\} \cup \{K \subset L, \ Z \not\subset K\} \in \mathfrak{F}_K,$$

Z_K is $\mathfrak{F}_K$-measurable for all $K \in \mathcal{K}$. For every $L \in \mathcal{K}$,

$$\{X \cap Z \cap L \neq \emptyset\} \cap \{Z \subset K\} = \{(X \cap K) \cap (Z_K \cap L) \neq \emptyset\} \cap \{Z \subset K\}.$$

Now $X \cap Z$ and $Z_K \cap L$ are measurable with respect to $\mathfrak{F}_K$. By Theorem 1.2.25, the intersection of these two random sets is also $\mathfrak{F}_K$-measurable. Finally, $\{Z \subset K\} \in \mathfrak{F}_K$ by the definition of the stopping set. □

The following result is similar to Theorem 2.6 and Proposition 2.7.

Proposition 2.29 (Generator of the stopping σ-algebra). *Let $\mathfrak{F}_K = \sigma(X \cap K)$ be a filtration generated by a random closed set X. Then for every stopping set Z, $\mathfrak{F}_Z = \sigma(X \cap Z)$.*

Proof. We outline the proof referring to Zuyev [631] for details. Given a stopping set Z_1 define the following σ-algebra

$$\mathfrak{F}_{Z_1-} = \sigma\Big(A_L \cap \{L \subset \operatorname{Int} Z_1\}, \ A_L \in \mathfrak{F}_L, \ L \in \mathcal{K}\Big).$$

The first step is to show that if Z and Z_1 are two stopping sets such that $Z \subset \operatorname{Int} Z_1$ a.s., then $\mathfrak{F}_Z \subset \mathfrak{F}_{Z_1-}$. Further, observe that for the natural filtration of X and any stopping set Z_1 one has

$$\mathfrak{F}_{Z_1-} = \sigma\Big(X \cap K, \ K \subset \operatorname{Int} Z_1, \ K \in \mathcal{K}\Big).$$

Choose a sequence of stopping sets $\{Z_n, n \geq 1\}$ such that $Z_n \downarrow Z$ and $Z \subset \operatorname{Int} Z_n$ for all n. Then

336   5 Random Sets and Random Functions

$$\mathfrak{F}_Z \subset \bigcap_{n \geq 1} \mathfrak{F}_{Z_n-} = \bigcap_{n \geq 1} \sigma\left(X \cap K, \ K \subset \text{Int } Z_n, \ K \in \mathcal{K}\right)$$
$$= \sigma\left(X \cap K, \ K \subset Z, \ K \in \mathcal{K}\right) = \sigma(X \cap Z).$$

Since Z is $\mathfrak{F}_Z$-measurable, it suffices to show that $(\{X \cap K \neq \emptyset\} \cap \{K \subset Z\}) \in \mathfrak{F}_Z$ for all K, that is

$$\{X \cap K \neq \emptyset\} \cap \{K \subset Z \subset L\} \in \mathfrak{F}_L$$

for any $L \in \mathcal{K}$. This is evident if $K \not\subset L$. Otherwise, $\{X \cap K \neq \emptyset\} \in \mathfrak{F}_L$ which finishes the proof. □

Set-indexed martingales

By adapting the definition of a martingale indexed by a partially ordered set from Kurtz [341] to the family of compact sets ordered by inclusion, a set-indexed martingale is defined as follows.

Definition 2.30 (Set-indexed martingale). A set-indexed random process ζ_K, $K \in \mathcal{K}$, is called a *martingale* if $\mathbf{E}(\zeta_{K_2}|\zeta_{K_1}) = \zeta_{K_1}$ a.s. for all $K_1, K_2 \in \mathcal{K}$ such that $K_1 \subset K_2$.

Under some uniform integrability condition it is possible to prove the optional sampling theorem that is formulated by replacing K_1 and K_2 in Definition 2.30 by stopping sets $Z_1 \subset Z_2$.

3 Semicontinuous random functions

3.1 Epigraphs of random functions and epiconvergence

Epiconvergence

In the following we assume that $\mathbb{E}$ is LCHS, although many concepts can be generalised for $\mathbb{E}$ being a general Polish space. If $f(x)$ is a lower semicontinuous function defined for $x \in \mathbb{E}$ with values in the extended real line $\bar{\mathbb{R}} = [-\infty, \infty]$, then

$$\text{epi } f = \{(x, t) \in \mathbb{E} \times \mathbb{R} : t \geq f(x)\}$$

is called the *epigraph* of f. By Proposition A.2, epi f is closed in the product topology on $\mathbb{E} \times \mathbb{R}$ if and only if f is lower semicontinuous.

The epigraph is an enormously influential concept in optimisation. If f is lower semicontinuous, then epi f contains all information necessary to evaluate $\inf_{x \in K} f(x)$ for any compact set K. Using epigraphs, the family of all lower semicontinuous functions is embedded into the family of closed subsets of $\mathbb{E} \times \mathbb{R}$. The convergence concepts for closed sets discussed in Appendix B can be used to define the convergence of lower semicontinuous functions using their epigraphs. The Fell topology is especially important if $\mathbb{E}$ is a locally compact space; in this case it is equivalent to the Painlevé–Kuratowski convergence. If $\mathbb{E}$ is a general Banach space, then the convergence of epigraphs is often considered in the Mosco sense.

Definition 3.1 (Epiconvergence). A sequence of lower semicontinuous functions $\{f_n, n \geq 1\}$ is said to *epiconverge* to f (notation $f_n \xrightarrow{\text{epi}} f$) if epi f_n converges to epi f as $n \to \infty$ in the space $\mathcal{F}$ of closed sets in the Painlevé–Kuratowski sense.

The limiting function f in Definition 3.1 may take infinite values. Note that the arithmetic sum is not continuous with respect to the epiconvergence, i.e. $f_n \xrightarrow{\text{epi}} f$ and $g_n \xrightarrow{\text{epi}} g$ does not necessarily imply the epiconvergence of $f_n + g_n$ to $f + g$.

Proposition 3.2 (Equivalent definitions of epiconvergence). *For a sequence of lower semicontinuous functions, the following statements are equivalent.*
(i) f_n epiconverges to f as $n \to \infty$.
(ii) *For all* $x \in \mathbb{E}$,
$$\liminf_{n \to \infty} f_n(x_n) \geq f(x)$$
for all sequences $x_n \to x$ *and*
$$\limsup_{n \to \infty} f_n(x_n) \leq f(x)$$
for at least one sequence $x_n \to x$.
(iii) *For all* $K \in \mathcal{K}$ *and* $G \in \mathcal{G}$,
$$\liminf_{n \to \infty} \left(\inf_{x \in K} f_n(x) \right) \geq \inf_{x \in K} f(x),$$
$$\limsup_{n \to \infty} \left(\inf_{x \in G} f_n(x) \right) \leq \inf_{x \in G} f(x).$$

The above properties are often alternatively used to define the epigraphical convergence. Another characterisation in terms of the convergence of excursion sets $\{f \leq t\}$ is also possible.

Proposition 3.3 (Epiconvergence in terms of excursion sets). *For a sequence of lower semicontinuous functions,* $f_n \xrightarrow{\text{epi}} f$ *if and only if the following two conditions hold:*
(1) $\limsup_{n \to \infty} \{f_n \leq t_n\} \subset \{f \leq t\}$ *for all sequences* $t_n \to t$;
(2) $\liminf_{n \to \infty} \{f_n \leq t_n\} \supset \{f \leq t\}$ *for some sequence* $t_n \to t$ *in which case this sequence can be chosen with* $t_n \downarrow t$.

The set of points that minimise f is denoted by
$$\operatorname{argmin} f = \{x \in \mathbb{E} : f(x) \leq \inf f < \infty\},$$
where $\inf f = \inf\{f(x) : x \in \mathbb{E}\}$. Note that dom f is the set of all $x \in \mathbb{E}$ such that $f(x)$ if finite, so that $\operatorname{argmin} f = \emptyset$ if dom $f = \emptyset$. The points which are nearly optimal or ε-optimal for some $\varepsilon > 0$ comprise the set

$$\varepsilon\text{-}\operatorname{argmin} f = \{x \in \mathbb{E} : f(x) \leq \inf f + \varepsilon < \infty\},$$

see Figure 3.1.

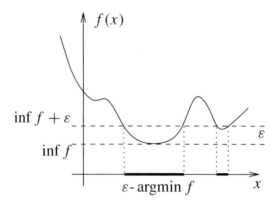

Figure 3.1. Optimal and nearly optimal points.

A sequence $\{f_n, n \geq 1\}$ is said to have a relatively compact sequence of ε-optimal points if, for any $\varepsilon > 0$, there exists a compact set K_ε and a sequence $\{a_n, n \geq 1\} \subset K_\varepsilon$ such that $f_n(a_n) < \inf f_n + \varepsilon$ for all $n \geq 1$. This is the case if there exists a compact set K such that $(\operatorname{dom} f_n) \subset K$ for all sufficiently large n.

Theorem 3.4 (Convergence of minimisers). *If a sequence of lower semicontinuous functions $\{f_n, n \geq 1\}$ epiconverges to f, then*

$$\limsup_{n \to \infty} (\inf f_n) \leq \inf f$$

with the equality holding if $\{f_n, n \geq 1\}$ has a relatively compact sequence of ε-optimal points. Furthermore,

$$\limsup_{n \to \infty}(\operatorname{argmin} f_n) \subset \operatorname{argmin} f. \tag{3.1}$$

If $\operatorname{argmin} f \neq \emptyset$, then

$$\operatorname{argmin} f = \bigcap_{\varepsilon > 0} \liminf_{n \to \infty} \left(\varepsilon\text{-}\operatorname{argmin} f_n \right)$$

if and only if $\inf f_n \to \inf f$ as $n \to \infty$.

Proposition D.2 together with (3.1) imply that the argmin functional is an upper semicontinuous multifunction on the family of lower semicontinuous functions. It should be noted that additional conditions (see Rockafellar and Wets [498, Prop. 3.42]) are required to ensure that $\operatorname{argmin} f_n$ converges to $\operatorname{argmin} f$ in the Fell topology.

It is easy to see that the uniform convergence of lower semicontinuous functions implies their epiconvergence. The epiconvergence of lower semicontinuous functions and their pointwise convergence do not imply each other. It is possible to show (see Dolecki, Salinetti and Wets [140]) that these two concepts coincide on any family $\mathcal{V}$ of lower semicontinuous functions which is *equi-lower semicontinuous*, i.e. for all $x \in \mathbb{E}$ and $\varepsilon > 0$ there exists a neighbourhood U of x such that $f(y) \geq \min(\varepsilon^{-1}, f(x) - \varepsilon)$ for all $y \in U$ and $f \in \mathcal{V}$.

Normal integrands

Consider a function $\zeta: \mathbb{E} \times \Omega \mapsto \bar{\mathbb{R}}$, where $\mathbb{E}$ is a LCHS space and $(\Omega, \mathfrak{F}, \mathbf{P})$ is a complete probability space. Such functions are sometimes called variational systems.

Definition 3.5 (Normal integrand). A function $\zeta(x, \omega)$, $x \in \mathbb{E}$, $\omega \in \Omega$, is called a *normal integrand* if its epigraph $X(\omega) = \text{epi } \zeta(\cdot, \omega)$ is a random closed set measurable with respect to $\mathfrak{F}$. A normal integrand is said to be *proper* if it does not take a value $-\infty$ and is not identically equal to $+\infty$.

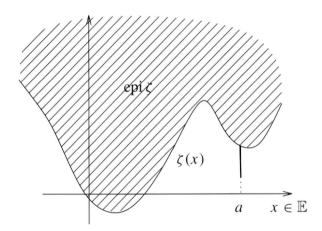

Figure 3.2. Normal integrand $\zeta(x)$, $x \in \mathbb{E}$, and its epigraph. Note that ζ is discontinuous at $x = a$, where $\zeta(a)$ is smaller than the corresponding left- and right-hand limits.

The closedness of the epigraph imposed in Definition 3.5 implies that ζ is lower semicontinuous with respect to x for almost all ω. To simplify notation we will usually write $\zeta(x)$ or ζ_x instead of $\zeta(x, \omega)$. If ξ is an $\mathbb{E}$-valued random element, then $\zeta(\xi)$ is a random variable, since

$$\{\omega: \zeta(\xi) \leq t\} = \{\omega: (\text{epi } \zeta) \cap (\{\xi\} \times (-\infty, t]) \neq \emptyset\} \in \mathfrak{F}, \quad t \in \mathbb{R}.$$

In particular, $\zeta(x)$ is a random variable for every $x \in \mathbb{E}$. However, if $\zeta(x)$ is a random variable for every $x \in \mathbb{E}$ and ζ is almost surely lower semicontinuous, then

ζ is not necessarily a normal integrand. In this case ζ is called a random lower semicontinuous function.

Proposition 3.6 (Joint measurability). *Let ζ be a random lower semicontinuous function. If $\zeta(x, \omega)$ is jointly measurable in (x, ω), then ζ is a normal integrand.*

Proof. For every measurable subset $A \subset \mathbb{E}$ and $t \in \mathbb{R}$,

$$\{(x, \omega) : (\text{epi } \zeta) \cap (A \times (-\infty, t]) \neq \emptyset\} = \{(x, \omega) : A \in K, \ \zeta(x, \omega) \leq t\}$$

is a measurable set in the product space, so that its projection on Ω is measurable by the completeness of the probability space, see Theorem E.5. □

Since $X = \text{epi } \zeta$ is a random closed subset of $\mathbb{E} \times \mathbb{R}$, its distribution is determined by the capacity functional $T(K)$ defined on compact subsets of $\mathbb{E} \times \mathbb{R}$. The finite-dimensional distributions of $\zeta(x)$, $x \in \mathbb{E}$, can be retrieved from the capacity functional of epi ζ, since

$$\mathbf{P}\{\zeta(x_1) > t_1, \ldots, \zeta(x_n) > t_n\} = 1 - T(\cup_{i=1}^{n}(\{x_i\} \times (-\infty, t_i])).$$

As in Section 1.6.1, $\mathfrak{S}_{\text{epi }\zeta}$ denotes the family of all $K \in \mathcal{K}(\mathbb{E} \times \mathbb{R})$ such that $T(K) = T(\text{Int } K)$.

The family of finite unions of sets of type $K_j \times [a_j, b_j]$ where $K_j \in \mathcal{K}(\mathbb{E})$ forms a separating class in $\mathbb{E} \times \mathbb{R}$. By the construction of the epigraph, epi ζ hits $K_j \times [a_j, b_j]$ if and only if epi ζ hits $K_j \times (-\infty, b_j]$. Therefore, it suffices to define the capacity functional on the finite unions of type $\cup_{j=1}^{m}(K_j \times (-\infty, t_j])$. The family of such sets will be denoted by $\mathcal{K}_e$. Without loss of generality assume that $t_1 < t_2 < \cdots < t_m$ and $K_1 \supset K_2 \supset \cdots \supset K_m$, see Figure 3.3.

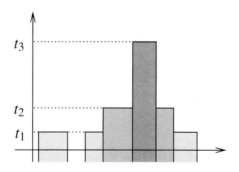

Figure 3.3. A set from $\mathcal{K}_e$.

A stochastic process ζ is *separable* if there exists an everywhere dense countable set $\mathbb{Q} \subset \mathbb{E}$ and a set Ω_0 of probability zero such that for every open set $G \subset \mathbb{E}$ and closed subset $F \subset \mathbb{R}$, the events $\{\zeta(x) \in F \text{ for all } x \in G \cap \mathbb{Q}\}$ and $\{\zeta(x) \in F \text{ for all } x \in G\}$ differ from each other at most on a subset of Ω_0, see Gihman and Skorohod [193]. The approach to stochastic processes based on their finite-dimensional distributions requires the separability to handle such properties like the continuity or the boundedness of stochastic processes. This is explained by the fact that the cylindrical σ-algebra is constructed without any topological assumptions on the index space, while these assumptions are being brought in later by means of the separability concept. In contrast, the epigraphical approach allows us to work with non-separable stochastic processes including those that may also have discontinuities of the second kind.

Example 3.7 (Non-separable process). Let ξ be a random element in $\mathbb{E} = \mathbb{R}^d$ with an absolutely continuous distribution and let $\zeta(x) = 0$ if $x = \xi$ and $\zeta(x) = 1$ otherwise, see Figure 3.4. Then ζ is not separable and its finite-dimensional distributions are indistinguishable with those of the function identically equal to 1. However, epi ζ is a non-trivial random closed set $X = (\{\xi\} \times [0, 1]) \cup (\mathbb{E} \times [1, \infty))$.

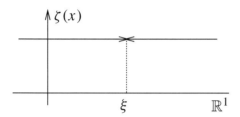

Figure 3.4. Non-separable stochastic process on $\mathbb{R}^1$.

For any normal integrand ζ, its epigraph epi ζ is a random closed set that possesses a Castaing representation, see Section 1.2.2. If ζ is a separable stochastic process, then it is possible to define constructively a Castaing representation of epi ζ. If $\mathbb{Q}$ is the separability set of ζ and $\mathbb{Q}_+$ is a countable dense subset of $[0, \infty)$, then $\{(q, \zeta(q) + a) : q \in \mathbb{Q}, a \in \mathbb{Q}_+\}$ provides a Castaing representation of epi ζ.

Definition 3.8 (Inner separability). A normal integrand is said to be *inner separable* if, to each compact set $K \in \mathfrak{S}_{\text{epi}\,\zeta}$ and $\varepsilon > 0$, there corresponds a finite set $I_\varepsilon \subset K$ such that $T(K) \leq T(I_\varepsilon) + \varepsilon$.

Proposition 3.9 (Separability and inner separability). *If ζ is a separable stochastic process, then the corresponding normal integrand is inner separable.*

Proof. It suffices to check the condition of Definition 3.8 for $K = \cup_{j=1}^m (K_j \times (-\infty, t_j])$ where $K_j \times (-\infty, t_j] \in \mathfrak{S}_{\text{epi}\,\zeta}$ for all j. For all $\varepsilon > 0$ there exist open sets $G_j \subset K_j$ and numbers $s_j < t_j$, $1 \leq j \leq m$, such that

$$T(\cup_j G_j \times (-\infty, s_j)) \geq T(\cup_j K_j \times (-\infty, t_j]) - \varepsilon/2.$$

By the separability property of ζ,

$$1 - T(\cup_j G_j \times (-\infty, s_j)) = \mathbf{P}(\cap_j \{\inf_{x \in G_j} \zeta(x) \geq s_j\})$$
$$= \mathbf{P}(\cap_j \{\inf_{x \in \mathbb{Q} \cap G_j} \zeta(x) \geq s_j\})$$
$$\geq \mathbf{P}(\cap_j \{\inf_{x \in D_j} \zeta(x) \geq s_j\}) - \varepsilon/2,$$

where $D_j \subset G_j$, $1 \leq j \leq m$, are finite sets. Then for $I = \{(x, s_j) : x \in D_j, 1 \leq j \leq m\}$, we have

$$T(I) \geq T(\cup_j G_j \times (-\infty, s_j)) - \varepsilon/2 \geq T(K) - \varepsilon. \qquad \square$$

The following proposition concerns the properties of normal integrands related to their level sets and minimisers.

Proposition 3.10 (Excursion sets and minimisers). *If ζ is a normal integrand, then*
(i) $\{\zeta \leq \alpha\} = \{x \in \mathbb{E} : \zeta(x) \leq \alpha\}$ *is a random closed set for every random variable $\alpha \in \mathbb{R}$;*
(ii) $\{\zeta \leq t\}$, $t \in \mathbb{R}$, *is an increasing set-valued process;*
(iii) $\inf \zeta$ *is a random variable with values in $\mathbb{R}$ and $\operatorname{argmin} \zeta$ and ε-$\operatorname{argmin} \zeta$ (for any $\varepsilon > 0$) are random closed subsets of $\mathbb{E}$.*

Proof.
(i) It suffices to note that $\{(\{\zeta \leq t\}) \cap K \neq \emptyset\}$ coincides with $\{(\operatorname{epi} \zeta) \cap (K \times (-\infty, t]) \neq \emptyset\}$ for every $K \in \mathcal{K}(\mathbb{E})$.
(ii) Apply (i) to $\alpha = t$.
(iii) Notice that $\alpha = \inf \zeta$ is a random variable and apply (i). $\qquad \square$

By Theorem 1.2.25(iii), the boundary $\partial \operatorname{epi} \zeta$ is a random closed set. Let $\partial^- \operatorname{epi} \zeta$ be the set of points $(x, t) \in \operatorname{epi} \zeta$ such that $(x, s) \notin \operatorname{epi} \zeta$ for all $s < t$, see Figure 3.5. Note that $\partial^- \operatorname{epi} \zeta$ is not necessarily closed.

Definition 3.11 (Sharp integrand). *If $\partial \operatorname{epi} \zeta = \operatorname{epi} \zeta$ a.s. and $\partial^- \operatorname{epi} \zeta$ is a locally finite subset of $\mathbb{E} \times \mathbb{R}$, then ζ is called a sharp integrand.*

Proposition 3.12. *If ζ is a sharp integrand, then $\partial^- \operatorname{epi} \zeta$ is a locally finite point process.*

Proof. It suffices to prove that $\partial^- \operatorname{epi} \zeta$ is measurable. Let $\mathcal{G}_0$ be a countable base of topology on $\mathbb{E}$ composed of relatively compact sets. For every $G \in \mathcal{G}_0$ consider the random closed set

$$X_G = \{(x, t) \in \mathbb{E} \times \mathbb{R} : \zeta(x) = t = \inf_{y \in \operatorname{cl} G} \zeta(y)\}.$$

Then $\partial^- \operatorname{epi} \zeta$ is the union of X_G for $G \in \mathcal{G}_0$, so is measurable. $\qquad \square$

Recall that the conjugate f° and the subdifferential of ∂f of a (convex) function f are defined in Appendix F. These definitions can be applied to a normal integrand ζ defined on $\mathbb{R}^d$.

3 Semicontinuous random functions 343

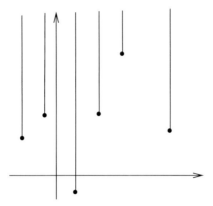

Figure 3.5. Sharp integrand $\zeta(x)$. The points of $\partial^- \text{epi}\,\zeta$ are shown as bold dots.

Theorem 3.13 (Conjugate and subdifferential of normal integrand).
(i) *If ζ is a normal integrand on $\mathbb{R}^d$, then its conjugate ζ° is a normal integrand.*
(ii) *If ζ is a proper normal integrand and ξ is an $\mathbb{E}$-valued random element such that $\zeta(\xi) < \infty$ a.s., then the subdifferential $\partial \zeta(\xi)$ is a random convex closed set (with possibly empty values).*

Proof.
(i) Let $\{(\xi_i, \alpha_i), i \geq 1\}$ be the Castaing representation of the random closed set epi ζ. Then
$$g_i(y) = \langle \xi_i, y \rangle - \alpha_i$$
is a normal integrand on $\Omega' = \{\omega : \text{epi}\,\zeta \neq \emptyset\}$ for every i, whence $\zeta^\circ(y) = \sup_i g_i(y)$ is a normal integrand on Ω'. For every $\omega \notin \Omega'$, we have $\zeta(x) = \infty$ for all x, so that $\zeta^\circ(y) = -\infty$ identically, i.e. epi $\zeta^\circ = \mathbb{R}^d \times \mathbb{R}$ in this case. Since Ω' is measurable, this finishes the proof of (i).
(ii) Let ζ be almost surely convex. Then
$$\partial \zeta(x) = \{y \in \mathbb{R}^d : \langle x, y \rangle - \zeta(x) \geq \langle x', y \rangle - \zeta(x') \text{ for all } x'\}$$
$$= \{y \in \mathbb{R}^d : \langle x, y \rangle - \zeta(x) \geq \zeta^\circ(y)\}$$
$$= \{y \in \mathbb{R}^d : \zeta^\circ(y) - \langle x, y \rangle \geq \zeta(x)\}.$$

Note that $\zeta^\circ(y) - \langle \xi, y \rangle$ is a normal integrand with respect to y. Therefore, $\partial \zeta(\xi)$ is the projection on $\mathbb{E} = \mathbb{R}^d$ of the random closed set obtained as the intersection of $\text{epi}(\zeta^\circ(\cdot) - \langle \xi, \cdot \rangle)$ and $\mathbb{E} \times [\zeta(\xi), \infty)$. It follows from the results of Section 1.2.3 that this projection is a random closed set in $\mathbb{E}$.

For a general ζ the statement follows from Theorem 1.2.25 concerning the measurability of limits for sequences of random closed sets. □

Weak epiconvergence

Definition 3.14 (Weak epiconvergence). A sequence $\{\zeta_n, n \geq 1\}$ of normal integrands *weakly epiconverges* (or epiconverges in distribution) to ζ if $X_n = \text{epi}\,\zeta_n$ converge weakly to $X = \text{epi}\,\zeta$ as random closed sets in $\mathbb{E} \times \mathbb{R}$.

Since $\mathbb{E}$ is LCHS, the weak epiconvergence can be formulated in terms of capacity functionals as

$$T_{\text{epi}\,\zeta_n}(K) \to T_{\text{epi}\,\zeta}(K) \quad \text{as } n \to \infty \tag{3.2}$$

for all $K \in \mathfrak{S}_{\text{epi}\,\zeta}$. By Corollary 1.6.9, it suffices to check (3.2) for all $K = \bigcup_{j=1}^m K_j \times (-\infty, t_j] \in \mathfrak{S}_{\text{epi}\,\zeta}$, where $K_1, \ldots, K_m$ belong to a separating class in $\mathbb{E}$ and $t_j \in \mathbb{R}$, $1 \leq j \leq m$, $m \geq 1$. Reformulating this fact for the stochastic processes ζ_n instead of their capacity functionals we come to the following conclusion.

Proposition 3.15 (Weak epiconvergence of normal integrands). *A sequence ζ_n, $n \geq 1$, of normal integrands weakly epiconverges to a normal integrand ζ if and only if*

$$\mathbf{P}\left\{\inf_{x \in K_i} \zeta_n(x) > t_i,\ i = 1, \ldots, m\right\} \to \mathbf{P}\left\{\inf_{x \in K_i} \zeta(x) > t_i,\ i = 1, \ldots, m\right\}$$

as $n \to \infty$ for all $m \geq 1$, $t_1, \ldots, t_m \in \mathbb{R}$ and $K_1, \ldots, K_m$ belonging to a separating class $\mathcal{A}$ on $\mathbb{E}$ and satisfying the continuity condition

$$\mathbf{P}\left\{\inf_{x \in K_i} \zeta(x) > t_i\right\} = \mathbf{P}\left\{\inf_{x \in \text{Int}\,K_i} \zeta(x) \geq t_i\right\}, \quad i = 1, \ldots, m.$$

Proposition 3.16 (Weak convergence for boundaries of epigraphs). *Let ζ and $\{\zeta_n, n \geq 1\}$ be proper normal integrands.*
  (i) *If $\partial\,\text{epi}\,\zeta_n$ weakly converges to $\partial\,\text{epi}\,\zeta$, then ζ_n weakly epiconverges to ζ.*
  (ii) *If $\mathbb{E}$ is locally connected and $\partial\,\text{epi}\,\zeta = \text{epi}\,\zeta$ a.s., then the weak epiconvergence of ζ_n to ζ implies that $\partial\,\text{epi}\,\zeta_n$ weakly converges to $\text{epi}\,\zeta = \partial\,\text{epi}\,\zeta$.*

Proof. By the continuous mapping theorem, it suffices to prove the conclusions for deterministic lower semicontinuous functions $\{f_n, n \geq 1\}$ that converge to a lower semicontinuous function f. For each $K \in \mathcal{K}$, $\inf_{x \in K} f(x) > t$ if and only if $\text{epi}\,f \cap (K \times (-\infty, t]) = \emptyset$; and for each relatively compact open set G, $\inf_{x \in G} f(x) < t$ if and only if

$$\text{epi}\,f \cap (G \times (-\infty, t)) \neq \emptyset.$$

(i) Assume that $\partial\,\text{epi}\,f_n$ converges to $\partial\,\text{epi}\,f$ in the Fell topology. For any $K \in \mathcal{K}$, $\inf_{x \in K} f(x) > t$ implies $(\partial\,\text{epi}\,f) \cap (K \times (-\infty, t]) = \emptyset$ and so $(\partial\,\text{epi}\,f_n) \cap (K \times (-\infty, t]) = \emptyset$ for all sufficiently large n, whence $\inf_{x \in K} f_n(x) > t$ (here it is important that the value $-\infty$ for f and f_n is excluded). Similarly, $\inf_{x \in G} f(x) < t$ implies $(\partial\,\text{epi}\,f_n) \cap (G \times (-\infty, t)) \neq \emptyset$, whence $\inf_{x \in G} f_n(x) < t$ for all sufficiently large n. By Proposition 3.2, f_n epiconverges to f.

(ii) Suppose that f_n epiconverges to f and $\partial \operatorname{epi} f = \operatorname{epi} f$. Since $\mathbb{E} \times \mathbb{R}$ is locally connected, taking a boundary is a lower semicontinuous mapping on the family of closed subsets of $\mathbb{E} \times \mathbb{R}$, see Proposition D.6(iii). Therefore,

$$\operatorname{epi} f = \partial \operatorname{epi} f \subset \liminf(\partial \operatorname{epi} f_n)$$
$$\subset \limsup(\partial \operatorname{epi} f_n) \subset \limsup(\operatorname{epi} f_n) = \operatorname{epi} f,$$

whence $\partial \operatorname{epi} f_n \xrightarrow{\text{PK}} \operatorname{epi} f$. □

Proposition 3.17 (Weak convergence of excursion sets).
(i) If ζ is a normal integrand, then the set of $t \in \mathbb{R}$ such that $\{\zeta \leq t\}$ almost surely coincides with $\operatorname{cl}(\{\zeta < t\})$ is dense in $\mathbb{R}$.
(ii) If ζ_n weakly epiconverges to ζ and $\{\zeta \leq t_i\} = \operatorname{cl}(\{\zeta < t_i\})$ a.s. for $t_1, \ldots, t_m$, then the m-tuple of random closed sets $(\{\zeta_n \leq t_1\}, \ldots, \{\zeta_n \leq t_m\})$ converges in distribution to $(\{\zeta \leq t_1\}, \ldots, \{\zeta \leq t_m\})$.
(iii) If, additionally to (ii), $\mathbb{E}$ is locally connected and $\partial\{\zeta \leq t_i\} = \{\zeta \leq t_i\}$ a.s. for all i (e.g. this is the case when ζ is a sharp integrand), then the m-tuple of random closed sets $(\partial\{\zeta_n \leq t_1\}, \ldots, \partial\{\zeta_n \leq t_m\})$ converges in distribution to $(\partial\{\zeta \leq t_1\}, \ldots, \partial\{\zeta \leq t_m\})$.

Proof.
(i) Let $\mathcal{G}_0$ be a countable base of $\mathcal{G}$. If $\{\zeta \leq t\} \neq \operatorname{cl}(\{\zeta < t\})$, then $\zeta(x) \leq t$ for some $x \notin \operatorname{cl}(\{\zeta < t\})$, whence $\inf_{y \in G} \zeta(y) \geq t$ for some $G \in \mathcal{G}_0$ with $x \in G$. Therefore, $\inf_{y \in G} \zeta(y) = t$ and

$$\left\{ t : \mathbf{P}\{\{\zeta \leq t\} = \operatorname{cl}(\{\zeta < t\})\} < 1 \right\} \subset \bigcup_{G \in \mathcal{G}_0} \left\{ t : \mathbf{P}\left\{ \inf_{x \in G} \zeta(x) = t \right\} > 0 \right\}.$$

The right-hand side is a countable union of at most countable sets, whence the set in the left-hand side is at most countable.

(ii, iii) If f_n epiconverges to f and $\{f \leq t\} = \operatorname{cl}(\{f < t\})$, then Proposition 3.3 implies that $\{f_n \leq t\}$ Painlevé–Kuratowski converges to $\{f \leq t\}$. If $\partial\{f \leq t\} = \{f \leq t\}$, then $\partial\{f_n \leq t\}$ converges to $\{f \leq t\}$ in the Painlevé–Kuratowski sense. Now all statements follow from the continuous mapping theorem. □

In general the weak epiconvergence of ζ_n to ζ does not follow from and does not imply the convergence of the corresponding finite-dimensional distributions. It is possible to enforce the relevant implications by imposing the uniform regularity conditions on the normal integrands using a kind of uniformity condition for the values of the capacity functionals on finite sets and sets of type $\cup_j (K_j \times (-\infty, t_j])$ (belonging to the family $\mathcal{K}_e$).

It was shown by Salinetti and Wets [512] that the weak epiconvergence implies the convergence of finite-dimensional distributions if and only if ζ_n are *equi-outer regular*, i.e. for every finite set $I \subset \mathbb{E} \times \mathbb{R}$ and every $\varepsilon > 0$ there exists a compact set $K \in \mathcal{K}_e \cap \mathfrak{S}_{\operatorname{epi} \zeta}$ such that $T_{\operatorname{epi} \zeta_n}(K) < T_{\operatorname{epi} \zeta_n}(I) + \varepsilon$. The convergence of finite-dimensional distributions implies the weak epiconvergence if the ζ_n's are *equi-inner separable*, i.e. for each $K \in \mathcal{K}_e$ and $\varepsilon > 0$, there exists a finite set $I \subset K$ such that $T_{\text{epi}\,\zeta_n}(K) < T_{\text{epi}\,\zeta_n}(I) + \varepsilon$ for all n.

Clearly, the epiconvergence is weaker than the conventional definition of the weak convergence of stochastic process in the uniform metric. But for the later convergence one should check some tightness conditions, while it is not necessary to deal with the tightness to prove the weak convergence of epigraphs.

Theorem 3.18 (Tightness for epigraphs). *Any sequence $\{\zeta_n, n \geq 1\}$ of normal integrands in a LCHS space is relative compact with respect to the weak epiconvergence, i.e. there exists a subsequence $\{\zeta_{n(k)}, k \geq 1\}$ which epiconverges in distribution.*

Proof. The family of closed subsets of $\mathbb{E} \times \mathbb{R}$ is compact in the Fell topology. The family of all epigraphs is closed in the Fell topology and so is also compact. This immediately implies the tightness of their probability distributions. □

Epigraphical representation of additive functionals

The epigraphical technique yields a representation of additive functionals acting on the space $\mathbf{L}^p$ of integrable random elements in $\mathbb{E} = \mathbb{R}^d$. A functional $W: \mathbf{L}^p(\Omega; \mathbb{R}^d) \mapsto \bar{\mathbb{R}}$ is said to be *additive* if

$$W(\mathbf{1}_A \xi + \mathbf{1}_{\Omega \setminus A} \xi) = W(\mathbf{1}_A \xi) + W(\mathbf{1}_{\Omega \setminus A} \xi)$$

for every $\xi \in \mathbf{L}^p$ and $A \in \mathfrak{F}$. The functional W is called *proper* if it never takes the value $-\infty$ and is not identically equal to $+\infty$.

Theorem 3.19. *Let $W: \mathbf{L}^p \mapsto \bar{\mathbb{R}}$ be an additive lower semicontinuous (with respect to $\mathbf{L}^p$-norm) proper functional, where $p \in [1, \infty]$. Then $W(\xi) = \mathbf{E}\zeta(\xi)$ for all ξ from $\mathbf{L}^p$, where ζ is a uniquely determined (up to a set of probability zero) proper normal integrand with $\zeta(0) = 0$ a.s.*

Proof. Since W is additive and proper, $W(0) = 0$. Let $\varXi$ be the set of pairs (ξ, α) with $\xi \in \mathbf{L}^p$ and $\alpha \in \mathbf{L}^1(\Omega; \mathbb{R})$ such that $W(\mathbf{1}_A \xi) \leq \mathbf{E}(\mathbf{1}_A \alpha)$ for all $A \in \mathfrak{F}$. If $(\xi, \alpha), (\xi', \alpha') \in \varXi$ and $B \in \mathfrak{F}$, then

$$W(\mathbf{1}_A(\mathbf{1}_B \xi + \mathbf{1}_{\Omega \setminus B} \xi')) = W(\mathbf{1}_{A \cap B} \xi) + W(\mathbf{1}_{A \setminus B} \xi')$$
$$\leq \mathbf{E}(\mathbf{1}_{A \cap B} \alpha) + \mathbf{E}(\mathbf{1}_{A \setminus B} \alpha')$$
$$= \mathbf{E}(\mathbf{1}_A(\mathbf{1}_B \alpha + \mathbf{1}_{\Omega \setminus B} \alpha')), \quad A \in \mathfrak{F}.$$

Therefore, $(\mathbf{1}_B \xi + \mathbf{1}_{\Omega \setminus B} \xi', \mathbf{1}_B \alpha + \mathbf{1}_{\Omega \setminus B} \alpha') \in \varXi$, i.e. $\varXi$ is a decomposable set of random elements, see Section 2.1.1. If $\{(\xi_n, \alpha_n), n \geq 1\}$ are elements of $\varXi$ that converge to (ξ, α) in $\mathbf{L}^p \times \mathbf{L}^1$, then the lower semicontinuity of W implies that

$$W(\mathbf{1}_A \xi) \leq \liminf_{n \to \infty} W(\mathbf{1}_A \xi_n) \leq \lim_{n \to \infty} \mathbf{E}(\mathbf{1}_A \xi_n) = \mathbf{E}(\mathbf{1}_A \xi), \quad A \in \mathfrak{F}.$$

Thus, $\varXi$ is a closed subset of $\mathbf{L}^p(\Omega; \mathbb{R}^d) \times \mathbf{L}^1(\Omega; \mathbb{R})$ which is also non-empty, since $(0, 0) \in \varXi$. By a variant of Theorem 2.1.6 for product spaces, there exists a random

3 Semicontinuous random functions 347

closed set Z in $\mathbb{E} \times \mathbb{R}$ such that $\varXi$ is the set of selections of Z. Let $\{(\xi_n, \alpha_n), n \geq 1\}$ be the Castaing representation of Z. Since $(\xi_n, \alpha_n + c) \in \varXi$ for every $c \geq 0$, it is easy to see that, on a subset $\Omega' \subset \Omega$ of probability one, $(x, t) \in Z$ implies that $\{x\} \times [t, \infty) \subset Z$.

Define a random function $\zeta(x)$ with values in the extended real line by

$$\zeta(x) = \begin{cases} \inf\{t : (x, t) \in Z\}, & \omega \in \Omega', \\ 0, & \omega \in \Omega \setminus \Omega'. \end{cases}$$

Then epi $\zeta = Z$ on Ω' and otherwise epi $\zeta = \mathbb{E} \times [0, \infty)$. Thus, epi ζ is a random closed set, i.e. ζ is a normal integrand. Let us prove that the constructed process ζ provides the required representation of W. Consider $\xi \in \mathbf{L}^p$ such that $W(\xi) < \infty$. If $A = \cup_{n=1}^{\infty} A_n$ with pairwise disjoint $A_n \in \mathfrak{F}, n \geq 1$, then

$$W(\mathbf{1}_A \xi) = \sum_{k=1}^{n} W(\mathbf{1}_{A_k} \xi) + W(\mathbf{1}_{\cup_{k>n} A_n} \xi).$$

Since $\liminf_{n\to\infty} W(\mathbf{1}_{\cup_{k>n} A_n} \xi) \geq W(0) = 0$,

$$W(\mathbf{1}_A \xi) \geq \limsup_{n\to\infty} \sum_{k=1}^{n} W(\mathbf{1}_{A_k} \xi).$$

Furthermore,

$$W(\mathbf{1}_A \xi) \leq \liminf_{n\to\infty} \sum_{k=1}^{n} W(\mathbf{1}_{A_k} \xi).$$

Therefore, $W(\mathbf{1}_A \xi) = \sum_{k=1}^{\infty} W(\mathbf{1}_{A_k} \xi)$, which means that $W(\mathbf{1}_A \xi)$ is a bounded signed measure on $\mathfrak{F}$, which is **P**-absolutely continuous. Let α be its Radon–Nikodym derivative with respect to **P**, so that $W(\mathbf{1}_A \xi) = \mathbf{E}(\mathbf{1}_A \alpha)$. Then $(\xi, \alpha) \in \varXi$, whence $(\xi, \alpha) \in Z$ and $\zeta(\xi) \leq \alpha$ a.s. Thus, $\mathbf{E}\zeta(\xi) \leq \mathbf{E}(\mathbf{1}_\Omega \alpha) = W(\xi)$.

Let $\mathbf{E}\zeta(\xi) < \infty$ for $\xi \in \mathbf{L}^p$. Choose a sequence $\{\alpha_n, n \geq 1\}$ of integrable random variables such that $\alpha_n \downarrow \zeta(\xi)$ a.s. Since $(\xi, \alpha_n) \in$ epi $\zeta = Z$ a.s., we have $(\xi, \alpha_n) \in \varXi$ and hence $W(\xi) \leq \mathbf{E}(\alpha_n) \downarrow \mathbf{E}\zeta(\xi)$. Thus,

$$W(\xi) \leq \mathbf{E}\zeta(\xi). \tag{3.3}$$

It remains to show that $\mathbf{E}\zeta(\xi)$ is defined (possibly being infinite) for every $\xi \in \mathbf{L}^p$. If $\mathbf{E}\zeta(\xi)$ is not defined, then $\mathbf{E}(\mathbf{1}_A \zeta(\xi)) = -\infty$ with $A = \{\omega : \zeta(\xi) < 0\}$. By (3.3), $W(\mathbf{1}_A \xi) = -\infty$ contrary to the assumption that W is proper. Therefore, $\mathbf{E}\zeta(\xi) = W(\xi)$ for all $\xi \in \mathbf{L}^p$.

It is possible to choose ζ in such a way that $\zeta(x)$ is a proper function on $\mathbb{E}$ for every ω. Since $W(0) = 0$ and $\zeta(0) < \infty$ a.s., $\zeta(x)$ can be replaced by $\zeta(x) - \zeta(0)$. Therefore, we can assume that $\zeta(0) = 0$ a.s. To show the uniqueness, it suffices to note that any two normal integrands ζ_1 and ζ_2 providing the representation for W will have identical epigraphs, whence ζ_1 and ζ_2 coincide almost everywhere. □

3.2 Stochastic optimisation

Convergence of minimisers

Stochastic optimisation in a broad sense deals with the convergence of minimisers and minimum values of a sequence of random functions. Consider a sequence $\{\zeta_n, n \geq 1\}$ of normal integrands defined on a LCHS space $\mathbb{E}$. Assume that ζ_n weakly epiconverges to ζ as $n \to \infty$. Then Theorem 3.4 implies that argmin ζ (respectively inf ζ) are stochastically larger (see Section 1.4.8) than lim sup(argmin ζ_n) (respectively lim sup(inf ζ_n)).

Theorem 3.20 (Weak convergence of infima). *If the normal integrands ζ_n, $n \geq 1$, weakly epiconverge to a normal integrand ζ, then*

$$\mathbf{P}\{\inf \zeta < t\} \leq \liminf_{n \to \infty} \mathbf{P}\{\inf \zeta_n < t\}$$

for all $t \in \mathbb{R}$. Furthermore, $\inf \zeta_n$ converges in distribution to $\inf \zeta$ if the sequence $\{\zeta_n, n \geq 1\}$ is equi-inf-compact, i.e. for every $t \in \mathbb{R}$ there exists a compact set K such that $\{\zeta_n \leq t\} \subset K$ a.s. for all $n \geq 1$.

Proof. The method of a single probability space yields a sequence ζ'_n, $n \geq 1$, of normal integrands having the same distribution as ζ_n and such that ζ'_n epiconverges almost surely to ζ', where the latter has the same distribution as ζ. The proof is finished by applying Theorem 3.4. □

The situation is especially simple if $\mathbb{E}$ is a compact space, since then the map $f \mapsto \inf f$ is continuous with respect to the epiconvergence, see Theorem 3.4.

Epiconvergence of averages

Let ζ be a proper normal integrand such that $\mathbf{E}\zeta(x)$ is well defined (but may be infinite). By Fatou's lemma $h(x) = \mathbf{E}\zeta(x)$ is lower semicontinuous. A sequence $\{\zeta_n, n \geq 1\}$ of i.i.d. normal integrands having the same distribution as ζ can be used to estimate h as

$$\eta_n(x) = \frac{1}{n} \sum_{i=1}^{n} \zeta_i(x), \quad x \in \mathbb{E}.$$

Then η_n is a proper normal integrand and, for all x, $\eta_n(x) \to h(x) = \mathbf{E}\zeta(x)$ a.s. as $n \to \infty$ by the strong law of large numbers. The following result shows that, under relatively mild conditions, η_n epiconverges almost surely to $h(x)$. This issue is central to optimisation problems since the epiconvergence may be used to obtain results about the convergence of minimum values and minimisers.

Theorem 3.21 (Epiconvergence of averages). *Assume that each $x_0 \in \mathbb{E}$ has an open neighbourhood G such that there exists an integrable random variable α with*

$$\inf_{x \in G} \zeta(x) \geq \alpha \quad \text{a.s.} \tag{3.4}$$

Then $\eta_n \xrightarrow{\text{epi}} h$ a.s. as $n \to \infty$.

Proof. Referring to Proposition 3.2, we prove that $\liminf \eta_n(x_n) \geq h(x_0)$ a.s. whenever $x_n \to x_0$ and $x_0 \in \mathbb{E}$. First, show this fact for the restriction of η_n and h onto an open neighbourhood G satisfying (3.4). If $\inf_{x \in G} \zeta(x)$ is infinite with positive probability, then the result evidently holds. Therefore, assume that $\zeta(x) < \infty$ a.s. for some $x \in G$.

For every $n \geq 1$, define a sequence of random functions $\{g_k^n(x), k \geq 1\}$ by induction as follows. First, put $g_0(x) = \alpha(x)$. If $g_k^n(x)$ is given, then put

$$\varphi_k^n(x) = \inf\left\{\rho(x, y) + |g_k^n(x) - r| : y \in G, r \geq \zeta_n(y)\right\}, \quad (3.5)$$

where ρ is a metric on $\mathbb{E}$. Note that $\varphi_k^n(x)$ determines a distance from $(x, g_k^n(x))$ to the epigraph of ζ_n restricted on G. Since ζ_n is not identically ∞, the function φ_k^n is finite. Define

$$g_{k+1}^n(x) = g_k^n(x) + \varphi_k^n(x).$$

By g_k, $k \geq 1$, we denote functions obtained by applying the above construction to the function ζ instead of ζ_n. Let us establish several useful properties of the sequence $\{g_k^n, k \geq 1\}$.

(a) g_k^n is measurable on $G \times \Omega$. Indeed, φ_k^n is measurable, since it is a distance from a point to a random closed set in the metric space $\mathbb{E} \times \mathbb{R}$ with the metric $\rho'((x_1, t_1), (x_2, t_2)) = \rho(x_1, x_2) + |t_1 - t_2|$, see Section 1.2.3.

(b) Each g_k^n is Lipschitz with a Lipschitz constant independent of n. First, g_0 is constant, so has a Lipschitz constant 0. Proceeding by induction, let c be a Lipschitz constant for g_k^n. It follows from (3.5) that φ_k^n is Lipschitz with constant $c + 1$, whence g_{k+1}^n has a Lipschitz constant $2c + 1$.

(c) $g_k^n(x) \geq \alpha$ and $g_k^n(x) \uparrow \zeta_n(x)$ as $k \to \infty$, which follows immediately from the construction noticing that $\varphi_k^n(x) \leq \zeta_n(x) - g_k^n(x)$.

Note that g_k and $\{g_k^n, n \geq 1\}$ are i.i.d. random functions. For x and k fixed, define

$$\gamma_k^n(x) = \frac{1}{n} \sum_{i=1}^n g_k^i(x).$$

The classical strong law of large numbers implies that

$$\gamma_k^n(x) \to \mathbf{E} g_k(x) \quad \text{a.s. as } n \to \infty \quad (3.6)$$

for each fixed x. Since all $\{g_k^n, n \geq 1\}$ have the same Lipschitz constant, it follows that $\mathbf{E} g_k$ and each γ_k^n are Lipschitz. Let $\{x_i, i \geq 1\}$ be a dense sequence in G. By the countability and the Lipschitz property (b), (3.6) almost surely holds for all $x \in G$ simultaneously.

Pick any $x_0 \in G$. Suppose that $h(x_0) = \mathbf{E}\zeta(x_0) < \infty$. Then (c) implies that for large k the value $\mathbf{E} g_k(x_0)$ is close to $h(x_0)$, i.e. $h(x_0) - \mathbf{E} g_k(x_0) < \varepsilon$. Since $\eta_n \geq \gamma_k^n$ always, it follows that

$$\liminf_{x_n \to x_0} \eta_n(x_n) \geq h(x_0) - \varepsilon \quad \text{a.s.} \tag{3.7}$$

Since ε is arbitrarily small, this concludes the proof. If $h(x_0) = \infty$, similar arguments are applicable with ε arbitrarily small replaced by $\mathbf{E} g_k(x_0)$ arbitrarily large. Since $\mathbb{E}$ is separable, it is covered by a countable number of open sets G such that (3.7) holds on every G. Then (3.7) holds almost surely for all $x_0 \in \mathbb{E}$ and all sequences $x_n \to x_0$.

We now verify that $\eta_n(y_n) \to h(x_0)$ a.s. for at least one sequence $y_n \to x_0$. Let $\{x_i, i \geq 1\}$ be a dense sequence in $\mathbb{E}$. The lower semicontinuity of h implies that, for each $x_0 \in \mathbb{E}$, a subsequence of $\{(x_i, h(x_i)), i \geq 1\}$ converges to $(x_0, h(x_0))$. By the strong law of large numbers, $\mathbf{E}\eta_n(x_i) \to h(x_i)$ a.s. as $n \to \infty$ simultaneously for all x_i. Hence it is possible to find a subsequence $\{y_n, n \geq 1\}$ of $\{x_i, i \geq 1\}$ such that $\eta_n(y_n) \to h(x_0)$ as $n \to \infty$. By Proposition 3.2, η_n almost surely epiconverges to h. □

If ζ is a non-negative normal integrand, then (3.4) automatically holds for $\alpha = 0$, so that the strong law of large numbers holds for non-negative integrands without any additional conditions.

Minimisation of expectations

Many stochastic optimisation problems can be written in the following form

$$J(x) = \mathbf{E} g(x, \theta) = \int_\Theta g(x, \theta) \mathbf{P}(\mathrm{d}\theta) \to \min, \quad x \in \mathbb{E}, \tag{3.8}$$

where θ is a random element in a measurable space Θ. This problem can also be interpreted within the framework of the *Bayesian decision theory*, where $\mathbb{E}$ is a decision space, $\theta \in \Theta$ is an unknown quantity affecting the decision process (state of nature) with the prior distribution $\mathbf{P}$ and $g(x, \theta)$ is the loss incurred when the chosen action is x and the true state of nature is θ.

Assume that $g(x, \theta)$ is a non-negative function jointly measurable in θ and x and lower semicontinuous with respect to x for $\mathbf{P}$-almost all θ. Sometimes, (3.8) can be approximated by a different problem:

$$J_n(x) = \frac{1}{n} \sum_{i=1}^{n} g(x, \theta_i) \to \min, \quad x \in \mathbb{E}, \tag{3.9}$$

where $\theta_1, \ldots, \theta_n$ is a sample of i.i.d. observations of θ distributed according to $\mathbf{P}$. Note that (3.9) is a particular case of the general setup, where

$$J_n(x, \omega) = \int_\Theta g(x, \theta) \mathbf{P}_n(\mathrm{d}\theta)(\omega)$$

for a sequence of random measures $\{\mathbf{P}_n(\omega), n \geq 1\}$ that converges to $\mathbf{P}$ in some sense as $n \to \infty$. Then (3.9) appears if $\mathbf{P}_n$ is the empirical probability measure.

3 Semicontinuous random functions

Theorem 3.22 (Convergence of expectations). *Let $\mathbb{E}$ be a Polish space and let g be a normal integrand that satisfies condition (3.4). Assume that Θ is a general topological space and*

$$\int f(\theta)\mathbf{P}_n(d\theta)(\omega) \to \int f(\theta)\mathbf{P}(d\theta) \quad \text{a.s. as } n \to \infty \tag{3.10}$$

for every measurable and $\mathbf{P}$-integrable function f on Θ. Then $J(x)$ is lower semicontinuous and J_n epiconverges to J a.s. as $n \to \infty$.

If $\mathbf{P}_n$ is an empirical measure, then (3.10) holds and Theorem 3.22 corresponds to Theorem 3.21.

The epiconvergence of J_n implies the convergence of minimisers if $\mathbb{E}$ is a compact space. If $\mathbb{E}$ is not compact, Theorem 3.4 implies that $\inf J_n \to \inf J$ if $\{J_n, n \geq 1\}$ almost surely has a relatively compact sequence of ε-optimal points.

Convergence of maximum likelihood estimators

Many statistical estimators appear as solutions of minimisation (or maximisation) problems. Since the epiconvergence is the weakest functional convergence which ensures the convergence of minimum points, it can be applied to prove strong consistency of estimators under minimal conditions.

Let $(\mathbb{E}, \mathfrak{F}, \mu)$ be a complete measurable space with a σ-finite measure μ. Consider a parametric family $\mathbf{P}_\theta$ of probability distributions on $\mathbb{E}$ which are absolutely continuous with respect to μ with the densities $f_\theta(x)$, $x \in \mathbb{E}$, parametrised by θ from a Polish space Θ.

Assumptions 3.23.
(i) f is jointly measurable with respect to x and θ.
(ii) For every $\theta \in \Theta$, f_θ is a probability density function with respect to μ.
(iii) For μ-almost all x, $f_\theta(x)$ is sup-compact as a function of θ, i.e. for each $r > 0$, the set $\{\theta : f_\theta(x) \geq r\}$ is compact in Θ.
(iv) $\theta_1 \neq \theta_2$ implies $\mu(\{x : f_{\theta_1}(x) \neq f_{\theta_2}(x)\}) > 0$.

Fix a $\theta_0 \in \Theta$ and assume that f_{θ_0} is a *probability density function* of a random element ξ. Consider a basic statistical problem of estimating θ_0 from a sample $x_1, \ldots, x_n$ of n realisations of ξ. The maximum likelihood approach suggests estimating θ_0 by the *maximum likelihood estimator* $\hat{\theta}_n$ defined as a maximiser of the likelihood function

$$L(x_1, \ldots, x_n | \theta) = \prod_{i=1}^{n} f_\theta(x_i), \quad \theta \in \Theta.$$

Define

$$b(x) = \sup_{\theta \in \Theta} f_\theta(x),$$

so that $b(x) = b_1(x)$, where

$$b_n(x_1, \ldots, x_n) = \sup_{\theta \in \Theta} L(x_1, \ldots, x_n | \theta), \quad n \geq 1.$$

Theorem 3.24 (Consistency of maximum likelihood estimators). *Let Assumptions 3.23 hold. Assume also that*

$$\mathbf{E}_{\theta_0}\left[\log \frac{b(\xi)}{f_{\theta_0}(\xi)}\right] < \infty, \qquad (3.11)$$

where $\mathbf{E}_{\theta_0}$ denotes the expectation with respect to $\mathbf{P}_{\theta_0}$. If $\varepsilon_n \downarrow 0$ as $n \to \infty$, then the following statements hold.

(i) There exists a sequence $\{\hat{\theta}_n, n \geq 1\}$ of ε_n-approximate maximum likelihood estimators, namely a sequence of measurable maps from $\mathbb{E}^n$ into Θ such that

$$L(x_1, \ldots, x_n | \hat{\theta}_n) \geq b_n(x_1, \ldots, x_n) - \varepsilon_n, \quad n \geq 1.$$

(ii) For every such sequence $\{\hat{\theta}_n, n \geq 1\}$, $\hat{\theta}_n \to \theta_0$ almost surely as $n \to \infty$.

Proof.
(i) By Assumptions 3.23(i),(iii), the functions b and b_n are measurable and finite. Assumption 3.23(iii) implies that f is upper semicontinuous with respect to θ, whence the likelihood function is upper semicontinuous too. For any fixed n and $\varepsilon > 0$ define

$$X_{n,\varepsilon} = \{\theta : L_n(x_1, \ldots, x_n | \theta) \geq b_n(x_1, \ldots, x_n) - \varepsilon\}$$

and

$$X_n = X_{n,0} = \{\theta : L_n(x_1, \ldots, x_n | \theta) = b_n(x_1, \ldots, x_n)\}.$$

By a dual to Proposition 3.10(iii) for upper semicontinuous functions, $X_{n,\varepsilon}$ and X_n are random closed subsets of Θ. By the fundamental selection theorem there exist selections $\hat{\theta}_{n,\varepsilon}$ and $\hat{\theta}_n$ of $X_{n,\varepsilon}$ and X_n respectively. This proves (i).
(ii) Define $g(x, \theta) = -\log f_\theta(x)$ for $x \in \Omega$, $\theta \in \Theta$ and put

$$\eta_n(\theta) = \frac{1}{n} \sum_{i=1}^{n} g(x_i, \theta) = -\frac{1}{n} \log L_n(x_1, \ldots, x_n | \theta).$$

Define a function with possibly infinite values by

$$\varphi(\theta) = \mathbf{E}_{\theta_0}\left[\log \frac{b(\xi)}{f_\theta(\xi)}\right], \qquad \theta \in \Theta.$$

Since it is possible to replace μ by $c\mu$ and the function f by f/b, assume that f takes values in $[0, 1]$, g takes values in $[0, \infty]$ and $b(x)$ is identically equal to 1. Using Jensen's inequality, it is easily seen that θ_0 is the unique minimiser of $\varphi(\theta)$, $\theta \in \Theta$.

The finding of the maximum likelihood estimator is equivalent to minimisation of η_n, which is a stochastic optimisation problem. By Theorem 3.21, η_n epiconverges to φ almost surely. Note that (3.11) is important to ensure that $\varphi(\theta_0) < \infty$, so that φ is not identically equal to ∞, i.e. $g(x, \theta)$ is a proper normal integrand with respect to θ. Now (ii) follows from (3.1) of Theorem 3.4. □

3.3 Epigraphs and extremal processes

Pointwise extremes

If ζ_1 and ζ_2 are two normal integrands, then

$$\operatorname{epi}\zeta_1 \cup \operatorname{epi}\zeta_2 = \operatorname{epi}\zeta$$

where $\zeta(x) = \min(\zeta_1(x), \zeta_2(x))$, $x \in \mathbb{E}$, is the pointwise minimum of ζ_1 and ζ_2. Therefore, a pointwise minimum of functions corresponds to the unions of their epigraphs. This observation allows us to apply results for unions of random closed sets for epigraphs of random functions in order to prove limit theorems for pointwise minima of random lower semicontinuous functions. In what follows we mostly discuss the dual statements for pointwise maxima of upper semicontinuous functions and unions of their hypographs, as is more typical in the literature devoted to the studies of maxima of random functions.

Assume that $\mathbb{E}$ is a LCHS space. Let $\mathrm{USC}(\mathbb{E})$ denote the family of upper semicontinuous functions on $\mathbb{E}$. The hypotopology on $\mathrm{USC}(\mathbb{E})$ is induced by the Fell topology on closed sets in $\mathbb{E} \times \mathbb{R}$. If f is a (random) upper semicontinuous function, then $f^\vee(K) = \sup_{x \in K} f(x)$ denotes the corresponding maxitive random capacity (or sup-measure).

Definition 3.25 (Max-infinite divisibility and max-stability). Let ζ be a random upper semicontinuous function on $\mathbb{E}$.

(i) ζ is called max-infinitely divisible if for each n there exist i.i.d. random upper semicontinuous functions $\eta_1, \ldots, \eta_n$ such that $\zeta(x)$, $x \in \mathbb{E}$, coincides in distribution with $\max_{1 \leq i \leq n} \eta_i(x)$, $x \in \mathbb{E}$.

(ii) ζ is max-stable if, for every $n \geq 2$, there are constants $a_n > 0$ and $b_n \in \mathbb{R}$ such that ζ coincides in distribution with $a_n \max_{1 \leq i \leq n} \zeta_i + b_n$ where $\zeta_1, \ldots, \zeta_n$ are i.i.d. with the same distribution as ζ.

Theorem 4.1.33 applied to the lattice of upper semicontinuous functions with the lattice operation being the intersection of hypographs (or pointwise minima of functions) yields the following result.

Theorem 3.26 (Max-infinite divisible functions). *A random upper semicontinuous function ζ which is not identically equal to ∞ is max-infinitely divisible if and only if $\zeta(x)$ coincides in distribution with $\max(h(x), \eta_i(x), i \geq 1)$, $x \in \mathbb{E}$, where*

(1) $h \colon \mathbb{E} \mapsto \bar{\mathbb{R}}$ is an upper semicontinuous function which is not identically equal to ∞ and satisfies

$$h(x) = \sup\{t : \mathbf{P}\{\zeta(x) \geq t\} = 1\};$$

(2) $\{\eta_i, i \geq 1\}$ are atoms of the Poisson point process on $\mathrm{USC}(\mathbb{E})$ with uniquely determined locally finite intensity measure ν on $\{f \in \mathrm{USC}(\mathbb{E}) : f \neq h, f \geq h\}$ (i.e. ν is finite on compact in the hypotopology sets) such that

$$\nu(\cup_{i=1}^n \{f : f^\vee(K_i) \geq x_i\}) = -\log \mathbf{P}(\cap_{i=1}^n \{\zeta^\vee(K_i) < x_i\})$$

for all $K_i \in \mathcal{K}$, $x_i > h^\vee(K_i)$, $i = 1, \ldots, n$ and $n \geq 1$.

354   5 Random Sets and Random Functions

Since pointwise maxima of random upper semicontinuous functions correspond to unions of their hypographs, it is possible to use limit theorems for unions of random closed sets in order to derive results for pointwise maxima of upper semicontinuous functions.

Semi-min-stable processes

A stochastic process $\zeta(t)$, $t \geq 0$, is said to be *semi-min-stable* of order α if, for each $n \geq 2$ and i.i.d. processes $\zeta, \zeta_1, \ldots, \zeta_n$, the process $n \min(\zeta_1(t), \ldots, \zeta_n(t))$, $t \geq 0$, coincides in distribution with $\zeta(n^\alpha t)$, $t \geq 0$. By considering the epigraph $X = \text{epi}\,\zeta$ as a random closed set in $\mathbb{R}^2$, it is easy to see that X is stable in the sense of Proposition 4.1.21 with $\mathbf{g} = (\alpha, 1)$. Examples of semi-min-stable processes can then be constructed using the scheme of Example 4.1.17 with Z being the epigraph of a deterministic lower semicontinuous function.

Continuous choice processes

Because of a number of applications to dynamic modelling of consumer choice it is interesting to study stochastic processes with values in $\text{USC}(\mathbb{E})$ or time-dependent sequences of random upper semicontinuous functions. Such an object is denoted by ζ_t, so that $\zeta_t = \zeta_t(x)$, $x \in \mathbb{E}$, is a random element in $\text{USC}(\mathbb{E})$ for every $t \geq 0$. Without loss of generality assume that all processes take non-negative values and so belong to the family Υ of non-negative upper semicontinuous functions punctured by removal of the function identically vanishing on $\mathbb{E}$. A sufficiently flexible model for random elements in Υ is provided by super-extremal processes.

Definition 3.27 (Super-extremal process). If $\Pi = \{(t_k, \eta_k), k \geq 1\}$ is a Poisson point process on $\mathbb{R}_+ \times \Upsilon$ with the intensity measure μ, then the *super-extremal process* is given by

$$\zeta_t = \sup_{t_k \leq t} \eta_k, \quad t > 0.$$

For every $B \in \mathfrak{B}(\mathbb{E})$, the process $\zeta_t^\vee(B)$, $t > 0$, is a classical univariate extremal process, see Resnick [481, p. 180]. By construction, ζ_t is max-infinitely divisible for every $t > 0$. It is shown in Resnick and Roy [482, Th. 3.1] that, for every $t > 0$, the defined ζ_t is a random element in Υ (i.e. hypo ζ_t is a random closed set) and there is a version of ζ_t which is continuous from the right and has limits from the left with respect to t (with respect to the hypotopology on Υ). The Markov property of ζ_t follows from the fact that $\zeta_t = \max(\zeta_s, \zeta_{s,t})$ for $s < t$, where

$$\zeta_{s,t} = \sup_{s < t_k \leq t} \eta_k$$

is independent of ζ_s.

The super-extremal process represents the evolution of random utilities where at each time t the utility for the choice x is given by $\zeta_t(x)$. At every time moment

t the customer choice aims to maximise the utility, so that the preferable options correspond to the argmax set

$$M_t = \{x \in \mathbb{E} : \zeta_t(x) = \zeta_t^\vee(\mathbb{E})\}$$

called the *choice process*, where $\zeta_t^\vee(\mathbb{E}) = \sup_{x \in \mathbb{E}} \zeta_t(x)$. Note that M_t can be empty if ζ_t is unbounded on $\mathbb{E}$ and does not take infinite values. This situation can be avoided if $\mathbb{E}$ is compact.

An analogue of Proposition 3.10(iii) immediately shows that M_t is a random closed set for every $t > 0$. Similar to Theorem 3.4, it is possible to show that the argmax functional is upper semicontinuous on USC($\mathbb{E}$). Since ζ_t is continuous from the right, M_t is upper semicontinuous from the right in the Fell topology.

Example 3.28 (Simple choice processes). Let $\{t_k\}$ form a stationary Poisson point process on $(0, \infty)$ and let $\eta_k(x) = \mathbf{1}_{X_k}(x)$, $k \geq 1$, where $\{X_k\}$ are random compact sets that constitute a Poisson point process on $\mathcal{K}(\mathbb{E})$. Then $M_t = \cup_{k:\, t_k \leq t} X_k$ is the increasing set-valued process, cf (1.3).

If $\eta_k(x) = s_k \mathbf{1}_{X_k}(x)$, where $Y = \{s_k\}$ is a union-stable Poisson point process on $[0, \infty)$ with intensity function $\lambda(x) = x^{-2}$ (see Example 1.16), then $M_t = X_n$, where $s_n = \max\{s_k : t_k \leq t\}$. Note that Y is bounded almost surely, so that $s_{n_0} = \max(Y)$. Then M_t converges in distribution to X_{n_0}.

From now on assume that $\mathbb{E}$ is compact, so that $\mathcal{F} = \mathcal{K}$. Let μ be a locally finite measure on $(0, \infty) \times \Upsilon$ such that $\mu(\{t\} \times \Upsilon) = 0$ and $\mu_t(\cdot) = \mu((0, t] \times \cdot)$ satisfies $\mu_t(\Upsilon) = \infty$ and

$$\mu_t(\{f : f^\vee(K) = \infty\}) = 0, \quad K \in \mathcal{K},$$

for all $t > 0$. Furthermore, assume that

$$c^\alpha \mu_t(cY) = \mu_t(Y), \quad c > 0, \quad (3.12)$$

for some $\alpha > 0$ and each Y from the family $\mathfrak{B}(\Upsilon)$ of the Borel subsets of Υ defined with respect to the hypotopology. If (3.12) holds, then ζ_t is said to have *max-stable components*; in particular, this implies that ζ_t itself is max-stable for every $t > 0$. By transforming $[0, \infty)$ it is possible to assume without loss of generality that $\alpha = 1$.

For $Y \in \mathfrak{B}(\Upsilon)$ put

$$\zeta_t[Y] = \sup_{t_k \leq t,\, \eta_k \in Y} \eta_k^\vee(\mathbb{E}).$$

Since $\{(t_k, \eta_k), k \geq 1\}$ is a Poisson point process, $\zeta_t[Y_i]$, $i = 1, \ldots, n$, are mutually independent for disjoint $Y_1, \ldots, Y_n$. Particular important subsets of Υ are

$$\Upsilon_K = \{f \in \Upsilon : (\operatorname{argmax} f) \cap K \neq \emptyset\},$$
$$\Upsilon^K = \{f \in \Upsilon : (\operatorname{argmax} f) \subset K\},$$
$$\Upsilon_1 = \{f \in \Upsilon : f^\vee(\mathbb{E}) \geq 1\}.$$

Lemma 3.29. *If ζ_t has max-stable components and $Y \in \mathcal{B}(\Upsilon)$ is a cone, i.e. $cY = Y$ for all $c > 0$, then $\zeta_t[Y]$ is does not have atoms and*

$$\mathbf{P}\{\zeta_t[Y] < s\} = \exp\{-s^{-1}\mu_t(Y \cap \Upsilon_1)\}, \quad t > 0.$$

Proof. The construction of the Poisson point process Π from Definition 3.27 implies that

$$\mathbf{P}\{\zeta_t[Y] < s\} = \exp\{-\mu_t(Y \cap \{f : f^\vee(\mathbb{E}) \geq s\})\}.$$

By (3.12) and the assumption on Y,

$$\mu_t(Y \cap \{f : f^\vee(\mathbb{E}) \geq s\}) = \mu_t(s(Y \cap \{f : f^\vee(\mathbb{E}) \geq 1\}))$$
$$= s^{-1}\mu_t(Y \cap \Upsilon_1). \qquad \square$$

Theorem 3.30 (Distribution of the choice process). *Let ζ_t be a super-extremal process with max-stable components in a compact space $\mathbb{E}$. Then*
 (i) $\zeta_t^\vee(\mathbb{E})$ and M_t are independent for each $t > 0$;
 (ii) the containment functional of M_t is given by

$$\mathbf{P}\{M_t \subset K\} = \frac{\mu_t(\Upsilon^K \cap \Upsilon_1)}{\mu_t(\Upsilon_1)}, \quad K \in \mathcal{K},$$

and is called the choice probability;
(iii) the capacity functional of M_t is

$$T_{M_t}(K) = \mathbf{P}\{M_t \cap K \neq \emptyset\} = \frac{\mu_t(\Upsilon_K \cap \Upsilon_1)}{\mu_t(\Upsilon_1)}, \quad K \in \mathcal{K}.$$

Proof. Note that

$$\mathbf{P}\{M_t \subset K\} = \mathbf{P}\left\{\zeta_t[\Upsilon^K] > \zeta_t[(\Upsilon^K)^c]\right\}$$

and $c\Upsilon^K = \Upsilon^K$ for every $K \subset \mathbb{E}$. Lemma 3.29 and the total probability formula imply that

$$\mathbf{P}\{M_t \subset K\} = \int_0^\infty \exp\{-s^{-1}\mu_t((\Upsilon^K)^c \cap \Upsilon_1)\} \, d\exp\{-s^{-1}\mu_t(\Upsilon^K \cap \Upsilon_1)\}$$
$$= \frac{\mu_t(\Upsilon^K \cap \Upsilon_1)}{\mu_t(\Upsilon_1)}$$

as required. The proof of (iii) is similar. To show (i), observe that

$$\zeta_t^\vee(\mathbb{E}) = \max(\zeta_t[\Upsilon^K], \zeta_t[(\Upsilon^K)^c]).$$

By a similar expression of $\mathbf{P}\{M_t \subset K\}$ with the integral taken from 0 to z we obtain

3 Semicontinuous random functions   357

$$\mathbf{P}\{M_t \subset K,\ \zeta_t^\vee(\mathbb{E}) \leq z\} = \mathbf{P}\left\{\zeta_t[\Upsilon^K] > \zeta_t[(\Upsilon^K)^c],\ \zeta_t^\vee(\mathbb{E}) \leq z\right\}$$
$$= \frac{\mu_t(\Upsilon^K \cap \Upsilon_1)}{\mu_t(\Upsilon_1)}(1 - \exp\{-z\mu_t(\Upsilon_1)\}).$$

Since the containment functional determines the distribution of M_t, this shows that M_t and $\zeta_t^\vee(\mathbb{E})$ are independent. □

Corollary 3.31. *If ζ_t has max-stable components and is time-homogeneous, i.e. $\mu_t(\cdot) = t\mu(\cdot)$, then the distribution of M_t does not depend on t and*

$$C(K) = \mathbf{P}\{M_t \subset K\} = \frac{\mu(\Upsilon^K \cap \Upsilon_1)}{\mu(\Upsilon_1)}, \quad K \in \mathcal{K}, \tag{3.13}$$

$$T(K) = \mathbf{P}\{M_t \cap K \neq \emptyset\} = \frac{\mu(\Upsilon_K \cap \Upsilon_1)}{\mu(\Upsilon_1)}, \quad K \in \mathcal{K}. \tag{3.14}$$

The following result establishes that the set-valued random process M_t, $t > 0$, is Markov and gives its transition probabilities.

Theorem 3.32 (Markov property of the choice process). *If ζ_t is a super-extremal process with max-stable components, then M_t, $t > 0$, is a Markov process with state space $\mathcal{K}$. For $0 < s < t$, $K, L \in \mathcal{K}$, the transition probabilities are determined*
(i) *in terms of the containment functional (choice probability) by*

$$\mathbf{P}\{M_t \subset K | M_s = L\} = \frac{\mu_{s,t}(\Upsilon^K \cap \Upsilon_1)}{\mu_t(\Upsilon_1)} + \mathbf{1}_{L \subset K}\frac{\mu_s(\Upsilon_1)}{\mu_t(\Upsilon_1)},$$

where $\mu_{s,t}(\cdot) = \mu_t(\cdot) - \mu_s(\cdot)$;
(ii) *in terms of the capacity functional by*

$$\mathbf{P}\{M_t \cap K \neq \emptyset | M_s = L\} = \frac{\mu_{s,t}(\Upsilon_K \cap \Upsilon_1)}{\mu_t(\Upsilon_1)} + \mathbf{1}_{L \cap K \neq \emptyset}\frac{\mu_s(\Upsilon_1)}{\mu_t(\Upsilon_1)}.$$

Corollary 3.33. *If ζ_t is a time-homogeneous super-extremal process with max-stable components, then*

$$\mathbf{P}\{M_t \subset K | M_s = L\} = t^{-1}\left((t-s)C(K) + s\mathbf{1}_{L \subset K}\right),$$

$$\mathbf{P}\{M_t \cap K \neq \emptyset | M_s = L\} = t^{-1}\left((t-s)T(K) + s\mathbf{1}_{L \cap K \neq \emptyset}\right),$$

where $C(K)$ and $T(K)$ are given by (3.13) and (3.14) respectively. The time-changed choice process M_{e^t} is a time-homogeneous Markov process.

Epiconvergence of support functions for polyhedral approximations of convex sets

The following example is related to the epiconvergence, unions of random sets and polygonal approximations of convex sets. Let F be a convex compact set in $\mathbb{R}^d$ with a

358   5 Random Sets and Random Functions

twice continuously differentiable boundary ∂F and let $\mathbf{n}(x)$ be the unit outer normal vector at $x \in \partial F$. Furthermore, let $\mathbf{P}$ be a probability measure on F with continuous density f which is non-vanishing on Int F. Consider i.i.d. random points $\xi_1, \ldots, \xi_n$ with distribution $\mathbf{P}$. Their convex hull

$$\Xi_n = \mathrm{co}(\xi_1, \ldots, \xi_n)$$

is a *random polyhedron* which approximates F as $n \to \infty$, see Figure 3.6. Define

$$\eta_n(u) = h(F, u) - h(\Xi_n, u), \qquad u \in \mathbb{S}^{d-1}.$$

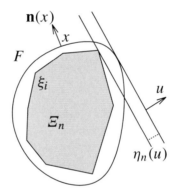

Figure 3.6. Polyhedral approximation of F.

Theorem 3.34 (Epiconvergence for polyhedral approximations).
(i) If f does not vanish identically on ∂F, then $n\eta_n$ weakly epiconverges to a sharp integrand η such that $\partial^- \mathrm{epi}\, \eta$ is a Poisson point process on $\mathbb{S}^{d-1} \times [0, \infty)$ with the intensity measure

$$\Lambda(K) = \int_{F_K} f(x) \mathcal{H}^{d-1}(\mathrm{d}x)\mathrm{d}t, \qquad K \subset \mathbb{S}^{d-1} \times [0, \infty),$$

where $\mathcal{H}^{d-1}$ is the $(d-1)$-dimensional Hausdorff measure and

$$F_K = \bigcup_{(u,t) \in K} \{(x, s) : x \in \partial F, \mathbf{n}(x) = u, s \in [0, t]\}.$$

(ii) Assume that $f(x) = 0$ for all $x \in \partial F$ and f is continuously differentiable in a neighbourhood of ∂F with $\langle f'(x), \mathbf{n}(x) \rangle$ not vanishing identically on ∂F, where f' is the vector of the partial derivatives of f. Then $\sqrt{n}\eta_n$ weakly epiconverges to a sharp integrand η such that $\partial^- \mathrm{epi}\, \eta$ is a Poisson point process on $\mathbb{S}^{d-1} \times [0, \infty)$ with the intensity measure

3 Semicontinuous random functions 359

$$\Lambda(K) = \int_{F_K} x \langle f'(x), \mathbf{n}(x) \rangle \mathcal{H}^{d-1}(dx) dt.$$

Proof.
(i) Since $\eta_n(u) = \min\{h(F, u) - \langle u, \xi_i \rangle : 1 \leq i \leq n\}$, the epigraph of η_n is the union of epigraphs of functions $h(F, u) - \langle u, \xi_i \rangle$ for $1 \leq i \leq n$, i.e.

$$H_n = \operatorname{epi} \eta_n = X_1 \cup \cdots \cup X_n,$$

where

$$X_i = \{(u, t) : u \in \mathbb{S}^{d-1}, t \geq 0, \langle u, \xi_i \rangle \geq h(F, u) - t\}, \quad i \geq 1,$$

are i.i.d. random closed sets. Fix a sequence $a_n \to 0$ of normalising constants. Then

$$a_n^{-1} \circ H_n = \{(u, a_n^{-1} t) : (u, t) \in H_n\} = \operatorname{epi}(a_n^{-1} \eta_n).$$

The weak convergence of $a_n^{-1} \circ H_n$ would follow from the pointwise convergence of its capacity functional on compact sets given by

$$\tilde{K} = \bigcup_{i=1}^{m} K_i \times [0, t_i], \tag{3.15}$$

where $K_1, \ldots, K_m$ are compact subsets of $\mathbb{S}^{d-1}$, $t_1, \ldots, t_m > 0$ and $m \geq 1$. It is possible to assume that $K_1, \ldots, K_m$ are regular closed with respect to the induced topology on the unit sphere and have disjoint interiors, since such sets form a separating class on $\mathbb{S}^{d-1}$. First, consider $\tilde{K} = K \times [0, x]$, where K is a regular closed subset of $\mathbb{S}^{d-1}$. Introduce the sets

$$M(K, t) = \{x \in F : \inf\{h(F, u) - \langle u, x \rangle : u \in K\} \leq t\},$$
$$N(K) = \{x \in \partial F : \mathbf{n}(x) \in K\},$$
$$N(K, t) = \{x - s\mathbf{n}(x) : x \in N(K), 0 \leq s \leq t\}.$$

Then

$$\mathbf{P}\{X_1 \cap t \circ \tilde{K} \neq \emptyset\} = \mathbf{P}(M(K, t)) = \int_{M(K, t)} f(u) du.$$

Note that for all $\varepsilon > 0$ and sufficiently small $t > 0$,

$$N(K, t) \subset M(K, t) \subset N(K^\varepsilon, t), \tag{3.16}$$

where $K^\varepsilon = \{v \in \mathbb{S}^{d-1} : B_\varepsilon(v) \cap K \neq \emptyset\}$. It can be shown that for a certain constant c and each sufficiently small positive t

$$\left| \int_{N(K,t)} f(x) dx - \int_{N(K)} \mathcal{H}^{d-1}(dx) \int_0^t f(x - s\mathbf{n}(x)) ds \right| \leq c \sup_{u \in N(K,t)} f(u) t^2.$$

Furthermore,

$$\int_{N(K)} \mathcal{H}^{d-1}(dx) \int_0^t f(x - s\mathbf{n}(x))ds$$

$$= t \int_{N(K)} f(x)\mathcal{H}^{d-1}(dx) - \frac{t^2}{2}\int_{N(K)} \langle f'(x), \mathbf{n}(x)\rangle \mathcal{H}^{d-1}(dx) + o(t^2)$$

as $t \to 0$. By the condition on f, $\mathcal{H}^{d-1}(\{v \in \partial F : f(v) \neq 0\}) > 0$. Therefore,

$$t^{-1} \int_{N(K,t)} f(x)dx \to \int_{N(K)} f(x)\mathcal{H}^{d-1}(dx) \quad \text{as } t \to 0.$$

It follows from (3.16) that

$$\int_{N(K)} f(x)\mathcal{H}^{d-1}(dx) \leq \lim_{t \to 0} t^{-1}\int_{M(K,t)} f(x)dx \leq \int_{N(K^\varepsilon)} f(x)\mathcal{H}^{d-1}(dx).$$

A similar inequality holds for K given by (3.15), i.e.

$$\sum_{i=1}^m t_i \int_{N(K_i)} f(x)\mathcal{H}^{d-1}(dx) \leq \lim_{s \to 0} s^{-1}\mathbf{P}\left\{X_1 \cap s \circ \tilde{K} \neq \emptyset\right\}$$

$$\leq \sum_{i=1}^m t_i \int_{N(K_i^\varepsilon)} f(x)\mathcal{H}^{d-1}(dx).$$

Theorem 4.2.3 implies that the random closed set $n \circ H_n$ (i.e. $a_n^{-1} \circ H_n$ for $a_n = n^{-1}$) converges weakly to the random set H having the capacity functional

$$T_H(K) = 1 - \exp\left\{-\int_{F_K} f(x)dx\,dt\right\}.$$

Note that $F_K = F_{\hat{K}}$, where $\hat{K} = \{(u, s) : s \leq t, (u, t) \in K\}$ and F_{K_1} and F_{K_2} are disjoint if $\hat{K}_1$ and $\hat{K}_2$ are disjoint. Therefore, $\Lambda(\hat{K})$ is a measure, whence $n\eta_n$ epiconverges to a sharp integrand η such that epi $\eta = H$ and ∂^-epi η is a Poisson point process with the intensity measure Λ.

(ii) In this case

$$\sup_{x \in N(K,t)} f(x) \to 0 \quad \text{as } t \to 0.$$

Hence

$$t^{-2}\int_{N(K,t)} f(x)\mathrm{d}x \to -\frac{1}{2}\int_{N(K)} \langle f'(x), \mathbf{n}(x)\rangle \mathcal{H}^{d-1}(\mathrm{d}x) \quad \text{as } t\to 0.$$

Thus, for $a_n = n^{-1/2}$, the weak epilimit of $a_n^{-1}\eta_n$ has the capacity functional

$$T_H(K) = 1 - \exp\left\{\int_{F_K} s\langle f'(x), \mathbf{n}(x)\rangle \mathrm{d}x\mathrm{d}s\right\}. \qquad \square$$

Example 3.35. If $\xi_1,\ldots,\xi_n$ are uniformly distributed on F, then the limiting capacity functional is given by

$$T_H(K) = 1 - \exp\{-\mu(F_K)/\operatorname{mes}(F)\},$$

where μ is the product of $\mathcal{H}^{d-1}$ and the Lebesgue measure on the line.

3.4 Increasing set-valued processes of excursion sets

It is possible to strengthen the epiconvergence by adding a uniformity requirement to it. Observe that if f is a lower semicontinuous function, then the family of its excursion sets $\{f \leq t\}$ parametrised by the level $t \in \mathbb{R}$ is an increasing set-valued process on $\mathbb{R}$ which is also right-continuous, i.e. $\{f \leq s\} \downarrow \{f \leq t\}$ if $s \downarrow t$. Discontinuities from the left are caused by local minima of f. If $f(x) = t$ is a local minimum, then $\{f \leq t\}$ contains x, while $\{f \leq s\}$ for $s < t$ does not hit a neighbourhood of x.

Let ρ be a metric on the family of closed sets in a LCHS space $\mathbb{E}$ that metrises the Fell topology, e.g. the Hausdorff–Busemann metric (B.1) or the Hausdorff metric if $\mathbb{E}$ is compact. Consider two lower semicontinuous functions f and g with values in $[0, 1]$. The corresponding *uniform metric* is defined as

$$\rho_U(f,g) = \sup_{0\leq t\leq 1} \rho(\{f\leq t\},\{g\leq t\}).$$

However, because of possible discontinuities, the uniform metric is too strong. Generalising the concept of convergence for numerical right-continuous functions with left limits, it is possible to use the *Skorohod distance* together with the Hausdorff (or Hausdorff–Busemann) metric for closed sets in order to define the DH-distance between lower semicontinuous functions.

Definition 3.36 (DH-distance and convergence). The *DH-distance* between lower semicontinuous functions f and g is defined as

$$\rho_{\mathrm{DH}}(f,g) = \inf_{\lambda\in\Sigma}\left[\rho_U(\lambda(f),g) + \sup_{0\leq t\leq 1}|\lambda(t)-t|\right],$$

where Σ is the family of continuous bijections of $[0, 1]$. We say that f_n *DH-converges* to f is $\rho_{\mathrm{DH}}(f_n,f) \to 0$ as $n\to\infty$.

It is easy to generalise the DH-convergence concept for functions with values in a general interval and then further to functions with possibly unbounded families of values by considering their arbitrary truncations. The DH-convergence is generally incomparable with classical definitions of convergence.

Example 3.37 (Convergence modes).
(i) Let $f_n(t) = \mathbf{1}_{[1/2,1]}(t)(2tn^{-1} + 1 - 2n^{-1})$. Then f_n uniformly converges to $\mathbf{1}_{[1/2,1]}$, but f_n does not DH-converge. The uniform convergence does not imply DH-convergence even if the limiting function is continuous.
(ii) The function $f_n(t) = \mathbf{1}_{[1/2+1/n,1]}(t) + \mathbf{1}_{[1/2,1/2+1/n]}(t)(nt - n/2)$ DH-converges to $\mathbf{1}_{[1/2,1]}(t)$, but does not admit a limit in the Skorohod space or in the uniform metric.
(iii) If $f_n(t) = t^n$, $0 \leq t \leq 1$, then f_n epiconverges to 0, but does not DH-converge, i.e. the epiconvergence does not imply the DH-convergence.

The following result follows from Proposition 3.3.

Proposition 3.38 (DH-convergence implies epiconvergence). *If lower semicontinuous functions f_n DH-converge to f, then the sequence $\{f_n, n \geq 1\}$ also epiconverges to f as $n \to \infty$.*

The *first passage* time
$$\tau_K = \inf\{t : \{f \leq t\} \cap K \neq \emptyset\}$$
of the process $\{f \leq t\}$ is directly related to the infimum of f, since $\tau_K \leq t$ if and only if $\inf_{x \in K} f(x) \leq t$. Therefore, the DH-convergence ensures the convergence of the first passage times.

Now consider the weak DH-convergence of a sequence $\{\zeta_n\}$ of normal integrands. We begin with the convergence of finite-dimensional distributions.

Theorem 3.39 (Finite-dimensional distributions for the excursion process). *Suppose that ζ is a random lower semicontinuous function such that the distribution of $\inf_{x \in K} \zeta(x)$ is atomless for each $K \in \mathcal{K}$. Then the finite-dimensional distributions of the set-valued process $\{\zeta_n \leq t\}$, $t \in [0, 1]$, converge to the finite-dimensional distributions of $\{\zeta \leq t\}$, $t \in [0, 1]$, if and only if ζ_n weakly epiconverges to ζ as $n \to \infty$.*

Proof. The one-dimensional distribution of $\{\zeta_n \leq t\}$ for $t \in [0, 1]$ is given by the corresponding capacity functional
$$\mathbf{P}\Big\{\{\zeta_n \leq t\} \cap K \neq \emptyset\Big\} = \mathbf{P}\Big\{\inf_{x \in K} \zeta_n \leq t\Big\}.$$
For the continuity property, note that
$$\mathbf{P}\Big\{\{\zeta \leq t\} \cap \operatorname{Int} K = \emptyset, \ \{\zeta \leq t\} \cap K \neq \emptyset\Big\} + \mathbf{P}\Big\{\inf_{x \in K} \zeta(x) = t\Big\}$$
$$= \mathbf{P}\Big\{(\operatorname{epi} \zeta) \cap F \neq \emptyset, \ (\operatorname{epi} \zeta) \cap \operatorname{Int} F = \emptyset\Big\},$$

where $F = K \times [0, t]$. The same holds for the capacity functional of the m-tuple of random sets $(\{\zeta_n \leq t_1\}, \ldots, \{\zeta_n \leq t_1\})$ for $t_1 < t_2 < \cdots < t_m$. Therefore, the weak epiconvergence of ζ_n is equivalent to the convergence of finite-dimensional distributions. □

In difference to Theorem 3.18, the convergence of finite-dimensional distributions of the process $\{\zeta_n \leq t\}$ does not suffice for the weak DH-convergence. One also needs to check the tightness condition described in Theorem 3.40. For a lower semicontinuous function f and $x \in \mathbb{E}$ define

$$\omega_\varepsilon(f, x) = f(x) - \inf\{f(y) : y \in B_\varepsilon(x)\},$$
$$\bar{\omega}_\varepsilon(f, x) = \inf\left\{f(y) : \inf f(B_\varepsilon(y)) > f(x)\right\} - f(x),$$

where the infimum over an empty set equals ∞. The value

$$\hat{\Delta}_\varepsilon(f) = \inf_{x \in \mathbb{E}} \max(\omega_\varepsilon(f, x), \bar{\omega}_\varepsilon(f, x)),$$

is related to the continuity modulus

$$\Delta_h(F) = \sup_{h \leq t \leq 1-h} \min(\rho(F_{t-h}, F_t), \rho(F_t, F_{t+h}))$$
$$+ \rho(F_0, F_h) + \rho(F_{1-h}, F_1)$$

for the set-valued function $F_t = \{f \leq t\}$, $t \in [0, 1]$. It is possible to show that for each $\varepsilon > 0$ and sufficiently small h, $\Delta_h(F) > \varepsilon$ implies $\hat{\Delta}_\varepsilon(f) < 2h$ and $\hat{\Delta}_\varepsilon(f) < h$ implies $\Delta_h(F) > \varepsilon$.

Theorem 3.40 (Tightness for DH-convergence). *A sequence $\{\zeta_n, n \geq 1\}$ of lower semicontinuous functions with values in $[0, 1]$ weakly DH-converges to a function ζ if the finite-dimensional distributions of $\{\zeta_n \leq t\}$ converge to the finite-dimensional distributions of $\{\zeta \leq t\}$ and, for each $\varepsilon > 0$,*

$$\limsup_{h \downarrow 0} \mathbf{P}\left\{\hat{\Delta}_\varepsilon(\zeta_n) \leq h\right\} = 0.$$

3.5 Strong law of large numbers for epigraphical sums

Operations with epigraphs

Let ζ_1 and ζ_2 be proper normal integrands defined on $\mathbb{E} = \mathbb{R}^d$. Two basic operations with functions, addition and pointwise minimum, can be reformulated for epigraphs, so that the pointwise minimum corresponds to the union of the epigraphs, i.e.

$$\text{epi}(\min(\zeta_1, \zeta_2)) = \text{epi}\,\zeta_1 \cup \text{epi}\,\zeta_2,$$

and the pointwise addition

364   5 Random Sets and Random Functions

$$\mathrm{epi}(\zeta_1+\zeta_2)=\{(x,t_1+t_2):\ (x,t_i)\in\mathrm{epi}\,\zeta_i,\ i=1,2\}$$

is obtained as a "vertical" sum of the epigraphs.

It is possible to define another operation with functions by taking Minkowski sums of their epigraphs as closed subsets of $\mathbb{R}^d\times\mathbb{R}$, so that

$$\mathrm{cl}(\mathrm{epi}\,\zeta_1+\mathrm{epi}\,\zeta_2)=\mathrm{cl}(\{(x_1+x_2,t_1+t_2):\ (x_i,t_i)\in\mathrm{epi}\,\zeta_i,\ i=1,2\})$$

is the epigraph of a lower semicontinuous function ζ called the *epigraphical sum* of ζ_1 and ζ_2 and denoted $\zeta=\zeta_1\oplus\zeta_2$. It is easy to see that

$$\zeta(x)=\inf\{\zeta_1(y)+\zeta_2(z):\ y+z=x\}. \qquad (3.17)$$

The scaling operation is defined similarly as $c\odot\zeta$ with

$$\mathrm{epi}(c\odot\zeta)=c\,\mathrm{epi}\,\zeta=\{(cx,ct):\ (x,t)\in\mathrm{epi}\,\zeta\},\quad c>0.$$

so that $(c\odot\zeta)(x)=c\zeta(c^{-1}x)$. These definitions show a great degree of similarity with the setup of Chapter 3 which deals with normalised Minkowski sums of random closed sets. However, the specific feature of integrands requires results for unbounded random closed sets.

Selection expectation for normal integrands

A normal integrand ζ is said to be *integrable* if epi ζ is an integrable random closed set. This is equivalent to the integrability of $\zeta(\xi)$ for an integrable random vector ξ in $\mathbb{R}^d$. In particular, it suffices to assume that $\zeta(x)$ is integrable for at least one $x\in\mathbb{R}^d$. If ζ is integrable, then epi ζ possesses a non-empty selection expectation $\mathbf{E}(\mathrm{epi}\,\zeta)$.

Proposition 3.41 (Selection expectation integrands). *If ζ is an integrable normal integrand, then* $\mathbf{E}(\mathrm{epi}\,\zeta)=\overline{\mathrm{co}}\,(\mathrm{epi}\,g)$ *where*

$$g(x)=\inf\{\mathbf{E}\zeta(\xi):\ \xi\in\mathbf{L}^1,\ \mathbf{E}\xi=x\},\quad x\in\mathbb{E}. \qquad (3.18)$$

Proof. For a lower semicontinuous function f define its strict epigraph by

$$\mathrm{epi}'\,f=\{(x,t):\ f(x)>t\}.$$

It suffices to show that the strict epigraph epi$'$ g coincides with the Aumann integral $\mathbf{E}_\mathrm{I}(\mathrm{epi}'\,\zeta)$, see Definition 2.1.13.

If $(x,t)\in\mathrm{epi}'\,g$, then there exists $\xi\in\mathbf{L}^1$ verifying $\mathbf{E}\xi=x$ and $\mathbf{E}\zeta(\xi)<t$. Define

$$\alpha=\zeta(\xi)+t-\mathbf{E}\zeta(\xi).$$

Then (ξ,α) is an integrable selection of the random (not necessarily closed) set epi$'$ ζ and $\mathbf{E}(\xi,\alpha)=(x,t)$, whence $(x,t)\in\mathbf{E}_\mathrm{I}\,\mathrm{epi}'\,\zeta$.

Conversely, if $(x,t)\in\mathbf{E}_\mathrm{I}(\mathrm{epi}'\,\zeta)$, then there are $\xi\in\mathbf{L}^1$ and an integrable random variable α such that $\mathbf{E}\xi=x$, $\mathbf{E}\alpha=t$ and $\zeta(\xi)<\alpha$ a.s. Then

$$g(x)\le\mathbf{E}\zeta(\xi)<\mathbf{E}\alpha=t. \qquad\square$$

If f is a lower semicontinuous function, then $\overline{\mathrm{co}}\,(\mathrm{epi}\,f)$ is the epigraph of a convex lower semicontinuous function called the *biconjugate* of f, which is the largest convex lower semicontinuous function which is smaller than f. Proposition 3.41 implies that $\mathbf{E}(\mathrm{epi}\,\zeta)$ is the epigraph of the biconjugate of $\mathbf{E}\zeta(x)$, $x \in \mathbb{E}$.

Law of large numbers and application to allocation problem

Theorem 3.1.21 implies that if $\{\zeta_n, n \geq 1\}$ are i.i.d. integrable proper normal integrands, then
$$\eta_n = n^{-1} \odot (\zeta_1 \oplus \cdots \oplus \zeta_n)$$
epiconverges almost surely to the function of g defined by (3.18). Note that η_n can be equivalently obtained as
$$\eta_n(x) = \inf\{n^{-1}\sum_{i=1}^{n}\zeta_i(x_i) : x_1, \ldots, x_n \in \mathbb{R}^d,\ n^{-1}\sum_{i=1}^{n}x_i = x\}.$$

This explains the close relationships with the *allocation problem* in mathematical economy, see Artstein and Hart [22], which is usually formulated for maximum instead of minimum and refers to the dual results for hypographs.

Consider a sequence $\{\zeta_n, n \geq 1\}$ of random production functions, which are i.i.d. random upper semicontinuous functions defined on $\mathbb{R}_+^d = [0, \infty)^d$ with values in $\mathbb{R}_+ = [0, \infty)$. The random variable $\zeta_i(x_i)$ determines the output of the ith firm if it has $x_i \in \mathbb{R}_+^d$ resources of d different types. Let $q \in \mathbb{R}_+^d$ be a vector with strictly positive coordinates, which represents the total initial resources to be allocated for n different firms with production functions $\zeta_1, \ldots, \zeta_n$ respectively. Consider the following optimisation (allocation) problem:
$$\frac{1}{n}\sum_{i=1}^{n}\zeta_i(x_i) \to \max \quad \text{subject to} \quad \frac{1}{n}\sum_{i=1}^{n}x_i \leq q, \qquad (3.19)$$
where the inequality in $\mathbb{R}_+^d$ is coordinatewise. The supremum in (3.19) is denoted by $v_n(q)$ (remember that $v_n(q)$ is a random variable so it also depends on ω). Note that $X_i = \mathrm{hypo}\,\zeta_i$, $i \geq 1$, form a sequence of i.i.d. random closed (unbounded) sets and
$$\mathbf{E}X_i = \mathrm{hypo}\,g, \qquad (3.20)$$
where g is the smallest concave function which is larger than $\mathbf{E}\zeta_i(x)$ for all $x \in \mathbb{R}_+^d$. If $\zeta_i = f$ is deterministic, then g is the smallest concave hull of f. Assume that $n \to \infty$, i.e. the number of firms is large.

Theorem 3.42 (Optimal allocations). *For any $q \in (0, \infty)^d$, the value $v_n(q)$ of the problem (3.19) converges almost surely as $n \to \infty$ to $v(q)$, which is the supremum of $\mathbf{E}g(\xi)$ subject to $\mathbf{E}\xi \leq q$, where g is defined by (3.20). The convergence is uniform for q from each bounded subset of $(0, \infty)^d$.*

Proof. It follows from (3.19) that $v_n(q)$ is the supremum of r such that (x,r) belongs to $n^{-1}(X_1+\cdots+X_n)$ and $x \leq q$. By Theorem 3.1.21, $n^{-1}(X_1+\cdots+X_n)$ converges in the Fell topology to the closed convex hull of $\mathbb{E}X_1$. The convergence of hypographs implies the convergence of maxima over compact sets, so that $v_n(q)$ converges uniformly for q in a bounded set to the supremum of r such that there exists $x \leq q$ with $(x,r) \in \overline{\mathrm{co}}\,(\mathbb{E}X_1)$. By (3.20) the latter is equal to $v(q)$. □

The concept of conditional selection expectation applied to a sequence of epigraphs leads to the following definition.

Definition 3.43 (Martingale integrand). A sequence $\{\zeta_n, n \geq 1\}$ of convex integrable proper normal integrands is called a *martingale integrand* if $\{\mathrm{epi}\,\zeta_n, n \geq 1\}$ is a multivalued martingale with closed convex values in $\mathbb{E} \times \mathbb{R}$.

Submartingale and supermartingale integrands are defined similarly which makes it possible to formulate for integrands a number of those results from Section 1.1 which are applicable for unbounded random closed sets.

3.6 Level sums of random upper semicontinuous functions

Deterministic level sums and convergence

Although it is possible to apply arithmetic addition or pointwise maxima to upper semicontinuous functions, there are several "non-traditional" operations which make sense in this setting. Here we consider the operation based on Minkowski addition of the excursion sets. For simplicity, assume that $\mathbb{E}$ is the Euclidean space $\mathbb{R}^d$, although many results hold for a general Banach space. Let Υ_0 denote the family of upper semicontinuous functions f with values in $[0,1]$ such that $\{f \geq t\}$ is compact for each $t \in (0,1]$ and $\{f = 1\} \neq \emptyset$. Furthermore, let $\{f > t-\}$ denote the closure of $\{x \in \mathbb{E} : f(x) > t\}$. The set $\{f > 0-\} = \mathrm{supp}\,f$ is called the *support* of f.

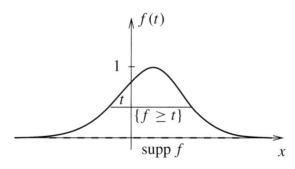

Figure 3.7. Upper semicontinuous function f and its upper excursion set.

A deterministic upper semicontinuous function $f \in \Upsilon_0$ can be represented as a "stacked" composition of its excursion sets $\{f \geq t\}$ for $0 < t \leq 1$, see Figure 3.7.

3 Semicontinuous random functions

Operations with functions can now be defined using set-theoretical operations with their excursion sets and then stacking them back together. For instance, the *level sum* of f and g is defined as

$$(f \boxplus g)(x) = \sup_{y+z=x} \min(f(y), g(z)), \quad x \in \mathbb{E},$$

cf. (3.17). It is easy to see that the excursion sets of $f \boxplus g$ are equal to Minkowski sums of the corresponding excursion sets of f and g, i.e.

$$\{(f \boxplus g) \geq t\} = \{f \geq t\} + \{g \geq t\}, \quad 0 < t \leq 1.$$

Multiplication by a real number $c \neq 0$ is defined as

$$(c \boxdot f)(x) = f(x/c).$$

The family Υ_0 can be equipped with various metrics. The following two metrics are especially important in the following:

$$\rho_H^\infty(f, g) = \sup_{0 < t \leq 1} \rho_H(\{f \geq t\}, \{g \geq t\}),$$

$$\rho_H^1(f, g) = \int_0^1 \rho_H(\{f \geq t\}, \{g \geq t\}) dt.$$

Clearly, $\rho_H^1(f, g) \leq \rho_H^\infty(f, g)$. It is shown by Puri and Ralescu [465] that the space Υ_0 with metric ρ_H^∞ is a complete non-separable metric space. In contrast, the metric ρ_H^1 turns Υ_0 into a separable metric space, see Klement, Puri and Ralescu [318]. It is possible also to define an analogue of the DH-metric from Definition 3.36.

The following fact states that the convergence in the uniform metric ρ_H^∞ follows from the pointwise convergence of the Hausdorff distances between the corresponding excursion sets and the supports.

Theorem 3.44 (ρ_H^∞-convergence in terms of excursion sets). *Let $\{f_n, n \geq 1\}$ be a sequence of functions from Υ_0. Then $\rho_H^\infty(f_n, f) \to 0$ as $n \to \infty$ for $f \in \Upsilon_0$ if and only if*

$$\rho_H(\{f_n \geq t\}, \{f \geq t\}) \to 0 \quad \text{as } n \to \infty \text{ for all } t \in (0, 1],$$

and

$$\rho_H(\{f_n > t-\}, \{f > t-\}) \to 0 \quad \text{as } n \to \infty \text{ for all } t \in [0, 1).$$

Proof. Necessity is clear, so let us prove sufficiency. Note that if $K_1 \subset K \subset K_2$ and $L_1 \subset L \subset L_2$ (all sets are compact), then

$$\rho_H(K, L) \leq \max(\rho_H(K_1, L_2), \rho_H(K_2, L_1)). \tag{3.21}$$

Fix $\varepsilon > 0$ arbitrarily small. Since the family of sets $\{f \geq t\}, 0 < t \leq 1$, is decreasing and all these sets are contained within the compact set $\operatorname{supp} f$, there is at most a finite number of points $0 < s_1 < \cdots < s_m < 1$ such that

$$\rho_H(\{f > s_i-\}, \{f \geq s_i\}) \geq \varepsilon.$$

Put $s_0 = 0$ and $s_{m+1} = 1$.

Fix any $i \in \{0, \ldots, m\}$. Since $\operatorname{supp} f$ is a compact set, we can partition the interval $(s_i, s_{i+1}]$ by points $s_i = t_0 < t_1 < \cdots < t_{n_i} = s_{i+1}$ such that

$$\rho_H(\{f > t_j-\}, \{f \geq t_{j+1}\}) < \varepsilon, \quad j = 0, \ldots, n_i - 1. \tag{3.22}$$

For any $j = 0, \ldots, n_i - 1$, by (3.21) and the triangle inequality

$$\sup_{t_j < t \leq t_{j+1}} \rho_H(\{f_n \geq t\}, \{f \geq t\})$$

$$\leq \max\left[\rho_H(\{f_n > t_j-\}, \{f \geq t_{j+1}\}), \rho_H(\{f_n \geq t_{j+1}\}, \{f > t_j-\})\right]$$

$$\leq \max\left[\rho_H(\{f_n > t_j-\}, \{f > t_j-\}), \rho_H(\{f_n \geq t_{j+1}\}, \{f \geq t_{j+1}\})\right]$$

$$+ \rho_H(\{f \geq t_{j+1}\}, \{f > t_j-\}).$$

The choice of partitioning points implies that the first term in the right-hand side converges to zero by the condition of the theorem and the second term is bounded by ε as chosen in (3.22).

Then, for any $i = 0, 1, \ldots, m$,

$$\sup_{s_i < t \leq s_{i+1}} \rho_H(\{f_n \geq t\}, \{f \geq t\}) \leq \max_{0 \leq j \leq n_i - 1} \sup_{t_j < t \leq t_{j+1}} \rho_H(\{f_n \geq t\}, \{f \geq t\})$$

converges to zero, which immediately yields $\rho_H^\infty(f_n, f) \to 0$ as $n \to \infty$. □

Random level sums

If ζ is a *random element* in Υ_0, then $\{\zeta \geq t\}$ is a random compact set for each $t \in (0, 1]$. Note that ζ is said to be *integrably bounded* if all random compact sets $\{\zeta \geq t\}$ are integrably bounded, i.e. $\mathbf{E}\|\{\zeta \geq t\}\| < \infty$ for all $0 < t \leq 1$. We call ζ *strongly integrable* if $\mathbf{E}\|\operatorname{supp}\zeta\| < \infty$. The expectation of ζ which complies with the above introduced operations can be defined by taking the selection expectations of the excursion sets of ζ and then stacking them together. This is possible because the selection expectation respects the inclusion relationship between random compact sets.

Assume that ζ is integrable and define

$$(\mathbf{E}_A \zeta)(x) = \sup\{t : x \in \mathbf{E}_A\{\zeta \geq t\}\}, \quad x \in \mathbb{E}. \tag{3.23}$$

It is easy to show that $\{\mathbf{E}_A \zeta \geq t\} = \mathbf{E}_A\{\zeta \geq t\}$ because the family $\mathbf{E}_A\{\zeta \geq t\}$ is monotone with respect to t and

3 Semicontinuous random functions

$$\mathbf{E}_A\{\zeta \geq t\} = \lim_{t_n \downarrow t} \mathbf{E}_A\{\zeta \geq t_n\}$$

by the convergence theorem for the selection expectation, see Section 2.1.2. In the following we assume that the basic probability space is non-atomic, so that all excursion sets of $\mathbf{E}_A \zeta$ are convex. An integrable random function ζ is strongly integrable if and only if $\operatorname{supp} \mathbf{E}_A \zeta$ is compact.

The strong law of large numbers for i.i.d. random upper semicontinuous functions $\zeta, \zeta_1, \zeta_2, \dots$ aims to establish the convergence of

$$\bar{\zeta}_n = n^{-1} \boxdot (\zeta_1 \boxplus \dots \boxplus \zeta_n)$$

to the expectation $\mathbf{E}_A \zeta$ in one of the metrics ρ_H^∞ or ρ_H^1. A simple inequality between these metrics entails that it suffices to prove it with respect to ρ_H^∞ only.

Theorem 3.45 (Strong law of large numbers for level sums). *Every sequence ζ_n, $n \geq 1$, of i.i.d. strongly integrable random upper semicontinuous functions satisfies the strong law of large numbers in ρ_H^∞ metric, i.e. $\rho_H^\infty(\bar{\zeta}_n, \mathbf{E}_A \zeta_1) \to 0$ a.s. as $n \to \infty$.*

Proof. It suffices to note that by the strong law of large numbers for random compact sets (Theorem 3.1.6) $\rho_H(\{\bar{\zeta}_n \geq t\}, \{\mathbf{E}\zeta_1 \geq t\}) \to 0$ (respectively $\rho_H(\{\bar{\zeta}_n > t-\}, \{\mathbf{E}\zeta_1 > t-\}) \to 0$) almost surely as $n \to \infty$ for each $t \in (0, 1]$ (respectively $t \in [0, 1)$). The proof is concluded by applying Theorem 3.44. □

3.7 Graphical convergence of random functions

Graphical convergence

As we have seen, random functions give rise to various random sets which appear as their epigraphs, hypographs or graphs. Apart from providing new interesting examples of random sets, this allows us to handle random functions in a new specific way. In this section we consider limit theorems for graphs of random functions. First, note that in limit theorems for stochastic processes two basic convergences are usually considered. This is the *uniform convergence* and the *Skorohod convergence*, see, e.g. Billingsley [70] and Skorohod [537]. However, often a sequence of functions does not converge in any of the topologies known from the classical theory of functional limit theorems. At the same time the pointwise convergence or epi- (hypo-) convergences are too weak to make conclusions about the limits of many interesting functionals.

Example 3.46 ("Non-traditional" convergence).
(i) Let $f_n(t) = nx \mathbf{1}_{[0,1/n]}(t) + \mathbf{1}_{(1/n,\infty)}(t)$, $0 \leq t \leq 1$, $n \geq 1$. Then $f_n(t)$ converges pointwise to $f(t) = \mathbf{1}_{t>0}$, but does not converge either uniformly or in the Skorohod sense.
(ii) $f_n(t) = nt$, $0 \leq t \leq 1$, "converges" to the vertical line as $n \to \infty$.

(iii) The sequence $f_n(t) = \sin nt$, $0 \leq t \leq 1$, $n \geq 1$, fills in the rectangle $[0, 1] \times [-1, 1]$, but does not converge pointwisely.

The convergence statements from Example 3.46 can be made precise in terms of the epiconvergence. An alternative is to formulate the convergence results for the completed graphs of functions considered as closed subsets of $[0, 1] \times \mathbb{R}$, see (3.35). While some closed sets in $[0, 1] \times \mathbb{R}$ are obtained as closed graphs of functions, many other do not admit such representations and are sometimes called "non-functional". Example 3.46(iii) provides a sequence of functions that converges to a rectangle. This calls for formulating the problem for a sequence of multifunctions $F_n(t) = \{\sin nt\}$ (they are actually single-valued) that converges to a set-valued function $F(t) = [-1, 1]$, $0 \leq t \leq 1$.

In many cases set-valued functions can be naturally used to replace single-valued functions. For instance, when considering the weak convergence of random variables, it may be helpful to replace jumps of the cumulative probability distribution function with segments filling the whole vertical intervals between its limits from the left and from the right.

Definition 3.47 (Graphical convergence). Let $F_n(t)$, $t \in \mathbb{E}$, $n \geq 1$, be a sequence of functions whose values are closed subsets of a topological space $\mathbb{E}'$. Assume that both $\mathbb{E}$ and $\mathbb{E}'$ are LCHS. The *graph* of F_n is a subset of $\mathbb{E} \times \mathbb{E}'$ defined as

$$\text{Graph } F_n = \{(t, x) : t \in \mathbb{E}, \ x \in F_n(t)\}.$$

The sequence $\{F_n, n \geq 1\}$ is said to *graphically converge* to F if $\text{cl}(\text{Graph } F_n)$ converge to $\text{cl}(\text{Graph } F)$ in the Painlevé–Kuratowski sense as subsets of $\mathbb{E} \times \mathbb{E}'$ with the product topology.

Note that F can be retrieved from its graph as

$$F(t) = \{x : (t, x) \in \text{Graph } F\}.$$

The graphical convergence of functions is in general not comparable with their pointwise convergence; it is easy to provide an example of a sequence that converges both pointwisely and graphically, but the limits differ. For instance, if $\mathbb{E} = [0, 1]$, $\mathbb{E}' = \mathbb{R}$ and $F_n(t) = \{f_n(t)\}$ are single-valued functions given by

$$f_n(t) = \begin{cases} nt, & 0 \leq t \leq 1/n, \\ 2 - nt, & 1/n < t \leq 2/n, \\ 0, & 2/n < t \leq 1, \end{cases}$$

then $f_n(t)$ converges pointwisely to zero, while its graphical limit is a set-valued function given by

$$F(t) = \begin{cases} [0, 1], & t = 0, \\ \{0\}, & 0 < t \leq 1. \end{cases}$$

If $\mathbb{E}'$ is a metric space, the uniform convergence with respect to the Hausdorff metric implies the graphical convergence.

Definition 3.48 (Graphical convergence in distribution). A sequence $\{Z_n, n \geq 1\}$ of random set-valued functions is said to *graphically converge in distribution* to a random closed set Z if $\mathrm{cl}(\mathrm{Graph}\, Z_n)$ weakly converges to Z as $n \to \infty$.

According to Definition 3.48, a sequence of random single-valued functions may converge to a random closed set which has no direct interpretation as a graph of a random single-valued function. However, the limiting random set Z can be interpreted as the graph of the random set-valued function $Z(t) = \{x : (t, x) \in Z\}$. In general spaces one has to define the weak convergence with respect to a topology that generates the Effros σ-algebra, see Theorem 1.2.7.

Random step-functions

Consider a situation when $\mathbb{E} = [0, 1]$ and $\mathbb{E}' = \mathbb{R}$, which means that all functions are defined on $[0, 1]$ and have values being subsets of the real line. Then all closed graphs become random closed subsets of $[0, 1] \times \mathbb{R}$. It is easy to see that the finite unions of rectangles $[s, t] \times [x, y]$ for (s, x), (t, y) from a countable dense set in $[0, 1] \times \mathbb{R}$ constitute a separating class $\mathcal{A}$ in $[0, 1] \times \mathbb{R}$. We start with a basic result concerning the graphical convergence of step-functions. Let $\{\alpha_n, n \geq 0\}$ be a sequence of i.i.d. random variables with distribution

$$\mathbf{P}(A) = \mathbf{P}\{\alpha_0 \in A\}, \qquad A \in \mathfrak{B}(\mathbb{R}). \tag{3.24}$$

Define a step-function

$$\zeta_n(t) = a_n^{-1} \alpha_{[nt]}, \qquad 0 \leq t \leq 1, \tag{3.25}$$

for a monotone sequence $\{a_n, n \geq 1\}$ of positive normalising constants satisfying either $a_n \downarrow 0$ or $a_n \uparrow \infty$, see Figure 3.8.

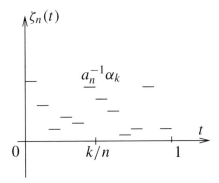

Figure 3.8. Step-function $\zeta_n(t)$, $0 \leq t \leq 1$.

Theorem 3.49 (Convergence of single-valued step-functions). *Random function ζ_n graphically converges in distribution to a non-trivial random set X if and only if there exists a dense set $\mathbb{Q}$ in $\mathbb{R}^2$ such that*

$$\lim_{n \to \infty} n\mathbf{P}(a_n[x, y]) = v([x, y]) \in [0, +\infty] \qquad (3.26)$$

for all $(x, y) \in \mathbb{Q}$ with $x \leq y$, where v is a measure on $\mathfrak{B}(\mathbb{R})$ such that $0 < v([x, y]) < +\infty$ for some $(x, y) \in \mathbb{Q}$.

Proof. Define $X_n = \mathrm{cl}(\mathrm{Graph}\,\zeta_n)$. Corollary 1.6.9 together with Theorem 1.6.5 imply that $X_n \xrightarrow{d} X$ if and only if $T_{X_n}(K) \to T_X(K)$ as $n \to \infty$ for each $K \in \mathcal{A} \cap \mathfrak{S}_X$. Let $K_\beta = [s + \beta, t - \beta] \times [x + \beta, y - \beta]$ for $\beta \geq 0$. The function

$$f(\beta) = Q_X(K_\beta) = 1 - T_X(K_\beta) = \mathbf{P}\{X \cap K_\beta = \emptyset\}$$

is monotone, so that $f(\beta)$ has arbitrary small points of continuity, where

$$T_X(K_\beta) = T_X(\mathrm{Int}\,K_\beta).$$

For any such $\beta \geq 0$, write $K_\beta = [\bar{s}, \bar{t}] \times [\bar{x}, \bar{y}]$. If $h_n = \mathbf{P}(a_n[\bar{x}, \bar{y}])$, then

$$(1 - h_n)^{[n\bar{t}]-[n\bar{s}]+3} \leq Q_{X_n}(K_\beta) \leq (1 - h_n)^{[n\bar{t}]-[n\bar{s}]-1}. \qquad (3.27)$$

Necessity. Assume that

$$\mathbf{P}\{X_n \cap K_\beta = \emptyset\} \to \mathbf{P}\{X \cap K_\beta = \emptyset\} \quad \text{as } n \to \infty.$$

If $h_n = \mathbf{P}(a_n[\bar{x}, \bar{y}]) \to 0$ as $n \to \infty$, then (3.27) yields

$$n(\bar{t} - \bar{s})h_n \to -\log Q(K_\beta) = v([\bar{x}, \bar{y}]) \in [0, +\infty],$$

and $Q(K_\beta) = \exp\{-(\bar{t} - \bar{s})v([\bar{x}, \bar{y}])\}$. If $h_n \geq \delta > 0$ for sufficiently large n, then $Q(K_\beta) = 0$ and $nh_n \to \infty$, i.e.

$$\lim_{n \to \infty} nh_n = \exp\{-(\bar{t} - \bar{s})v([\bar{x}, \bar{y}])\}$$

with $v([\bar{x}, \bar{y}]) = \infty$. The same conclusion holds if $h_{n_k} \to 0$ for a subsequence $\{n_k,\ k \geq 1\}$, but $h_n \not\to 0$.

The set $\mathbb{Q}$ of all pairs $(\bar{x}, \bar{y})$ such that $\lim_{n \to \infty} n\mathbf{P}(a_n[\bar{x}, \bar{y}])$ exists is dense in $\mathbb{R}^2$. There is one pair $(\bar{x}, \bar{y}) \in \mathbb{Q}$ that yields a non-trivial limit, since otherwise $Q(K_\beta)$ is either zero or one and the limiting distribution is degenerated by Theorem 1.1.27.

Sufficiency. Let $Q(K) = \exp\{-(\mathrm{mes}_1 \otimes v)(K)\}$, where $K \in \mathcal{K}([0, 1] \times \mathbb{R})$. Then $1 - Q(K) = T(K)$ is a capacity functional of a random set X. Since $0 < Q([0, 1] \times [x, y]) < 1$ for some x, y, the limiting random closed set is non-degenerated.

From (3.27) we deduce that $Q_{X_n}([s, t] \times [x, y]) \to Q(K)$ if $(s, x), (t, y) \in \mathbb{Q}$. Inequality (3.27) can be generalised for $K = \bigcup_{j=1}^m ([s_j, t_j] \times [x_j, y_j])$ and used to prove that also in this case $Q_{X_n}(K) \to Q(K)$ as $n \to \infty$. $\square$

3 Semicontinuous random functions 373

The limiting random closed set X in Theorem 3.49 has the capacity functional

$$T_X(K) = 1 - \exp\{-(\text{mes}_1 \otimes \nu)(K)\}, \qquad (3.28)$$

where mes_1 is the Lebesgue measure in $\mathbb{R}^1$. By Theorem 4.1.6, X is infinitely divisible for unions. Indeed, X is a Poisson random set in $[0, 1] \times \mathbb{R}$ with intensity $\text{mes}_1 \otimes \nu$, see Definition 1.8.8. If the total mass of ν is infinite, then X has an infinite number of points. It is easily seen that if ν is supported by the whole line, then both the epi- and hypo-limits of ζ_n are degenerated, while ζ_n may have a non-trivial graphical limit in distribution.

Proposition 3.50. *Let X be a random closed set with the capacity functional (3.28).*
(i) X has no fixed points if and only if $\nu([x, y]) < \infty$ for all finite x and y.
(ii) If X has no fixed points, then X is the limit of step-functions with $\mathbf{P}(\{0\}) = 0$ and $a_n \to 0$.
(iii) The set of fixed points of X is

$$F_X = \{(t, y) : 0 \le t \le 1, \ y \in \mathbb{R}, \ \nu(\{y\}) = \infty\}.$$

Proof.
(i) follows from Lemma 4.1.8, since X is infinitely divisible for unions.
(ii) If $x < 0 < y$, then $\mathbf{P}(a_n[x, y]) \to \mathbf{P}(\{0\})$. Furthermore, $n\mathbf{P}(a_n[x, y])$ converges to $\nu([x, y])$ being finite by the condition. This implies $\mathbf{P}(\{0\}) = 0$. Clearly, $a_n \to \infty$ is impossible, since in this case $[0, 1] \times \{0\}$ are fixed points of the limiting random closed set.
(iii) It follows from (3.28) that $(x, y) \in F_X$ if and only if $\nu(\{y\}) = \infty$, which proves the conclusion. □

Graphical convergence of set-valued processes

Consider a generalisation of the previous scheme for set-valued processes. Let $\{X_n, n \ge 0\}$ be a sequence of i.i.d. random closed sets in $[0, 1] \times \mathbb{R}^d$. For $a, b \ge 0$, $t \ge 0$ and $x \in \mathbb{R}^d$ denote

$$(b, a) \circ (t, x) = (bt, ax)$$

and similarly $(b, a) \circ F$ for $F \subset [0, 1] \times \mathbb{R}^d$, cf (4.2.6). Define the set-valued process

$$Z_n(t) = (n^{-1}, a_n^{-1}) \circ \left(X_{[nt]} + ([nt], 0)\right), \quad 0 \le t \le 1. \qquad (3.29)$$

The projection of X_n onto its $\mathbb{R}^d$-coordinate is denoted by

$$\tilde{X}_n = \{x \in \mathbb{R}^d : ([0, 1] \times \{x\}) \cap X_n \ne \emptyset\}.$$

Since for every $K \in \mathcal{K}(\mathbb{R}^d)$

$$\mathbf{P}\left\{(a_n^{-1} \cup_{i=1}^n \tilde{X}_i) \cap K = \emptyset\right\} = \mathbf{P}\left\{\text{cl}(\text{Graph } Z_n) \cap ([0, 1] \times K) = \emptyset\right\},$$

374   5 Random Sets and Random Functions

the graphical convergence in distribution of Z_n implies the weak convergence of the normalised unions for the sequence $\{\tilde{X}_n, n \geq 1\}$. The following theorem states that under some conditions the converse statement holds.

Theorem 3.51. *The set-valued process Z_n defined by (3.29) graphically converges in distribution if*
(1) *$Y_n = a_n^{-1} \cup_{i=1}^{n} \tilde{X}_i$ converges in distribution as $n \to \infty$ and*
(2) *there exists $\alpha \neq 0$ such that, for every $t \geq 0$,*

$$\lim_{n \to \infty} \frac{a_{[nt]}}{a_n} = t^{-1/\alpha}. \tag{3.30}$$

Then $nT_{\tilde{X}_1}(a_n K)$ converges as $n \to \infty$ to a completely alternating upper semicontinuous capacity $\Psi(K)$ for for all $K \in \mathcal{K} \cap \mathfrak{S}_\Psi$ and Z_n graphically converges in distribution to a random closed subset Z in $[0, 1] \times \mathbb{R}^d$ with the capacity functional

$$T_Z(K) = 1 - \exp\left\{-\int_0^1 \Psi(\{x : (t, x) \in K\}) dt\right\}. \tag{3.31}$$

The random closed set Z can be represented as

$$Z = \bigcup_{i \geq 1} (\{t_i\} \times F_i), \tag{3.32}$$

where $\{t_i, i \geq 1\}$ are i.i.d. random variables uniformly distributed on $[0, 1]$ and $\{F_i, i \geq 1\}$ is a Poisson point process on the family $\mathcal{F}$ of closed subsets of $\mathbb{R}^d$ with intensity measure Λ satisfying $\Lambda(\mathcal{F}_K) = \Psi(K)$.

Proof. The family $\mathcal{A}$ of sets $\cup_{i=1}^{m}[s_i, t_i] \times K_i$ where $m \geq 1, 0 \leq s_1 < t_1 \leq \cdots \leq s_n < t_n \leq 1$, and $K_1, \ldots, K_m \in \mathcal{K}(\mathbb{R}^d)$ is a separating class in $[0, 1] \times \mathbb{R}^d$. If $m = 1$, then

$$\left(\mathbf{P}\left\{a_n^{-1}\tilde{X}_1 \cap K = \emptyset\right\}\right)^{[nt]-[ns]+3} \leq \mathbf{P}\{\mathrm{cl}(\mathrm{Graph}\, Z_n) \cap ([s, t] \times K) = \emptyset\}$$

$$\leq \left(\mathbf{P}\left\{a_n^{-1}\tilde{X}_1 \cap K = \emptyset\right\}\right)^{[nt]-[ns]-1}.$$

Therefore,

$$\left|\mathbf{P}\{\mathrm{cl}(\mathrm{Graph}\, Z_n) \cap ([s, t] \times K) = \emptyset\}\right.$$

$$\left. - \mathbf{P}\left\{\left(a_n^{-1} \cup_{i=1}^{[n(t-s)]} \tilde{X}_i\right) \cap K = \emptyset\right\}\right| \to 0$$

as $n \to \infty$. Furthermore,

$$\mathbf{P}\left\{\left(a_n^{-1} \cup_{i=1}^{[n(t-s)]} \tilde{X}_i\right) \cap K = \emptyset\right\} = \mathbf{P}\left\{(a_n^{-1}a_{[n(t-s)]})Y_{[n(t-s)]} \cap K = \emptyset\right\},$$

where $Y_{[n(t-s)]} \xrightarrow{d} Y$ as $n \to \infty$. By (3.30),

$$\mathbf{P}\left\{a_n^{-1} \cup_{i=1}^{[n(t-s)]} \tilde{X}_i \cap K = \emptyset\right\} \to \mathbf{P}\left\{(t-s)^{-1/\alpha} Y \cap K = \emptyset\right\}$$
$$= Q_Y((t-s)^{1/\alpha} K) = \exp\{-\Psi((t-s)^{1/\alpha} K)\}.$$

By Theorem 4.2.8, Y is union-stable with parameter α, whence

$$\exp\{-\Psi((t-s)^{1/\alpha} K)\} = \exp\{-(t-s)\Psi(K)\}.$$

By similar arguments using approximations from below and above we obtain

$$\mathbf{P}\left\{\mathrm{cl}(\mathrm{Graph}\, Z_n) \cap (\cup_{i=1}^m [s_i, t_i] \times K_i) \neq \emptyset\right\} \to 1 - \prod_{i=1}^m Q_Y((t_i - s_i)^{1/\alpha} K_i)$$
$$= 1 - \exp\{-\sum_{i=1}^m (t_i - s_i)\Psi(K_i)\}.$$

The right-hand side can be extended to a capacity functional on $\mathcal{K}([0, 1] \times \mathbb{R}^d)$ which is determined by (3.31). By comparison of the capacity functionals it is easy to check the applicability of the construction given by (3.32). □

Note that both (1) and (2) in Theorem 3.51 hold if $\{X_n, n \geq 1\}$ satisfies the conditions of Theorem 4.2.1 (for $\alpha < 0$) or its dual for $\alpha > 0$.

Theorem 3.51 provides also an alternative way to derive Theorem 3.49 for the case of X_n being singletons in $\mathbb{R}$. If in this case $a_n \to 0$ and $\mathbf{P}(\{0\}) = 0$ in (3.24), then Theorem 4.2.9 implies that condition (2) of Theorem 3.51 can be omitted. Then Ψ is a measure that is denoted by ν. If Z is a union-stable random set without fixed points, Corollary 4.1.13 yields $\alpha > 0$ and

$$\nu([x, y]) = \theta(y) - \theta(x) \quad (3.33)$$

with

$$\theta(x) = \mathbf{1}_{x>0} c_1 x^\alpha - \mathbf{1}_{x<0} c_2 |x|^\alpha,$$

$c_1, c_2 \geq 0$ and $c_1 + c_2 > 0$.

Weakly dependent sequences and linearly interpolated step-functions

Assume that the sequence $\{X_n, n \geq 1\}$ is *m-dependent*, i.e. there exists $m \geq 1$ such that X_i, $i \leq n$, and X_k, $k \geq n + m + 1$ are independent for every $n \geq 1$. Then Theorem 3.51 also holds, i.e. the graphical convergence of Z_n in distribution implies that $a_n^{-1} \circ \cup_{i=1}^n \tilde{X}_i$ converges in distribution as $n \to \infty$ and the converse holds if the sequence of normalising constants satisfies (3.30).

This can be used to prove a limit theorem for linearly interpolated step-functions constructed by a sequence $\{\alpha_n, n \geq 0\}$ of i.i.d. random variables with distribution (3.24). For any set $F \subset \mathbb{R}$ write $F^\sim = \mathrm{co}(F \cup \{0\})$ for the convex hull of the union of F and the origin. If $K \in \mathcal{K}([0, 1] \times \mathbb{R})$ define $K^\sim = \mathrm{co}(K \cup K')$, where $K' = \mathrm{proj}_{[0,1]} K$ is the projection of K onto $[0, 1]$.

Proposition 3.52. *Let $a_n \to 0$ and let*

$$\lim_{n\to\infty} n\mathbf{P}(a_n[x, y]^{\sim}) = v([x, y]^{\sim}) \tag{3.34}$$

for (x, y) from a dense set $\mathbb{Q} \subset \mathbb{R}^2$, where v is a measure on $\mathfrak{B}(\mathbb{R})$ such that $v([x, y]) < \infty$ for at least one $(x, y) \in \mathbb{Q}$. Assume that $\alpha_0 > 0$ a.s. or $\alpha_0 < 0$ a.s. Then

$$\zeta_n(t) = a_n^{-1}(\alpha_{[nt]} + (nt - [nt])(\alpha_{[nt]+1} - \alpha_{[nt]})), \quad 0 \le t \le 1,$$

graphically converges in distribution to a random closed set Z with the capacity functional

$$T_Z(K) = 1 - \exp\{-(\mathrm{mes}_1 \otimes v)(K^{\sim})\}, \quad K \in \mathcal{K}([0, 1] \times \mathbb{R}).$$

Proof. Let X_k be the segment in $[0, 1] \times \mathbb{R}$ with the end-points being $(0, \alpha_k)$ and $(1, \alpha_{k+1})$. Then Graph ζ_n = Graph Z_n, where Z_n is the set-valued process defined by (3.29). The sequence $\{X_n, n \ge 1\}$ defined above is 1-dependent. By the necessary conditions in the limit theorem for extremes of random variables (see Galambos [186, Th. 2.4.3, 2.4.4]), (3.34) implies (3.30). It suffices to prove that $Y_n = a_n^{-1} \circ \cup_{i=1}^n \tilde{X}_i$ converges in distribution as $n \to \infty$. Note that $\tilde{X}_i$ is the segment with the end-points α_{i-1} and α_i. If K is a compact subset of $\mathbb{R}$, then

$$\mathbf{P}\{Y_n \cap K = \emptyset\} = \prod_{i=1}^{n+1} \mathbf{P}\{\alpha_i \le a_n \inf K\} + \prod_{i=1}^{n+1} \mathbf{P}\{\alpha_i \ge a_n \sup K\}.$$

Assume that $\alpha_0 \ge 0$ a.s. (the case $\alpha_0 \le 0$ is treated similarly). Then the first summand in the right-hand side tends to zero. By (3.34), $\mathbf{P}\{Y_n \cap K = \emptyset\} \to \exp\{-v(K^{\sim})\}$ with $K^{\sim} = [0, \sup K]$. An application of Theorem 3.51 (modified for weakly dependent random closed sets) finishes the proof. □

As shown by Lyashenko [365], the conditions of Proposition 3.52 are also necessary. The limiting random closed set Z in Proposition 3.52 is a sharp integrand, so that $\partial^- Z$ is a Poisson point process with the intensity measure Λ. A similar result holds for the case $a_n \to \infty$. Further generalisations are possible for random functions obtained as partial linear interpolators of the step function defined in (3.25).

Graphically continuous functionals

The maximum and minimum of a function over a closed subset of $[0, 1]$ are continuous functionals with respect to the graphical convergence. However, a number of other interesting functionals are not continuous in the Fell topology that generates the graphical convergence of functions. Fortunately, it is possible to define "smooth" versions of many geometric functionals which are continuous with respect to the graphical convergence. Fix an $\varepsilon > 0$. For $K \in \mathcal{K}([0, 1] \times \mathbb{R})$ and $F \in \mathcal{F}([0, 1] \times \mathbb{R})$ define

$$\text{Len}_\varepsilon(F) = \frac{1}{2\varepsilon}\text{mes}_2(F^\varepsilon),$$
$$\text{Card}_\varepsilon(F) = \frac{1}{\pi\varepsilon^2}\text{mes}_2(F^\varepsilon),$$
$$\text{Ran}_\varepsilon(F) = \text{mes}_1(\text{pr}_\mathbb{R} F^\varepsilon) - 2\varepsilon,$$

where $\text{pr}_\mathbb{R}$ is the projection of the second coordinate. If ε is small, then $\text{Len}_\varepsilon(F)$ (respectively $\text{Card}_\varepsilon(F)$) approximates the perimeter (respectively cardinality) of F. Corollary E.13(ii) implies that these ε-versions of conventional geometric functionals are continuous with respect to the graphical convergence.

M_2-topology

Another concept of the convergence of graphs can be introduced as follows. For a Polish space $\mathbb{E}$, consider a family $\mathbb{K}_\mathbb{E}$ of functions $f : [0, 1] \mapsto \mathbb{E}$ that are continuous from the right and have left limits everywhere on $[0, 1]$. Let Γ_f be the *completed graph* of f, i.e. the set of all points $(t, x) \in [0, 1] \times \mathbb{E}$ such that x belongs to the segment with end-points $f(t - 0)$ and $f(t)$. Note that Γ_f is a continuous curve in $[0, 1] \times \mathbb{E}$. A sequence $\{f_n, n \geq 1\} \subset \mathbb{K}_\mathbb{E}$ is said to converge to f if

$$\rho_H(\Gamma_f, \Gamma_{f_n}) \to 0 \quad \text{as } n \to \infty. \tag{3.35}$$

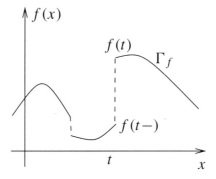

Figure 3.9. A function and its completed graph.

This topology appears under the name of M_2 topology in Skorohod [537]. If $\mathbb{E}$ is a real line, then (3.35) is equivalent to the convergence of maximum and minimum of f_n over any segment $[s, t] \subset [0, 1]$. Therefore, (3.35) holds if and only if f_n both epi- and hypo-converges to f as $n \to \infty$. Although Theorem 3.18 and its variant for the hypoconvergence imply that any sequence of random functions from $\mathbb{K}_\mathbb{R}$ is tight in this topology, the limit may be non-functional. A tightness condition that guarantees the weak convergence to a random function in $\mathbb{K}_\mathbb{R}$ can be formulated as follows.

Theorem 3.53. *A sequence ζ_n, $n \geq 1$, of random functions from $K_\mathbb{E}$ converges in distribution to a random function ζ in the M_2 topology if the finite dimensional distributions of $\zeta_n(t)$ converge to those of $\zeta(t)$ for t from a dense set in $[0, 1]$ and*

$$\lim_{c \downarrow 0} \limsup_{n \to \infty} \mathbf{P}\{\Delta(c, \zeta_n) > \varepsilon\} = 0,$$

for all $\varepsilon > 0$, where, for every $f \in K_\mathbb{E}$,

$$\Delta(c, f) = \sup_{t \in [0,1],\, t' \in [t_c^* - c/2, t_c^*],\, t'' \in [t_c, t_c + c/2]} \rho_H(f(t), \mathrm{co}\{f(t'), f(t'')\}),$$

$t_c = \max(0, t - c)$ *and* $t_c^* = \min(1, t + c)$. *These conditions are also necessary if $\mathbb{E}$ contains a subset which can be mapped continuously onto a line segment.*

Notes to Chapter 5

Section 1.1. The studies of multivalued martingales in the Euclidean space were initiated by van Cutsem [113, 114] and Neveu [425] who investigated their convergence properties. Hiai and Umegaki [255] extended the theory to integrably bounded set-valued martingales in a Banach space. Theorem 1.3 is a synthesis of results of Papageorgiou [446] and Hess [242]. The fundamental convergence theorems (Theorems 1.5 and 1.6) for integrably bounded multivalued martingales are taken from Hiai and Umegaki [255]. For compact-valued martingales, it is possible to replace the uniform integrability condition with the so-called terminal uniform integrability.

Papageorgiou [446, Th. 3.1] showed that the separability assumption of $\mathbb{E}^*$ in Theorem 1.6 can be dropped; the proof refers to the Castaing representation of multivalued martingales described in Theorem 1.12(ii). The Radon–Nikodym property of $\mathbb{E}$ can also be replaced by the condition that $X_n \subset Y$ for every $n \geq 1$, where Y is a weakly compact integrably bounded random set, see Papageorgiou [446, Th. 3.2]. It is shown by Choukari-Dini [100] that $\mathbf{E}(\mathbf{1}_A X_n)$ Mosco converges to $\mathbf{E}(\mathbf{1}_A X_\infty)$ for every $A \in \mathfrak{F}$ under the conditions of Theorem 1.6. Under the same conditions, X_n converges almost surely to X_∞ in the Mosco sense, see Li and Ogura [352]. If all σ-algebras in the filtration are countably generated and one of the conditions (ii) or (iii) of Theorem 2.1.24 holds, then Theorem 1.6 holds for multivalued submartingales, see Li and Ogura [352].

The convergence theorem for multivalued martingales in $\mathbb{R}^d$ with uniformly integrable norms goes back to van Cutsem [113] and Neveu [425]. An alternative proof of the convergence results by Papageorgiou [442] refers to Theorem 1.3(iii) which says that taking support functions preserves martingale properties. This fact yields the scalar convergence of the sequence $\{X_n, n \geq 1\}$, which corresponds to the pointwise convergence of their support functions. Other types of convergence can be obtained by requiring some conditions on the sequence of support functions. For example, it is possible to deduce a convergence theorem for submartingales by assuming that the support functions are equi-lower semicontinuous, see

Papageorgiou [442, Th. 2.3, 2.4]. To ensure the boundedness of the support function it is usually assumed that $X_n \subset Y$ for an integrably bounded random weakly compact convex set Y. Without loss of generality Y can be chosen to be symmetric, since by the Krein–Smulian theorem $\overline{\text{co}}\,(Y \cup \check{Y})$ is weakly compact. Therefore, $|h(X_n, u)| \leq h(Y, u)$ for every $u \in \mathbb{E}^*$. This argument was used to prove a convergence theorem for multivalued supermartingales in Papageorgiou [444].

Theorem 1.8 was proved by Papageorgiou [446]. The Mosco convergence of supermartingales was studied by Li and Ogura [352] and also in the context of fuzzy random sets in Li and Ogura [353]. Couvreux and Hess [107] obtained a Lévy type martingale convergence theorem for random closed sets with unbounded values and under sufficiently weak conditions.

Subsequent extensions followed a number of different routes. First, many results have been extended for reversed martingales (see Hiai and Umegaki [255]) defined for time parameter $n \leq 0$ and applied to derive strong laws of large numbers by Castaing and Ezzaki [89]. These results can be extended for supermartingales and submartingales. Ziat [626] obtained convergence results of multivalued reversed martingales.

As shown by Hess [242], it is possible to drop the boundedness condition for multivalued supermartingales in Theorem 1.9, which extends earlier results by Choukari-Dini [100].

Theorems 1.12 and 1.13 concerning the existence of martingale selections and the Castaing representation for multivalued martingales were proved by Hess [242] in this generality, although a number of other results existed earlier, see Luu [362]. Theorem 1.13 can be generalised for non-reflexive spaces given that $X_n \cap B_r(0)$ is almost surely weakly compact for all $n \geq 1$ and $r > 0$, see Hess [242]. A result similar to the Castaing representation for submartingales was obtained by Li and Ogura [352]. Further results on representations and convergence of multivalued martingales have been published by Wang and Xue [597]. Multivalued martingales in continuous time were investigated by Dong and Wang [141].

de Korvin and Kleyle [325, 326] proved a convergence theorem for integrably bounded supermartingales with uniformly integrable norms and applied it to decision-making.

The optional sampling theorem for multivalued martingales in Euclidean spaces was proved in Aló, de Korvin and Roberts [7] under some restrictions of the uniform integrability type. The presented optional sampling theorem (Theorem 1.15) is due to Papageorgiou [448]. Wang [595] introduced the concept of essential closure (or closed convex hull) for a family of random closed sets and set-valued processes in order to extend the optional sampling theorem for integrable (but not necessarily integrably bounded) multivalued martingales with closed convex values in a Banach space.

Single-valued pramarts and mils were surveyed by Egghe [159]. Multivalued generalisations are due to Luu [361], Bagchi [41], Choukari-Dini [100, 101] and Papageorgiou [446, 448]. Further results have been obtained by Krupa [332] and Avgerinos and Papageorgiou [33]. Results from the vast Chinese literature on multivalued martingales are summarised in Zhang, Wang and Gao [624]. Recent results on multivalued and fuzzy-valued martingales are presented by Li, Ogura and Kreinovich [354].

Multivalued martingales appear as Radon–Nikodym derivatives of multivalued measures (see Theorem 1.35) with respect to an increasing family of σ-algebras. The convergence theorems for multivalued martingales can be applied to show that the estimator $\mathbf{E}(\theta|\xi_1,\ldots,\xi_n)$ of a set-valued parameter θ based on a sample $\xi_1,\ldots,\xi_n$ converges almost surely to $\mathbf{E}(\theta|\mathfrak{F}_\infty)$ with the σ-algebra $\mathfrak{F}_\infty$ generated by all observations.

It is possible to define multivalued martingales in general metric spaces using the Herer expectation of random sets, see Definition 2.3.7. Several results on such martingales can be found in Herer [234, 236].

Section 1.2. Set-valued Markov process have been considered by Xu [617] and Gao and Zhang [187]. However, little is known about them beyond the representation theorems in terms of Markov selections. Set-valued Markov processes in a discrete space $\mathbb{E}$ are widely studied in probability theory, see Harris [222] and Liggett [357]. Set-valued stationary processes were studied by Wang and Wang [596].

It is possible to derive a central limit theorem for increasing processes defined by partial Minkowski sums of square integrable random compact sets. Schreiber [524] derived a large deviation principle for set-valued union processes. Example 1.20 is taken from Schreiber [524]. Krupa [333] considered a problem of finding the optimal stopping time for a sequence of random sets that maximises the selection expectation, which is a multivalued analogue of Snell's optimisation problem.

Non-trivial growth models that lead to random fractal sets are widely used in physics. Physical applications of growth models are discussed by Vicsek [574], see also the collection of papers [541] edited by H.E. Stanley and N. Ostrowsky, where further references, simulated and real pictures and discussions can be found.

One of the first stochastic models of spread on the lattice is due to Richardson, see also Durrett and Liggett [150]. Growth models that describe the spread of fires using increasing processes of random sets are discussed by Vorob'ev [587]. Stoyan and Lippmann [545] surveyed continuous models that can be used to describe the growth of cancer. Growth models related to the Boolean model of random sets (see Molchanov [406]) were considered by Cressie and Hulting [111]. Such processes may be used to describe the growth of cells or crystals in space.

Theorem 1.21 on random differential inclusions was proved by Papageorgiou [445], who generalised earlier results by Nowak [438]. Deterministic differential inclusions are discussed in detail by Aubin and Cellina [29].

Stochastic integrals of set-valued processes were introduced by Kisielewicz and Sosulski [314, 315] and Kim and Kim [302], where a number of results concerning selections of the integrals and the corresponding set-valued stochastic inclusions can be found. The same approach can be used to define the set-valued integral $\int_0^t X_s \, ds$ and an integral with respect to a Poisson measure. Set-valued stochastic differential inclusions were studied by Kree [331]. Jung and Kim [283] investigated set-valued integrals in case the set-valued process has intervals as possible values. Applications to controlled processes are discussed in Kisielewicz [314].

Ransford [474] defined a family of set-valued processes (called subholomorphic) on the Riemann sphere by assuming regularity properties of their paths and a predictability-like assumption. The family of such processes is closed under a wide variety of operations. In particular, it is possible to derive an inequality for the number of downcrossings, which leads to a number of convergence results.

Real-valued shot-noise processes were introduced in the 1960s, mostly in relation to modelling of queueing systems and physical phenomena, see Heinrich and Schmidt [232] for a survey and mathematical results concerning limit theorems for shot-noise processes. Typically, the response function f takes values in the real line; in many applications the marked point process of events is assumed to be Poisson. The Minkowski shot-noise process is presented here for the first time, while the union shot-noise was studied by Heinrich and Molchanov [231], where Theorem 1.27 is proved. It should be noted that all results can be generalised for N being Brillinger mixing (see Heinrich and Schmidt [232]) point processes and can be formulated in terms of the second order factorial cumulant measures. These cumulant measures vanish in the Poisson case.

It is possible to generalise Theorem 1.27 for multivalued functions F which are regularly varying as considered in Appendix H. The proof relies on the inversion theorem for regularly varying multifunctions (see Theorem H.5). The details can be found in Heinrich and Molchanov [231].

Section 1.3. Stochastic processes on random open convex domains and their finite-dimensional distributions have been considered by Dynkin and Fitzsimmons [153], see also Hu [266]. These papers pay particular attention to applications of these results to the construction to Markov processes.

The Tietze extension theorem for continuous functions (see Kuratowski [337, p. 127]) is generalised for random functions with random domains by Zhdanok [625], Brown and Schreiber [78] and Beg and Shahzad [57]. Measurability issues for random functions with random domains have been studied by Bocşan [74].

Cross [112] provides a comprehensive account of the theory of deterministic multivalued linear operators. Random multivalued operators have been studied by Itoh [273]. Further results are due to Beg and Shahzad [58, 59] and Tarafdar et al. [557].

Random multivalued operators with stochastic domain were introduced by Engl [162]. Theorem 1.31 and a number of further results on multivalued operators with stochastic domains are proved by Papageorgiou [445]. Other classical results on fixed points also have their set-valued counterparts that are often formulated using selections. A major restriction is the separability requirement on X, which is necessary to apply Theorem 1.31. It is possible to prove a stochastic variant of the Riesz representation theorem for random functionals defined on a random linear subspace X of a Hilbert space, see Papageorgiou [445, Th. 5.1]. An application of random multivalued functions with stochastic domains to differential equations and inclusions is described by Kandilakis and Papageorgiou [291].

Section 2.1. Theorem 2.1 was proved by Bulinskaya [84]. Theorem 2.2 is a famous result that was first derived heuristically by Rice [487]. Level crossings of a stationary (Gaussian) process are discussed in detail by Cramér and Leadbetter [108] and Leadbetter, Lindgren and Rootzen [346]. Further results and references concerning the Rice formula can be found in [346, Ch. 7]. A multidimensional analogue of the Rice formula is due to Radchenko [471]. Under certain conditions on the covariance, the point set of upcrossings converges in distribution to the Poisson point process, see Aldous [6].

Basic results on geometric properties of random fields can be found in a monograph by Adler [4]. The expected value of the Euler–Poincaré characteristic given by (2.2) is derived in Adler [4, Th. 5.4.1] as a corollary of a general result that gives the expected value of the so-called differential geometric characteristic of the level set, see Adler [4, Th. 5.3.1]. See also the recent survey by Adler [5].

Further advances are due to Worsley [612, 613] who discussed a number of applications in image analysis and astronomy (where results in two and three dimensions are of particular importance). The relevant random field may describe an image, so that particular high values (or upper excursions) signpost "active" areas that are important, for instance, in brain imaging. In astronomy such random fields determine cosmic background radiation where fluctuations may give some insight into the formation of the universe. Results for random fields with distributions derived from the Gaussian one (χ^2, F or t) can be found in Worsley [614]. It is also possible to relax the conditions on W by allowing piecewise smooth boundaries in two or three dimensions. Related properties of random surfaces were considered by Wschebor [615]. Nott and Wilson [436] considered parameter estimation problems for excursion sets of Gaussian random fields.

Hitting times of sets by Markov processes are widely studied in the potential theory for Markov processes, see Blumenthal and Getoor [73]. The q-capacities are discussed in Bertoin [65], where Proposition 2.4 also originates from. If, for some $q > 0$, two transient Markov processes share the same capacities given by (2.3) or the hitting probabilities $\mathbf{P}^x\{\tau_K < \infty\}$ of the two processes coincide for all x and $K \in \mathcal{K}$, then the processes are time changes of each other, see Fitzsimmons [174, 175] for exact formulations and further references. This conclusion also holds for 0-capacities of symmetric processes, see [175]. Glover and Rao [204] applied the Choquet theorem in this context.

The hitting times for diffusion processes and the corresponding capacities are discussed by Itô and McKean [272]. Intersections of paths of Brownian motions are studied by Khoshnevisan [301]. The capacity functional of type (2.3) determined by hitting times of a stochastic process was considered by Le Yan [345]. The construction can be extended for stochastic processes that are right-continuous and have left limits at every time moment.

Section 2.2. Theorem 2.6 and the related results are reproduced from Molchanov [393]. The concept of a random (not necessarily closed) set is one of the key concepts in Dellacherie [131], where random sets are used to study measurability, predictability and further related concepts for stochastic processes on the positive half-line. Theorem 2.8 also holds for not necessarily closed sets. Azema [37] discusses in detail random sets on the line.

Section 2.3. The concept of a strong Markov set is due to Hoffman-Jørgensen [261] who used this name in order to emphasise that the Markov property holds at random times. It was also called a Markov set by Krylov and Yushkevitch [335]. These random sets were thoroughly studied by Maisonneuve and Meyer [374, 389]. Later on the name regenerative set became more widely used. Regenerative sets form a special case of general regenerative systems studied by Maisonneuve [372]. Definition 2.10 is taken from Molchanov [393, 396].

Definition 2.11 is due to Maisonneuve [373] who proved Proposition 2.12 that establishes a relationship between regenerative and strong Markov random sets, see also Fitzsimmons, Fristedt and Maisonneuve [176]. They called random sets satisfying Definition 2.10 renewal sets.

A classification of regenerative phenomena according to the elements of the Laplace exponent (or the cumulant of the subordinator) is due to Kingman [309]. The explicit construction of the local time is described in Kingman [311]. It is a particular case of the general definition of the occupation density (see Bertoin [65, Sec. 5.1]) applied to the age process. A thorough study of various definitions of Markov random sets (using semilinear processes, intrinsic definition and the definition as the range of a subordinator) is due to Horowitz [265].

Although it is not straightforward to extend the described concepts for stationary Markov sets, it is also possible to show that such sets correspond to the image of a subordinator, see Taksar [554, 555] and Fitzsimmons and Taksar [178] for results in this area.

Theorem 2.17 is proved by Hawkes [226]. It is related to earlier results by Kesten [298]. Theorem 2.19 and the subsequent example are due to Hawkes [224].

Intersections of standard regenerative phenomena were studied by Kendall [293] and Kingman [309] using multiplication of their p-functions. Theorem 2.18 was proved by Fitzsimmons, Fristedt and Maisonneuve [176]. A thorough study of intersections of regenerative sets has been undertaken by Bertoin [67]. Fristedt [184] examined the weak convergence, intersections and the infinite divisibility for intersections of regenerative sets within various settings: for subsets of non-negative integers, all integers, the positive half-line and the whole line. It is shown by Molchanov [396] how to find the distribution of the intersection of independent strong Markov sets X and Y by solving an integral equation for the function $\chi(t) = \mathbf{P}\{X \cap (Y + t) \neq \emptyset\}$.

Regenerative embedding of Markov sets was studied by Bertoin [66] who also discussed related concepts of thinning and thickening of regenerative sets. Geometric properties of regenerative sets, e.g. their fractal dimension, are discussed in Bertoin [68].

Knight [319] characterised level sets of the so-called quasi-diffusions (or gap diffusions) in terms of their local times. The corresponding problem for diffusion processes is still open, see Itô and McKean [272, p. 217].

Kallenberg [289] studied symmetric interval partitions (or exchangeable random sets) that are the finite-interval counterpart of regenerative sets. If $b_1 \geq b_2 \geq b_3 \geq \cdots > 0$ are numbers that sum up to 1, then the exchangeable random sets X on the line has the complement X^c, where the intervals of lengths $b_1, b_2, \ldots$ occur in random order, i.e. for every $\varepsilon > 0$, the intervals lengths greater than ε enumerated from left to right form a finite exchangeable sequence. Then X is the closure of the range of the process

$$\zeta_t = \sum_k b_k \mathbf{1}_{\tau_k \leq t},$$

where $\{t_k, k \geq 1\}$ are i.i.d. random variables uniformly distributed on $[0, 1]$.

Section 2.4. This section follows Zuyev [631], who applied the concept of a stopping set to obtain rather general results for the case when the basic filtration is generated by a Poisson point process. As shown in Zuyev [631], the optional sampling theorem holds for all set-indexed martingales obtained as likelihood ratios. Propositions 2.28 and 2.29 have been proved in Zuyev [631] generalising similar results for stopping times by Molchanov [393]. The underlying theory of martingales on partially ordered sets was developed by Kurtz [341], see also Edgar and Sucheston [157].

Set-indexed martingales have been thoroughly studied by Ivanoff and Merzbach [274]. Their definition of the stopping set is dual to Definition 2.27 and requires that $\{K \subset Z\} \in \mathfrak{F}_K$ for all compact K. This definition is more convenient to apply for stopping sets whose values belong to some predetermined family of closed sets since then the family of stopping sets becomes closed under intersection. This establishes close links with the studies of multiparametric martingales, where stopping sets also naturally appear. It is possible to consider stopping sets with closed but not necessarily compact values. The predictability and progressive measurability aspects have been studied in Ivanoff, Merzbach and Schiopu-Kratina [275]. Stopping sets have been used to formulate strong Markov property of random fields by Rozanov [505] and Evstigneev [164], see also Kinateder [304] and Balan [42].

Set-indexed Brownian motion was investigated by Pyke [468] and Bass and Pyke [51].

Section 3.1. The epiconvergence of lower semicontinuous functions is a well-known concept described by Attouch [26], Dal Maso [116] and Rockafellar and Wets [499]. Aubin and Frankowska [30] presented this concept within the unified treatment of set-valued analysis. This is an extremely influential concept in optimisation and modern variational calculus. In the calculus of variations it is sometimes called the Γ-convergence, while in the studies of extremes it appears under the name of inf-vague convergence as opposed to the sup-vague convergence for sequences of upper semicontinuous functions and their hypographs studied by Vervaat [572]. Propositions 3.2 and 3.3 are proved in Rockafellar and Wets [499, Ch. 7,14], see also Attouch [26]. Further results mentioned in Section 3.1 stem from Rockafellar and Wets [498] and Salinetti and Wets [512]. Theorem 3.4 is due to Rockafellar and Wets [498] and Attouch [26]. Some topological generalisations for $\mathbb{E}$ being a Polish space are due to Beer [56].

An interpretation of epigraphs of stochastic processes as random closed sets was developed by Salinetti and Wets [512, 513]. They introduced the concepts of the (equi-) inner

384   5 Random Sets and Random Functions

separability and the equi-outer regularity for random lower semicontinuous functions. Normal integrands are often called random lower semicontinuous functions. Propositions 3.16 and 3.17 go back to Norberg [431] (however, formulated for upper semicontinuous functions and their hypographs). Theorem 3.13 (with part (ii) formulated for convex integrands) goes back to Rockafellar [496, pp. 181-183]. Later on it was generalised by Hess [244] for integrands defined on a Banach space with a strongly separable dual. Epiderivatives are discussed by Rockafellar and Wets [497, 499] and Aubin and Frankowska [30]. It is clearly possible to define epiderivatives for normal integrands using the almost sure convergence. An alternative definition of epiderivative in distribution is possible using the weak epiconvergence of normal integrands.

The weak convergence of sharp integrands is closely related to the weak convergence conditions for point processes discussed in Section 1.8.1, see Norberg [431, Prop. 2.3].

Theorem 3.19 is proved by Hiai [251, 254], but is reformulated using the probabilistic terminology.

It should be noted that the duals for the introduced concepts are possible for upper semicontinuous functions, their hypographs and lower excursions.

Section 3.2. The convergence of minimum for convex normal integrands was considered by Ch.J. Geyer (unpublished report). Anisimov and Seilhamer [11] investigated the convergence of minimisers in a setup when the limiting process has only a unique minimum point.

The epiconvergence of averages of i.i.d. lower semicontinuous functions (as formulated in Theorem 3.21) was studied by Attouch and Wets [27] assuming the existence of a quadratic minorant, in which case also bounds for the Hausdorff distance between the epigraphs are given. Hess [245, Th. 5.1] obtained this result for a general (not necessarily complete) metric space $\mathbb{E}$ and non-negative integrands. Theorem 3.21 as formulated in Section 3.2 goes back to Artstein and Wets [25, Th. 2.3], where also uniformity problems have been addressed and the case of a general Polish space $\mathbb{E}$ has been considered. This result was mentioned also as a possible extension by Hess [245]. Theorem 3.22 is due to Zervos [623] who proved it for a Souslin space $\mathbb{E}$ and generalised it for a general topological space $\mathbb{E}$ and a separable metric space Θ assuming that $\mathbf{P}_n$ converges narrowly to $\mathbf{P}$ under a relaxed condition on the integrands. Earlier results by Berger and Salinetti [64] concern the case when both $\mathbb{E}$ and Θ are Euclidean spaces in view of applications of the epiconvergence technique to Bayes decision theory. The average in (3.9) can be weighted, which is often useful in the framework of Monte Carlo importance sampling, see Berger and Salinetti [64].

Applications of the epiconvergence technique to stochastic optimisation are described in [25, 64, 509, 512] among a number of other references. Applications to convergence of estimators have been pioneered by J. Pfanzagl and further studied by Dupačová and Wets [149], Hoffman-Jørgensen [262, Ch. 13] and Hess [245]. Dupačová and Wets [149] discussed applications of epiderivatives to the asymptotic analysis of estimators. Further general results concerning asymptotics of solutions of stochastic programming problems have been obtained by King and Rockafellar [306]. They are formulated in terms of derivatives of multivalued functions. The large deviation technique is exploited by Kaniovski, King and Wets [292]. Stability issues in stochastic programming are surveyed by Schultz [527].

The convergence of zero sets, $\{x : \zeta_n(x) = 0\}$, is studied by Anisimov and Pflug [10]. For this, one typically requires that ζ_n converges weakly uniformly or the bands constructed around the graph of ζ_n converge as random closed sets.

Section 3.3. Norberg [431] initiated studies of extremes for random semicontinuous processes from the point of view of convergence of the associated random closed sets. He mostly worked with upper semicontinuous processes, their hypographs and associated sup-measures. A generalisation of Theorem 3.26 for max-infinitely divisible capacities is given in Norberg [430]. This theorem yields a characterisation of max-infinitely divisible random vectors as a particular case, see Balkema and Resnick [47]. Giné, Hahn and Vatan [201] modified the representation given by Theorem 3.26 (for semicontinuous processes) for the case of sample continuous processes and obtained spectral representation for max-stable sample continuous processes. The approach based on limit theorems for unions of random sets was developed by Molchanov [398, Sec. 8.3]. The application to polygonal approximations of convex sets was adapted from Molchanov [404]. Semi-min-stable processes have been introduced and characterised by Penrose [452] without using the random sets interpretation.

Super-extremal processes and their applications to continuous choice models have been considered by Resnick and Roy [482, 483], where most of the presented results originated from (however, notation is generally different and the statement of Theorem 3.32 has been corrected). It is possible to define lattice valued extremal processes which would serve as a generalisation of time-dependent processes with values in the family of upper semicontinuous functions.

Section 3.4. A representation of level sets of upper semicontinuous functions as a set-valued function from the Skorohod space goes back to Colubi et al. [102] and Kim [303]. The convergence results and properties of DH-convergence presented in this section are new. It is possible to generalise them for lower semicontinuous functions with arbitrary real values.

Section 3.5. A law of large numbers for epigraphs was obtained by Artstein and Hart [22] who also considered applications to allocation problem as described in Section 3.5. The optimality of allocations related to random closed sets was further studied by Papageorgiou [441, 443]. Generalisations of the law of large numbers for epigraphs are possible for $\mathbb{E}$ being a general Banach space, see Hess [243], King and Wets [307] and Krupa [332]. Castaing and Ezzaki [89] showed that the strong law of large numbers for epigraphs can be naturally derived from the Mosco convergence of reversed integrand martingales. Results on martingale integrands can be found in Hess [242] and Krupa [332]. The ergodic theorem for integrands is proved by Choirat, Hess and Seri [97]

Section 3.6. Random upper semicontinuous functions appear under different names in various settings. For instance, they are called random fuzzy sets or fuzzy random variables, see Puri and Ralescu [461, 465], or random grey-scale images, see Serra [532] and Molchanov [407]. The approach based on level sets decomposition and subsequent stacking them together is popular in image processing, where its generalisations give rise to the so-called stack filters, see Wendt, Cole and Callagher [608] and Maragos and Schafer [376].

Random fuzzy sets (or random upper semicontinuous functions) have been extensively studied by Puri and Ralescu [461, 465] and Klement, Puri and Ralescu [318]. The proofs given in Section 3.6 are taken from Molchanov [409]. The simple fact given by Theorem 3.44 was apparently overlooked in the previous papers that dealt with fuzzy random variables. A more complicated proof for the strong law of large numbers with respect to the uniform metric $\rho_{\mathrm{H}}^{\infty}$ is obtained by Colubi et al. [103] using approximation of upper semicontinuous functions by functions with simple excursion sets derived in López-Díaz and Gil [359]. A strong law of large numbers for upper semicontinuous random functions under exchangeability conditions was proved by Terán [562].

The definition of Gaussian random sets from Section 3.2.2 can be easily generalised for excursion sets of random upper semicontinuous functions. The corresponding theorem for Gaussian random sets was generalised for upper semicontinuous random functions by Puri and Ralescu [463]. Li, Ogura, Proske and Puri [355] proved a central limit theorem counterpart of Theorem 3.44.

Distances between random fuzzy sets and various concepts of expectations are considered by Näther [422]. A natural generalisation of the conditional selection expectation (see Section 2.1.6) for the case of random upper semicontinuous functions leads to the concept of a martingale. The corresponding convergence theorems were proved by Puri and Ralescu [466] and Stojaković [543]. Dominated convergence theorems for expectations of sequences of random upper semicontinuous functions in the metrics ρ_H^1 and ρ_H^∞ are proved by Klement, Puri and Ralescu [318, 465]. It is possible to consider level sums of capacities or non-additive measures, so that $(\varphi \boxplus \psi)(K)$ equals the supremum of $\min(\varphi(K_1), \psi(K_2))$ for $K_1 + K_2 \subset K$.

Formula (3.23) can be used for other expectations from Chapter 2. However, additional care is needed since other expectations do not necessarily respect the monotonicity relationship between random sets, see Section 2.3.4.

Section 3.7. Several topologies on the space of functions have been defined by Skorohod [537]. One particular topology (called J_2-topology) has later become widely used in the studies of stochastic processes under the name of the D-topology, see Billingsley [70]. The D-convergence of stochastic processes has been defined first for the parameter space $[0, 1]$ and then generalised for more general parameters spaces, see e.g. Bass and Pyke [52] and Lindvall [358]. It should be noted that other topologies defined by Skorohod [537] are more intrinsically related to graphs of random functions. One example is the M_2 topology that appears in Theorem 3.53 proved in Skorohod [537]. This topology is discussed in detail by Whitt [609]. The idea of M_2-topology was extended to set-indexed random functions by Bass and Pyke [52]. Kisyński [316] showed that the Skorohod D-topology can be generated by the Hausdorff distance between the graphs of functions $(\zeta(t-), \zeta(t))$ taking values in the product space $\mathbb{E} \times \mathbb{E}$ if ζ is a $\mathbb{E}$-valued function.

The graphical convergence of set-valued functions is considered in Rockafellar and Wets [499, Sec. 5E]. The formal definition of the graphical convergence in distribution of random set-valued functions is apparently new, although it has been studied extensively by Lyashenko [365, 367, 369] who obtained necessary and sufficient conditions for the graphical convergence in distribution for single-valued random functions and $a_n \to 0$. Theorem 3.49 goes back to Lyashenko [367] while its generalisation for set-valued processes was formulated in Lyashenko [370]. The representation (3.33) was obtained in Lyashenko [368] directly without using a characterisation theorem for union-stable random closed sets.

The present formulation of Theorem 3.51 is new. Lyashenko [370] presented further results concerning the case when random closed sets X_n in (3.29) are given by graphs of random functions. The results are formulated using regular variation ideas in an implicit form. Random functions obtained by a partial interpolation of step-functions are considered in Lyashenko [365]. It was assumed that the step-functions are interpolated for points from a fixed set of integers. The current presentation of these results based on set-valued processes is new.

Appendices

A Topological spaces and linear spaces

Sets

We use standard set-theoretic notation for union and intersection, $A \setminus B$ denotes the set-theoretic difference of A and B, $A \triangle B = (A \setminus B) \cup (B \setminus A)$ is the symmetric difference. The same sign is used for strict and non-strict inclusions, so that $A \subset B$ allows for $A = B$. Further, $A_n \uparrow A$ (respectively $A_n \downarrow A$) means that A_n is a non-decreasing (respectively non-increasing) sequence of sets with $A = \cup A_n$ (respectively $A = \cap A_n$). The set of all integers is denoted by $\mathbb{Z}$. The integer part of a real number x is denoted by $[x]$.

Topological spaces

An arbitrary set $\mathbb{E}$ can be made a topological space by choosing a *topology* $\mathcal{G}$, which is a family of open sets, so that $\mathcal{G}$ is closed under arbitrary unions and finite intersections. A subfamily $\mathcal{G}_0 \subset \mathcal{G}$ is called the *base* of topology if each open set $G \in \mathcal{G}$ can be represented as a union of sets from $\mathcal{G}_0$. For instance, if $\mathbb{E} = \mathbb{R} = (-\infty, +\infty)$ is the real line, then the base of the standard topology is given by all open intervals and a countable base is formed by intervals with rational end-points. A *sub-base* of a topology is a family of sets such that their finite intersections form the base of the topology.

Open sets from $\mathcal{G}$ yield *closed* sets as their complements. The family of closed sets is denoted by $\mathcal{F}$, so that $\mathcal{F} = \{G^c : G \in \mathcal{G}\}$, where $G^c = \mathbb{E} \setminus G$ denotes the complement of G. We write $\mathcal{G}(\mathbb{E})$ and $\mathcal{F}(\mathbb{E})$ to denote the space of open sets and closed sets of the particular carrier space $\mathbb{E}$ and omit $\mathbb{E}$ where no ambiguity occurs.

If $\mathbb{E}' \subset \mathbb{E}$ then the *induced* (or relative) topology on $\mathbb{E}'$ is given by intersections $G \cap \mathbb{E}'$ for all $G \in \mathbb{E}$. If $\mathbb{E}$ and $\mathbb{E}'$ are two topological spaces, then their *product space* $\mathbb{E} \times \mathbb{E}'$ consists of all pairs (x, y) for $x \in \mathbb{E}$ and $y \in \mathbb{E}'$. The topology on $\mathbb{E} \times \mathbb{E}'$ has the base given by $G \times G'$ for $G \in \mathcal{G}(\mathbb{E})$ and $G' \in \mathcal{G}(\mathbb{E}')$.

Let A be an arbitrary subset of $\mathbb{E}$. The intersection of all closed sets that contain A is denoted by cl A or $\overline{A}$ and is called the *closure* of A. An open set U such that $U \supset A$ is said to be a *neighbourhood* of A. A point $x \in A$ is said to be an *interior* point of A if $U \subset A$ for some neighbourhood U of x. The set of all interior points of A is denoted by Int A and called the *interior* of A. A set A is said to be *regular closed* if A coincides with the closure of its interior, i.e. $A = \text{cl}(\text{Int } A)$. The *boundary* of A equals the set-theoretic difference between the closure and interior of A, i.e. $\partial A = (\text{cl } A) \setminus (\text{Int } A)$. Equivalently, $\partial A = (\text{cl } A) \cap (\text{cl } A^c)$, i.e. the boundary of A consists of the limiting points for both A and its complement. A closed set A is said to be *perfect* if A does not have isolated points, i.e. every $x \in A$ is a limit of a sequence $\{x_n, n \geq 1\}$ such that $x_n \neq x$ for all $n \geq 1$. A sequence $\{x_n, n \geq 1\}$ is said to converge to x as $n \to \infty$ if every neighbourhood of x contains all x_n with $n \geq n_0$ for some n_0. A set is closed if and only if it contains the limit for every convergent sequence of its points.

A set $K \subset \mathbb{E}$ is *compact* if each open covering of K admits a finite subcovering, i.e. $K \subset \cup_{i \in I} G_i$ for any open sets $\{G_i, i \in I\}$ with an arbitrary family of subscripts I implies $K \subset G_{i_1} \cup \cdots \cup G_{i_n}$ for a finite set of subscripts $\{i_1, \ldots, i_n\} \subset I$. The family of all compact sets is denoted by $\mathcal{K}$ or $\mathcal{K}(\mathbb{E})$. Letters F, G and K (with or without indices) are typically used to denote generic closed, open and compact subsets of $\mathbb{E}$. The *empty* set is both open and closed and also compact. The family of all *non-empty* closed (respectively open, compact) sets is denoted by $\mathcal{F}'$ (respectively $\mathcal{G}', \mathcal{K}'$).

A class $\mathcal{A}$ of subsets of $\mathbb{E}$ is said to be *separating* if, whenever $K \subset G$ for $K \in \mathcal{K}$ and $G \in \mathcal{G}$, there exists an $A \in \mathcal{A}$ such that $K \subset A \subset G$.

If $\mathbb{E}$ is itself a compact set, then $\mathbb{E}$ is called a *compact* space. Furthermore, $\mathbb{E}$ is called *locally compact* if each point $x \in \mathbb{E}$ has a neighbourhood with compact closure. If $\mathbb{E}$ can be represented as a countable union of compact sets, $\mathbb{E}$ is said to be σ-*compact*. A set $B \subset \mathbb{E}$ is said to be *relatively compact* if $\text{cl}(B)$ is a compact set. If $\mathcal{M}$ is any family of sets, then $\mathcal{M}_k$ denotes the family of relatively compact sets from $\mathcal{M}$, for example $\mathcal{G}_k$ denotes the family of relatively compact open sets.

A locally compact space can be made compact by adding one additional point located "at infinity" and adjusting appropriately the topology. This construction is called one-point (or Aleksandrov) *compactification*. The corresponding open sets are sets from $\mathcal{G}(\mathbb{E})$ and the added point $\{\infty\}$ has neighbourhoods which are complements to compact sets.

If the topology on $\mathbb{E}$ has a countable base (in this case $\mathbb{E}$ is called *second countable*), then the compactness property of $K \in \mathcal{F}$ is equivalent to the fact that every sequence $\{x_n, n \geq 1\} \subset K$ admits a convergent subsequence. The existence of a countable base implies that $\mathbb{E}$ is *separable*, i.e. $\mathbb{E} = \text{cl } \mathbb{Q}$ for a countable set $\mathbb{Q} \subset \mathbb{E}$.

A topological space $\mathbb{E}$ is said to be *Hausdorff* if each two disjoint points of $\mathbb{E}$ have disjoint open neighbourhoods. We often assume this property, which however is not automatically valid for general topological spaces. If all singletons are closed (which is a weaker requirement), then $\mathbb{E}$ is said to be T_1-space. A weaker condition which singles out T_0-spaces requires that for each pair of different points there exists an open set which contains one point and does not contain the other one. Every compact set in a Hausdorff space is closed, i.e. $\mathcal{K} \subset \mathcal{F}$. This might not be the case if the space is not Hausdorff, where compact sets are usually called *quasicompact*.

The *saturation*, sat A, of a set $A \subset \mathbb{E}$ is the intersection of all open sets that contain A. Set A is called *saturated* if $A = \text{sat} A$. All sets are saturated if and only if $\mathbb{E}$ is a T_1-space.

A locally compact Hausdorff second countable space is said to be *LCHS* space. Sometimes these spaces are called *semi-compact*. The following well-known proposition says that in LCHS spaces compact sets can be approximated by open sets and open sets can be approximated from below by compact sets.

Proposition A.1. *If $\mathbb{E}$ is a LCHS space, then the following statements hold.*
  (i) *Each compact set K has a sequence of neighbourhoods $\{G_n, n \geq 1\}$ such that $G_n \subset G$ for some n and each open set $G \subset K$.*
  (ii) *For each $G \in \mathcal{G}$ there exists a sequence of relatively compact open sets $\{G_n, n \geq 1\}$ such that $\text{cl } G_n \in \mathcal{K}$, $\text{cl } G_n \subset G_{n+1}$ for all n and $G = \cup_{n \geq 1} G_n$.*
  (iii) *If $K \in \mathcal{K}$ and $F \in \mathcal{F}$ are disjoint, then K and F have disjoint neighbourhoods.*

Sometimes, we need a stronger property than Proposition A.1(iii), which requires that every two disjoint closed sets have disjoint open neighbourhoods. Then the space $\mathbb{E}$ is called *normal*.

A function $f : \mathbb{E} \mapsto \mathbb{E}'$ which maps $\mathbb{E}$ into another topological space $\mathbb{E}'$ is *continuous* if, for each $G \in \mathcal{G}(\mathbb{E}')$, the *inverse image*

$$f^{-1}(G) = \{x \in \mathbb{E} : f(x) \in G\}$$

is an open set in $\mathbb{E}$.

Algebras and σ-algebras

A family of sets is called an *algebra* if this family contains $\emptyset$ and is closed under taking complements and finite unions. An algebra $\mathfrak{F}$ is called a σ-*algebra* if it is closed under countable unions. If $\mathcal{M}$ is any family of sets, then $\sigma(\mathcal{M})$ denotes the smallest σ-algebra generated by $\mathcal{M}$.

One particular case of this construction is worth special attention. The minimal σ-algebra which contains the family $\mathcal{G}$ of all open sets is called the *Borel σ-algebra* on $\mathbb{E}$ and denoted by $\mathfrak{B}(\mathbb{E})$ or, shortly, $\mathfrak{B}$ if no ambiguity occurs. It is easy to see that $\mathfrak{B}$ contains all closed sets and can be equivalently defined as the minimal σ-algebra generated by the family of all closed sets, so that $\mathfrak{B} = \sigma(\mathcal{G}) = \sigma(\mathcal{F})$. Furthermore, $\mathfrak{B}_k$ denotes the family of relatively compact Borel sets.

If $\mathfrak{F}$ is a σ-algebra on $\mathbb{E}$ and $\mathfrak{F}'$ is a σ-algebra on $\mathbb{E}'$, then a function $f : \mathbb{E} \mapsto \mathbb{E}'$ is called $(\mathfrak{F}, \mathfrak{F}')$-*measurable* if $f^{-1}(A) \in \mathfrak{F}$ for every $A \in \mathfrak{F}'$. A function $f : \mathbb{E} \mapsto \mathbb{R}$ is called *Borel* if it is $(\mathfrak{B}(\mathbb{E}), \mathfrak{B}(\mathbb{R}))$-measurable.

A *paving* of $\mathbb{E}$ is any class $\mathcal{E}$ of subsets of $\mathbb{E}$ which includes the empty set. If $\mathcal{E}$ is a paving, then $\mathcal{E}_\sigma$ (respectively $\mathcal{E}_\delta$) denotes the class of countable unions (respectively intersections) of sets from $\mathcal{E}$. An $\mathcal{E}$-*analytic* (analytic over $\mathcal{E}$) set is a set which can be represented as $\cup_{(n_k)} (Y_{n_1} \cap Y_{n_1 n_2} \cap \cdots)$, where the outer union is taken over all possible sequences $(n_k) = (n_1, n_2, \ldots)$ of non-negative integers and $Y_{n_1 \ldots n_k} \in \mathcal{E}$ for every k-tuple $n_1, \ldots, n_k$ and $k \geq 1$. A set is called *analytic* if it is a subset of a metrisable space and is analytic over the class of closed subsets of this space. The family of analytic sets is richer than the Borel σ-algebra on the corresponding space.

Metric spaces

The topological properties of $\mathbb{E}$ are especially simple if $\mathbb{E}$ is a metric (or metrisable) space. A *metric* on $\mathbb{E}$ is a non-negative function $\rho(x, y)$, $x, y \in \mathbb{E}$, such that
(1) $\rho(x, y) = 0$ implies $x = y$ (full identification property);
(2) $\rho(x, y) = \rho(y, x)$ (symmetry);
(3) $\rho(x, z) \leq \rho(x, y) + \rho(y, z)$ (triangle inequality).

A sequence of points $\{x_n, n \geq 1\}$ in $\mathbb{E}$ is a *Cauchy sequence*, if, for every $\varepsilon > 0$, there exists an $n \geq 1$ such that $\rho(x_k, x_m) < \varepsilon$ for every $k, m \geq n$. A metric space is called *complete* if every Cauchy sequence is convergent. Complete separable metric spaces are called *Polish*. Continuous images of Polish spaces are called *Souslin* spaces. All analytic sets in a Polish space can be characterised as continuous images of another Polish space. A metric space is compact if and only if it is complete and totally bounded, i.e. for any $\varepsilon > 0$ the space can be covered by a finite number of balls of radius ε. A *ball* of radius $r \geq 0$ centred at $x \in \mathbb{E}$ is denoted by

$$B_r(x) = \{y \in \mathbb{E} : \rho(x, y) \leq r\}.$$

A set G in a metric space is open if and only if, for each $x \in G$, there exists $r > 0$ such that $B_r(x) \subset G$. Furthermore, x_n converges to x as $n \to \infty$ if and only if $\rho(x_n, x) \to 0$ as $n \to \infty$. A *separable* metric space always has a countable base, so is also second countable.

A *distance* from a point x to a non-empty set A is defined as

$$\rho(x, A) = \inf\{\rho(x, y) : y \in A\}.$$

The union of $B_r(x)$ for $x \in A$ is said to be the *r-envelope* (or parallel set) of A and denoted by A^r. see Figure A.1. The r-envelope of A is alternatively defined as

$$A^r = \{x \in \mathbb{E} : B_r(x) \cap A \neq \emptyset\} = \{x \in \mathbb{E} : \rho(x, A) \leq r\}. \quad \text{(A.1)}$$

The *open r-envelope* of A is defined by

$$A^{r-} = \{x \in \mathbb{E} : \rho(x, A) < r\} \quad \text{(A.2)}$$

and the *inner parallel* set by

$$A^{-r} = \{x \in \mathbb{E} : B_r(x) \subset A\}.$$

The *diameter* of A is defined by

$$\mathrm{diam}(A) = \sup\{\rho(x, y) : x, y \in A\},$$

A Topological spaces and linear spaces

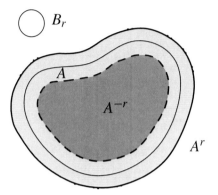

Figure A.1. Outer and inner parallel sets.

and A is called *bounded* if $\text{diam}(A)$ is finite. If $\mathbb{E}$ is σ-compact, then all bounded sets are relatively compact. By $\text{nc}(A)$ we denote the Kuratowski measure of non-compactness of $A \subset \mathbb{E}$, which is equal to the smallest $\varepsilon > 0$ such that A admits a finite covering of sets of diameter less than or equal to ε.

A function f which maps a metric space $(\mathbb{E}, \rho)$ into another metric space $(\mathbb{E}', \rho')$ is said to be *Lipschitz* if there exists a constant $c > 0$ such that $\rho'(f(x), f(y)) \leq c\rho(x, y)$ for every $x, y \in \mathbb{E}$. A subset $F \subset \mathbb{E}$ is said to be *d-rectifiable* if F is an image of a bounded subset of $\mathbb{R}^d$ under a Lipschitz mapping.

Semicontinuity

A function $f : \mathbb{E} \mapsto \bar{\mathbb{R}}$ defined on a topological space $\mathbb{E}$ with values in the extended real line $\bar{\mathbb{R}} = [-\infty, \infty]$ is called *upper semicontinuous* at $x \in \mathbb{E}$ if

$$\limsup_{y \to x} f(y) \leq f(x),$$

and *lower semicontinuous* if

$$\liminf_{y \to x} f(y) \geq f(x).$$

Furthermore, f is said to be upper (lower) semicontinuous if it is upper (lower) semicontinuous at every $x \in \mathbb{E}$. The family of all upper semicontinuous functions on $\mathbb{E}$ is denoted by $\text{USC}(\mathbb{E})$. It is easy to see that the *indicator function*

$$\mathbf{1}_A(x) = \mathbf{1}_{x \in A} = \begin{cases} 1, & x \in A, \\ 0, & \text{otherwise}, \end{cases}$$

is upper semicontinuous if and only if A is closed. The following proposition introduces a number of important concepts related to semicontinuous functions, establishes their equivalence and fixes notation.

Proposition A.2 (Semicontinuous functions). Let $f : \mathbb{E} \mapsto \mathbb{R}$ be a real-valued function. Then the following statements are equivalent.
(U1) f is upper semicontinuous.
(U2) The hypograph
$$\text{hypo } f = \{(x,t) : t \leq f(x)\}$$
is closed in $\mathbb{E} \times \mathbb{R}$.
(U3) For each $t \in \mathbb{R}$, the upper excursion set
$$\{f \geq t\} = \{x \in \mathbb{E} : f(x) \geq t\}$$
is closed.
(U4) $f^{-1}((-\infty, t))$ is open for all $t \in \mathbb{R}$.
Furthermore, all statements from the following group are equivalent
(L1) f is lower semicontinuous.
(L2) The epigraph
$$\text{epi } f = \{(x,t) : t \geq f(x)\}$$
is closed in $\mathbb{E} \times \mathbb{R}$.
(L3) For each $t \in \mathbb{R}$, the lower excursion set
$$\{f \leq t\} = \{x \in \mathbb{E} : f(x) \leq t\}$$
is closed.
(L4) $f^{-1}((t, \infty))$ is open for all $t \in \mathbb{R}$.

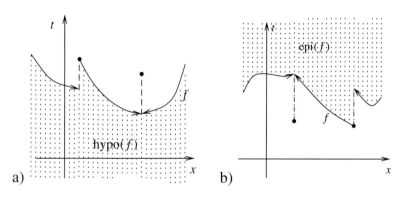

Figure A.2. Examples of upper semicontinuous (a) and lower semicontinuous (b) functions.

It is possible to characterise hypographs of upper semicontinuous functions as closed subsets of the product space $\mathbb{E} \times \mathbb{I}$, where $\mathbb{I}$ is the extended real line topologised in such a way that only non-trivial open sets are $(x, \infty]$, see Vervaat [572]. The studies of random closed sets in non-Hausdorff spaces are motivated by the fact that the corresponding product space is not Hausdorff.

A Topological spaces and linear spaces

Projective systems

Let $\{A_n, n \geq 1\}$ be a sequence of non-empty sets and let $\{u_{m,n}\}$ be a two-parametric sequence of functions such that $u_{m,n} : A_n \mapsto A_m$ for all $m \leq n$. Assume that $u_{m,m}$ is the identity map on A_m for every $m \geq 1$ and $u_{m,p}$ is equal to the superposition of $u_{m,n}$ and $u_{n,p}$ for every $m \leq n \leq p$. The sequence $\{A_n, n \geq 1\}$ together with the functions $\{u_{m,n}\}$ is called a *projective system*.

Let $A = A_1 \times A_2 \times \cdots$ be the Cartesian product of the $\{A_n, n \geq 1\}$. By proj_n denote the *projection* from A onto A_n. The subset $A' \subset A$ defined by

$$A' = \{x = (x_1, x_2, \ldots) : \operatorname{proj}_m(x) = u_{m,n}(\operatorname{proj}_n(x)), \ m \leq n\}$$

is called the *projective limit* of the projective system defined above. The definition of the projective limit for a sequence of sets is sometimes useful when dealing with unbounded random closed sets. The following two results describe two important cases when the projective limit is not empty.

Proposition A.3. *If $\{A_n, n \geq 1\}$ are compact topological spaces and $\{u_{m,n}\}$ are continuous for every $m \leq n$, then the projective limit A' is non-empty and compact.*

Theorem A.4 (Mittag–Leffler's theorem). *If $\{A_n, n \geq 1\}$ are complete metric spaces, $\{u_{m,n}\}$ are uniformly continuous for every $m \leq n$ and $u_{n,n+1}(A_{n+1})$ is dense in A_n for every $n \geq 1$, then the projective limit A' is non-empty and $\operatorname{proj}_n(A')$ is dense in A_n for every $n \geq 1$.*

Linear normed spaces

Assume that $\mathbb{E}$ is a *linear normed* space. The linear structure entails $\mathbb{E}$ with two operations: addition and multiplication by scalars (real numbers). The corresponding norm $\|\cdot\|$ satisfies the identification property and the triangle inequality, so that $\rho(x, y) = \|x - y\|$ is a metric on $\mathbb{E}$. If $\mathbb{E}$ is complete with respect to this norm, $\mathbb{E}$ is called *Banach* space. The *norm* of a set $A \subset \mathbb{E}$ is defined by

$$\|A\| = \sup\{\|x\| : x \in A\}. \tag{A.3}$$

Sets with a finite norm are bounded. Clearly, $\|A\| \leq \operatorname{diam}(A) \leq 2\|A\|$.

The *closed linear hull* of a set $A \subset \mathbb{E}$ is the closure of all finite linear combinations of points from A. A subset $A \subset \mathbb{E}$ is called a *linear subspace* if A is closed with respect to addition and multiplication by constants.

A set $F \subset \mathbb{E}$ is called *convex* if $tx + (1-t)y \in F$ for every $x, y \in F$ and $t \in (0, 1)$. The family of closed convex subsets of $\mathbb{E}$ is denoted by $\operatorname{co} \mathcal{F}$. The prefix "co" denotes the convex sets from the chosen family, for instance, $\operatorname{co} \mathcal{K}$ is the family of convex compact sets, $\operatorname{co} \mathcal{K}'$ is the family of non-empty convex compact sets, etc. The space $\mathbb{E}$ is called *locally convex* if each point $x \in \mathbb{E}$ has a convex neighbourhood. The empty set is always regarded as being convex.

If $t_1, \ldots, t_n$ are non-negative numbers that sum to 1, then $t_1 x_1 + \cdots + t_n x_n$ is called a convex combination of $x_1, \ldots, x_n \in \mathbb{E}$. The *convex hull*, co A, of $A \subset \mathbb{E}$ is the set of all finite convex combinations of points from F. If A is closed, its convex hull is not necessarily closed. The *closed convex hull* $\overline{\text{co}}\, A$ is the closure of co A. Then $\overline{\text{co}}\, A$ equals the intersection of all convex closed sets that contain A.

A function $u: \mathbb{E} \mapsto \mathbb{R}$ is said to be a *linear functional* if $u(\alpha x + \beta y) = \alpha u(x) + \beta u(y)$ for all $x, y \in \mathbb{E}$ and real numbers α, β. The family of all linear continuous functionals is denoted by $\mathbb{E}^*$ and is called the *dual* space to $\mathbb{E}$. A generic element of $\mathbb{E}^*$ is often denoted by u. We often write $\langle x, u \rangle$ instead of $u(x)$. The norm of u in $\mathbb{E}^*$ is given by
$$\|u\| = \sup\{\langle x, u \rangle : \|x\| \leq 1\}.$$
The dual space $\mathbb{E}^*$ with this norm is also a linear normed space. A linear space $\mathbb{E}$ is called *reflexive* if the second dual space $(\mathbb{E}^*)^*$ is isomorphic to $\mathbb{E}$, so that there is a bijection between $(\mathbb{E}^*)^*$ and $\mathbb{E}$ which preserves the linear operations and the norm.

The norm on $\mathbb{E}$ generates the corresponding metric and topology. The convergence in this topology is called *strong* convergence. A sequence $\{x_n, n \geq 1\} \subset \mathbb{E}$ is said to converge *weakly* to $x \in \mathbb{E}$ if $\langle x_n, u \rangle \to \langle x, u \rangle$ for every $u \in \mathbb{E}^*$. Both strong and weak topologies generate the corresponding concepts of the closedness, compactness, etc. For example, a set is called strong compact (or compact) if it is compact in the strong topology, a set is called weakly closed if it is closed with respect to the weak topology, etc. The prefix "s–" usually denotes concepts that are related to the strong topology, while "w–" denotes the corresponding concepts for the weak topology, e.g. w–$\lim x_n$ is the weak limit of $\{x_n, n \geq 1\}$.

A function $x \mapsto Ax$ from $\mathbb{E}$ into another linear space $\mathbb{E}'$ is called a *linear operator* if $A(x + y) = Ax + Ay$ and $A(cx) = cAx$ for all $x, y \in \mathbb{E}$ and $c \in \mathbb{R}$. The norm of A is defined by $\|A\| = \sup\{\|Ax\| : \|x\| \leq 1\}$, where the norm $\|Ax\|$ is taken in $\mathbb{E}'$. A linear operator has a finite norm if and only if it is continuous. A linear operator A is called an *isometry* if $\|Ax\| = \|x\|$ for all $x \in \mathbb{E}$.

A linear space $\mathbb{E}$ is called a *Hilbert* space if it is complete and is equipped with an inner product $\langle \cdot, \cdot \rangle$ which is a real valued function defined on $\mathbb{E} \times \mathbb{E}$ such that, for all $x, y, z \in \mathbb{E}$ and real α, β,
(1) $\langle x, y \rangle = \langle y, x \rangle$;
(2) $\langle \alpha x + \beta y, z \rangle = \alpha \langle x, z \rangle + \beta \langle y, z \rangle$;
(3) $\langle x, x \rangle \geq 0$ with equality if and only if $x = 0$.
The corresponding norm is given by $\|x\|^2 = \langle x, x \rangle$. The Riesz representation theorem says that every continuous functional u on Hilbert space $\mathbb{E}$ can be represented as $u(y) = \langle x, y \rangle$ for a unique $x \in \mathbb{E}$. Thus, the space $\mathbb{E}^*$ is isomorphic to $\mathbb{E}$.

Cones

A *cone* is a set $\mathbb{C}$ endowed with an addition and a scalar multiplication. The addition is supposed to be associative and commutative and there is a neutral element 0. For the scalar multiplication the usual associative and distributive properties hold. The cancellation law, stating that $x + y = x + z$ implies $y = z$, however does not hold in general. A (possibly non-convex) cone can be also defined as a subset $\mathbb{C} \subset \mathbb{E}$ such that $\mathbb{C}$ is closed under multiplication by positive scalars.

An *order* on $\mathbb{C}$ is a reflexive transitive relation $\leq$ such that $x \leq y$ implies $x + z \leq y + z$ and $cx \leq cy$ for all $x, y, z \in \mathbb{C}$ and $c \geq 0$. A *sublinear* (respectively *superlinear*) functional on $\mathbb{C}$ is a mapping $p \colon \mathbb{C} \mapsto (-\infty, \infty]$ such that $p(cx) = cp(x)$ and $p(x + y)$ is smaller (respectively greater) than or equal to $p(x) + p(y)$ for every $x, y \in \mathbb{C}$ and $c \geq 0$. The following Hahn–Banach type sandwich theorem is the basis for the duality theory of ordered cones, see Roth [504].

Theorem A.5 (Sandwich theorem). *Let $\mathbb{C}$ be an ordered cone and let p be a sublinear and q a superlinear functional such that $q(x) \leq p(y)$ whenever $x \leq y$ for $x, y \in \mathbb{C}$. Then there exists a monotone linear functional f such that $q \leq f \leq p$.*

Euclidean space

The d-dimensional *Euclidean* space $\mathbb{R}^d = \mathbb{R} \times \cdots \times \mathbb{R}$ consists of d-tuples $x = (x_1, \ldots, x_d)$ (d-dimensional vectors) with the standard coordinatewise addition and the norm

$$\|x\| = \sqrt{x_1^2 + \cdots + x_d^2}.$$

The *origin* 0 has all zero coordinates $0 = (0, \ldots, 0)$. The space $\mathbb{R}^d$ is not compact, but locally compact and also σ-compact, since it can be represented as the union of compact balls $B_n(0)$. In $\mathbb{R}^d$, we write shortly B_r instead of $B_r(0)$ for a ball centred at the origin. Compact sets in $\mathbb{R}^d$ can be characterised as bounded closed sets. The *unit sphere* is denoted by

$$\mathbb{S}^{d-1} = \{u \in \mathbb{R}^d \colon \|u\| = 1\}.$$

The space $\mathbb{R}^d$ is a Hilbert space with the inner product

$$\langle x, y \rangle = \sum_{i=1}^{d} x_i y_i.$$

Linear operators that map $\mathbb{R}^d$ into $\mathbb{R}^d$ are given by matrices. Among them very important are *rigid motions* in $\mathbb{R}^d$: translations, $x \mapsto x + a$, and rotations, $x \mapsto \mathbf{w}x$.

It is possible to define a partial order on $\mathbb{R}^d$. A point $x = (x_1, \ldots, x_d)$ is said to be *lexicographically* smaller than $y = (y_1, \ldots, y_d)$ (notation $x \leq y$) if $x_1 \leq y_1$, or $x_1 = y_1$ and $x_2 \leq y_2$, or $x_1 = y_1, x_2 = y_2$ and $x_3 \leq y_3$, etc.

A *hyperplane* (or $(d-1)$-dimensional affine subspace of $\mathbb{R}^d$) is defined as

$$\mathbb{H}_u(t) = \{x \in \mathbb{R}^d \colon \langle x, u \rangle = t\}$$

for some $u \in \mathbb{S}^{d-1}$ and $t > 0$. Then u is the corresponding normal vector and t is the distance between the hyperplane and the origin. The hyperplane $\mathbb{H}_u(t)$ bounds the closed half-space

$$\mathbb{H}_u^-(t) = \{x \in \mathbb{R}^d \colon \langle x, u \rangle \leq t\}.$$

Minkowski operations

Vector operations with points in a general linear normed space $\mathbb{E}$ induce operations with subsets of $\mathbb{E}$. For any $A \subset \mathbb{E}$ its *dilation* by a real number c (or homothety) is defined by

$$cA = \{cx : x \in A\}.$$

In particular, for $c = -1$ we obtain the *reflection* of A with respect to zero:

$$\check{A} = -A = \{-x : x \in A\}.$$

A set A is said to be *centrally symmetric* if $A = \check{A}$.

For $A, B \subset \mathbb{E}$ define

$$A \oplus B = \{x + y : x \in A, y \in B\}.$$

Then $A \oplus B$ is called the *Minkowski sum* of A and B, see Figure A.3. It is clear that this operation is commutative and associative. By agreement the sum is empty if at least one summand is empty. If A is convex, then $A \oplus A = 2A$, while there are non-convex sets such that $A \oplus A$ is strictly larger than $2A$, for example, this is the case for a two-point set A. The set $\frac{1}{2}(A \oplus \check{A})$ is called the *central symmetrisation* of A and $A \oplus \check{A}$ is called the *difference body* for A. If no ambiguity occurs, we will not use the special sign for the Minkowski addition and write $A + B$ instead of $A \oplus B$ and $A_1 + \cdots + A_n$ instead of $A_1 \oplus \cdots \oplus A_n$. For $x \in \mathbb{E}$ we always write $x + A$ instead of $\{x\} \oplus A$.

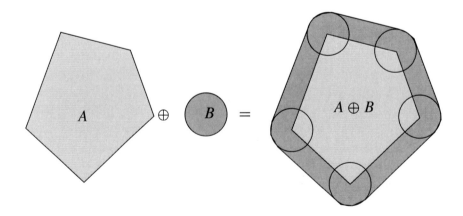

Figure A.3. Minkowski sum of a polygon and a circle.

If both A and B are compact, then $A \oplus B$ is a compact set. If at least one summand is compact and the other is closed, then the sum is closed. However, care should be taken when adding two closed non-compact sets, since their Minkowski sum is not necessarily closed. For example, if $\mathbb{E}=\mathbb{R}^d$, $A=\{(x_1,x_2): x_1>0, x_2=1/x_1\}$ and $B=\{(x_1,0): x_1\leq 0\}$, then $A\oplus B$ is not closed.

The *Minkowski addition* is not an invertible operation in general, so that for A and B it may be impossible to find a set X such that $A\oplus X=B$. For example, if B is a triangle (or any polygon) and A is a ball, then $A\oplus X=B$ is impossible for any X, since $A\oplus X$ smooths out all vertices as shown in Figure A.3. The *Minkowski difference* of two sets is defined by

$$A\ominus B=\{x\in\mathbb{E}: x+\check{B}\subset A\}.$$

This subtraction operation is however not the exact dual to the Minkowski addition, since the inclusion $(A\ominus B)\oplus B\subset A$ can be strict. If A is bounded, then $A\ominus\check{A}=\{0\}$, while this is not generally applicable to unbounded sets.

Some properties of the Minkowski addition and subtraction are listed below. They are formulated for three arbitrary sets $A,B,C\subset\mathbb{E}$. The proofs are easy and can be found in Matheron [381, Sec. 1.5] as well as in a number of other texts devoted to mathematical morphology.

$$(A\ominus B)\ominus C=A\ominus(B\oplus C), \qquad (A\ominus C)\oplus B\subset(A\oplus B)\ominus C,$$
$$A\oplus(B\cup C)=(A\oplus B)\cup(A\oplus C), \qquad A\oplus(B\cap C)\subset(A\oplus B)\cap(A\oplus C),$$
$$A\ominus(B\cup C)=(A\ominus B)\cap(A\ominus C), \qquad A\ominus(B\cap C)\supset(A\ominus B)\cup(A\ominus C),$$
$$(B\cap C)\ominus A=(B\ominus A)\cap(C\ominus A), \qquad (B\cup C)\ominus A\supset(B\ominus A)\cup(C\ominus C).$$

In image processing and mathematical morphology (see Dougherty [146], Heijmans [228], Serra [532])

$$A\oplus\check{B}=\{x: (x+B)\cap A\neq\emptyset\}$$

is called a *dilation* of A by B and

$$A\ominus\check{B}=\{x: (x+B)\subset A\}$$

is called the *erosion* of A by B. In this context the set B is called the *structuring element*. Combinations of dilations and erosions are used to create effective filtering procedures for binary images represented by sets in $\mathbb{R}^d$. For instance, the combination of dilation and erosion $(A\oplus\check{B})\ominus\check{B}$ is called the *closing*, while the reversed combination $(A\ominus\check{B})\oplus\check{B}$ is called the *opening*. Both closing and opening are idempotent operations with the closing being extensive. i.e. $(A\oplus\check{B})\ominus\check{B}\supset A$, and the opening being anti-extensive, i.e. $(A\ominus\check{B})\oplus\check{B}\subset A$, see Matheron [381, Ch. 7].

There are further definitions of difference of sets, which are better adjusted for convex sets, but are less popular that the Minkowski subtraction. Let us describe one such concept called the *Demyanov difference* [507]. For a compact convex set K

$$H(K,u)=\{x\in K: \langle x,u\rangle=\sup_{x\in K}\langle x,u\rangle\}, \quad u\in\mathbb{S}^{d-1},$$

is called the support set (or the face) of K. The Demyanov difference of two convex compact sets K and L is defined as the closed convex hull of all conventional differences $H(K, u) - H(L, u)$ taken for all $u \in \mathbb{S}^{d-1}$ such that $H(K, u)$ and $H(L, u)$ are singletons.

B Space of closed sets

Fell topology

The space $\mathcal{F}$ of closed sets of a topological space is one of the standard objects in set theory and general topology. There are a number of ways to endow the family $\mathcal{F}$ with a structure of a topological space, see Beer [56], Lucchetti and Torre [360] and Sonntag and Zălinescu [539]. Below we describe only several possible topologies on $\mathcal{F}$ leaving the reader to explore the wealth of topologies in the cited references. Note also that the space of closed sets can also be described as a full locally convex cone, see Roth [503].

Topologies on $\mathcal{F}$ are often introduced by describing their sub-bases. The following notation are useful for this purpose. For each $A \subset \mathbb{E}$,

$$\mathcal{F}_A = \{F \in \mathcal{F} : F \cap A \neq \emptyset\}$$

denotes the family of closed sets which have non-empty intersection with A;

$$\mathcal{F}^A = \{F \in \mathcal{F} : F \cap A = \emptyset\}$$

is the family of closed sets which miss A.

Definition B.1. (Vietoris and Fell topologies)
  (i) The *Vietoris* topology has as a sub-base sets $\mathcal{F}_G$ for all $G \in \mathcal{G}$ and $\mathcal{F}^F$ for all $F \in \mathcal{F}$, see Michael [390].
  (ii) The *Fell* topology has a sub-base which consists of $\mathcal{F}_G$ for all $G \in \mathcal{G}$ and $\mathcal{F}^K$ for all $K \in \mathcal{K}$, see Fell [169] and Matheron [381].

The Fell topology is weaker than the Vietoris topology, so that all sequences that converge in the Vietoris topology do converge in the Fell topology. If $\mathbb{E}$ is Hausdorff, then both topologies agree on the family $\mathcal{F}'$ of non-empty closed sets if and only if $\mathbb{E}$ is compact.

The Vietoris topology is also called *exponential* topology, see Kuratowski [337, § 17] and [338, § 42] where related topological results are discussed. For instance, if $\mathbb{E}$ is compact, so is the space $\mathcal{F}$ with the Vietoris topology (and only if provided $\mathbb{E}$ is a T_1-space). The Fell topology is sometimes called the *vague* topology. It can be modified for non-Hausdorff spaces by replacing compact sets with the saturated sets, see Vervaat [572].

Since
$$\mathcal{F}^{K_1} \cap \mathcal{F}^{K_2} = \mathcal{F}^{K_1 \cup K_2},$$
intersections of open sets from the sub-base described in Definition B.1(ii) are given by

$$\mathcal{F}^K_{G_1,\ldots,G_n} = \mathcal{F}^K \cap \mathcal{F}_{G_1} \cap \cdots \cap \mathcal{F}_{G_n}$$
$$= \{F \in \mathcal{F} : F \cap K = \emptyset, F \cap G_1 \neq \emptyset, \ldots, F \cap G_n \neq \emptyset\}$$

for $K \in \mathcal{K}$ and $G_1, \ldots, G_n \in \mathcal{G}, n \geq 0$ (so that none of the G's are present if $n = 0$). Since the family of such sets is closed under finite intersections, they constitute a base of the Fell topology on $\mathcal{F}$. The following results have been proved in Beer [56, Sec. 5.1]. Some of them are contained in Matheron [381, Sec. 1.2] for the case of locally compact $\mathbb{E}$.

Theorem B.2 (Properties of the Fell topology).
  (i) If $\mathbb{E}$ is a Hausdorff space, then $\mathcal{F}$ is compact in the Fell topology and $\mathcal{F}_K$ is compact in $\mathcal{F}$ for every $K \in \mathcal{K}$.
 (ii) If $\mathbb{E}$ is a locally compact Hausdorff space, then $\mathcal{F}$ is a compact Hausdorff space and $\mathcal{F}'$ is a locally compact Hausdorff space.
(iii) If $\mathbb{E}$ is Hausdorff, then $\mathbb{E}$ is locally compact second countable if and only if $\mathcal{F}'$ is locally compact and second countable in the Fell topology (or, equivalently, $\mathcal{F}'$ is a Polish space).
(iv) If $\mathbb{E}$ is Hausdorff and second countable, then $\mathcal{F}'$ is compact in the Fell topology if and only if $\mathbb{E}$ is compact.

If $\mathbb{E}$ is a LCHS space, then the Fell topology is metrisable, see (B.1). If F_n converges to F in the Fell topology, we write $F_n \xrightarrow{F} F$. Note that $F_n \xrightarrow{F} \emptyset$ if and only $F_n \cap K = \emptyset$ for all $K \in \mathcal{K}$ and sufficiently large n, i.e. F_n "escapes" from all compact sets.

Proposition B.3 (see Salinetti and Wets [511]). Let $\mathbb{E}$ be a Polish locally compact space with metric ρ. A sequence of closed sets $\{F_n, n \geq 1\}$ converges to F in the Fell topology if and only if

$$(F_n \setminus F^{r-}) \cup (F \setminus F_n^{r-}) \xrightarrow{F} \emptyset$$

for all $r > 0$, where F^{r-} is the open r-envelope of F, see (A.2).

Painlevé–Kuratowski convergence

Definition B.4 (Lower and upper limits of a sequence of sets). Let $\{A_n, n \geq 1\}$ be a sequence of (not necessarily closed) subsets of $\mathbb{E}$. The lower limit, $\liminf A_n$, consists of points x such that $x_n \to x$ for $x_n \in A_n, n \geq 1$. The upper limit is the set $\limsup A_n$ of all points $x \in \mathbb{E}$ such that $x_{n(k)} \to x$ for $x_{n(k)} \in A_{n(k)}$ and a subsequence $\{n(k), k \geq 1\}$.

The above defined limits differ from the lower and upper set-theoretic limits of a sequence of sets:

$$\text{Liminf} A_n = \bigcup_{n=1}^{\infty} \bigcap_{m \geq n} A_m, \quad \text{Limsup} A_n = \bigcap_{n=1}^{\infty} \bigcup_{m \geq n} A_m,$$

so that the following inclusions hold:

$$\text{Liminf} A_n \subset \liminf A_n, \quad \text{Limsup} A_n \subset \limsup A_n.$$

Definition B.5 (Painlevé–Kuratowski convergence). A sequence $\{A_n, n \geq 1\}$ of subsets of $\mathbb{E}$ is said to converge to A in the Painlevé–Kuratowski sense if $A = \limsup A_n = \liminf A_n$. In this case we write $A_n \xrightarrow{PK} A$ or PK-$\lim A_n = A$.

The following results proved in Beer [56, Th. 5.2.6, 5.2.10] establish relationships between the Painlevé–Kuratowski convergence and the convergence in the Fell topology.

Theorem B.6 (Fell topology and Painlevé–Kuratowski convergence). Let $F \in \mathcal{F}$ and let $\{F_n, n \geq 1\}$ be a sequence of closed sets in a Hausdorff space $\mathbb{E}$.
(i) $F_n \xrightarrow{F} F$ if $F_n \xrightarrow{PK} F$ (and only if in case $\mathbb{E}$ is locally compact).
(ii) If $\mathbb{E}$ is separable, then $F_n \xrightarrow{PK} F$ if and only if $F_n \xrightarrow{F} F$.

The equivalence of the Fell topology and the Painlevé–Kuratowski convergence can be reformulated as follows.

Corollary B.7 (see Matheron [381]). Let $\mathbb{E}$ be LCHS. A sequence $\{F_n, n \geq 1\}$ of closed sets converges to F in the Painlevé–Kuratowski sense if and only if the following two conditions hold:

(F1) if $K \cap F = \emptyset$, $K \in \mathcal{K}$, then $K \cap F_n = \emptyset$ eventually (for all sufficiently large n);
(F2) if $G \cap F \neq \emptyset$ for $G \in \mathcal{G}$, then $G \cap F_n \neq \emptyset$ eventually.

This characterisation of the Painlevé–Kuratowski convergence is helpful to establish the continuity of maps on the space $\mathcal{F}$, see Proposition D.5. However, many other interesting operations with sets are only semicontinuous, see Appendix D.

Example B.8. If $F_n = \{1/n\}$, then $F_n \xrightarrow{PK} F = \{0\}$, but $F_n \cap F = \emptyset$, whence $F_n \cap F$ does not converge to $F \cap F = \{0\}$. Thus, the intersection of sets is not a continuous operation.

Closed sets in Polish spaces

In infinite-dimensional linear spaces the Painlevé–Kuratowski convergence and the Fell topology do not have nice properties and Definition B.5 has to be modified in order to formulate a useful convergence concept. Recall that a normed linear space $\mathbb{E}$ can be equipped with the strong topology generated by the norm and the weak topology induced by the convergence of all linear continuous functionals, see Appendix A.

Definition B.9 (Mosco convergence). A sequence $\{F_n, n \geq 1\}$ of weakly closed sets in a normed linear space $\mathbb{E}$ is said to *Mosco* converge to F (notation $F_n \xrightarrow{M} F$) if
$$\text{w-lim sup } F_n \subset F \subset \text{s-lim inf } F_n,$$
i.e. each $x \in F$ is a strong limit of a sequence $\{x_n, n \geq 1\}$ with $x_n \in F_n, n \geq 1$, and for each subsequence $\{n(k), k \geq 1\}$ and $x_{n(k)}$ from $F_{n(k)}, k \geq 1$, the weak convergence $x_{n(k)} \to x$ implies $x \in F$.

Let w-$\mathcal{F}$ designate the family of weakly closed sets. It is possible to specify the sub-base of the Mosco topology as the families w-$\mathcal{F}_G$ and w-$\mathcal{F}^K$ for strongly open sets G and weakly compact sets K.

Proposition B.10 (see Hess [242]). Let $\{F_n, n \geq 1\}$ be a sequence of closed sets and let $r_k \to \infty$ be an increasing sequence of positive real numbers. Assume that $(F_n \cap B_{r_k}(0)) \xrightarrow{M} F'_k$ for every $k \geq 1$. Then F_n Mosco converges to the (necessarily) closed set $F = \cup_k F'_k$.

The following convergence concept is particularly useful for closed sets in general Polish spaces. Remember that $\rho(x, F)$ is the distance function of F, i.e. the minimum distance from x to F. The distance function of the empty set $F = \emptyset$ identically equals infinity.

Definition B.11 (Wijsman convergence). The *Wijsman* topology on $\mathcal{F}$ is the topology which is determined by the pointwise convergence of the distance functions, so that a sequence of closed sets $\{F_n, n \geq 1\}$ converges to F in the Wijsman topology (notation $F_n \xrightarrow{W} F$) if $\rho(x, F_n) \to \rho(x, F)$ for every $x \in \mathbb{E}$.

Theorem B.12 (see Beer [56, Th. 2.5.4]). If $\mathbb{E}$ is a Polish space, then $\mathcal{F}'$ equipped with the Wijsman topology is also Polish.

Theorem B.13 (see Francaviglia, Lechicki and Levi [180]). If $\{F_n, n \geq 1\}$ is a sequence of closed sets, then
(i) $F_n \xrightarrow{W} F$ implies $F_n \xrightarrow{PK} F$.
(ii) $F \supset \limsup F_n$ if $\liminf \rho(x, F_n) \geq \rho(x, F)$ for all $x \in \mathbb{E}$.
(iii) $F \subset \liminf F_n$ if and only if $\limsup \rho(x, F_n) \leq \rho(x, y)$ for all $x \in \mathbb{E}$.
(iv) The Wijsman and Painlevé–Kuratowski convergence are identical if closed balls in $\mathbb{E}$ are compact.
(v) If the distance function $\rho(x, F_n)$ converges pointwisely to $\rho(x, F)$, then it converges uniformly for x from every compact set K.

It is possible to show that co $\mathcal{F}'$ is a closed subset of $\mathcal{F}'$ in the Wijsman topology. If $\mathbb{E}$ is a separable reflexive Banach space with a Fréchet differentiable norm, then a sequence of closed convex sets converges in Mosco sense if and only if it converges in the Wijsman topology. The space $\mathcal{F}$ is compact in the Wijsman topology if and only if each closed ball in $\mathbb{E}$ is compact, see Lechicki and Levi [347]. The Wijsman topology is metrisable if $\mathbb{E}$ is separable.

The difference between the Wijsman and Painlevé–Kuratowski convergences can be easily explained for sequences of closed sets that converge to an empty set. If F_n converges to $\emptyset$ in the Wijsman topology, F_n eventually escapes from every bounded set, while $F_n \xrightarrow{PK} \emptyset$ means that F_n eventually escapes from every compact set. The Attouch–Wets topology is also related to the convergence of the distance functions.

Definition B.14 (Attouch–Wets convergence). A sequence $\{F_n, n \geq 1\}$ of closed sets is said to converge to F in the *Attouch–Wets* topology if

$$\sup_{x \in B} |\rho(x, F_n) - \rho(x, F)| \to 0 \quad \text{as } n \to \infty$$

for every bounded set B.

The Attouch–Wets topology can be strengthened by assuming that $\rho(x, F_n)$ converges to $\rho(x, F)$ uniformly over all $x \in \mathbb{E}$. This concept leads to the Hausdorff metric between closed sets discussed in Appendix C. Then F_n converges to F in the *Hausdorff metric* (notation $F \xrightarrow{H} F$, ρ_H-$\lim F_n = F$ or $\rho_H(F_n, F) \to 0$) if

$$\rho_H(F_n, F) = \sup_{x \in \mathbb{E}} |\rho(x, F_n) - \rho(x, F)| \to 0 \quad \text{as } n \to \infty.$$

This is a quite restrictive concept for closed non-compact sets, since, for example, bounded closed sets may not converge to unbounded. A variant of this metric

$$\rho_{HB}(F_n, F) = \sup_{x \in \mathbb{E}} e^{-\rho(0,x)} |\rho(x, F_n) - \rho(x, F)| \tag{B.1}$$

called the *Hausdorff–Buseman* metric metrises the Fell topology.

A sequence of (not necessarily closed) sets $\{A_n, n \geq 1\}$ in a Banach space is said to converge *scalarly* to A if $h(A_n, u) \to h(A, u)$ for every $u \in \mathbb{E}^*$, where $h(A_n, u)$ is the support function of A_n, see (F.1). The family of all weakly compact convex sets is a Polish space with respect to the scalar topology. Sometimes (see Papageorgiou [447]) this mode of convergence is unfortunately called a weak convergence, which causes misunderstanding when discussing the weak convergence of random closed sets.

C Compact sets and the Hausdorff metric

Myopic topology

The family $\mathcal{K}$ of compact subsets of $\mathbb{E}$ can be equipped with topologies induced by the topologies on the space $\mathcal{F}$ of closed sets described in Appendix B (assuming that $\mathbb{E}$ is Hausdorff so that compact sets are closed). However, it is more appropriate to endow $\mathcal{K}$ with a topology and a convergence specific to the fact that the sets are compact.

C Compact sets and the Hausdorff metric

Definition C.1 (Myopic topology). The *myopic* (or narrow) topology on $\mathcal{K}$ has the sub-base that consists of

$$\mathcal{K}^F = \{K \in \mathcal{K} : K \cap F = \emptyset\}, \quad F \in \mathcal{F},$$

and

$$\mathcal{K}_G = \{K \in \mathcal{K} : K \cap G \neq \emptyset\}, \quad G \in \mathcal{G}.$$

By comparing the above families with the sub-base of the Fell topology on $\mathcal{F}$ it is easy to see that the myopic topology is stronger than the topology induced on $\mathcal{K}$ by the Fell topology. For example, $K_n = \{0, n\}$ converges in the Fell topology to $\{0\}$, but does not converge in $\mathcal{K}$. This explains the fact that $\mathcal{K}$ with the myopic topology is not a compact space, but only locally compact.

Theorem C.2 (see Matheron [381]). *Let $\mathbb{E}$ be a locally compact Hausdorff space.*
(i) A set $\mathcal{V} \subset \mathcal{K}$ is compact in $\mathcal{K}$ if and only if $\mathcal{V}$ is closed in the Fell topology and there exists $K_0 \in \mathcal{K}$ such that $K \subset K_0$ for all $K \in \mathcal{V}$.
(ii) $\mathcal{K}$ is locally compact in the myopic topology.
(iii) A sequence $\{K_n, n \geq 1\}$ of compact sets myopically converges to $K \in \mathcal{K}$ if $K_n \xrightarrow{PK} K$ and there exists $K_0 \in \mathcal{K}$ such that $K_n \subset K_0$ for all $n \geq 1$.

It is possible to see that $\emptyset$ is an isolated point in $\mathcal{K}$, i.e. no sequence of non-empty compact sets converges to $\emptyset$. If $\mathbb{E}$ is separable (so that $\mathbb{E} = \mathrm{cl}\,\mathbb{Q}$ for a countable set $\mathbb{Q}$), then $\mathcal{K}$ is separable. A countable dense set in $\mathcal{K}$ consists of all finite sets from $\mathbb{Q}$.

Proposition C.3. *Let $\{K_n, n \geq 1\}$ be a sequence of compact sets. If $K_n \downarrow K$ then K_n myopically converges to K. If $K_n \uparrow A$ and $\mathrm{cl}(A) \in \mathcal{K}$, then K_n myopically converges to $\mathrm{cl}(A)$.*

Hausdorff metric

If $\mathbb{E}$ is a metric space, then the myopic topology on the family $\mathcal{K}'$ of non-empty compact sets is metrisable by the *Hausdorff metric*.

Definition C.4 (Hausdorff metric). Let $\mathbb{E}$ be equipped with metric ρ. For each two non-empty compact sets K and L define the *Hausdorff* metric (or Hausdorff distance) between K and L by

$$\rho_H(K, L) = \max\left(\sup_{x \in K} \rho(x, L), \sup_{y \in L} \rho(y, K)\right).$$

The Hausdorff distance can be defined for any pair of bounded sets in a metric space. The following well-known result clarifies the structure of open sets in the topology generated by the Hausdorff metric on $\mathcal{K}$ and the corresponding Borel σ-algebra, see Castaing and Valadier [91, Th. II-6, II-10].

Theorem C.5 (Topology and σ-algebra generated by Hausdorff metric). *Let $\mathbb{E}$ be a metric space.*

(i) For each $G \in \mathcal{G}$, the sets $\{K \in \mathcal{K} : K \subset G\}$ and $\{K \in \mathcal{K} : K \cap G \neq \emptyset\}$ are open in the topology generated by the Hausdorff metric.
(ii) For each $K_0 \in \mathcal{K}$, a basis of neighbourhoods of K_0 in $(\mathcal{K}, \rho_H)$ consists of the sets $\{K \in \mathcal{K} : K \subset G, K \cap G_1 \neq \emptyset, \ldots, K \cap G_n \neq \emptyset\}$ for all $n \geq 0$ and open sets $G, G_1, \ldots, G_n$ such that $G \supset K_0$.
(iii) The Borel σ-algebra $\mathfrak{B}(\mathcal{K})$ generated on $\mathcal{K}$ by the Hausdorff metric coincides with both of the following σ-algebras:
(1) generated by $\{K \in \mathcal{K} : K \subset G\}$ for $G \in \mathcal{G}$;
(2) generated by $\{K \in \mathcal{K} : K \cap G \neq \emptyset\}$ for $G \in \mathcal{G}$.

Proof.
(i) If $K_0 \subset G$, then, by the compactness of K_0,

$$\varepsilon = \inf\{\rho(x, y) : x \in K_0, y \in G^c\} > 0.$$

Then $K \subset G$ for each compact set K with $\rho_H(K, K_0) < \varepsilon$, whence the set $\{K : K \subset G\}$ is open. If $K_0 \cap G \neq \emptyset$, then there exists an open ball centred at $x \in K_0 \cap G$ and contained in G. Therefore $\rho_H(K, K_0) < \varepsilon$ implies $K \cap G \neq \emptyset$, so that $\{K : K \cap G \neq \emptyset\}$ is open.

(ii) Fix a compact set K_0 and $\varepsilon > 0$. Let $G_1, \ldots, G_n$ be open balls of radius $\varepsilon/2$ which cover K_0 and let $G = \{x : \rho(x, K) < \varepsilon\}$. If $K \subset G$ and K hits each of $G_1, \ldots, G_n$, then $\rho_H(K, K_0) < \varepsilon$. This immediately implies (ii).

(iii) By (i), the σ-algebras (1) and (2) are contained in $\mathfrak{B}(\mathcal{K})$. The reverse inclusion would follow from (ii) if we prove that (1) and (2) define identical σ-algebras. Note that $G = \cup_n F_n$ with $F_n = \{x : \rho(x, G^c) \geq n^{-1}\}$. Then

$$\{K : K \cap G \neq \emptyset\} = \bigcup_n \{K : K \cap F_n \neq \emptyset\} = \bigcup_n (\mathcal{K} \setminus \{K : K \subset F_n^c\}).$$

This implies that (2) is included in (1).

Let $G_n = \{x : \rho(x, G^c) < n^{-1}\}$. Then $G^c = \cap_n G_n$ and $\mathcal{K} \setminus \{K : K \subset G\} = \cap_n \{K : K \cap G_n \neq \emptyset\}$. Therefore, (1) is included in (2), so that the σ-algebras (1) and (2) are identical. $\square$

Corollary C.6 (Myopic topology and Hausdorff metric). *The topology on $\mathcal{K}'$ generated by the Hausdorff metric on $\mathcal{K}'$ is equivalent to the myopic topology on $\mathcal{K}'$.*

Taking into account Corollary C.6, we write $\rho_H\text{-}\lim K_n = K$ to denote the myopic convergence of K_n to K in a metric space. It is possible to formulate Corollary C.6 for the whole family $\mathcal{K} = \mathcal{K}' \cup \{\emptyset\}$ if we allow the Hausdorff metric to take infinite values and put $\rho_H(\emptyset, K) = \infty$ for every non-empty K.

Proposition C.7 (Alternative definitions of Hausdorff metric).
(i) For every $K, L \in \mathcal{K}'$,

$$\rho_H(K, L) = \max\{d_H(K, L), d_H(L, K)\}, \tag{C.1}$$

where

$$d_H(K, L) = \inf\{\varepsilon > 0 : K \subset L^\varepsilon\} \tag{C.2}$$

and L^ε is the ε-envelope of L, see (A.1).

(ii) *The Hausdorff distance between K and L equals the uniform distance between their distance functions, i.e.*

$$\rho_H(K, L) = \sup_{x \in \mathbb{E}} |\rho(x, K) - \rho(x, L)|.$$

The following result is widely known for $\mathbb{E}$ being a locally compact space (see Schneider [520, Th. 1.8.2]), while the proof in this generality goes back to Fedorchuk and Filippov [168].

Theorem C.8 (Completeness of $\mathcal{K}'$). *If $\mathbb{E}$ is a complete metric space, then the space $\mathcal{K}'$ of non-empty compact sets is complete in the Hausdorff metric.*

Proof. Let $\{x_n\}$ be a sequence of points from $\mathbb{E}$. Then either $\{x_n\}$ has a convergent subsequence or there exist $\delta > 0$ and a subsequence $\{x_{n(k)}\}$ such that $\rho(x_{n(j)}, x_{n(k)}) \geq \delta$ for $j \neq k$. Indeed, define a sequence $\gamma^0 = \{x_n^0\}$ by $x_n^0 = x_n$ for all $n \geq 1$. Proceed by induction. If $A_n = B_{1/k}(x_n^{k-1}) \cap \gamma^{k-1}$ is infinite for at least one n, then γ^k is a subsequence of γ^{k-1} which consists of points from A_n. If all A_n are finite, then by taking a point from every non-empty $A_n \setminus \bigcup_{j=1}^{n-1} A_j$ we can find a subsequence such that distances between every two points are not less than $\delta = 1/k$. If it is possible to continue this construction until infinity, we get sequences $\gamma^0, \gamma^1, \ldots$ such that $\{x_n^n\}$ is a Cauchy sequence which has a limit, since $\mathbb{E}$ is complete.

Let $\{K_n\}$ be a Cauchy sequence in $\mathcal{K}'$. Put $F_0 = \mathrm{cl}(\bigcup_{n=1}^\infty K_n)$. Each of the sets K_n is separable, so that F_0 is separable and is second countable (because $\mathbb{E}$ is a metric space). Therefore, every open cover of F_0 contains a countable subcover. In view of this, the compactness of F_0 is equivalent to the property that every sequence of points of F_0 has a convergent subsequence.

It suffices to prove that there is no sequence $\{x_n\} \subset F_0$ such that $\rho(x_i, x_j) \geq \delta$ for some $\delta > 0$ and all $i \neq j$. Assume that such a sequence exists. Fix $N \geq 1$ such that $\rho_H(K_m, K_n) \leq \delta/5$ for all $m, n \geq N$. Then $\bigcup_{n=N}^\infty K_n$ is a subset of the closed $\delta/5$-envelope of K_N, so that

$$F_0 \subset K_1 \cup \cdots \cup K_{N-1} \cup (K_N)^{\delta/5}.$$

For each $i \geq 1$ put $y_i = x_i$ if $x_i \in K_1 \cup \cdots \cup K_{N-1}$ and otherwise let y_i be any point of $K_N \cap B_{2\delta/5}(x_i)$. For $i \neq j$,

$$\rho(y_i, y_j) \geq \rho(x_i, x_j) - \rho(x_i, y_i) - \rho(x_j, y_j) > \delta - 2\delta/5 - 2\delta/5 = \delta/5,$$

which means that $\{y_i\}$ does not have a convergent subsequence, contrary to the compactness of $K_1 \cup \cdots \cup K_N$.

For every $n \geq 1$, the set $F_n = \mathrm{cl}(\bigcup_{i=n}^\infty K_i)$ is compact, since F_n is a closed subset of F_0. The decreasing sequence of compact sets $\{F_n\}$ converges in the Hausdorff metric to $K = \bigcap_n F_n$. Hence, for any given $\varepsilon > 0$ there exists n_0 such that $K_n \subset K^\varepsilon$

for every $n \geq n_0$. Since $\{K_n\}$ is a Cauchy sequence, there exists $N \geq n_0$ such that $K_n \subset K_m^\varepsilon$ for every $m, n \geq N$. Thus, $F_n \subset K_m^\varepsilon$, so that $K \subset K_m^\varepsilon$ for every $m \geq N$. This implies $\rho_H(K, K_m) \leq \varepsilon$ for every $m \geq N$, so that K_n converges to K in the Hausdorff metric. □

The following theorem easily follows from the compactness of $\mathcal{F}$, see Theorem B.2.

Theorem C.9 (Relative compactness in $\mathcal{K}'$). *If $\mathbb{E}$ is locally compact, then from each bounded sequence in $\mathcal{K}'$ one can select a convergent subsequence.*

The following result taken from Hansen and Hulse [219] provides a useful condition for the convergence in the Hausdorff metric.

Proposition C.10. *Let $\{K_n, n \geq 1\} \subset \mathcal{V}$, where $\mathcal{V}$ is a closed subset of $\mathcal{K}$. If $\mathrm{cl}(\cup_n K_n)$ is compact and $\{d_H(K_n, V), n \geq 1\}$ converges for each $V \in \mathcal{V}$, then $\{K_n, n \geq 1\}$ converges in the Hausdorff metric.*

Assume that $\mathbb{E}$ is a linear normed space and restrict attention to the family $\mathrm{co}\,\mathcal{K}'$ of convex bodies (non-empty compact convex sets). The statement (ii) of the following theorem is the famous *Blaschke selection theorem* (originally formulated for $\mathbb{E}$ being the Euclidean space $\mathbb{R}^d$).

Theorem C.11 (Relative compactness in $\mathrm{co}\,\mathcal{K}'$). *Let $\mathbb{E}$ be a linear normed space. Then*
(i) $\mathrm{co}\,\mathcal{K}'$ is a closed subset of $\mathcal{K}$.
(ii) If $\mathbb{E}$ is locally compact, then every bounded sequence of convex bodies has a subsequence converging to a convex body.

Proof.
(i) It suffices to show that $\mathcal{K} \setminus \mathrm{co}\,\mathcal{K}'$ is open. Let K be a non-convex compact set. Then there are points $x, y \in K$ and $t \in (0, 1)$ such that $B_\varepsilon(z) \cap K = \emptyset$ for some $\varepsilon > 0$, where $z = tx + (1 - t)y$. Consider an arbitrary $K' \in \mathcal{K}$ such that $\rho_H(K, K') < \varepsilon/2$. Then there are points $x', y' \in K'$ such that $\rho(x, x') < \varepsilon/2$ and $\rho(y, y') < \varepsilon/2$, so that $z' = tx' + (1 - t)y'$ satisfies $\rho(z, z') < \varepsilon/2$. If $z' \in K'$, then there is a point $w \in K$ such that $\rho(w, z') < \varepsilon/2$, which leads to a contradiction $\rho(w, z) < \varepsilon$. Thus every such K' is not convex.
(ii) follows from (i) and Theorem C.9. □

The space $\mathcal{K}$ can be metrised using a number of different metrics. A family of metrics useful in image analysis is obtained by considering $\mathbf{L}^p$ distances between the distance functions.

Definition C.12 (Baddeley's delta-metric). Suppose that $\mathbb{E}$ is equipped with a Radon measure ν which satisfies

$$\inf_{x \in \mathbb{E}} \nu(B_r(x)) > 0$$

for any fixed $r > 0$. Let $w \colon [0, \infty] \mapsto [0, \infty]$ be any bounded concave function with $w(0) = 0$. For $p \in [1, \infty)$ and $K, L \in \mathcal{K}'$ define

$$\Delta_w^p(K, L) = \left(\int_{\mathbb{E}} |w(\rho(x, K)) - w(\rho(x, L))| \nu(\mathrm{d}x) \right)^{1/p}.$$

This metric has been introduced by Baddeley [38], who proved that it generates the myopic topology under the assumption that $\mathbb{E}$ is compact or w is eventually constant, i.e. $w(t) = c$ for all sufficiently large t. Further metrics on the family of compact convex sets can be defined using $\mathbf{L}^p$ distances between their support functions, see Vitale [578].

Convexification

For compact sets in the Euclidean space $\mathbb{R}^d$, it is possible to provide a useful bound for the Hausdorff distance between Minkowski sums of sets and their convex hulls. The corresponding result is known under the name of *Shapley–Folkman–Starr theorem* (or Shapley-Folkman theorem). The proof given below is adapted from Arrow and Hahn [13]. For a compact set K denote its radius by

$$\mathrm{rad}(K) = \inf_x \sup_{y \in K} \|x - y\| = \inf_x \|K - x\|.$$

Clearly, $\mathrm{rad}(K) \leq \|K\|$ and $\mathrm{rad}(K) = \frac{1}{2}\mathrm{diam}(K)$.

Theorem C.13 (Shapley–Folkman–Starr). *For each $K_1, \ldots, K_n \in \mathcal{K}$,*

$$\rho_H(K_1 + \cdots + K_n, \mathrm{co}(K_1 + \cdots + K_n)) \leq \sqrt{d} \max_{1 \leq i \leq n} \mathrm{rad}(K_i). \tag{C.3}$$

Lemma C.14. *For all $K_1, \ldots, K_n \in \mathcal{K}$,*

$$\rho_H(K_1 + \cdots + K_n, \mathrm{co}(K_1 + \cdots + K_n))^2 \leq \sum_{i=1}^{n} (\mathrm{rad}(K_i))^2. \tag{C.4}$$

Proof. We proceed by induction on n. If $n = 1$, then each $x \in \mathrm{co}(K)$ can be represented as $x = \alpha_1 y_1 + \cdots + \alpha_m y_m$ for some $y_1, \ldots, y_m \in K$ and non-negative coefficients $\alpha_1, \ldots, \alpha_m$ with $\alpha_1 + \cdots + \alpha_m = 1$. Let $\mathrm{rad}(K) = \|K - x^*\|$. It is easily seen that

$$0 = \sum_{i=1}^{m} \alpha_i \langle x - y_i, x - x^* \rangle.$$

It is impossible that $\langle x - y_i, x - x^* \rangle > 0$ for all y_i because their weighted sum vanishes. Pick $y = y_i$ such that $\langle x - y, x - x^* \rangle \leq 0$. For such y,

$$(\text{rad}(K))^2 \geq \|x^* - y\|^2 = \|(x - x^*) - (x - y)\|^2$$
$$= \|x - y\|^2 + \|x - x^*\|^2 - 2\langle x - x^*, x - y\rangle$$
$$\geq \|x - y\|^2.$$

Since x is arbitrary, we obtain (C.4) for $n = 1$.

Next, suppose that (C.4) holds for n sets; we will prove it for $n + 1$ sets $K_1, \ldots, K_{n+1}$. Each $x \in \text{co}(K_1 + \cdots + K_{n+1})$ can be represented as the sum $x = x' + x''$ where $x' \in \text{co}(K_1 + \cdots + K_n)$ and $x'' \in \text{co}(K_{n+1})$. By the induction hypothesis, there is $y' \in K_1 + \cdots + K_n$ such that

$$\|x' - y'\|^2 \leq \sum_{i=1}^n (\text{rad}(K_i))^2.$$

Choose $z = z^*$ to minimise $\|x - y' - z\|$ for $z \in \text{co}(K_{n+1})$. Then

$$\|x - y' - z^*\|^2 \leq \|x - y' - x''\|^2 = \|x' - y'\|^2 \leq \sum_{i=1}^n (\text{rad}(K_i))^2.$$

For each $z \in \text{co}(K_{n+1})$ and $0 < t \leq 1$,

$$\|x - y' - z^*\| \leq \|x - y' - (tz + (1 - t)z^*)\|^2$$
$$= \|x - y' - z - t(z - z^*)\|^2$$
$$= \|x - y' - z^*\|^2 - 2t\langle x - y' - z, z - z^*\rangle + t^2\|z - z^*\|^2.$$

Letting $t \downarrow 0$ yields

$$\langle x - y' - z, z - z^*\rangle \leq 0 \tag{C.5}$$

for all $z \in \text{co}(K_{n+1})$. Choose a finite subset $K'_{n+1} \subset K_{n+1}$ and $y'' \in K'_{n+1}$ such that

$$\|y'' - z^*\| \leq (\text{rad}(K'_{n+1}))^2 \leq (\text{rad}(K_{n+1}))^2.$$

Note that $z = z^* + t(y'' - z^*) \in \text{co}(K_{n+1})$ for all sufficiently small $|t|$. Substituting such z into (C.5) we get

$$t\langle x - y' - z^*, y'' - z^*\rangle \leq 0.$$

Since t can be of either sign,

$$\langle x - y' - z^*, y'' - z^*\rangle = 0.$$

Let $y = y' + y''$. Then

$$\|x - y\|^2 = \|(x - y' - z^*) - (y'' - z^*)\|^2$$
$$= \|x - y' - z^*\|^2 - 2\langle x - y' - z^*, y'' - z^*\rangle + \|y'' - z^*\|^2$$
$$\leq \sum_{i=1}^{n+1} (\text{rad}(K_i))^2. \qquad \square$$

Proof of Theorem C.13.. If $n \leq d$, then by (C.4),

$$\rho_H(K_1 + \cdots + K_n, \mathrm{co}(K_1 + \cdots + K_n))^2 \leq \sum (\mathrm{rad}(K_i))^2$$
$$\leq d(\max_{1 \leq i \leq n} \mathrm{rad}(K_i))^2,$$

whence (C.3) immediately follows. If $n \geq d$, decompose

$$x = x' + x'' \in \mathrm{co}(K_1 + \cdots + K_n)$$

so that x' belongs to a convex hull of the sum of at most d sets from $K_1, \ldots, K_n$ and x'' belongs to the sum of all other sets. Such a decomposition exists by [13, Th. 8, Appendix B]. Then Theorem C.13 follows from Lemma C.14 applied to x' and the chosen d sets from $K_1, \ldots, K_n$. □

It is also possible to tighten the bound in (C.3) as follows, see Arrow and Hahn [13, p. 399]. Let

$$\mathrm{rad}_i(K) = \sup_{x \in \mathrm{co}(K)} \inf_{x \in \mathrm{co}(L),\, L \subset K} \mathrm{rad}(L) \qquad (\mathrm{C}.6)$$

denote the inner radius of $K \in \mathcal{K}$. Note that $\mathrm{rad}_i(K) = 0$ if K is convex.

Theorem C.15. *For every $K_1, \ldots, K_n \in \mathcal{K}$,*

$$\rho_H(K_1 + \cdots + K_n, \mathrm{co}(K_1 + \cdots + K_n)) \leq \sqrt{d} \max_{1 \leq i \leq n} \mathrm{rad}_i(K_i).$$

D Multifunctions and semicontinuity

Consider a topological space $\mathbb{E}$ and a set $\mathbb{V}$. A function $F: \mathbb{V} \mapsto \mathcal{F}$ on $\mathbb{V}$ with values in $\mathcal{F}(\mathbb{E})$ is called a closed *set-valued* function or closed-valued *multifunction*. The set $\mathbb{V}$ is often the space Ω of elementary events, the real line or $\mathbb{E}$ itself. The *graph* of F is a subset of $\mathbb{V} \times \mathbb{E}$ defined by

$$\mathrm{Graph}\, F = \{(v, x) : x \in F(v)\}.$$

The *effective domain* of F is

$$\mathrm{dom}\, F = \{v \in \mathbb{V} : F(v) \neq \emptyset\}.$$

The *inverse* of F is a multifunction

$$F^-(A) = \{v \in \mathbb{V} : F(v) \cap A \neq \emptyset\}, \quad A \subset \mathbb{E},$$

which acts on subsets of $\mathbb{E}$.

Assume that both $\mathbb{V}$ and $\mathbb{E}$ are second countable Hausdorff topological spaces. Extending the concept of semicontinuous functions with values in the real line, it is possible to define semicontinuous functions with values in $\mathcal{F}$.

Definition D.1 (Semicontinuous multifunctions). A closed-valued multifunction $F\colon \mathbb{V} \mapsto \mathcal{F}(\mathbb{E})$ is said to be
  (i) upper semicontinuous if $F^{-}(K)$ is closed in $\mathbb{V}$ for all $K \in \mathcal{K}(\mathbb{E})$;
  (ii) lower semicontinuous if $F^{-}(G)$ is open in $\mathbb{V}$ for all $G \in \mathcal{G}(\mathbb{E})$.

In view of Theorem B.6, it is possible to reformulate Definition D.1 in terms of the Painlevé–Kuratowski convergence of sequences of sets.

Proposition D.2. *A multifunction F is lower (respectively upper) semicontinuous if and only if, for every $v \in \mathbb{V}$,*

$$\text{PK-}\liminf_{v' \to v} F(v') \supset F(v) \qquad (\text{respectively } \text{PK-}\limsup_{v' \to v} F(v') \subset F(v) \,).$$

A function $f\colon \mathbb{V} \mapsto \bar{\mathbb{R}} = [-\infty, \infty]$ is upper semicontinuous if and only if $F(v) = [-\infty, f(v)]$ is an upper semicontinuous multifunction and f is lower semicontinuous if and only if $F(v) = [f(v), \infty]$ is a lower semicontinuous multifunction.

Proposition D.3 (see Rockafellar and Wets [498]). *A closed-valued multifunction F is upper semicontinuous if and only if* Graph F *is a closed subset of* $\mathbb{V} \times \mathbb{E}$.

Semicontinuous multifunctions are $(\mathfrak{B}(\mathbb{V}), \mathfrak{B}(\mathcal{F}))$-measurable, where $\mathfrak{B}(\mathbb{V})$ is the Borel σ-algebra on $\mathbb{V}$ and $\mathfrak{B}(\mathcal{F})$ is the Borel σ-algebra generated by the Fell topology.

Definition D.4 (Continuous multifunction). A closed-valued multifunction F is called *continuous* if it is both upper and lower semicontinuous, i.e. F is continuous in the Fell topology.

Applied for bounded single-valued functions, Definition D.4 coincides with the conventional continuity definition, whereas for unbounded functions, it may lead to a different concept. For instance, the function $F(v) = \{v^{-1}\}$ becomes continuous on $\mathbb{V} = \mathbb{R}$ if we put $F(0) = \emptyset$.

It is often useful to put $\mathbb{V} = \mathcal{F}(\mathbb{E})$ or $\mathbb{V} = \mathcal{F}(\mathbb{E}) \times \mathcal{F}(\mathbb{E})$, so that the corresponding closed-valued multifunctions map $\mathcal{F}$ into itself or act on pairs of closed sets. It is useful to list the continuity properties for several such transformations, see Matheron [381].

Proposition D.5 (Continuous maps). *The following maps are continuous.*
  (i) $(F, F') \mapsto (F \cup F')$ *from* $\mathcal{F} \times \mathcal{F}$ *into* $\mathcal{F}$.
  (ii) $F \mapsto cF$ *from* $\mathcal{F}$ *into* $\mathcal{F}$ *and* $K \mapsto cK$ *from* $\mathcal{K}$ *into* $\mathcal{K}$, *where c is any real number.*
  (iii) *The map* $(F, K) \mapsto (F \oplus K)$ *from* $\mathcal{F} \times \mathcal{K}$ *into* $\mathcal{F}$ *and* $(K, L) \mapsto (K \oplus L)$ *from* $\mathcal{K} \times \mathcal{K}$ *into* $\mathcal{K}$.
  (iv) $F \mapsto \check{F}$ *from* $\mathcal{F}$ *into* $\mathcal{F}$ *and* $K \mapsto \check{K}$ *from* $\mathcal{K}$ *into* $\mathcal{K}$.

Proposition D.6 (Semicontinuous maps).
(i) $(F, F') \mapsto F \cap F'$ from $\mathcal{F} \times \mathcal{F}$ into $\mathcal{F}$ is upper semicontinuous.
(ii) $F \mapsto \mathrm{cl}(F^c)$ from $\mathcal{F}$ into $\mathcal{F}$ is lower semicontinuous.
(iii) If $\mathbb{E}$ is locally connected (so that each point has a connected neighbourhood), then $F \mapsto \partial F$ (boundary of F) is a lower semicontinuous map from $\mathcal{F}$ into $\mathcal{F}$.
(iv) $F \mapsto \overline{\mathrm{co}}\,(F)$ (the closed convex hull of F) from $\mathcal{F}$ into $\mathcal{F}$ is lower semicontinuous.
(v) $(F, F') \mapsto \mathrm{cl}(F \oplus F')$ from $\mathcal{F} \times \mathcal{F} \mapsto \mathcal{F}$ is lower semicontinuous (assuming $\mathbb{E}$ is a linear space).
(vi) Closing $(F, K) \mapsto (F \oplus \check{K}) \ominus \check{K}$ and opening $(F, K) \mapsto (F \ominus \check{K}) \oplus \check{K}$ are upper semicontinuous maps from $\mathcal{F} \times \mathcal{K}'$ into $\mathcal{F}$.

It is possible to define and characterise semicontinuous functions on $\mathcal{K}$ as it has been done for the space $\mathcal{F}$. The following useful result (see Matheron [381, Prop. 1.4.2]) deals with *increasing* real-valued functionals $T : \mathcal{K} \mapsto \mathbb{R}$. Note that T is increasing if $T(K_1) \leq T(K_2)$ for $K_1 \subset K_2$.

Proposition D.7 (Semicontinuity of increasing functionals). *Let $\mathbb{E}$ be LCHS. An increasing map $T : \mathcal{K} \mapsto \mathbb{R}$ is upper semicontinuous if and only if $T(K_n) \downarrow T(K)$ as $K_n \downarrow K$ and $n \to \infty$.*

Proof. Clearly, $K_n \downarrow K$ yields $T(K_n) \downarrow T(K)$ if T is upper semicontinuous. For the converse statement, assume that K_n myopically converges to K. Then $K = \cap_n \tilde{K}_n$, where $\tilde{K}_n$ is the closure of $\cup_{i \geq n} K_i$. By Theorem C.2(iii), the sequence $\{K_n\}$ is bounded, so that $\tilde{K}_n$ is compact. Then $\tilde{K}_n \downarrow K$, whence $T(\tilde{K}_n) \downarrow T(K)$. By the monotonicity of T,

$$\limsup T(K_n) \leq \limsup T(\tilde{K}_n) = T(K)\,. \qquad \square$$

Some maps are semicontinuous on $\mathcal{K}$ but not on $\mathcal{F}$. For instance, if $\mathbb{E} = \mathbb{R}^d$, then the *Lebesgue measure* $\mathrm{mes}(F)$ is not upper semicontinuous on $\mathcal{F}$, since there exists $F_n \downarrow F$ such that $\mathrm{mes}(F_n)$ does not converge to $\mathrm{mes}(F)$. The following result implies, in particular, that the Lebesgue measure is upper semicontinuous on $\mathcal{K}$.

Proposition D.8 (Semicontinuity of a measure). *Let $\mathbb{E}$ be a LCHS space. Then any σ-finite measure μ is upper semicontinuous on $\mathcal{K}$.*

Proof. By Proposition D.7, it suffices to assume that $K_n \downarrow K$. Then the result follows from the dominated convergence theorem. $\qquad \square$

A wealth of material on multifunctions in Euclidean spaces can be found in Rockafellar and Wets [499], Aubin and Frankowska [30] and Hiriart-Urruty and Lemaréchal [258].

E Measures, probabilities and capacities

Measures

Consider an abstract set Ω with a σ-algebra $\mathfrak{F}$. A *measure* μ is a function defined on $\mathfrak{F}$ with values in $[0, \infty]$ which is σ-additive, i.e.

$$\mu(\cup_{n=1}^{\infty} A_n) = \sum_{n=1}^{\infty} \mu(A_n),$$

where $\{A_n, n \geq 1\}$ is any sequence of disjoint sets from $\mathfrak{F}$. It is possible to consider measures with values in more general spaces than $\mathbb{R}$, for example vector measures or set-valued measures. If μ and λ are two measures on spaces $(\Omega_1, \mathfrak{F}_1)$ and $(\Omega_2, \mathfrak{F}_2)$, then $\mu \otimes \lambda$ denotes the product-measure on $\Omega_1 \times \Omega_2$ with the product σ-algebra $\mathfrak{F}_1 \otimes \mathfrak{F}_2$ generated by $A_1 \times A_2$ for $A_1 \in \mathfrak{F}_1$ and $A_2 \in \mathfrak{F}_2$. If $f: \Omega \mapsto \Omega'$ is a measurable map on $(\Omega, \mathfrak{F}, \mu)$, then $\mu' = f(\mu)$ (or $f \circ \mu$) is the *image* of μ, which is a measure on the σ-algebra on Ω' generated by $\{A \subset \Omega' : f^{-1}(A) \in \mathfrak{F}\}$ and $\mu'(A) = \mu(f^{-1}(A))$.

A measure μ is called *finite* if $\mu(\Omega) < \infty$; μ is σ-finite if Ω can be represented as a countable union of sets of finite measure; μ is *locally finite* if every point $x \in \mathbb{E}$ has a neighbourhood of finite measure. The *Lebesgue integral* of a real-valued function $f: \Omega \mapsto \mathbb{R}$ with respect to μ is denoted by $\int f(\omega) \mu(d\omega)$ or $\int f d\mu$. The *Lebesgue measure* on $\mathbb{R}^d$ is usually denoted by mes or mes$_d$ to specify the dimension. The measure of value 1 concentrated at a point $x \in \mathbb{E}$ is called the *Dirac* measure and denoted by δ_x, so that $\delta_x(A) = 1$ if $x \in A$ and $\delta_x(A) = 0$ otherwise.

A measure μ is *absolutely continuous* with respect to another measure λ defined on the same σ-algebra (notation $\mu \ll \lambda$) if $\mu(A) = 0$ for every $A \in \mathfrak{F}$ such that $\lambda(A) = 0$. The *Radon–Nikodym derivative* (if it exists) is denoted by $d\mu/d\lambda$, so that

$$\mu(A) = \int_A \frac{d\mu}{d\lambda}(x) \lambda(dx).$$

Assume now that $\Omega = \mathbb{E}$ is a topological space. The *support* of a measure μ (notation supp μ) on a topological space $\mathbb{E}$ is the set of all points $x \in \mathbb{E}$ such that $\mu(G) > 0$ for each open set G that contains x. A measure μ on a topological space $\mathbb{E}$ with a σ-algebra $\mathfrak{F}$ is called *Radon* if μ is locally finite and $\mu(B)$ for each $B \in \mathfrak{F}$, can be approximated from below by $\mu(K)$ with K being a compact subset of B. Every locally finite measure on a Polish space is Radon.

Carathéodory's construction

Here we recall *Carathéodory's construction* described in Mattila [382, p. 54 ff.] or Federer [167, pp. 169–170] that produces a measure from a rather general functional on a family of sets. Let $\varphi: \mathcal{M} \mapsto \mathbb{R}$ be a map defined on a family $\mathcal{M}$ of subsets of a metric space $\mathbb{E}$, such that

E Measures, probabilities and capacities

(1) for every $\delta > 0$ there are $A_1, A_2, \ldots \in \mathcal{M}$ such that $\bigcup_{n=1}^{\infty} A_n = \mathbb{E}$ and diam$(A_n) < \delta$;
(2) for every $\delta > 0$ there is $A \in \mathcal{M}$ such that $\varphi(A) \leq \delta$ and diam$(A) \leq \delta$.

For each $\delta > 0$ define the approximating outer measure

$$\bar{\varphi}_\delta(A) = \inf \left\{ \sum_{n=1}^{\infty} \varphi(A_n) \right\}, \quad A \subset \mathbb{E}, \tag{E.1}$$

where the infimum is taken over all countable coverings $\{A_n, n \geq 1\}$ of A by sets $A_n \in \mathcal{M}$ with diam$(A_n) < \delta$. The limit

$$\bar{\varphi}(A) = \sup_{\delta > 0} \bar{\varphi}_\delta(A) = \lim_{\delta \downarrow 0} \bar{\varphi}_\delta(A) \tag{E.2}$$

is a measure on Borel sets. If $\mathcal{M}$ consists of Borel sets then $\bar{\varphi}$ is Borel regular. The process of deriving $\bar{\varphi}$ from φ is called *Carathéodory's construction* and $\bar{\varphi}$ is called Carathéodory's extension of φ. If φ is a restriction of a Radon measure μ onto a family $\mathcal{M}$ that contains all open balls, then $\bar{\varphi} = \mu$, see Baddeley and Molchanov [39]. Furthermore, if $\mathcal{M}'$ is a subclass of $\mathcal{M} \subset \mathcal{G}$ such that $\mathcal{M}'$ is closed under finite unions and all elements of $\mathcal{M}$ are (possibly uncountable) unions of elements of $\mathcal{M}'$ then the Carathéodory' extensions of φ using $\mathcal{M}$ and $\mathcal{M}'$ coincide.

Example E.1 (Hausdorff measures). For $\alpha \in [0, d]$ let

$$\varphi(A) = 2^{-\alpha} \varkappa_\alpha (\text{diam } A)^\alpha$$

where

$$\varkappa_\alpha = \frac{\Gamma(1/2)^\alpha}{\Gamma(\alpha/2 + 1)} = \frac{2\pi^{\alpha/2}}{\alpha \Gamma(\alpha/2)}. \tag{E.3}$$

Note that $\varkappa_k$ is the volume of the k-dimensional unit ball, when $\alpha = k$ is an integer. If $\mathcal{M}$ is the class of all closed sets, or open sets, or convex sets of $\mathbb{R}^d$ then $\bar{\varphi}$ is the α-dimensional *Hausdorff* measure $\mathcal{H}^\alpha$ in $\mathbb{R}^d$. In the case $\alpha = d$ we obtain the Lebesgue measure.

For a set A, the infimum of the values of $\alpha \geq 0$ with $\mathcal{H}^\alpha(A) = 0$ is called the *Hausdorff dimension* of A and is denoted by $\dim_H(A)$.

The measure $\mathcal{H}^{d-1}$ is the surface area for "sufficiently smooth" (rectifiable) surfaces. In particular, the surface area measure on the unit sphere is denoted by ν_d and

$$\omega_d = \nu_d(\mathbb{S}^{d-1}) = \mathcal{H}^{d-1}(\mathbb{S}^{d-1}) = d\varkappa_d = \frac{2\pi^{d/2}}{\Gamma(d/2)}$$

is the surface area of the unit sphere in $\mathbb{R}^d$. The measure ν_d is the unique (up to proportionality) measure on $\mathbb{S}^{d-1}$ which is invariant under rotations; it is called the *Haar* measure.

Example E.2 (Counting measure). Let $\varphi(A)$ be identically equal to 1. Then $\bar{\varphi}$ is the *counting* measure, i.e. $\bar{\varphi}(A) = N(A) = \text{card}(A)$ is the cardinality of A if A is finite.

Projective systems and measures

Let $\{A_n, n \geq 1\}$ be a sequence of Hausdorff spaces. Let $u_{m,n}\colon A_n \mapsto A_m$, $m \leq n$, be a family of projection maps, so that $\{A_n\}$ and $\{u_{m,n}\}$ form a projective system. Assume that the projective system has a limit A', see Appendix A. A decreasing projective system of measures is a sequence of Radon measures μ_n on A_n, $n \geq 1$, such that $u_{m,n}(\mu_n) \leq \mu_m$ for $m \leq n$. An exact projective system of measures is a decreasing projective system with the additional condition that $u_{m,n}(\mu_n) = \mu_m$ for $m \leq n$. The following theorem is adapted from Schwartz [529, Th. 21] and formulated for the case of a countably indexed projective system.

Theorem E.3 (Projective systems of measures).
  (i) Given a decreasing projective system of finite Radon measures $\{\mu_n, n \geq 1\}$ on $\{A_n, n \geq 1\}$, there exists a Radon measure μ on A' such that $\mathrm{proj}_n(\mu) \leq \mu_n$ for every $n \geq 1$.
  (ii) Given an exact projective system of Radon probability measures $\{\mu_n, n \geq 1\}$ on $\{A_n, n \geq 1\}$, there exists a Radon measure μ on A' such that $\mathrm{proj}_n(\mu) = \mu_n$ for every $n \geq 1$.

Example E.4. Let $\{A_n, n \geq 1\}$ be an increasing sequence of measurable sets and let μ_n be a probability measure on A_n so that $\mu_n(A) = \mu_m(A)$ whenever A is a measurable subset of A_m. Put $\tilde{A}_n = A_1 \times \cdots \times A_n$ and $\tilde{\mu}_n = \mu_1 \times \cdots \times \mu_n$. Then $\{\tilde{A}_n, n \geq 1\}$ and $\{\tilde{\mu}_n, n \geq 1\}$ form an exact projective system with respect to the projection maps $u_{m,n}$. Theorem E.3 establishes the existence of the projective limit μ which can be interpreted as a measure on $\cup_{n \geq 1} A_n$ such that its restriction onto every A_n coincides with μ_n.

Probability measures

If $\mu(\Omega) = 1$, then μ is said to be a *probability* measure, and then we usually write $\mathbf{P}$ instead of μ. The triplet $(\Omega, \mathfrak{F}, \mathbf{P})$ is called the *probability space*. The σ-algebra $\mathfrak{F}$ and also the probability space are called *complete* if, for every set $A \in \mathfrak{F}$ with $\mathbf{P}(A) = 0$, all subsets of A are contained in $\mathfrak{F}$. The abbreviation a.s. indicates that a certain property holds almost surely (i.e. with probability 1) with respect to the underlying probability measure.

A *random variable* is a real-valued Borel measurable function ξ defined on Ω. The cumulative distribution function of ξ is $F(x) = \mathbf{P}\{\xi \leq x\}$. If $A \in \mathfrak{F}$, then the corresponding indicator random variable is denoted by $\mathbf{1}_A$. The expectation of ξ is denoted by $\mathbf{E}\xi$; ξ is said to be *integrable* if it has a finite expectation. A sequence of random variables $\{\xi_n, n \geq 1\}$ is said to be *uniformly* integrable if

$$\sup_{n \geq 1} \mathbf{E}(\mathbf{1}_{|\xi_n| > c} |\xi_n|) \to 0 \quad \text{as } c \uparrow \infty.$$

An $\mathbb{E}$-valued *random element* ξ is a $(\mathfrak{F}, \mathfrak{B}(\mathbb{E}))$-measurable map from Ω into $\mathbb{E}$. This map induces a probability measure on the family of Borel subsets of $\mathbb{E}$, so that

$\mathbf{P}(B) = \mathbf{P}\{\xi \in B\}$ for $B \in \mathfrak{B}(\mathbb{E})$. Two random elements ξ and η are said to be identically distributed (notation $\xi \stackrel{d}{\sim} \eta$) if $\mathbf{P}\{\xi \in B\} = \mathbf{P}\{\eta \in B\}$ for all $B \in \mathfrak{B}(\mathbb{E})$.

If $\mathbb{E} = \mathbb{R}^d$, then $\xi = (\xi_1, \ldots, \xi_d)$ is a random *vector*. If $\mathbb{E}$ is a Banach space and $\mathbf{E}\|\xi\|$ is finite, then it is possible to define the *Bochner expectation* (or the Bochner integral), $\mathbf{E}\xi$, by approximating ξ with simple random elements. Alternatively, $\mathbf{E}\xi$ can be defined as a (unique) element $x \in \mathbb{E}$ such that $\mathbf{E}\langle\xi, u\rangle = \langle x, u\rangle$ for every u from the dual space $\mathbb{E}^*$. If $A \in \mathfrak{F}$, then $\mathbf{E}(\mathbf{1}_A \xi)$ is the Bochner integral of ξ restricted onto the set A.

A sequence $\{\xi_n, n \geq 1\}$ is said to be an *i.i.d.* sequence if the random elements $\xi_1, \xi_2, \ldots$ are independent and share the same distribution.

The following useful theorem (Dellacherie [131, I-T32]) concerns measurability of projections.

Theorem E.5 (Measurability of projections). *Let* $(\Omega, \mathfrak{F}, \mathbf{P})$ *be a complete probability space and let* $\mathbb{E}$ *be a LCHS space with its Borel* σ*-algebra* $\mathfrak{B}$. *For every* A *from the product* σ*-algebra* $\mathfrak{F} \otimes \mathfrak{B}$, *the projection of* A *onto* Ω *is* $\mathfrak{F}$*-measurable.*

Weak convergence

A sequence of probability measures $\{\mathbf{P}_n, n \geq 1\}$ on $\mathfrak{B}(\mathbb{E})$ *weakly* converges to $\mathbf{P}$ if $\int f\,d\mathbf{P}_n \to \int f\,d\mathbf{P}$ for every bounded, continuous real function f. Equivalently, a sequence of random elements $\{\xi_n, n \geq 1\}$ with distributions $\{\mathbf{P}_n, n \geq 1\}$ is said to converge weakly (or in distribution) to ξ (notation $\xi_n \stackrel{d}{\to} \xi$) if $\mathbf{E}f(\xi_n) \to \mathbf{E}f(\xi)$ for every bounded continuous function f. A set $A \in \mathfrak{B}(\mathbb{E})$ is said to be a $\mathbf{P}$-*continuity* set if $\mathbf{P}(\partial A) = 0$.

Theorem E.6 (see Billingsley [70]). *The following conditions are equivalent.*

(i) $\mathbf{P}_n$ *weakly converges to* $\mathbf{P}$ *(i.e.* $\xi_n \stackrel{d}{\to} \xi$*).*
(ii) $\lim_n \int f\,d\mathbf{P}_n = \int f\,d\mathbf{P}$ *(i.e.* $\lim_n \mathbf{E}f(\xi_n) = \mathbf{E}f(\xi)$*) for all bounded, uniformly continuous real function* f.
(iii) $\limsup_n \mathbf{P}_n(F) \leq \mathbf{P}(F)$ *(i.e.* $\limsup_n \mathbf{P}\{\xi_n \in F\} \leq \mathbf{P}\{\xi \in F\}$*) for all closed* F.
(iv) $\liminf_n \mathbf{P}_n(G) \geq \mathbf{P}(G)$ *(i.e.* $\liminf_n \mathbf{P}\{\xi_n \in G\} \geq \mathbf{P}\{\xi \in G\}$*) for all open* G.
(v) $\lim_n \mathbf{P}_n(A) = \mathbf{P}(A)$ *(i.e.* $\lim_n \mathbf{P}\{\xi_n \in A\} = \mathbf{P}\{\xi \in A\}$*) for all* $\mathbf{P}$*-continuity sets* A.

If $\mathbb{E}$ is a Polish space and $\mathbf{P}_n$ weakly converges to $\mathbf{P}$, then it is possible to construct $\mathbb{E}$-valued random elements $\{\xi_n, n \geq 1\}$ and ξ with distributions $\{\mathbf{P}_n, n \geq 1\}$ and ξ defined on the same probability space such that ξ_n converges to ξ in probability, i.e. $\mathbf{P}\{\rho(\xi_n, \xi) \geq \varepsilon\} \to 0$ for every $\varepsilon > 0$. This useful tool is due to A.V. Skorohod and is known under the name of a single probability space technique. This technique can be extended as follows.

Proposition E.7 (see Kallenberg [288]). *Let* $f, f_1, f_2, \ldots$ *be measurable maps between Polish spaces* $\mathbb{E}$ *and* $\mathbb{E}'$ *and let* $\xi, \xi_1, \xi_2, \ldots$ *be random elements in* $\mathbb{E}$ *satisfying*

$f_n(\xi_n) \xrightarrow{d} f(\xi)$. Then there exist, on a suitable probability space, random elements $\eta \stackrel{d}{\sim} \xi$ and $\eta_n \stackrel{d}{\sim} \xi_n$, $n \geq 1$, such that $f_n(\eta_n) \to f(\eta)$ a.s.

A subclass $\mathcal{V}$ of $\mathfrak{B}(\mathbb{E})$ is a *convergence determining* class if the convergence $\mathbf{P}_n(A) \to \mathbf{P}(A)$ for all **P**-continuity sets $A \in \mathcal{V}$ entails the weak convergence of $\mathbf{P}_n$ to **P**. In separable spaces, the finite intersections of spheres constitute a convergence-determining class. Further, $\mathcal{V}$ is convergence-determining if $\mathcal{V}$ is closed under finite intersections and each open set in $\mathbb{E}$ is at most a countable union of elements of $\mathcal{V}$. Each convergence-determining class is also a *determining* class. The latter means any two probability measures are identical if they agree on $\mathcal{V}$. The class of closed sets is a determining class as is any other family that generates $\mathfrak{B}(\mathbb{E})$.

Capacities

Let $(\mathbb{E}, \mathcal{E})$ be a *paved space* for which the paving $\mathcal{E}$ is closed under finite unions and intersections.

Definition E.8 (Choquet capacity). A function φ defined on all subsets of $\mathbb{E}$ with values in the extended real line is called a *Choquet capacity* (or Choquet $\mathcal{E}$-capacity) if it satisfies the following properties.
(1) φ is increasing, i.e. $\varphi(A_1) \leq \varphi(A_2)$ if $A_1 \subset A_2$.
(2) $\varphi(A_n) \uparrow \varphi(A)$ for each increasing sequence of sets $A_n \uparrow A$.
(3) $A_n \in \mathcal{E}$, $n \geq 1$, and $A_n \downarrow A$ implies $\varphi(A_n) \downarrow \varphi(A)$.

A set $A \subset \mathbb{E}$ is called *capacitable* (or φ-capacitable) if

$$\varphi(A) = \sup\{\varphi(B) : B \subset A, \ B \in \mathcal{E}_\delta\}.$$

Note that $\mathcal{E}_\delta$ (respectively $\mathcal{E}_\sigma$) is the family of all countable intersections (respectively unions) of sets from $\mathcal{E}$, and $\mathcal{E}_{\sigma\delta}$ is the family of all countable intersections of sets from $\mathcal{E}_\sigma$. The following principal result is due to Choquet [98], see also Meyer [388, Th. III.19]. The idea of the proof is to show that every set from $\mathcal{E}_{\sigma\delta}$ is capacitable and then use a representation of analytic sets as projections of $(\mathcal{E}' \times \mathcal{E})_{\sigma\delta}$-sets, where $(\mathbb{E}', \mathcal{E}')$ is an auxiliary paved space.

Theorem E.9 (Choquet capacitability theorem). *If φ is a Choquet $\mathcal{E}$-capacity, then every $\mathcal{E}$-analytic set is capacitable.*

Among many applications of this capacitability theorem one can mention the proof of the measurability of the first hitting times and a number of other results from the general theory of stochastic processes, see Dellacherie [131].

A fundamental example of a Choquet capacity can be obtained as follows. Let $(\Omega, \mathfrak{F}, \mu)$ be a complete finite measure space. Then $\mu^*(A) = \inf\{\mu(B) : A \subset B, \ B \in \mathfrak{F}\}$ is the corresponding outer measure defined for all $A \subset \mathbb{E}$. Let $\mathcal{E}$ be the family of finite unions of subsets of $\mathbb{R}_+ \times \Omega$ of the form $C \times B$, where $C \in \mathcal{K}(\mathbb{R}_+)$ and $B \in \mathfrak{F}$. If $A \subset \mathbb{R}_+ \times \Omega$, then $\pi(A)$ denotes the projection of A on Ω. Then $\varphi(A) = \mu^*(\pi(A))$ is a Choquet $\mathcal{E}$-capacity on the product space.

Definition E.10 (Strongly subadditive capacity). A function φ on $\mathcal{E}$ with values in either $[-\infty, \infty)$ or $(-\infty, \infty]$ is called *strongly subadditive* if φ increasing and

$$\varphi(A \cup B) + \varphi(A \cap B) \leq \varphi(A) + \varphi(B) \tag{E.4}$$

for all $A, B \in \mathcal{E}$.

If φ is increasing, then (E.4) is equivalent to

$$\varphi(A \cup B \cup C) + \varphi(C) \leq \varphi(A \cup C) + \varphi(B \cup C) \tag{E.5}$$

for all $A, B, C \in \mathcal{E}$. Under the assumption that φ is increasing, (E.4) is also equivalent to

$$\varphi(\cup_{i=1}^{m} B_i) + \sum_{i=1}^{n} \varphi(A_i) \leq \varphi(\cup_{i=1}^{m} A_i) + \sum_{i=1}^{n} \varphi(B_i) \tag{E.6}$$

for $A_i \subset B_i$, $i = 1, \ldots, n$. If (E.6) holds for countable sequences, then φ is called *countably strongly subadditive*.

Let φ be a non-negative strongly subadditive set function on $\mathcal{E}$ and let $\hat{\mathcal{E}}$ be a subclass of $\mathcal{E}_\sigma$, closed under finite intersections and countable unions. Define

$$\varphi_*(A) = \sup\{\varphi(B) : B \subset A, B \in \mathcal{E}\}, \quad A \in \hat{\mathcal{E}},$$
$$\varphi^*(A) = \inf\{\varphi_*(B) : A \subset B, B \in \hat{\mathcal{E}}\}, \quad A \subset \mathbb{E},$$

where $\inf \emptyset = +\infty$. The following result describes the conditions which guarantee that the extension is well defined.

Theorem E.11 (see Doob [143, Appendix II]). *Let φ be a non-negative strongly subadditive set function on $\mathcal{E}$. If*
(1) $\varphi(A_n) \uparrow \varphi(A)$ *for all* $A_n, A \in \mathcal{E}$, $A_n \uparrow A$;
(2) $\varphi(A_n) \downarrow \varphi^*(A)$ *for all* $A_n \in \mathcal{E}$, $A_n \downarrow A$,
*then φ^* is an extension of φ, it does not depend on the choice of $\hat{\mathcal{E}}$ and φ^* is a Choquet $\mathcal{E}$-capacity, which is countably strongly subadditive on the family of all subsets of $\mathbb{E}$.
If $\varphi^*(A) < \infty$, then A is capacitable if and only if for all $\varepsilon > 0$ there exist $A'_\varepsilon \in \mathcal{E}_\delta$ and $A''_\varepsilon \in \hat{\mathcal{E}}$ such that $A'_\varepsilon \subset A \subset A''_\varepsilon$ and $\varphi^*(A''_\varepsilon) < \varphi^*(A'_\varepsilon) + \varepsilon$.*

Capacities on $\mathcal{K}$

Assume that $\mathbb{E}$ is a LCHS space. A particularly important paving $\mathcal{E} = \mathcal{K}$ consists of all compact subsets of $\mathbb{E}$. Let $\varphi \colon \mathcal{K} \mapsto [0, \infty]$ be a capacity that satisfies the following conditions:

(S1) φ is strongly subadditive on $\mathcal{K}$;
(S2) if $K_n \downarrow K$ with $K \in \mathcal{K}$ and $\{K_n, n \geq 1\} \subset \mathcal{K}$, then $\varphi(K_n) \downarrow \varphi(K)$.

Note that (S2) implies that φ is upper semicontinuous on $\mathcal{K}$ with the myopic topology, see Proposition D.7. If (S2) holds, then φ is called a (topological) precapacity on $\mathbb{E}$. Let $\hat{\mathcal{E}}$ be the family $\mathcal{G}$ of all open subsets which is a subfamily of $\mathcal{E}_\sigma = \mathcal{K}_\sigma$. Theorem E.11 provides an extension of φ:

$$\varphi^*(G) = \sup\{\varphi(K) : K \subset G,\ K \in \mathcal{K}\}, \quad G \in \mathcal{G},$$
$$\varphi^*(A) = \inf\{\varphi^*(G) : A \subset G,\ G \in \mathcal{G}\}, \quad A \subset \mathbb{E}.$$

All analytic sets are capacitable and $\varphi^*(K) = \varphi(K)$ for every $K \in \mathcal{K}$.

The following result provides a continuity condition for a capacity.

Proposition E.12 (Continuity of capacity). *Let $\mathcal{K}^\varepsilon = \{K^\varepsilon : K \in \mathcal{K}\}$ be the family of ε-envelopes of compact sets for some fixed $\varepsilon > 0$. If φ is a Choquet $\mathcal{K}^\varepsilon$-capacity and $\varphi(K) = \varphi(\mathrm{Int}\, K)$ for all $K \in \mathcal{K}^\varepsilon$, then φ is continuous on $\mathcal{K}^\varepsilon$ in the Hausdorff metric.*

Proof. If $K, K_1, K_2, \ldots \in \mathcal{K}^\varepsilon$ and $\rho_\mathrm{H}(K_n, K) \to 0$, then $K^{-\delta_n} \subset K_n \subset K^{\delta_n}$ for a sequence $\delta_n \downarrow 0$, where $K^{-\delta_n}$ is the inner parallel set. By Definition E.8, $\varphi(K^{-\delta_n}) \uparrow \varphi(\mathrm{Int}\, K)$ and $\varphi(K^{\delta_n}) \downarrow \varphi(K)$. Therefore, φ is continuous. $\square$

Propositions D.8 and E.12 yield the following useful fact.

Corollary E.13 (Continuity of measure). *Let $\mathbb{E}$ be a LCHS space.*
(i) *If μ is a σ-finite measure such that $\mu(K) = \mu(\mathrm{Int}\, K)$ for every $K \in \mathcal{K}^\varepsilon$ for some $\varepsilon > 0$, then μ is continuous on $\mathcal{K}^\varepsilon$ in the Hausdorff metric.*
(ii) *If μ is a finite Radon measure, then μ is continuous with respect to the Fell topology on $\mathcal{F}^\varepsilon = \{F^\varepsilon : F \in \mathcal{F}\}$.*

Vague topology

We also use the term *capacity* for any function φ on $\mathcal{K}$ with values in $[0, \infty]$ satisfying the outer continuity condition (S2) above and such that $\varphi(\emptyset) = 0$. The family of all such capacities can be endowed with the *vague* topology generated by the families $\{\varphi : \varphi(K) < t\}$ and $\{\varphi : \varphi(G) > t\}$ where $K \in \mathcal{K}$, $G \in \mathcal{G}$ and $t > 0$. The *narrow* topology is generated by the families $\{\varphi : \varphi(F) < t\}$ and $\{\varphi : \varphi(G) > t\}$ for $F \in \mathcal{F}$, $G \in \mathcal{G}$ and $t > 0$. Then φ_n converges to φ vaguely if and only if

$$\limsup_n \varphi_n(K) \leq \varphi(K), \quad K \in \mathcal{K},$$
$$\liminf_n \varphi_n(G) \geq \varphi(G), \quad G \in \mathcal{G}.$$

The relative compactness for various families of capacities, upper semicontinuous functions and closed sets is studied in O'Brien and Watson [439].

Theorem E.14 (see Norberg [430]). *The family of all capacities in a LCHS space is itself a Polish space in the vague topology.*

Sup-measures

An important particular family of capacities is provided by sup-measures, see Vervaat [572]. Write $\wedge$ instead of the minimum and $\vee$ instead of the maximum of numbers from the extended real line $\bar{\mathbb{R}}$. If φ is a function on $\mathcal{G}$ with values in the extended real line, then its *sup-derivative* is the function $d^\wedge \varphi : \mathbb{E} \mapsto \bar{\mathbb{R}}$ defined by

$$d^\wedge \varphi(x) = \bigwedge_{G \ni x} \varphi(G), \quad x \in \mathbb{E}.$$

The *sup-integral* of a function $f : \mathbb{E} \mapsto \bar{\mathbb{R}}$ is the function $f^\vee : \mathcal{G} \mapsto \bar{\mathbb{R}}$ given by

$$f^\vee(G) = \bigvee_{t \in G} f(t), \quad G \in \mathcal{G}.$$

A function $\varphi : \mathcal{G} \mapsto \bar{\mathbb{R}}$ is called a *sup-measure* if $\varphi(\emptyset) = 0$ and for all families $\{G_j, j \in J\}$ of open sets

$$\varphi(\cup_{i \in J} G_j) = \bigvee_{j \in J} \varphi(G_j).$$

The vague topology induced on the family of measures is called the vague topology for measures; the vague topology induced on the family of sup-measures is called the *sup-vague* topology and is equivalent to the Fell convergence of the hypographs of the corresponding functions. A closed set F gives rise to the indicator sup-measure $\varphi(K) = \mathbf{1}_{K \cap F \neq \emptyset}$, so that the relative vague topology on the family of such sup-measures coincides with the Fell topology on $\mathcal{F}$, see Norberg [430].

The family of all sup-measures can be endowed with a topology using the following construction. Let $\mathcal{B}$ be a family of subsets of $\mathbb{E}$ with the only requirement that $\emptyset \in \mathcal{B}$. The sup-$\mathcal{B}$-topology is the smallest topology that makes the function $\varphi \mapsto \varphi(A)$ upper semicontinuous for every $A \in \mathcal{B}$ and lower semicontinuous for every $A \in \mathcal{G}$. If $\mathcal{B} = \mathcal{F}$, then the sup-$\mathcal{F}$-topology is called the *sup-narrow* topology; the choice $\mathcal{B} = \mathcal{K}$ yields the sup-vague topology.

If $f \in \mathrm{USC}(\mathbb{E})$ is an upper semicontinuous function, then its sup-integral $f^\vee$ is a sup-measure. The sup-$\mathcal{B}$-topology on the family of sup-measures induces the sup-$\mathcal{B}$-topology on $\mathrm{USC}(\mathbb{E})$ defined as the smallest topology that makes $f \mapsto f^\vee(A)$ upper semicontinuous for every $A \in \mathcal{B}$ and lower semicontinuous for every open set A. Then $f_n \to f$ in this topology if and only if

$$\limsup_n f_n^\vee(B) \le f^\vee(B), \quad B \in \mathcal{B},$$
$$\liminf_n f_n^\vee(G) \ge f^\vee(G), \quad G \in \mathcal{G}.$$

An *extremal process* is a random element with values in the family of sup-measures. It is often useful to be able to extend such a random element from its values on a base $\mathcal{G}_0$ of the topology $\mathcal{G}$. Assume that $\mathbb{E}$ is locally compact. If $\zeta(G)$,

$G \in \mathcal{G}_0$, is a stochastic process defined on $\mathcal{G}_0$ such that $\zeta(\cup_{n\geq 1} G_n) = \vee_i \zeta(G_i)$ a.s. for each sequence $\{G_n, n \geq 1\} \subset \mathcal{G}_0$ with $\cup_{n\geq 1} G_n \in \mathcal{G}_0$, then there exists an extremal process $\zeta^*(G)$ defined for $G \in \mathcal{G}$ such that $M(G) = \zeta(G)$ a.s. for every $G \in \mathcal{G}_0$ separately, see Vervaat [572].

Potentials and capacities

Let $k(x, y): \mathbb{R}^d \times \mathbb{R}^d \mapsto [0, \infty]$ be a lower semicontinuous function which is said to be a *kernel*. Then

$$U_\mu^k(x) = \int_K k(x, y) \mu(dy)$$

denotes the *potential* of μ. Assume that k satisfies the *maximum principle*, i.e. $U_\mu^k(x) \leq a$ for all $x \in \operatorname{supp} \mu$ implies this inequality everywhere on $\mathbb{R}^d$. Then the functional

$$C(K) = \sup \left\{ \mu(K): U_\mu^k(x) \leq 1, \, x \in \operatorname{supp} \mu \right\}, \quad K \in \mathcal{K}, \tag{E.7}$$

is a completely alternating Choquet capacity on $\mathcal{K}$, see Choquet [98] and Landkof [343]. Then $C(K) = \mu(K)$ for a measure μ called the *equilibrium measure* and satisfying $U_\mu^k(x) \leq 1$ for $x \in \operatorname{supp} \mu$ and $U_\mu^k(x) \geq 1$ up to a set of zero capacity (approximately everywhere) on K, see Landkof [343, Th. 2.3]. The existence of the equilibrium measure implies that the capacity C is *subadditive* and *upper semicontinuous*, see Landkof [343, p. 141]. It is possible to show that $C(K)$ can be obtained as

$$C(K) = \operatorname{cap}_k(K) = \left[\inf \{ E_k(\mu) : \mu(K) = 1, \, \mu(K^c) = 0 \} \right]^{-1},$$

where the infimum is taken over Radon measures μ and

$$E_k(\mu) = \iint k(x, y) \mu(dx) \mu(dy)$$

denotes the *energy* of μ.

An important family of kernels that produce completely alternating capacities is the family of *Riesz* kernels $k(x, y) = k_{d,\gamma} \|x - y\|^{\gamma - d}$ with

$$k_{d,\gamma} = \frac{\pi^{\gamma - d/2} \Gamma((d - \gamma)/2)}{\Gamma(\gamma/2)}.$$

The corresponding $C = \operatorname{cap}_k$ is called the *Riesz capacity* and is often denoted as $\operatorname{cap}_\gamma$. It is known (see Landkof [343, p. 143]) that $\operatorname{cap}_\gamma$ satisfies the maximum principle and therefore is completely alternating in the case $0 < \gamma \leq 2$ for $d \geq 3$ and $0 < \gamma < 2$ for $d = 2$. In this case the equilibrium potential of μ satisfies $U_\mu^k(x) \leq 1$ everywhere and $U_\mu^k(x) = 1$ approximately everywhere on K.

Note also that the sup-measure $\varphi(K) = \sup_{x \in K} f(x)$ can be obtained using (E.7) with $k(x, y) = 1/\max(f(x), f(y))$.

F Convex sets

Support functions

The *support function* of a closed subset F of a linear space $\mathbb{E}$ is defined on the dual space $\mathbb{E}^*$ as

$$h(F, u) = \sup\{\langle x, u \rangle : x \in F\}, \quad u \in \mathbb{E}^*. \tag{F.1}$$

The support function may take infinite values if F is unbounded. The weak* topology on $\mathbb{E}^*$ is the topology of weak convergence of linear functionals. The support function of a compact set can be characterised as a weak*-continuous function h on the unit ball B_1^* in $\mathbb{E}^*$, which is *subadditive*, i.e.

$$h(u + v) \leq h(u) + h(v) \quad \text{for all } u, v \in B_1^* \text{ with } u + v \in B_1^*; \tag{F.2}$$

and *positively homogeneous*, i.e.

$$h(cu) = ch(u), \quad c > 0, \ u, cu \in B_1^*. \tag{F.3}$$

Both (F.2) and (F.3) are summarised by saying that h is *sublinear*. The family of support functions of compact convex sets is a closed cone in the space of weak*-continuous functions on B_1^*. A singleton $\{x\}$ has as the support function $h(\{x\}, u) = \langle x, u \rangle$, which is linear. Conversely, if $h \colon B_1^* \mapsto \mathbb{R}$ is linear, then h is a support function of a singleton.

Theorem F.1. *Let $\mathbb{E}$ be a linear normed space. For each bounded closed set F, its support function is Lipschitz with Lipschitz constant $\|F\|$, that is*

$$\|h(F, u_1) - h(F, u_2)\| \leq \|F\| \|u_1 - u_2\|, \quad u_1, u_2 \in \mathbb{E}^*. \tag{F.4}$$

Proof. For each u and v

$$h(F, u + v) \leq h(F, u) + h(F, v) \leq h(F, u) + \|F\| \|v\|,$$

whence the Lipschitz property immediately follows. □

Convex sets in $\mathbb{R}^d$

For the rest of Appendix F, assume that $\mathbb{E} = \mathbb{R}^d$. Figure F.1 shows the geometrical meaning of the support function in the planar case. For every closed set F, its support function $h(F, u)$ is defined for all $u \in \mathbb{R}^d$, while (F.3) implies that the support function can be uniquely extended from its values on the unit sphere $\mathbb{S}^{d-1}$. An important theorem establishes that every real-valued sublinear function on $\mathbb{R}^d$ is a support function of the unique convex compact set, see Schneider [520, Th. 1.7.1].

The *supporting plane* of F is defined by

$$\mathbb{H}(F, u) = \{x \in \mathbb{R}^d : \langle x, u \rangle = h(F, u)\}.$$

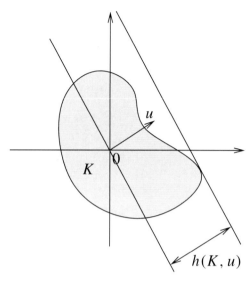

Figure F.1. Support function of a planar compact set.

Every non-empty closed convex set in $\mathbb{R}^d$ is the intersection of its supporting half-spaces and so is uniquely determined by its support function. Furthermore, $H(F, u) = \mathbb{H}(F, u) \cap F$ is called the *support set* of F.

The function
$$w(K, u) = h(K, u) + h(K, -u)$$
is called the *width function* of $K \in \mathcal{K}$. Its mean value

$$\mathsf{b}(K) = \frac{2}{\omega_d} \int_{\mathbb{S}^{d-1}} h(K, u) \mathcal{H}^{d-1}(\mathrm{d}u), \tag{F.5}$$

is called the *mean width* $\mathsf{b}(K)$. The vector-valued integral

$$\mathsf{s}(K) = \frac{1}{\varkappa_d} \int_{\mathbb{S}^{d-1}} h(K, u) u \mathcal{H}^{d-1}(\mathrm{d}u) \tag{F.6}$$

defines the *Steiner point* $\mathsf{s}(K)$ of K. For every non-empty convex compact set K, we have $\mathsf{s}(K) \in K$. Moreover, the Steiner point belongs to the relative interior of K taken in the topology induced on the smallest affine subspace that contains K.

If $\mathbb{E}$ is a linear space, a function $f : \mathbb{E} \mapsto (-\infty, \infty]$ is *convex* if f is not identically equal to infinity and if

$$f((1-t)x + ty) \le (1-t)f(x) + tf(y)$$

for all $x, y \in \mathbb{E}$ and for $0 \le t \le 1$. A function f is convex if and only if its epigraph epi f is a non-trivial convex subset of $\mathbb{E} \times (-\infty, \infty]$.

For convex functions on $\mathbb{R}^d$ the concepts of gradient and differential have natural extensions. If f is a convex function, then

$$\partial f(x) = \{v \in \mathbb{R}^d : f(y) \geq f(x) + \langle v, y - x \rangle \text{ for all } y \in \mathbb{R}^d\}$$

is a convex closed set called the *subdifferential* of f at x; each element of $\partial f(x)$ is called a *subgradient* of f. If $f(u) = h(F, u)$ is a support function of a convex closed set F which is distinct from the whole space, then $\partial f(0) = \partial h(F, 0) = F$. If $K \in \text{co}\,\mathcal{K}$ and $u \in \mathbb{R}^d \setminus \{0\}$, then the subdifferential $\partial h(K, u)$ of the support function at u coincides with the support set $H(K, u)$.

Consider $K \in \text{co}\,\mathcal{K}$ such that $0 \in \text{Int}\,K$. The set

$$K^\circ = \{x \in \mathbb{R}^d : \langle x, y \rangle \leq 1, \ y \in K\}$$

is called the *polar* set to K (in a general linear space, the polar is a subset of the dual space). If f is a convex function on $\mathbb{R}^d$, then its *conjugate* function is defined by

$$f^\circ(v) = \sup\{\langle v, x \rangle - f(x) : x \in \mathbb{R}^d\}, \quad v \in \mathbb{R}^d.$$

The mapping $f \mapsto f^\circ$ is called the Legendre–Fenchel transform, see Rockafellar and Wets [499, Sec. 11A].

Intrinsic volumes

While intersections of convex sets are convex, this is clearly not the case for their unions. The family of finite unions of compact convex sets in $\mathbb{R}^d$ is called the *convex ring* and is denoted by $\mathcal{R}$. The *extended* convex ring $\bar{\mathcal{R}}$ consists of all sets $F \in \mathcal{F}$ such that $F \cap K \in \mathcal{R}$ for every $K \in \text{co}\,\mathcal{K}$. A numerical function φ on $\mathcal{R}$ is called a *valuation* if

$$\varphi(K \cup L) + \varphi(K \cap L) = \varphi(K) + \varphi(L), \quad K, L \in \mathcal{R}.$$

A particularly important valuation assigns 1 to every convex non-empty set. It extends uniquely to a valuation on the convex ring; this extension is called the *Euler–Poincaré characteristic* and denoted by $\chi(K)$ for $K \in \mathcal{R}$.

Important functionals on the family co $\mathcal{K}'$ of non-empty convex compact sets in $\mathbb{R}^d$ are the *intrinsic volumes* $\mathbf{V}_j(K)$, $0 \leq j \leq d$, which can be defined by means of the *Steiner formula*

$$\text{mes}(K^r) = \sum_{j=0}^{d} r^{d-j} \varkappa_{d-j} \mathbf{V}_j(K), \tag{F.7}$$

where K^r is the r-envelope of K and $\varkappa_{d-j}$ is the volume of the $(d-j)$-dimensional unit ball given by (E.3), see Schneider [520, p. 210] and Schneider and Weil [522, pp. 38–39]. Then $\mathbf{V}_0(K) = 1$, $\mathbf{V}_d(K) = \mathbf{V}(K) = \text{mes}(K)$ is the volume or the d-dimensional Lebesgue measure, $\mathbf{V}_{d-1}(K)$ equals half of the surface area of K and $\mathbf{V}_1(K)$ is proportional to the mean width of K, more precisely,

$$V_1(K) = \omega_d b(K)/(2\varkappa_{d-1}). \tag{F.8}$$

Other functionals $V_{d-2}(K), \ldots, V_2(K)$ can be found as integrals of curvature functions (if K is smooth). The intrinsic volumes are equal up to proportionality constants to the so-called quermassintegrals, see Leichtweiss [348].

It is possible to write down a local variant of the Steiner formula. For every $x \notin K$ and $K \in \text{co}\,\mathcal{K}'$ let $p(x, K)$ be the uniquely determined nearest point to x in K. The local parallel set to K is the set $M_r(K, B \times A)$ of all points $x \in \mathbb{R}^d$ such that $\rho(x, K) \leq r$, $p(x, K)$ belongs to $B \in \mathfrak{B}(\mathbb{R}^d)$ and the unit normal vector to K at $p(x, K)$ belongs to a Borel set A in $\mathbb{S}^{d-1}$. The *local* Steiner formula establishes that $\text{mes}(M_r(K, B \times A))$ is a polynomial given by

$$\text{mes}(M_r(K, B \times A)) = \frac{1}{d} \sum_{i=0}^{d} r^{d-i} \binom{n}{i} \Theta_i(K, B \times A),$$

where $\Theta_i(K, \cdot)$ can be extended to a measure on $\mathfrak{B}(\mathbb{R}^d \times \mathbb{S}^{d-1})$ called the ith *generalised curvature* measure. In particular, $C_i(K, B) = \Theta_i(K, B \times \mathbb{S}^{d-1})$, $B \in \mathfrak{B}(\mathbb{R}^d)$, is called the *curvature* measure and $S_i(K, A) = \Theta_i(K, \mathbb{R}^d \times A)$, $A \in \mathfrak{B}(\mathbb{S}^{d-1})$, is the *area* measure of K. Being normalised by $(d\varkappa_{d-i})^{-1}\binom{n}{i}$, their total values are equal to the value of the corresponding intrinsic volume $V_i(K)$.

The intrinsic volumes can be extended to valuations on the convex ring, see also Schneider [520] and Weil [603]. If K is the union of convex sets $K_1, \ldots, K_n$, then the *additive* extension

$$V_j(K) = \sum_i V_j(K_i) - \sum_{i_1 < i_2} V_j(K_{i_1} \cap K_{i_2}) + \cdots$$
$$+ (-1)^{n+1} V_j(K_1 \cap \cdots \cap K_n)$$

is defined by the usual inclusion-exclusion formula. Then V_d and V_{d-1} retain their geometrical meanings and remain positive, while other functionals can become negative, in particular $V_0(K) = \chi(K)$ is the Euler–Poincaré characteristic of K. The curvature can also be extended by additivity for $K \in \mathcal{R}$. A further extension to general compact sets is described in Hug, Last and Weil [269].

The *positive* extension of the intrinsic volumes onto the convex ring can be defined as follows, see Weil [603]. For $x \in \mathbb{R}^d$ and K from the convex ring we call a point $y \in K$ a metric projection of x onto K if there is a neighbourhood U of y such that $\|x - z\| > \|x - y\|$ for all $z \in U \cap K$, $z \neq y$, see Figure F.2.

Let $\bar{c}_r(K, x)$ be the number of projections y of x with $\|y - x\| \leq r$ for $r > 0$. The expansion

$$\int_{\mathbb{R}^d} \bar{c}_r(K, x) dx = \sum_{j=0}^{d} r^{d-j} b_{d-j} \bar{V}_j(K)$$

defines the positive extensions of the intrinsic volumes. If K is convex, then the left-hand side equals the volume of K^r, whence $\bar{V}_j = V_j$ for convex compact sets.

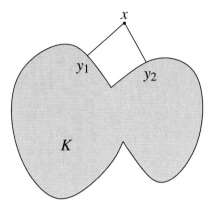

Figure F.2. Two metric projections of x onto K.

An important theorem of convex geometry yields an expression of the volume for Minkowski combinations of convex compact sets $K_1, \ldots, K_n$ as

$$V(\lambda_1 K_1 + \cdots + \lambda_n K_n) = \sum_{i_1=1}^{m} \cdots \sum_{i_d=1}^{m} \lambda_{i_1} \cdots \lambda_{i_d} V(K_{i_1}, \ldots, K_{i_d}) \qquad (F.9)$$

valid for all $\lambda_1, \ldots, \lambda_n \geq 0$. The coefficients $V(K_{i_1}, \ldots, K_{i_d})$ are called *mixed volumes*.

A substantial amount of information on convex sets and integral geometry can be found in a monograph by Schneider [520] and in the handbook by Gruber and Wills [214]. Geometric measure theory and rectifiable sets are treated in detail by Federer [167] and Mattila [382].

G Semigroups and harmonic analysis

Let S be an arbitrary abelian *semigroup*, i.e. S is a set equipped with a commutative and associative binary operation called addition and denoted $+$. Assume the existence of a neutral element denoted by 0.

Definition G.1 (Semicharacter). A *semicharacter* on S is a function $\chi : S \mapsto [-1, 1]$ satisfying $\chi(0) = 1$ and $\chi(s + t) = \chi(s)\chi(t)$ for all $s, t \in S$.

The set of all semicharacters on S is denoted $\hat{S}$, which itself is an abelian semigroup under pointwise multiplication and with the neutral element being the semicharacter identically equal to 1.

Definition G.2 (Positive and negative definiteness).
(i) A real-valued function $f: \mathsf{S} \mapsto \mathbb{R}$ is called *positive definite* (notation $f \in \mathsf{P}(\mathsf{S})$) if f is bounded and the matrix $\|f(s_i + s_j)\|_{i,j=1,\ldots,n}$ is positive semidefinite for every $n \geq 1$ and for each n-tuple $s_1, \ldots, s_n \in \mathsf{S}$.
(ii) A function $f: \mathsf{S} \mapsto \mathbb{R}_+ = [0, \infty)$ is called *negative definite* (notation $f \in \mathsf{N}(\mathsf{S})$) if the matrix $\|f(s_i) + f(s_j) - f(s_i + s_j)\|_{i,j=1,\ldots,n}$ is positive semidefinite for every $n \geq 1$ and for each n-tuple $s_1, \ldots, s_n \in \mathsf{S}$.

The following results provide representations of positive definite and negative definite functions on semigroups.

Theorem G.3 (Representaion of positive definite functions). *For every $f \in \mathsf{P}(\mathsf{S})$ there exists one and only one positive Radon measure μ on $\hat{\mathsf{S}}$ such that*

$$f(s) = \int_{\hat{\mathsf{S}}} \chi(s) \mu(d\chi), \quad s \in \mathsf{S}.$$

Theorem G.4 (Representation of negative definite functions). *Let $f \in \mathsf{N}(\mathsf{S})$. Then*

$$f(s) = c + h(s) + \int_{\hat{\mathsf{S}} \setminus \{1\}} (1 - \chi(s)) \mu(d\chi), \quad (G.1)$$

where f uniquely determines: a non-negative constant c, a function $h: \mathsf{S} \mapsto \mathbb{R}_+$ satisfying $h(s + t) = h(s) + h(t)$ for all $s, t \in \mathsf{S}$ and a non-negative Radon measure μ on $\hat{\mathsf{S}} \setminus \{1\}$ with the property

$$\int_{\hat{\mathsf{S}} \setminus \{1\}} (1 - \chi(s)) \mu(d\chi) < \infty, \quad s \in \mathsf{S}.$$

The measure μ in (G.1) is said to be the *Lévy–Khinchin* measure of f. Note that h vanishes if f is bounded.

For a function $f: \mathsf{S} \mapsto \mathbb{R}$ introduce the successive differences as

$$\Delta_{s_1} f(s) = f(s) - f(s + s_1),$$
$$\Delta_{s_n} \cdots \Delta_{s_1} f(s) = \Delta_{s_{n-1}} \cdots \Delta_{s_1} f(s) - \Delta_{s_{n-1}} \cdots \Delta_{s_1} f(s + s_n), \quad n \geq 2,$$

where $s, s_1, \ldots, s_n \in \mathsf{S}$.

Definition G.5 (Complete alternation and monotonicity). A function $f: \mathsf{S} \mapsto \mathbb{R}_+$ is said to be *completely alternating* (notation $f \in \mathsf{A}(\mathsf{S})$) if $\Delta_{s_n} \cdots \Delta_{s_1} f(s) \leq 0$ and *completely monotone* (notation $f \in \mathsf{M}(\mathsf{S})$) if $\Delta_{s_n} \cdots \Delta_{s_1} f(s) \geq 0$ for all $n \geq 1$ and $s, s_1, \ldots, s_n \in \mathsf{S}$.

Theorem G.6 ($\mathsf{M}(\mathsf{S})$ and $\mathsf{A}(\mathsf{S})$ as cones). *$\mathsf{M}(\mathsf{S})$ is an extreme sub-cone of the family $\mathsf{P}(\mathsf{S})$ of positive definite functions on S; $\mathsf{A}(\mathsf{S})$ is an extreme sub-cone of the family $\mathsf{N}(\mathsf{S})$ of all negative definite functions.*

The following is a useful characterisation of completely alternating functions.

Theorem G.7 (Completely alternating functions). *Each function $f \in \mathsf{A}(\mathsf{S})$ admits the representation (G.1) with the Lévy–Khinchin measure μ concentrated on the set $(\hat{\mathsf{S}} \setminus \{1\})_+$ of non-negative semicharacters from $\hat{\mathsf{S}} \setminus \{1\}$.*

Theorem G.8. *Let f be a non-negative function on S. Then $f \in \mathsf{A}(\mathsf{S})$ if and only if $e^{-tf} \in \mathsf{M}(\mathsf{S})$ for all $t > 0$.*

A function $f \in \mathsf{P}_1(\mathsf{S}) = \{f \in \mathsf{P}(\mathsf{S}) : f(0) = 1\}$ is called *infinitely divisible* if and only if for each $n \geq 1$ there exists $f_n \in \mathsf{P}_1(\mathsf{S})$ such that $f = (f_n)^n$. The family of all infinitely divisible functions is denoted by $\mathsf{P}^i(\mathsf{S})$. Let $\mathsf{N}_\infty(\mathsf{S})$ be the closure of $\{f \in \mathsf{N}(\mathsf{S}) : f(0) = 0\}$ with respect to the pointwise convergence. Note that functions from $\mathsf{N}_\infty(\mathsf{S})$ may attain infinite values. The following theorem characterises logarithms of infinitely divisible functions.

Theorem G.9 (Infinitely divisible functions). *Let f be a non-negative function in $\mathsf{P}_1(\mathsf{S})$ and let $\varphi = -\log f$. Then the following conditions are equivalent.*
(i) $f \in \mathsf{P}^i(\mathsf{S})$.
(ii) $\varphi \in \mathsf{N}_\infty(\mathsf{S})$.
(iii) $e^{-t\varphi} \in \mathsf{P}_1(\mathsf{S})$ for all $t > 0$.
The same results hold if P_1 is replaced by M_1 (the family of all completely monotone functions, $f(0) = 1$), P^i by M^i (infinitely divisible functions from M_1) and N_∞ by A_∞ (completely alternating functions which may assume infinite values).

An important example of infinitely divisible functions is related to random elements in semigroups. If ξ is a random element in S, then ξ is said to be infinitely divisible if ξ coincides in distribution with $\xi_1 + \cdots + \xi_n$ for each $n \geq 2$, where $\xi_1, \ldots, \xi_n$ are i.i.d. random elements in S. Then

$$f(\chi) = \mathbf{E}\chi(\xi), \quad \chi \in \hat{\mathsf{S}},$$

is called the *Laplace transform* of ξ. It is easy to check that

$$\Delta_{\chi_n} \cdots \Delta_{\chi_1} \mathbf{E}\chi(\xi) = \mathbf{E}[\chi(\xi)(1 - \chi_1(\xi)) \cdots (1 - \chi_n(\xi))] \geq 0,$$

so that the Laplace transform is completely monotone on $\hat{\mathsf{S}}$, i.e. $f \in \mathsf{M}(\hat{\mathsf{S}})$. The infinite divisibility property of ξ implies that its Laplace transform is infinitely divisible on $\hat{\mathsf{S}}$, i.e. $\mathbf{E}\chi(\xi) \in \mathsf{M}^i(\hat{\mathsf{S}}) \subset \mathsf{P}^i(\hat{\mathsf{S}})$. Theorem G.9 yields

$$\mathbf{E}\chi(\xi) = e^{-\varphi(\chi)}, \quad \chi \in \hat{\mathsf{S}}, \qquad (G.2)$$

where $\varphi \in \mathsf{A}_\infty(\hat{\mathsf{S}})$.

Assume that S is *idempotent*, that is $s + s = s$ for every $s \in \mathsf{S}$. It is possible to define an ordering on S by $s \leq t$ if $s + t = t$, so that $s + t = s \vee t$. Every semicharacter χ on S is a $\{0, 1\}$-valued function. For each $\chi \in \hat{\mathsf{S}}$, the set $I = \chi^{-1}(\{1\})$ is a semigroup which is hereditary on the left, i.e. for $s, t \in \mathsf{S}$ with $s \leq t$

and $t \in I$ we have $s \in I$. Conversely, if $I \subset S$ is a semigroup and hereditary on the left, then $\mathbf{1}_I$ is a semicharacter. Therefore, $\hat{S}$ is isomorphic with the set $\mathcal{I}$ of semigroups which are hereditary on the left, considered as a semigroup under intersection. Every function $f \in P(S)$ is non-negative, decreasing and bounded, while each $f \in N(S)$ is increasing and satisfies $f(0) \leq f(s)$. The following theorem provides the representations of positive definite and negative definite functions on idempotent semigroups.

Theorem G.10 (Positive and negative definite functions on imdepotent semigroup). *Let* S *be an idempotent semigroup. For* $f \in P(S)$ *there is a unique Radon measure* μ *on the semicharacters* $\hat{S}$ *(or, equivalently on the family* $\mathcal{I}$ *of hereditary on the left semigroups) such that*

$$f(s) = \mu(\{\chi \in \hat{S} : \chi(s) = 1\}) = \mu(\{I \in \mathcal{I} : s \in I\}), \quad s \in S.$$

For $f \in N(S)$ *there is a unique Radon measure* μ *on* $\hat{S} \setminus \{1\} = \mathcal{I} \setminus \{S\}$ *such that*

$$\begin{aligned} f(s) &= f(0) + \mu(\{\chi \in \hat{S} : \chi(s) = 0\}) \\ &= f(0) + \mu(\{I \in \mathcal{I} \setminus \{S\} : s \notin I\}), \quad s \in S. \end{aligned}$$

The semicharacters on $\hat{S}$ can be identified with elements of S. Therefore, if ξ is an infinitely divisible random element in an idempotent semigroup, then Theorem G.10 (applied to the semigroup $\hat{S}$) and (G.2) imply

$$\mathbf{E}\mathbf{1}_{\xi \in I} = \mathbf{P}\{\xi \in I\} = \exp\{-\mu(S \setminus I)\}, \tag{G.3}$$

where μ is concentrated on $S \setminus \{0\}$.

Most of the results presented above are taken from Berg, Christensen and Ressel [60]. A comprehensive account on harmonic analysis on semigroups can be found in the monograph Berg, Christensen and Ressel [61].

H Regular variation

Numerical regularly varying functions

A measurable function $f : \mathbb{R}_+ = [0, \infty) \mapsto \mathbb{R}_+$ is *regularly varying* with the exponent α if, for each $x > 0$,

$$\lim_{t \to \infty} \frac{f(tx)}{f(t)} = x^\alpha, \tag{H.1}$$

see Seneta [530] and Bingham, Goldie and Teugels [71]. We then write $f \in \mathrm{RV}_\alpha$. In this case (H.1) holds uniformly for $x \in [a, b]$, $0 < a < b < \infty$. A function L is said to be *slowly* varying if L satisfies (H.1) with $\alpha = 0$. Each $f \in RV_\alpha$ can be written as

$$f(x) = x^\alpha L(x) \tag{H.2}$$

for a slowly varying function L. A slowly varying function L admits the representation

$$L(x) = \exp\left\{\eta(x) + \int_b^x t^{-1}\varepsilon(t)dt\right\}, \quad x \geq b, \tag{H.3}$$

for a certain $b \geq 0$. Here $\eta(x)$ is bounded on $[b, \infty)$ and admits a finite limit as $x \to \infty$ and $\varepsilon(t)$ is a continuous function such that $\varepsilon(t) \to 0$ as $t \to \infty$.

Let us mention several useful properties of regularly varying functions (all proofs can be found in Seneta [530]). The letters f and L stand for arbitrary regular varying and slowly varying functions.

1. For each $\alpha > 0$

$$x^\alpha L(x) \to \infty \quad \text{and} \quad x^{-\alpha} L(x) \to 0 \quad \text{as } x \to \infty.$$

2. If $f \in RV_\alpha$, then there exists a function $f^- \in RV_{1/\alpha}$ (called the *asymptotic inverse* function for f) such that

$$f(f^-(x)) \sim x \quad \text{and} \quad f^-(f(x)) \sim x \quad \text{as } x \to \infty.$$

3. Let L be slowly varying. If $f(x) = x^\alpha L(x)$ is non-decreasing on $[a, \infty)$ for some $a > 0$ and $\alpha > 0$, then

$$f^-(x) = \inf\{y: f(y) \geq x, y \in [a, \infty)\}$$

is the asymptotic inverse function for f. This fact is called the *inverse theorem* for univariate regularly varying functions.

Multivariate regularly varying functions

For the generalisation to multivariate functions we follow Yakimiv [618], see also de Haan and Resnick [216, 217] and Resnick [480]. Let $\mathbb{C}$ be a cone in $\mathbb{R}^d$ and let $\mathbb{C}' = \mathbb{C} \setminus \{0\}$. A measurable function $f: \mathbb{C} \mapsto \mathbb{R}_+$ is said to be regularly varying with exponent α if, for any fixed $e \in \mathbb{C}'$,

$$\lim_{t \to \infty} \left|\frac{f(tx)}{f(te)} - \varphi(x)\right| = 0, \quad x \in \mathbb{C}'. \tag{H.4}$$

The function φ is *homogeneous* with exponent α, i.e.

$$\varphi(tx) = t^\alpha \varphi(x), \quad t > 0, \, x \in S. \tag{H.5}$$

Condition (H.4) is too weak to ensure desirable properties of regular varying *multivariate* functions. The function f is said to belong to $RV_\alpha^u(\mathbb{C})$ if (H.4) is valid uniformly on $\mathbb{S}^{d-1} \cap \mathbb{C}$, i.e.

$$\lim_{t \to \infty} \sup_{x \in \mathbb{S}^{d-1} \cap \mathbb{C}} \left| \frac{f(tx)}{f(te)} - \varphi(x) \right| = 0. \tag{H.6}$$

It is possible to show that (H.4) and (H.6) are equivalent if and only if $d = 1$.

Let $\mathrm{RV}_0^{\mathrm{u}}(\mathbb{C})$ be the class of slowly varying functions on $\mathbb{C}$ such that (H.6) holds with the function φ identically equal to 1. Furthermore, let $\mathrm{Hom}_\alpha(\mathbb{C})$ be the class of all continuous functions which satisfy (H.5). It was proved in Yakimiv [618] that $f \in \mathrm{RV}_\alpha^{\mathrm{u}}(\mathbb{C})$ if and only if

$$f = L\varphi \tag{H.7}$$

for $L \in \mathrm{RV}_0^{\mathrm{u}}(\mathbb{C})$ and $\varphi \in \mathrm{Hom}_\alpha(\mathbb{C})$. It was proved in de Haan and Resnick [217] that for each $L \in \mathrm{RV}_0^{\mathrm{u}}(\mathbb{C})$ and $c > 0$, $\varepsilon > 0$ there exists t_0 such that

$$(1-\varepsilon)\|x\|^{-\varepsilon} \le \frac{L(tx)}{L(te)} \le (1+\varepsilon)\|x\|^{\varepsilon} \tag{H.8}$$

for all $t \ge t_0$ and $\|x\| \ge c$.

Multivalued regular variation

Regular varying *multifunctions* have been introduced in Molchanov [398, Ch. 6], where further details and proofs can be found. Let $\mathbb{C}$ be a regular closed cone in $\mathbb{R}^m$, $\mathbb{C}' = \mathbb{C} \setminus \{0\}$ and let $M: \mathbb{C} \mapsto \mathcal{F}$ be a function on $\mathbb{C}$ with values in the family $\mathcal{F}$ of closed subsets of $\mathbb{R}^d$. Assume that $M(0) = \{0\}$ and M is measurable, i.e. for each compact set K the set $\{u \in \mathbb{C}: M(u) \cap K \ne \emptyset\}$ is measurable. The dimensions d and m are not supposed to be equal.

Definition H.1 (Regularly varying multifunction). A multifunction $M: \mathbb{C} \mapsto \mathcal{F}$ is said to be regularly varying with the limiting function Φ and index α if

$$\mathrm{PK\text{-}}\lim_{t \to \infty} \frac{M(tu)}{g(t)} = \Phi(u) \tag{H.9}$$

for every sequence $u_t \in \mathbb{C}'$, such that $u_t \to u \ne 0$ as $t \to \infty$, where $\Phi(u)$ is a non-trivial closed subset of $\mathbb{R}^d$, $\Phi(u) \ne \{0\}$ for $u \ne 0$ and $g: (0, \infty) \mapsto (0, \infty)$ is a numerical univariate regularly varying function of index α. We then write $M \in \mathrm{RV}_{\alpha,g,\Phi}^{\mathrm{u}}(\mathbb{C}; \mathcal{F})$ or, briefly, $M \in \mathrm{RV}_\alpha^{\mathrm{u}}(\mathbb{C}; \mathcal{F})$. If M takes only compact values and (H.9) holds in the Hausdorff metric, i.e.

$$\rho_{\mathrm{H}}\text{-}\lim_{t \to \infty} \frac{M(tu_t)}{g(t)} = \Phi(u), \tag{H.10}$$

then M is said to belong to $\mathrm{RV}_{\alpha,g,\Phi}^{\mathrm{u}}(\mathbb{C}; \mathcal{K})$.

Definition H.1 complies with the definitions of multivariate regular varying functions. Indeed, a function $h: \mathbb{C} \mapsto \mathbb{R}^1$ belongs to $\mathrm{RV}_\alpha^{\mathrm{u}}(\mathbb{C})$ if and only if $M(u) = \{h(u)\} \in \mathrm{RV}_\alpha^{\mathrm{u}}(\mathbb{C}; \mathcal{K})$. The limiting multifunction $\Phi(u)$ in (H.9) is *homogeneous*, i.e.

$$\Phi(su) = s^\alpha \Phi(u), \quad s > 0, \ u \in \mathbb{R}^m. \tag{H.11}$$

It is obvious that continuous homogeneous functions satisfying (H.11) are regularly varying.

Theorem H.2 (see Molchanov [398]). *If $M \in \mathrm{RV}_\alpha^\mathrm{u}(\mathbb{C}; \mathcal{F})$ (respectively $M \in \mathrm{RV}_\alpha^\mathrm{u}(\mathbb{C}; \mathcal{K})$), then the function Φ from (H.9) (respectively (H.10)) is continuous in the Fell topology (respectively in the Hausdorff metric) on $\mathbb{C}'$.*

Example H.3. Let $h_i : \mathbb{R}^m \mapsto \mathbb{R}^1$, $1 \leq i \leq d$, be regularly varying numerical multivariate functions from $\mathrm{RV}_{\alpha,\varphi}^\mathrm{u}(\mathbb{C})$ such that

$$\lim_{t \to \infty} \frac{h_i(tu)}{g(t)} = \varphi_i(u), \quad 1 \leq i \leq d, \ u \in \mathbb{C}', \tag{H.12}$$

for a univariate function $g \in \mathrm{RV}_\alpha$. Then

$$M(u) = \{(h_1(u), \ldots, h_d(u))\} \in \mathrm{RV}_{\alpha,g,\Phi}^\mathrm{u}(\mathbb{C}; \mathcal{K})$$

where $\Phi(u) = \{(\varphi_1(u), \ldots, \varphi_d(u))\}$ is a singleton for each u. The function M maps $\mathbb{R}^m$ into the class of single-point subsets of $\mathbb{R}^d$.

The following lemma shows that some set-theoretic operations preserve the regular variation property. It easily follows from the continuity of these operations with respect to the Hausdorff metric.

Lemma H.4. *Let $M_i \in \mathrm{RV}_\alpha^\mathrm{u}(\mathbb{C}; \mathcal{K})$, $c_i > 0$, $1 \leq i \leq p$. Then the functions*

$$c_1 M_1 \cup \cdots \cup c_p M_p,$$
$$\mathrm{co}(M^{(1)}),$$
$$c_1 M_1 \oplus \cdots \oplus c_p M_p$$

belong to the class $\mathrm{RV}_\alpha^\mathrm{u}(\mathbb{C}; \mathcal{K})$.

Inversion theorems

If $M : \mathbb{C} \mapsto \mathcal{F}$ is a multifunction defined on $\mathbb{C}$, then

$$M^-(K) = \{u \in \mathbb{C} : M(u) \cap K \neq \emptyset\} \tag{H.13}$$

is said to be the *inverse* for M. The following theorem is an analogue of the inversion theorem for numerical univariate regularly varying functions.

Theorem H.5 (Inversion theorem). *Let $M \in \mathrm{RV}_{\alpha,g,\Phi}^\mathrm{u}(\mathbb{C}; \mathcal{F})$ with $\alpha > 0$. Assume that for all $u_0 \in \mathbb{C}'$ and $\varepsilon > 0$ there exists $\delta > 0$ such that*

$$\Phi(u_0)^\delta \subset \bigcup_{u \in B_\varepsilon(u_0)} \frac{M(tu)}{g(t)} \tag{H.14}$$

for all sufficiently large t. Then, for every fixed $a > 0$,

$$\text{PK-}\lim_{t\to\infty} \frac{M^-(tK_t)}{g^-(t)} \cap \mathbb{C}_a = \Phi^-(K) \cap \mathbb{C}_a \tag{H.15}$$

if $\rho_H(K_t, K) \to 0$ and $0 \notin K$. Here $\mathbb{C}_a = \{u \in \mathbb{C} : \|u\| \geq a\}$, $\gamma = 1/\alpha$, g^- is the asymptotic inverse function to g and

$$\Phi^-(K) = \{u \in \mathbb{C} : \Phi(u) \cap K \neq \emptyset\}. \tag{H.16}$$

If $M \in \text{RV}^u_{\alpha,g,\Phi}(\mathbb{C}; \mathcal{K})$ and $0 \notin \Phi(u)$ for all $u \in \mathbb{C}'$, then the Painlevé–Kuratowski limit in (H.15) can be replaced with the limit in the Hausdorff metric. If

$$\rho_H\text{-}\lim_{t\to\infty} \frac{M(tu_t)}{g(t)} = \{0\}, \tag{H.17}$$

whenever $u_t \to 0$ as $t \to \infty$, then (H.15) holds for $a = 0$.

The closed sets K, K_t, $t > 0$, in Theorem H.5 are allowed to be non-compact, provided $\rho_H(K, K_t) \to 0$ as $t \to \infty$.

The following propositions reformulate (H.14) for particular functions M.

Proposition H.6. *Let $M(u) = \{(h_1(u), \ldots, h_d(u))\}$ be the function from Example H.3, where $h_i \in \text{RV}^u_\alpha(\mathbb{C})$, $1 \leq i \leq d$, are continuous functions. If the function g in (H.12) is continuous, then (H.14) holds.*

Proposition H.7. *Let $M(u) = g(\|u\|)F(e_u)$, where $F: \mathbb{S}^{m-1} \mapsto \mathcal{F}$ is a multifunction on the unit sphere and $e_u = u/\|u\|$. Then (H.17) holds if F is bounded on S and $g(tx_t)/g(t) \to 0$ as $x_t \to 0$ and $t \to \infty$.*

Note also that all functions from Lemma H.4 satisfy the conditions (H.14) in case all their components M_i, $1 \leq i \leq p$, satisfy the same condition. Without (H.14), statement (H.15) should be replaced with

$$\text{PK-}\lim_{t\to\infty} \frac{M^-(tK_t)}{g^-(t)} \cap \mathbb{C}_a \subset \Phi^-(K) \cap \mathbb{C}_a.$$

Now consider a particular case of Theorem H.5, which, in fact, is the *inversion theorem* for multivariate regularly varying functions.

Theorem H.8. *Let $h \in \text{RV}^u_{\alpha,\varphi}(\mathbb{C})$ be a continuous regularly varying numerical function with $\alpha > 0$. Assume that the corresponding normalising function g in (H.12) is continuous. Define for any fixed $a > 0$*

$$M^-(x) = \{u \in \mathbb{C} : \|u\| \geq a, \ h(u) \geq x\}, \quad x > 0.$$

Then $M^- \in \text{RV}^u_{\gamma,g^-,\Phi^-}((0,\infty), \mathcal{F})$, where g^- is the asymptotically inverse function for g, $\gamma = 1/\alpha$ and $\Phi^-(x) = \{u \in \mathbb{C} : \varphi(u) \geq x\}$.

Integrals of regularly varying functions

The following result concerns asymptotic properties of an integral, whose domain of integration is a regularly varying multifunction.

Theorem H.9. *Let $L \in RV_0^u(\mathbb{C})$ be a slowly varying function with non-negative values and let $\varphi \colon \mathbb{C} \mapsto \mathbb{R}^1$ be a continuous homogeneous function of exponent $\alpha - d$ with $\alpha < 0$, where $\mathbb{C}$ is a cone in $\mathbb{R}^d$. Furthermore, let*

$$M \colon \mathbb{R}^1 \mapsto \mathcal{F}(\mathbb{C}') = \{F \in \mathcal{F} \colon F \subset \mathbb{C}'\}$$

be a multivalued function, whose values are closed subsets of $\mathbb{C}' = \mathbb{C} \setminus \{0\}$. Suppose that

$$\inf\left\{\varepsilon > 0 \colon D^{-\varepsilon} \cap K \subset \frac{M(t)}{g(t)} \cap K \subset D^{\varepsilon} \cap K\right\} \to 0 \quad \text{as } t \to \infty \quad (H.18)$$

holds for a regular closed set D missing the origin and for every compact set K, where $g \in RV_\gamma(\mathbb{R}_+)$ with $\gamma > 0$. Then, for every $e \in \mathbb{C}'$,

$$\int_{M(t)} \varphi(u) L(u) du \sim L(g(s)e) g(s)^\alpha \int_D \varphi(u) du \quad \text{as } s \to \infty.$$

Note that (H.18) is more restrictive than

$$\text{PK-}\lim_{t \to \infty} \frac{M(t)}{g(t)} = D.$$

Nevertheless, for convex-valued multifunctions these conditions are equivalent.

Corollary H.10. *Let $M \in RV_\alpha^u(\mathbb{C}; \mathcal{K})$ be a convex-valued multifunction and let the functions φ and L satisfy the conditions of Theorem H.9. Then*

$$H(v) = \int_{M(v)} \varphi(u) L(u) du$$

is a regularly varying multivariate function from RV_α^u. In particular, this holds if $H(v)$ is the Lebesgue measure of $M(v)$.

References

The bibliography in this book is created using the BibTeX-database *Random Sets and Related Topics* maintained by the author. The full bibliography is downloadable from
http://liinwww.ira.uka.de/bibliography/Math/random.closed.sets.html
The bibliography for this particular book is available from
http://www.cx.unibe.ch/~ilya/rsbook/rsbib.html

1. ABID, M. (1978). Un théorème ergodique pour des processes sous-additifs et sur-stationnaires. *C. R. Acad. Sci., Paris, Ser. I* **287**, 149–152.
2. ADAMSKI, W. (1977). Capacitylike set functions and upper envelopes of measures. *Math. Ann.* **229**, 237–244.
3. ADLER, A., ROSALSKY, A. and TAYLOR, R. L. (1991). A weak law for normed weighted sums of random elements in Rademacher type p Banach spaces. *J. Multivariate Anal.* **37**, 259–268.
4. ADLER, R. J. (1981). *The Geometry of Random Fields*. Wiley, New York.
5. ADLER, R. J. (2000). On excursion sets, tube formulas and maxima of random fields. *Ann. Appl. Probab.* **10**, 1–74.
6. ALDOUS, D. (1989). *Probability Approximations via the Poisson Clumping Heuristic*. Springer, New York.
7. ALÓ, R. A., DE KORVIN, A. and ROBERTS, C. (1979). The optional sampling theorem for convex set-valued martingales. *J. Reine Angew. Math.* **310**, 1–6.
8. ANDERSON, T. W. (1955). The integral of a symmetric unimodal function over a symmetric convex set and some probability inequalities. *Proc. Amer. Math. Soc.* **6**, 170–176.
9. ANGER, B. and LEMBCKE, J. (1985). Infinitely divisible subadditive capacities as upper envelopes of measures. *Z. Wahrsch. verw. Gebiete* **68**, 403–414.
10. ANISIMOV, V. V. and PFLUG, G. C. (2000). Z-theorems: limits of stochastic equations. *Bernoulli* **6**, 917–938.
11. ANISIMOV, V. V. and SEILHAMER, A. V. (1995). Asymptotic properties of extreme sets of random fields. *Theory Probab. Math. Statist.* **51**, 29–38.
12. ARAUJO, A. and GINÉ, E. (1980). *The Central Limit Theorem for Real and Banach Valued Random Variables*. Wiley, New York.
13. ARROW, K. J. and HAHN, F. H. (1971). *General Competitive Analysis*. Holden-Day, San Francisco.
14. ARTSTEIN, Z. (1972). Set-valued measures. *Trans. Amer. Math. Soc.* **165**, 103–125.

15. ARTSTEIN, Z. (1980). Discrete and continuous bang-bang and facial spaces or: look for the extreme points. *SIAM Rev.* **22**, 172–185.
16. ARTSTEIN, Z. (1983). Distributions of random sets and random selections. *Israel J. Math.* **46**, 313–324.
17. ARTSTEIN, Z. (1984). Convergence of sums of random sets. In *Stochastic Geometry, Geometric Statistics, Stereology*, edited by R. V. Ambartzumian and W. Weil, 34–42. Teubner, Leipzig. Teubner Texte zur Mathematik, B.65.
18. ARTSTEIN, Z. (1984). Limit laws for multifunctions applied to an optimization problem. In *Multifunctions and Integrands*, edited by G. Salinetti, vol. 1091 of *Lect. Notes Math.*, 66–79. Springer, Berlin.
19. ARTSTEIN, Z. (1998). Relaxed multifunctions and Young measures. *Set-Valued Anal.* **6**, 237–255.
20. ARTSTEIN, Z. and BURNS, J. A. (1975). Integration of compact set-valued functions. *Pacific J. Math.* **58**, 297–307.
21. ARTSTEIN, Z. and HANSEN, J. C. (1985). Convexification in limit laws of random sets in Banach spaces. *Ann. Probab.* **13**, 307–309.
22. ARTSTEIN, Z. and HART, S. (1981). Law of large numbers for random sets and allocation processes. *Math. Oper. Res.* **6**, 485–492.
23. ARTSTEIN, Z. and VITALE, R. A. (1975). A strong law of large numbers for random compact sets. *Ann. Probab.* **3**, 879–882.
24. ARTSTEIN, Z. and WETS, R. J.-B. (1988). Approximating the integral of a multifunction. *J. Multivariate Anal.* **24**, 285–308.
25. ARTSTEIN, Z. and WETS, R. J.-B. (1995). Consistency of minimizers and the SLLN for stochastic programs. *J. Convex Anal.* **2**, 1–17.
26. ATTOUCH, H. (1984). *Variational Convergence for Functions and Operators*. Pitman, Boston.
27. ATTOUCH, H. and WETS, R. J.-B. (1990). Epigraphical processes: law of large numbers for random LSC functions. *Sém. Anal. Convexe* **20**(Exp. No.13), 29 pp.
28. AUBIN, J.-P. (1999). *Mutational and Morphological Analysis. Tools for Shape Evolution and Morphogenesis*. Birkhäuser, Boston.
29. AUBIN, J.-P. and CELLINA, A. (1984). *Differential Inclusions*. Springer, Berlin.
30. AUBIN, J.-P. and FRANKOWSKA, H. (1990). *Set-Valued Analysis*, vol. 2 of *System and Control, Foundation and Applications*. Birkhäuser, Boston.
31. AUMANN, R. J. (1965). Integrals of set-valued functions. *J. Math. Anal. Appl.* **12**, 1–12.
32. AUMANN, R. J. and SHAPLEY, L. S. (1974). *Values of Non-Atomic Games*. Princeton University Press, Princeton, NJ.
33. AVGERINOS, E. P. and PAPAGEORGIOU, N. S. (1999). Almost sure convergence and decomposition of multivalued random processes. *Rocky Mountain J. Math.* **29**, 401–435.
34. AYALA, G., FERRÁNDIZ, J. and MONTES, F. (1991). Random set and coverage measure. *Adv. Appl. Probab.* **23**, 972–974.
35. AYALA, G. and MONTES, F. (1997). Random closed sets and random processes. *Rend. Circ. Mat. Palermo (2)* **50**, 35–41.
36. AYALA, G. and SIMÓ, A. (1995). Bivariate random closed sets and nerve fibre degeneration. *Adv. Appl. Probab.* **27**, 293–305.
37. AZEMA, J. (1985). Sur les fermes aleatoires. *Lect. Notes Math.* **1123**, 397–495.
38. BADDELEY, A. J. (1992). Errors in binary images and an L^p version of the Hausdorff metric. *Nieuw Archief voor Wiskunde* **10**, 157–183.
39. BADDELEY, A. J. and MOLCHANOV, I. S. (1997). On the expected measure of a random set. In *Advances in Theory and Applications of Random Sets*, edited by D. Jeulin,

3–20, Singapore. Proceedings of the International Symposium held in Fontainebleau, France (9-11 October 1996), World Scientific.
40. BADDELEY, A. J. and MOLCHANOV, I. S. (1998). Averaging of random sets based on their distance functions. *J. Math. Imaging and Vision* **8**, 79–92.
41. BAGCHI, S. N. (1985). On a.s. convergence of multivalued asymptotic martingales. *Ann. Inst. H.Poincaré, Sect. B, Prob. et Stat.* **21**, 313–321.
42. BALAN, R. M. (2001). A strong Markov property for set-indexed processes. *Statist. Probab. Lett.* **53**, 219–226.
43. BALDER, E. J. (1984). A unifying note on Fatou's lemma in several dimensions. *Math. Oper. Res.* **9**, 267–275.
44. BALDER, E. J. (1988). Fatou's lemma in infinite dimensions. *J. Math. Anal. Appl.* **136**, 450–465.
45. BALDER, E. J. and HESS, C. (1995). Fatou's lemma for multifunctions with unbounded values. *Math. Oper. Res.* **20**, 175–188.
46. BALDER, E. J. and HESS, C. (1996). Two generalizations of Komlós theorem with lower-closure-type applications. *J. Convex Anal.* **3**, 25–44.
47. BALKEMA, A. A. and RESNICK, S. I. (1977). Max-infinite divisibility. *J. Appl. Probab.* **14**, 309–319.
48. BARBATI, A., BEER, G. and HESS, C. (1994). The Hausdorff metric topology, the Attouch-Wets topology, and the measurability of set-valued functions. *J. Convex Anal.* **1**, 107–119.
49. BARBATI, A. and HESS, C. (1998). The largest class of closed convex valued multifunctions for which Effros measurability and scalar measurability coincide. *Set-Valued Anal.* **6**, 209–236.
50. BARNETT, V. (1976). The orderings of multivariate data. *J. R. Statist. Soc. Ser. A* **139**, 318–354.
51. BASS, R. F. and PYKE, R. (1984). The existence of set-indexed Lévy processes. *Z. Wahrsch. verw. Gebiete* **66**, 157–172.
52. BASS, R. F. and PYKE, R. (1985). The space $\mathcal{D}(A)$ and weak convergence of set-indexed processes. *Ann. Probab.* **13**, 860–884.
53. BAUDIN, M. (1984). Multidimensional point processes and random closed sets. *J. Appl. Probab.* **21**, 173–178.
54. BEDDOW, J. K. and MELLOY, T. P. (1980). *Testing and Characterization of Powder and Fine Particles*. Heyden & Sons, London.
55. BEDNARSKI, T. (1981). On solutions of minimax test problems for special capacities. *Z. Wahrsch. verw. Gebiete* **58**, 397–405.
56. BEER, G. (1993). *Topologies on Closed and Closed Convex Sets*. Kluwer, Dordrecht.
57. BEG, I. and SHAHZAD, N. (1995). Random extension theorems. *J. Math. Anal. Appl.* **196**, 43–52.
58. BEG, I. and SHAHZAD, N. (1996). On random approximation and coincidence point theorems for multivalued operators. *Nonlinear Anal.* **26**, 1035–1041.
59. BEG, I. and SHAHZAD, N. (1997). Measurable selections: in random approximations and fixed point theory. *Stochastic Anal. Appl.* **15**, 19–29.
60. BERG, C., CHRISTENSEN, J. P. R. and RESSEL, P. (1976). Positive definite functions on abelian semigroups. *Math. Ann.* **223**, 253–272.
61. BERG, C., CHRISTENSEN, J. P. R. and RESSEL, P. (1984). *Harmonic Analysis on Semigroups*. Springer, Berlin.
62. BERGER, J. and BERLINER, L. M. (1986). Robust Bayes and empirical Bayes analysis with ϵ-contaminated priors. *Ann. Statist.* **14**, 461–486.

References

63. BERGER, J. O. (1984). The robust Bayesian viewpoint (with discussion). In *Robustness of Bayesian Analyses*, edited by J. Kadane, vol. 4 of *Stud. Bayesian Econometrics*, 63–144. North-Holland, Amsterdam.
64. BERGER, J. O. and SALINETTI, G. (1995). Approximations of Bayes decision problems: the epigraphical approach. *Ann. Oper. Res.* **56**, 1–13.
65. BERTOIN, J. (1996). *Lévy Processes*. Cambridge University Press, Cambridge.
66. BERTOIN, J. (1997). Renerative embedding of Markov sets. *Probab. Theor. Relat. Fields* **108**, 559–571.
67. BERTOIN, J. (1999). Intersection of independent regerative sets. *Probab. Theor. Relat. Fields* **114**, 97–121.
68. BERTOIN, J. (1999). Subordinators: examples and applications. In *Lectures on Probability Theory and Statistics*, edited by P. Bernard, vol. 1717 of *Lect. Notes Math.*, 1–91. Springer, Berlin.
69. BICKEL, P. J. and YAHAV, J. A. (1965). Renewal theory in the plane. *Ann. Math. Statist.* **36**, 946–955.
70. BILLINGSLEY, P. (1968). *Convergence of Probability Measures*. Wiley, New York.
71. BINGHAM, N. H., GOLDIE, C. M. and TEUGELS, J. L. (1987). *Regular Variation*. Cambridge University Press, Cambridge.
72. BIRGÉ, L. (1977). Tests minimax robustes. *Asterisque* **43-44**, 87–133.
73. BLUMENTHAL, R. M. and GETOOR, R. K. (1968). *Markov processes and potential theory*. Academic Press, New York.
74. BOCŞAN, G. (1986). *Random Sets and Related Topics*, vol. 27 of *Monografii Matematice*. Universitatea din Timişoara, Timişoara.
75. BÖHM, S. and SCHMIDT, V. (2003). Palm representation and approximation of the covariance of random closed sets. *Adv. Appl. Probab.* **35**, 295–302.
76. BRÄKER, H. and HSING, T. (1998). On the area and perimeter of a random convex hull in a bounded convex set. *Probab. Theor. Relat. Fields* **111**, 517–550.
77. BRONOWSKI, J. and NEYMAN, J. (1945). The variance of the measure of a two-dimensional random set. *Ann. Math. Statist.* **16**, 330–341.
78. BROWN, L. and SCHREIBER, B. M. (1989). Approximation and extension of random functions. *Monatsh. Math.* **107**, 111–123.
79. BROZIUS, H. (1989). Convergence in mean of some characteristics of the convex hull. *Adv. Appl. Probab.* **21**, 526–542.
80. BROZIUS, H. and DE HAAN, L. (1987). On limiting laws for the convex hull of a sample. *J. Appl. Probab.* **24**, 852–862.
81. BRU, B., HEINICH, H. and LOOTGIETER, J.-C. (1993). Distances de Lévy et extensions des theoremes de la limite centrale et de Glivenko-Cantelli. *Publ. Inst. Stat. Univ. Paris* **37**(3-4), 29–42.
82. BUJA, A. (1986). On the Huber-Strassen theorem. *Probab. Theor. Relat. Fields* **73**, 149–152.
83. BULDYGIN, V. V. and SOLNTSEV, S. A. (1991). Convergence of sums of independent random vectors with operator normalizations. *Theory Probab. Appl.* **36**, 346–351.
84. BULINSKAYA, E. V. (1961). On mean number of crossings of a level by a stationary Gaussian process. *Theory Probab. Appl.* **6**, 474–478. In Russian.
85. BYRNE, C. L. (1978). Remarks on the set-valued integrals of Debreu and Aumann. *J. Math. Anal. Appl.* **62**, 243–246.
86. CARNAL, H. (1970). Die konvexe Hülle von n rotationssymmetrisch verteilten Punkten. *Z. Wahrsch. verw. Gebiete* **15**, 168–179.

87. CASCOS FERNÁNDEZ, I. and MOLCHANOV, I. (2003). A stochastic order for random vectors and random sets based on the Aumann expectation. *Statist. Probab. Lett.* **63**, 295–305.
88. CASSELS, J. W. S. (1975). Measures of the non-convexity of sets and the Shapley-Folkman-Starr theorem. *Math. Proc. Cambridge Philos. Soc.* **78**, 433–436.
89. CASTAING, C. and EZZAKI, F. (1997). SLLN for convex random sets and random lower semicontinuous integrands. *Atti Sem. Mat. Fis. Univ. Modena* **45**, 527–553.
90. CASTAING, C., EZZAKI, F. and HESS, C. (1997). Convergence of conditional expectations for unbounded closed convex sets. *Studia Math.* **124**, 133–148.
91. CASTAING, C. and VALADIER, M. (1977). *Convex Analysis and Measurable Multifunctions*, vol. 580 of *Lect. Notes Math.*. Springer, Berlin.
92. CERF, R. (1999). Large deviations for sums of i.i.d. random compact sets. *Proc. Amer. Math. Soc.* **127**, 2431–2436.
93. CERF, R. and MARICONDA, C. (1994). Oriented measures and bang-bang principle. *C. R. Acad. Sci., Paris, Ser. I* **318**, 629–631.
94. CHATEAUNEUF, A. and JAFFRAY, J.-Y. (1989). Some characterizations of lower probabilities and other monotone capacities through the use of Möbius inversion. *Math. Soc. Sciences* **17**, 263–283.
95. CHATEAUNEUF, A., KAST, R. and LAPIED, A. (1996). Choquet pricing for financial markets with frictions. *Math. Finance* **6**, 323–330.
96. CHATTERJI, S. D. (1964). A note on the convergence of Banach-space valued martingales. *Anal. Math.* **153**, 142–149.
97. CHOIRAT, C., HESS, C. and SERI, R. (2003). A functional version of the Birkhoff ergodic theorem for a normal integrand: a variational approach. *Ann. Probab.* **31**, 63–92.
98. CHOQUET, G. (195354). Theory of capacities. *Ann. Inst. Fourier* **5**, 131–295.
99. CHOQUET, G. (1957). Potentiels sur un ensemble de capacités nulles. *C. R. Acad. Sci., Paris, Ser. I* **244**, 1707–1710.
100. CHOUKAIRI-DINI, A. (1989). M-convergence des martingales (asymptotique) multivoques. épi-martingales. *C. R. Acad. Sci., Paris, Ser. I* **309**, 889–892.
101. CHOUKARI-DINI, A. (1996). On almost sure convergence of vector valued pramarts and multivalued pramarts. *J. Convex Anal.* **3**, 245–254.
102. COLUBI, A., DOMÍNGUEZ-MENCHERO, J. S., LÓPEZ-DÍAZ, M. and RALESCU, D. (2002). A $D_E[0, 1]$ representation of random upper semicontinuous functions. *Proc. Amer. Math. Soc.* **130**, 3237–3242.
103. COLUBI, A., LÓPEZ-DÍAZ, M., DOMÍNGUEZ-MENCHERO, J. S. and GIL, M. A. (1999). A generalized strong law of large numbers. *Probab. Theor. Relat. Fields* **114**, 401–417.
104. DE COOMAN, G. and AEYELS, D. (1999). Supremum preserving upper probabilities. *Inform. Sci.* **118**, 173–212.
105. COSTÉ, A. (1975). Sur les multimeasures à valeurs fermées bornées d'un espace de Banach. *C. R. Acad. Sci., Paris, Ser. I* **280**, 567–570.
106. COUSO, I., MONTES, S. and GIL, P. (2002). Stochastic convergence, uniform integrability and convergence in mean on fuzzy measure spaces. *Fuzzy Sets and Systems* **129**, 95–104.
107. COUVREUX, J. and HESS, C. (1999). A Lévy type martingale convergence theorem for random sets with unbounded values. *J. Theoret. Probab.* **12**, 933–969.
108. CRAMÉR, H. and LEADBETTER, M. R. (1967). *Stationary and Related Stochastic Processes*. Wiley, New York.

109. CRAUEL, H. (2002). *Random Probability Measures on Polish Spaces*. Taylor & Francis, London and New York.
110. CRESSIE, N. A. C. (1979). A central limit theorem for random sets. *Z. Wahrsch. verw. Gebiete* **49**, 37–47.
111. CRESSIE, N. A. C. and HULTING, F. L. (1992). A spatial statistical analysis of tumor growth. *J. Amer. Statist. Assoc.* **87**, 272–283.
112. CROSS, R. (1998). *Multivalued Linear Operators*. Marcel Dekker, New York.
113. VAN CUTSEM, B. (1969). Martingales de multiapplications à valeurs convexes compactes. *C. R. Acad. Sci., Paris, Ser. I* **269**, 429–432.
114. VAN CUTSEM, B. (1972). Martingales de convexes fermés aléatoires en dimension finie. *Ann. Inst. H.Poincaré, Sect. B, Prob. et Stat.* **8**, 365–385.
115. DAFFER, P. Z. and TAYLOR, R. L. (1982). Tightness and strong laws of large numbers in Banach spaces. *Bull. Inst. Math. Acad. Sinica* **10**, 252–263.
116. DAL MASO, G. (1993). *An Introduction to Γ-Convergence*. Birkhäuser, Boston.
117. DALEY, D. J. and VERE-JONES, D. (1988). *An Introduction to the Theory of Point Processes*. Springer, New York.
118. DAVIS, R. A., MULROW, E. and RESNICK, S. I. (1987). The convex hull of a random sample in $\mathbf{R}^2$. *Comm. Statist. Stochastic Models* **3**(1), 1–27.
119. DAVIS, R. A., MULROW, E. and RESNICK, S. I. (1988). Almost sure limit sets of random samples in $\mathbf{R}^d$. *Adv. Appl. Probab.* **20**, 573–599.
120. DAVYDOV, Y., PAULAUSKAS, V. and RAČKAUSKAS, A. (2000). More on p-stable convex sets in Banach spaces. *J. Theoret. Probab.* **13**, 39–64.
121. DAVYDOV, Y. and THILLY, E. (1999). Réarrangements convexes de processus stochastiques. *C. R. Acad. Sci., Paris, Ser. I* **329**, 1087–1090.
122. DAVYDOV, Y. and VERSHIK, A. M. (1998). Réarrangements convexes des marches aléatoires. *Ann. Inst. H.Poincaré, Sect. B, Prob. et Stat.* **34**, 73–95.
123. DAVYDOV, Y. and ZITIKIS, R. (2003). Generalized Lorenz curves and convexifications of stochastic processes. *J. Appl. Probab.* **40**, 906–925.
124. DEBREU, G. (1967). Integration of correspondences. In *Proc. Fifth Berkeley Symp. Math. Statist. and Probability*, vol. 2, 351–372. Univ. of California Press.
125. DEBREU, G. and SCHMEIDLER, D. (1972). The Radon-Nikodym derivative of a correspondence. In *Proc. Sixth Berkeley Symp. Math. Statist. and Probability*, vol. 2, 41–56. Univ. of California Press.
126. DELBAEN, F. (1974). Convex games and extreme points. *J. Math. Anal. Appl.* **45**, 210–233.
127. DELBAEN, F. (2002). Coherent risk measures on general probability spaces. In *Advances in Finance and Stochastics*, edited by K. Sandmann and P. J. Schönbucher, 1–37. Springer, Berlin.
128. DELFOUR, M. C. and ZOLÉSIO, J.-P. (1994). Shape analysis via oriented distance functions. *J. Funct. Anal.* **123**, 129–201.
129. DELLACHERIE, C. (1969). Ensembles aléatoires I, II. In *Séminaire de Probabilités, III*, edited by P. A. Meyer, vol. 88 of *Lect. Notes Math.*, 97–136. Springer, Berlin.
130. DELLACHERIE, C. (1971). Quelques commentaires sur les prolongement de capacités. In *Séminair de Probabilités V*, edited by M. Karoubi and P. A. Meyer, vol. 191 of *Lect. Notes Math.*, 77–81. Springer, Berlin.
131. DELLACHERIE, C. (1972). *Capacités et processus stochastiques*. Springer, Berlin.
132. DELLACHERIE, C. and MEYER, P.-A. (1978). *Probabilities and Potential*. North-Holland, Amsterdam.
133. DEMPSTER, A. P. (1967). Upper and lower probabilities induced by a multivalued mapping. *Ann. Math. Statist.* **38**, 325–329.

134. DEMPSTER, A. P. (1968). A generalization of Bayesian inference. *J. R. Statist. Soc. Ser. B* **30**, 205–247.
135. DENNEBERG, D. (1994). *Non-Additive Measure and Integral*. Kluwer, Dordrecht.
136. DENTCHEVA, D. (1998). Differentiable selections and Castaing representations of multifunctions. *J. Math. Anal. Appl.* **223**, 371–396.
137. DEROBERTIS, L. and HARTIGAN, J. A. (1981). Bayesian inference using intervals of measures. *Ann. Statist.* **9**, 235–244.
138. DHAENE, J., DENUIT, M., GOOVAERTS, M. J., KAAS, R. and VYNCKE, D. (2002). The concept of comonotonicity in actuarial science and finance: applications. *Insurance Math. Econom.* **31**, 133–161.
139. DHAENE, J., DENUIT, M., GOOVAERTS, M. J., KAAS, R. and VYNCKE, D. (2002). The concept of comonotonicity in actuarial science and finance: theory. *Insurance Math. Econom.* **31**, 3–33.
140. DOLECKI, S., SALINETTI, G. and WETS, R. J.-B. (1983). Convergence of functions: equi-semicontinuity. *Trans. Amer. Math. Soc.* **276**, 409–429.
141. DONG, W. and WANG, Z. P. (1998). On representation and regularity of continuous parameter multivalued martingales. *Proc. Amer. Math. Soc.* **126**, 1799–1810.
142. DOOB, J. L. (1953). *Stochastic Processes*. Wiley, New York.
143. DOOB, J. L. (1984). *Classical Potential Theory and its Probabilistic Counterparts*. Springer, Berlin.
144. DOSS, S. (1949). Sur la moyenne d'un élément aléatoire dans un espace distancié. *Bull. Sci. Math.* **73**, 48–72.
145. DOSS, S. (1962). Moyennes conditionneles et martingales dans un espace métrique. *C. R. Acad. Sci., Paris, Ser. I* **254**, 3630–3632.
146. DOUGHERTY, E. R., ed. (1993). *Mathematical Morphology in Image Processing*. Marcel Dekker, New York.
147. DOZZI, M., MERZBACH, E. and SCHMIDT, V. (2001). Limit theorems for sums of random fuzzy sets. *J. Math. Anal. Appl.* **259**, 554–565.
148. DUNFORD, N. and SCHWARTZ, J. T. (1958). *Linear Operators. Part I: General Theory*. Interscience Publishers, New York.
149. DUPAČOVÁ, J. and WETS, R. J.-B. (1988). Asymptotic behaviour of statistical estimators and of optimal solutions of stochastic optimization problems. *Ann. Statist.* **16**, 1517–1549.
150. DURRETT, R. and LIGGETT, T. M. (1981). The shape of limit set in Richardson's growth model. *Ann. Probab.* **9**, 186–193.
151. DYCKERHOFF, R. and MOSLER, K. (1993). Stochastic dominance with nonadditive probabilities. *ZOR — Methods and Models of Operation Research* **37**, 231–256.
152. DYNKIN, E. B. and EVSTIGNEEV, I. V. (1976). Regular conditional expectations of correspondences. *Theory Probab. Appl.* **21**, 325–338.
153. DYNKIN, E. B. and FITZSIMMONS, P. J. (1987). Stochastic processes on random domains. *Ann. Inst. H.Poincaré, Sect. B, Prob. et Stat.* **23**, 379–396.
154. EDDY, W. F. (1980). The distribution of the convex hull of a Gaussian sample. *J. Appl. Probab.* **17**, 686–695.
155. EDDY, W. F. (1984). Set-valued orderings for bivariate data. In *Stochastic Geometry, Geometric Statistics, Stereology*, edited by R. Ambartzumian and W. Weil, 79–90, Leipzig. Teubner. Teubner Texte zur Mathematik, B.65.
156. EDDY, W. F. and GALE, J. D. (1981). The convex hull of a spherically symmetric sample. *Adv. Appl. Probab.* **13**, 751–763.
157. EDGAR, G. and SUCHESTON, L. (1992). *Stopping Times and Directed Processes*. Cambridge University Press, Cambridge.

158. EFRON, B. (1965). The convex hull of a random set of points. *Biometrika* **52**, 331–343.
159. EGGHE, L. (1984). *Stopping Time Techniques for Analysist and Probabilists*, vol. 100 of *London Math. Soc. Lect. Notes Ser.*. Cambridge Univ. Press, Cambridge.
160. EL AMRI, K. and HESS, C. (2000). On the Pettis integral of closed valued multifunctions. *Set-Valued Anal.* **8**, 329–360.
161. EMERY, M. and MOKOBODZKI, G. (1991). Sur le barycentre d'une probabilité dans une variété. In *Sèminaire de Probabilités XXV*, edited by J. Azéma, P. A. Meyer and M. Yor, vol. 1485 of *Lect. Notes Math.*, 220–233. Springer, Berlin.
162. ENGL, H. W. (1978). Random fixed point theorems for multivalued mappings. *Pacific J. Math.* **76**, 351–360.
163. EVANS, S. N. and PERES, Y. (1998). Eventual intersection for sequences of Lévy processes. *Electron. Commun. in Probab.* **3**, 21–27.
164. EVSTIGNEEV, I. V. (1988). Stochastic extremal problems and the strong Markov property of random fields. *Russian Math. Surveys* **43**(2), 1–49.
165. FALCONER, K. J. (1990). *Fractal Geometry*. Wiley, Chichester.
166. FARO, R., NAVARRO, J. A. and SANCHO, J. (1995). On the concept of differential of a measure. *Arch. Math.* **64**, 58–68.
167. FEDERER, H. (1969). *Geometric Measure Theory*. Springer, Heidelberg.
168. FEDORCHUK, V. and FILIPPOV, V. (1988). *General Topology. Basic Constructions*. Moscow University, Moscow. In Russian.
169. FELL, J. M. G. (1962). A Hausdorff topology for the closed subsets of a locally compact non-Hausdorff space. *Proc. Amer. Math. Soc.* **13**, 472–476.
170. FENG, Y. (2001). Sums of independent fuzzy random variables. *Fuzzy Sets and Systems* **123**, 11–18.
171. FINE, T. L. (1988). Lower probability models for uncertainty and nondeterministic processes. *J. Statist. Plan. Inf.* **20**, 389–411.
172. FISHER, L. (1969). Limiting sets and convex hulls of samples from product measures. *Ann. Math. Statist.* **40**, 1824–1832.
173. RAYNAUD DE FITTE, P. (1997). Théoreme ergodique ponctuel et lois fortes des grand nombres pour des points aléatoires d'un espace métrique à courbure négative. *Ann. Probab.* **25**, 738–766.
174. FITZSIMMONS, P. J. (1987). On the identification of Markov processes by the distribution of hitting times. In *Seminar on Stochastic Processes (Cahrlottesville, 1986)*, edited by E. Çinlar, K. L. Chung, R. K. Getoor and J. Glover, vol. 13 of *Progress in Probability and Statistics*, 15–19. Birkhäuser, Boston, MA.
175. FITZSIMMONS, P. J. (1999). Markov processes with equal capacities. *J. Theoret. Probab.* **12**, 271–292.
176. FITZSIMMONS, P. J., FRISTEDT, B. and MAISONNEUVE, B. (1985). Intersections and limits of regenerative sets. *Z. Wahrsch. verw. Gebiete* **70**, 157–173.
177. FITZSIMMONS, P. J. and KANDA, M. (1992). On Choquet's dichotomy of capacity for Markov processes. *Ann. Probab.* **20**, 342–349.
178. FITZSIMMONS, P. J. and TAKSAR, M. (1988). Stationary regenerative sets and subordinators. *Ann. Probab.* **16**, 1308–1332.
179. FORTET, R. and KAMBOUZIA, M. (1975). Ensembles aléatoires induits par une répartition ponctuelle aléatoire. *C. R. Acad. Sci., Paris, Ser. I* **280**, 1447–1450.
180. FRANCAVIGLIA, A., LECHICKI, A. and LEVI, S. (1985). Quasi-uniformization of hyperspaces and convergence of nets of semicontinuous multifunctions. *J. Math. Anal. Appl.* **112**, 347–370.
181. FRÉCHET, M. (1948). Les éléments aléatoires de nature quelconque dans un espace distancié. *Ann. Inst. H.Poincaré, Sect. B, Prob. et Stat.* **10**, 235–310.

182. FRIEL, N. and MOLCHANOV, I. S. (1998). A class of error metrics for grey-scale image comparison. In *Mathematical Modelling and Estimation Techniques in Computer Vision*, edited by J. L. D. F. Prêteux and E. R. Dougherty, vol. 3457 of *Proceedings of SPIE*, 194–201, San Diego, California. SPIE.
183. FRIEL, N. and MOLCHANOV, I. S. (1999). A new thresholding technique based on random sets. *Pattern Recognition* **32**, 1507–1517.
184. FRISTEDT, B. (1996). Intersections and limits of regenerative sets. In *Random Discrete Structures*, edited by D. Aldous and R. Pemantle, vol. 76 of *The IMA Volumes in Mathematics and its Applications*, 121–151. Springer, New York.
185. GÄHLER, S. and MURPHY, G. (1981). A metric characterization of normed linear spaces. *Math. Nachr.* **102**, 297–309.
186. GALAMBOS, J. (1978). *The Asymptotic Theory of Extreme Order Statistics*. Wiley, New York.
187. GAO, Y. and ZHANG, W. X. (1994). Theory of selection operators on hyperspaces and multivalued stochastic processes. *Sci. China Ser. A* **37**, 897–908.
188. GERRITSE, B. (1996). Varadhan's theorem for capacities. *Comment. Math. Univ. Carolin.* **37**, 667–690.
189. GERRITSE, G. (1986). Supremum self-decomposable random vectors. *Probab. Theor. Relat. Fields* **72**, 17–33.
190. GHIRARDATO, P. (1997). On independence for non-additive measures, with a Fubini theorem. *J. Econom. Theory* **73**, 261–291.
191. GHOUSSOUB, N. and STEELE, J. M. (1980). Vector valued subadditive processes and applications. *Ann. Probab.* **8**, 83–95.
192. GIERZ, G., HOFMANN, K. H., KEIMEL, K., LAWSON, J. D., MISLOVE, M. and SCOTT, D. S. (1980). *A Compendium of Continuous Lattices*. Springer, Berlin.
193. GIHMAN, I. I. and SKOROHOD, A. V. (1969). *Introduction to the Theory of Stochastic Processes*. W.B. Saunders, Philadelphia.
194. GILBOA, I. and LEHRER, E. (1991). Global games. *Intern J. Game Theory* **20**, 129–147.
195. GILBOA, I. and SCHMEIDLER, D. (1993). Updating ambiguous beliefs. *J. Econ. Th.* **59**, 33–49.
196. GILBOA, I. and SCHMEIDLER, D. (1994). Additive representation of non-additive measures and the Choquet integral. *Ann. Oper. Res.* **52**, 43–65.
197. GILBOA, I. and SCHMEIDLER, D. (1995). Canonical representation of set functions. *Math. Oper. Res.* **20**, 197–212.
198. GINÉ, E. and HAHN, M. G. (1985). Characterization and domains of attraction of p-stable compact sets. *Ann. Probab.* **13**, 447–468.
199. GINÉ, E. and HAHN, M. G. (1985). The Lévy-Hinčin representation for random compact convex subsets which are infinitely divisible under Minkowski addition. *Z. Wahrsch. verw. Gebiete* **70**, 271–287.
200. GINÉ, E. and HAHN, M. G. (1985). M-infinitely divisible random sets. *Lect. Notes Math.* **1153**, 226–248.
201. GINÉ, E., HAHN, M. G. and VATAN, P. (1990). Max-infinitely divisible and max-stable sample continuous processes. *Probab. Theor. Relat. Fields* **87**, 139–165.
202. GINÉ, E., HAHN, M. G. and ZINN, J. (1983). Limit theorems for random sets: application of probability in Banach space results. In *Probability in Banach spaces, IV (Oberwolfach, 1982)*, edited by A. Beck and K. Jacobs, vol. 990 of *Lect. Notes Math.*, 112–135. Springer, Berlin.
203. GIROTTO, B. and HOLZER, S. (2000). Weak convergence of bounded, motonote set functions in an abstract setting. *Real Anal. Exchange* **26**, 157–176.

204. GLOVER, J. and RAO, M. (1996). Condenser potentials. *Astérisque* **236**, 125–131.
205. GODET-THOBIE, C. (1980). Some results about multimeasures and their selectors. In *Measure Theory*, edited by D. Kölzow, vol. 794 of *Lect. Notes Math.*, 112–116. Springer, Berlin.
206. GOODEY, P. and WEIL, W. (1993). Zonoids and generalizations. In *Handbook of Convex Geometry*, edited by P. M. Gruber and J. M. Wills, 1299–1326. North-Holland, Amsterdam.
207. GOODMAN, I. R. and NGUYEN, H. T. (1985). *Uncertainty Models for Knowledge-Based Systems*. North-Holland, Amsterdam.
208. GRAF, S. (1980). A Radon-Nikodym theorem for capacities. *J. Reine Angew. Math.* **320**, 192–214.
209. GRAF, S. (1995). On Bandt's tangential distribution for self-similar measures. *Monatsh. Math.* **120**, 223–246.
210. GRENANDER, U. (1963). *Probabilities on Algebraic Structures*. Wiley, New York.
211. GRIZE, Y. L. and FINE, T. L. (1987). Continuous lower probability-based models for stationary processes with bounded and divergent time averages. *Ann. Probab.* **15**, 783–803.
212. GROENEBOOM, P. (1988). Limit theorems for convex hulls. *Probab. Theor. Relat. Fields* **79**, 327–368.
213. GRUBER, P. M. (1993). Aspects of approximation of convex bodies. In Gruber and Wills [214], 319–345.
214. GRUBER, P. M. and WILLS, J. M., eds. (1993). *Handbook of Convex Geometry. Vol. A, B.*. North-Holland, Amsterdam.
215. DE HAAN, L. (1984). A spectral representation for max-stable processes. *Ann. Probab.* **12**, 1194–1204.
216. DE HAAN, L. and RESNICK, S. I. (1977). Limit theorems for multivariate sample extremes. *Z. Wahrsch. verw. Gebiete* **40**, 317–337.
217. DE HAAN, L. and RESNICK, S. I. (1987). On regular variation of probability densities. *Stochastic Process. Appl.* **25**, 83–93.
218. HALMOS, P. R. and VAUGHAN, H. E. (1950). The marriage problem. *Am. J. of Math.* **72**, 214–215.
219. HANSEN, J. C. and HULSE, P. (2000). Subadditive ergodic theorems for random sets in infinite dimensions. *Statist. Probab. Lett.* **50**, 409–416.
220. HARDING, J., MARINACCI, M., NGUYEN, N. T. and WANG, T. (1997). Local Radon-Nikodym derivatives of set functions. *Int. J. Uncertainty, Fuzziness and Knowledge-Based Systems* **5**, 379–394.
221. HARRIS, T. E. (1968). Counting measures, monotone random set functions. *Z. Wahrsch. verw. Gebiete* **10**, 102–119.
222. HARRIS, T. E. (1976). On a class of set-valued Markov processes. *Ann. Probab.* **4**, 175–199.
223. HART, S. and KOHLBERG, E. (1974). Equally distributed correspondences. *J. Math. Econ.* **1**, 167–174.
224. HAWKES, J. (1977). Intersections of Markov random sets. *Z. Wahrsch. verw. Gebiete* **37**, 243–251.
225. HAWKES, J. (1981). Trees generated by a simple branching process. *J. London Math. Soc.* **24**, 373–384.
226. HAWKES, J. (1998). Exact capacity results for stable processes. *Probab. Theor. Relat. Fields* **112**, 1–11.
227. HAZOD, W. (1997). Remarks on convergence of types theorems on finite dimensional vector spaces. *Publ. Math. Debrecen* **50**, 197–219.

228. HEIJMANS, H. J. A. M. (1994). *Morphological Image Operators*. Academic Press, Boston.
229. HEIJMANS, H. J. A. M. and MOLCHANOV, I. S. (1998). Morphology on convolution lattices with applications to the slope transform and random set theory. *J. Math. Imaging and Vision* **8**, 199–214.
230. HEILPERN, S. (2002). Using Choquet integral in economics. *Statist. Papers* **43**, 53–73.
231. HEINRICH, L. and MOLCHANOV, I. S. (1994). Some limit theorems for extremal and union shot-noise processes. *Math. Nachr.* **168**, 139–159.
232. HEINRICH, L. and SCHMIDT, V. (1985). Normal convergence of multidimensional shot noise and rates of this convergence. *Adv. Appl. Probab.* **17**, 709–730.
233. HENGARTNER, W. and THEODORESCU, R. (1973). *Concentration Functions*. Academic Press, New York, London.
234. HERER, W. (1991). Mathematical expectation and martingales of random subsets of a metric space. *Probab. Math. Statist.* **11**, 291–304.
235. HERER, W. (1992). Mathematical expectation and strong law of large numbers for random variables with values in a metric space of negative curvature. *Probab. Math. Statist.* **13**, 59–70.
236. HERER, W. (1997). Martingales of random subsets of a metric space of negative curvature. *Set-Valued Anal.* **5**, 147–157.
237. HESS, C. (1979). Théorème ergodique et loi forte des grands nombres pour des ensembles aléatoires. *C. R. Acad. Sci., Paris, Ser. I* **288**, 519–522.
238. HESS, C. (1983). Loi de probabilité des ensembles aléatoires à valeurs fermées dans un espace métrique séparable. *C. R. Acad. Sci., Paris, Ser. I* **296**, 883–886.
239. HESS, C. (1985). Loi forte des grand nombres pour des ensembles aléatoires non bornés à valeurs dans un espace de Banach séparable. *C. R. Acad. Sci., Paris, Ser. I* **300**, 177–180.
240. HESS, C. (1986). Quelques résultats sur la mesurabilité des multifonctions à valeurs dans un espace métrique séparable. *Séminaire d'analyse convexe. Montpelier* **16**(Exp. No. 1), 43 pp.
241. HESS, C. (1991). Convergence of conditional expectations for unbounded random sets, integrands and integral functionals. *Math. Oper. Res.* **16**, 627–649.
242. HESS, C. (1991). On multivalued martingales whose values may be unbounded: martingale selectors and Mosco convergence. *J. Multivariate Anal.* **39**, 175–201.
243. HESS, C. (1994). Multivalued strong laws of large numbers in the slice topology. Application to integrands. *Set-Valued Anal.* **2**, 183–205.
244. HESS, C. (1995). On the measurability of the conjugate and the subdifferential of a normal integrand. *J. Convex Anal.* **2**, 153–165.
245. HESS, C. (1996). Epi-convergence of sequences of normal integrands and strong consistency of the maximum likelihood estimator. *Ann. Statist.* **24**, 1298–1315.
246. HESS, C. (1999). Conditional expectation and martingales of random sets. *Pattern Recognition* **32**, 1543–1567.
247. HESS, C. (1999). The distribution of unbounded random sets and the multivalued strong law of large numbers in nonreflexive Banach spaces. *J. Convex Anal.* **6**, 163–182.
248. HESS, C. (2002). Set-valued integration and set-valued probability theory: An overview. In *Handbook of Measure Theory*, edited by E. Pap, chap. 14, 617–673. Elsevier.
249. HESS, C. and ZIAT, H. (2002). Théorème de Komlós pour des multifonctions intégrables au sens de Pettis et applications. *Ann. Sci. Math. Québec* **26**, 181–198.
250. HIAI, F. (1978). Radon-Nikodym theorem for set-valued measures. *J. Multivariate Anal.* **8**, 96–118.

251. HIAI, F. (1979). Representation of additive functionals on vector-valued normed Köthe spaces. *Kodai Math. J.* **2**, 300–313.
252. HIAI, F. (1984). Strong laws of large numbers for multivalued random variables. In *Multifunctions and Integrands*, edited by G. Salinetti, vol. 1091 of *Lect. Notes Math.*, 160–172. Springer.
253. HIAI, F. (1985). Convergence of conditional expectations and strong laws of large numbers for multivalued random variables. *Trans. Amer. Math. Soc.* **291**, 613–627.
254. HIAI, F. (1985). Multivalued conditional expectations, multivalued Radon-Nikodym theorem, integral representation of additive operators, and multivalued strong laws of large numbers. Unpublished paper.
255. HIAI, F. and UMEGAKI, H. (1977). Integrals, conditional expectations, and martingales of multivalued functions. *J. Multivariate Anal.* **7**, 149–182.
256. HILDENBRAND, W. and MERTENS, J.-F. (1971). On Fatou's lemma in several dimensions. *Z. Wahrsch. verw. Gebiete* **17**, 151–155.
257. HIMMELBERG, C. (1974). Measurable relations. *Fund. Math.* **87**, 53–72.
258. HIRIART-URRUTY, J.-B. and LEMARÉCHAL, C. (1993). *Convex Analysis and Minimization Algorithms*, vol. 1 and 2. Springer, Berlin.
259. HOBOLTH, A. and VEDEL JENSEN, E. B. (2000). Modeling stochastic changes in curve shape, with application to cancer diagnostics. *Adv. Appl. Probab.* **32**, 344–362.
260. HOEFFDING, W. (1953). On the distribution of the expected values of the order statistics. *Ann. Math. Statist.* **24**, 93–100.
261. HOFFMAN-JØRGENSEN, J. (1969). Markov sets. *Math. Scand.* **24**, 145–166.
262. HOFFMAN-JØRGENSEN, J. (1994). *Probability with a View towards Statistics*, vol. 1 and 2. Chapman & Hall, New York.
263. HOFFMAN-JØRGENSEN, J. (1998). Convergence in law of random elements and random sets. In *High Dimensional Probability*, edited by E. Eberlein, M. Hahn and M. Talagrand, vol. 43 of *Progress in Probability*, 151–189. Birkhäuser, Basel.
264. HÖRMANDER, L. (1954). Sur la fonction d'appui des ensembles convexes dans un espace localement convexe. *Ark. Mat.* **3**, 181–186.
265. HOROWITZ, J. (1972). Semilinear Markov processes, subordinators and renewal theory. *Z. Wahrsch. verw. Gebiete* **24**, 167–193.
266. HU, K. Y. (1988). A generalization of Kolmogorov's extension theorem and an application to the construction of stochastic processes with random time domains. *Ann. Probab.* **16**, 222–230.
267. HUBER, P. J. (1981). *Robust Statistics*. Wiley, New York.
268. HUBER, P. J. and STRASSEN, V. (1973). Minimax tests and the Neyman-Pearson lemma for capacities. *Ann. Statist.* **1, 2**, 251–263, 223–224.
269. HUG, D., LAST, G. and WEIL, W. (2004). A local Steiner-type formula for general closed sets and applications. *Math. Z.* **246**, 237–272.
270. HUNEYCUTT, J. E. (1971). On an abstract Stieltjes measure. *Ann. Inst. Fourier* **21**, 143–154.
271. IOFFE, A. D. and TIHOMIROV, V. M. (1979). *Theory of Extremal Problems*. North-Holland, Amsterdam.
272. ITÔ, K. and MCKEAN, H. P. (1965). *Diffusion Processes and their Sample Paths*. Springer, Berlin.
273. ITOH, S. (1979). Measurability of condensing multivalued mappings and random fixed point theorems. *Kodai Math. J.* **2**, 293–299.
274. IVANOFF, B. G. and MERZBACH, E. (1995). Stopping and set-indexed local martingales. *Stochastic Process. Appl.* **57**, 83–98.

275. IVANOFF, B. G., MERZBACH, E. and SCHIOPU-KRATINA, I. (1995). Lattices of random sets and progressivity. *Statist. Probab. Lett.* **22**, 97–102.
276. JACOBS, M. Q. (1969). On the approximation of integrals of multivalued functions. *SIAM J. Control Optim.* **7**, 158–177.
277. JAFFRAY, J.-Y. (1992). Bayesian updating and belief functions. *IEEE Trans Syst. Man Cybernetics* **22**, 1144–1152.
278. JAFFRAY, J.-Y. (1997). On the maximum of conditional entropy for upper/lower probabilities generated by random sets. In *Applications and Theory of Random Sets*, edited by J. Goutsias, R. Mahler and H. T. Nguyen, vol. 97 of *The IMA Volumes in Mathematics and its Applications*, 107–127, Berlin. Springer.
279. JAIN, N. C. and MARKUS, M. B. (1975). Central limit theorems for $C(S)$-valued random variables. *J. Funct. Anal.* **19**, 216–231.
280. JANG, L. C. and KWON, J. S. (2000). On the representation of Choquet integrals of set-valued functions, and null sets. *Fuzzy Sets and Systems* **112**, 233–239.
281. JONASSON, J. (1998). Infinite divisibility of random objects in locally compact positive convex cones. *J. Multivariate Anal.* **65**, 129–138.
282. JONASSON, J. (1998). On positive random objects. *J. Theoret. Probab.* **11**, 81–125.
283. JUNG, E. J. and KIM, J. H. (2003). On set-valued stochastic integrals. *Set-Valued Anal.* **21**, 401–418.
284. JUREK, Z. J. and MASON, J. D. (1993). *Operator-Limit Distributions in Probability Theory*. Wiley, New York.
285. KADANE, J. B. and WASSERMAN, L. (1996). Symmetric, coherent, Choquet capacities. *Ann. Statist.* **24**, 1250–1264.
286. KALLENBERG, O. (1973). Characterization and convergence of random measures and point processes. *Z. Wahrsch. verw. Gebiete* **27**, 9–21.
287. KALLENBERG, O. (1983). *Random Measures*. Akademie-Verlag/Academic Press, Berlin/New York, 3rd edn.
288. KALLENBERG, O. (1996). Improved criteria for distributional convergence of point processes. *Stochastic Process. Appl.* **64**, 93–102.
289. KALLENBERG, O. (2001). Local hitting and conditioning in symmetric interval partitions. *Stochastic Process. Appl.* **94**, 241–270.
290. KAMAE, T., KRENGEL, U. and O'BRIEN, G. L. (1977). Stochastic inequalitites on partially ordered spaces. *Ann. Probab.* **5**, 899–912.
291. KANDILAKIS, D. A. and PAPAGEORGIOU, N. S. (1990). Properties of measurable multifunctions with stochastic domain and their applications. *Math. Jap.* **35**, 629–643.
292. KANIOVSKI, Y. M., KING, A. J. and WETS, R. J.-B. (1995). Probabilistic bounds (via large deviations) for the solutions of stochastic programming problems. *Ann. Oper. Res.* **56**, 189–208.
293. KENDALL, D. G. (1968). Delphic semigroups, infinitely divisible regenerative phenomena, and the arithmetic of p-functions. *Z. Wahrsch. verw. Gebiete* **9**, 163–195.
294. KENDALL, D. G. (1973). On the non-occurrence of a regenerative phenomenon in given interval. In *Stochastic Analysis*, edited by E. F. Harding and D. G. Kendall, 294–308. Wiley, Chichester etc.
295. KENDALL, D. G. (1974). Foundations of a theory of random sets. In *Stochastic Geometry*, edited by E. F. Harding and D. G. Kendall, 322–376. Wiley, New York.
296. KENDALL, M. G. and MORAN, P. A. P. (1963). *Geometrical Probability*. Charles Griffin, London.
297. KENDALL, W. S. (2000). Stationary countable dense random sets. *Adv. Appl. Probab.* **32**, 86–100.

298. KESTEN, H. (1969). *Hitting Probabilities of Single Points for Processes with Stationary Independent Increments*, vol. 93 of *Memoirs of the American Mathematical Society*. American Mathematical Society, Providence, R.I.
299. KHAN, M. A. and MAJUMDAR, M. (1986). Weak sequential convergence in $L_1(\mu, X)$ and an approximate version of Fatou's lemma. *J. Math. Anal. Appl.* **114**, 569–573.
300. KHAN, M. A. and SUN, Y. (1996). Integrals of set-valued functions with a countable range. *Math. Oper. Res.* **21**, 946–954.
301. KHOSHNEVISAN, D. (2003). Intersections of Brownian motions. *Expos. Math.* **21**, 97–114.
302. KIM, B. K. and KIM, J. H. (1999). Stochastic integrals of set-valued processes and fuzzy processes. *J. Math. Anal. Appl.* **236**, 480–502.
303. KIM, Y. K. (2001). Compactness and convexity on the space of fuzzy sets. *J. Math. Anal. Appl.* **264**, 122–132.
304. KINATEDER, K. K. J. (2000). Strong Markov properties for Markov random fields. *J. Theoret. Probab.* **13**, 1101–1114.
305. KING, A. J. (1989). Generalized delta theorems for multivalued mappings and measurable selections. *Math. Oper. Res.* **14**, 720–736.
306. KING, A. J. and ROCKAFELLAR, R. T. (1993). Asymptotic theory for solutions in statistical estimation and stochastic programming. *Math. Oper. Res.* **18**, 148–162.
307. KING, A. J. and WETS, R. J.-B. (1991). Epi-consistency of convex stochastic programs. *Stochastics Stoch. Rep.* **34**, 83–92.
308. KINGMAN, J. F. C. (1964). The stochastic theory of regenerative events. *Z. Wahrsch. verw. Gebiete* **2**, 180–224.
309. KINGMAN, J. F. C. (1972). *Regenerative Phenomena*. Wiley, London.
310. KINGMAN, J. F. C. (1973). Homecomings of Markov processes. *Adv. Appl. Probab.* **5**, 66–102.
311. KINGMAN, J. F. C. (1973). An intrinsic description of local time. *Bull. London Math. Soc.* **6**, 725–731.
312. KINGMAN, J. F. C. (1973). Subadditive ergodic theory. *Ann. Probab.* **1**, 883–909.
313. KINOSHITA, K. and RESNICK, S. I. (1991). Convergence of random samples in $\mathbf{R}^d$. *Ann. Probab.* **19**(4), 1640–1663.
314. KISIELEWICZ, M. (1997). Set-valued stochastic integrals and stochastic inclusions. *Stochastic Anal. Appl.* **15**, 783–800.
315. KISIELEWICZ, M. and SOSULSKI, W. (1995). Set-valued stochastic integrals over martingale measures and stochastic inclusions. *Discuss. Math. Algebra Stochastic Methods* **15**, 179–188.
316. KISYNSKI, J. (1990). Metrization of $D_E[0, 1]$ by Hausdorff distances between graphs. *Ann. Pol. Math.* **51**, 195–203.
317. KLEI, H.-A. (1988). A compactness criterion in $L^1(E)$ and Radon-Nikodym theorems for multifunctions. *Bull. Sci. Math.* **112**, 305–324.
318. KLEMENT, E. P., PURI, M. L. and RALESCU, D. A. (1986). Limit theorems for fuzzy random variables. *Proc. R. Soc. London A* **407**, 171–182.
319. KNIGHT, F. B. (1981). Characterization of the Lévy measures of inverse local times of gap diffusions. In *Seminar on Stochastic Processes, 1981*, vol. 1 of *Progr. Probab. Statist.*, 53–78, Boston. Birkhäuser.
320. KOCH, K., OHSER, J. and SCHLADITZ, K. (2003). Spectral theory for random closed sets and estimating the covariance via frequency space. *Adv. Appl. Probab.* **35**, 603–613.
321. KOLMOGOROV, A. N. (1950). *Foundations of the Theory of Probability*. Chelsea, New York.

322. KOLMOGOROV, A. N. and LEONTOVITCH, M. A. (1992). On computing the mean Brownian area. In *Selected works of A. N. Kolmogorov, Volume II: Probability and mathematical statistics*, edited by A. N. Shiryaev, vol. 26 of *Mathematics and its applications (Soviet series)*, 128–138. Kluwer, Dordrecht, Boston, London.
323. KOMLÓS, J. (1967). A generalisation of a problem by Steinhaus. *Acta Math. Hungar.* **18**, 217–229.
324. KÖNIG, H. (1997). *Measure and Integration*. Springer, Berlin.
325. DE KORVIN, A. and KLEYLE, B. (1984). Goal uncertainty in a generalised information system: convergence properties of the estimated expected utilities. *Stochastic Anal. Appl.* **2**, 437–457.
326. DE KORVIN, A. and KLEYLE, B. (1985). A convergence theorem for convex set valued supermartingales. *Stochastic Anal. Appl.* **3**, 433–445.
327. KOSHEVOY, G. A. and MOSLER, K. (1998). Lift znoids, random convex hulls and the variability of random vectors. *Bernoulli* **4**, 377–399.
328. KÖTHE, G. (1969). *Topological Vector Spaces. I.* Springer, Berlin.
329. KRAMOSIL, I. (1999). Measure-theoretic approach to the inversion problem for belief functions. *Fuzzy Sets and Systems* **102**, 363–369.
330. KRÄTSCHMER, V. (2003). When fuzzy measures are upper envelopes of probability measures. *Fuzzy Sets and Systems* **138**, 455–468.
331. KREE, P. (1982). Diffusion equations for multivalued stochastic differential equations. *J. Funct. Anal.* **49**, 73–90.
332. KRUPA, G. (1998). *Limit Theorems for Random Sets*. Ph.D. thesis, University of Utrecht, Utrecht, The Netherlands.
333. KRUPA, G. (2003). Snell's optimization problem for sequences of convex compact valued random sets. *Probab. Math. Statist.* **23**, 77–91.
334. KRUSE, R. (1987). On the variance of random sets. *J. Math. Anal. Appl.* **122**, 469–473.
335. KRYLOV, N. V. and YUSHKEVITCH, A. A. (1964). Markov random sets. *Theory Probab. Appl.* **9**, 738–743. In Russian.
336. KUDO, H. (1953). Dependent experiments and sufficient statistics. *Natural Science Report. Ochanomizu University* **4**, 905–927.
337. KURATOWSKI, K. (1966). *Topology I*. Academic Press, New York.
338. KURATOWSKI, K. (1968). *Topology II*. Academic Press, New York.
339. KURATOWSKI, K. and RYLL-NARDZEWSKI, C. (1965). A general theorem on selectors. *Bull. Polon. Sci. Sér. Sci. Math. Astronom. Phys.* **13**, 397–403.
340. KURTZ, T. G. (1974). Point processes and completely monotone set functions. *Z. Wahrsch. verw. Gebiete* **31**, 57–67.
341. KURTZ, T. G. (1980). The optional sampling theorem for martingales indexed by directed sets. *Ann. Probab.* **8**, 675–681.
342. LACHOUT, P. (1995). On multifunction transforms of probability measures. *Ann. Oper. Res.* **56**, 241–249.
343. LANDKOF, N. S. (1972). *Foundations of Modern Potential Theory*. Springer, Berlin.
344. LANTUÉJOUL, C. (2002). *Geostatistical Simulation*. Springer, Berlin.
345. LE JAN, Y. (1983). Quasi-continuous functions and Hunt processes. *J. Math. Soc. Japan* **35**, 37–42.
346. LEADBETTER, M. R., LINDGREN, G. and ROOTZEN, H. (1986). *Extremes and Related Properties of Random Sequences and Processes*. Springer, Berlin.
347. LECHICKI, A. and LEVI, S. (1987). Wijsman convergence in the hyperspace of a metric space. *Boll. Un. Mat. Ital. B (7)* **1**, 439–451.
348. LEICHTWEISS, K. (1980). *Konvexe Mengen*. VEB Deutscher Verlag der Wissenschaften, Berlin.

349. LÉVY, P. (1992). *Processus Stochastiques et Mouvement Brownien.* Édition Jacques Gabay, Sceaux. Reprint of the second (1965) edition.
350. LEWIS, T., OWENS, R. and BADDELEY, A. J. (1999). Averaging feature maps. *Pattern Recognition* **32**, 1615–1630.
351. LI, Q. D. and LEE, E. S. (1995). On random α-cuts. *J. Math. Anal. Appl.* **190**, 546–558.
352. LI, S. and OGURA, Y. (1998). Convergence of set valued sub- and supermartingales in the Kuratowski–Mosco sense. *Ann. Probab.* **26**, 1384–1402.
353. LI, S. and OGURA, Y. (1999). Convergence of set valued and fuzzy-valued martingales. *Fuzzy Sets and Systems* **101**, 453–461.
354. LI, S., OGURA, Y. and KREINOVICH, V. (2002). *Limit Theorems and Applications of Set-Valued and Fuzzy Set-Valued Random Variables.* Kluwer, Dordrecht.
355. LI, S., OGURA, Y., PROSKE, F. N. and PURI, M. L. (2003). Central limit theorem for generalized set-valued random variables. *J. Math. Anal. Appl.* **285**, 250–263.
356. LIGGETT, T. M. (1985). An improved subadditive ergodic theorem. *Ann. Probab.* **13**, 1279–1285.
357. LIGGETT, T. M. (1985). *Interacting Particle Systems.* Springer, New York.
358. LINDVALL, T. (1973). Weak convergence of probability measures and random functions on the $\mathcal{D}[0, \infty)$. *J. Appl. Probab.* **10**, 109–121.
359. LÓPEZ-DÍAZ, M. and GIL, M. A. (1998). Reversing the order of integration in iterated expectations of fuzzy random variables, and statistical applications. *J. Statist. Plan. Inf.* **74**, 11–29.
360. LUCCHETTI, R. and TORRE, A. (1994). Classical set convergences and topologies. *Set-Valued Anal.* **2**, 219–240.
361. LUU, D. Q. (1984). Applications of set-valued Radon–Nikodym theorems to convergence of multivalued L^1-amarts. *Math. Scand.* **54**, 101–113.
362. LUU, D. Q. (1985). Quelques résultats de représentation des amarts uniformes multivoques dans les espaces de Banach. *C. R. Acad. Sci., Paris, Ser. I* **300**, 63–65.
363. LUU, D. Q. (1986). Representation theorem for multivalued (regular) L^1-amarts. *Math. Scand.* **58**, 5–22.
364. LYASHENKO, N. N. (1982). Limit theorems for sums of independent compact random subsets of Euclidean space. *J. Soviet Math.* **20**, 2187–2196.
365. LYASHENKO, N. N. (1983). Geometric convergence of random processes and statistics of random sets. *Soviet Math.* **27**(11), 89–100.
366. LYASHENKO, N. N. (1983). Statistics of random compacts in the Euclidean space. *J. Soviet Math.* **21**, 76–92.
367. LYASHENKO, N. N. (1983). Weak convergence of step processes in a space of closed sets. *Zapiski Nauch. Seminarov LOMI* **130**, 122–129. In Russian.
368. LYASHENKO, N. N. (1984). Estimation of parameters of Poisson random sets. *Zap. Nauchn. Sem. Leningrad. Otdel. Mat. Inst. Steklov. (LOMI)* **136**, 121–141.
369. LYASHENKO, N. N. (1987). Graphs of random processes as random sets. *Theory Probab. Appl.* **31**, 72–80.
370. LYASHENKO, N. N. (1989). Geometric limits for noises with arbitrary elementary components. In *Statistics and Control of Random Processes*, edited by A. N. Shiryaev, 121–135. Nauka, Moscow. In Russian.
371. MAHLER, R. P. S. (1997). Random sets in information fusion. an overview. In *Applications and Theory of Random Sets*, edited by J. Goutsias, R. Mahler and H. T. Nguyen, vol. 97 of *The IMA Volumes in Mathematics and its Applications*, 129–164, Berlin. Springer.
372. MAISONNEUVE, B. (1974). *Systèmes Régénératifs*, vol. 15 of *Astérisque*. Société Mathématique de France.

373. MAISONNEUVE, B. (1983). Ensembles régénératifs de la droite. *Z. Wahrsch. verw. Gebiete* **63**, 501–510.
374. MAISONNEUVE, B. and MEYER, P. A. (1974). Ensembles aléatoires markoviens homogènes. In *Séminaire de Probabilités VIII*, edited by C. Dellacherie, P. A. Meyer and M. Weil, vol. 381 of *Lect. Notes Math.*, 172–261. Springer, Berlin.
375. MANCHAM, A. and MOLCHANOV, I. S. (1996). Stochastic models of randomly perturbed images and related estimation problems. In *Image Fusion and Shape Variability Techniques*, edited by K. V. Mardia and C. A. Gill, 44–49, Leeds. Leeds University Press.
376. MARAGOS, P. and SCHAFER, R. W. (1987). Morphological filters – part II: Their relations to median, order-statistics, and stack filters. *IEEE Trans. Acoustic, Speech and Signal Proc.* **35**, 1170–1184.
377. MARINACCI, M. (1999). Limit laws for non-additive probabilities and their frequentist interpretation. *J. Econ. Th.* **84**, 145–195.
378. MARINACCI, M. (1999). Upper probabilities and additivity. *Sankhyā: The Indian J. of Statist. Ser. A* **61**, 358–361.
379. MARTELLOTTI, A. and SAMBUCINI, A. R. (2001). On the comparison of Aumann and Bochner integrals. *J. Math. Anal. Appl.* **260**, 6–17.
380. MASE, S. (1979). Random compact sets which are infinitely divisible with respect to Minkowski addition. *Adv. Appl. Probab.* **11**, 834–850.
381. MATHERON, G. (1975). *Random Sets and Integral Geometry*. Wiley, New York.
382. MATTILA, P. (1995). *Geometry of Sets and Measures in Euclidean Spaces*. Cambridge University Press, Cambridge.
383. MATTILA, P. (1995). Tangent measures, densities, and singular integrals. In *Fractal Geometry and Stochastics*, edited by C. Bandt, S. Graf and M. Zähle, 43–52. Birkhäuser, Basel.
384. MCBETH, D. and RESNICK, S. J. (1994). Stability of random sets generated by multivariate samples. *Comm. Statist. Stochastic Models* **10**, 549–574.
385. MCCLURE, D. E. and VITALE, R. A. (1975). Polygonal approximation of plane convex bodies. *J. Math. Anal. Appl.* **51**, 326–358.
386. MEAYA, K. (1997). Caractérisation d'ensembles aléatoires gaussiens. *Afrika Mat.* **8**, 39–59.
387. MEESTER, R. and ROY, R. (1996). *Continuum Percolation*. Cambridge University Press, New York.
388. MEYER, P.-A. (1966). *Probability and Potentials*. Waltman, London.
389. MEYER, P.-A. (1970). Ensembles régénératifs, d'après Hoffman-Jørgensen. In *Séminaire de Probabilités IV*, vol. 124 of *Lect. Notes Math.*, 133–150. Springer, Berlin.
390. MICHAEL, E. (1951). Topologies on spaces of subsets. *Trans. Amer. Math. Soc.* **71**, 152–182.
391. MOLCHANOV, I. (1984). A generalization of the Choquet theorem for random sets with a given class of realizations. *Theory Probab. Math. Statist.* **28**, 99–106.
392. MOLCHANOV, I. (1984). Labelled random sets. *Theory Probab. Math. Statist.* **29**, 113–119.
393. MOLCHANOV, I. S. (1985). The structure of strict Markov labelled random closed sets. *Ukrainian Math. J.* **37**, 63–68.
394. MOLCHANOV, I. S. (1989). On convergence of empirical accompanying functionals of stationary random sets. *Theory Probab. Math. Statist.* **38**, 107–109.
395. MOLCHANOV, I. S. (1993). Characterization of random closed sets stable with respect to union. *Theory Probab. Math. Statist.* **46**, 111–116.
396. MOLCHANOV, I. S. (1993). Intersections and shift functions of strong Markov random closed sets. *Probab. Math. Statist.* **14**(2), 265–279.

397. MOLCHANOV, I. S. (1993). Limit theorems for convex hulls of random sets. *Adv. Appl. Probab.* **25**, 395–414.
398. MOLCHANOV, I. S. (1993). *Limit Theorems for Unions of Random Closed Sets*, vol. 1561 of *Lect. Notes Math.*. Springer, Berlin.
399. MOLCHANOV, I. S. (1993). Limit theorems for unions of random sets with multiplicative normalization. *Theory Probab. Appl.* **38**(3), 541–547.
400. MOLCHANOV, I. S. (1993). On distributions of random closed sets and expected convex hulls. *Statist. Probab. Lett.* **17**, 253–257.
401. MOLCHANOV, I. S. (1993). On regularly varying multivalued functions. In *Stability Problems for Stochastic Models*, edited by V. V. Kalashnikov and V. M. Zolotarev, vol. 1546 of *Lect. Notes Math.*, 121–129. Springer, Berlin.
402. MOLCHANOV, I. S. (1993). Strong law of large numbers for unions of random closed sets. *Stochastic Process. Appl.* **46**(2), 199–212.
403. MOLCHANOV, I. S. (1994). On probability metrics in the space of distributions of random closed sets. *J. of Math. Sciences* **72**, 2934–2940.
404. MOLCHANOV, I. S. (1995). On the convergence of random processes generated by polyhedral approximations of compact convex sets. *Theory Probab. Appl.* **40**, 383–390.
405. MOLCHANOV, I. S. (1997). Statistical problems for random sets. In *Applications and Theory of Random Sets*, edited by J. Goutsias, R. Mahler and H. T. Nguyen, vol. 97 of *The IMA Volumes in Mathematics and its Applications*, 27–45, Berlin. Springer.
406. MOLCHANOV, I. S. (1997). *Statistics of the Boolean Model for Practitioners and Mathematicians*. Wiley, Chichester.
407. MOLCHANOV, I. S. (1998). Grey-scale images and random sets. In *Mathematical Morphology and its Applications to Image and Signal Processing*, edited by H. J. A. M. Heijmans and J. B. T. M. Roerdink, vol. 12 of *Computational Imaging and Vision*, 247–257. Kluwer, Dordrecht.
408. MOLCHANOV, I. S. (1998). Random sets in view of image filtering applications. In *Nonlinear Filters for Image Processing*, edited by E. R. Dougherty and J. Astola, chap. 10, 419–447. SPIE, New York.
409. MOLCHANOV, I. S. (1999). On strong laws of large numbers for random upper semicontinuous functions. *J. Math. Anal. Appl.* **235**, 349–355.
410. MOLCHANOV, I. S., OMEY, E. and KOZAROVITZKY, E. (1995). An elementary renewal theorem for random convex compact sets. *Adv. Appl. Probab.* **27**, 931–942.
411. MOLCHANOV, I. S. and TERÁN, P. (2003). Distance transforms for real-valued functions. *J. Math. Anal. Appl.* **278**, 472–484.
412. MÖNCH, G. (1971). Verallgemeinerung eines Satzes von A. Rényi. *Studia Sci. Math. Hungar.* **6**, 81–90.
413. MOORE, M. (1984). On the estimation of a convex set. *Ann. Statist.* **12**, 1090–1099.
414. MORI, S. (1997). Random sets in data fusion. multi-object state-estimation as a foundation of data fusion theory. In *Applications and Theory of Random Sets*, edited by J. Goutsias, R. Mahler and H. T. Nguyen, vol. 97 of *The IMA Volumes in Mathematics and its Applications*, 185–207, Berlin. Springer.
415. MOSLER, K. (2002). *Multivariate Dispersion, Central Regions and Depth. The Lift Zonoid Approach*, vol. 165 of *Lect. Notes Statist.*. Springer, Berlin.
416. MOURIER, E. (1955). L-random elements and L^*-random elements in Banach spaces. In *Proc. Third Berekeley Symp. Math. Statist. and Probability*, vol. 2, 231–242. Univ. of California Press.
417. MÜLLER, A. (1997). Integral probability metrics and their generating classes of functions. *Adv. Appl. Probab.* **29**, 429–443.

418. MÜLLER, A. and STOYAN, D. (2002). *Comparison Methods for Stochastic Models and Risks.* Wiley, Chichester.
419. MUROFUSHI, T. (2003). A note on upper and lower Sugeno integrals. *Fuzzy Sets and Systems* **138**, 551–558.
420. MUROFUSHI, T. and SUGENO, M. (1991). A theory of fuzzy measures: representations, the Choquet integral, and null sets. *J. Math. Anal. Appl.* **159**, 532–549.
421. NAGEL, W. and WEISS, V. (2003). Limits of sequences of stationary planar tessellations. *Adv. Appl. Probab.* **35**, 123–138.
422. NÄTHER, W. (2000). On random fuzzy variables of second order and their application to linear statistical inference with fuzzy data. *Metrika* **51**, 201–222.
423. VON NEUMANN, J. (1949). On rings of operators. Reduction theory. *Ann. Math.* **50**, 401–485.
424. NEVEU, J. (1965). *Mathematical Foundations of the Calculus of Probability.* Holden-Day Inc., San Francisco, Calif.
425. NEVEU, J. (1972). Convergence presque sûre de martingales multivoques. *Ann. Inst. H.Poincaré, Sect. B, Prob. et Stat.* **8**, 1–7.
426. NGUYEN, H. T. (1978). On random sets and belief functions. *J. Math. Anal. Appl.* **65**, 531–542.
427. NGUYEN, H. T. (1979). Some mathematical tools for linguistic probabilities. *Fuzzy Sets and Systems* **2**, 53–65.
428. NGUYEN, H. T. and NGUYEN, N. T. (1998). A negative version of Choquet theorem for Polish spaces. *East-West J. Math.* **1**, 61–71.
429. NORBERG, T. (1984). Convergence and existence of random set distributions. *Ann. Probab.* **12**, 726–732.
430. NORBERG, T. (1986). Random capacities and their distributions. *Probab. Theor. Relat. Fields* **73**, 281–297.
431. NORBERG, T. (1987). Semicontinuous processes in multi-dimensional extreme-value theory. *Stochastic Process. Appl.* **25**, 27–55.
432. NORBERG, T. (1989). Existence theorems for measures on continuous posets, with applications to random set theory. *Math. Scand.* **64**, 15–51.
433. NORBERG, T. (1992). On the existence of ordered couplings of random sets — with applications. *Israel J. Math.* **77**, 241–264.
434. NORBERG, T. (1997). On the convergence of probability measures on continuous posets. In *Probability and Lattices*, edited by W. Vervaat and H. Holwerda, vol. 110 of *CWI Tracts*, 57–92. CWI, Amsterdam.
435. NORBERG, T. and VERVAAT, W. (1997). Capacities on non-Hausdorff spaces. In *Probability and Lattices*, edited by W. Vervaat and H. Holwerda, vol. 110 of *CWI Tracts*, 133–150. CWI, Amsterdam.
436. NOTT, D. J. and WILSON, R. J. (1997). Parameter estimation for excursion set texture models. *Signal Processing* **63**, 199–201.
437. NOTT, D. J. and WILSON, R. J. (2000). Multi-phase image modelling with excursion sets. *Signal Processing* **80**, 125–139.
438. NOWAK, A. (1986). Applications of random fixed point theorems in the theory of generalised random differential equations. *Bull. Acad. Sci. Pol. Sci. Ser. Math.* **34**, 487–494.
439. O'BRIEN, G. L. and WATSON, S. (1998). Relative compactness for capacities, measures, upper semicontinuous functions and closed sets. *J. Theoret. Probab.* **11**, 577–588.
440. PANCHEVA, E. (1988). Max-stability. *Theory Probab. Appl.* **33**, 167–170.
441. PAPAGEORGIOU, N. S. (1985). On the efficiency and optimality of allocations. *J. Math. Anal. Appl.* **105**, 113–135.

442. PAPAGEORGIOU, N. S. (1985). On the theory of Banach space valued multifunctions I, II. *J. Multivariate Anal.* **17**, 185–206, 207–227.
443. PAPAGEORGIOU, N. S. (1986). On the efficiency and optimality of allocations II. *SIAM J. Control Optim.* **24**, 452–479.
444. PAPAGEORGIOU, N. S. (1987). A convergence theorem for set-valued supermartingales with values in a separable Banach space. *Stochastic Anal. Appl.* **5**, 405–422.
445. PAPAGEORGIOU, N. S. (1988). On measurable multifunctions with stochastic domains. *J. Austral. Math. Soc. Ser. A* **45**, 204–216.
446. PAPAGEORGIOU, N. S. (1991). Convergence and representation theorem for set-valued random processes. *Probab. Math. Statist.* **11**, 253–269.
447. PAPAGEORGIOU, N. S. (1992). Convergence theorems for set-valued martingales and semimartingales. *Anal. Math.* **18**, 283–293.
448. PAPAGEORGIOU, N. S. (1995). On the conditional expectation and convergence properties of random sets. *Trans. Amer. Math. Soc.* **347**, 2495–2515.
449. PAPAMARCOU, A. and FINE, T. L. (1986). A note on undominated lower probabilities. *Ann. Probab.* **14**, 710–723.
450. PAPAMARCOU, A. and FINE, T. L. (1991). Stationarity and almost sure divergence of time averages in interval-valued probability. *J. Theoret. Probab.* **4**, 239–260.
451. PEMANTLE, R., PERES, Y. and SHAPIRO, J. W. (1996). The trace of spatial Brownian motion is capacity-equivalent to the unit square. *Probab. Theor. Relat. Fields* **106**, 379–399.
452. PENROSE, M. D. (1992). Semi-min-stable processes. *Ann. Probab.* **20**, 1450–1463.
453. PERES, Y. (1996). Intersection-equivalence of Brownian paths and certain branching processes. *Comm. Math. Phys.* **177**, 417–434.
454. PERES, Y. (1996). Remarks on intersection-equivalence and capacity-equivalence. *Ann. Inst. H.Poincaré, Sect. B, Prob. et Stat.* **64**, 339–347.
455. PERES, Y. (2001). An invitation to sample paths of Brownian motion. Lecture Notes.
456. PHILIPPE, F., DEBS, G. and JAFFRAY, J.-Y. (1999). Decision making with monotone lower probabilities of infinite order. *Math. Oper. Res.* **24**, 767–784.
457. PICARD, J. (1994). Barycentres et martingales sur une varété. *Ann. Inst. H.Poincaré, Sect. B, Prob. et Stat.* **30**, 647–702.
458. PICK, R. (1987). Expectation in metric spaces. *Studia Scientiarium Mathematicarum Hungarica* **22**, 347–350.
459. PITMAN, J. and YOR, M. (1996). Random discrete distributions derived from self-similar random sets. *Electron. J. Probab.* **1**, 1–28.
460. PUCCI, P. and VITILARO, G. (1984). A representation theorem for Aumann integrals. *J. Math. Anal. Appl.* **102**, 86–101.
461. PURI, M. L. and RALESCU, D. A. (1983). Differentials of fuzzy functions. *J. Math. Anal. Appl.* **91**, 552–558.
462. PURI, M. L. and RALESCU, D. A. (1983). Strong law of large numbers for Banach space-valued random sets. *Ann. Probab.* **11**, 222–224.
463. PURI, M. L. and RALESCU, D. A. (1985). The concept of normality for fuzzy random variables. *Ann. Probab.* **13**, 1373–1379.
464. PURI, M. L. and RALESCU, D. A. (1985). Limit theorems for random compact sets in Banach space. *Math. Proc. Cambridge Philos. Soc.* **97**, 151–158.
465. PURI, M. L. and RALESCU, D. A. (1986). Fuzzy random variables. *J. Math. Anal. Appl.* **114**, 409–422.
466. PURI, M. L. and RALESCU, D. A. (1991). Convergence theorem for fuzzy martingales. *J. Math. Anal. Appl.* **160**, 107–122.

467. PURI, M. L., RALESCU, D. A. and RALESCU, S. S. (1987). Gaussian random sets in Banach space. *Theory Probab. Appl.* **31**, 598–601.
468. PYKE, R. (1983). The Haar-function construction of Brownian motion indexed by sets. *Z. Wahrsch. verw. Gebiete* **64**, 523–539.
469. RACHEV, S. T. (1986). Lévy-Prokhorov distance in a space of semicontinuous set functions. *J. Soviet Math.* **34**, 112–118.
470. RACHEV, S. T. (1991). *Probability Metrics and the Stability of Stochastic Models.* Wiley, Chichester.
471. RADCHENKO, A. N. (1985). Measurability of a geometric measure of a level set of a random function. *Theory Probab. Math. Statist.* **31**, 131–140.
472. RÅDSTRÖM, H. (1952). An embedding theorem for spaces of convex sets. *Proc. Amer. Math. Soc.* **3**, 165–169.
473. RALESCU, D. and ADAMS, G. (1980). The fuzzy integral. *J. Math. Anal. Appl.* **75**, 562–570.
474. RANSFORD, T. J. (1990). Holomorphic, subharmonic and subholomorphic processes. *Proc. London Math. Soc.(3)* **61**, 138–188.
475. RANSFORD, T. J. (1990). Predictable sets and set-valued processes. *Lect. Notes Math.* **1426**, 41–45.
476. RATAJ, J. (2004). On set covariance and three-point sets. *Czechoslovak Math. J.* **54**, 205–214.
477. REISS, R.-D. (1989). *Approximate Distributions of Order Statistics.* Springer, Berlin.
478. RÉNYI, A. and SULANKE, R. (1963). Über die konvexe Hülle von n zufällig gewällten Punkten. I. *Z. Wahrsch. verw. Gebiete* **2**, 75–84.
479. REPOVŠ, D. and SEMENOV, P. V. (1998). *Continuous Selections of Multivalued Mappings.* Nijhoff, Dordrecht.
480. RESNICK, S. I. (1986). Point processes, regular variation and weak convergence. *Adv. Appl. Probab.* **18**, 66–138.
481. RESNICK, S. I. (1987). *Extreme Values, Regular Variation and Point Processes.* Springer, Berlin.
482. RESNICK, S. I. and ROY, R. (1994). Super-extremal processes and the argmax process. *J. Appl. Probab.* **31**, 958–978.
483. RESNICK, S. I. and ROY, R. (1994). Superextremal processes, max-stability and dynamic continuous choice. *Ann. Appl. Probab.* **4**, 791–811.
484. RESNICK, S. I. and TOMKINS, R. (1973). Almost sure stability of maxima. *J. Appl. Probab.* **10**, 387–401.
485. RÉTI, T. and CZINEGE, I. (1989). Shape characterization of particles via generalised Fourier analysis. *J. Microscopy* **156**, 15–32.
486. REVUZ, A. (1955). Fonctions croissantes et mesures sur les espaces topologiques ordonnés. *Ann. Inst. Fourier* **6**, 187–268.
487. RICE, S. O. (1945). Mathematical analysis of random noise. *Bell Syst. Techn. J.* **24**, 46–156.
488. RICHTER, H. (1963). Verallgemeinerung eines in der Statistik benötigten Satzes der Maßtheorie. *Math. Ann.* **150**, 85–90 and 440–441.
489. RIEDER, H. (1977). Least favourable pairs for special capacities. *Ann. Statist.* **5**, 909–921.
490. RIPLEY, B. D. (1976). The foundations of stochastic geometry. *Ann. Probab.* **4**, 995–998.
491. RIPLEY, B. D. (1976). Locally finite random sets: foundations for point process theory. *Ann. Probab.* **4**, 983–994.

492. RIPLEY, B. D. (1981). *Spatial Statistics.* Wiley, New York.
493. RIPLEY, B. D. and RASSON, J.-P. (1977). Finding the edge of a Poisson forest. *J. Appl. Probab.* **14**, 483–491.
494. ROBBINS, H. E. (1944). On the measure of a random set. I. *Ann. Math. Statist.* **15**, 70–74.
495. ROBBINS, H. E. (1945). On the measure of a random set. II. *Ann. Math. Statist.* **16**, 342–347.
496. ROCKAFELLAR, R. T. (1976). Integral functionals, normal integrands and measurable selections. In *Nonlinear Operators and the Calaculus of Variations*, edited by J. P. Gossez, E. J. Lami Dozo, J. Mawhin and L. Waelbroeck, vol. 543 of *Lect. Notes Math.*, 157–207, Berlin. Springer.
497. ROCKAFELLAR, R. T. (1980). Generalized directional derivatives and sugradients of nonconvex functions. *Canad. J. Math.* **32**, 331–355.
498. ROCKAFELLAR, R. T. and WETS, R. J.-B. (1984). Variational systems, an introduction. In *Multifunctions and Integrands*, edited by G. Salinetti, vol. 1091 of *Lect. Notes Math.*, 1–54. Springer, Berlin.
499. ROCKAFELLAR, R. T. and WETS, R. J.-B. (1998). *Variational Analysis.* Springer, Berlin.
500. ROSENFELD, A. and PFALZ, J. L. (1968). Distance functions on digital pictures. *Pattern Recognition* **1**, 33–61.
501. ROSS, D. (1986). Random sets without separability. *Ann. Probab.* **14**, 1064–1069.
502. ROSS, D. (1990). Selectionable distributions for a random set. *Math. Proc. Cambridge Philos. Soc.* **108**, 405–408.
503. ROTH, W. (1996). Integral type linear functional on ordered cones. *Trans. Amer. Math. Soc.* **348**, 5065–5085.
504. ROTH, W. (2000). Hahn-Banach type theorems for locally convex cones. *J. Austral. Math. Soc. Ser. A* **68**, 104–125.
505. ROZANOV, Y. A. (1982). *Markov Random Fields.* Springer, New York.
506. RUBIN, R. H. and VITALE, R. A. (1980). Asymptotic distribution of symmetric statistic. *Ann. Statist.* **8**, 165–170.
507. RUBINOV, A. M. and AKHUNDOV, I. S. (1992). Difference of compact sets in the sense of Demyanov and its application to nonsmooth analysis. *Optimization* **23**, 179–188.
508. SAINTE-BEUVE, M.-F. (1978). Some topological properties of vector measures with bounded variation and its applications. *Ann. Mat. Pura Appl. (4)* **116**, 317–379.
509. SALINETTI, G. (1987). Stochastic optimization and stochastic processes: the epigraphical approach. *Math. Res.* **35**, 344–354.
510. SALINETTI, G., VERVAAT, W. and WETS, R. J.-B. (1986). On the convergence in probability of random sets (measurable multifunctions). *Math. Oper. Res.* **11**, 420–422.
511. SALINETTI, G. and WETS, R. J.-B. (1981). On the convergence of closed-valued measurable multifunctions. *Trans. Amer. Math. Soc.* **266**, 275–289.
512. SALINETTI, G. and WETS, R. J.-B. (1986). On the convergence in distribution of measurable multifunctions (random sets), normal integrands, stochastic processes and stochastic infima. *Math. Oper. Res.* **11**, 385–419.
513. SALINETTI, G. and WETS, R. J.-B. (1990). Random semicontinuous functions. In *Lectures in Applied Mathematics and Informatics*, 330–353. Manchester Univ. Press, Manchester.
514. SCHLATHER, M. (2002). Models for stationary max-stable random fields. *Extremes* **5**, 33–44.
515. SCHMEIDLER, D. (1970). Fatou's lemma in several dimensions. *Proc. Amer. Math. Soc.* **24**, 300–306.

516. SCHMEIDLER, D. (1986). Integral representation without additivity. *Trans. Amer. Math. Soc.* **97**, 255–261.
517. SCHMEIDLER, D. (1989). Subjective probability and expected utility without additivity. *Econometrica* **57**, 571–587.
518. SCHMITT, M. and MATTIOLI, J. (1993). *Morphologie Mathématique*. Masson, Paris.
519. SCHNEIDER, R. (1988). Random approximations of convex sets. *J. Microscopy* **151**, 211–227.
520. SCHNEIDER, R. (1993). *Convex Bodies. The Brunn–Minkowski Theory*. Cambridge University Press, Cambridge.
521. SCHNEIDER, R. and WEIL, W. (1983). Zonoids and related topics. In *Convexity and its Applications*, edited by P. M. Gruber and J. M. Wills, 296–317. Birkhäuser, Basel.
522. SCHNEIDER, R. and WEIL, W. (1992). *Integralgeometrie*. B. G. Teubner, Stuttgart.
523. SCHNEIDER, R. and WEIL, W. (2000). *Stochastische Geometrie*. Teubner, Leipzig.
524. SCHREIBER, T. (2000). Large deviation principle for set-valued unions process. *Probab. Math. Statist.* **20**, 273–285.
525. SCHREIBER, T. (2002). Variance asymptotics and central limit theorems for volumes of unions of random closed sets. *Adv. Appl. Probab.* **34**, 520–539.
526. SCHREIBER, T. (2003). Limit theorems for certain functionals of unions of random closed sets. *Theory Probab. Appl.* **47**, 79–90.
527. SCHULTZ, R. (2000). Some aspects of stability in stochastic programming. *Ann. Oper. Res.* **100**, 55–84.
528. SCHÜRGER, K. (1983). Ergodic theorems for subadditive superstationary families of convex compact random sets. *Z. Wahrsch. verw. Gebiete* **62**, 125–135.
529. SCHWARTZ, L. (1973). *Radon Measures on Arbitrary Topological Spaces and Cylindrical Measures*. Oxford University Press, Bombay.
530. SENETA, E. (1976). *Regularly Varying Functions*, vol. 508 of *Lect. Notes Math.*. Springer, Berlin.
531. SERFLING, R. (1980). *Approximation Theorems of Mathematical Statistics*. Wiley, New York.
532. SERRA, J. (1982). *Image Analysis and Mathematical Morphology*. Academic Press, London.
533. SHAFER, G. (1976). *Mathematical Theory of Evidence*. Princeton University Press.
534. SHAFER, G. (1979). Allocations of probability. *Ann. Probab.* **7**, 827–839.
535. SHAPLEY, L. S. (1971). Cores of convex games. *Internat. J. Game Theory* **1**, 12–26.
536. SHEPHARD, G. C. (1974). Combinatorial properties of associated zonotopes. *Canad. J. Math.* **26**, 302–321.
537. SKOROHOD, A. V. (1956). Limit theorems for stochastic processes. *Theory Probab. Appl.* **1**, 261–290.
538. SMALL, C. G. (1990). A survey on multidimensional medians. *Internat. Statist. Rev.* **58**, 263–277.
539. SONNTAG, Y. and ZĂLINESCU, C. (1994). Set convergences: A survey and a classification. *Set-Valued Anal.* **2**, 329–356.
540. STAM, A. J. (1984). Expectation and variance of the volume covered by a large number of independent random sets. *Comp. Math.* **52**, 57–83.
541. STANLEY, H. E. and OSTROWSKY, N., eds. (1986). *On Growth and Form*. Nijhoff, Dordrecht.
542. STICH, W. J. A. (1988). An integral for nonmeasurable correspondence and the Shapley-integral. *Manuscripta Math.* **61**, 215–221.
543. STOJAKOVIĆ, M. (1994). Fuzzy random variables, expectations, and martingales. *J. Math. Anal. Appl.* **184**, 594–606.

544. STOYAN, D., KENDALL, W. S. and MECKE, J. (1995). *Stochastic Geometry and its Applications*. Wiley, Chichester, 2nd edn.
545. STOYAN, D. and LIPPMANN, G. (1993). Models of stochastic geometry — a survey. *Z. Oper. Res.* **38**, 235–260.
546. STOYAN, D. and STOYAN, H. (1980). On some partial orderings of random closed sets. *Math. Operationsforsch. Statist. Ser. Optimization* **11**, 145–154.
547. STOYAN, D. and STOYAN, H. (1994). *Fractals, Random Shapes and Point Fields*. Wiley, Chichester.
548. STRAKA, F. and ŠTĚPÁN, J. (1989). Random sets in [0,1]. In *Information Theory, Statistical Decision Functions, Random Processes, Trans. 10th Prague Conf., Prague / Czech., 1986, Vol. B*, 349–356.
549. STRASSEN, V. (1964). Messfehler und Information. *Z. Wahrsch. verw. Gebiete* **2**, 273–305.
550. STRASSEN, V. (1965). The existence of probability measures with given marginals. *Ann. Math. Statist.* **36**, 423–439.
551. SUGENO, M. (1974). *Theory of Fuzzy Integrals and its Applications*. Ph.D. thesis, Tokyo Institute of Technology, Tokyo.
552. SUGENO, M., NARUKAWA, Y. and MUROFUSHI, T. (1998). Choquet integral and fuzzy measures on locally compact space. *Fuzzy Sets and Systems* **99**, 205–211.
553. SVERDRUP-THYGESON, H. (1981). Strong law of large numbers for mesures of central tendency and dispersion of random variables in compact metric spaces. *Ann. Statist.* **9**, 141–145.
554. TAKSAR, M. I. (1980). Regenerative sets on real line. In *Seminar on Probability, XIV*, edited by J. Azéma and M. Yor, vol. 784 of *Lect. Notes Math.*, 437–474. Springer, Berlin.
555. TAKSAR, M. I. (1987). Stationary Markov sets. In *Séminaire de Probabilités, XXI*, edited by J. Azéma, P.-A. Meyer and M. Yor, vol. 1247 of *Lect. Notes Math.*, 303–340. Springer, Berlin.
556. TALAGRAND, M. (1978). Capacités invariantes extrémales. *Ann. Inst. Fourier* **28**, 79–146.
557. TARAFDAR, E., WATSON, P. and YUAN, X.-Z. (1997). The measurability of Carathéodory set-valued mappings and random fixed point theorems. *Acta Math. Hungar.* **74**, 309–319.
558. TAYLOR, R. L. (1978). *Stochastic Convergence of Weighted Sums of Random Elements in Linear Spaces*, vol. 672 of *Lect. Notes Math.*. Springer, Berlin.
559. TAYLOR, R. L. and INOUE, H. (1985). Convergence of weighted sums of random sets. *Stochastic Anal. Appl.* **3**, 379–396.
560. TAYLOR, R. L. and INOUE, H. (1985). A strong law of large numbers for random sets in Banach spaces. *Bull. Inst. Math. Acad. Sinica* **13**, 403–409.
561. TAYLOR, R. L. and INOUE, H. (1997). Laws of large numbers for random sets. In *Random Sets: Theory and Applications*, edited by J. Goutsias, R. P. S. Mahler and H. T. Nguyen, 347–360. Springer, New York.
562. TERÁN, P. (2003). A strong law of large numbers for random upper semicontinuous functions under exchangeability conditions. *Statist. Probab. Lett.* **65**, 251–258.
563. THOMA, H. M. (1991). Belief function computation. In *Conditional Logic in Expert Systems*, edited by I. R. Goodman, M. M. Gupta, H. T. Nguyen and G. S. Rogers, 269–308. Elsevier, North Holland, Amsterdam.
564. TORQUATO, S. (2002). *Random Heterogeneous Materials*. Springer, New York.
565. TRADER, D. A. (1981). *Infinitely Divisible Random Sets*. Ph.D. thesis, Carnegie-Mellon University.

566. UEMURA, T. (1993). A law of large numbers for random sets. *Fuzzy Sets and Systems* **59**, 181–188.
567. UHL, JR., J. J. (1969). The range of a vector-valued measure. *Proc. Amer. Math. Soc.* **23**, 158–163.
568. VAKHANIYA, N. N., TARIELADZE, V. I. and CHOBANYAN, S. A. (1987). *Probability Distributions on Banach spaces*. D. Reidel Publ. Co., Dordrecht.
569. VALADIER, M. (1971). Multi-applications mesurables à valeurs convexex compactes. *J. Math. Pures Appl.* **50**, 265–292.
570. VALADIER, M. (1980). On conditional expectation of random sets. *Ann. Mat. Pura Appl. (4)* **126**, 81–91.
571. VALADIER, M. (1980). Sur l'espérance conditionelle multivoque non convexe. *Ann. Inst. H.Poincaré, Sect. B, Prob. et Stat.* **16**, 109–116.
572. VERVAAT, W. (1997). Random upper semicontinuous functions and extremal processes. In *Probability and Lattices*, edited by W. Vervaat and H. Holwerda, vol. 110 of *CWI Tracts*, 1–56. CWI, Amsterdam.
573. VERVAAT, W. and HOLWERDA, H., eds. (1997). *Probability and Lattices*, vol. 110 of *CWI Tracts*. CWI, Amsterdam.
574. VICSEK, T. (1989). *Fractal Growth Phenomena*. World Scientific, Singapore.
575. VILKOV, B. N. (1995). Asymptotics of random convex broken lines. *Zap. Nauchn. Sem. S.-Peterburg. Otdel. Mat. Inst. Steklov. (POMI)* **223**, 263–279. In Russian.
576. VITALE, R. A. (1983). Some developments in the theory of random sets. *Bull. Inst. Intern. Statist.* **50**, 863–871.
577. VITALE, R. A. (1984). On Gaussian random sets. In *Stochastic Geometry, Geometric Statistics, Stereology*, edited by R. V. Ambartzumian and W. Weil, 222–224. Teubner, Leipzig. Teubner Texte zur Mathematik, B.65.
578. VITALE, R. A. (1985). L_p metrics for compact, convex sets. *J. Approx. Theory* **45**, 280–287.
579. VITALE, R. A. (1987). Expected convex hulls, order statistics, and Banach space probabilities. *Acta Appl. Math.* **9**, 97–102.
580. VITALE, R. A. (1987). Symmetric statistics and random shape. In *Proceedings of the 1st World Congress of the Bernoulli Society. Vol.1. Probability theory and applications*, edited by Y. A. Prohorov and V. V. Sazonov, 595–600, Utrecht. VNU Science Press.
581. VITALE, R. A. (1988). An alternate formulation of mean value for random geometric figures. *J. Microscopy* **151**, 197–204.
582. VITALE, R. A. (1990). The Brunn–Minkowski inequality for random sets. *J. Multivariate Anal.* **33**, 286–293.
583. VITALE, R. A. (1991). Expected absolute random determinants and zonoids. *Ann. Appl. Probab.* **1**, 293–300.
584. VITALE, R. A. (1991). The translative expectation of a random set. *J. Math. Anal. Appl.* **160**, 556–562.
585. VITALE, R. A. (1994). Stochastic smoothing of convex bodies: two examples. *Rend. Circ. Mat. Palermo (2)* **35**, 315–322.
586. VOROB'EV, O. Y. (1984). *Srednemernoje Modelirovanie (Mean-Measure Modelling)*. Nauka, Moscow. In Russian.
587. VOROB'EV, O. Y. (1996). Random set models of fire spread. *Fire Technology* **32**, 137–173.
588. WAGNER, D. H. (1977). Survey of measurable selection theorem. *SIAM J. Control Optim.* **15**, 859–903.

589. WAGNER, D. H. (1979). Survey of measurable selection theorem: an update. In *Measure Theory*, edited by D. Kölzow, vol. 794 of *Lect. Notes Math.*, 176–219. Springer, Berlin.
590. WALLEY, P. (1987). Belief function representations of statistical evidence. *Ann. Statist.* **15**, 1439–1465.
591. WALLEY, P. (1991). *Statistical reasoning with Imprecise Probabilities*. Chapman and Hall, London.
592. WALLEY, P. and MORAL, S. (1999). Upper probabilities based only on the likelihood function. *J. R. Statist. Soc. Ser. B* **61**, 831–847.
593. WANG, G. and LI, X. (2000). On the weak convergence of sequences of fuzzy measures and metric of fuzzy measures. *Fuzzy Sets and Systems* **112**, 217–222.
594. WANG, R. (1998). Some properties of sums of independent random sets. *Northeast. Math. J.* **14**, 203–210.
595. WANG, R. (2001). Essentiual (convex) closure of a family of random sets and its applications. *J. Math. Anal. Appl.* **262**, 667–687.
596. WANG, R. and WANG, Z. (1997). Set-valued stationary processes. *J. Multivariate Anal.* **63**, 180–198.
597. WANG, Z. P. and XUE, X. H. (1994). On convergence and closedness of multivalued martingales. *Trans. Amer. Math. Soc.* **341**, 807–827.
598. WASSERMAN, L. A. (1990). Belief functions and statistical inference. *Canad. J. Statist.* **18**, 183–196.
599. WASSERMAN, L. A. (1990). Prior envelopes based on belief functions. *Ann. Statist.* **18**(1), 454–464.
600. WASSERMAN, L. A. and KADANE, J. B. (1990). Bayes' theorem for Choquet capacities. *Ann. Statist.* **18**, 1328–1339.
601. WASSERMAN, L. A. and KADANE, J. B. (1992). Symmetric upper probabilities. *Ann. Statist.* **20**, 1720–1736.
602. WEIL, W. (1982). An application of the central limit theorem for Banach-space-valued random variables to the theory of random sets. *Z. Wahrsch. verw. Gebiete* **60**, 203–208.
603. WEIL, W. (1983). Stereology: A survey for geometers. In *Convexity and Its Applications*, edited by P. M. Gruber and J. M. Wills, 360–412. Birkhäuser, Basel.
604. WEIL, W. (1995). The estimation of mean shape and mean particle number in overlapping particle systems in the plane. *Adv. Appl. Probab.* **27**, 102–119.
605. WEIL, W. and WIEACKER, J. A. (1984). Densities for stationary random sets and point processes. *Adv. Appl. Probab.* **16**, 324–346.
606. WEIL, W. and WIEACKER, J. A. (1987). A representation theorem for random sets. *Probab. Math. Statist.* **6**, 147–151.
607. WEIL, W. and WIEACKER, J. A. (1993). Stochastic geometry. In *Handbook of Convex Geometry*, edited by P. M. Gruber and J. M. Wills, 1393–1438. Elsevier Sci. Publ., North-Holland, Amsterdam.
608. WENDT, P. D., COYLE, E. J. and CALLAGHER, N. C. (1986). Stack filters. *IEEE Trans. Acoustic, Speech and Signal Proc.* **34**, 898–911.
609. WHITT, W. (2002). *Stochastic-Process Limits*. Springer, New York.
610. WICHURA, M. (1970). On the construction of almost uniformly convergent random variables with given weakly convergent image laws. *Ann. Math. Statist.* **41**, 284–291.
611. WOLFENSON, M. and FINE, T. L. (1982). Bayes-like decision making with upper and lower probabilities. *J. Amer. Statist. Assoc.* **77**, 80–88.
612. WORSLEY, K. J. (1995). Boundary corrections for the expected Euler characteristic of excursion sets of random fields, with an application to astrophysics. *Adv. Appl. Probab.* **27**, 943–959.

613. WORSLEY, K. J. (1995). Estimating the number of peaks in a random field using the Hadwiger characteristic of excursion sets, with applications to medical images. *Ann. Statist.* **23**, 640–669.
614. WORSLEY, K. J. (1995). Local maxima and the expected Euler characteristic of excursion sets of χ^2, F and t fields. *Adv. Appl. Probab.* **26**, 13–42.
615. WSCHEBOR, M. (1985). *Surface Aléatoires*, vol. 1147 of *Lect. Notes Math.*. Springer, Berlin.
616. XIA, A. (2004). Point processes characterized by their one dimensional distributions. *Probab. Theor. Relat. Fields* **128**, 467–474.
617. XU, M. (1996). Set-valued Markov processes and their representation theorem. *Northeast. Math. J.* **12**, 171–182.
618. YAKIMIV, A. L. (1981). Multivariate Tauberian theorems and their applications to the branching Bellman-Harris processes. *Math. USSR-Sb.* **115**, 453–467. In Russian.
619. YANNELIS, N. C. (1988). Fatou's lemma in infinite-dimensional spaces. *Proc. Amer. Math. Soc.* **102**, 303–310.
620. YURACHKIVSKY, A. P. (2002). A functional central limit theorem for the measure of a domain covered by a flow of random sets. *Theory Probab. Math. Statist.* **67**, 151–160. In Ukrainian.
621. ZADEH, L. A. (1987). *Fuzzy Sets and Applications: Selected Papers*. Wiley, New York.
622. ZADEH, L. A. (1999). Fuzzy sets as a basis for a theory of possibility. *Fuzzy Sets and Systems* **100**, 9–34.
623. ZERVOS, M. (1999). On the epiconvergence of stochastic optimization problems. *Math. Oper. Res.* **24**, 495–508.
624. ZHANG, W.-X., WANG, P.-W. and GAO, Y. (1996). *Set-Valued Stochastic Processes*. Science Publ. Co., Bejing. In Chinese.
625. ZHDANOK, T. A. (1983). Extension by continuity of a random function on a random set. *Theory of Random Processes* **11**, 39–41. in Russian.
626. ZIAT, H. (1997). Martingales renversées et loi forte des grand nombres multivoque pour la topologie de Wijsman. *Ann. Sci. Math. Québec* **21**, 191–201.
627. ZIEZOLD, H. (1977). On expected figures and a strong law of large numbers for random elements in quasi-metric spaces. In *Trans. 7th Prague Conf. Inf. Th., Statist. Dec. Func., Random Processes (Prague, 1974).*, vol. A, 591–602, Dordrect. Reidel.
628. ZOLOTAREV, V. M. (1979). Ideal metrics in the problems of probability theory. *Austral. J. Statist.* **21**, 193–208.
629. ZOLOTAREV, V. M. (1991). Asymptotic expansions with probability 1 of sums of independent random variables. *Theory Probab. Appl.* **36**, 800–802.
630. ZOLOTAREV, V. M. (1997). *Modern Theory of Summation of Independent Random Variables*. VSP, Utrecht.
631. ZUYEV, S. (1999). Stopping sets: Gamma-type results and hitting properties. *Adv. Appl. Probab.* **31**, 355–366.

List of Notation

The notation list is ordered by the following groups: Latin letters (lowercase, uppercase, boldface, blackboard style, calligraphic, serif); Greek letters; Gothic letters; mathematical operators; symbols; abbreviations.

Letters F, G and K usually denote a general closed, open and compact set; X, Y, etc. are random sets; t, s are time points; x, y, z are points in space; ξ, ζ are random elements or random functions. Spaces, lattices, cones are usually denoted by blackboard style, families of sets by calligraphic and σ-algebras by Gothic letters.

a_n (normalising constant) 262
$a_n(K)$ (adjusted normalising constants) 262
$d^\wedge \varphi$ (sup-derivative) 419
f^{-1} (inverse image) 389
$f^\vee$ (sup-integral) 419
$h(F, u)$ (support function) 421
$h(K, u)$ (support function) 198
$k(x, y)$ (kernel) 420
p_X (coverage function) 23, 176
$\langle x, u \rangle$ (linear functional) 394
$v_K(x)$ (function $T(xK)$) 262
x_t^- (backward recurrence process) 326
x_t^+ (forward recurrence process) 326

A^r (r-envelope) 390
A^{-r} (inner r-envelope) 390
A^{r-} (open r-envelope) 390
B_1 (unit ball) 198
B_r (ball centred at origin) 395
$B_r(x)$ (ball centred at x) 390
B_1^* (unit ball in dual space) 28, 200
C_X (containment functional) 22, 279

F^- (inverse multifunction) 409
X^* (inversion) 251
F_X (fixed points of X) 243
$H(F, u)$ (support set) 422
$H(K)$ (renewal function) 227
H_X (invariant translations) 255
I_X (inclusion functional) 22
L_X (directions u with a.s. finite $h(X, u)$) 103
M_t (choice process) 355
Q_X (avoidance functional) 22
S_K^+ (positive directions) 227, 293
S_n (partial sum of i.i.d. sets) 199, 209
T, $T(K)$, $T_X(K)$ (capacity functional) 4
T^* (extension of T) 9
U_μ^k (potential of μ) 420
X (random set) 1
X^b (isolated or right-limit points) 326
X_∞ (limit of martingale) 305
X^- (inverse of X) 2, 26
$\bar{X}_n$ (Minkowski average) 214
$\|X\|$ (norm of X) 3, 150

List of Notation

Y_n (closed convex hull of union) 278
Z_n (intersection of random half-spaces) 292
Z_n (union of n i.i.d. random sets) 262
$Z(p)$ (fractal set generated by dyadic cubes) 61

$\mathbf{d}$ (distance function) 179
$\bar{\mathbf{d}}$ (mean distance function) 180
$\mathbf{E}_A \zeta$ (level selection expectation) 368
$\mathbf{E}X$ (selection expectation) 151
$\mathbf{E}_A X$ (selection expectation) 151
$\mathbf{E}_B X$ (Debreu expectation) 156
$\mathbf{E}_{DA} X$ (distance average) 180
$\mathbf{E}_F X$ (Fréchet expectation) 184
$\mathbf{E}_H X$ (Herer expectation) 187
$\mathbf{E}_I X$ (Aumann integral) 151
$\mathbf{E}\xi$ (Bochner expectation) 146
$\mathbf{E}_D \xi$ (Doss expectation) 186
$\mathbf{E}(\xi|\mathfrak{H})$ (conditional expectation) 170
$\mathbf{E}X^\natural$ (reduced expectation) 161
$\mathbf{E}(X|\mathfrak{H})$ (conditional expectation) 170
$\mathbf{E}^\mathfrak{H} X$ ($\mathfrak{H}$-selection expectation) 161
$\mathbf{E}_T X$ (translative expectation) 163
$\mathbf{E}_V X$ (Vorob'ev expectation) 177
$\mathbf{L}^p$ (p-integrable random elements) 146
$\mathbf{n}(u)$ (unit normal vector) 358
$\mathbf{P}$ (probability measure) 414
$\mathbf{s}(\cdot)$ (Steiner point) 36, 422

$\mathbb{C}$ (cone) 394
$\mathbb{C}_S$, $\mathbb{C}_S(t)$ (cone generated by S) 276
$\mathbb{E}$ (space) 1
$\mathbb{E}^*$ (dual space) 27, 394
$\mathbb{H}(F, u)$ (supporting plane) 421
$\mathbb{H}_u(t)$ (hyperplane) 395
$\mathbb{H}_u^-(t)$ (half-space) 395
$\mathbb{I}$ (top of lattice) 258
$\mathbb{L}$ (lattice) 183, 258
$\mathbb{L}$ (poset) 42
$\mathbb{L}_\xi$ (support of $\xi \in \mathbb{L}$) 259
$\mathbb{M}$ (locally finite measures) 115
$\mathbb{P}$ (subfamily of all probability measures) 124
$\mathbb{Q}$ (separant) 53
$\mathbb{R}$ (real line) 387
$\bar{\mathbb{R}}$ (extended line) 410
$\mathbb{R}^d$ (Euclidean space) 1, 395
$\mathbb{S}^{d-1}$ (sphere in $\mathbb{R}^d$) 198, 395

$\mathbb{Y}$ (linearising Banach space) 174
$\mathbb{Z}$ (integers) 387
$\mathbb{Z}^d$ (integer grid in $\mathbb{R}^d$) 114

$\mathcal{A}$ (separating class) 19
$\mathcal{A}_0$ (pre-separating class) 19
$\mathcal{C}(L)$ (convex closed sets with infinite on L support function) 104
$\mathcal{C}_X$ (admissible realisations) 282
$\mathcal{E}$ (algebra of sets in $\mathbb{E}$) 124
$\mathcal{E}$ (paving) 389, 416
$\mathcal{F}$ (closed sets) 1, 387
$\mathcal{F}^K_{K_1,\ldots,K_n}$ (closed sets hitting $K_1,\ldots,K_n$ and missing K) 5
$\mathcal{F}'$ (non-empty closed sets) 179, 388
$\mathcal{F}_A$ (closed sets hitting A) 398
$\mathcal{F}^A$ (closed sets missing A) 398
$\mathcal{G}$ (open sets) 1, 387
$\mathcal{G}_0$ (base of topology) 387
$\mathcal{H}^\alpha$ (Hausdorff measure) 413
$\mathcal{H}^{d-1}$ (($d-1$)-dimensional Hausdorff measure) 115
$\mathcal{J}(\mathbb{L})$, $\mathcal{J}_0(\mathbb{L})$ (increasing functions, sets) 67
$\mathcal{K}$ (compact sets) 1, 388
$\mathcal{K}_0$ (closures of relatively compact sets from $\mathcal{G}_0$) 20
$\mathrm{co}\,\mathcal{K}_X$ (convex compact sets containing F_X) 280
$\mathcal{K}'$ (non-empty compact sets) 388
$\mathcal{K}_G$ (compact sets hitting G) 403
$\mathcal{K}^F$ (compact sets missing F) 403
$\mathcal{K}^X$ (compact sets missing F_X) 244
$\mathcal{L}$ (Scott open filters) 42, 258
ℓ (sup-generating family in $\mathbb{L}$) 183
$\mathcal{L}_x$ (Lawson duality map) 42
$\mathcal{M}$ (subfamily of compact sets) 294
$\mathcal{N}$ (counting measures) 106
$\mathcal{P}$ (all subsets of $\mathbb{E}$) 9
$\mathcal{R}$ (convex ring) 423
$\bar{\mathcal{R}}$ (extended convex ring) 423
$\mathcal{V}$ (compact sets with $\liminf T(xK) = 0$) 262
$\mathcal{T}$ (trapping system) 99
$\mathcal{Z}$ (subfamily of $\mathcal{F}$) 100, 174
$\mathcal{S}_\mathfrak{H}(X)$ ($\mathfrak{H}$-measurable selections) 33
$\mathcal{S}(\nu)$ (selectionable distributions) 34
$\mathcal{S}(X)$ (measurable selections) 26, 146
$\mathcal{S}^p(X)$ (p-integrable selections) 146

List of Notation   465

A (completely alternating) 426
A($\mathcal{D}$), A$_\cup$($\mathcal{D}$) (completely alternating) 7
A$_\cap$($\mathcal{D}$) (completely $\cap$-alternating) 7
b (mean width) 422
C (continuous functions) 198
M (completely monotone) 426
M($\mathcal{D}$), M$_\cap$($\mathcal{D}$) (completely monotone) 7
M$_\cup$($\mathcal{D}$) (completely $\cup$-monotone) 7
N (negative definite) 426
P (positive definite) 426
S (semigroup) 425
$\hat{\text{S}}$ (semicharacters) 425
V (volume) 201
V$_j$ (intrinsic volume) 423

χ (Euler–Poincaré characteristic) 423
$\Delta_{K_n} \cdots \Delta_{K_1}$ (nth-order difference) 5
$\Delta_{s_n} \cdots \Delta_{s_1}$ (nth-order difference) 426
Δ_K (1st-order difference) 5
Δ_w^p (Δ-metric) 407
δ_x (Dirac measure) 412
$(\varepsilon_u, \delta_u)$ (adjunction) 183
Γ_X (covariance of $h(X, \cdot)$) 214
$\gamma(x)$ (variogram) 24
$\mathfrak{H}$ (σ-algebra) 12, 170
$\varkappa_d$ (volume of unit ball) 203
Λ (intensity measure) 110, 246
$\mu(X)$ (measure of X) 59
∇_K (1st-order difference) 7
$\nabla_{K_n} \cdots \nabla_{K_1}$ (nth-order difference) 7
ω_d (surface area of $\mathbb{S}^{d-1}$) 413
$(\Omega, \mathfrak{F}, \mathbf{P})$ (probability space) 1, 414
Φ (Laplace exponent) 331
φ (set-function) 7
$\bar{\varphi}$ (Carathéodory's extension) 119
φ^- (outer extension of φ) 19
φ^* (extension of set-function) 417
$\tilde{\varphi}$ (dual functional) 8
φ^0 (inner extension of φ) 19
Π_Λ (Poisson process) 109, 246
Ψ (capacity related to T) 244
Ψ_Z (limiting capacity for unions) 263
ρ_H (Hausdorff metric) 198, 404
ρ_H^1 ($\mathbf{L}^1$-Hausdorff metric) 367
ρ_H^∞ (uniform Hausdorff metric) 367
$\rho(x, A)$ (distance from point to set) 390

$\sigma(\cdot)$ (generated σ-algebra) 389
$\sigma(L)$ (σ-algebra on $\mathcal{C}(L)$) 104
$\sigma(\mathbb{L})$ (σ-algebra on $\mathbb{L}$) 42
τ (stopping time) 311
Θ (limiting capacity) 271
Υ (non-negative upper semicontinuous functions) 354
υ (deterministic capacity) 120
υ (weighting function) 296
Υ_0 (upper semicontinuous functions with compact support) 366
$\Xi_\lambda(x)$ (set-valued shot-noise) 317
ξ_X (linearisation of X) 174
ζ (Gaussian function on $\mathbb{S}^{d-1}$) 214
$\zeta(G)$ (hitting process) 97

$\mathfrak{B}$ (Borel sets) 389
$\mathfrak{B}(\mathcal{F})$ (Borel σ-algebra in $\mathcal{F}$) 2
$\mathfrak{B}_k$ (relative compact Borel sets) 389
$\mathfrak{d}$ (metric in $\mathbb{Y}$) 174
$\mathfrak{F}$ (σ-algebra) 412
$\mathfrak{F}_K$ (set-indexed filtration) 334
$\mathfrak{F}_n$ (filtration) 303
$\mathfrak{F}_t$ (filtration) 325
$\mathfrak{F}_\tau$ (stopping σ-algebra) 326
$\mathfrak{F}_X$ (σ-algebra generated by X) 27
$\mathfrak{J}$ (family of finite sets) 23
$\mathfrak{L}$ (Lévy metric) 94
m (probability metric) 294
m(ξ, η) (probability metric) 93
p (Prokhorov metric) 32
p$_\mathrm{H}$ (Prokhorov–Hausdorff metric) 35
u (uniform probability metric) 94
u$_\upsilon$ (weighted uniform metric) 296
$\mathfrak{S}_\varphi$ (continuity sets for φ) 19
$\mathfrak{S}_{T_X}$, $\mathfrak{S}_X$ (continuity sets) 85

$\mathbf{1}_A$ (indicator function) 391
argmax (maximiser) 81
argmin (minimiser) 39, 337
cap$_\gamma$ (Riesz capacity) 420
cap$_k$ (k-capacity) 420
card (cardinality) 25
$\overline{\mathrm{co}}$ (convex closed) 394
cl A, $\overline{A}$ (closure of A) 388
co (convex) 393
diam (diameter) 390
dim$_\mathrm{H}$ (Hausdorff dimension) 413
dom (domain) 409

466 List of Notation

epi (epigraph) 336, 392
epi$'$ (strict epigraph) 364
$\text{cl}(F;\mathcal{T})$ ($\mathcal{T}$-closure of F) 99
$\text{cl}(F;U)$ (U-closure of F) 183
Graph (graph) 27, 370, 409
Hom_α (homogeneous) 430
hypo (hypograph) 392
Int (interior) 388
mes (Lebesgue measure) 201, 411
nc (measure of non-compactness) 391
$\text{OFilt}(\mathbb{L})$ (Scott open filters) 42
$\text{Pois}_+(\Lambda)$ (compound Poisson set) 222
proj_n (projection on coordinate n) 393
proj_X (metric projection on X) 40
rad (radius) 407
RV_α (regularly varying) 428
$RV_{\beta,g,\Theta}(\mathcal{A})$ (regularly varying capacity) 270
RV_α^u (regularly varying multivariate) 429
$RV_{\alpha,g,\Phi}^u$ (regularly varying multivalued) 430
$\text{Scott}(\mathbb{L})$ (Scott topology) 42
supp f (support of function) 366
supp μ (support of measure) 106, 115, 412
USC (upper semicontinuous functions) 391
w– (weak) 394
s– (strong) 394

$\check{\ }$ (reflection of set) 396
$\circ$ (scaling in $\mathbb{R}^d$) 265
$\cdot^c$ (complement) 387
$\stackrel{d}{\sim}$ (equality in distribution) 4, 415
$\stackrel{\text{epi}}{\to}$ (epiconvergence) 337

$[\cdot]$ (integer part) 387
$\ll$ (absolute continuity) 412
$\ll$ (way below relation) 42
$\|\cdot\|$ (norm) 393
$\|\cdot\|_\infty$ (uniform norm) 146, 214
$\|\cdot\|_p$ ($\mathbf{L}^p$-norm) 146
$\odot$ (epigraphical rescaling) 364
$\ominus$ (Minkowski difference) 397
$\oplus$ (epigraphical sum) 364
$\oplus, +$ (Minkowski sum) 396
$\otimes$ (product-measure) 412
∂A (boundary of A) 388
$\partial^-\text{epi}$ (strict lower boundary) 342
$\partial f(x)$ (subdifferential) 423
K^o (polar set) 423
$\langle\cdot,\cdot\rangle$ (scalar product) 394
$\boxdot$ (level rescaling) 367
$\boxplus$ (level sum) 367
$\stackrel{F}{\to}$ (Fell convergence) 399
$\stackrel{H}{\to}$ (Hausdorff metric convergence) 402
$\stackrel{M}{\to}$ (Mosco convergence) 401
$\stackrel{PK}{\to}$ (Painlevé–Kuratowski convergence) 400
$\stackrel{W}{\to}$ (Wijsman convergence) 401
$\triangle$ (symmetric difference) 387
$\uparrow x$ (upper set) 42
$\vee D$ (supremum of $D \subset \mathbb{L}$) 42

a.s. (almost surely) 414
CLT (central limit theorem) 214
i.i.d. (independent identically distributed) 415
i.o. (infinitely often) 91, 273
LCHS (locally compact Hausdorff separable) 389
SLLN (strong law of large numbers) 198

Name Index

Abid, M. 230
Adams, G. 126, 142
Adamski, W. 144
Adler, A. 206
Adler, R.J. 381
Aeyels, D. 142
Aldous, D. 381
Aló, R.A. 379
Anderson, T.W. 202, 238
Anger, B. 139, 144
Anisimov, V.V. 384
Araujo, A. 214, 221, 222
Arrow, K.J. 237, 407, 409
Artstein, Z. 34, 35, 137, 154, 191, 192, 200, 236–238, 240, 365, 384, 385
Attouch, H. 383, 384
Aubin, J.-P. 32, 36, 136, 137, 194, 320, 380, 383, 384, 411
Aumann, R.J. 143, 191, 192, 238
Avgerinos, E.P. 379
Ayala, G. 141, 142, 193
Azema, J. 382

Baddeley, A.J. 138, 140, 142, 190, 193, 407, 413
Bagchi, S.N. 379
Balan, R.M. 383
Balder, E.J. 167, 192, 240
Balkema, A.A. 283, 299, 385
Barbati, A. 136
Barnett, V. 302
Bass, R.F. 383, 386
Baudin, M. 141
Beddow, J.K. 193

Bednarski, T. 144
Beer, G. 29, 136, 383, 398–401
Beg, I. 381
Berg, C. 16, 135, 237, 244, 428
Berger, J.O. 144, 384
Berliner, L.M. 144
Bertoin, J. 59, 138, 382, 383
Bickel, P.J. 240
Billingsley, P. 32, 35, 86–88, 90, 104, 118, 141, 369, 386, 415
Bingham, N.H. 428
Birgé, L. 139
Blumenthal, R.M. 382
Bocşan, Gh. 381
Böhm, S. 136
Bräker, H. 91
Bronowski, J. 138
Brown, L. 381
Brozius, H. 302
Bru, B. 193
Buja, A. 144
Buldygin, V.V. 200
Bulinskaya, E.V. 381
Burns, J.A. 191
Byrne, C.L. 155, 191

Callagher, N.C. 385
Carnal, H. 301
Cascos, I. 139
Cassels, J.W.S. 198, 238
Castaing, C. 26, 32, 136, 192, 239, 379, 385, 403
Cellina, A. 380
Cerf, R. 191, 240

Name Index

Chateauneuf, A.   136, 139, 144
Chatterji, S.D.   303, 305
Chobanyan, S.A.   30, 146, 303
Choirat, Ch.   385
Choquet, G.   13, 135, 139, 140, 143, 144, 416, 420
Choukari-Dini, A.   378, 379
Christensen, J.P.R.   16, 135, 237, 244, 428
Cole, E.J.   385
Colubi, A.   385
de Cooman, G.   142
Costé, A.   192
Couso, I.   142
Couvreux, J.   379
Cramér, H.   381
Crauel, H.   136, 138
Cressie, N.A.C.   239, 380
Cross, R.   381
van Cutsem, B.   192, 378
Czinege, I.   193

Daffer, P.Z.   207, 208
Daley, D.J.   121, 141, 142
Dal Maso, G.   383
Davis, R.A.   140, 274, 300–302
Davydov, Yu.A.   239, 240
Debbs, G.   143
Debreu, G.   136, 191, 192, 238
Delbaen, F.   139, 142
Delfour, M.C.   193
Dellacherie, C.   VII, 72, 135, 137, 326–329, 382, 415, 416
Dempster, A.P.   143
Denneberg, D.   72, 139, 142
Dentcheva, D.   36, 137, 316
DeRobertis, L.   144
Dhaene, J.   139
Dolecki, S.   339
Dong, W.   379
Doob, J.L.   54, 417
Doss, S.   193, 194
Dougherty, E.R.   VIII, 397
Dozzi, M.   240
Dunford, N.   154, 155, 165
Dupačová, J.   140, 384
Durrett, R.   380
Dyckerhoff, R.   143
Dynkin, E.B.   138, 192, 320, 381

Eddy, W.F.   300, 302
Edgar, G.   383
Efron, B.   301
Egghe, L.   379
El Amri, K.   192
Emery, M.   194
Engl, H.W.   136, 138, 381
Evans, S.N.   142
Evstigneev, I.V.   192, 383
Ezzaki, F.   192, 239, 379, 385

Falconer, K.J.   138
Faro, R.   80, 140
Federer, H.   77, 412, 425
Fedorchuk, V.V.   405
Fell, J.M.G.   398
Feller, W.   138
Feng, Y.   240
Ferrandiz, J.   142
Filippov, V.V.   405
Fine, T.L.   130, 139, 143, 144
Fisher, L.   300
Fitzsimmons, P.J.   138–140, 320, 330, 331, 381, 382
Fortet, R.   135
Francaviglia, A.   401
Frankowska, H.   32, 36, 136, 137, 320, 383, 384, 411
Fréchet, M.   193
Friel, N.   141, 193
Fristedt, B.   331, 334, 382

Gähler, S.   194
Galambos, J.   241, 242, 265, 266, 268, 283, 285, 299, 376
Gale, J.D.   300
Gao, Y.   36, 137, 140, 379, 380
Gerritse, B.   127, 142
Gerritse, G.   299, 300
Getoor, R.K.   382
Geyer, Ch.J.   384
Ghirardato, P.   139
Ghoussoub, N.   230
Gierz, G.   137, 280
Gihman, I.I.   54, 55, 327, 341
Gil, M.A.   385
Gil, P.   142
Gilboa, I.   125, 139, 142, 143

Giné, E. 163, 191, 214, 217, 221–223, 238, 239, 283, 301, 385
Girotto, B. 142
Glover, J. 382
Godet-Thobie, C. 164, 192
Goldie, C.M. 428
Goodey, P. 238
Goodman, I.R. 134
Graf, S. 74, 126, 131, 139, 140
Grenander, U. VII
Grize, Y.L. 143
Groeneboom, P. 91, 301
Gruber, P.M. 140, 425

de Haan, L. 299, 301, 302, 429, 430
Hahn, F.H. 237, 407, 409
Hahn, M.G. 163, 191, 217, 223, 238, 239, 283, 301, 385
Halmos, P.R. 33, 137
Hansen, J.C. 200, 238, 240, 406
Harding, J. 139
Harris, T.E. 142, 380
Hart, S. 33, 137, 191, 238, 240, 365, 385
Hartigan, J.A. 144
Hawkes, J. 138, 332, 382
Hazod, W. 254
Heijmans, H.J.A.M. VIII, 182, 183, 193, 397
Heilpern, S. 144
Heinich, H. 193
Heinrich, L. 318, 380, 381
Hengartner, W. 294, 295
Herer, W. 189, 193, 194, 379
Hess, Ch. 29, 136, 137, 167, 191–193, 212, 238–240, 308, 310, 378, 379, 384, 385, 401
Hiai, F. 91, 136, 150, 165, 169, 172, 173, 191, 192, 212, 238, 378, 379, 384
Hildenbrand, W. 192
Himmelberg, C. 26, 136
Hiriart-Urruty, J.-B. 104, 411
Hobolth, A. 193
Hoeffding, W. 160
Hoffman-Jørgensen, J. 140, 330, 382, 384
Holwerda, H. 137
Holzer, S. 142
Hörmander, L. 157
Horowitz, J. 382
Hsing, T. 91

Hu, K.Y. 381
Huber, P.J. 130–132, 139, 144
Hug, D. 424
Hulse, P. 240, 406
Hulting, F.L. 380
Huneycutt, J.E. 135, 143

Inoue, H. 208, 238, 239
Ioffe, A.D. 191
Itô, K. 253, 323, 334, 382, 383
Itoh, S. 381
Ivanoff, B.G. 383

Jacobs, M.Q. 192
Jaffray, J.-Y. 129, 136, 139, 143, 144
Jain, N.C. 214, 218
Jang, L.C. 142
Jonasson, J. 193, 240
Jung, E.J. 380
Jurek, Z.J. 299

Kadane, J.B. 131, 132, 139, 142, 144
Kallenberg, O. 112, 141, 142, 383, 415
Kamae, T. 139
Kambouzia, M. 135
Kanda, M. 139
Kandilakis, D.A. 381
Kaniovski, Y.M. 384
Kast, R. 144
Kendall, D.G. VIII, 135, 137, 138, 141, 382
Kendall, M.G. 138
Kendall, W.S. VIII, 84, 109, 115, 122, 123, 137, 138, 141, 142
Kesten, H. 138, 331, 333, 382
Khan, M.A. 191, 192
Khoshnevisan, D. 382
Kim, B.K. 380
Kim, J.H. 380
Kim, Y.K. 385
Kinateder, K.K.J. 383
King, A.J. 135, 140, 384, 385
Kingman, J.F.C. 57, 138, 230, 331, 382
Kinoshita, K. 275, 276, 300, 301
Kisielewicz, M. 380
Kisyński, J. 386
Klee, H.-A. 159
Klement, E.P. 367, 385, 386
Kleyle, B. 311, 379

470 Name Index

Knight, F.B. 334, 383
Koch, K. 25, 136
Kohlberg, E. 33, 137, 191
Kolmogorov, A.N. VII, 138
Komlós, J. 226
König, H. 142
de Korvin, A. 311, 379
Koshevoi, G. 238
Köthe, G. 197
Kozarovitzky, E. 226, 228, 240
Kramosil, I. 143
Krätschmer, V. 144
Kree, P. 380
Kreinovich, V. 239, 379
Krengel, U. 139
Krupa, G. 240, 379, 380, 385
Kruse, R. 194
Krylov, N.V. 330, 382
Kudo, H. 191
Kuratowski, K. 32, 136, 381, 398
Kurtz, T.G. 141, 336, 383
Kwon, J.S. 142

Landkof, N.S. 135, 420
Lapied, A. 144
Last, G. 424
Lantuéjoul, Ch. 24
Le Yan, Y. 382
Leadbetter, M.R. 241, 254, 299, 381
Lechicki, A. 401
Lee, E.S. 193
Lehrer, E. 143
Leichtweiss, K. 424
Lemaréchal, C. 104, 411
Lembcke, J. 139, 144
Leontovich, M. 138
Levi, S. 401
Lévy, P. 323
Lewis, T. 190
Li, Q.D. 193
Li, S. 171, 191, 239, 378, 379, 385
Li, X. 142
Liggett, T.M. 230, 380
Lindgren, G. 241, 254, 299, 381
Lindvall, T. 386
Lippmann, G. 380
Lootgieter, J.-C. 193
López-Díaz, M. 385
Lucchetti, R. 398

Luu, D.Q. 165, 192, 310, 379
Lyashenko, N.N. 20, 140, 238–240, 376, 386

Mahler, R.P.S. 140
Maisonneuve, B. 331, 382
Majumdar, M. 192
Mancham, A. 193
Maragos, P. 385
Mariconda, C. 191
Marinacci, M. 142, 144
Markus, M.B. 214, 218
Martellotti, A. 191
Mase, S. 239
Mason, J.D. 299
Matheron, G. VII, VIII, XI, 13, 24, 66, 134–138, 192, 238, 244, 248, 299, 397–400, 403, 410, 411
Mattila, P. 60, 61, 140, 412, 425
Mattioli, J. 138
McBeth, D. 300
McClure, D.E. 91
McKean, H.P. 253, 323, 334, 382, 383
Meaya, K. 239
Mecke, J. VIII, 84, 109, 115, 122, 123, 141, 142
Meester, R. 65
Mellow, T.P. 193
Mertens, J.-F. 192
Merzbach, E. 240, 383
Meyer, P.-A. 137, 382, 416
Michael, E. 398
Mokobodzki, G. 194
Molchanov, I. 85, 115, 122, 138–142, 192–194, 226, 228, 240, 263, 290, 299–302, 380–383, 385, 413, 430, 431
Mönch, G. 141
Montes, F. 142, 193
Montes, S. 142
Moore, M. 302
Moral, S. 135
Moran, P.A.P. 138
Mori, S. 140
Mosler, K. 143, 238
Mourier, E. 199
Müller, A. 97, 139
Mulrow, E. 140, 274, 300–302
Murofushi, T. 124, 126, 127, 142
Murphy, G. 194

Nagel, W.   300
Narukawa, Y.   124, 126, 142
Näther, W.   386
Navarro, J.A.   80, 140
Neveu, J.   16, 192, 378
Neyman, J.   138
Nguyen, H.T.   41, 134, 135, 137, 143
Nguyen, N.T.   41, 135, 137
Norberg, T.   11, 18, 43, 44, 48, 118, 135, 137, 139–142, 244, 246, 270, 299–301, 384, 385, 418, 419
Nott, D.J.   141, 381
Novikov, P.S.   137
Nowak, A.   380

O'Brien, G.L.   139, 418
Ogura, Y.   171, 191, 239, 378, 379, 385
Ohser, J.   25, 136
Okon, T.   193
Omey, E.   226, 228, 240
Ostrowsky, N.   380
Owens, R.   190

Pancheva, E.   300
Papageorgiou, N.S.   136, 155, 191, 192, 307, 310, 312, 321, 322, 378–381, 385, 402
Papamarcou, A.   130, 143, 144
Paulauskas, V.   239
Pemantle, R.   62, 138
Penrose, M.D.   300, 385
Peres, Y.   62, 138, 142, 323
Pfalz, J.L.   193
Pfanzagl, J.   384
Pflug, G.Ch.   384
Philippe, F.   143
Picard, J.   194
Pick, R.   194
Pitman, J.   138
Proske, F.N.   385
Pucci, P.   192
Puri, M.L.   198, 238, 239, 367, 385, 386
Pyke, R.   383, 386

Rachev, S.T.   93, 141, 293, 294, 302
Račkauskas, A.   239
Radchenko, A.N.   381
Rådström, H.   156, 199

Ralescu, D.A.   126, 142, 198, 238, 239, 367, 385, 386
Ralescu, S.S.   239
Ransford, T.J.   136, 380
Rao, M.   382
Rasson, J.-P.   302
Rataj, J.   136
Raynaud de Fitte, P.   194
Reiss, R.-D.   299
Rényi, A.   91, 301
Repoš, D.   137
Resnick, S.I.   140, 274–276, 283, 286, 299–302, 354, 385, 429, 430
Ressel, P.   16, 135, 237, 244, 428
Réti, T.   193
Revuz, A.   135, 143
Rice, S.O.   381
Richter, H.   191, 238
Rieder, H.   144
Ripley, B.D.   141, 302
Robbins, H.E.   138
Roberts, C.   379
Rockafellar, R.T.   136, 137, 338, 383, 384, 386, 410, 411, 423
Rokhlin, V.A.   137
Rootzen, H.   241, 254, 299, 381
Rosalsky, A.   206
Rosenfeld, A.   193
Ross, D.   137
Roth, W.   395, 398
Roy, R.   65, 354, 385
Rozanov, Y.A.   383
Rubin, R.H.   234
Ryll-Nardzewski, C.   32, 136

Sainte-Beuve, M.F.   192
Salinetti, G.   20, 88, 137, 140, 339, 345, 383, 384, 399
Sambucini, A.R.   191
Sancho, J.   80, 140
Schafer, R.W.   385
Schiopu-Kratina, I.   383
Schladitz, K.   25, 136
Schlather, M.   299
Schmeidler, D.   125, 126, 139, 142, 143, 192
Schmidt, V.   136, 240, 318, 380
Schmitt, M.   138

Schneider, R.   VIII, 81, 91, 140, 200, 228, 238, 301, 405, 421, 423–425
Schreiber, B.M.   381
Schreiber, T.   301, 314, 380
Schultz, R.   384
Schürger, K.   240
Schwartz, J.T.   154, 155, 165
Schwartz, L.   414
Seilhamer, A.V.   384
Semenov, P.V.   137
Seneta, E.   242, 428, 429
Serfling, R.   206, 234
Seri, R.   385
Serra, J.   VII, 385, 397
Shafer, G.   127, 129, 136, 143
Shahzad, N.   381
Shapiro, J.W.   62, 138
Shapley, L.S.   125, 142, 143
Shephard, G.C.   206
Simó, A.   141
Skorohod, A.V.   54, 55, 327, 341, 369, 377, 386, 415
Small, C.G.   182
Solntsev, S.A.   200
Sonntag, Y.   398
Sosulski, W.   380
Stam, A.J.   302
Stanley, H.E.   380
Steele, J.M.   230
Štěpán, J.   42, 137
Stich, W.J.A.   191
Stojaković, M.   386
Stoyan, D.   VIII, 84, 109, 115, 122, 123, 139, 141, 142, 193, 380
Stoyan, H.   139, 141, 193
Straka, F.   42, 137
Strassen, V.   70, 130, 139, 144
Sucheston, L.   383
Sugeno, M.   124, 126, 142
Sulanke, R.   91, 301
Sun, Y.   191
Sverdrup-Thygeson, H.   193

Taksar, M.   330, 382
Talagrand, M.   135, 138
Tarafdar, E.   381
Tarieladze, V.I.   30, 146, 303
Taylor, R.L.   206–208, 238, 239
Terán, P.   193, 385

Teugels, J.L.   428
Theodorescu, R.   294, 295
Thilly, E.   240
Thoma, H.M.   136
Tihomirov, V.M.   191
Tomkins, R.   274, 286
Torquato, S.   135
Torre, A.   398
Trader, D.A.   141, 299, 301

Uemura, T.   213, 239
Umegaki, H.   136, 150, 172, 173, 191, 192, 378, 379

Vakhaniya, N.N.   30, 145, 303
Valadier, M.   26, 32, 136, 173, 192, 403
Vatan, P.   283, 301, 385
Vaughan, H.E.   33, 137
Vedel Jensen, E.B.   193
Vere-Jones, D.   121, 141, 142
Vershik, A.M.   240
Vervaat, W.   137, 140–142, 276, 383, 392, 398, 419, 420
Vicsek, T.   380
Vilkov, B.N.   240
Vitale, R.A.   91, 141, 191, 192, 234, 238–240, 407
Vitiliaro, G.   192
von Neumann, J.   137, 167
Vorob'ev, O.Yu.   193, 380

Wagner, D.H.   136
Walley, P.   128, 135, 139, 143, 144
Wang, G.   142
Wang, P.-W.   379
Wang, R.   192, 240, 379, 380
Wang, Z.P.   379, 380
Wasserman, L.A.   129, 131, 132, 134, 139, 142–144
Watson, S.   418
Weil, W.   VIII, 142, 238, 239, 423, 424
Weiss, V.   300
Wendt, P.D.   385
Wets, R.J.-B.   20, 88, 137, 140, 154, 192, 338, 339, 345, 383–386, 399, 410, 411, 423
Whitt, W.   386
Wichura, M.   90
Wieacker, J.   VIII

Wieacker, J.A.   142
Wills, J.M.   425
Wilson, R.J.   141, 381
Wolfenson, M.   139, 143
Worsley, K.J.   323, 381
Wschebor, M.   381

Xia, A.   141
Xu, M.   380
Xue, X.H.   379

Yahav, J.A.   240
Yakimiv, A.L.   429, 430
Yannelis, N.C.   159, 192
Yor, M.   138
Yurachkivsky, A.P.   301

Yushkevitch, A.A.   330, 382

Zadeh, L.A.   134, 135
Zălinescu, C.   398
Zervos, M.   384
Zhang, W.X.   36, 137, 140, 379, 380
Zhdanok, T.A.   381
Ziat, H.   240, 379
Ziezold, H.   193
Zinn, J.   217, 238, 239
Zitikis, R.   240
Zolésio, J.P.   193
Zolotarev, V.M.   95, 216, 293–295, 300, 302
Zuyev, S.   335, 383

Subject Index

adjunction 183
age process *see* backward recurrence process
algebra 389
allocation
  efficient 321
  optimal 235, 321, 365
allocation problem 240, 365, 385
α-cut 193
alternating renewal process 58, 66, 329
area measure 81, 200, 424
argmin functional 338
Aumann expectation *see* selection expectation
Aumann integral 151, 157, 165, 166, 191, 238
  closedness 158
  convexity in $\mathbb{R}^d$ 152
Aumann–Pettis integral 192
avoidance functional 22, 106, 277

backward recurrence process 326, 330
ball 390, 395
  in dual space 421
  volume of 413
Banach space 393
  of type p 197, 206, 212
bang-bang principle 191
Bartlett spectrum 25
barycentre 194
Bayes risk 132
belief function 127, 143
  Bayesian updating 128
  condensable 128
  extension 127
  likelihood based 129
  updating 128
  vacuous 128
Besicovitch covering theorem 77, 78
binary distance transform 193
Blaschke
  expectation 200
  selection theorem 406
  sum 200
Bochner expectation 146, 156, 186, 188, 192, 415
Bochner integral *see* Bochner expectation
Boolean model 115, 238, 380
Borel σ-algebra 389, 390
boundary 388
branching process 62
broken line 234, 240
Brownian motion 62
Brunn–Minkowski inequality 192, 200, 238
  for random sets 201
Bulinskaya's theorem 322

c-trap 101
Campbell theorem 110, 318
  capacity version 121
Cantor set 62
capacitability theorem 10, 416
capacity 9, 117, 417, 418
  2-alternating 132
  absolutely continuous 74
  C-additive 65, 138, 247

476 Subject Index

Choquet   *see* Choquet capacity
completely alternating   9, 420
completely monotone   103
continuity of   418
countably strongly additive   417
dichotomous   76
differentiable   79
indicator   79
max-infinitely divisible   385
maxitive   11, 252
  complete alternation   11
  upper semicontinuity   11
minitive   272
Newton   252, 296
random   *see* random capacity
regularly varying   270, 300
Riesz   252, 299, 420
strongly subadditive   417
subadditive   420
upper semicontinuous   420
vague convergence   418
capacity functional   4, 7, 10, 22, 70, 94, 130, 132, 244, 344
  and hitting time   325
  as probability measure   10
  complete alternation property   7
  conditional   129
  continuous in Hausdorff metric   96
  equalised   76, 77, 139
  extension   9, 98
  maxitive   135
  monotonicity   4
  non-additive   6
  of the limit for unions   263
  of the union   242
  of union-infinite-divisible random set   244
  of union-stable random set   248
  on finite sets   52, 63, 138, 345
  on open sets   40, 44
  rotation invariant   50
  semicontinuity   4
  translation invariant   50, 107
capacity functionals
  pointwise convergence   86, 87, 264, 270, 359
  uniform convergence   96, 265
Carathéodory's extension   77, 119, 413
  of random capacity   119

cardinality   377
Castaing representation   32, 94, 150, 309, 328
Cauchy distribution   288
Cauchy sequence   390
central limit theorem   239
centroid   193
Chapman–Kolmogorov equation   330
character   236
choice probability   356
choice process   355
  capacity functional   356
  containment functional   356
  Markov property   357
  transition probability   357
Choquet
  capacitability theorem   416
  capacity   416, 420
  $\mathcal{E}$-capacity   416, 417
Choquet integral   70, 126, 139
  comonotonic additivity   72
  derivative of   82
  indefinite   73, 75
  lower   72
  properties   71
  upper   72
Choquet theorem   10, 41, 48
  harmonic analysis proof   18
  measure-theoretic proof   13, 141
Choquet–Kendall–Matheron theorem   *see* Choquet theorem
class
  convergence determining   87, 104, 416
  determining   416
  pre-separating   19, 112, 270
  separating   19, 86, 98, 118, 388
  countable   118
closed set-valued function   *see* multifunction
closing   397
closure   388
coalition   124
compactification   388
compactly uniform integrable sequence   207
completed graph   377
conditional expectation   170
  convergence   174
  properties   171

Subject Index   477

cone   394
  asymptotic   167
  full locally convex   398
  ordered   395
  polar   167
conjugate   342
containment functional   22, 70, 75, 102, 104, 125, 127, 279, 280
  on convex compact sets   103
  orthogonal sum   128
  pointwise convergence   104
continuity family   85
continuity set   19, 340
convergence
  in distribution   *see* weak convergence
  of minimisers   338
  of types   254
  strong   394
  weak   *see* weak convergence, 394
convergence of sets
  lower and upper limits   399
  Mosco   401
  Painlevé–Kuratowski   400, 410
  scalar   402
  Wijsman   401
convex cone   156
convex hull   358, 394
  closed   394
convex hulls   314
  volumes of   301
convex rearrangement   234
convex ring   423
  extended   112, 423
covariance function   23
  exponential   24
covariance matrix   160
coverage function   23, 55, 134, 176, 177, 243
cumulant   58, 331
curvature   91
curvature measure   424

D-convergence   386
D-topology   361, 386
Debreu expectation   156
decision
  Bayesian theory   350
  making   311
  space   350

decomposable set   148
delta theorem   140
Δ-metric   179, 193, 407
Dempster rule of combination   128
Demyanov difference   397
derivative of capacity   80
DH-convergence   361, 385
  tightness condition   363
  weak   362
DH-distance   361
diameter   390
difference
  body   396
  set-theoretic   387
  symmetric   387
differential inclusion   315
  random   315, 380
    existence of solution   315
  stochastic   316
dilation
  by a number   396
  by set   397
distance average   180, 189, 190
distance function   27, 88, 179, 401, 405
  indicator   179
  mean   180
  metric   179
  signed   179, 180
  square   179, 182
distribution selectionable   34, 69
Doss expectation   187, 190
  in metric space   186
  of bounded random set   187

effective domain   409
Effros σ-algebra   2, 26, 84, 134
elementary renewal theorem   226
  for random sets   227, 228, 240
  multivariate   226
empirical probability measure   350
energy of measure   420
entropy condition   214
envelope   48, 390, 405
  open   390
epiconvergence   337, 370, 383
  of averages   348
  weak   344, 346, 348, 358, 362
epiderivative   384
epigraph   213, 336, 353, 392

478   Subject Index

strict   364
unions of   353
epigraphical sum   364
equilibrium measure   420
ergodic theorem   230, 240
   pointwise   230
erosion   397
Euclidean space   1, 395
Euler–Poincaré characteristic   323, 324, 423, 424
evaluation   183
exact programme   236
excursion set   176, 322, 337, 361, 366
   lower   392
   upper   392
expectation
   lower   131
   upper   131
expected utility   143
extremal process   313, 419
extreme sub-cone   426
extreme values   241

Fatou's lemma   158, 159, 165, 192
   approximate   167
   finite dimensional   167
   infinite-dimensional   168
Fell topology   2, 398
   base of   399
   continuity of measure   418
   convergence   399
   generated σ-algebra   410
   metrisability   399
   properties of   399
filter   42
filtration   303
   natural   325, 328–330
   set-indexed   334
finite-dimensional distributions   52
first passage time   362
fixed point   161, 243
forward recurrence process   326
   transition probability   330
Fourier transform   25
Fréchet
   expectation   186, 193
   mean   184
   variance   184, 224
function

asymptotic inverse   429, 432
biconjugate   365
Borel   389
completely alternating   426, 427
completely monotone   334, 426
conjugate   423
continuous   389
convex   422
homogeneous   251, 433
indicator   *see* indicator function
infinitely divisible   427
Lipschitz   391
lower semicontinuous   361, 391
measurable   389
negative definite   426
positive definite   135, 426
positively homogeneous   421
regularly varying   428, 429, 432
   multivariate   429, 433
semicontinuous   410
set-valued   *see* multifunction
slowly varying   216, 429, 430
strong incidence   100
   random   101
subadditive   421
sublinear   157, 421
support   *see* support function
unimodal   202
upper semicontinuous   353, 366, 391
   support of   366
weak incidence   100
   random   101
functional
   accompanying   *see* capacity functional
   additive   346
      epigraphical representation   346
   alternating of infinite order   *see* functional, completely alternating
   capacity   *see* capacity functional
   completely alternating   7, 14, 73
   completely monotone   7, 125, 127, 320
   concave   9
   containment   *see* containment functional
   continuous   233, 394
   dual   7, 22
   inclusion   *see* inclusion functional
   increasing   411
   infinitely alternating   *see* functional, completely alternating

Subject Index

linear 394
  continuous 394, 400
Lipschitz 219
maxitive *see* capacity, maxitive
positively linear 219
proper 346
strictly decreasing 272
strongly subadditive *see* functional concave
strongly superadditive *see* functional convex
subadditive 6
sublinear 71, 395
superlinear 71, 395
upper semicontinuous 73
functions
  comonotonic 72
  equi-lower semicontinuous 339, 378
fundamental measurability theorem 26, 114, 326
fundamental selection theorem 32, 328
fuzzy random variable 385
fuzzy set 134
  membership function 134

game *see* non-additive measure
Γ-convergence 383
gauge function 228
Gaussian process 322
Gaussian random field 323
geometric covariogram 24
germ-grain model 114
graph 409
graphical convergence 370, 386
  continuous functionals 377
  in distribution 371, 376
growth model 380

$\mathfrak{H}$-atom 173
half-space 395
  random 103, 292
harmonic analysis 16
Hausdorff dimension 12, 60, 61, 123, 323, 331, 413
Hausdorff distance *see* Hausdorff metric
Hausdorff measure 115, 123, 358, 413
Hausdorff metric 21, 402, 403
  completeness 405
  continuity of measure 418

convergence 406
  for random sets distributions 35
Hausdorff space 388, 399, 414
Hausdorff–Busemann metric 361, 402
Herer expectation 187, 188, 379
Hilbert space 184, 198, 394
hitting functional 22
hitting probability 140
hitting process 97–100, 141
  extension 98
  finite-dimensional distributions 101
Hoeffding theorem 160
homothety 396
horizontal integral 142
Hörmander embedding theorem 157, 199
hyperplane 395
hypograph 353, 365, 392
hypotopology 353

inclusion 387
inclusion functional 22, 55, 320
  multiple integral of 60
  of random open set 63
indicator 3
indicator function 176, 391
  first-order stationary 51
inf-vague convergence 383
infimum, measurability of 39
inner extension 19
inner radius 198, 409
inner separability 341
integrable selections 146, 151
  decomposability 148
  on atomic and non-atomic spaces 151
  weak compactness 155
integral *see* named integrals
integral functional 150
intensity function 109
intensity measure 109, 110, 246, 295
interior 388
intrinsic density 122
intrinsic volume 160, 234, 423, 424
  additive extension 424
  positive extension 424
inverse image 174, 389
inverse multifunction *see* multifunction, inverse of, 409
inversion theorem 429, 431

for multivariate regular varying functions 432
isometry 394
isoperimetric inequality 203
isotropic rotation 203, 204

K-convergence 226
Kakutani fixed point theorem 320
kernel 252, 420
   Riesz 252, 420
Khinchin lemma 254
Komlós theorem 226
Korolyuk's theorem 121
Krein–Smulian theorem 379
Krickeberg's decomposition theorem 308, 310
Kudo–Aumann integral 191
Kuratowski measure of non-compactness 41, 391

Laplace exponent 331, 333
Laplace transform 236, 427
large deviation 232, 240
   principle 233, 314
   theory 127
lattice 42, 183
   complete 42
   continuous 42
   of closed sets 48
law of iterated logarithm 223, 239
Lawson duality 42
LCHS space 1, 389
Lebesgue integral 71
Lebesgue measure 201, 411–413, 423
Legendre–Fenchel transform 423
level set 322
level sum 367
Lévy measure 222, 246, 331, 332
Lévy metric 294
Lévy–Khinchin measure
   on a lattice 259, 260
Lévy–Khinchin representation 221
Lévy–Khinchin theorem 239
lexicographical minimum 149
lexicographical order 395
lift zonoid 238
likelihood
   function 129, 351
   region 129

linear hull 393
linear operator 394
linearisation 175
local time 331
Lorenz curve 240
loss 350
lower expectation 131
lower probability 129
lower semicontinuous function
   random 340
L^p-norm 146
Lyapunov's theorem 152, 153, 191

M_2-topology 386
martingale 303
   set-indexed 336, 383
martingale in the limit 312
martingale integrand 366, 385
martingale selection 309, 310
max-stable 299
maximum likelihood estimator 351
   consistency 352
maximum principle 420
mean distance function 180, 193
mean ergodic theorem 231
mean width 159, 233, 422, 423
   expected 160
measurable selection *see* selection
measure 412
   absolutely continuous 412
   completely random 221
   counting 105, 106, 413
      locally finite 105
      random 106
   Dirac 412
   energy of 420
   feasible 124
   finite 412
   finitely additive 191
   fuzzy 124
   generalised curvature 424
   Haar 413
   Hausdorff *see* Hausdorff measure
   image of 412
   Lebesgue *see* Lebesgue measure
   Lévy–Khinchin 426, 427
   locally finite 412
   multivalued 164, 165
      absolutely continuous 164

Subject Index   481

integral with respect to   165
   selection of   164
   variation of   164
non-additive   *see* non-additive measure
outer   416
potential of   420
Radon   412, 418, 426
semicontinuity of   411
set-valued   412
σ-finite   176, 412
spectral   221
support of   106, 115, 412
measures
   vague topology   419
membership function   193
metric   390
   Hausdorff   *see* Hausdorff metric
metric entropy   215, 218
minimisation of expectations   350
Minkowski addition   *see* Minkowski sum
Minkowski average   213
Minkowski combination   425
Minkowski difference   397
Minkowski sum   195, 314, 396
   closedness   396
   operator-normalised   200
   strong law of large numbers   199
   volume of   240
   weighted   156
Mittag–Leffler's theorem   393
mixed volume   233, 425
Möbius inverse   76
Möbius inversion   25, 136
Mosco convergence   210, 211
Mosco topology   239, 401
multifunction   25, 315, 370, 409
   Borel measurable   41
   continuous   410
      examples of   410
   effective domain of   409
   Effros measurable   26
   graph of   27, 370, 409, 410
   homogeneous   288, 319, 430
   inverse of   319, 409, 431
   P-a.s. semi-differentiable   90
   regularly varying   292, 430
   semicontinuous   410
      examples of   411
   strongly measurable   26

   weakly measurable   26
multivalued amart   310, 312
multivalued function   *see* multifunction
multivalued martingale   304, 310, 366
   as closure of martingale selections   311
   Castaing representation   311
   convergence   305
   integrably bounded   304
   Mosco convergence   307
   optional sampling theorem   *see* optional sampling theorem
   reversed   379
   uniformly integrably bounded   307
multivalued measure   164
multivalued operator
   fixed point   320
   random   320, 381
   with stochastic domain   321, 381
multivalued pramart   312
multivalued quasi-martingale   312
multivalued submartingale   304, 316
multivalued supermartingale   304, 310
   convergence   307
   in Banach space   308
multivariate distribution
   characterisation of   160
multivariate quantile   302
myopic topology   21, 403, 418
   convergence   403
   properties of   403

n-point coverage probabilities   23
neighbourhood   388
neutral element   425
Newton capacity   123
Neyman–Pearson lemma for capacities   133
Neyman–Pearson test   133
non-additive measure   124, 142
   coherent   124
   convex   124
   core of   124
   decomposable   124
   dual   124
   equalised   124
   Jordan decomposition   125
   outer   127
   weakly symmetric   124
non-closed random set   41

norm   393
  composition   125
  total variation   125
norm of set   *see* set, norm of
normal integrand   339, 340, 348, 384
  conjugate of   343
  inner separable   341
  integrable   364
  non-negative   350
  proper   339, 346, 363
  selection expectation   364
  sharp   342, 358, 376, 384
  subdifferential   343
  weak convergence   *see* epiconvergence, weak, 344
normal integrands
  convergence of finite-dimensional distributions   345
  equi-inner separable   345
  equi-outer regular   345
normal vector   358
null-array of random sets   269

opening   397
optional sampling theorem
  for multivalued martingales   311, 379
order statistic   160
origin   395
outer extension   19

p-function   57
  standard   57, 329
parallel set   390
  inner   390
partially ordered set   42
paving   389, 416, 417
payoff   124
perimeter   160, 377
  expected   160, 203
Pettis integral   192
plane supporting   421
plausibility function   127
point
  exposed   197
  interior   388
point process   106, 141, 342
  marked   110
  on $\mathcal{K}$ 112
  orderly   121

parametric measure   121
  simple   106, 107, 109, 121
  stationary   106, 317
  superposition   300
  thinning   109
  weak convergence   111
Poisson point process   66, 80, 109, 246, 261, 286, 295, 318, 353, 354, 358, 374, 376
  capacity functional   109, 251
  on co $\mathcal{K}$ 222
  stationary   109, 251
Poisson-rescaled random set   252
polar set   423
Polish space   25, 390
  of capacities   418
  of closed sets   401
poset   *see* partially ordered set
positive convex cone   236
possibility measure   135
potential   420
pramart   379
precapacity   117, 418
  extension of   418
prevision   139
  coherent   140
probability density function   351
probability generating function   244
probability measure   414
  weak convergence   84, 415
probability metric   93, 97, 294
  compound   94
  homogeneous   296
  ideal   296
  integral   97
  Lévy   93–95
  Prokhorov   93
  regular   296
  simple   94
  uniform   94
probability space   1, 146, 414
  complete   414
progressive measurability   326
projection   393, 424
projection theorem   415
projective limit   309, 393
projective system   393, 414
  exact   414
Prokhorov metric   32, 35

Prokhorov theorem   35
Prokhorov–Hausdorff metric   35
pseudometric   174

quasi-diffusion   383
quermassintegral   424

radius-vector expectation   182
radius-vector function   105, 182, 276
  expected   182
Radon–Nikodym derivative   412
  for capacities   73
  of multivalued measure   379
Radon–Nikodym property   158, 164, 305
Radon–Nikodym theorem
  for capacities   75
  for multivalued measures   307
Rådström embedding theorem   156, 199
random ball   3, 86, 289, 291
random Borel set   41, 322
random capacity   117, 353
  continuity set   117
  extension   118
  indicator   119, 121
  integrable   120
  intensity   120
  intensity measure of   121
  intrinsic density   122
  parametric measure of   121
  stationary   122
  weak convergence   118
random closed set   1, 2, 339
  a.s. continuous   56
  adapted   334, 335
  additive union-stable   256
  affine union-stable   254
  approximable   30
  boundary of   37
  closed complement of   37
  closed convex hull of   37
  concentration function   294
  conditional distribution   12
  convex   see random convex closed set
  exchangeable   383
  first-order stationary   50
  fixed point of   see fixed point
  g-invariant   49
  Hausdorff approximable   154
  homogeneous at infinity   256
  in extended convex ring   112
  in Polish space   27, 40
  infinite divisible for unions   242, 373
  integrable   151, 364
  integrably bounded   150, 199, 305
  integrably bounded in $\mathbb{R}^d$ 159
  intrinsic density   122
  isotropic   49, 203
  kernel of   105
  locally finite   106, 107
    stationary on the line   107
  marked   141
  minimal σ-algebra   27
  natural filtration   see natural filtration
  non-approximable   31
  non-trivial   243
  norm of   3
  P-continuous   55
  Pettis integrable   240
  Poisson   373
  Poisson rescaled   252
  quantile   177
  quantile of   176
  quasi-stationary   51
  reduced representation   151, 201
  regenerative   330
  regular closed   63, 96
  self-similar   51, 248, 332
  semi-Markov   66, 247
  separable   53, 63, 128, 138
  simple   30, 154, 156
  star-shaped   105
  stationary   49, 122, 137, 322
  stochastic order   67, 68
  strong Markov   see strong Markov random set
  surface measure of   115
  union-infinitely-divisible   see random closed set, infinite divisible for unions
  union-stable   see union-stable random closed set
  variance   194
  Wijsman approximable   31
random closed sets
  a.s. convergence   90, 92
  capacity equivalent   62
  convergence in probability   92
  identically distributed   4
  independent   12

Subject Index

intersection-equivalent 62
Minkowski sum 37
relative compactness of distributions 85
unions 314
weak convergence *see* weak convergence
random compact set 21, 23, 151, 335
  a.s. convergence 91
  compound Poisson 222
  convergence in probability 93
  convex *see* random convex compact set
  covariance function 214
  Gaussian 219, 220, 239
  Hausdorff approximable 31
  isotropic 228
  M-infinitely divisible 222
  p-stable 220, 239
  square integrable 214
  truncation 224
random convex closed set 64, 102, 343
  convex-stable 281
    characterisation of 283
  infinitely divisible for convex hulls 279
  integrably bounded 157
  strictly convex-stable 281
    characterisation of 282
  unbounded 103
random convex compact set 88, 102, 305
random convex hull 203
random element 93, 146, 174, 414
  conditional expectation of 170
  expectation *see* Bochner expectation
  Fréchet expectation 184
  Gaussian 160
  in a cone 237
  in semigroup 427
    infinitely divisible 427, 428
  infinite divisible in a semi-lattice 258
  integrable 146, 170
  self-decomposable 262, 300
  tight sequence 89
random field 381, 383
random fractal set 380
random fuzzy set 386
random grey-scale image 385
random indicator function 42
random interval 328
random matrix 205
random measure 115, 120

counting 121
random open set 63, 319
  convex 64
  inclusion functional of 63
random polyhedron 90, 358
random rotation 291
random sample
  almost sure stability 300
  convergence 286, 300
random segment 204
random set
  Borel 167
  closed *see* random closed set
  discrete 193
  graph-measurable 41, 326
  open *see* random open set
  optional 328
  Poisson 109, 286
  T-closed 101
random set-valued function 371
random translation 163
random triangle 3, 289, 291
random upper semicontinuous function 368, 385
  dominated convergence 386
  integrably bounded 368
  max-infinitely divisible 353
  max-stable 353
  strong law of large numbers 369
  strongly integrable 368, 369
random variable 160, 414
  expectation of 414
  integrable 414
  max-stable 241
random vector 90, 146, 415
  Gaussian 160
  lift zonoid of 205
  zonoid of 205
rate function 233, 314
reduced representation 13, 44
reflection 396
regenerative embedding 333
regenerative event 57, 66, 138, 329
  avoidance functional 58
  instantaneous 58
  stable 58
  standard 57
relative compact sequence of ε-optimal points 338

Subject Index 485

relaxed programme   236
renewal function   226
   containment   227
   hitting   229
   inclusion   229
residual lifetime process   *see* forward recurrence process
response function   317
   multivalued   317
Rice's formula   323
rigid motion   395
Robbin's formula weighted   122
Robbins' theorem   59, 176
   capacity version   120
robust statistics   132
rounding   204

sandwich theorem   71, 395
saturation   389
sausage   325
scalar convergence   378
scheme of series   270
Scott topology   42
   second countable   40, 43, 101
second spectral moment   323
selection   26, 94, 130, 145
   adapted   316
   existence   31
   $\mathfrak{F}_X$-measurable   33, 149
   generalised   89
   integrable   94, 146, 304
      existence   149
      properties of   146
      set of   *see* integrable selections
   Lipschitz   316
   Markov   313
   of set-valued process   312
   stationary   313
selection expectation   94, 139, 151, 156, 157, 184, 188, 199
   conditional   *see* conditional expectation
   convergence   165
   convexity   153
   dominated convergence theorem   166, 169
   limit of averages   214
   monotone convergence   169
   of segment   205
   reduced   161, 209

selection operator   36, 88
semi-lattice   42, 258
semi-min-stable process   300
semicharacter   16, 425, 427, 428
semigroup   9, 16, 425
   idempotent   16, 427, 428
semiring   44
separant   53
separating class   344
sequence m-dependent   375
sequence of random closed sets
   relatively stable   275
   stable   275
sequence of random variables
   relatively stable   275
set
   analytic   390, 418
   bounded   391, 393
   capacitable   416
   centrally symmetric   396
   closed   387, 388, 391
   compact   388
   convex   393, 396
      closed   393, 394
   decomposable   148, 305, 346
   increasing   67
   inverse   251
   irreducible   49
   locally finite   106
   M-infinitely divisible   196
   norm of   393
   of fixed points   243
   open   387
   P-continuous   415
   parallel   390
      inner   390
   perfect   388
   quasicompact   389
   radius of   407
   rectifiable   391
   regular closed   28, 388
   relatively compact   19, 388, 391
   saturated   48, 389
   Scott open   42
   separating   42
   star-shaped   182, 276
   stopping   *see* stopping set
   support function of   *see* support function
   $\mathcal{T}$-closed   99, 100

totally bounded   390
upper   42, 67
weakly ball-compact   168
set-indexed process   334
set-valued   *see* multivalued
set-valued function   *see* multifunction
set-valued martingale   *see* multivalued martingale
set-valued process   312, 317, 373
  adapted   303
  finite-dimensional distributions   362
  Gaussian   239
  increasing   313, 361
  Markov   313, 357
  second-order stationary   313
  stochastic integral of   380
  strictly stationary   313
    ergodic theorem   313
  subholomorphic   380
  union   380
Shapley–Folkman theorem   *see* Shapley–Folkman–Starr theorem
Shapley–Folkman–Starr theorem   195, 218, 407
shot-noise process   317, 380
  Minkowski   317
    weak convergence   318
  union   317
σ-algebra   389
  Borel   *see* Borel σ-algebra
  complete   414
  completion   327
  stopping   335
singleton   161, 421
  random   2, 10
    convergence   86
Skorohod convergence   369
Skorohod distance   361
Skorohod space   385
slice topology   29, 239
Snell's optimisation problem   380
space
  Banach   *see* Banach space
  compact   388, 390
  dual   27, 394
  Euclidean   *see* Euclidean space
  Hausdorff   *see* Hausdorff space
  Hilbert   *see* Hilbert space
  LCHS   389

locally compact   388, 399
locally compact Hausdorff second countable   389
locally connected   411
locally convex   393
metric   390, 403
  complete   390
  Doss-convex   189
  of negative curvature   194
  separable   390
non-Hausdorff   392
of closed sets   145, 398
  topology   *see* topology on $\mathcal{F}$
paved   416
Polish   *see* Polish space
product   387, 392
reflexive   394
second countable   388, 390
separable   388
σ-compact   388
sober   49
Souslin   390
T_1 388, 389
T_0 388
topological   387
space law   138
spatial median   182
sphere   395, 421
  surface area of   413
stack filter   385
Steiner formula   423
  local   424
Steiner point   36, 163, 422
  generalised   36
step-function random   371
  linearly interpolated   375
stochastic control   317
stochastic integral set-valued   315
stochastic optimisation   348, 350, 352
stochastic order   67
stochastic process
  Gaussian   318
  max-stable   299
  semi-min-stable   354, 385
  separable   340, 341
  with random open domain   319
    finite-dimensional distributions   320
stopping set   335
stopping σ-algebra   311, 326

Subject Index

stopping time 311, 312, 326, 328
  optimal 380
strong decomposition property 74
strong law of large numbers
  Mosco convergence 210
  Painlevé–Kuratowski convergence 209
  Wijsman convergence 212
strong law of large numbers for Minkowski sums 199
strong Markov process
  level set 329, 330
strong Markov random set 329, 330
  embedding 333
  intersection of 333
  stable 332, 333
  weak convergence 333
structuring element 397
sub-probability measure 10
sub-σ-algebra 161
subdifferential 342, 423
subgradient 423
subordinator 58, 331
  occupation measure 332
  stable 331, 332
successive differences 5, 7, 68
Sugeno integral 126
sup-derivative 419
sup-generating family 183
sup-integral 97, 419
sup-measure 11, 353, 384, 419, 420
  derivative of 81
  random 119
sup-vague convergence 383
super-extremal process 354, 385
  with max-stable components 355, 356
superstationary sequence 230
support estimation 302
support function 27, 39, 157, 198, 421
  covariance 214
  Gaussian 219, 239
  Lipschitz property 421
  subdifferential 423
support set 422, 423
surface area 413
symmetric interval partition 383
symmetric order 139
symmetrisation 396

$\mathcal{T}$-closure 99, 101

tangent cone 37
tessellation 300
three series theorem 225, 240
Tietze extension theorem 381
tight sequence of random sets 207
top of lattice 258
topologies for capacities 418
topology 387
  base of 387, 419
  decreasing 68
  exponential 398
  induced 387
  narrow *see* myopic topology, 418
  scalar 402
  strong 394, 400
  sub-base of 387
  sup-narrow 419
  sup-vague 419
  vague 111, 117, 398, 418, 419
  weak 159, 394, 400
  weak* 421
topology on $\mathcal{F}$
  Attouch–Wets 402
  Fell *see* Fell topology
  Vietoris 398
  Wijsman *see* Wijsman topology
translative expectation 163, 202
translative integral formula 81
trap 100
  c-trap 101
trapping space 101
trapping system 99, 100, 141
triangular array 230
  subadditive 230
  superstationary 230
two series theorem 225
types convergence 254

U-closure 183
U-expectation 184
U-statistic 206, 234
unambiguous event 131
unanimity game 125, 128
uniform convergence 369
uniform integrability 414
  terminal 378
uniform metric 294
  generalised 296
union-infinite-divisibility 242

characterisation 244
union-stable random closed set 80, 247, 295, 386
   characterisation 248
unit ball, volume of 203
upper expectation 131
   symmetric 131
upper level set *see* excursion set
upper probability 75, 129
   regular 131
utility function 321

vague convergence 79
valuation 423
variational system 339
volume 201
von Neumann selection theorem 167
Vorob'ev expectation 177
   as minimiser 177
   generalisation 178
Vorob'ev median 178

Wald's formula 240
Wald's identity 311
weak convergence 84, 86, 169, 371
   and Lévy metric 95
   of random convex compact sets 104
width function 422
Wiener process 51, 160, 316
   zero set 323, 331, 333
Wijsman topology 29, 401
   metrisability 401
   properties of 401

zonoid 204, 238
zonotope 204, 205

Printed by Publishers' Graphics LLC